Networks in Action

Business Choices and Telecommunications Decisions

The Wadsworth Series in Management Information Systems

Networks in Action

Business Choices and Telecommunications Decisions

Peter G. W. Keen
International Center for
Information Technologies
and Fordham University

J. Michael Cummins
Emory University

Wadsworth Publishing Company
Belmont, California
A Division of Wadsworth, Inc.

Publisher: *Kathy Shields*
Development Editor: *Alan Venable*
Editorial Assistant: *Tamara Huggins*
Production Editor: *Carol Carreon Lombardi*
Managing Designer: *Cloyce Wall*
Print Buyer: *Diana Spence*
Art Editor: *Donna Kalal*

Permissions Editor: *Peggy Meehan*
Designer: *Harry Voigt*
Copy Editor: *Alan Titche*
Technical Illustrator: *Lotus Art*
Cover Designer: *Harry Voigt*
Compositor: *Weimer Graphics*
Printer: *Maple-Vail Book Manufacturing*

 This book is printed on acid-free recycled paper.

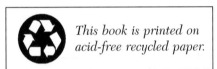 **International Thomson Publishing**
The trademark ITP is used under license.

Printed in the United States of America

2 3 4 5 6 7 8 9 10—98 97 96 95 94

Library of Congress Cataloging-in-Publication Data

Keen, Peter G. W.
    Networks in action : business choices and telecommunications decisions / Peter
Keen, J. Michael Cummins.
    p. cm.—(The Wadsworth series in management information systems)
    Includes bibliographical references and index.
    ISBN 0-534-17766-2
      1. Business—Communication systems. 2. Telecommunication systems. 3. Computer
networks. 4. Data transmission systems. I. Cummins, J. Michael, 1945– . II. Title.
    III. Series.
HF5541.T4K443   1994
651.8—dc20   93-27745

Contents

3   **The Telecommunications Decision Sequence** 97

Part Two Network Building Blocks 151

4 Terminals and Workstations 152

5 Transmission Links   196

16 Total Quality Operations 581

17 Managing Costs 618

Part Five   The Telecommunications Career Agenda 643

18   Telecommunications Management and the Organization 644

19   Telecommunications and Your Career 671

Preface

We wrote *Networks in Action* to fill a major and well-known void in colleges: the lack of a book that brings together the technical and business aspects of telecommunications with a primary business and organizational focus and yet does not neglect the often very complex details of the ever- and fast-changing technology. Previously, there have been a number of first-rate books on the nuts and bolts of the technology and some on the exciting business opportunities of telecommunications, but very few that fuse the two. That is what we set out to do.

We chose our title carefully: *Networks in Action* is remorselessly practical. We want to help our three intended audiences learn how to mesh business opportunities with technical choices:

▸ *To the business student,* we offer a sound, comprehensive, detailed, but nonintimidating explanation of the technology that is literally reshaping the nature of business and organization.

▸ *To the MIS student* we provide a needed solid grounding in telecommunications, a topic that is generally covered only peripherally in MIS textbooks, which are centered on computers, not telecommunications.

▸ *To the network professional and student,* we offer practical planning and decision tools plus rich examples of telecommunications in action that can help them not only extend their technical knowledge (in a field where it is notoriously hard to keep up to date) but also extend their understanding of how to relate what they know to broader business trends and opportunities.

Decision Sequences and Dialogues

We have built *Networks in Action* around a four-step telecommunications decision framework. The four-step sequence is the product of our work as researchers, educators, and consultants; and each of these three areas of our experience was equally influential in our development of these frameworks. The four decision steps are: (1) choose the business opportunity, (2) define the business criteria for technical decisions, (3) design the telecommunications resource, and (4) make the economic case. This decision sequence involves three parties: the business client, the telecommunications designer, and the technical implementer. Fundamentally, *Networks in Action* is about their dialogue, with each step in the decision sequence providing them a shared blueprint and language.

Throughout the book, we highlight the role of the client, the designer, and the implementer, with the designer our primary focus. The two main dialogues are the business-centered one between client and designer, which addresses business opportunities and their translation into criteria for technical design; and the client–implementer technology-centered dialogue, which encompasses design, operations, and management. In this way, we bring together naturally the topics that are so often kept entirely separate in textbooks: business planning, technical design, and delivery. This is what, we believe, students need for effective careers in a world increasingly dependent on telecommunications in core elements of service and coordination. Certainly, what we see senior executives and information services managers searching for is the new generation that can translate comfortably between business and technology.

Theory into Practice: The Minicases

Each framework in the decision sequence draws on the best of current research and theory and has been applied in a wide range of companies (some of which used the frameworks as the entire base for developing their telecommunications strategies). Our frameworks are proven in concept and in practice. That is important to us; we want *Networks in Action* to help students build their careers, and to do this we must offer them tools that are grounded in the best of theory and proven in the best of practice.

The frameworks have also been derived from literally hundreds of case studies. Because we believe very strongly that students need—and enjoy—practical illustrations that help them make sense of technical concepts, we do not use "toy" problems or examples but draw on these experiences from companies around the world. Two entire chapters of *Networks in Action*, Chapters 9 and 13, are cases; each of the other seventeen chapters includes four minicases. The book thus contains 74 cases, each with assignments and questions for discussion.

These cases cover the world: one of our main priorities from the start was to provide a global perspective on a technology that is one of the key driving forces of global communication and coordination. The cases thus cover Mexico's and Hungary's privatization of telecommunications; Hong Kong and Singapore as exemplars in using telecommunications as an explicit weapon for economic development; the international networks of Texas Instruments, Analog Devices, Volkswagen, and Digital Equipment; examples from companies in every continent and most major industries, ranging from cases on airlines such as United Airlines and British Airways; banks such as the Royal Bank of Canada, Bank of the West in California, and the Industrial Bank of Japan; car manufacturers including Chrysler; insurance companies like Progressive and Travelers; transportation firms like Conrail; telecommunications equipment and service providers like PictureTel; and public sector organizations, including the IRS, the State of California, and Maryland's welfare agency.

We chose the cases to focus on specific business or technical issues and in particular to show how companies have implemented leading-edge technology developments not covered in many books. Each case highlights a technology: very small aperture satellite terminals, image processing, frame relay, SONET, local area–wide area network interconnection, data compression, videoconferencing, and many others. Although we wanted our book to have a strong business focus, we also made sure it addresses the technology in detail and depth.

Integrating Business and Technical Components

A central underlying theme in how we explore the technology is, we believe, uncommon in textbooks but vital for business implementation: integration of network and computing components across the entire enterprise. Many books focus on the technical components of a firm's telecommunications resource, but far less on how they fit together. Our analysis focuses on their role as building blocks of the integrated business resource: (1) the terminals that access services, (2) the transmission links that connect the terminal to the service, (3) the transmission methods that turn the electronic signals into meaningful messages, (4) the nodes and switches that manage the flow of communications traffic, and (5) the architecture and standards that provide the blueprint for all the other building blocks to work together both efficiently and effectively.

One of our other main priorities in writing *Networks in Action* has been to make the technology itself concrete and relevant to business decisions rather than abstract and conceptual. We thus focus on how technical standards that are defined by committees arrive in the marketplace as proven products and how technological innovations move from concept to practice. We devote an entire chapter to the domestic and international telecommunications industry and also provide an appendix that describes specific telecommunications products available in the market as of mid-1993. We strongly believe that too few books stress the role of the industry as a key force in innovation. Out in the "real world," telecommunications designers and implementers are wrestling with the challenges of how to assess vendors, when to move with the leading edge and yet avoid being on the "bleeding edge," and how to make sense of the many and often conflicting claims about technical innovations. We hope that students will get a good sense of the industry in action and learn how to view the technical concepts of telecommunications innovation realistically and imaginatively, so that they can contribute in their careers to managing telecommunications, not just knowing about it.

The central focus of *Networks in Action* is not telecommunications but the management of telecommunications as a business resource. We include chapters on managing network operations, making the economic case for investing in telecommunications, and organizing the telecommunications resource. We do not know of any other textbook that explicitly views telecommunications in the multiple perspectives of business opportunity, technology, industry, organization, and economics. Yet they all must be addressed for telecommunications to be made as effective a business resource as finance, product development, or human resource management.

The Structure of *Networks in Action*

Networks in Action consists of five parts, which alternate between business opportunity and technical choices.

Part I, *Telecommunications, the Opportunity and Challenge,* presents the overall decision sequence. It includes a succinct but comprehensive summary of the technical network building blocks. Its aim is to give the student a solid understanding of the whole before looking at the parts. We have found in our teaching that although many textbooks facilitate understanding of transmission links or LAN protocols, most leave readers struggling to grasp these elements in the context of the overall enterprise network resources.

Part II, *Network Building Blocks,* contains the five chapters that describe the technology of telecommunications—in action. We do not address the theoretical foundation of telecommunications, because that is the specialized domain of the telecommunications engineer. We emphasize how the technical elements work together, through choice of standards, architecture, and tools for interoperability. Together, Parts I and II provide students with a thorough grounding in the business and the technology of most impact on organizations' efficiency and effectiveness.

Part III, *Linking Business Choices and Technical Decisions,* examines in depth the business side of telecommunications, the proven high-payoff opportunities: the three dimensions of business functionality—reach, range, and responsiveness—of the telecommunications service platform; and making the economic case for investing in telecommunications infrastructures. It ends with a full-chapter case showing how a large European firm applied the decision sequence to evolve an entirely new telecommunications capability at the core of its business strategy.

Part IV, *Managing the Telecommunications Resource,* addresses the telecommunications industry and how to assess trends and suppliers, network design, total quality network operations, and management of costs. The framework for design focuses on the trade-offs among capability, reliability, flexibility, and cost that must take into account changing business and changing technology. The chapter on the industry covers the domestic long-distance and local marketplace, international service and equipment providers. The chapter on network operations applies principles of total quality management. The chapter on costs highlights the practical concerns of telecommunications managers and their clients.

Part V, *The Telecommunications Career Agenda,* looks forward to the organizational issues and skills students need to understand and address to be effective in their own careers. It describes the evolution of telecommunications in the organization and the main challenges telecommunications managers now face, especially in supporting local area network users and building new technical expertise. It ends with an analysis of four career "quadrants" that require differing mixes of technical and business skills.

Each chapter has cases. Parts II and III each end with a chapter that is entirely cases (Part II) or a single case (Part III). Within the chapters, new terms are highlighted in bold and included in the Glossary. The text also includes

- ▶ Summary and Review Questions to ensure understanding of the material in the chapter;
- ▶ Assignments and questions for class discussion that include open-ended questions that ask for opinions;
- ▶ A complete Glossary;
- ▶ A comprehensive Bibliography, plus selected recommended references; and
- ▶ A unique Product Summary Appendix, which briefly describes products available on the market as of mid-1993, including prices and features of each. The aim here is to give students a concrete sense of the products that embody the more abstract principles and theory of telecommunications.

| | | | | | | | ## About the Authors

The authors met when both were teaching at Stanford University. Our backgrounds are both complementary and different. Mike Cummins is an economist, who worked for the Federal Communications Commission and INTELSAT. He spearheaded the University of Miami's innovative MBA program in telecommunications management and worked closely with Motorola on developing education programs built on the principles of total quality management. Peter Keen has moved between the worlds of academia and business, acting as a translator; wrote one of the earliest books on the linkage between business and telecommunications, *Competing in Time: Using Telecommunica-*

tions for Business Advantage; and has constantly spanned in his writing, teaching, and consulting the worlds of technology, business, and human resources. We hope and believe that we have brought all of our disparate experiences together in a book that provides an integrated perspective on the integration of business opportunities and technical choices.

Acknowledgments

We would like to thank the following colleagues for their valuable input during the development of this project: Dennis Adams, University of Houston; Lynda Armbruster, Rancho Santiago Community College; Carl Clavadetscher, California State Polytechnic University; George Fowler, Texas A&M University; Bezalel Gavish, Vanderbilt University; Joan E. Hoopes, Marist College; Thomas Housel, University of Southern California, Los Angeles; Kenneth L. Marr, Hofstra University; G. Premkumar, Iowa State University; Satya Prakash Saraswat, Bentley College; Robert A. Schultheis, Southern Illinois University at Edwardsville; Harold T. Smith, Brigham Young University; Gholamreza Torkzadeh, University of Texas at El Paso; David Van Over, University of Idaho; and Carol E. Young, Georgia State University.

A note from Michael Cummins

In any undertaking of this magnitude, the strong support of family and friends is absolutely essential for the strength to persevere. To my wife, Ann, a very special thanks for her steadfast encouragement and loving support throughout the long and arduous process. To my daughter, Amy, thanks for her patient understanding in putting up with my frequent absences and lack of attention so that I could work on the book. My son, Scott, was mercifully spared by being away at college. There were numerous times over the more than three years of writing this book when I needed special encouragement just to keep going and not abandon the project altogether. My mother was there during those dark hours, as she always is, with unwavering faith that her son could and would succeed. I also want to thank my mother- and father-in-law, Mary Marshall and Harrison Vaughan, and my special friends Tom and Sherry Barrat, Eric and Susy Korth, and Tom and Grace Will for their continued encouragement, interest, and support. In the end, even the destruction of our home by Hurricane Andrew did not halt the writing of this book. My only regret is that my father is not alive to see the fruits of my labor. He would have been proud.

J. Michael Cummins

A note from Peter Keen

This is my seventh book, so that my family and friends already know what's involved in the lengthy, complex, and generally neurotic process by which A Great Idea ends up as something you can hold and read, albeit a year or so later than scheduled. As always, my wife, Lynda, has protected my time, contributed ideas, and worked with me on the practical business projects that, I hope and believe, give this book its balance of ideas and examples and its real-world richness. A million thanks to her is a million too few.

My long cooperative association with MCI Communications has helped me keep up to date with the fast-moving technology, brought me into contact with senior telecommunications managers across the world, and funded much of my research. I've tried to make sure this book is not influenced by my relationship with and liking for MCI, but this is a great company. I have benefited in so many ways from working with it, beginning from my friendship with Bill McGowan, whose death took away the person who created or at least guaranteed the tele-communications revolution. Bill surely will be ranked among the dozen or so greatest business persons of the past fifty years; he was fun, too, and brilliant. I owe special thanks to John Zimmerman of MCI, who has for years been the coordinator of the generous research support MCI has provided me.

Peter G.W. Keen

From both authors

At Wadsworth, rather than give thanks, we should probably offer medals: to Alan Venable, for patience in the face of procrastination and muddled syntax; to Carol Carreon for humor in the face of procrastination and missing artwork, and to Rhonda Gray, for not screaming in the face of telephone tag. The biggest thanks we have to offer, though, are to Frank Ruggirello, who generated the series of which this book is part and played a constant role in supporting, challenging, stimulating, and improving it.

To our families

One

Telecommunications as a Business Resource

1

Telecommunications: The Opportunity and the Challenge

Objectives

- See the technology of telecommunications in its wider business context
- Learn about the nature of an effective dialogue between business and technical experts
- Briefly analyze several short real-world minicases that demonstrate the impact of telecommunications on competition and innovation

Chapter Overview

How Telecommunications Is Reshaping Business

The Explosive Pace of Technological Change

The Need for New Management Processes

Innovation and Globalization

Networks: The Core Technology

Minicase 1-1 THEi: Reinventing a Company

Minicase 1-2 Shearson Lehman: Streamlining Workflows

Minicase 1-3 California State Government: Telecommuting

Minicase 1-4 "Telestroika": Hungary's National Telecommunications Policy

| | | | | | |

Chapter Overview

This introductory chapter gives you an overview of both the business opportunities that telecommunications has created for firms and the main components of the technology itself. Thus, rather than beginning with "fundamentals" of the technology, we focus instead on the wider *business* context of the technology. This approach reflects our aim of helping you balance the business issues that drive telecommunications and the technical ones that are crucial to turning the choices into action.

The chapter is nontechnical, though it introduces and defines some core technical concepts, particularly ones that address the issue that is driving modern telecommunications: the efforts to ensure that the many separate types of computer and telecommunications resources in use today can be interlinked instead of remaining separate services requiring separate equipment and delivery infrastructures.

The chapter also highlights the need for and nature of the dialogue among client, designer, and implementer that is the organizing focus of *Networks in Action*. The client is the manager or group that has a business need or sees an opportunity that can be met through telecommunications. The designer translates the opportunity into the technical blueprint that the implementer uses to build and operate the telecommunications capability. The chapter also discusses the business opportunity of telecommunications in domestic industries and in developed and developing countries alike.

| | | | | | |

How Telecommunications Is Reshaping Business

Telecommunications is the electronic transmission of information, whether it be voice phone calls, computer data, photographs, music, documents, or television broadcasts. **Transmission** is the movement of signals along a communications link and is the core of telecommunications; it provides a connection between sender and receiver and manages the routing of the communication. The telephone system, for example, transmits your voice along a complex set of lines and exchanges. An automated teller machine transmits to your bank's computer your request to withdraw cash from your checking account. CBS transmits its programs to your television antenna.

As a business resource, telecommunications is already vital to the efficiency and effectiveness of public and private organizations and can only become more so in the coming decade. By making it practical to transmit information at high speeds, telecommunications has transformed the basics of entire industries. Obvious everyday examples are the automated teller machines (ATMs) that provide an electronic bank branch that stays open nearly 24 hours a day instead of a bricks and mortar one with much shorter hours, and the airline reservation

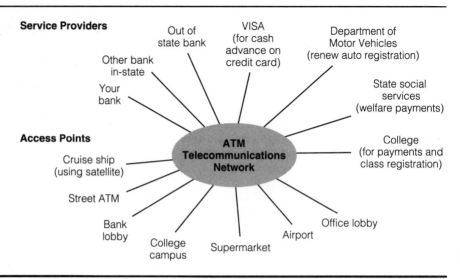

Figure 1–1

**The Power of
Telecommunications:
Bank Automated
Teller Machines**

The telecommunications network for a multibank national ATM network (such as Cirrus, Monec, or Most) allows you to use your bank's ATM card at any location that links to the network. The telecommunications network is depicted as a cloud to indicate that it is something a user neither sees nor needs to know about; you merely insert your card and enter some information to get cash.

The diagram shows both some services that are available and some typical points of access to the network. Note that access need not occur from a bank and that services are not restricted to authorizing a withdrawal, making a deposit, or finding out your balance. The functions this telecommunications network provides are essential to users in that when the ATM network is "down," so is the bank.

systems that allow travel agents to book flights on hundreds of airlines in **real-time**. Real-time means that as soon as the booking is entered by hitting a key on a **terminal** that accesses the system, the relevant computer records, which may be thousands of miles away, are immediately updated. The word *terminal* was originally used for the devices that are the end points of the telecommunications service; today, personal computers, phones, bar code readers, and television sets are all forms of terminals. Figure 1-1 shows a very simplified diagram of an ATM network shared by many banks that allows you to use, say, your First American Bank of Transylvania bank card at an ATM in Boston, Little Rock, or Los Angeles. The telecommunications network handles the transmission of the transaction and the routing of it to your own bank's computer

ATMs are so well-established that we take them for granted and can easily overlook how they transformed the nature of banking in less than ten years.[1] A

---

[1] In the late 1970s, a number of studies showed that ATMs were not cost-effective on the basis of known volumes of transactions. By 1983, for a bank not to have ATMs was like its not issuing checkbooks. Gary Hector (1988) describes the different strategies used by Citibank and Bank of

less common but almost surely as far-reaching example of the transformation of an industry through telecommunications is the electronic claims processing systems that reduce the paper and time involved in handling five billion medical insurance claims a year in the United States, costing $8 billion to administer. These systems cut insurers' costs by 50 percent or more and speed up payment by weeks; it costs around $5 to process a paper claim but just 3–10 cents if the claim is sent directly to a computer that processes it and makes the payment electronically.[2] It costs 29 cents just for the stamp on a mailed document. Industry experts have good empirical evidence to suggest that the savings from electronic claims processing will be $10 billion a year by 1995.[3]

In July 1992, 15 health care providers announced a phased plan that will streamline the entire claims and payment process by 1996. All the technology is already in use and has been proven by leading health care insurers and hospitals. The plan is a response to a challenge by the head of the federal Department of Health and Human Services to ensure industrywide use of electronic claims processing and cut administrative costs, which amount to one-quarter of the $800 billion spent on health care in the United States each year. Figure 1-2 shows an electronic claims processing system that links doctors' offices to insurers, eliminating paper and delays.

EBS (electronic benefits system) may be as common an acronym in 1995 as ATM is today. It is also increasingly difficult to identify any industry whose *basics* are likely to remain unaffected by telecommunications.

| | | | | | | |

The Explosive Pace of Technological Change

The rate of change and its effects in the technology of telecommunications is explosive. Four factors, which occur individually and/or in combination, pace the use of telecommunications as a business and social resource: (1) competition, (2) speed of transmission, (3) access tools, and (4) interoperability.

America in 1978; "Citicorp was investing—recklessly it appeared—in a scheme to put two automated teller machines in each of its retail branches. . . . The common wisdom in the industry was that Citicorp was betting big on a system that would never pay off. Bank of America's studies concluded that [ATMs] would never pay for themselves" (p. 97). In 1981, "worried that Bank of America was too far behind, [its new CEO] ordered the bank to install more than one machine a day" (p. 147).

[2] These figures come from the federal Health Care Financing Administration, which operates Medicare. HCFA also estimates that it costs $47 to replace a check that gets lost in the mail.

[3] One executive who has played a leading role in mobilizing the health care industry for electronic claims processing, Joseph Brophy of Travelers Insurance, is quoted as saying, "I truly believe we can have an automated claims processing environment within two years, and I believe that system will save $10 billion in paper claims administration annually." (*New York Times*, 1992) Most of the figures on the actual benefits hospitals have already received, as reported in the trade press, suggest that his estimates may be conservative.

Figure 1–2

Electronic Insurance Claims Processing Network

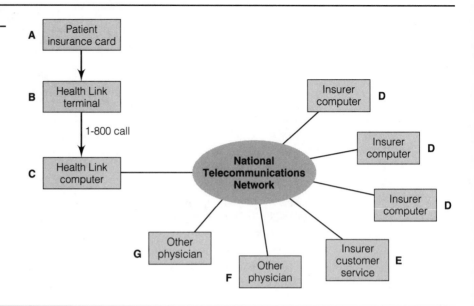

Health Link was introduced in 1991 by a Princeton, New Jersey firm that offers thousands of doctors, insurers, and health providers a simple service that reduces paper, complexity, and time. *Computerworld* summarizes it as making "the process of filing insurance claims as paperless and speedy as using a credit card to buy a new suit and tie."

A doctor's office staff uses a small and unobtrusive computer terminal, almost identical to the device used to scan a credit card for authorization in a store. The card here is the patient's insurance card (A in the diagram). The terminal (B) dials a 1-800 number that connects to a national telecommunications network to link to the Health Link computer (C). This processes the claim request, identifying any special information needed for the relevant insurer and formatting the data so that it can be directly processed through the network by the insurer's computer system (D). The insurer's system confirms coverage and immediately authorizes payment after the doctor has finished. If the patient is not covered for the treatment, the Health Link system automatically dials the insurer's customer service unit (E) for help. The physician may also refer the patient to other physicians on the patient's health plan (F, G).

Health Link has cut claims processing costs for its users by about half and processing time by two-thirds.

Competition

Up to the 1980s, telecommunications historically was treated as a "natural monopoly," part of a nation's social infrastructure. A **PTT**—Poste Telegraphique et Telephonique, the generic term for the government-controlled agency that operates a nation's telecommunications monopoly—operated the telephone system, with limited licensing of other service providers in such areas as telex. The United States has never had a PTT, but AT&T was the equivalent; its chairman offered in the 1920s that it be subject to federal and state regulation in return for its being allowed to operate as a monopoly. When the United States first deregulated the long-distance phone industry in 1982, and the United Kingdom liberalized it (allowed partial competition) a few years later, they stood alone. In most countries, companies had no choice of equipment, services, or prices beyond what was offered by the PTT. The **divestiture** of AT&T in 1984, the outcome of an antitrust suit, created competition in long distance and split up AT&T's local phone services into seven regional phone monopolies.

Now, monopolies are the exception, not the norm. Sweden, the Netherlands, New Zealand, Spain, Mexico, Germany, and many other nations are recognizing the importance of low-cost, high-quality telecommunications in a world of global operations and competition. The result has been a drop in prices of, typically, 30 percent and a flood of innovation in products and suppliers. For example, a U.S. leader in the use of one of the most advanced high-speed telecommunications transmission methods is a firm called WilTel, a subsidiary of an oil company that was set up to dispose of unused oil pipelines. Instead, it installed **fiber-optic** cables in them and became a long-distance telecommunications provider.[4] Fiber optics send light signals along glass cables, thinner than a human hair, via laser beams that pulse signals at sub-sub-subsecond speeds. This permits high communications transmission rates and the corresponding ability to move volumes of information cheaply and rapidly, both of which were unimaginable ten years ago. WilTel preempted the traditional providers of long-distance computer-to-computer communications, such as AT&T, MCI, and Sprint, by exploiting its ownership of the pipelines and rights of way, fiber-optics technology, and new transmission methodology.

Competition in the long-distance telecommunications market has cut the average price of a phone call by over 40 percent in under ten years. Lack of competition in the local market, which is three times bigger, has resulted in just a 10 percent cut. The competition between AT&T, MCI, and Sprint, the three major long-distance providers, has created a flood of discounts, new products such as personal 1-800 toll-free numbers, and aggressive technical innovation. New technology creates new competition. In 1991, for instance, more cellular phones were put into service than regular residential phone lines: 2.5 million versus 1.9 million.

Competition has at last come to the local phone companies. Cable television firms can offer phone call transmission, and phone companies can offer entertainment.[5] A new race is on to capture customers and preempt the

---

[4]The impact of fiber optics on telecommunications can be illustrated by the change in capacity of the transatlantic submarine cables (prefixed by "TAT") laid down for international telephone calls. TAT-1 was made operational in 1958 and could handle 72 calls simultaneously. TAT-5 (1970) could carry 2,112 calls, and TAT-7 (1983) 10,500 calls. These cables were coaxial; subsequent cables used fiber optics. Thus TAT-8 (1988) carried 40,000 calls and TAT-9 (1991) 80,000.

[5]Time Warner launched an experiment in December 1991 to deliver entertainment to thousands of homes in Queens, New York, through its fiber-optics network. It can deliver 54 channels of pay-per-view plus 96 channels of regular programming, with capacity left over for other services, including phone calls. The local phone companies have the capacity to add video through their existing networks. Bell Atlantic's 1991 Annual Report states: "We were particularly pleased by the

marketplace. Mobile communications, cable TV, fiber into the home, satellites, personal communication networks, and electronic publishing are a little like Hollywood in the 1920s; It is difficult to see who will win and where the marketplace will move. But the movement will be at least as fast in terms of technical innovation, products, and cost cutting as that in the long-distance telephone market between 1984 and 1993.

Speed of Transmission

The core of telecommunications is how quickly and cheaply information can be moved. A famous epigram by science fiction writer Arthur Clarke captures the progress of telecommunications in this area; he said that in any society there is a point at which technology is indistinguishable from magic. Much of what is happening in transmission in the 1990s is magic in this sense; almost nothing you can predict is implausible. The telecommunications planners of the 1970s hoped to achieve by the end of this century a transmission speed that has already been exceeded by a factor of almost 40,000 because of new telecommunications transmission media. Transmission media are the various types of cables and radio, and light waves that carry the transmission signal. Originally, telecommunications meant the movement of electrical signals along copper wires; now it also means the movement of light signals and radio signals. Fiber-optic cables occupy the same space as copper cables but carry information up to 100,000 times faster. Radio waves do not need any wires but move information via satellites 22,300 miles above the earth or through line-of-sight microwave relays. (Line of sight means that no physical blockage stands between sender and receiver.)

The same information may be sent over a variety of transmission media, all of which are equivalent in terms of the basic techniques of telecommunications. A phone call may travel by cable, or radio waves (with or without a satellite). Figure 1-3 shows an unusual mode of transmission: radio signals beamed into the sky until they bounce off the trail of ion dust from a meteor at a height of 52–73 miles. Tin cans and string, fiber optics, satellites, meteors, and even barbed-wire fences can all be used to carry a transmission signal.

The main advantage of fiber optics is massive speed at low cost, but it requires installing new cables or replacing existing ones. The advantage of wireless communications is that it provides mobility; cellular phones can be used in cars, for example. The main impediments to deploying wireless communications are the high cost (over $100 million) of launching satellites, many unresolved issues of regulation and allocation of radio frequencies by the Federal Communications Commission (FCC),[6] and the greater level of interference from noise, weather, and other sources.

[U.S.] Court of Appeals' ruling that permits us to enter the information services market [and] the FCC's proposal to allow telephone companies to offer video dial-tone services." (pp. 4–5)

[6] Allocating space on the airwaves and determining how licenses will be awarded is both a contentious political issue and one with immense economic impacts. The process is also very slow. Some economists estimate that the 11-year delay in the FCC's approval of cellular communications cost U.S. vendors $86 billion in revenues.

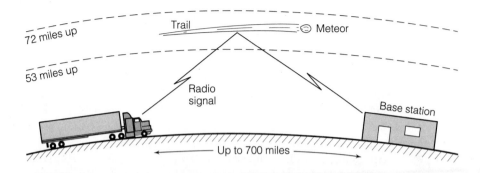

Figure 1–3

Tracking Trucks by Bouncing Radio Waves Off Meteor Trails

Transtrack Inc. helps transportation companies reduce the cost of locating their vehicles. A radio signal is constantly broadcast from each of five company bases into the sky. The signal bounces off meteor dust trails in the atmosphere and returns to earth to be received by the target truck. The truck's unit picks up the message and sends back a signal that indicates its location. On average, it takes ten minutes of signaling for the base station to find a suitable dust trail. As quoted in *The Economist* (1989), "To human eyes, a meteor is beautiful. To a radio wave, it's just another thing to bounce off."

The main advantage of the copper phone lines in use everywhere today is simply that they are in place. Replacing them with fiber optics is mainly a question of costs, and it will be at least three decades before every house has fiber. Moreover, the early 1990s has seen major innovations in transmission over existing phone lines. Your local phone company is racing to head off cable television providers and others to prevent them from capturing their core residential and business market; they are offering high-speed transmission that can even provide pay-per-view video with no replacement of the installed wires.[7] One of the primary targets of the cable companies is large businesses and long-distance carriers that must currently route their calls via the local phone company; close to 50 cents of every dollar of revenue MCI gets from long distance is paid to the local exchange carrier as an access charge.[8] This all adds up to guaranteed innovation and competition because fiber optics, satellite communication, and high-speed transmission through copper wire cut costs, make wider uses of telecommunications cost-effective, and create new uses. Consider, for example, the transmission of a document. Before long-distance competition cut phone rates, faxing a document was very expensive and took several minutes. Now, it takes just seconds along a standard phone line. However, it is still impractical to send a full-color photograph by fax. Fiber optics make it practical, as does

[7]The FCC, which governs U.S. interstate and international telecommunications regulation, approved local telephone companies' practice of offering video dial tone early in 1992. Bell Atlantic announced its entry into the home-video-by-phone market in October 1992.

[8]In early 1992, the largest cable operator in the United States, Tele-Communications, Inc., bought 49 percent of Telport Communications Group, which has built a $100 million business by providing companies ways to services that bypass the local phone company.

satellite communication, but this capability is confined to use by businesses that lease the necessary high-speed transmission services. However, we are already seeing local phone companies offering this capability for an installation cost of under $500 and for only pennies per minute.

The National Research and Education Network, approved by President Bush in 1992, will be able to transmit in a single second 100,000 typed pages, or 1,000 high-resolution photographs, or 10,000 three-dimensional X-rays. In 1983, sending a database of a thousand figures, such as a spreadsheet or price list, took from five seconds (a special leased line) to a minute (over "POTS"— plain old telephone system). Today, it takes under a second over a leased data communications link and around 10 seconds on a standard "dial-up" phone line, at a cost that is at least a tenth of that just a decade ago. T1, the standard transmission link used by telephone companies and large businesses, can transmit *Networks in Action* in about five seconds. No end is in sight for continued acceleration of the pace of technical change in every area of telecommunications transmission.

Moving signals along a transmission link is one of the two basic elements of transmission. The other—routing the signal from originator to receiver—is needed because it is economically senseless to even think about providing a direct connection from every sending device to every receiving one. Instead, signals move through the network through a series of switches. As the term suggests, a switch is a device that receives a message and directs it to the next transmission link.

In the telecommunications industry, the national phone system is termed the "public switched network." Public means open to any subscriber, and switched means calls are routed from sender to receiver via equipment at nodes in the network. This process is depicted in Figure 1-4, which shows in highly simplified form how a phone call from Long Island to Los Angeles moves through AT&T's long-distance network.

Historically, the electrical signals transmitted along the phone line were processed by switches that were **electromechanical devices,** which have moving parts driven by electricity. Your phone call connected you to the local exchange, where a device responded to the signals and mechanically moved connectors to establish the circuit to complete the call. Now, more elements in the switching process are handled by computers, which operate at speeds measured in microseconds (millionths of a second), nanoseconds (billionths), and even picoseconds (trillionths). As a result, switches have become faster and cheaper, enabling them to handle the faster transmission media of fiber optics and wireless communications. Together transmission links and switching technology form the core of modern telecommunications—and its cost base as well. The investment in the public long-distance phone network is estimated at over $250 billion. Japan has calculated that it will cost at least $215 billion to upgrade its public system to fiber optics, a process that will take over 25 years.

Access Tools   Much of the innovation in the use of telecommunications stems from an increase in the places from and tools by which people can access services. Until

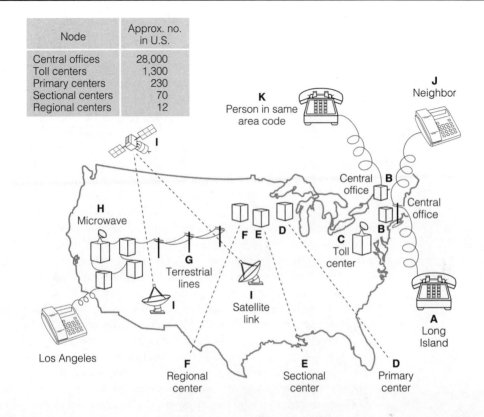

Node	Approx. no. in U.S.
Central offices	28,000
Toll centers	1,300
Primary centers	230
Sectional centers	70
Regional centers	12

Figure 1–4

The National Phone System

In a call via AT&T from Long Island to Los Angeles, the telephone (A) is connected by copper wires to the first node, the local phone company's central office (B), which switches the call to a toll center (C). From there it is switched to a primary center (D), a sectional center (E), and a regional center (F), either along terrestrial cable trunk lines (G), microwave radio (H), or wireless satellite link (I). The centers concentrate traffic onto high-speed communication links, with the switches selecting the route on the basis of cost and congestion. A local phone call will be switched directly from the central office if the subscriber is connected to it—to a neighbor of the caller, for instance (J), or to another central office if the person called is within the same area code (K). Other switches and transmission links handle international traffic; these are not shown.

the 1970s, the tools were limited to telephones, telex machines, radios, and very expensive and slow facsimile machines; now they include personal computers, mobile phones, paging devices, portable fax machines, bar code readers, cash registers, and even interactive television. POTS now offers push-button **voice messaging**, in which you select from a menu of options by pressing, say, a "1" or "8" in response to recorded voice instructions; a computer receives the instructions and responds to your selection. Personal computers can dial up over the phone line a host of information services. ATMs can be used to collect welfare payments or renew car registration and are even on aircraft carriers. Telecommunications is fundamentally about ease of access and convenience. Its major impact on society is that it removes time and location as barriers to service and coordination.

Interoperability

The more services and facilities to which an access tool can link, the more valuable and convenient it will be. The telephone is valuable because it is easy to use and can access other states and countries. Imagine what it would be like if you had to use twelve different types of phone—one for connecting with AT&T customers, one for MCI, one for Sprint, one for calling the UK, and so on. What if your AT&T phone could call rotary phones only but not push-button ones?

Historically, the field of information technology—which comprises access tools, computers, telecommunications, and information stores—was dominated by exactly that level of incompatibility. Incompatibility means that this device cannot directly link to that device, this computer cannot use that software, or this personal computer cannot use that telecommunications facility. In 1981, the New York corporate treasurer's office of a leading multinational firm had 17 different computer terminals in daily use. One was for Citibank's electronic cash management system, one was for Chase's, another was for Midland Bank (of the United Kingdom), another was for Reuters foreign exchange quotes, still another was for foreign exchange trading, and so on. Even today, most such companies have a mass of incompatible computers and corresponding multiple and incompatible telecommunications systems.

Ending incompatibility is the major goal in the information technology field: providers, users, and the many consortia and committees are trying to work together to reduce incompatibility and increase what is often termed **interoperability**, the working together of the elements of a telecommunications resource. The wider term **integration** refers to the working together of all the elements of an information technology resource: hardware, software, and telecommunications.

The move toward interoperability is the fourth key force driving the explosive technological pace of innovation in telecommunications. The starting point was to interconnect countries' phone systems, each of which had different voltages, signals, and equipment. Now, you can routinely dial just about any country, and switching equipment handles the needed conversions. The next step was to enable communication to, from, and between computers. Here, every major vendor developed its own telecommunications procedures and techniques, greatly limiting interconnection.

The situation in telecommunications in 1982, the pivotal date when the divestiture of AT&T was announced, was thus one in which the technology base for voice communications—spoken messages sent from and to telephones—was very separate from data communications, the transmission and processing of computer-generated information. One of the goals of industry leaders was voice/ data integration—the ability to handle either type of traffic through the same telecommunications facilities. The business community of telecommunications users was more interested in **open systems**, ones that were not vendor-dependent, so that they could interconnect the many different types of computer and communications facilities they operated. Open systems are contrasted with **proprietary systems**, which require vendor-specific products.

The move toward integration of just about everything is now the driver of information technology. Integration poses massive technical challenges, and many vested interests are slowing it. That said, every year sees more and more progress in interoperability. This allows a firm like Wal-Mart to integrate its entire supply chain. Its stores' point-of-sale registers can link by satellite to its head office computers, which in turn link directly to suppliers' computers, which link to shippers, banks, insurers, and other parties.

Change as the Norm in Technology

These four forces are all moving rapidly. Competition ensures innovation in products and decrease in prices. Developments in transmission stimulate competition, drastically cut prices, and increase the type of information that can be transmitted and hence the range of practical uses of telecommunications. For instance, 1992 saw widespread pilot projects using teleradiology—the transmission of high-resolution images of X-rays from small hospitals to medical centers, where experts interpret them and review them over a telecommunications link with the relevant physician. This process was completely impractical as late as 1990; transmission speeds, let alone costs, blocked the opportunity.

Access tools open up new uses of telecommunications, mainly because it can cost only a little to add a new service to an existing base. ATMs are becoming the access point for registering at college or, handling state benefit payments. Telephones contact computers now, not just other phones. Finally, interoperability of information technology systems allows interconnection of business processes and eliminates many procedures, documents, locations, and people that can slow service and reduce coordination. That is why a few banks can now process a mortgage application over the phone in 15 minutes instead of 30–60 days, why Federal Express can tell a customer within half an hour where a package is along its complex delivery chain, and why **electronic data interchange (EDI)** has cut the number of staff in purchasing departments from hundreds to around 20. EDI is the transmission of electronic documents in a form that allows another firm's computer to convert the sender's message format (for, say, a purchase order) into its own formats.[9] Figure 1-5 shows a

---

[9] Here are a few other examples of EDI's benefits: Service Merchandising Corporation reduced its cost per order from $50 to $12, saving $7.5 million a year; RJ Reynolds saved $5–10 million a year in labor, staff, and lead times; Westinghouse and Portland General reduced the elapsed time for an order from 15 days to one-half day and its cost from $90 to $10; and Design Inc. cut its replenishment cycle from 14 days to three days by using Levi-Strauss's Levilink system. Levi-Strauss estimates that before the introduction of Levilink, 70 percent of its business data was manually keyed into a computer; those same data were subsequently manually reentered into at least one other computer. (Data from Keen, 1991)

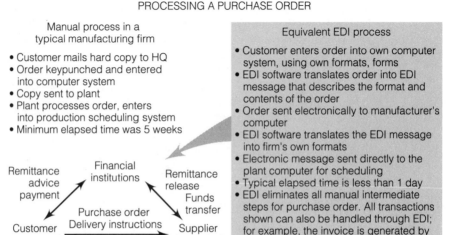

PROCESSING A PURCHASE ORDER

Manual process in a
typical manufacturing firm

- Customer mails hard copy to HQ
- Order keypunched and entered
  into computer system
- Copy sent to plant
- Plant processes order, enters
  into production scheduling system
- Minimum elapsed time was 5 weeks

Equivalent EDI process

- Customer enters order into own computer
  system, using own formats, forms
- EDI software translates order into EDI
  message that describes the format and
  contents of the order
- Order sent electronically to manufacturer's
  computer
- EDI software translates the EDI message
  into firm's own formats
- Electronic message sent directly to the
  plant computer for scheduling
- Typical elapsed time is less than 1 day
- EDI eliminates all manual intermediate
  steps for purchase order. All transactions
  shown can also be handled through EDI;
  for example, the invoice is generated by
  the manufacturer's computer and sent
  immediately and electronically to the
  customer's computer, which processes
  the payment electronically and sends it
  via EDI to its bank.

Remittance
advice
payment

Financial
institutions

Remittance
release
Funds
transfer

Purchase order
Delivery instructions

Customer
computer/
workstation

Supplier
computer/
workstation

Invoice
Dispatch advice
Delivery schedule

Shipping
notice

Bill of
lading

Transportation
carrier

Figure 1–5

**Electronic Data
Interchange**

manufacturing firm's processes before and after EDI and indicates the degree
of streamlining of business processes that telecommunications makes practical.

| | | | | | | | # The Need for New Management Processes

The mixture of technical innovation, competition, and variety of applications
makes telecommunications (1) an important business resource, (2) a complex,
specialized, and volatile technology, and (3) a field that is highly interdependent
with the computers and information systems that process business transactions
and access and manage the relevant databases. An ongoing problem organizations report is that few professionals or managers are comfortable in all three of
these areas. Business experts do not understand a technology whose jargon
makes computer terminology look simple. Telecommunications specialists are
often unfamiliar with business needs and opportunities. Neither they nor their
peers in information systems can be expected to keep up with each others' areas

of expertise, and each is often ignorant of terms and key developments in the other's field. Organizations everywhere are talking about the need for people who can build a new dialogue that can bring these people together.[10]

Networks in Action is about that dialogue and how to build it. It focuses on two main issues:

▸   *Business decisions:* the practical business and organizational opportunities of telecommunications

▸   *Technical choices* of building blocks needed to create and operate the relevant telecommunications resource

The book is built around a telecommunications business decision sequence that has four steps, each of which applies one of four core decision frameworks developed for *Networks in Action:*

▸   *Identify the business opportunity*

▸   *Define the business criteria* for the technical design of the telecommunications platform needed to turn opportunity into action

▸   *Identify the network design priorities* and trade-offs between them

▸   *Make the economic case* for the investment

Figure 1-6 summarizes this decision sequence.

The Need for Hybrid Designers

Business decisions and technical choices obviously must mesh; they often do not. For decades, the two worlds of business and technology have been marked by a wide gap in vocabulary, knowledge, attitudes, and priorities. Business people criticize "techies" for their jargon and lack of understanding of business needs and operations: "All they do is throw acronyms at you. They don't listen and they don't understand business." Skilled technical professionals express similar frustration about the lack of involvement by "users" and their frequent abdication of responsibility: "They won't make up their minds. All they care about is cost. They seem to think dial tone comes from God."

Firms are looking for "hybrids"—people who understand the business context *and* the technology. Such people cannot be expected to be both business experts and technical experts at the same time, but if they are comfortable with both business and technical issues and the links between them, they will know how to draw on specialized expertise and translate the two languages.

Consider these extracts from an article on telecommunications management (Kronstadt 1990):

▸   "To make sure his group is on the right track, Duffy [director of telecommunications and computing services for ICI Americas] has put together a team of 'account representatives' that work directly with business units."

---

[10] *Information Week* (1991) interviewed five heads of information services and the CEOs of their companies; every one of them identified trust, communication, and "teaching both sides to speak a common language" as the key to their relationships.

Figure 1–6	**1. Identify the Business Opportunity**
The Telecommunications Business Decision Sequence	Run the business better
	Gain an edge in existing markets
	Create product and market innovation
	2. Define the Business Criteria for the Telecommunications Platform
	Reach: the locations/people/organizations to which we must link
	Range: the variety of information and transactions we must be able to share directly and automatically across business functions and processes
	Responsiveness: the level of service we guarantee, in terms of speed, reliability, and security
	3. Identify the Network Design Priorities and Trade-Offs
	Capability: the technical applications and services the platform provides
	Flexibility: how easily it can accommodate change
	Certification: how reliable it is and what service it guarantees
	Cost: fixed, variable development, and operations
	4. Make the Economic Case
	Improve profit management and alerting systems
	Improve traditional costs: labor, real estate, information technology
	Provide premium quality without adding a cost premium
	Provide premium customer service without adding a cost premium
	Improve long-term business infrastructure costs; improve revenues

▸ "We're not trying just to provide technical services. We want to be project managers, consultants, facilitators, negotiators and catalysts."

▸ "The whole thing typically depends on two or three people in high-level staff positions who can blend technical knowledge with business vision. . . . Even at the largest companies, you could count the pathfinders like this on the fingers of one hand. . . . Telecommunications is changing now as fast as computers did in the 1980s and some young turks are finding themselves stopped cold by old-line information systems professionals' attitudes."

▸ "The tragedy of telecommunications has been the failure of its managers to be able to articulate the opportunities to the board of directors."

▸ "One thing seems certain: Whatever decision is made today about any aspect of networks, it will have to be re-examined tomorrow in the light of new technologies and business trends."

These comments highlight the vital need to be able to "articulate the opportunities" of telecommunications to business managers; that is the aim of Steps 1 and 2 in the telecommunications business decision sequence. The style the sequence helps you develop requires an ability to "blend technical knowledge with business vision," to be one of the "pathfinders" who can be counted on the fingers of one hand. It requires the ability to apply a technology that is changing as fast as computers did in the 1980s—or perhaps even faster, for in mid-1990, when the previously quoted comments were made, telecommunications was in a relatively placid period of change compared to now. Today, the skills of a consultant, catalyst, or facilitator able to explain business issues must be complemented by the ability both to explain technical issues and to identify the business implications of radical new technical developments. "Old-line" attitudes must give way to young turks' fresh thinking.

The style of thinking that the decision sequence helps you develop meshes business and technology in a way that very few telecommunications professionals have been able to do. Obviously, each company has its own planning methods, and your job will focus on specific areas of telecommunications technology and/or use. The hybrid's style will be of great value to you in your career regardless of your exact job because the ability to span business and technology rests on dialogue.

Three Voices: The Client, Designer, and Implementer

Three main parties participate in the dialogue:

▸ The *client*, who has a business need

▸ The telecommunications *designer*, who translates that need into the technical blueprint (often termed the "architecture") for implementation

▸ The *implementer*, who chooses the specific technical components to implement the design

The dialogue moves from *what* to *how:* What is the business target of opportunity, and how can that target best be met?

Networks in Action focuses primarily on the pivotal role of the designer—the bridge between the world of the client, who may know almost nothing about the technology, and that of the implementer, who may be just as unfamiliar with the client's business and its objectives. The book is not about marketing, finance, manufacturing, or distribution but looks instead at where and how telecommunications is a major resource in these and other business areas. Examples are point-of-sale technology, the core of retailing operations and a key factor in the rise and fall of industry competitors; **image processing**, which is an equally vital force in the war against paper consumption; and electronic data interchange, which is rapidly becoming central to streamlining transactions between firms. (Image processing is the scanning of documents and their storage in electronic form for transmission, access, and processing by computers.) These and many other uses of telecommunications will all be discussed in terms of both business opportunity and technical choices, so that the client and designer can work together practically and creatively.

The technology will be described at the level of detail that enables the designer and implementer to work together efficiently. The designer need not know the detailed engineering of a specific technical component of the network and will assume that it has been correctly handled by the provider. The implementer must have a far more detailed level of knowledge that is outside the scope of this book.

The Cost of Silence Across the Business/ Telecommunications Gap

The topics of this book are relevant to clients, implementers, and designers. The main theme of *Networks in Action* is building effective dialogues: between client and designer, designer and implementer, and (to a lesser degree) client and implementer. The client and implementer can significantly expand both their abilities to contribute to the dialogue and their understanding of the context of their work by using *Networks in Action.* Ignorance of the business implications of telecommunications caused a number of airline, insurance, banking, and retailing firms to fall badly behind their competitors. Kmart, for instance, fell four years behind Wal-Mart in the use of point-of-sale technology and lost its position as the industry leader; it cost Kmart over $1 billion to catch up, and many observers saw this as a fight for survival.

In each industry in which telecommunications has become a core element of basic operations, quality, and customer service, at least one of the old leaders fails to keep up with the new demands. One explanation for this phenomenon is that it is very difficult for a senior management team to challenge the assumptions that have made the firm successful. If those assumptions did not depend heavily on telecommunications, it is very difficult to include the topic in business planning. One of the major opportunities for the telecommunications designer is to highlight telecommunications as a key business force.

Any expert in 1980 who was predicting success in the airlines industry in 1990 would have picked Delta, with its first-rate service, and Eastern Airlines, with its excellent route structure. He or she would have identified American Airlines as a likely loser. Similarly, the predicted winners in retailing looked like Sears and Kmart. USAA and BancOne of Ohio were relatively unknown players in insurance and banking, well behind better known and more traditional firms. Now, like American Airlines, they are way ahead of most of their old rivals, because of their senior executives' ability to mesh information technology into their thinking and to plan.

In the same way that lack of business understanding of telecommunications has often led to loss of competitive strength, ignorance of business has led many skilled technical managers to come up with elegant systems that missed the market. There could be a telecommunications Fiasco Hall of Fame of ideas that ought to work but have failed to attract customers; these ideas are solutions looking for a problem. Home banking is the most notorious example. The success of ATMs and the wide diffusion of personal computers (PCs) made it "obvious" that people would respond favorably to home banking and would make payments via telecommunications, balance their checkbooks, handle their own investments, and transfer funds between accounts.

The peak for home banking occurred in 1983, when 40,000 new users signed up for the systems provided by Chemical Bank, Citibank, and others, a figure that was a factor of ten too small to break even. In 1984, 30,000 users signed up, and by 1988 Chemical and Citibank had dropped their service. The problem was not one of technology; home banking used standard personal computers and dial-up phone lines, and the software was fairly simple. The enthusiasts had overlooked how people use banks. First, around 85 percent of ATM transactions are cash withdrawals; home banking cannot dispense cash. Writing

Figure 1–7

**The Business/
Telecommunications
Career Map**

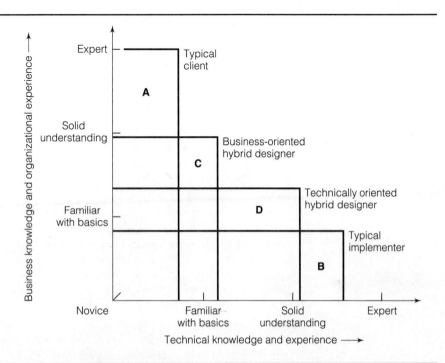

out a check and putting it in an envelope is simpler for most of us than typing payee numbers and other information into a personal computer. It may indeed be easier to balance your checkbook and statement via PC—if you are one of those few who ever balance it anyway—but most people do not have multiple accounts and thus don't need to transfer money, and most personal computers used at home have no telecommunications capability.

Home banking has been successful in some markets. Union Bank of Switzerland, for instance, has targeted it to small businesses, adding such features as international electronic payments and cash-flow analysis. Many Australian farmers love home banking because they live up to 300 miles from a bank branch.

Technology is not in and of itself a guarantee of business success. The need is to combine insight into business with both knowledge of telecommunications and the ability to connect the technology to business goals. That takes special skills and that is the role of the hybrid designer.

*A Business/
Telecommunications
Career Map*

Figure 1-7 provides a way of describing the nature of hybrids. It maps each party in the business/telecommunications dialogue onto the two dimensions of business expertise and technical expertise, ranging from novice to expert.

The wider or taller the rectangle, the broader the individual's expertise. For instance, two typical career paths are represented by Person A, who might be a corporate banker or a sales representative, and Person B, a computer programmer or a telecommunications specialist. They have moved along a career ladder, each new job and promotion building on the previous jobs. Each individual is an expert in his or her own field but knows very little about others' worlds. Person C is less knowledgeable about business than is A but is more at home with technology. Similarly, C is less knowledgeable about technology but far more comfortable with business than is B. Person C is a business-oriented hybrid, whereas person D is a technology-oriented hybrid.

Dialogue between A and B is limited. The hybrids, C and D, extend the possibilities for a meaningful exchange of ideas. They both understand business, but whereas A has more business expertise, C has stronger technical expertise. It is very difficult in business today to identify business functions that do not depend heavily on hybrids to ensure that there is not a wide gap of mutual understanding and expertise. Point-of-sale retailing needs a first-rate understanding of merchandising and superb technical knowledge of advanced data communications. Similarly, airline distribution personnel need business skills in marketing and technical skills in network design and operation. When all firms in an industry have access to the same technology—as is the case for airlines, retailing, banking, and others—the evidence is that it is management and dialogue that make the difference.

In Figure 1-7, Person A (typically the client) and Person B (typically the implementer) is each an expert in his or her own field, but they need a bridge between them. That means either that they must extend their own range of expertise by learning about the key aspects of telecommunications technology (for Person A) or of the business relevance of telecommunications (for B), or that they must talk with C or D, who is a designer. Thus although *Networks in Action* is focused on the designer in the business/telecommunications dialogue, it is also relevant to the client and implementer.

Innovation and Globalization

Telecommunications and Business Innovation

Well-known technologies that indicate how deeply telecommunications affects us all in our daily lives are telephones, facsimile (fax) machines, automated teller machines, satellite dishes, credit card authorizations, and telemarketing. Figure 1-8 lists some of the recent innovative uses of telecommunications to provide entirely new products, services, and ways of working, communicating, and operating.

Together with computers (most obviously personal computers), telecommunications changes the status quo. Just 20 years ago, many of the services listed in Figure 1-8 bordered on fantasy. Today, it is routine for long-distance phone companies to send tens or even hundreds of thousands of phone calls simultaneously along a strand of optical fiber. (When AT&T's monopoly was ended in 1984, the number of calls that could be carried along a phone line simultaneously ranged from 50 to 600.)

Figure 1–8 **Eight Proven Telecommunications Innovations of the 1990s**	**1. Interactive education by satellite**

1. Interactive education by satellite

Maine: 47th in percentage of high school graduates going on to college, 50th in adult education; 2/3 of population too far away from one of seven college campuses to attend classes. Maine's Interactive TV system has over 4,000 students, with many graduates and enrollments growing at 12% a year
Access tool: large-screen television sets in schools and even fire stations
Enabler: satellite transmission, with data compression reducing the amount of information to be sent
Impact: for less than the cost of a dormitory, Maine has created a whole new education infrastructure; major impact on women in rural Maine who dropped out of high school and have no job training

2. ATMs for welfare payments

Maryland: Uses existing interbank ATM network (called Most) for food stamps and welfare payments via benefit card
Access tool: ATM and point of sale register in supermarkets (for processing food stamps electronically, eliminating physical coupons)
Enabler: in-place, low-cost ATM infrastructure; switching technology connecting to Maryland and Federal computers (for food stamps)
Impact: major cost savings for the state and for supermarkets handling food stamps; reductions in thefts and losses of checks

3. Tax refunds authorized on the spot

Virginia: uses on-line systems to enter and check key tax form data, approve refund, and send message to central system to issue the check
Access tool: workstations in 40 tax offices, where citizens can bring in their returns and get requests for refund authorized on the spot; 1,800 other workstations in government offices
Enabler: low-cost private network
Impact: saves $50 million a year, with no staff increase in three years despite 7% in growth volume a year

4. A mortgage in 15 minutes

Citibank: replaces a paper-dominated process that took 30–60 days with a 15-minute session over the phone
Access tool: telephone (in real estate dealer's office or at home); workstations in Citibank with access to credit bureau, bank records, and so on
Enabler: 800 numbers; powerful personal computerlike workstations for on-line processing by customer service agent
Impact: a radical change in the basics of service for the mid-1990s

Manufacturers today routinely handle orders and deliveries with no paper, delays, or people in the middle. Companies such as General Motors, Sears, and Wal-Mart require their suppliers to link to them through EDI instead of mail. Transactions that once took weeks now take days instead. The $25 billion that apparel retailers once spent annually on administration is being halved through the use of EDI and related **quick response (QR)** systems in which computers

Figure 1–8	**5. $1 trillion a day traded in futures and options**
(continued)	**Tokyo:** trading in Chicago from Tokyo on 24-hour GLOBEX system **Access tool:** personal computers **Enabler:** low-cost global telecommunications links; satellite and high-speed fiber optics **Impact:** complete globalization around the clock, even when the Chicago exchange is "closed"
	6. Personal 800 numbers
	Sprint: offer of an 800 number for family use: students phoning home, elderly parents and so on; small business use, too **Access tool:** standard phone **Enabler:** falling costs of high-capacity wide area network transmission **Impact:** over half the traffic on the long-distance phone network at peak hours is now 800 calls; allows small businesses to go national
	7. Paperless aircraft carrier
	U.S. Navy: the Navy's paperless-ship project stores documents in image form **Access tool:** personal computer **Enabler:** compact disk storage, ultra-highspeed local area network **Impact:** the 40 tons of basic documents needed for daily operations can be stored on a few dozen optical disks; time spent in finding documents is cut by 70%; frees up space for bunks, storage, weapons
	8. Worldwide delivery in 48 hours with no warehouses
	Laura Ashley: this retailer, with 547 stores in 26 countries, has eliminated all its warehouses and uses Federal Express to handle all distribution, with a guarantee of delivery anywhere Laura Ashley operates within 48 hours. **Access tools:** personal computers in Laura Ashley locations that are linked to Federal Express's systems **Enabler:** Federal Express worldwide network and use of mobile communications **Impact:** goods can move straight from the supplier in, say, Hong Kong, to Vienna without having to go through Laura Ashley's UK distribution headquarters

and telecommunications are used to streamline the entire chain of transactions, from point of sale to retailer to manufacturer to shipper to warehouse to store.

Routinely, the CEO of a company can know yesterday's sales to the penny and be alerted to events and trends now, instead of in several months, when it may be too late to do anything about them. According to the chief executive of Frito-Lay, one of the leaders in using hand-held computers and telecommunications to move information at the start and end of the business day to and from its field sales force and head office, the result was that for the first time in his career he had the information to act instead of react (Jordan, 1991). Top retailers such as Wal-Mart, Toys R Us, and Dillards get information from stores every two hours via telecommunications. They can spot today's problems today and can spot a trend in under a week. The rest of the industry hears about the trend after it has already happened.[11]

Many industries across the world now literally depend on telecommunications operations. The international banking network SWIFT, which handles transfers of funds between international banks, electronically processes a trillion dollars of transactions *each day*. That is almost a billion dollars a minute! The city of London processes over $120 billion of foreign exchange deals a day; currency traders live by telecommunications. The computerized reservation systems for American Airlines and United Airlines each handle close to 2,000 transactions a second. Omaha, Nebraska, handles 100 million "800" calls a year, fully justifying its claim to be the 800-number capital of the United States.

These business innovations create both new opportunities and new management challenges. When the optical fiber link is cut, thousands of phone calls die. When the manufacturer's purchasing systems cannot link to suppliers for half a day, production may stop dead. "Telecommunications" is synonymous with "service" in more and more aspects of both everyday business and our personal lives.

The New Gross Information Product

Telecommunications is a relatively new factor in international business competition. It is also a relatively new part of how we all live. The power of telecommunications in everyday life became especially apparent during the Gulf War, the attempted 1991 coup in Russia, and in the Tiananmen Square massacre in China. During World War II, people watched cinema newsreels that were a month old. During the Vietnam War, television news was only some 24 hours behind the events it showed. During the Gulf War, news was in real-time. Intelligence officers watching on CNN saw where Scud missiles landed and immediately either adjusted their aim or instituted countermeasures. The new tools of journalism included phones that fit in a suitcase, portable satellite earth stations, fax machines to send Arabic documents to experts for immediate translation, and high-resolution photographs transmitted via phone lines. None of this is "gee whiz" any more than is ordering a pizza by fax, or receiving 110 channels of TV via satellite, or being able to direct-dial Tokyo, Moscow, or Rio de Janeiro. We all live in an age in which telecommunications is now the equivalent of electricity, cars, and airplanes 90 years ago.

Telecommunications has become so crucial to business operations that it is widely recognized as one of the main forces for economic growth of cities, states, and countries and for the competitive survival of companies. In the 1980s, world expenditures on telecommunications grew by 600 percent; even with the recession of the early 1990s, it is on pace to grow by another 500 percent in this decade. Across the world, there is a strong correlation between a country's per capita gross domestic product and the number of telephones per

---

[11] "In December 1986, for example, Toys R Us tried out scooters—skateboards with handles—with a trial order of 10,000. They sold out in two days, a trend the computers immediately spotted and acted on. Last year, Toys R Us sold over 1 million scooters. When the computer conflicts with his own judgement, [the CEO] will go with the computers." (*Forbes,* February 22, 1988).

thousand population. That is a major explanation of why India launched its Mission Better Telecommunications in 1987 and will spend close to $40 billion by 2000 on its goal of doubling its entire communications systems in five years.

To be modern, a city or country must have modern telecommunications. The world is moving more and more business transactions **on-line**—that is, the data needed to answer a question or carry out a transaction can be accessed immediately by a computer, instead of having to be located in a file or a book and then made available. We are seeing the emergence of what may be termed the gross information product, and it dwarfs the gross domestic product. The total *physical* exports of the United States and Germany, the two world coleaders in trade (Japan is in third place), each amount to just over $400 billion a year. That is less than a week of London's international *electronic* trade in the form of foreign exchange, securities, funds transfers, and credit card transactions. It is well under half a day's electronic financial transactions handled in New York City, which amount to $1.5 trillion every day.

In international trade and finance, many cities are exploiting telecommunications as an explicit element of their economic policy. Singapore and Hong Kong, for instance, are using EDI to eliminate the masses of paper documents that add about 7 percent to the cost of international shipments. Government documents that once took several days to obtain and process are now electronically handled in under 15 minutes and routed to the relevant port, shipping company's office, or freight forwarder. EDI helped move Singapore from tenth to first in the world in terms of volume of shipments. Substituting telecommunications for paper is rapidly becoming a new cornerstone for business across the world.

Telecommunications and International Development

Telecommunications disparities among countries become disparities in both economic well-being and competitiveness. Sweden has 615 phone lines for every 1,000 people; the United States has 409, West Germany 403, and Japan 367. India has 4, China 5, and Mexico 47. Every 1 percent increase in the number of phones in a developing country is estimated to increase per capita income by 3 percent. The relationship is a causal one: Improve telecommunications and a business has new opportunities, but impede telecommunications and those opportunities are blocked. The *Financial Times* of May, 1990 commented that telecommunications traffic is now a far better measure for gauging economic activity than are more traditional metrics of imports and exports and gross national product.

The takeoff of the Pacific Rim is as much a reflection of its investments in telecommunications infrastructures as of reliance on cheap and very skilled labor; one without the other would not have led to growth in South Korea, Indonesia, Singapore, Hong Kong, and Taiwan at a rate that will double their GNPs in this decade. In order to attract trading services, Singapore set its rates for international phone calls at a third those of other Asian countries. Its 1987 National Information Technology Plan states: "The information communications system is the backbone of the information economy. We must continue to have the best information communication system in the world in order to maintain an advantage in the information age."

Telecommunications in Developing Countries

Other areas of the world recognize the growing link between telecommunications and economic growth. As Latin America replaces decades of military dictatorships with a new era of economic liberalization, telecommunications has moved from being a supremely inefficient government muddle to an entrepreneurial business resource. Mexico, with the fastest growing stock market in the world in 1991, will invest over $5 billion in telecommunications in partnership with U.S. Southwestern Bell and France Telecom. By contrast, in Argentina people have been employed full-time quite literally to spend their day dialing out to get a dial tone; the reported bribe for getting a phone in Venezuela is $5,000. Each of these countries is now aggressively abandoning the view of telecommunications as a government bureaucracy and is looking for partners that will leapfrog it into the communications era.

In Eastern Europe, Hungary deregulated telecommunications even before Mikhail Gorbachev began his astonishing reforms in the former Soviet Union. At the time of German unification, the former East Germany's phone system was using the same technology as when Hitler was in power in the 1940s. By 1994, Germany may well have one of the most advanced telecommunications system in the world; it has already committed well over $60 billion to the three main infrastructures of modern life: education, roads, and telecommunications. Other countries in Eastern Europe are finding that telecommunications is a critical element in their efforts to build a new economy. Poland, for instance, built a new banking system and stock market but quickly found that clearing checks, moving funds, and making trades rests on telecommunications as well as on privatization and entrepreneurship. The race is on to make sure that lack of a communications infrastructure does not block growth.

The situation in Africa is much bleaker in every way. The 470 million people in black Africa have the purchasing power of Belgium, whose population is just 10 million. A computer that costs $75,000 in the United States sells for $280,000 in Nigeria. Humidity, high temperatures, lack of skilled staff, government paranoia about the free flow of information, concentration of population in a few large cities and many villages, and almost complete inaccessibility to loans, investment capital, and foreign exchange have left Africa outside the information economy. The World Bank is trying to change this. In 1990, it sent a team to 17 African countries to get approval to install a **VSAT (Very Small Aperture Terminals)** network; VSATs are the tiny satellite earth station dishes that brought news from Saudi Arabia to CNN in real-time during the Gulf War. The World Bank's aim is to bring African governmental agencies, airlines, universities, and businesses into the new global information economy.

South Africa is moving, too. F. W. de Clerk's initiatives to end apartheid have brought the country back into the mainstream of international business. Nigeria, the strongest opponent of apartheid and the country that spearheaded the 1984 sanctions against South Africa, welcomed de Clerk as a hero in 1992. In the same year Singapore signed major economic agreements with South Africa, signalled by direct flights to Johannesburg on Singapore Airlines. In South Africa, businesses looking to be international players have loudly complained about the costs and regulations by Telekom, the government monopoly. In 1992, Telekom became a privatized company. More importantly, it and a consortium of large businesses announced plans in mid-1992 to launch a satellite whose footprint (the area on earth that can pick up the signal) will cover Southern Africa, not just the Republic of South Africa. As one CEO stated in a conversation with one of the authors of *Networks in Action,* "Sanctions blocked us. Now sanctions have ended, but without telecommunications we will be just as blocked."

The 15th century began the "globalization" of our world society, starting with the Portuguese exploration of Africa. In the final decade of the 20th century, globalization does not mean sending out ships, missionaries, and soldiers, but rather sending up to a satellite a signal that returns to an earth station. Remember: GNPs of countries and their exports are measured in billions of dollars a year, and the electronic movements of money are not included in GNP, and what we call here the gross information product is measured in billions of dollars daily, not annually.

| | | | | | | | # Networks: The Core Technology

The basic element of a telecommunications resource is a **network**—a set of devices that share a directory and can thus directly access each other. The directory is analogous to a standard telephone directory, except that it provides an address for each component of the network. If a device is not included in the directory, it is not part of the network. Directories must be updated when components and locations change, in the same way that if you wish to add a new phone to the public telephone system, you must have a phone number; having a handset is not enough.

The main distinctions between networks are public versus private and local area versus wide area. A **public network** is available to a wide range of subscribers on a pay-as-you-go basis; the most obvious and widely used example is the public phone system. A **private network** is one in which the transmission facility is leased by a firm for its own use. A **local area network** links personal computers and printers within the same building or building complex; A **wide area network** covers a broad geographic area.

Examples of business networks are a bank's wide area network that connects ATMs to central computers that handle check-cashing and deposits; a local area network that connects, say, the personal computers in a company's legal department; and an international private network in which a firm leases telecommunications services for its own (hence private) use.

There are many other types of network, and a typical network may link thousands of devices and include thousands of additional components. Networks of connected networks can create new ways of doing business; leading retailers like Wal-Mart, Toys R Us, and Dillards, for instance, have linked their own stores' point-of-sale workstations and local (in-store) networks to their corporate cross-country network to provide up-to-the-hour data on stocks and purchases. The network links directly to suppliers for automated and increasingly paperless purchasing, invoicing, and payments through EDI. Many of the suppliers' networks in turn link to manufacturers' networks. This "quick response" technology has transformed the competitive dynamics of retailing. During the 1991 recession, when retailing slumped badly, Wal-Mart, the Gap, Limited, and Toys R Us reported sales gains of up to 30 percent. Their networked logistical systems are a major explanation of their success; they have better information, faster reaction times, and more coordinated distribution than their competitors.

Many of the most innovative and far-reaching business opportunities of telecommunications come from linking networks across companies and even industries. So, too, do many of the most complex technical problems. Generally, a specific network is built around a limited set of technical building blocks and is designed to meet a targeted range of needs and types of traffic. Adding new devices into the network is relatively simple if they fit into the original technical design. Linking networks of very different technical components is not so easy and may require special-purpose devices and software.

Summary

Telecommunications has reshaped the basics of many industries. An airline without a reservation system is simply not an airline, nor a bank without ATMs a bank. We can expect that telecommunications will continue to change the very basics of industries.

The pace of innovative technical change in telecommunications is explosive, driven, and often compounded by (1) competition, (2) improvements in speed of transmission, (3) new types of access tools, and (4) increased interoperability among services.

Historically, telecommunications has been a monopoly, regulated and operated by a PTT, a quasi-government agency. The United States deregulated long-distance telecommunications in 1982 through the divestiture of AT&T, with local telecommunications remaining both regulated and largely a monopoly of the seven Regional Bell Operating Companies created in the divestiture. Soon after, the United Kingdom liberalized telecommunications, allowing limited competition. In 1992, the United States began to open competition in the local phone market, with cable television companies being allowed to provide phone services, and phone companies allowed to enter the information services and entertainment markets for the first time.

Now, more and more PTTs are either being privatized or opened up to competition. Competition in telecommunications leads to rapid price cuts; costs of long-distance phone calls in the United States have dropped 40 percent since 1984, when deregulation took effect. Competition also fuels product innovation.

The key to telecommunications is how quickly and cheaply information can be transmitted. It is in this area that telecommunications technology has made thousandfold improvements in only a decade, with innovations emerging every few months. In particular, fiber optics provides such speed that 100,000 typed pages can be sent in a second.

Tools that access telecommunications-based services also fuel innovation. The telephone now connects to computers as well as to other phones. Personal computers used in offices or homes open up new points of access and create new types of service. ATMs can link not only to banks but also to computers that process welfare payments or renew driving licenses.

Interoperability of previously separate services, information, and equipment allows leading retailers to streamline and interconnect their entire supply chain—from the point-of-sale registers in each store, to computers in the head office, to suppliers', banks', and shippers' systems.

The business importance of telecommunications creates the new management challenges that are the main topic of this book. Technological innovations open up business opportunities, and business innovation increasingly rests on telecommunications. The new need is to bridge the gap between the businessperson who has never needed to understand telecommunications and the telecommunications specialist who has focused his or her attention on the technology and not the wider context of its business use.

Companies are searching for hybrids, people who understand both the business context and the technology of telecommunications and can help build a dialogue among three parties: (1) the client, who has a business need or is looking for a business opportunity; (2) the telecommunications designer, who translates that need or opportunity into a technical blueprint, often termed an architecture; and (3) the implementer, who chooses the specific technical components needed to implement the design and operate the service. This book mainly focuses on the role of the designer, who is the bridge between the business world of the client and the technical world of the implementer.

Such a dialogue is essential to avoid the common situation in which a firm loses its competitive position in the marketplace because its business managers are unaware of the implications of telecommunications technology and its uses. It is also needed to avoid innovations that seem likely to succeed but fail because the technical designers did not understand the customers and/or the marketplace.

The link between telecommunications and business innovation is crucial and widespread, as evidenced by ATMs, electronic data interchange, electronic trading, and customer-supplier computer-to-computer transactions. We are seeing the emergence of what can be termed the gross information product, which dwarfs the gross domestic product.

Internationally, telecommunications expenditures are growing at a rate of 500–600 percent in this and the past decade. Telecommunications has become key to developing countries' efforts to improve their economic infrastructures; telecommunications traffic is now the most reliable metric of a nation's economic activity. In Latin America, the Pacific Rim, and Eastern Europe, upgrading the telecommunications infrastructure is seen as an urgent priority; Africa lags badly in this regard.

Selected References

The purpose of the references provided here is to draw your attention as a student or telecommunications professional to books worth buying or finding in your library or magazines and journals worth subscribing to. References are part of a telecommunications designer's tool kit, enriching what you learn in class or on the job. They are tools to help you develop the style of thinking and communication that is the central theme of this book.

References cited in this chapter begin on page 727.

Books

The following books provide discussions of telecommunications-related topics and include many concrete examples. Titles marked "B" address only business and ignore technology; "BT" indicates that both business and telecommunications technology are covered; "T" indicates a relatively strong focus on technology, but not at a level that is beyond the understanding of a nontechnical student; "P" indicates that the main focus is on public policy and/or economics; "G" denotes a book aimed at a general audience; and "S" identifies a scholarly book.

Together, these books constitute a valuable minilibrary for the telecommunications designer. Many—perhaps most—clients will be familiar with the books that aim at a general business audience and thus provide both a common context for dialogue and plenty of examples. The more scholarly books are listed here because they provide sound, in-depth coverage of their topics.

Antonelli, Christiano (ed.), *The Economics of Information Networks* (Amsterdam: North-Holland, 1992). (P,S)

Bradley, Steven P., and J. A. Haussman (eds.), *Future Competition in Telecommunications* (Cambridge, MA: Harvard Business School Press, 1989). (P,S)

Coll, Steve, *The Deal of the Century: The Breakup of AT&T* (New York: Simon & Schuster, 1986). (B,G)

Crandall, Robert W., and K. Flamm (eds.), *Changing the Rules: Technological Change, International Competition, and Regulation in Telecommunications* (Washington, DC: The Brookings Institution, 1991). (PT,S)

Davis, Stan, and Bill Davidson, *20/20 Vision* (New York: Simon & Schuster, 1991). (B,G)

Hyman, Leonard, R. C. Toole, and R. M. Avellis, *The New Telecommunications Industry: Evolution and Organization* New York: Public Utilities Reports, Inc., 1987). (PT,S)

Johansen, Robert, et al., *Leading Business Teams: How Teams Can Use Technology and Group Process Tools to Enhance Performance* (Reading, MA: Addison-Wesley, 1990). (BT,G)

Keen, Peter G. W., *Competing in Time: Using Telecommunications for Competitive Advantage*, 2d ed., (Cambridge, MA: Ballinger, 1988). (BT,G)

Palvia, Shailendra, P. Palvia, and R. M. Zigli, *The Global Issues of Information Technology Management* (Harrisburg, PA: Ida Group Publishing, 1992). (PT,S)

Pierce, John R., and A. M. Noll, *Signals: The Science of Telecommunications* (New York: Scientific American Library, 1990). (T,S)

Stalk, George, Jr., and T. M. Hout, *Competing Against Time* (New York: The Free Press, 1990). (B,G)

Toffler, Alvin, *Power Shift* (New York: Bantam Books, 1990). (B,G)

Winston, Brian, *Understanding Media* (Cambridge, MA: Harvard University Press, 1986). (PT,S)

Magazines and Journals

The following publications are useful and timely sources of information on telecommunications, especially trends in the technology and the industry and new applications (those marked° are strongly recommended):

Business Communications Review: a very technical bimonthly journal that focuses on the telecommunications industry. It is one of the few telecommunications publications that covers both voice and data communications rather than focusing on one or the other.

°*CIO:* a well-written and easy-to-read monthly publication that covers information technology in general, with plenty of examples from business. Each year it publishes an issue that focuses on telecommunications.

°*Communications International:* highly technical monthly publication that is indispensable in its news and analyses of PTTs and international standards.

Data Communications: a monthly publication that is very technical, but well worth reading (or skimming, even by nontechnical people) because it avoids fads and hype and explains key technical developments thoroughly and accurately.

Network Computing: a monthly publication that focuses on personal computers and local area networks at a very technical level of detail; mainly useful for implementers, rather than designers, but covers aspects of PC/telecommunications interdependencies that most of the telecommunications trade press ignores.

°*Network World:* a weekly trade periodical that is indispensable in its coverage of just about every aspect of networks in action, including in-depth analyses of technology and easy-to-read and timely business examples.

Telecommunications: a monthly publication with in-depth analysis of emerging technology and standards.

Review Questions

1. Identify five services you use at least once a week that depend entirely on telecommunications. Identify (a) the access tool (for example, ATM, personal computer, phone), (b) the service provider, and (c) why the service would be impractical without telecommunications. If your service provider did not offer the applicable telecommunications-dependent facility, would you drop it and give your business to another one that does?

2. Questions 2–6 check your understanding of the role of switches in a network. In the following diagram, how many lines are needed to connect each of the five phones to every other phone?

3. How many lines are needed to connect each of ten phones to every other phone?

4. Is the increase in the number of lines needed to connect two, three, five, and ten phones linear (that is, a constant increase)? Does doubling the number of receivers mean doubling the number of lines needed? Or is it nonlinear, meaning that if you double the lines you must more than double the connections?

5. If the phone line from each of the ten receivers is connected to a switch, how many lines are needed so that each phone can link to every other phone?

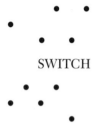

SWITCH

6. Refer to Figure 1-4, which shows the public switched network. Which of the following things shown in that diagram are switches?
   a. the phone receiver (A)
   b. the terrestrial trunk line (G)
   c. the toll center (C)
   d. the central office (B)

7. Which of the following activities involve voice communications, which are data communications, which use both (explain which is used where), and which do not use any telecommunications?
   a. using an ATM to make a cash withdrawal outside your bank branch
   b. using a personal computer to run a spreadsheet program
   c. using a personal computer to send a copy of a term paper electronically to your professor's personal computer
   d. leaving a message on a friend's answering machine

   e. cashing a check in a supermarket
   f. using an ATM in a supermarket
   g. making a plane reservation by calling the airline
   h. ordering goods from a mail-order firm
   i. being stopped for speeding by a police officer who asks for your driver's license

8. Which telephone company provides your long-distance phone service? Local phone service? Are either of these a PTT? Why or why not?

9. Which of the following is a local area network and which a wide area network?
   a. a university's network that links its libraries, dorms, and classrooms
   b. a bank's ATM network
   c. Federal Express's mobile communications that link delivery vans to its central databases and computers
   d. MCI's long-distance phone network
   e. a network linking your university to other universities across the state
   f. a cellular telephone system

10. Refer back to Figure 1-3:
   a. What is the transmission medium?
   b. How is the signal switched from the base station to the truck and from the truck to the base station?
   c. Could this system operate using fiber optics? Why or why not?
   d. Transtrack claims this system is cheaper than using satellites? On what might Transtrack base this claim?

Assignments and Questions for Discussion

1. Refer to Figure 1-1, which shows a bank ATM network.
   a. Using this as a guide, draw the access points and service providers for a major airline reservation system, showing the network as a cloud. An obvious access point is a travel agent's reservation terminal. Could a personal computer in the home be an access point? If so, include it in your diagram.

b. Could the network also handle car rental reservations? Hotels? If so, show them on your diagram, and add other potential service providers.

c. What are the consequences of the network being "down" for a day?

d. American Airlines and United Airlines were forced by the U.S. government to change the way in which flights are displayed on the screen in answer to a travel agent's query for flights from, say, Dallas to New York on Thursday, June 10, between 9 and 11 A.M. What do you think was the reason for the charge of "bias"—that it was unfair that the first airline to appear on the screen was American Airlines because travel agents often chose the first airline they saw there— and why do you think other airlines continue to charge that the two airlines, which together handle close to 80 percent of travel agent reservations, have an unfair and "predatory" advantage?

e. What information must be stored and accessed on-line in order to make a reservation and for the travel agent to issue a ticket?

f. What additional competitive opportunities might this information provide an airline?

g. Frequent flier programs ask passengers to provide their unique frequent flier number. American Airlines has its frequent flier information database linked through the network to its reservation system. Its major competitors did not do this; their databases are on computers that do not link to the reservation network. What, if any, advantage could American Airlines gain from this?

2. Refer to Figure 1-3, which shows Transtrack's truck tracking system.

a. Satellites orbit the earth 22,300 miles up in the sky. If the trucks had a small earth station antenna and the base station had a large dish to send and receive signals, would this setup be any different in basic concept from Transtrack's cheaper system?

b. Transtrack's service uses a wireless form of transmission: VHF radio. Which of the following devices are wireless, and which use physical cables and wires? Do any of them use both? What difference does it make to the customer and the provider?

1) national television broadcasts
2) cellular radio
3) cable television
4) car phones
5) long-distance phone calls
6) automated teller machines
7) standard personal computers
8) fax machines

3. The long-distance telephone industry in the United States is deregulated. Local phone service was kept as a regulated monopoly.

a. What are the main arguments you see in support of competition in the long-distance market?

b. What are the main arguments for keeping local services a monopoly?

c. What are the arguments against competition in long distance?

d. What are the arguments for competition in the local telephone industry?

4. Refer to Figure 1-2, which shows Health Link's electronic insurance claims processing network.

a. Outline the process a doctor must use to create and file a claim if such a system did not exist. The patient arrives for his or her appointment and provides the name of the insurance company and the patient identification number. Show the steps involved subsequently, including any filling out of forms, mailing, and so on.

b. Compare these steps with those required to use Health Link. What are the savings you see, in terms of time, cost, and accuracy?

c. In what ways does Health Link's use of a personal computerlike terminal that accesses a remote computer provide advantages over the doctor's office staff phoning the insurance company directly to check coverage?

Minicase 1-1

THEi

Reinventing a Company

THEi (Thorn Home Electronics International) operates 3,000 electronics rental and retail stores in 19 countries. Headquartered in the United Kingdom, THEi has a market share in Great Britain that exceeds 50 percent. THEi's U.S. subsidiary, Rent-A-Center, has grown from 180 stores in 1987 to well over 1,000 in 1991, and its market share has quadrupled from 3 percent to 12 percent.

THEi's three networks are simple from a technical perspective. Information flows on a daily basis to and from the stores; as we will see in later chapters, many large retailers have gone well beyond this, with stores sending data on an hourly basis and central computers directly initiating orders to suppliers, handling payments on-line, and so on.

The components of both the network and computer processing and information systems are fairly standard. Telecommunications is used to move volumes of information on a daily basis and to access smaller amounts of on-line information. A store manager can use the network to access head office inventory databases. AT&T, MCI, Sprint, and the Regional Bell Operating Companies in the United States provide many facilities that meet these needs, as do British Telecom and Mercury in Great Britain. Other European companies offer comparable services through public data networks.

THEi's success story grew out of a disaster. In 1985, Jim Maxmin was appointed chairman and CEO of a newly reconstructed organization. The electronics rental industry in Great Britain was in decline; most TV sets there had been rented until the early 1980s, but the globalization of electronics manufacturing had dropped prices and made TVs and other types of electronics affordable commodities. THEi's rental units were rapidly losing subscribers, with price and availability dominating consumers' choice between purchasing and renting.

Maxmin is a specialist in rescuing companies in trouble who had previously held firefighting posts in several British firms. After rebuilding THEi, he moved on to what a British newspaper described as "the job of hauling [the new firm] out of the mire." Maxmin certainly did just that for THEi, doubling profits in under three years. He left the firm the international leader in its marketplace.

He saw the business problem as fundamentally one of organization. To compete in a commodity market, THEi would have to shift from administration to marketing and customer service. "What we came up with was that you can actually get brand loyalty and high levels of repeat purchase [in individual stores] through outstanding customer service and personal contact." This may sound like common sense, but the barriers to action were immense. Store managers had almost no authority. Decisions on pricing, equipment, credit, and budgets were made by headquarters. Turnover of store managers was around 40 percent per year. They had few performance incentives, nothing much by which to differentiate their stores from competitors', and little meaningful interaction with customers.

Maxmin's philosophy was to turn THEi from a single company of 2,000 stores in 1985 into 2,000 individual businesses units—"one times 2,000." Each store would be independently run, and each manager would make the decisions and be given incentives and opportunities to build his or her own clientele, handle inventory and pricing, and manage credit.

How this works in practice is illustrated by a Rent-A-Center store in the United States. Gregg Allen, its young manager, maintains a map showing information on rentals and demographics. He uses this to make decisions—his own decisions—such as blanketing a target area with brochures, suggesting a VCR rental to a well-established customer who rents a television, or making a sales call to a customer whose rent-to-own plan is close to completion. He has run several successful special events, such as a Las Vegas Night and free weeks of rentals as prizes.

Allen contrasts his job today with his previous experiences at several Rent-A-Center stores. The old organization "had a lot of good ideas and theories. But they weren't really spelled out and laid down. A manager very often had only a vague idea of what he could and couldn't do." As a result, store managers did little to promote their products. "THEi has delivered more authority into the manager's hands." Allen adds that "the computer is really what puts the manager in control." Computers and telecommunications were a key enabler of Maxmin's practical vision of an empowered, "informated" (instead of automated) organization. Maxmin made several immediate moves that eliminated as many as five layers of management at the head office. He wanted store managers to feel they had the authority to deal with any problem, instead of having to refer matters up the decision-making chain.

Maxmin's concern was that this decentralization of authority needed to be accompanied by adequate information and support. Each store "had to be able to handle the complexity that had been the role of middle management." As part of meeting this need, THEi built a telecommunications network that linked its London data center's large IBM mainframe computer to each of its UK stores. All the applications were aimed at helping the stores make sound basic decisions about inventory, credit, pricing, and so on and to track customer rental contracts, equipment upgrades, termination dates, and the like. The stores could use the information to build special programs for individual customers and to manage the long-term customer relationship; as a result, the average contract life per customer has increased by four months—to seven years, an increase of 5 percent.

The stores send data via telecommunications back to THEi's London headquarters daily, providing an audit capability and helping the firm's smaller, leaner, more responsive management team oversee the company's overall operations and performance. Each store has a point-of-sale terminal that automatically collects sales and customer information during the day; this data is "squirted" directly to the London mainframe computer, using public network links. Corporate management focuses on "the big picture" in analyzing national trends, financial performance, and the like.

There are now three separate networks: one for the UK, one for the U.S., and one for 13 European countries. These send to and receive data from London; Wichita, Kansas; and Copenhagen, Denmark, respectively. The U.S. and Danish operations send data to London daily.

Maxmin emphasizes that the turnaround of THEi was based on organizational transformation, not technology. He points to several failures in introducing new systems, instances in which some conflict existed between organizational and technical issues. One example was an automated service application that management used to berate and control inefficient service staff, violating its own aims of "informating" instead of automating and of ensuring local authority instead of central control. Such "false starts" reflect, according to Maxmin, the realities of organizational change; he calculates that it will take at least ten years to complete the process.

THEi has invested heavily in training facilities in all the countries in which it operates. It emphasizes personal goal setting and individual responsibility, significantly upgrades the quality of its store managers, provides substantial performance incentives, emphasizes ongoing training, and makes a real effort to retain its managers.

Questions for Discussion

1. Assume that Maxmin had taken over THEi in 1965 (instead of 1985) and had tried to implement the same organizational strategy. In 1965, the following technical resources would not have been available to him: cost-effective international data communications, low-cost personal computers, point-of-sale terminals, and database management systems.

    a. Could he have carried out his 1985 strategy?

    b. How could he have implemented his philosophy of local authority?

2. Maxmin has cut the number of managers and staff in the head office and thus reduced controls on store managers. What information does he need to ensure that his "one times 2,000" is a set of 2,000 effective profit centers and decision makers, and not 2,000 soon-to-fail mom-and-pop stores? How can he balance centralized oversight and coordination with decentralized decision making?

3. In 1988, Maxmin relocated to Boston, Massachusetts, to oversee the operations of Rent-A-Center. He remained CEO of THEi worldwide. What impacts do you think his move had on the organization, on management of the UK and other European operations, and on Maxmin's own mode of management? Would your answer be the same if the date were 1965 and the telecommunications base were not in place?

Minicase 1-2

Shearson Lehman

Streamlining Workflows

Many newer applications of telecommunications exploit two relatively recent developments in information technology: local area network (LANs) and image processing systems. Local area networks link devices within a building; image processing systems scan documents and store them on optical disks, which use the same technology as compact disks that store music or video. The stored documents can be accessed from a personal computer through an LAN.

Shearson Lehman applied imaging processing through LANs in mid-1991 to streamline applications from its 400 U.S. branches for approving clients' suitability for options trading. This is a high-risk investment strategy, and a financial consultant in the Shearson branch is required to interview potential customers and appraise each's financial stability. Then the relevant forms are sent to Client Documentation Services in New York, where they are hand-checked for completeness and accuracy and the data are manually entered into a mainframe computer via a terminal. The compliance department then makes the decision on acceptance or rejection.

By law, each client must receive a written response from Shearson Lehman within 15 days. The firm has been hard pushed to meet the deadline. The complex paper flow and checking process meant that forms were often sent back through the system to repeat or correct an earlier step—and, of course, forms got lost. Shearson's more than 400 branches have 8,500 financial consultants; all the applications they handle are sent to New York for approval.

The new image processing system, which will take several years to deploy across the organization, has already cut turnaround times for processing applications from an average of 13 days to under five. Around 20 branches are linked to it, with more being added regularly. The new process begins in the branch, where the initial acceptance is made. The broker then faxes the forms to New York, saving several days over mailing. The faxes are never printed out on paper, however; instead, a standard personal computer accepts 16 faxes simultaneously using fax cards that plug into the machine. The forms are sent via the LAN into the image processing system, which is provided by Wang Labs, the computer manufacturer. This OPEN/Image system routes forms to a "virtual mailroom." The term *virtual* here is contrasted with *physical*; a piece of computer software that acts as if it were a physical mailroom gives the forms a unique electronic identification number, stores them in the mainframe, and immediately sends them (again through the LAN) to the options reviewer's desktop personal computer. The forms are displayed, together with any relevant background information from the mainframe. If the reviewer needs more information from the branch, he or she sends an electronic mail message—a "messenger wire"—that is sent to the branch financial consultant's PC over the firm's wide area network.

When the reviewer is satisfied, a single click on an icon (a small picture on the screen that indicates options for the PC user) sends it to the data entry department. This step still requires some manual entry, but there is no longer any paper to be shipped, and the data entry operator sees a high-resolution display of the form, with highlights indicating which data should be keyed into the mainframe computer. The main reason for not eliminating this step is that as yet the Wang image processing software is not fully compatible with the complex IBM computer terminals. It will take time to migrate from them; thus they remain both an essential tool and a stumbling block on the path to full integration.

The options reviewers and data entry operators share the same local area network, but the compliance department has a different one and is located on another floor. The electronic image of the application form is therefore sent from the first LAN to the second one via the wide area network. Modifications can be sent to and from each LAN. Only when the compliance department issues its approval is the information sent to the mainframe, which prints out a letter to be sent to the client. The forms can be annotated on the PC screen.

The benefits of the system have been dramatic. The reduction of a week or more in turnaround time is an obvious one that both improves customer service and reduces work loads. Another is the reduction in errors: Any long stream of paper processing is vulnerable to a mistake or oversight somewhere along the line. If each step—filling out the form, checking it, mailing it, opening the mail, sorting it, and so on—has a 95 percent chance of being correctly handled, it takes only 14 steps for the likelihood of a completely error-free and delay-free processing of an application to fall below 50 percent.

The director of client operations at Shearson Lehman comments that "the worse thing you can do to a customer is go back and ask for new forms because the originals are missing." The imaging system eliminates the physical movement of documents between different work areas and buildings; documents can be tracked at all stages, and if a branch calls in, it can be told exactly where a form is in the workflow. The branches have needed such feedback for a long time; it is frustrating for them not to know where things stand or to be unable to keep a customer informed.

One unexpected benefit of the system has been a new perspective on workflows and the resultant chance to redefine them. The previous process lacked consistency; individual reviewers handled forms in their own way, workflows had to fit the floor plan, and there were many inefficiencies. Imaging and network movement of documents has standardized the workflow; most reviewers find this an improvement, though some concerns about loss of flexibility and control existed at first. A manager comments: "[It] changed the way they work, and they were a little anxious. . . . There was something almost mystical about that piece of paper, but I've seen a tremendous change in attitudes already. Now, users are beginning to find out that they have more control with imaging."

Although more and more organizations are implementing image processing systems, this is a "first" in the industry. Regulators are cautious about any innovations that open up risks of either altering data in ways that cannot be detected or losing security. One manager predicts that "eventually the industry will accept this technology, but it's going to be a long haul. We're the first to do it, so obviously we have to deal with the regulators." Cultural hurdles are a constant concern. Shearson Lehman committed substantial resources to a three-phased training program and to getting support and feedback from reviewers.

The process for developing and implementing the system began at a fairly senior level. Shearson Lehman is part of American Express, which has a Technology Planning Group that looks at emerging technologies and makes presentations about their opportunities to nontechnical business unit managers, who have full autonomy over investment decisions concerning technology. The sponsor for imaging in Shearson Lehman was the director of Client Operations, an executive vice president. He and his team of business managers saw a major opportunity to improve customer service and internal processing efficiency. The team that drove the project was composed of:

1. *Clients:* business managers, including ones at a fairly senior level, from Client Documentation Services

2. *Designers:* technical specialists from American Express's Technical Planning's Securities Information Group, which handles Shearson's brokerage processing and hence understands its business as well as the relevant technology

3. *Implementers:* specialists from Amex, Shearson, and Wang

One member of the design team commented about the mix of skills: "We learned a lot from [Wang] about imaging application development, and they learned from us about the business." She added that throughout the design and implementation process, business and technical issues were highly interdependent. So, too, were information systems development and telecommunications issues. The imaging system required developing a large number of programs, which had to be custom-tailored to Shearson's needs. The telecommunications development needed to link two remote local area networks to each other via the wide area network and to link the branch offices to New York, again via the WAN. This was a mix of technical building blocks, with several gaps and blockages. The new technology must coexist with the old IBM systems and equipment.

This minicase is very representative of what is going on in many large and medium-sized firms.

Image processing tackles the ever-present problem of paper-dominated workflows. The proven benefits are so great that one day every firm will be using imaging as routinely as they now use word processing. (The comparison recalls the recency of word processing; most firms were using typewriters just ten years ago.) Today, most firms process paper physically; ten years from now, image processing will be the norm. Image processing depends on networking; as here, the original paper document is converted to a digital image as early in the process as practical; from then on, it is moved electronically. Local area networks are the obvious key infrastructure for imaging because most documents are processed in part or completely by a specific department.

Questions for Discussion

1. What applications can you see for applying a combination of image processing and local area networks to streamline your college or university's administration? Where might this improve student services? (If your school has already done this, change the tense of this and subsequent questions—from "where might this improve" to "where has this improved?", for example.)

2. If such a system were in place, what new types of services would become practical that would significantly improve speed of service, responsiveness, and cost?

3. What are the likely barriers to implementing such an innovation? Include organizational, economic, and financial issues in your analysis.

Minicase 1-3

California State Government

Telecommuting

In early 1990, the state of California completed a two-year pilot project in which just under 200 workers "telecommuted"—worked from their homes most of the week, instead of going into the office. The aims of the experiment were twofold: to increase the individual productivity of the telecommuters and to address longer-term issues of traffic congestion, air pollution, and overcrowding in the Sacramento area.

The director of the project reported that the gains in productivity and direct cost savings amounted to 300 percent of the investment. The California legislature subsequently passed a bill requiring private companies to set up "telework centers"; these are satellite-equipped office complexes specifically designed for telecommuting.

The potential and problems of telecommuting have been discussed widely in many organizations, and a growing number have carried out experiments. A smaller number have institutionalized the process. Advocates of telecommuting claim that productivity gains average 20 percent; that it offers opportunities for companies to cut travel costs, meet the personal needs of employees (especially in single-parent families), and reduce office costs; and that it dramatically reduces the time and frustration of sitting in traffic jams. As in the California example, city and state governments are urging companies to support "flexible work options" for environmental reasons and to reduce the need for increasingly expensive investments in infrastructures.

Here are a few examples of experiences and conclusions:

1. "One of the stereotypes that prevails is that telecommuting is just a perk for employees. The manager envisions employees sitting at home and maybe sneaking off to watch a soap opera and asks the legitimate question, "Why should we do this?" —a flexible-work options consultant.

2. "It needs a bit of a catalyst." As a result of the sudden jump in oil prices in 1991 created by the Persian Gulf crisis, many firms encouraged staff to work from home to reduce travel. "Lo and behold, they realized there were significant reductions in expenses and improvements in productivity." —a senior manager in the regulatory policy analysis unit of Northern Telecom.

3. "Industry isn't moving to telecommuting as an [alternative] to working in the office, but as [a way of enhancing] the ability to get your work done. If you can still take care of your daily business, regardless of where you are, then the office becomes less sanctified." —information services director in Control Data Corporation.

4. "The office, in most cases, is a lousy place to work. One thing that most workers tell us is that by getting away from the office they get away from distractions that keep them from working well."

5. "Today's telecommuters are much more likely to be autonomous business professionals and executives with whom monitoring [of performance] is done by results. You don't monitor a white-collar worker. They either make their objectives or lose their jobs. Clerical jobs and data-entry jobs where monitoring is much more of an issue and could become more oppressive have fallen away from telecommuting."

Technical Background

The telecommunications options for telecommuting range from just a telephone so that people can talk to you and vice versa, to a telephone and a personal computer that can dial up the firm's electronic mail system to handle messages, all the way up to having a satellite dish on your roof so that you can access large-scale databases, send and receive information at high speeds, and even hold electronic videoconferences. The PC can be a standard desktop machine, a portable laptop, or a top-of-the-line equivalent of a mainframe computer. (Electronic mail sends messages to a computer, where they are stored and forwarded when the person to whom they are addressed next logs on to the mail service; unlike a fax, which is sent to a specific phone number and machine, electronic mail is sent to the person, no matter where he or she is.)

The questions for discussion at the end of this minicase ask you to think through the impacts of increasing the capacity of the telecommunications capability. Here are a few key points to guide your thinking:

1. The standard dial-up telephone line has limited transmission speed. Typically, a personal computer sends and receives data over the telephone system at speeds of between 600 and 800 characters ("4", "m", and "$", for examples) a second. A standard high-speed leased data line transmits information at 7,000 characters a second and up.

2. The approximate sizes of messages and transactions in characters (or their equivalent) are:

A credit-card authorization request	125
A one-page electronic mail message	600
High-resolution facsimile (1 page)	12,500
One second of full-motion video	1,250,000

Questions for Discussion

1. Identify targets of opportunity for a large urban firm to use telecommuting for each of the following personnel:
   a. middle-level white-collar workers in its human resource department
   b. field sales reps
   c. data entry clerks.

2. The simplest telecommunications option for telecommuting is a telephone and a PC. What are the advantages, problems, and/or limitations of this simple setup for each of the three categories of worker listed in the previous question?

3. Many additional telecommunications facilities can be added to the simple setup, such as electronic mail, access to corporate databases, voice mail (like electronic mail, but recorded voice messages that are stored and forwarded, instead of written ones), or direct links to processing systems (such as for a sales rep to input orders directly). What are the likely advantages, problems, and/or limitations of these and any other telecommunica-

tions options you see, for both the telecommuter and the company? Make sure you think broadly, and include such issues as security, training, relocation of the individual, and cost.

4. Is telecommuting really likely to take off, or is it just a gimmick or a naive space-age idea? What factors will block it? What factors could make it take off?

Minicase 1-4

"Telestroika"

Hungary's National Telecommunications Policy

It is now widely recognized that a country's telecommunications infrastructure is a major factor in its economic growth (or lack of it). This realization has driven the liberalization or privatization of the PTT in country after country, first in Europe, and then across all continents.

Hungary began to upgrade completely its telecommunications capabilities far earlier than other Eastern European nations, and indeed well ahead of most developed countries elsewhere. In 1989, it announced a ten-year plan that included divestiture of the postal, broadcasting, and telecommunications administrations; joint ventures with foreign suppliers; tax incentives; and relaxation in laws concerning foreign investments and foreign ownership.

Hungary had already begun developing a mobile cellular network using the old analog (as opposed to digital) technology (see Chapter 2), with help in the form of a loan from the World Bank. In the United States mobile cellular telecommunications are wireless and are used in car phones. Hungary plans to have in place a digital mobile cellular system by 1995; the existing telephone system is analog, as it largely is in all countries.

The director of the telecommunications department of the old PTT stated in 1989 that "We believe no economic development can be achieved in our country without an efficient telecommunications infrastructure." At that time, Hungary had under eight phone lines per 100 people. In 1990, the ratio had improved to around 11 lines per 100 people—900,000 phone lines for a population of 10 million. The goal for the year 2000 is 30 lines per 100 people. (Canada leads the world in phones, with 20 million for a population of 25 million. Sweden has 62 lines per 100 people, and the United States has 41 per 100 people (1988 figures).

As in Japan, Germany, and many other countries, Hungary's telecommunications deregulation distinguishes among types of service, with varying degrees of competition allowed. Basic services, such as telex and standard telephone service, remain almost fully regulated. Public data networking and mobile communications are partially deregulated through licensing and limited competition. Value-added services—such as electronic funds transfers or access to databases or reservation systems, all of which offer more than just transmission of data—are fully open for competition. Most countries strictly regulate who can offer basic and who can offer value-added services.

Progress to date in Hungary has been rapid. The mobile cellular system was launched in 1991 through a partnership of the new Hungarian Radiotelephone Company and US West, one of the seven Regional Bell Operating Companies. The first 1,000 cellular connections were wildly oversubscribed; pent-up demand for phones is huge. In 1990, there was only one digital exchange in the country, supplied by Northern Telecom. Digital exchanges with a capacity of 100,000 phone lines were put in operation in late 1991. A leading German manufacturer has established a joint venture with a Hungarian company to produce 300,000 lines of telephone exchanges a year in 1992. The PTT's plans assume that 30–40 percent of equipment will be imported and offers generous tax incentives for joint ventures. The PTT has been entirely reorganized and has begun to create customer service centers in selected cities.

To keep that progress in perspective, it is worth noting that the goal for 2000 is that waiting time for installing a phone will be down to "less than one year." At the end of 1990, the waiting list for telephones contained 600,000 requests: only 900,000 phones were in use. The goal for the annual rate of growth in installations in the ten-year plan is 3.5 times faster than that achieved between 1987 and 1989.

The 1991–1994 priorities for the new Hungarian Telecommunications Company include:

1. Meeting business needs by building a digital network that can also take over much of the existing analog traffic.

2. Providing an "acceptable" level of basic service on the analog network.

3. Accelerating investment in major geographical areas that can provide a high rate of return.

4. Improving service to rural and underdeveloped areas of the country; today, 1,500 locations still have operators making the connections manually with plugs.

5. Increasing the number and availability of pay phones.

The equipment market has been liberalized and is highly entrepreneurial, and a wide variety of equipment is available in retail stores.

Financing is a major problem for Hungary. Foreign debt is $20 billion; the estimated cost of the ten-year plan is $6–$7 billion. The government's strategy is a mix of self-financing from increased telecommunications revenues, tax incentives to foreign firms, and joint ventures. The World Bank is expected to supplement its 1987 loan of $70 million by lending another $200 million.

The impacts to date of Hungary's Telestroika have been encouraging. The number of lines in use grew 13% in 1992 and 21% in 1993. New regional organizations have sprung up, together with a public company called Antenna Hungary, which offers VSAT services and is aggressively looking to enter the digital mobile communications market. First Pest Telephone (EPT) offers subscribers in Budapest links to 29 remote exchanges. Muszertechnika has built an equity partnership with Ericsson, the Swedish manufacturer of switches, installing 40,000 lines in 1991, 150,000 in 1992, and an estimated 200,000 in 1993. Regional groups, analogous to the U.S. Bell Operating companies, are positioning to capture business and to challenge the long-distance monopoly of Magav, the privatized national provider.

However, Hungary's 10 lines per thousand population (1993) still leaves it second or third from the bottom across the whole of Europe.

Technical Background

Although analog transmission has been the basis for voice (phone) traffic across the world, it is now obsolete and is being replaced by digital communications. Indeed, today *digital* and *telecommunications* are virtually equivalent terms.

Mobile cellular communications are wireless; no cables are needed, and transmission occurs via radio signals. Satellites operate on the same principle.

Questions for Discussion

Note: The following questions can best be answered by considering the economic effects of depreciation and construction lead times.

1. Until 1994, the Hungarian PTT plans to use analog systems exclusively for the extension of existing systems. Does this make sense? Why not move aggressively to conversion to digital systems?

2. Hungary's major new investments focus on mobile cellular networks. What advantages might this offer for businesses' growth? For social development? Instead of mobile communications, why not make use of high-speed physical links that offer far higher speeds and lower costs the highest priority?

3. Hungary was the first country in the old Eastern Bloc to set up a stock exchange. To date, it trades only a few securities. Will the telecommunications improvements help this market grow? If so, how?

2

Telecommunications Network Fundamentals

Objectives

- To learn the core terms and concepts of telecommunications technology
- To understand the nature and role of architecture and standards in developing an effective network capability
- To analyze several minicases that illustrate typical networks

Chapter Overview

| | | | | |

Chapter Overview

This chapter is a self-contained tutorial on the fundamentals of a firm's telecommunications resource. It does not go into much detail; that is provided in Part Two. The goal here is to introduce just the basic set of concepts and terms you need to know to get a clear overview of telecommunications as a whole. Of necessity, telecommunications is jargon-loaded; the rapid changes in technology create entirely new tools and techniques for which no words exist, and the need for extreme precision in defining and identifying procedures and products makes terms like *IEEE 802.5, hot-insertable cards, skinny stack, APPN,* and *V.21 bis* essential, even though often confusing. This chapter is about the fundamentals that underlie these and other terms.

The chapter describes the five main building blocks of a telecommunications network: terminals, transmission links, nodes and switches, transmission methods and procedures, and architecture and standards.

| | | | | |

The Technical Foundation of Telecommunications

Digital Communications: The Enabler of Modern Technology

Information technology (IT) includes computers, database management, and modern telecommunications. IT is built on a simple base: the ability to represent any form of information as a combination of representations of only two conditions—the presence or absence of an electrical signal. An on signal is indicated by a 1 and an off signal by 0. Each 0 or 1 is called a **bit**, which is the fundamental unit of coding for computers. **Digital coding** is the use of bits to represent information. Essential in representing information are **standards**—precisely-defined agreements on formats, procedures, and interfaces that permit designers and users of information technology, products, and services to develop systems with the assurance that they will work together with any others based on the same standards.

The **ASCII** (American Standard Code for Information Interchange) code is one of the oldest standards for digital information. The letter "c" in ASCII is represented by "0010011," for instance. Personal computer software for word processing generally offers a facility for creating or reading an ASCII file. There are thousands of standards relating to coordinating the elements of a telecommunications resource, such as linking a personal computer to a local area network, sending data through the switches and transmission links of a wide area network, routing a phone call, and so on. Standards provide the basis for both manufacturers (for designing their products) and users to link many varieties of equipment and services).

Figure 2-1 illustrates digital coding. It shows how varying an electrical signal is used to represent zeros and ones in the ASCII code. This highly simplified illustration shows a transmission link, such as a phone line, sending a 0 as a

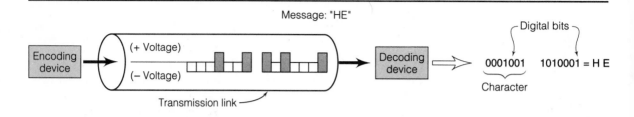

Figure 2–1

Example of digital coding using ASCII.

positive voltage and a 1 as a negative voltage; there are many techniques for transmitting digital data that maximize the amount of information that can be accurately sent along a transmission link. All that is needed is a mechanism to distinguish a 0 from a 1. Of course, the receiving device must also use ASCII to recognize and decode incoming messages.

Digital transmission, which sends bits as pulses in a high-speed stream, replaces the older mode of analog transmission, which is the base for the world's existing telephone systems and is still widely in use. **Analog transmission** replicates the speaker's voice by continuously varying the electrical signal to correspond with the variation in the sound wave (see Figure 2-2).

Digital transmission allows faster speeds, greater accuracy, improved error checking, more accurate storage and retrieval of information, and many other advantages. It also merges telecommunications and computing.

The Components of a Network

The five main building blocks of any network—terminals and workstations, transmission links, transmission methods, nodes and switches, and network architecture and standards—are introduced in the following subsections.

Terminals and Workstations These devices, which include personal computers, **intelligent workstations**, barcode readers, telephones, mainframe computers, and mobile phones, are the access points to the network. (The term intelligent workstation is increasingly used as a general description of a personal computer with built-in telecommunications capability that manages many aspects of a transaction through its own software, instead of relying on a computer at the other end of the telecommunications link.) An ATM and the Health Link electronic claims processing terminal (Figure 1-2) are both intelligent workstations.

Transmission Links These means of conveying a signal range from cables to fiber optics to satellites. These media are essentially equivalent, though they obviously have different costs, features, and operating requirements; a phone call may be sent via a physical cable, a microwave relay, or a transatlantic satellite.

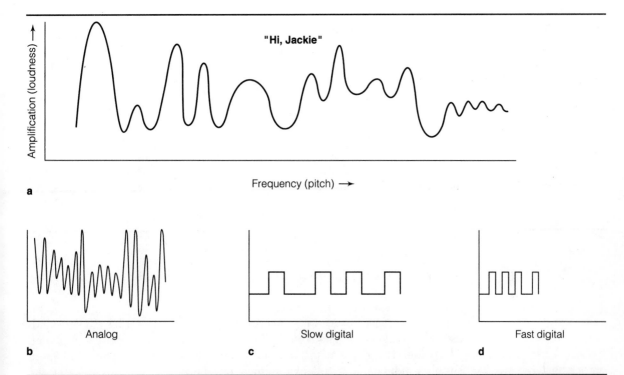

"Hi, Jackie"

Amplification (loudness) →

Frequency (pitch) →

a

Analog

b

Slow digital

c

Fast digital

d

Figure 2–2

Analog and Digital Transmission Compared

In contrast to digital transmission, in which the signal has only two states (0, 1), the signal in analog transmission varies as the analog of the sound of a voice (graph a). If the frequency of the analog signal is increased (as when an audiotape is fast-forwarded, making the voice squeaky and high-pitched), the signal becomes blurred and unrecognizable (b). In contrast, if a digital signal (c) is speeded up, each discrete bit remains distinct and recognizable (d).

Transmission Methods The same transmission medium (satellite, fiber optics, and so on) may send and receive information through very different techniques and **protocols**—procedures for establishing a telecommunications link, interpreting how the data is represented and coded, how it is transmitted, and how errors are to be handled.

Nodes and Switches A *switch* is a generic term for a device that manages the coordination and routing of traffic through a network. A **node** is a point to which a group of devices and transmission lines connect; it will usually contain a switch or a series of switches. Hundreds or even hundreds of thousands of devices may at any one time be sending messages across a network via the nodes, making hardware and software technology, cost-efficiency, and reliability major challenges to the network implementer. The diagram of the public-switched telephone network (Figure 1-4 on p. 11) showed just a few of the nodes that contain powerful switches.

Figure 2–3	Building Blocks	Examples
The Five Main Building Blocks of a Telecommunications Network	Terminals	Personal computers, telephones, satellite earth stations, cellular phones, computers, point-of-sale cash registers
	Transmission links	Twisted pair telephone wire, coaxial cable, ultrahigh frequency radio, line-of-sight microwave, fiber optic cables, infrared
	Transmission methods	Circuit switching, packet switching, analog, digital
	Nodes and switches	Bridges, routers, private branch exchange (PBX), multiplexers, gateways, concentrators, front-end processors
	Architecture and standards	Transmission protocols (X.25, Ethernet, token ring); message format standards (X.400, X12); network architectures (SNA, OSI); network management standards

Network Architecture Standards These network building blocks define the interfaces between equipment and applications and the procedures for sending and receiving information. An **interface** is the point of interconnection between two devices, such as a printer and a personal computer.

Once the specifications for how two devices interconnect have been standardized, suppliers need not know the internal details of each others' equipment. The wider the range of established standards, the more open—the more supplier-independent—telecommunications becomes. Standards are at the heart of the telecommunications field today and define the limits of the practical versus the theoretical.

The **architecture** is the master blueprint of standards for a network. The technical architecture—in effect the telecommunications strategy—is built on a list of standards that ensure interoperability of equipment, information, and services to meet specific business needs.

The five main building blocks of any telecommunications network are summarized pictorially in Figure 2-3. The concepts of standards and architectures may be difficult to grasp. For example, the term *standards* often sounds like *regulations* to business people. "Our standard is XYZ" then is interpreted as "You *must* use XYZ" and thus as an effort by telecommunications bureaucrats to impose their authority on decentralized business units. "Architecture" is associated in most peoples' minds with fixed buildings, so what, then, is an "integrated, open architecture"? The use of the terms *architecture, integration,* and *standards* too often sounds like centralization and rigidity when it is intended to create flexibility and decentralization of decisions on the uses of telecommunications. For the designer to understand these terms is not at all the same as explaining them in a way the client can understand. This is one of the main challenges this book attempts to address.

Types of Networks The two main distinctions between networks are their geographic span and whether they are private or public. Geographic span ranges from a local area (within a building or a campus) to a wide area (a region, a country, or global),

Type of Access	Wide Area Networks		Local Area Networks	
	International	Domestic	Metropolitan/campus	Local area
Public	International direct dial phone system; satellites used by television broadcasters	AT&T, MCI and Sprint long distance; individual nations' public data networks	Cellular phone networks (within city or region); regional Bell operating company cell relay and frame relay services	Not applicable: LANs are intraorganizational facilities owned by the company
Semipublic value added networks	Airline international message service (SITA); European automotive manufacturer/ supplier network (ODETTE)	On-line information services (Prodigy, CompuServe, Dow Jones); banking shared access ATM network (Monec, Cirrus)	Not applicable: public providers span metropolitan areas via public services	Not applicable: VANs are interorganizational shared services
Private Networks	Multinational firms' leased circuits linking international locations	VSAT network used by individual company	Leased lines linking, for example, a university's dorms, medical center, libraries, and multiple campuses.	Departmental systems linking personal computers, file servers, printers.

Figure 2-4

Types of Networks

with emerging intermediate variations. A private network is a service offered by telecommunications providers on a fixed-price, leased basis to firms or other users. Public network services are priced on a pay-as-you-go basis and are used by a wide range of subscribers; the standard telephone service is a public network, for instance. Figure 2-4 summarizes the resulting types of network, with examples. The five most widely used types of network are:

Local Area Network A local area network (LAN) links together devices that are located close to each other, typically within a single building. Campus networks and **metropolitan area networks** extend the limits; these networks (or linked collections of LANs) are localized but cover a bigger area than a LAN.[1] LANs are purchased and operated by the user organization, so they are not referred to as private facilities, even though they are obviously not a public service; the term *private* is used only to refer to services offered by external telecommunications providers.

Figure 2-5 shows an example of a local area network and the typical devices LANs connect. This system consists of 15 Macs and three PCs at Blue Cross

---

[1]The term *global LAN* sounds like an oxmoryon—like English cuisine, jumbo shrimp, rap music, and military intelligence—but many experts believe that LANs will completely take over from WANs and that there is no limit to the number of LANs that can be connected together to create what is termed an "enterprise" network.

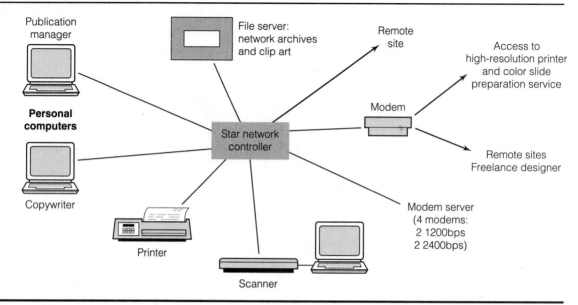

Figure 2–5

Blue Cross and Blue Shield: A LAN for Electronic and Desktop Publishing

headquarters plus additional Macs at remote sites. The printers and scanners use AppleTalk in a star configuration. Since implementation, the LAN has halved the cost of producing publications and reduced lead time from over a week to two days. Remote servers allow access from staff or freelancer homes, and the data compression feature allows faster transmission of text and graphic files.

Wide Area Private Network A wide area network spans a large geographic area. A WAN may be a voice network, a data network, or one that combines voice, data, and any other type of information, such as video. WANs typically connect very large numbers of devices and carry many types of traffic. The central wide area network that links a firm's distributed operations and that may link to outside WANs (suppliers, banks, public networks, and so on) is often termed the **backbone** network. Some firms operate private voice networks as well as data networks.

Figure 2-6 shows two possible backbone designs for a wide area network containing ten main nodes. Depending on the mix of traffic in terms of volume and type (engineering design documents, electronic mail, fax, voice, and so on), either design may be better than the other. Cheaper is not necessarily better; nor is bigger necessarily better.

Figure 2–6a

Wide Area Network Design Options Option A: 10 nodes, 16 links, leased-line costs of $1.3 million per month

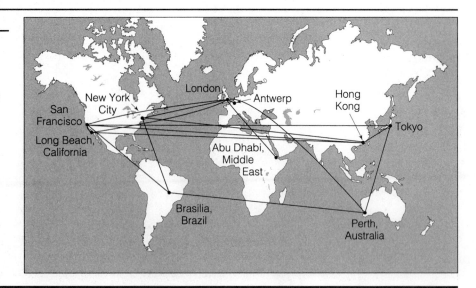

Public Voice Network The public voice network is the telephone system, which increasingly is being adapted to handle many other types of traffic. In the United States, the long-distance part of the public voice network is operated by AT&T, and the local parts by the seven Regional Bell Operating Companies (RBOCs), your "local phone company." RBOCs are required to allow long-distance service providers, such as AT&T, MCI, Sprint, and others—called **alternative operator services**—to connect to the public network. Alternative operator services generally provide special-service, low-cost transmission, niching their products; some, for example, offer hotels or schools bulk discounts for the right to handle calls by guests and students.

Public Data Networks (PDNs) Public data networks are an alternative to private networks and in many countries are the only available or legal option. The main reasons for this are the long history of telecommunications as a monopoly in every country until recently (that is, the view that telecommunications is a social infrastructure like roads), and the lack of money, technology, or economies of scale to make private networks practical. The advantages of using a PDN are that the infrastructure is already in place and that the heavy costs of investment and operation are amortized over many users.

Value-added Networks (VANs) Value-added networks are semiprivate networks that offer more than just transmission: They also include software and/or standards that handle its messages and transactions. Many, for example, are industry-specific, such as LIMNET VAN used by international insurance brokers and EDI networks used by retailers and their suppliers.

Figure 2–6b

Option B: 10 nodes, 11 links, leased-line costs of $0.8 million per month

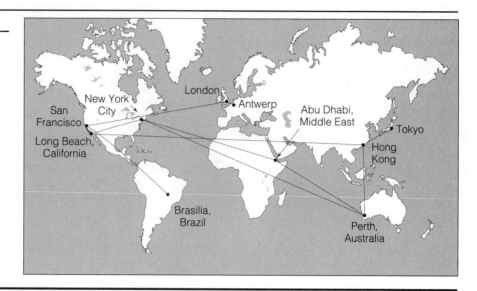

Another type of VAN is the virtual private network (VPN), which combines the guaranteed availability of private networks and the immediate accessibility of public networks. Table 2-1 provides examples of several leading VANs.

Uses of Networks

A large company will typically use a variety of both LANs, private WANs, the public phone system, VANs, and PDNs. Telecommunications suppliers are creating variants of each of these; some are even offering a global LAN. The Regional Bell Operating Companies are aggressively offering high-speed public network services that previously were confined to private networks; the most important of these is **switched megabit data services (SMDS)**, which is ideal for intermittent and occasional transmission of video and such images as X-rays and design documents. Switched megabit data services are high-speed (hence megabit) telecommunications that can be used on demand.

As a result of technology such as SMDS, cable TV companies and local phone-service providers are now positioning to enter each others' markets. Bell Atlantic, the most innovative of the RBOCs, is targeting entertainment as a growth area; Southwestern Bell has bought a cable TV firm; and several cable TV companies are experimenting with joint ventures with phone companies to add phone calls to their high-capacity, underutilized cables. US West has acquired part of Time Warner.

The various types of network that can be linked and the services and information they can share depend on the extent to which they are based on an architecture, a master plan of standards that integrates their very different technologies. Linking incompatible LANs is a major growth area today. Linking LANs to WANs to LANs is *the* headache for most telecommunications managers in large organizations.

Table 2–1	Company	Proportion of market*	Position/strategy
Leading Value-added Network (VAN) Service Providers	**BT GNS**	26%	Very aggressive; has exploited London's central position (literally) in world's electronic markets. Greenwich Mean Time provides it with widest time zone span from U.S. to Asia when markets are open.
	Sprint	24%	International services are a growing priority as U.S. long-distance market share remains static.
	GEIS	16%	Early VAN provider; owned by General Electric. Focused on vertical and niche markets, such as security, fire alarms, and maintenance systems for apartment buildings.
	IBM IN	13%	Operates one of the world's largest business networks built on its own SNA. Largest provider of systems integration services. Merged IN with Sears Technology Services to create Advantis in 1992. VANS are not a core business but an extra weapon in other areas of competition.
	AT&T Easy Line/Istel	8%	Strategy similar to BT: acquired Istel (Great Britain's equivalent to Tymnet).
	Infonet	8%	The leader in VAN technology. Owned by a consortium of foreign PTTs and MCI. Innovations include Infolan, which connects LANS around the world, and international EDI services.
	Compuserve	3%	Provides consumers and businesses with VAN services: information, shipping, electronic mail, bulletin boards. United States only.

*World market currently totals $2 billion
Data from Yankee Group, 1992.

LANs and WANs are built on entirely different technology bases. LANs developed as part of the personal computer revolution, and WANs evolved from long-distance telephony. PCs developed their own operating systems that were completely independent of those used on mainframe computers.

Transmission Media   A transmission medium is whatever wire or radio signal is used to carry coded information. Carrying a signal across a distance is central to telecommunications. Modern telecommunications exploits the electromagnetic spectrum using a carrier signal, such as a radio wave or an electrical signal along a cable, to move information. The **electromagnetic spectrum** is the invisible radio and infrared rays and visible light whose frequencies range from low to ultra-high,— from the sound we can hear (low frequency) through AM radio, then FM, then

UHF television, radar, microwave, infrared, and finally visible light. The higher a signal's frequency (cycles per second, or hertz), the more information that can be encoded onto it. High fidelity sound, for instance, operates at around 10^4 cycles per second, which gives it a frequency range of 3 to 30 kilohertz. Figure 2-7 shows the telecommunications transmission media associated with each frequency range.

Any message can be sent through *any* telecommunications medium that can carry the signal. When you make a transatlantic telephone call, you do not know in advance if it will travel by satellite or cable. In practice, of course, each medium has its own characteristics, costs, advantages, and disadvantages. The one-quarter-second delay created by the time it takes a signal to travel up to and back from a satellite 22,000 miles in the sky, for instance, is only an irritant on the phone.

Bandwidth

Bandwidth is the basic measure of the carrying capacity of a transmission link. It is the range of usable frequencies—the width of the band—and is computed by taking the difference between the highest and lowest frequencies. Frequencies are expressed in **hertz**, and its multiples of 1,000: **kilohertz** (thousands), **megahertz** (millions), **gigahertz** (billions), and even **terahertz** (trillions). Figure 2-8 summarizes the technical basics of bandwidth.

Fiber optics operate in the near infrared region of the electromagnetic spectrum, at 10^{14} cycles a second; the frequencies in this region are 30 to 400 terahertz (trillions of hertz). The copper wires used in the predigital era (when telecommunications meant telephones and telex machines) operated at a frequency of 300 to 3,300 hertz. This means that a fiber-optic link that replaces a copper cable can carry thousands of times more traffic. The physical length of the link is the same, and indeed the fiber optic cable may physically replace the copper one, but the bandwidth is much larger.

A microwave radio signal can range from 2 to 25 gigahertz. This 20-billion-cycle frequency range can be split up into different **channels**—specified parts of the transmission signal. A radio channel operating at 30 megahertz (30 million cycles of usable frequency) can in turn be split up into 6,000 voice channels, each of which uses 5,000 hertz. A basic principle of modern telecommunications is to share high-speed transmission links. It would not make sense to dedicate—to reserve for one user only—the 30 megahertz to a single speaker because the human voice has a frequency of under 7,000 hertz (roughly 200 to 7,000 hertz). Figure 2-9 illustrates how a signal is shared among many channels; the principle is the same for fiber, cable, satellite, or radio.

Satellites receive and send radio signals from a geosynchronous orbit 22,300 miles above the earth. **Geosynchronous** means that satellites remain apparently stationary directly overhead; at this altitude they maintain their position relative to the earth as it rotates. Satellites operate along two main frequencies: the lower-frequency **C band** (4 and 6 gigahertz) and the higher-frequency (11 and 14 gigahertz) **Ku band,** which is digital. The higher the frequency, the shorter the wavelength of the signal, which means that a smaller satellite earth station can receive that signal—and a single rain drop can distort it. **Very small aper-**

Electro-magnetic spectrum	Frequency (cycles per second)	Frequency Range	Communication Service
	— 10 —		
	— 10^2 —	Extremely low frequency (ELF) Hertz	Telegraph, teletypewriter
	— 10^3 —	Voice frequency (VF) 300 Hz to 3 kHz	Telephone circuit
	— 10^4 —	Very low frequency (VLF) 3 to 30 kHz	High fidelity
	— 10^5 —	Low frequency (LF) 30 to 300 kHz	Fixed, maritime mobile, navigational radio broadcast
Radio	— 10^6 —	Medium frequency (MF) 300 kHz to 3 MHz	Land and maritime mobile radio, radio broadcast
	— 10^7 —	High frequency (HF) 3 to 30 MHz	Fixed, mobile, maritime and aeronautical mobile, radio broadcast, amateur radio
	— 10^8 —	Very high frequency (VHF) 30 to 300 MHz	Fixed, mobile, maritime and aeronautical mobile, amateur radio, television broadcast, radio location and navigation, meteorological communication
	— 10^9 —	Ultrahigh frequency (UHF) 300 MHz to 3 GHz	Television, military, long-range radar
	— 10^{10} —	Superhigh frequency (SHF) 3 to 30 GHz	Fixed, mobile, radio location and navigational, space and satellite communication, microwave systems
	— 10^{11} —	Extremely high frequency (EHF) 30 to 300 GHz	Waveguides, radio astronomy, radar, radiometry
	— 10^{12} —	Far infrared region 300 GHz to 3 THz	
Infrared	— 10^{13} —	Mid-infrared region 3 to 30 THz	
	— 10^{14} —	Near infrared region 30 to 400 THz	Optical fibers
Visible		Visible	Heliograph, signal flags

Figure 2-7

The Electromagnetic Spectrum

1 kilohertz = 1000 hertz; 1 megahertz = 1,000,000 (10^6) hertz;
1 gigahertz = 1,000,000,000 (10^9) hertz;
1 terahertz = 1,000,000,000,000 (10^{12}) hertz

ture terminals (VSATs)—satellite earth stations that are less than six feet in diameter—are widely used by retailers because their low cost allows the smallest store to have the same communications capabilities as the biggest.[2]

---

[2]Think of VSATs as digital TV sets that can receive voice and data communications instead of programs. Magnavox offers a portable, mobile VSAT receiver with attached console that weighs 50 pounds and can be taken as carry-on luggage on an airplane.

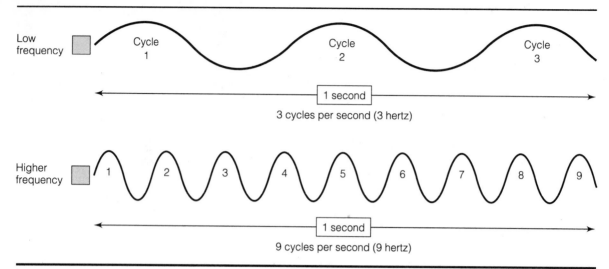

Low
frequency

Cycle 1   Cycle 2   Cycle 3

1 second

3 cycles per second (3 hertz)

Higher
frequency

1   2   3   4   5   6   7   8   9

1 second

9 cycles per second (9 hertz)

Figure 2–8

The Technical Basics of Bandwidth

The carrier signal sent by a transmitting device has the form of a sine wave. Bandwidth is the range of usable frequencies, delimited by the highest and lowest frequencies of a signal. Increasing the cycles per second increases the bandwidth.

The footprint of a satellite—the circle within which the signal from the satellite can be picked up by an earth station—can be as much as a third of the earth's surface. An earth station is a dish-shaped receiver commonly seen on top of buildings and in people's backyards. It may be as large as 100 feet in diameter or as small as 3 feet, with even tinier receivers coming onto the market. Figure 2-10 summarizes satellite technology and providers.

Transmission Speed

The speed at which information is sent along a transmission link is usually stated in rates: bits per second, kilobits per second, megabits per second, or gigabits per second.

Bits Per Second (bps) Bits per second is the standard measure of digital communications; the bits are individual zeros and ones. A typical voice-grade telephone line transmits at 1,200 bps up to 9,600 bps.

Kilobits Per Second (kbps) Thousands of bits per second. 9,600 bps may thus be expressed as 9.6 kbps (pronounced "nine dot six"). Common high-speed transmission rates are 56 kbps and 64 kbps.

These rates can handle most types of data both efficiently and at low cost.

Megabits Per Second (mbps) Millions of bits per second. The T1 link, which has become the standard communications offering from providers such as AT&T, has a total capacity of 1.544 mbps, typically shared among users in 56-

Figure 2–9

Channels: Sharing a Signal

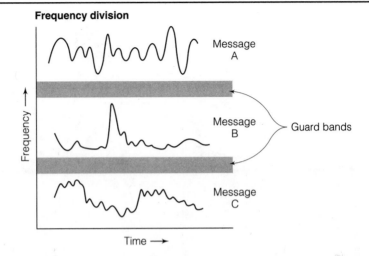

Frequency division

Message A

Message B

Guard bands

Message C

Frequency →

Time →

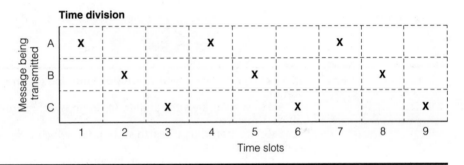

Time division

Message being transmitted

	1	2	3	4	5	6	7	8	9
A	X			X			X		
B		X			X			X	
C			X			X			X

Time slots

In frequency division sharing, signals of different frequencies, separated by guard bands to avoid interference between channels, are transmitted simultaneously. Messages A, B, and C in this case are analogous to cellular phones, FM radio, and fax machines, respectively. In time division sharing, each message uses all the frequency, but is transmitted only during rotating time slots.

kbps channels. Broadband communication uses high-frequency transmission to send large volumes of traffic over long distances.

Gigabits Per Second (gbps) Billions of bits per second. Gigabits per second represent the new frontier of telecommunications and is possible only through fiber optics. The emerging new fiber public network standard, **SONET** (synchronous optical network), provides speeds of up to 2.4 gbps. SONET is the base of the next generation of advanced networks. To date, the highest reported speed attained over fiber over distance is 28 gbps.

Figure 2–10a

**Satellite
Communication**

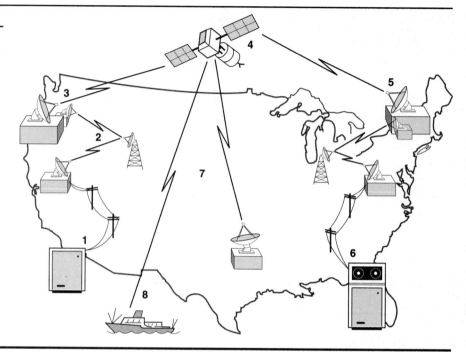

PC links to network; signal travels by land line (1) and microwave (2) to earth station (3).
Earth station links to satellite (4) in geosynchronous orbit 22,300 miles above the earth.
Satellite transmits to earth station (5), which links to remote host computer (6).
Signal can also be picked up by any other earth station within the satellite's footprint (7, 8).

Figure 2-11 summarizes the typical transmission speeds in the hierarchy of telecommunications links from ultrafast local to international long-distance transmission.

*Message Speed
Versus Transmission
Speed*

Transmission speed in and of itself does not determine message speed. Different types of data require more digital bits to represent their contents. For instance, a one-page typed memo amounts to about 15,000 bits. At 1,200 bps, this takes around 12 seconds to transmit. A high-resolution image of that same page, capturing its exact format, shading, and print fonts, requires over 100,000 bits. Sending this over a 1,200 bps line takes 90 seconds.

All digital communications data and *all* computer data are coded in bits. Examples of the bits needed to code different types of communication are:

▸   A credit card authorization request: 1,000 bits

▸   An electronic mail message: 5,000–15,000 bits

Figure 2–10b

(continued)

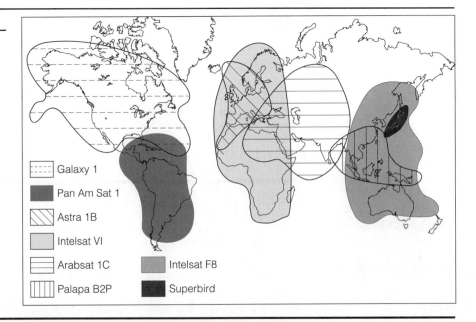

Galaxy 1

Pan Am Sat 1

Astra 1B

Intelsat VI

Arabsat 1C

Palapa B2P

Intelsat F8

Superbird

CNN uses eight satellites to get news in and out, including four of the most powerful ringing the earth: Astra, Palapa, Arabsat, and Superbird. The "footprints" of the satellites cover most of the world, allowing CNN to reach virtually everywhere.

▸ Digital voice (1 second): 56,000 bits.

▸ Full motion video, (television quality, one second): 10,000,000 bits without data compression

Because digital voice communication requires 56,000 bits to code, 56 kbps has become a standard for data communications; it provides an efficient base for transmitting voice and data together.

Data compression techniques can reduce the number of bits needed for coding a message. This is especially useful in handling high-resolution documents and video; coding color takes many bits, and full-motion video demands 30 frames a second. Digital **HDTV (high definition television)**—television transmission that provides a high-resolution picture with over 1,000 lines on the screen versus today's 525 lines—would need transmission speeds of several gigabits per second without data compression.

Data compression essentially screens the image, and instead of sending "red dot shade 109, red dot shade 109, red dot shade 109" it transmits "next 1,200 dots all red shade 109." This can reduce the number of bits to be transmitted for video by up to a factor of 200. Data compression is dramatically improving costs and efficiency of data communications and is expected to transform the economics of broadcasting and cable television.

	Type of Network	Bandwidth	Source
Figure 2–11	Intra-facility LAN	Gigabits	Company-owned cabling (fiber, coaxial)
Typical Transmission Speeds	Campus LAN	16-100 Mbps	Company-owned facility (fiber, microwave)
	Metropolitan	45 Mbps	Leased circuits (local carrier alternative access provider)
	Regional	1.5-45 Mb	Leased circuits (RBOCs, interexchange carriers, T1, T3)
	Nationwide	56-1.5 Mb	Leased circuits (interexchange carriers)
	International	9.6 kbps-64 kbps	PTT, VAN services

The differences in speed are mainly created by distance. Signals fade, electrical power is erratic, and outside objects interfere (an airplane overhead or a fluorescent lamp being switched on). Additional equipment such as repeaters, adapters, and switches must be installed, slowing the traffic flow.

The more detailed the information and the more accurately a video format and color must be represented, the greater the bandwidth needed. In addition, the faster that computers create information that must be transmitted across a network, the greater the required bandwidth of the link. The chip revolution of the 1980s transformed the processing of raw data. The bandwidth revolution that began toward the end of the 1980s promises to make that transformation look slow. With fiber optics in the home, a gigabit per second is not unlikely within the first decade of the coming millennium.

Terminals

Dumb Versus Intelligent Terminals

Terminals are input/output devices that interface directly with the user and originate or terminate the communications message being transmitted. They convert data, text, speech, or images into electronic signals for transmission across the network and then convert the signals back again to data, text, speech, or images at their destination in the network. Terminal devices include video display units, personal computers, telephones, electronic image scanners, fax machines, computer printers, and burglar alarms, among many others.

Terminals have evolved from single-purpose devices (such as the telegraph key) designed to send or receive one type of information from point A to point B, to multipurpose units that are capable both of sending and receiving a wide range of information types to multiple users simultaneously and accessing a range of services on the network. The reader is of course aware of the dazzling variety of features now common in the ordinary household telephone. There is much yet to come, including pay-per-view movies.

On the data communications side, the personal computer (PC) has evolved from a stand-alone data processing system that runs word processing and spreadsheet software to a dumb terminal accessing a mainframe computer, to an intelligent workstation that is part terminal and part computer connected to a sophisticated network of similar devices. That intelligence may be used to handle error-checking, format displays on the screen, compress data, or provide security.

As terminals become more intelligent, they become capable of handling many more types and quantities of information, which leads to new opportunities and tradeoffs and a choice concerning how much information processing is done in the terminal itself.

Operating Systems   Sharing data and carrying out transactions depend on the computer's **operating system**—the complex software that manages all aspects of a computer's applications, such as basic utilities, file management, running software applications, and error-handling. The operating system determines the capabilities of a computer, especially the variety of software it can run. **Network operating systems (NOS)** are a relatively recent development in which the software manages the functions of a LAN. A NOS controls the operations of the network, coordinating the communication between the devices on it and determining which devices can communicate via the network and how. Table 2-2 on page 62 summarizes the functions of a network operating system, with examples.

Operating systems have replaced hardware as the main battleground among computer vendors. In particular, Microsoft, Apple, and IBM have fought vigorously (and often viciously) to establish *their* operating system as both the base for consumer uses of personal computers and the cornerstone for businesses' technical architectures. Even though computer operating systems are not directly concerned with telecommunications, in practice the two are highly interrelated, not so much at the level of transmission of data, but in terms of how the operating system handles higher level functions.

Higher level functions go beyond the simple movement and processing of bits to include data management, security, transactions, and specific network standards and protocols. Higher level functions are especially complex and important in a high-volume transaction processing environment, such as banking or point of sale; in those cases, special-purpose communications management software is often required to ensure reliability, efficiency, and security. The boundaries among telecommunications, the operating system, and software application are then blurred, making it very difficult to change one without affecting the others. The most complex operating systems are those used on the mainframe computers that have historically been the workhorses for large transaction processing systems. Even though more and more computing functions are being distributed to intelligent workstations, these mainframe operating systems remain central to many firms' operations and are a major constraint on telecommunications options.

Table 2–2

Leading Network Operating Systems

Features	Novell 3.11	Microsoft Manager	Banyan VINES 5.0	Santa Cruz Unix
Memory required (megabytes)	4	8	8	2
Storage supported (gigabytes)	32,000	4	4	Unlimited
Global directory available?	No	No	Yes	Yes
Computer-to-computer communication?	No	No	No	Yes
Connects to IBM network management system?	Yes	Yes	No	Yes
Supports appropriate software?				
OSI network management	Yes	Yes	Yes	Yes
Macintosh System 7	Yes	Yes	Yes	Yes
IBM OS/2	Yes	Yes	No	Yes
Unix	No	Yes	Yes	Yes

Even with the far simpler and more standardized operating systems used on personal computers, the boundaries between them and telecommunications are often blurred as well. One difficulty that has resulted from greater intelligence and processing capability in PCs is incompatibility among PC operating systems. For example, the IBM DOS operating system is not compatible with the Apple Macintosh operating system.

The **UNIX** operating system has evolved independent of both Apple and IBM. UNIX offers particular features that differentiate it from most other operating systems, including better exploitation of hardware, greater **portability**, the ability to move UNIX-based software between hardware environments, and **multitasking** (running many software programs simultaneously). 1993 saw the introduction of Microsoft's NT operating system, which competes directly with IBM's flagship OS/2. New operating systems are expected from IBM and Apple's joint venture, Taligent, and others in 1994 and 1995.

It is common within a firm or department to find Macs; UNIX-based workstations; DOS- and Windows-based desktop, notebook, and laptop PCs; and PS/2s—all of them incompatible. Consequently, these powerful stand-alone computers can communicate directly with each other only in a limited fashion: through the use of various conversion devices in the network. Much of Novell's success in the LAN software market through its NetWare NOS reflects its early commitment to interconnecting these different types of personal computers.

Although the main trends in computing and telecommunications move applications off the mainframe and onto personal computers, the transaction base often remains on mainframes. The use of operating systems like UNIX that are compatible with multiple types of central processors and computer systems can eliminate this incompatibility. However, existing incompatibilities may prevent switching from one operating system to another. For example, application programs and databases built on computers that run on, say, IBM's MVS operating system or Digital Equipment Corporation's NAS cannot be directly converted to run under UNIX. In addition, there is no perfect or universal operating system, and it will be many years before any standard software or data resource can be transferred to any other system, regardless of its manufacturer and operating system.

Other Types of Terminals

Although workstations are the most important type of terminal from a business perspective, there are many others, including barcode scanners, machine tools, digital phones, sensors, and alarms. Every terminal should provide transparent, easy-to-use access. Transparent simply means that the user does not even know the interface is there, any more than you "see" the electricity that runs your toaster through the standard three-pin plug. Terminals use a wide range of general- and special-purpose connectors, including adapters, cables, and sockets. **Adapters** are small boxes or boards that have microprocessors in them; a typical use of adapters is to connect a PC to a specific type of LAN.

Terminal technologies are the main forces driving innovation in the use of both telecommunications and computers. With dumb terminals, the intelligence is entirely in the central computer; most recent developments in computing exploit intelligence in the workstation so that functions can be shared. **Distributed processing** shares the workload in computing and in communication between the local PC and the remote mainframe. **Client/server computing** extends this sharing to create a dynamic workflow between a powerful client application that needs data or software and a server that provides them. Two types of software that manage the linkages between client and server are **application program interfaces (APIs)**, which pass between two software elements the information needed for them to work together, and **remote program calls (RPCs)**, in which one software service triggers the activation of another. **Cooperative processing** extends client/server computing; complex transaction processing systems are built up out of software on different hardware units that work together across the network.[3]

Figure 2-12 on page 64 summarizes trends in distributed systems. The technical details of each of these systems are very complex, but the logic is the same: Exploit the power of low-cost, high-power electronics to shift intelligence as far out in the network as practical. Each system relies on telecommunications to coordinate the process of local and remote computer cooperation. This trend has removed the bottleneck of large mainframe computers as the key element in a complex business network; it has also brought computers and telecommunications into almost total interdependency.

---

[3]The term *client/server computing* is replacing *cooperative processing* as the general term for the new generation of distributed systems. They are essentially the same in their aims and principles. The distinction is useful, however, as a reminder that LAN-based client/server applications are far simpler to design and implement than enterprise-wide ones.

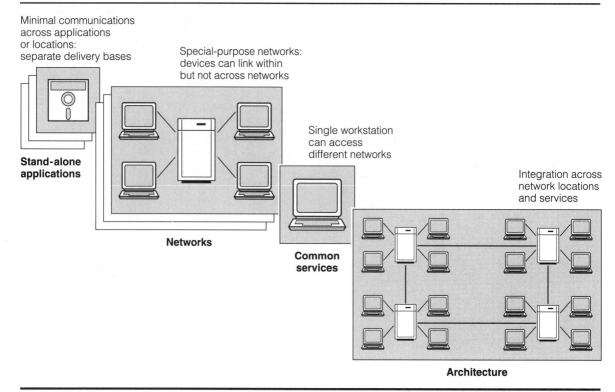

Figure 2–12

Trends in Distributed Systems

Transmission Links

Transmission links are the paths between the terminal devices that carry the coded electronic signal from origin to destination. The medium used to create the path may be some form of cable (coaxial cable, twisted wire pair, optical fiber, and so on), or through-the-air transmission (microwave radio, satellite, and so on). **Coaxial cable**, or coax, is an insulated wire surrounded by a second wire that insulates it and sheathed by a solid outer insulation; this is the black wire used to connect VCRs to TV sets. **Twisted wire pair** is the standard two-wire phone cable. The properties of the various media (that is, cable, satellite, microwave) determine when it is cost-effective and more efficient to use one type or the other (or a combination) for a given transmission. Figure 2-13 shows just a few of the many types of cables, plugs, and sockets in common use.

Transmission links can be set up to carry information in analog or digital form. For transmission over long distances it is often efficient (and in fact nec-

Figure 2–13

Cables, Plugs, and Sockets

Cables

Plugs

Sockets

Only a very few "standard" devices are pictured. The vast multitude of devices is the source of incompatibility.

essary with satellites and microwave radio) to superimpose the digital signal onto an analog signal carrier for the long-distance haul, and then unload the digital signal from the analog carrier at the distant end. A device called a **modem** (modulator-demodulator) is used to do the loading and unloading of the digital signal onto the analog carrier. Modems modulate the digital pulse to convert it to an analog wave; they also establish the speed of transmission in bits per second. Fiber optics is fully digital; a laser device pulses light at rapid speeds and has only the two conditions (pulse and no pulse) in contrast to the continuous signal of, say, a radio broadcast. There is thus no need for a modem in fiber optics.

Transmission links can vary considerably in their information carrying capacity. This capacity is a function of two factors: (1) the *strength of the signal* being transmitted, relative to the noise in the system; and (2) the *bandwidth of the link.*

The strength of the signal of a radio station is defined by the number of watts at which it broadcasts. The strength of an electrical signal along a cable depends on voltage. As a signal moves through its transmission medium, it fades and may get distorted. We experience this routinely with AM and FM radio stations' broadcasts. **Repeaters** are devices placed at intervals along the link to check, regenerate, and boost the signal. In the fiber optics TAT-9 submarine cable that links Great Britain and France to New Jersey, the repeaters are placed every 41 miles; TAT-7, the coaxial cable it replaced, had repeaters every six miles.

Noise

Noise is a key engineering concept in telecommunications. Indeed, the theoretical foundation of modern telecommunications is built on Claude Shannon's theorem that first defined "signal to noise ratio." **Noise** is a distortion that makes a telecommunications signal either fuzzy or inaccurate. In everyday life, we encounter noise on telephone lines and in television reception. A car radio becomes noisy when you drive under a high-voltage electrical transmission line, for instance.

Some transmission media such as phone lines are inherently noisy. Fiber optics is not only the fastest available medium today but the least noisy. Weather conditions can create noise in satellite radio transmissions. The need to check for and compensate for noise depends on both the medium and the nature of the information being transmitted.

Transmission errors are routinely caused by noise. The level of noise on a phone call rarely prevents effective conversation, but in data communications incorrect transmission of a single bit can change the value of a funds transfer from $20 million to $2 million or $200 million. A voltage change that generates a spark of what is termed *impulse noise,* for instance, might last just 1/100 of a second; that would be enough to distort close to 100 bits on a 9.6-Kbps line. A significant problem for many of the emerging ultra-high-speed transmission methods is that the network software is too slow to recognize and process an error; the data arrives at gigasecond (billionth of a second) speeds but the soft-

Figure 2–14

**Common Causes of
Electrical Problems**

The ideal alternating current: sine wave

TRANSIENT: Motor-driven device interference
(copying machine, elevator) or radio interferences

SPIKE: Sudden increase in voltage when loads
drop off a circuit (coffee machine or heater
turned off): typically lasts for just a millisecond

OVER VOLTAGE: Spikes several seconds long
typically caused by termination of heavy loads
on electrical utility

UNDER VOLTAGE: Defined as 20% or more
reduction lasting over 15 milliseconds; occurs
when loads suddenly increase or when utility cuts
power in brownout: 90% of office power problems

OUTAGE: Blackout of utility, fiber line cut,
switch failure

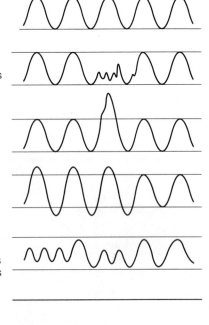

ware operates at microsecond (millionth of a second) speeds. Thus there can be a thousandfold difference between fiber optics (gigaseconds), and computer hardware and software (microseconds), and between them and disk access and storage (milliseconds).

The designer of a business networking application need not be concerned with distortion and error correction and will assume that they are properly provided for, but a telecommunications provider and the implementer of the service will give these factors constant attention. A large part of the telecommunications equipment industry is devoted to handling noise and other sources of errors. The telecommunications engineering profession has generated ingenious ways of detecting and correcting errors, many of which depend on complex mathematical techniques for analyzing a signal.

For LANs and digital WANs, **bit error rates**—the basic measure of quality of transmission—are shown as the number of errors per million or billion bits. One error per million bits is barely acceptable for today's business networks. Fiber optics is immune to electrical interference; as a result it has a bit error rate of around 1 in 10^{-9} (one error per billion bits). This is a thousandfold less than that for metal wire.

Clearly, electrical problems constitute a significant source of noise. Figure

Figure 2–15

Multiplexers

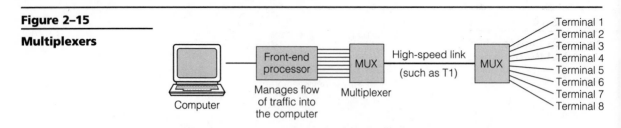

Multiplexers share a high-speed link among many slower devices in their function as aggregators and distributors of signals in networks.

2-14 shows the common causes of electrical problems. Ensuring reliable power supply is a central element in the implementer's work.

Nodes and Switches

Switches are the electronic equivalent of the airport hub in the airline system. Modern electronic switches receive signals arriving at a node (a functional unit connecting transmission lines) in the network and redirect these signals through the network to their final destination. They also serve to increase the utilization and efficiency of transmission links through **multiplexers**—devices that aggregate and bundle small streams of slow-speed traffic, place them on high-capacity trunk lines (this activity is called multiplexing), and then disaggregate the small traffic streams again at the distant end of the link for delivery to the final destination (see Figure 2-15).

In frequency division multiplexing (FDM), the high-speed link is divided into frequency bands (eight of them in Figure 2-15) separated by guardbands. Terminal 1 uses frequency band A, terminal 2 uses B, and so on. In time division multiplexing (TDM), each terminal is allotted a time slice (a fraction of a second) to send or receive a frame (or a blank, if inactive). To keep traffic flowing, the high-speed link must be faster (in Figure 2-15, eight times faster) than each terminal's data flow. In statistical multiplexing, 18 terminals can use a line that is fast enough for eight terminals in TDM. The multiplexer determines which terminals are active, adds a terminal identifier to the frame, and sends traffic from eight of the 18 terminals. Fast packet multiplexing is similar to statistical multiplexing, but the voice or data message is first divided into packets of 128 bytes; the multiplexer processes whole packets rather than characters (as in TDM).

Switches are substitutes for additional direct line connections; they improve the efficiency of existing links by allowing the sharing of links by different traffic streams. Today's intelligent switches can route both voice and data traffic, and many are capable of reformatting messages to ensure compatibility as data traf-

Figure 2–16

Switches

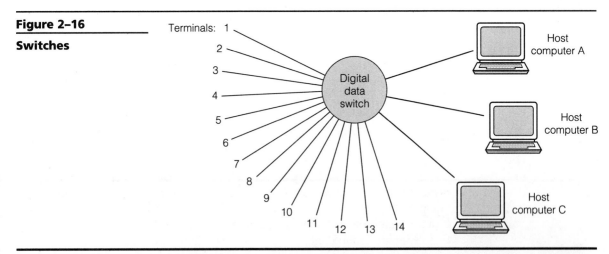

Switches are computers that process a stream of telecommunications signals instead of, say, accounting data or sales transactions. In this example, a digital data switch (or digital cross-connect switch, as it is often called) manages queues of incoming and outgoing messages, routes traffic to and from host computers, assigns priority, and manages the electronic equivalent of traffic jams. Additional features include security and accounting.

fic is routed across different types of interconnected networks. Large networks, such as the public telephone network, contain a number of sophisticated high-capacity switches connected in a **hierarchical configuration;** a given message or phone call may be routed through several switches before reaching its final destination as shown in Figure 1-4 (page 11).

Switching equipment—essentially a type of computer optimized to handle specific types of communications traffic—exploits microelectronics. Switches were large and very expensive in the days when there was only one U.S. telephone company, AT&T. AT&T made its own switches, which could cost over $1 billion to design and produce. The public network used these switches in central offices, to which circuits connected. **PBX (private branch exchange)** switches that companies could install on their own premises were available to large firms, who also used multiplexers in their private networks. Figure 2-16 shows examples of modern switches, whose range of features gets ever broader as the technology expands.

Many types of switch (voice, data, or both) are designed to handle particular types of network traffic in order to maximize throughput. Although such switches have more speed, versatility, and power, they include many features of bridges and routers, which operate on a store-and-forward basis. This adds "latency"—a delay of many milliseconds that reduces transmission link efficiency. Switches can speed up connection and transmission.

The classic switch is the phone company's central office switch, which establishes connections for calls, routes them, and handles billing information. The impact of switching technology is apparent in the speed with which a caller gets a dial tone and hears the signal indicating that the phone is ringing at the other end. Nondigital switches handle this job through electromechanical relays; digital switches use hardware that operates in microseconds.

Cabling and Switches

One of the unfortunate realities of complex networks (especially LANs) is that they create a jumble of cables and switches—unfortunate in the sense that many companies have a morass of cables, boxes, switching devices, and the like that are difficult to keep track of and to change. When an employee is transferred, it can take literally weeks to move his or her telephone and personal computer.

This level of complexity alarms telecommunications managers, and they need solutions quickly. The fastest growing part of the telecommunications equipment market is thus for such devices as **smart hubs**—single boxes that replace a multitude of separate hardware components (see Figure 2-17). This helps simplify cabling, interconnect incompatible systems, and provide network management diagnostics and many other tools needed to make it practical to operate an orderly multitechnology telecommunications environment. Proteon, Inc. introduced its new intelligent wiring hub in early 1992. The hub supports a range of types of cable in a single box that is just 19 inches wide. A major selling point is that it is **"hot swappable,"** which means users can swap equipment modules without disrupting network operations; previously, such a change would require stopping transmission.

Cisco Systems Inc., another fast-rising company, supplies **routers**—key switching devices for linking local area networks that use the same network operating system; many of them now handle multiple protocols. Many manufacturers also produce **bridges**, which also connect similar local area networks; unlike routers, however, they do not select the fastest or least congested route.

Intelligence in Switches

The degree of intelligence in a switch is a key strategic element of the network, and more sophisticated switches include capabilities for error handling, accounting and control, message format conversion, analog-to-digital conversion, prevention of unauthorized access, and fault detection and diagnosis. Switches can run the gamut from high-capacity computers directing thousands of telephone calls a minute to simple routing or bridging devices passing data traffic from one network to another.

Service providers (such as AT&T, MCI, Sprint, and value-added networks) increasingly build intelligence into their networks, both to enhance the range of features they can offer and to reduce the complexity of operations for users. MCI's enhanced voice service, for example, allows companies to link to it without expensive **customer premises equipment (CPE)**—the switches that interface the customer's locations to the network (see Figure 2-18). Over 3,000 manufacturers produce CPE, which includes telephone bandsets; key stations that route calls, provide hunting features, and have "hold" buttons, intercoms, and other features; switches; and data communications equipment.

The overall goal of modern switching, bridging, multiplexing, and routing technology is to create a seamless service that exploits both the intelligence of

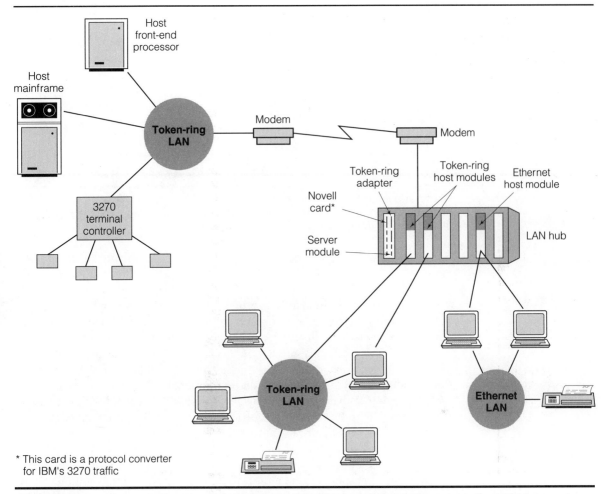

* This card is a protocol converter
for IBM's 3270 traffic

Figure 2–17

A Smart Hub

A smart hub is a piece of hardware with slots for connecting adapters and hardware modules for routing data from heterogeneous LANs. A single box thus replaces many boxes and the numerous cables between them.

computer hardware and software and the speed of raw transmission. It will be several years, though—and perhaps even a decade—before incompatibility among the elements of a complex business network ceases to be a constraint.

Transmission Methods and Procedures

Protocols

Health Link (Figure 1-2 on page 6) provides a service that electronically links doctors to insurers' computers. The main reason doctors connect to Health

Figure 2–18

Customer Premises Equipment (CPE)

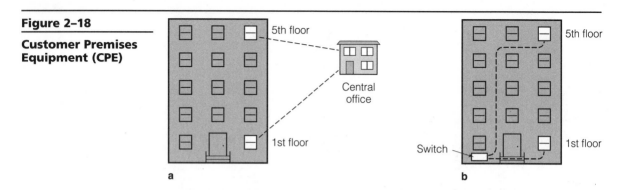

a

b

Without a switch on a company's premises (a), all internal calls are routed via the phone company's central office. With a switch (b), calls can be routed inside the building.

Link's computer, which then links to the relevant health-care insurer's computer, rather than linking directly to the insurer, is that all of the providers use different formats, standards, and architectures. Health Link translates the incoming message into the relevant formats. All the companies, however, use the telecommunications standard adopted by the value-added network that links them. This standard, called **X.25,** is perhaps the most well-known and widely used data communications standard in the world. X.25 is a standard protocol for packet-switched networks and is the base for most public data networks, most value-added networks, and almost all international networks.

Although terminals, transmissions links, and switches are the basic building blocks of the information highway infrastructure, they constitute only the skeleton of the network. Something more is needed to ensure that information being sent across the network gets from origin to destination intact and without error.

This discussion suggests the need for orderly rules and procedures designed to efficiently transport information through the network. Telecommunications protocols are the electronic equivalent of interpersonal greetings used to initiate any type of conversation. People's protocols typically start with an oral greeting such as, "Hello, I'm Mike," followed by a handshake and a few other exchanges that establish the identities of the conversants and set the tone for the subsequent exchange of information.

Because humans are so intelligent and adaptable, interpersonal protocols need not be precise and can vary somewhat from conversation to conversation, like "Hi" for "Hello." However, telecommunications protocols must be absolutely precise and complete right down to the individual bit, voltage, and any other relevant parameters, in order to ensure proper interpretation by electronic devices in the network.

The basis of a protocol is a frame that envelops the user's data and ensures that the data is transmitted as accurately and efficiently as possible; it includes a *header* that precedes the data and a *trailer* that follows it. Protocols are built in

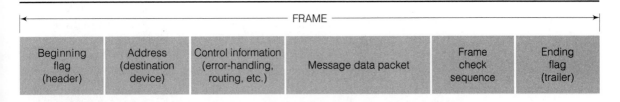

Figure 2–19

Link Control: Using a Frame to Regulate the Flow of the Digital Bit Stream

layers; the lowest layer establishes the electrical connections, whereas the highest ones interpret the format and meaning of the 0-1 bits.

A complete telecommunications protocol for sending information in digital form should include six basic functions:

1. *initiation of communication* through accessing the link or network
2. *character identification and grouping* to distinguish protocol signals from information signals
3. *message or conversation identification* to determine when one ends and another begins
4. *link control* to ensure the connection and to regulate the flow of data
5. *error detection and correction*
6. *termination* of the conversation or session.

The link control part of the protocol is concerned with regulating the flow across the network of information coded as a digital bit stream. The serial stream of 0s and 1s sent by a terminal over a transmission link has neither meaning nor organization. Data link control protocols structure the data in a frame (Figure 2-19); they also structure the traffic, establish and manage the physical communications path, and detect errors. Further, different parts of the network system often operate at different speeds, and terminal devices may not be able to process the signal at the rate at which it is being received, even with the use of buffers—storage devices that hold a message in the equivalent of a "pending" tray until the device can catch up with the processing. Consequently, one of the major functions of the link control protocol is to regulate the transmitting terminal by signaling it to suspend or resume transmitting, depending on the status of the receiving terminal.

Duplex and Half-duplex Transmission

The manner in which an information signal flows across the network must also be defined. Signals can be carried both directions simultaneously along the same transmission link (**full duplex transmission**), or they may travel only one way at a time (**half-duplex** or two-way alternate transmission). A full duplex transmission uses two parallel links—one for transmissions from A to B and the other for transmissions from B to A—just as traffic on a two-lane highway travels in the opposite direction in each lane. The half-duplex method is analogous to a one-lane bridge: Traffic flows in one direction across the bridge for a certain period of time, and then the traffic flowing in the other direction passes over the same bridge. Figure 2-20 shows the basics of duplex versus half-duplex transmission.

Circuit Switching Versus Packet Switching

Networks typically use either a circuit-switching or packet switching method of transmission. The choice determines the efficiency with which messages move through the network by maximizing the use of the available transmission capacity and by minimizing bottlenecks. In **circuit switching**, the signal is routed through the various nodes and switches along a fixed path that connects the point of origin to the destination point and is dedicated—used exclusively and not shared—for the duration of the call, session, or message. The transmitting terminal initiates the communication by signaling the network and requesting a link (a circuit) to the destination terminal. The network switches search for available links and lock in a specific transmission path through the network that is fixed for the duration of the communication session.

Because circuit switching creates a dedicated transmission path for the duration of the session, the network is inefficiently used during pauses in transmission. Historically, circuit switching was best suited to high-volume transactions that involved a highly interactive communication sequence that keeps the circuit busy, such as in the phone system. You have to spend time on setting up each call—on dialing and waiting a second or so for the person at the other end to pick up the phone. This overhead is small, especially if you then talk for 20 minutes, but it would be excessive if you had to repeat it for every sentence during your call. Circuit switching has been the base for telephony and for IBM's main telecommunications architectures.

Packet switching is the well-established alternative to the circuit switching method of transmission. Information to be sent is divided up into small blocks called **packets**, each of which is sent separately and travels over various routes before being reassembled in correct order at the final destination. Each packet is first coded with a destination address and sequence number and is then transmitted individually through the network. As a packet reaches a node or switch in the network, the destination address on the packet is read, and the packet is sent on to another node on the way to the final destination. Depending upon traffic volume and availability of facilities, different packets from the same message can follow different routes through the network, with all packets ultimately arriving at the final destination point, but not necessarily in the order in which they were sent. Upon final arrival, the packets are reassembled in correct order, and the information is decoded and forwarded to the terminal.

Packet switching is an efficient use of transmission facilities because links are not dedicated for a particular session's duration; this permits packets from other sessions to share the same transmission link, rather than having the link sit idle during pauses in information flow. Public data networks rely on packet

Figure 2–20

Duplex Versus Half-duplex Transmission

Step 1

DATA

Computer A

Computer B

Step 2

ACK

Computer A

Computer B

a

DATA | DATA

ACK | ACK

Computer A

Computer B

b

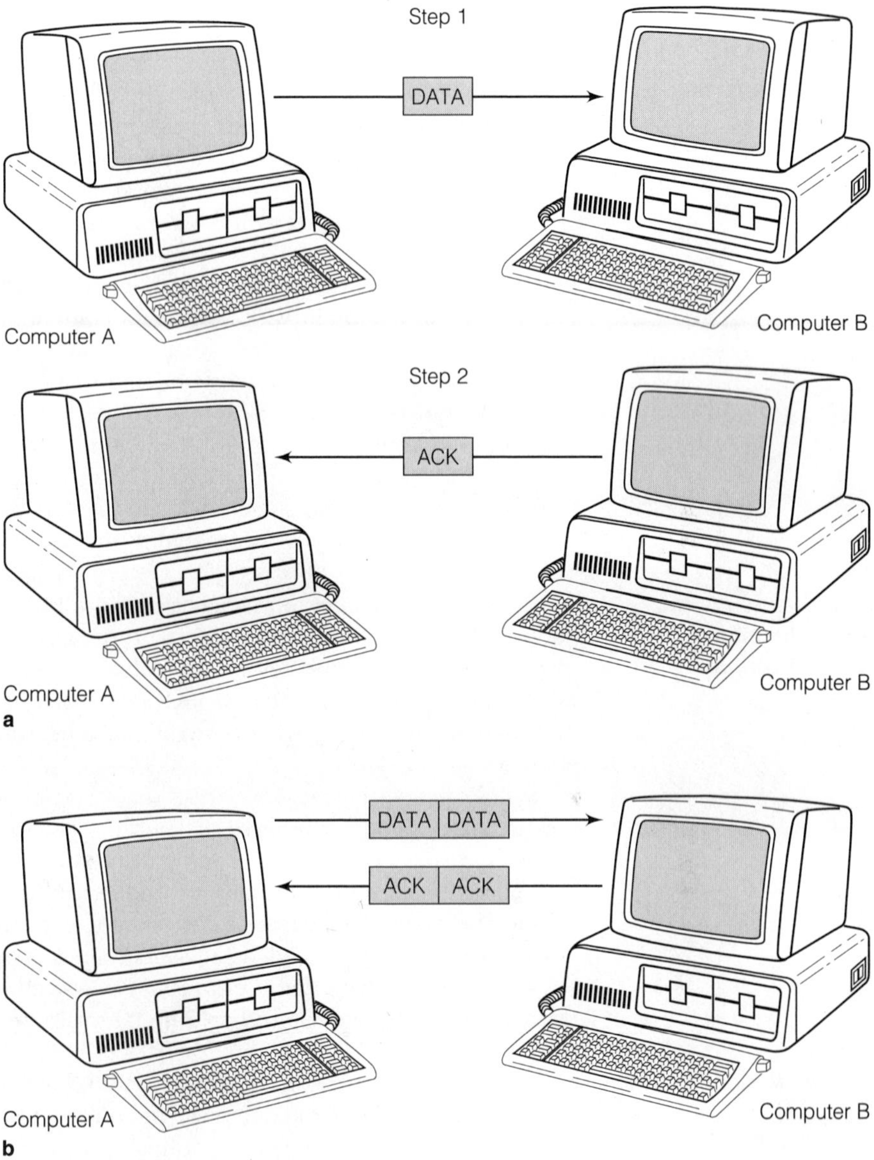

"ACK" = receipt of data acknowledged

Half-duplex transmission (a) is one way only; the protocol transmits a block of data (step 1) and then waits for an acknowledgment (step 2). Full duplex transmission (b) is two-way and has no propagation delay.

switching as the most cost-effective way to share a large telecommunications transmission link among hundreds of thousands of users.

New techniques are dramatically improving the speed of message flow, which has historically been slow because of the overhead time required to assemble and disassemble packets. Fast packet switching is a generic term for recent developments in digital transmission methods that in effect announce the next generation of advanced networks.

With traditional packet switching, the data is checked for accuracy at each node. The emerging technology of **frame relay** cuts out this delay by checking only at the entry and exit points of the network. Frame relay is one of the fastest moving developments in wide area data communications for large organizations.

Emerging just as fast but a little behind frame relay in development is **asynchronous transfer mode**, which is expected to become the base for the next generation of local area networks. Asynchronous transfer mode technology will also be a core element in public networks. ATM uses a method called cell relay to pulse data as very short packages, at rates far faster than has been possible with X.25 and the relative slow switching equipment and software of the 1980s.

Synchronous and Asynchronous Transmission

Asynchronous transmission means that there is no regular timing relationship between the transmitting and receiving terminals (that is, they are not synchronized); characters are sent either individually or in small blocks, with start/stop information included in each character or block of characters. The start/stop information tells the receiving terminal that a new character is on its way or has just been completed.

Synchronous transmission means that the transmitting and receiving terminals are synchronized in time, so that certain specified and fixed subsecond time intervals are designated for data and others for control information; this enables continuous streams of information to be sent efficiently in large blocks, minimizing the amount of capacity needed for control or protocol information. Start/stop bits are not needed for synchronous transmissions because the receiving terminal knows that if bits are received within a particular time interval they are data bits, and that other time intervals are reserved for control bits.

Protocol Converters

Protocol converters are pieces of equipment that use hardware and/or software to enable data to flow from one piece of equipment to another in a manner that is transparent to the user. They modify the bit stream passing through them to ensure that it conforms to the rules and procedures of the new network or piece of equipment that it is entering. Protocol conversion is increasingly a built-in feature of many switches, especially smart hubs but also intelligent workstations. They generally handle only a limited range of protocols, obviously focusing on the ones most widely used and that thus offer the largest market.

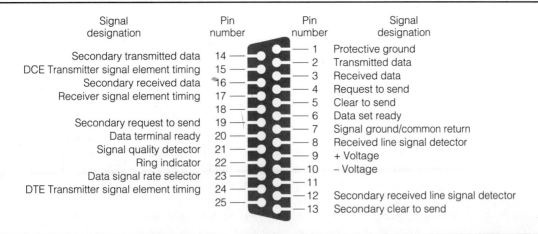

Signal designation	Pin number		Pin number	Signal designation
Secondary transmitted data	14		1	Protective ground
DCE Transmitter signal element timing	15		2	Transmitted data
Secondary received data	16		3	Received data
Receiver signal element timing	17		4	Request to send
	18		5	Clear to send
	19		6	Data set ready
Secondary request to send	19		7	Signal ground/common return
Data terminal ready	20		8	Received line signal detector
Signal quality detector	21		9	+ Voltage
Ring indicator	22		10	– Voltage
Data signal rate selector	23		11	
DTE Transmitter signal element timing	24		12	Secondary received line signal detector
	25		13	Secondary clear to send

Figure 2–21

The RS-232 Interface

This example of a technical standard defines the purpose, electrical signals, and timing in its function as an interface that connects a computer to a peripheral device (such as a modem, printer, or mouse). Although 25 wires are in the cable, most are not used in any given device interfaced by RS-232.

| | | | | | | |

Integration and Standards

Network Architecture and Standards

Integration—the combination of telecommunications, computing, and information management—is the central driving force today across the entire field of information technology. Integration requires compatibility—the ability of a wide variety of components to operate together. Historically, incompatibility has been the norm, partly because of differences among vendors' products, partly because of the extremely rapid pace of technological change, and partly because of the lack of any clear forum for ensuring commonality of standards. Standards, whether purely technical or business-centered, are the key to integration.

One example of a technical standard that you may already be familiar with (and perhaps frustrated by) in using a personal computer is the **RS-232** interface (Figure 2-21). This is a standard for connecting a computer to an external device such as a printer. RS-232 defines the electrical features and timing of signals for a 25-pin interface cable. Just as an engineer designing a fax machine takes it for granted that it will operate at 115 volts and use a standard phone jack, an engineer designing a new printer can be sure that, with an RS-232 interface, the printer can be connected to a standard PC. The engineer need not know *how* either the 115-volt electrical system or the PC works, only that a specific interface is required.

There are literally thousands of standards. Some specify physical interfaces between equipment (cables, sockets, and electrical signals); others address "higher level" message formats, procedures, network features, types of transmission, and the like. A standard is not a "rule"; nor do devices that adopt the same standard have to be at all the same in internal design and features, any more than photocopiers that share the U.S. standard for paper of 8½ by 11 inches are the same; they offer different speeds and options such as color copies, reduction and enlargement, collating, and double-sided copying within the same standard for paper size. It is in this sense that standards increase rather than decrease users' options and autonomy.

The Architecture as Blueprint

A telecommunications architecture is a blueprint, a set of design principles, and a choice of standards that translate business function into technical form to meet a client's requirements for the efficient and effective movement of information. It is the electronic equivalent of a master plan for a city, which lays out major roadways, zoning boundaries, and areas for future growth; it does not describe the details of houses to be located in the city, but it imposes standards for size, construction, and safety codes.

An efficient architecture is generally based on the ability to integrate services, applications, data, and networks in order to eliminate duplication, incompatibility, and redundancy while taking full advantage of economies of scale. In turn, standards are the key to ensuring both seamless (invisible and uninterrupted) interfaces between various network components and compatibility among the same type of equipment (for example, terminals or printers) manufactured by different vendors.

Open Systems

In the context of the standards, open means supplier-independent; open standards are thus independent of equipment, software, and suppliers. (Proprietary standards, by contrast, are vendor-dependent.) **Open systems** facilitate integration because in an ideal standards environment, all equipment and services are compatible and work together, and thus technical choices for the network are not constrained by vendor, vintage, or type. This is not yet the situation in telecommunications, although there has been significant progress in the past decade in first defining and then implementing open standards.

The time gap between the definition of a standard and its practical implementation in products generally ranges from five to 15 years. It is particularly difficult to establish open standards in telecommunications and information systems because historically the industry has been dominated by proprietary systems and the corresponding incompatibility among products made by different vendors. Most early communications systems were built for specific, stand-alone applications, and little thought was given to the need either for integration and interchangeability of components made by various vendors or future interconnection with other systems and networks.

A standard is considered de facto open if it meets the following criteria:

1.   It is fully defined.

2.   It has proved to be stable in products.

	Layer	Name	Functions
Figure 2–22	7	Application	Creates user message (e-mail), data base inquiry
The Layered Architecture of the OSI Reference Model	6	Presentation	Formats and converts data
	5	Session	Establishes and terminates communications session
	4	Transport	Manages end-to-end control of the communication
	3	Network	Routes the data to the correct destination
	2	Data link	Transmits the data to and from each node
	1	Physical	Connects nodes electronically and physically

In such a layered architecture, the application layer assumes that the engineering for lower layers is correctly implemented. Thus designers need not be concerned with the details of other layers. Each of the major standards and protocols fits into a specific layer (for example, X.25 is in Layer 2).

3. Its interfaces are published, and thus any vendor can ensure that its products conform to the interfaces.

4. No single party controls its development and/or use.

The Open Systems Interconnection (OSI) Reference Model

The **Open Systems Interconnection (OSI)** architecture model has been the most focused effort to end incompatibility. Developed under the auspices of the International Standards Organization (ISO), OSI was called a "reference" model; it was intended to guide the development of products, not to define a product in itself. Its purpose is to establish complete and consistent communication rules that permit the exchange of information between dissimilar systems. The OSI model has in many ways helped create the intellectual base for modern data communications. Although it has not been completely implemented, it has helped provide a framework for defining several key standards.

The OSI model is known as a layered architecture (or modular) approach to communications standards. Each of its seven layers represents a necessary element in the electronic communication process, with the application and user interface elements at the top and the physical connection standards at the bottom (Figure 2-22).

In further recognition of the need for open system standards, the **Integrated Services Digital Network (ISDN)** standard was developed in the late 1970s. 64 kbps is the unit of capacity provided by ISDN, a blueprint for the public networks of the mid-1990s that combines voice, data, text, and image on the same phone line in the home or office. ISDN is being implemented more slowly in the United States than in many other countries, partly because of the cost of replacing old analog facilities and partly because the pace of technical innovation in telecommunications makes it already obsolescent. That said, ISDN costs are around 10 percent of those for fiber installation, keeping ISDN an attractive option for many business and consumer applications for the mid- to late 1990s. In 1992 ISDN gained major new momentum as the RBOCs began to see it as a potential source of new revenues and as they and the long-distance carriers worked to test the interoperability of their ISDN services.

ISDN was developed by the Consultative Committee on International Telegraphy and Telephony (CCITT). CCITT is an international standards body with over 150 countries represented on it. It is one of four units of the International Telecommunications Union founded in 1865. ISDN is a worldwide standard for public telecommunications systems that is fully digital and brings together in a single transmission facility every type of information (voice, data, image, and video). Because the standard is worldwide any U.S. network engineered to meet ISDN standards should be compatible with any other ISDN system developed anywhere in the world. The first successful comprehensive test of ISDN across all major long-distance providers was made in late 1992, 15 years after the standard was first defined.

Although completely open systems and the universal adoption of the OSI standards are supported by virtually all factions in the industry, the world is moving rather slowly toward this goal. It is one thing to develop a set of standards, but quite another to achieve widespread implementation of such standards. No matter how detailed the standard specifications, they are open to interpretation, which means that even a good-faith effort on the part of manufacturers to meet the standards may not ensure compatibility. A case in point is a software package developed for the IBM PC that does not run properly (or at all) on a so-called "IBM compatible PC."

Another difficulty in implementation is that the development of technology often moves faster than the formal standards-setting process. Even as the basic ISDN standard is still being implemented, a new **Broadband ISDN** technology is emerging. **Broadband ISDN (BISDN)** is a second-generation ISDN that has more capacity than ISDN and is designed to meet the growing telecommunications needs of large users. Meanwhile, SONET (a fiber optic-based standard) is already in use and is leapfrogging ISDN and BISDN with speeds of up to 2.4 gbps.

Network Topologies

Network **topology** refers to the physical structure of the transmission links. Most networks are classified using the following descriptions: star, bus, mesh, ring, or hierarchical.

In the **star configuration** (Figure 2-23a), all nodes connect directly to a central hub, and all communications between the various nodes are routed through this hub, making it easy to coordinate and manage the network. The **bus topology** (Figure 2-23b) involves individual nodes connected directly to a common transmission medium (typically a cable or wire) and generally is associated with high-speed transmission over short distances (within a building or department).

A fully interconnected **mesh network** provides a direct connection between every pair of nodes in the network. However, a full mesh network is seldom implemented because of concern over excessive cost. Generally, the major nodes are directly connected together, and the minor nodes are connected to one or more locations, depending upon the exact communication requirements.

Figure 2–23a

Network Topologies

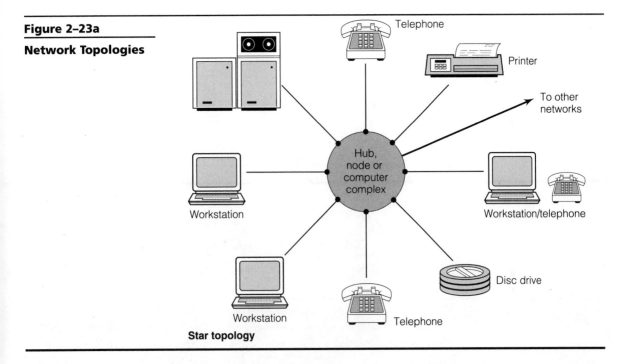

Star topology

The mesh topology provides a high level of reliability due to the redundancy available through the multiple paths terminating at each major node. **Redundancy** means that equipment and/or transmission links that are not needed for operation are provided and are available immediately if there is any problem; if one node is down, alternative routes are available.

The **ring topology** (Figure 2-23c) consists of nodes connected to a common circular or oval loop medium (cable or wire). The ring system is analogous to a circular rapid-transit system in which the train stops at each station, either takes on or lets off passengers, and then moves to the next station, continuing endlessly around the loop. The ring network operates the same way: Each node on the ring sequentially undergoes **polling**—the process in which software or hardware queries a device on the network to determine whether it has traffic to send or is ready to receive traffic. Ring networks are usually used when the network nodes are relatively close together (same building or department).

A **hierarchical network** is similar to a mesh network but has fewer direct connections between the various nodes. The public telephone network is an example of a hierarchical network that efficiently interconnects various-sized traffic streams to a large number of nodes while providing a high level of reliability and service quality (largely due to the alternate routing capability inherent in the hierarchical network topology).

Figure 2-24 (page 84) shows a network in greater detail than is typically

Figure 2–23b

(continued)

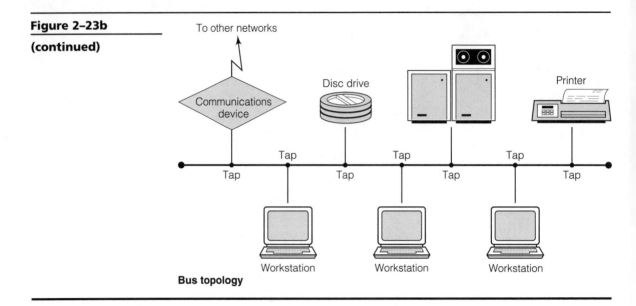

Bus topology

provided by vendor's ads or telecommunications textbooks. The figure is provided to show just how much is missing from the diagrams that show overall links, capabilities, switches, and the like—and what the implementer must address.

LANs and WANs: Competitive and Cooperative Technologies

The five network building blocks are generic; they apply to LANs and WANs, voice and data.

There are many specific differences in the technology used in LANs versus the public voice network and wide area data networks. More important than the technical differences from the client's, designer's, and implementer's perspectives is the historical shift from wide area networks as the core of a firm's communication resource to (first) local area networks as the driver and (now) to interoperability of all types of network as the key requirement. LANs are a very recent development within the telecommunications field. Before the mid-1980s, the industry was dominated by the public analog voice network, wide area public data networks, and private leased lines.

LAN technology was a response to the needs for high-capability telecommunications facilities to enable PCs to share information.

Figure 2–23c

(continued)

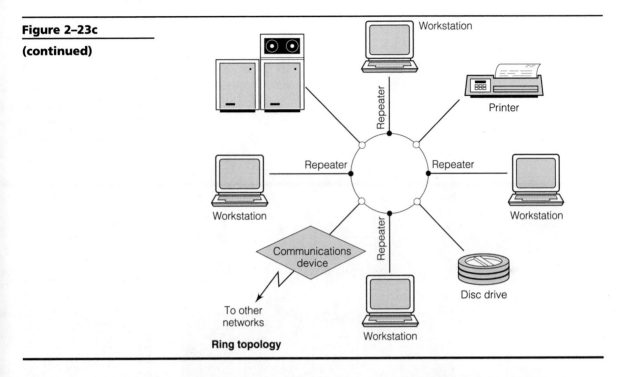

Ring topology

LANs became the driver of business innovations in the use of personal computers in the late 1980s. In terms of raw transmission capability and cost-effectiveness, they rapidly outpaced WAN technology. They provided low-cost transmission—with no usage charges by the phone company—and they were digital, matching high-speed digital transmission to high-performance digital computers.

For this reason, the most topical aspects of telecommunications mainly relate to LANs today, in terms of innovations and standards—and of problems as well. The volatility and pace of change in the technology outpaced firms' base of skills and experience; the need to link LANs across types of technology and locations meant that a new generation of technology has emerged to provide interoperability between local and wide area facilities; the need to ensure reliability and security of operations has meant that LAN operating systems must do more than just link PCs; the multiplicity of competing technologies and standards increases the complexity of network planning; and the increasingly blurred distinctions between telecommunications and computing escalate complexity, problems in keeping up to date, and choice of standards and vendors.

For all these reasons, the telecommunications implementer must have an in-depth knowledge of LAN technology far more than of WAN technology, and technical textbooks on telecommunications increasingly focus on LANs. In this

Figure 2–24

A Detailed Schematic Diagram of a Network

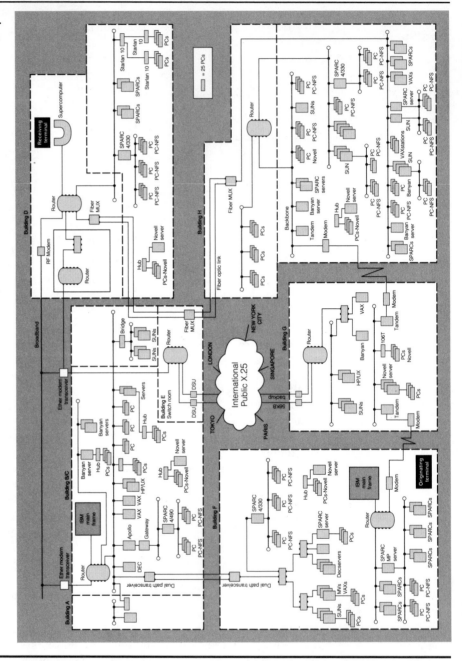

This diagram reveals the complexity of a typical large firm's network. Try tracking the route of a message from an originating terminal in Building F (lower left) to a receiving terminal in Building D (upper right).

book you will see many examples of the use of LANs, but far less in-depth discussion of their technology. The designer's emphasis must be on how the building blocks of the network work together to create a business resource. In that context, WANs and LANs are increasingly interdependent. Many of the problems implementers are wrestling with today relate to integration and interoperability. LANs now drive the need for innovations in WANs, and vice versa. *Networks in Action* thus tries to address both of these areas together, rather than one at the expense of the other. That said, from the implementer's viewpoint it is the LAN that dominates most aspects of telecommunications usage for the simple reason that the terminals that request services, access information, and share communications today are almost always connected to a LAN, which in turn is connected to a WAN.

Summary

An organization's telecommunications resource combines five technical building blocks: (1) terminals and workstations, (2) transmission links, (3) transmission methods, (4) nodes and switches, and (5) standards and architecture.

The two main distinctions between types of network are private versus public and local area versus wide area. Private networks use facilities leased from the provider of the transmission service; public networks are available to subscribers on a pay-as-you-go basis. Virtual private networks and switched megabit data services combine aspects of private networks delivered through the public network. Value-added networks offer specialized services that are generally used by consortia of firms that need to communicate with each other. Local area networks link devices within a single building or group of buildings. Wide area networks span large geographic distances. Campus and metropolitan networks handle intermediate areas.

Terminals are the access devices for telecommunications services. They include telephones; dumb terminals; intelligent workstations; barcode scanners; point-of-sale registers; machine tools; and many other devices. Telecommunications and computers are increasingly interdependent, and computer operating systems greatly constrain telecommunications options.

Transmission is the electronic movement of information. Transmission media include cables, microwave radio, and fiber optics. All information, ranging from voice phone calls, computer data, and images and video, can be coded in digital 0-1 form and transmitted over any media that can carry the signal.

The basic measure of transmission speed is bits per second. Originally, all transmission was analog, in which an electrical signal paralleled the sound wave of the speaker's voice. Today, digital communication is becoming the norm; in it, information is coded in 0-1 bit form. All computer data is digital. Digital information can be stored, manipulated, corrected, and pulsed at vary rapid speeds. The main barriers to shifting to fully digital communication are the existing investments in and reliance on analog facilities and equipment.

The capacity of a transmission link is determined by its bandwidth, which is the range of usable frequencies of the signal. Frequencies are measured in cycles per second, with the difference between the lowest and highest frequencies indicating the bandwidth. Cycles per second are defined in hertz, kilohertz (thousands), megahertz (millions), and gigahertz (billions).

All transmission media exploit the electromagnetic spectrum—from very low frequency electrical signals sent along copper wires, to high, very high, and ultrahigh frequency radio waves, to laser beams sent over optical fiber cable. Fiber optics provide transmission speeds measured in millions and even billions of bits per second.

Transmission methods vary widely. They rely on protocols—agreed on and precisely defined formats and procedures that have six main functions: (1) initiate the communications (2) identify and group characters, distinguishing protocol signals from information signals; (3) recognize and synchronize the beginning and ending of a message; (4) ensure the connection and routing of a message through link control; (5) detect and correct errors in transmission; and (6) terminate the communication.

Transmission may be duplex, in which signals travel in both directions simultaneously, or half-duplex, in which data travels in one direction at a time. Circuit switching dedicates the entire capacity of the channel to the devices sending and receiving messages to and from each other. Packet switching breaks messages up into short packets that move through the telecommunications link as if on a conveyor belt. Packet switching shares the link among many users, maximizing its capacity. Fast packet-switching techniques are rapidly increasing the speed of throughput, with frame relay and asynchronous transfer mode in early implementation as of late 1992.

Transmission may be asynchronous, in which a start/stop bit indicates the beginning or end of a character so that the two devices sending and receiving it need not be synchronized, or synchronous, in which they must be.

Switches are a generic term for equipment that handles the routing of traffic across the network. They range from the bridges and routers that link local area networks to large-scale multiplexers, private branch exchange, and smart hubs that coordinate many types of traffic that may use a variety of different protocols.

Standards are central to modern telecommunications. They define interfaces between equipment, software, or services. Proprietary standards are vendor-dependent; open standards are vendor-independent. Historically, proprietary systems have been the norm, but user demand is driving a shift toward open systems. The Open Systems Interconnection model has been the main blueprint guiding progress, although the rapid pace of change in technology has meant that in many instances, standard-setting organizations cannot keep up, and de facto standards emerge from users' and/or vendors' choices. The ISDN (International Services Digital Network) standard has been overtaken by broadband ISDN, frame relay, and asynchronous transfer mode, even as it is being tested.

Major standards-setting organizations include ANSI (American National Standards Institute), ISO (International Standards Organization), and CCITT (Consultative Committee on International Telegraphy and Telephony). Important and widely used standards include X.25, the base for packet-switched public data networks and RS-232, a standard for connecting a PC to a peripheral device.

The master blueprint for a firm's network strategy is called an architecture. Each architecture is built on a set of standards. The designer must determine the degree of integration the architecture must permit and the range of open systems it will include. Integration is the long-term goal of modern information technology.

Network topologies refer to the physical arrangement of the devices to be interlinked. The star configuration, which has a central hub through which all devices communicate with each other, has been the traditional configuration for airline reservation systems and bank ATM networks, which rely on large, central mainframe computers. A bus network simply hooks devices onto a length of cable. A mesh network connects all nodes to each other. A ring network links nodes into a circle. A hierarchical network is a mesh network but with fewer interconnections.

Review Questions

1. Briefly describe how digital coding can represent a voice conversation, a video image, and a one-page office memo.

2. What are some of the advantages of digital coding and transmission compared with analog coding and transmission?

3. What do barcode readers, personal computers, mainframe computers, and mobile phones have in common?

4. If a network contains workstations connected via copper wire transmission links, isn't this sufficient for communication to take place? Explain.

5. Why include switches in a network configuration; wouldn't it be better simply to directly connect all nodes communicating over the network?

6. Why is an architecture important to the development and evolution of a company's telecommunications capability?

7. Draw a simple matrix and label it with the main categories that define types of networks.

    a. Can all types of networks be classified into one of the four cells in the matrix, or are there types of networks that fall outside this classification scheme? Explain.

    b. True or false: There is only one type of public network that is used for data applications. Explain.

    c. True or false: Large companies typically limit themselves to one type of network that most closely matches their particular communications requirements. Explain.

8. Briefly explain the concept of a virtual private network and some of the advantages it has over a true dedicated private network.

9. What is the common standard shared by most public data networks, most value-added networks, and almost all international networks?

10. Explain how the phrase "We must lower our overhead" applies to telecommunications protocols.

11. What determines the information carrying capacity of a transmission link? Explain.

12. What are the factors in the evolution of communications satellites that have enabled the use of smaller and less expensive earth stations to transmit and receive information.

13. Explain the following statement: Transmission speed does not determine message or information speed.

14. List several of the important implications of having more intelligence incorporated into terminals that access a network.

15. What are the two factors that determine the information carrying capacity of a transmission link?

16. Why is "noise" a key concept in telecommunications?

17. Briefly explain the role of nodes, switches, and multiplexers in a network by using an airline transportation analogy.

18. List some of the functions performed by modern network switches that give them the designation "intelligent switches."

19. Describe the various components of a protocol system by making an analogy between it and a conversation.

20. Briefly explain the difference between a circuit-switching system and a packet-switching system.

21. What are the major differences between regular and fast packet-switching technologies that enable significantly reduced transmission time for the latter?

22. Briefly explain the difference between synchronous and asynchronous transmission methods.

23. What single element is the key to integration of telecommunications, computing, and information management?

24. What is the relationship between open systems and suppliers/vendors of equipment?

25. What are the main types of network topology? What advantages does each offer in terms of cost, efficiency, and technology?

Assignments and Questions for Discussion

1. Terminals/workstations, transmissions links, and switches constitute the physical skeleton of the network. Describe what else is needed to bring the telecommunications skeleton to life, and discuss its importance to the overall network concept.

2. Explain the importance of fast-packet technologies to the evolution of networks in supporting business activity.

3. Discuss the relationship between an open systems architecture and the integration of telecommunications, computing, and information management.

4. Explain why computer and network operating systems have replaced hardware as the main battleground among computer vendors.

5. Discuss the meaning of the following statement: "The power of the personal computer and its networking capability means freedom and flexibility and brings computers and telecommunications into almost total interdependency."

Minicase 2-1

Terra International

Reaching the Boonies Through VSAT

Terra International Inc. serves growers and dealers in the farming industry. It operates 300 retail stores, manufacturing plants, and wholesale offices in the United States, almost all of which are in rural locations. Terra provides on-line real-time processing for many transactions, including payments, inventory management, shipping, bill of lading preparation, and analysis of soil tests.

This business obviously requires a comprehensive telecommunications capability. Until the early 1990s, Terra relied on a network of leased lines connecting personal computers at each location to its IBM mainframe computer in Sioux City, Iowa. The point-of-sale network was provided by a wide range of telephone companies, many of them small and local and each with their own pricing schemes, state regulations, and capabilities. When any network outage occurred, Terra had to rely on several different companies to solve the problem. The analog transmission facilities were error-prone, and costs were high.

Today, Terra has no terrestrial leased lines. Instead, it uses AT&T's Tridom VSAT network (see Figure 2-25). Some 220 remote locations each has a VSAT dish on its roof. The Sioux City computer is linked to Tridom's Atlanta hub, which has highly sophisticated equipment and technical expertise. The hub is linked to backup facilities in New York and Los Angeles. Overall network availability is over 99.5 percent.

The network uses a well-established protocol for satellites, called Aloha (Hawaiian for "hello"), that is well-suited for handling short interactive transactions. Transactions that require sending a long stream of data, such as a computer file transfer, are automatically switched from Aloha to a multiplexer. The shared hub has a 9.2-meter antenna that collects signals for the satellite. A key element of the system is a software package called Clearlink Network Control System that handles many aspects of network control, such as fault determination, diagnostics, collection of performance statistics, configuration management, capacity planning, and change management. Without these, the 99.5 percent availability would drop at least 5 percent.

Terra has cut its communications costs by 25 percent and has greatly improved its service and reliability.

Questions for Discussion

1. Terra's system has a 99.5 percent availability. Assuming it operates nonstop, how many hours a week is the system unavailable? If the availability dropped to 94.5 percent, how many hours would be lost per week?

2. Where are the main switches in the Terra system?

89

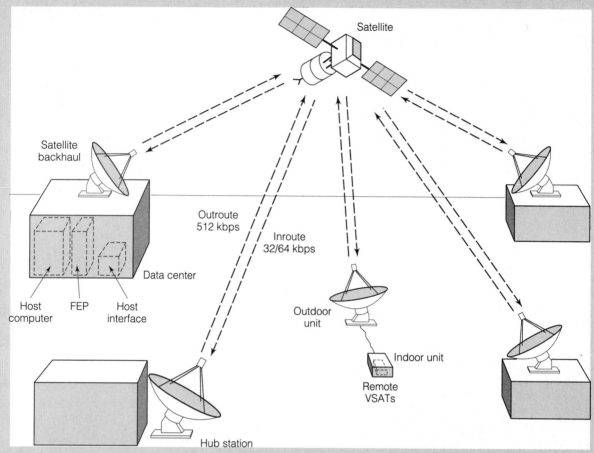

Figure 2–25

AT&T's Tridom VSAT Network

Each VSAT location has a small antenna plus an outdoor unit (ODU) and an indoor unit (IDU). The IDUs convert into packets data from cluster controllers (small, special-purpose switches) that interface personal computers to the system. The packets are sent to the ODU, which modulates them onto a radio frequency carrier. The signal is sent by satellite to the Atlanta or backup hub and then to the mainframe computer.

The VSAT receivers are 1.8 meters or 2.4 meters in diameter. They include special anti-icing features, for obvious reasons. The hub sends data at 512 kbps, using the 14.0–14.5 Ghz frequency; this is the uplink. It receives data from the VSAT stations at 32 or 64 kbps in the 11.7–12.2 Ghz downlink.

3. If Terra opens a new store, distribution center, or office, what must happen for the new location to be added to the network? Assume that the old terrestrial network were in place; would there then be any differences in what must happen to add the location to the network?

4. The following figures are the number of bits in a typical message:

a. A credit card authorization request: 1,000 bits

b. An electronic mail message: 5,000–15,000 bits

c. Digital voice (1 second): 56,000 bits

d. Full-motion video (television quality, one second): 10,000,000 bits

Identify which types of message can be efficiently handled by the network and which cannot be. Note that the central hub can send data at 512 kbps, whereas each location's uplink to the satellite operates at 32 kbps or 64 kbps.

5. If Terra operated a terrestrial network that transmitted data at 9.6 kbps, which types of message could be handled and which could not?

Minicase 2-2

Dun and Bradstreet

Extending the WAN to the LAN

Dun and Bradstreet was described by *Business Week* in 1986 as "a money machine." That machine was built on D&B's massive network that delivers to customers credit, financial, and marketing data. D&B decided in the 1970s that the key to electronic information services was to own the network and add more and more traffic to it, including other firms' databases. No company has made more money from electronic information services than D&B. It has 20 different divisions worldwide, including Nielsen (market research and television ratings), Moody's Investor Services (finance), and Donnelly (directories). When it sold the Official Airlines Guide for $750 million, that sum exceeded the profits of the entire airline industry that paid D&B to include their own data—schedules—in the guide.

The technology of the 1980s pointed toward a proprietary wide area network, large mainframes, and centralized databases. This was the classic private network. D&B added off-line delivery of information through mailings of floppy disks and optical storage media. The main access mechanism remained the personal computer/workstation.

In 1992, D&B began to extend this network to include local area networks in order to streamline the delivery process. Instead of 20 users in a given client location each accessing data and storing it on a hard disk, D&B planned to download the database to a LAN file server on the customer premises, thereby eliminating the need to access D&B's mainframe computers and reducing telecommunications costs for both D&B and the customer. (The file server contains the databases that individual personal computers access through the LAN; D&B manages the data and updates it on the file server.) This requires a fast workstation that supports the main PC operating systems customers use: UNIX, DOS, Windows, and OS/2.

The concept is easy. LAN servers can store gigabytes of data, making it virtually certain that D&B will distribute its data. D&B itself began in late 1991 to downsize its internal computing operations, moving data off mainframes and onto desktop systems linked to LANs. The emerging technologies of client/server computing provide the base for more distribution of data.

The implementation is much more difficult and it will be several years before D&B converts to the new process. Security and integrity of data are key issues. LANs may be just one element of distribution; metropolitan area networks (MANs) may allow D&B to distribute data to more than just a single site—to a city, office complex, or multisite company. (MANs are an intermediate type of network between LAN and WAN.) The advantage of the centralized system is that D&B can manage its data center and databases relatively simply. There is a master version of the database; access to it is made through a well-defined and well-designed access point. Updates are made in a single location.

Distributing data to client LANs means multiple versions of databases, multiple access mechanisms, complex client/server software, complex network management demands, complex data administration for D&B—and, almost certainly, a new infusion of profit for a money machine.

Questions for Discussion

1. Explain how Dun & Bradstreet made money from adding more and more information databases to its inventory and building up traffic on its network.

2. Discuss the advantage to D&B of downloading databases to LAN servers, rather than continuing to manage access from its central mainframe computers.

3. You have just been hired as director of LAN information services for D&B, and your first assignment is to prepare a brief outline of the major issues associated with managing an information service in a distributed processing, LAN environment. Produce such an outline.

Minicase 2-3

Saving New York State Millions

Building a Massive Network

As in most large states, New York's government agencies have built up a large number and wide variety of networks on a case-by-case basis. In the early 1990s, the state's telecommunications agency designed and implemented Empire Net, a backbone network that carries traffic between offices across the state. The logic was to replace multiple slower-speed leased circuits with a high-capacity shared resource. Most backbone networks for a large company have five to ten nodes; Empire Net has 153, as well as 450 T1 circuits, each operating at 1.5 Mbps. Nine thousand tail circuits connect state offices to the backbone.

Empire Net is a starting point, not an end point. As its project manager commented in late 1991, "We've only scratched the surface on this network for state business. We think the performance and cost of the network will make it attractive for other state business." Its largest initial user, the New York State Lottery, had been paying $430 a month for a tail circuit; Empire Net reduces this to $200.

Empire Net is a data network. The tail circuits that connect to it combine data and voice; the local loop transmits both over twisted-pair copper wire. The multiplexer at each state office transmits this "voice-over-data" traffic to a central office multiplexer at a NYNEX (the Regional Bell Operating Company) central office. There, the voice traffic is routed into the telco's public network, and the data traffic is routed into Empire Net.

Many firms cooperated in the development of Empire Net, which has many "largest" labels attached to it: It is one of the largest networks in the country, and probably the largest voice-over-data network. According to one consultant, "The project management was incredibly complex. When you get a network with this many nodes, the routing algorithms get weird. All sorts of things start to happen that you just don't see in 10-node corporate networks."

It may be complex, but it works.

Questions for Discussion

1. Why does the Empire Network have so many more nodes than a typical large company network, and what impact does this have on the design and management of the network?

2. What is the point of a voice-over-data system that combines the two streams initially and then splits them up, feeding one to the Regional Bell Operating Company and the other to Empire Net?

3. Describe some of the difficulties that are created by routing traffic over a network with many nodes but are not present in a corporate network with only a few nodes.

Minicase 2-4

International Telecommunications

The View from Volkswagen

Volkswagen AG is the fourth largest car manufacturer in the world and operates in just about every country. It regards telecommunications as the key element in "building a global patchwork of individual plants and sales organizations and making these bodies effective." Recent new pieces of the patchwork include VW's acquisitions in the Czech Republic and plants in Shanghai and Chanchung, China. It is important to VW to integrate new companies and partners quickly into the infrastructure. "If we do not accomplish this, we will not be able to integrate the business process and then we will not reach the economies of scale, which is the primary reason we bring companies together." (All the comments quoted here are from interviews with VW managers in December 1991.)

At the end of 1991, VW was restructuring its entire corporate backbone network, mainly to ensure better backup availability. VW planned to use 2 Mbps links across the backbone and slower backup links. Finalizing the strategy depended less on technology than on tariffs: "In Germany, we are charged a factor of 20 times higher for higher-speed lines—if they are available at all—than our partners in the U.S. or Japan."

This is just one of the hurdles. Satellite connections are the only way for VW to integrate its Czech Republic plants with operations in Wolfsburg, Germany, even though the distance is only a few hundred miles; the highest speed available before 1992

was 9.6 kbps. VW is outsourcing its Chinese computers and communications: "We would have to deal with the Chinese PTT and local authorities there. They probably only speak one Chinese dialect, so this would make it very difficult to coordinate the business from here or even from our offices in Shanghai or Chanchung. Therefore, our China links will most certainly be established via a service provider."

Across most of Europe, "it is still very much the PTTs that are setting the scene because of their traditional monopolies." VW must fight to get crossborder services. In Spain, only 9.6-kbps links were available; VW lobbied to get 64 kbps. It needs 2 Mbps to link its different Spanish locations. The charge for a connection from Wolfsburg to Paris is equivalent to one from New York to Mexico. In the United States, VW has benefited from its VSAT dealer network in terms of cost and efficiency. "However, regulatory and cost issues do not allow the use of VSAT in Europe as we do it in the U.S."

Where it is practical to do so in terms of regulation and cost, VW exploits technology aggressively through voice/data integration. It has a 384-kbps Wolfsburg-to-U.S. link that is 70 percent voice; voice is compressed to 32 kbps. When the network is reconfigured, this will improve to 8 kbps. The extra bandwidth freed up will be used to carry data applications, such as CAD/CAM.

VW had discussed the possibility of outsourcing

some network operations with a number of companies. In the United States, MCI is the service provider for its satellite link to Mexico and Canada and across the Atlantic. "But in Europe, we are greatly restricted because of the lack of competition and the lack of responsiveness from too many of the PTTs."

Questions for Discussion

1. Explain why a company's desire and willingness to create a high-speed international network linking its many plants and divisions is not sufficient to make it happen.

2. Does it make any sense to have a 2 Mbps link between several nodes in the network but only 9.6-kbps links between other nodes due to the limited availability of high-speed circuits in some countries? Isn't the capability of the network determined by its slowest, lowest-capacity link? Discuss.

3. Prices for the same service in different countries can vary by a factor of 20 or more. The differences in the cost of providing the service cannot vary by that much, so what accounts for the dramatic price differences?

3

The Telecommunications Decision Sequence

Objectives

- To learn the four core frameworks for the telecommunications decision sequence used in this book
- To understand the telecommunications/business decision sequence that uses these frameworks
- To review a range of examples that both illustrate business opportunities and show the technical building blocks they use
- To apply the decision sequence through a concrete case illustration

Chapter Overview

| | | | | | |

Chapter Overview

This chapter brings together the topics of Chapters 1 and 2 and completes Part I of *Networks in Action*. Chapter 1 addressed the business opportunities of telecommunications and the nature of an effective dialog among client, designer, and implementer; Chapter 2 addressed the technical fundamentals of telecommunications. This chapter fits these two pieces together, linking business opportunities and technical choices and presenting a telecommunications/business decision sequence that has the following four steps:

▸ Bring together a team to define the business opportunity

▸ Identify the business criteria for the telecommunications platform needed to deliver the required services

▸ Assess network design priorities and trade-offs

▸ Make the economic case for the investment

At the end of the telecommunications decision sequence, the implementer can select the specific products and equipment needed to turn the opportunity into action with a clear statement of the business priorities, the business criteria for technical choices, and the economic priorities to drive those choices.

Each step in the telecommunications decision sequence applies one of the four core frameworks developed in *Networks in Action*. Together, they provide a systematic approach to linking business opportunities and technical choices. This is not a planning "methodology" or a cookbook, nor will you find it practiced as such in any company; there is no standard established methodology for telecommunications planning. That said, the decision sequence has been derived from studies of the best of current practice, and all four core frameworks have been applied in a range of organizations.[1] The frameworks are a vehicle to help you create a style of thinking and analysis that matches what companies look for in their telecommunications planners and designers.

It has not been at all easy for firms to find people who can span business and technology in this way. The technology itself provides little (if any) clear guidance about how to identify business opportunities to apply it. Knowing about each of the basic building blocks of telecommunications discussed in Chapter 2 suggests some potential applications—for instance, the very concept of a local area network suggests ways in which departments in a company can link personal computers or share databases—but knowing this does not in itself indicate

---

[1] Earlier versions of several of the frameworks appear in Peter G. W. Keen, *Competing in Time: Using Telecommunications for Competitive Advantage*, 1986; *Shaping the Future*, 1990; and Eric R. Clemons, Peter Keen, and Steven Kimborough, *Network Design Variables*, 1986. The best test of frameworks is that organizations apply them and report that they work. Companies that have applied some of these frameworks in major projects include Ford of Europe, The Royal Bank of Canada, IBM, United Airlines, British Airways, Citibank Latino, Glaxo, and about 50 other organizations in Europe, Latin America, and North America. They are thus not "concepts" but guides to action that have both an academic and a real-world foundation.

where exactly to focus the technology. Equally, knowing that fiber optics offers the chance to move masses of information quickly does not in itself say anything about what information to move and why.

Similarly, although the business topics discussed in Chapter 1 highlight some ways in which telecommunications may open up major business opportunities, they do not give any sense of which technical building blocks to use, how to design the network to turn opportunity into action, or whether the opportunity is practical (or one in which the technical problems, risks, or costs make it unrealistic).

Chapters 1 and 2 provided a broad understanding of both the business context of telecommunications and the core technology. Chapter 3 provides the broad understanding of how to bring this knowledge together, with a detailed case illustration that applies the decision sequence.

The Four Core Frameworks for the Telecommunications/Business Decision Sequence

The four core frameworks used in the telecommunications decision sequence (see Figure 3-1) are a template for learning and, more importantly, for applying that knowledge to, first, the cases and assignments in this book, and, later, to a job as a client, designer, or implementer. The four steps in the decision sequence are as follows:

▶   *Step 1: Define the Business Opportunity.* The telecommunications business opportunity checklist (core framework 1) identifies three practical and proven high-payoff strategies for exploiting telecommunications: (1) run the business better, (2) gain an edge in existing markets, and (3) create market and organizational innovation. The checklist is based on an analysis of over 300 real-life cases. The logic of the checklist is to focus on *proven* generic uses of telecommunications across private- and public-sector organizations and across different industries, sizes of company, and countries. This focus helps clients understand telecommunications as a business opportunity and provides a simple structure for brainstorming and thinking both innovatively and practically. In addition, it provides a framework for explaining telecommunications in business terms.

▶   *Step 2: Identify the Business Criteria for the Platform.* The telecommunications services platform map (core framework 2) defines three dimensions of business capability needed in the telecommunications delivery base, which in this book is termed the telecommunications services platform. The business dimensions the platform addresses are:

1.   Its reach: What locations, people, and/or other organizations need to be connected to the telecommunications services?

Step	Core Framework	Roles		
		Client	Designer	Implementer
1. Define the business opportunity	Telecommunications business opportunity checklist	List and brainstorm	Suggest ideas to review; highlight key options	Check for realism
2. Identify the business criteria for the telecommunications platform	Telecommunications services platform map	Confirm priorities	Lead the review	Check for technical implications
3. Identify the network design priorities and trade-offs	Network design variables worksheet	Review trade-offs and priorities	Translate implications of priorities and options	Lead, suggest, and review options
4. Make the economic case	Quality profit engineering framework	Make the business justification	Suggest ideas to review; highlight key options	Guarantee realism of plans

Figure 3-1

The Telecommunications/ Business Decision Sequence

2. Its range: What information and transactions must be shared directly and automatically across business functions and processes?
3. Its responsiveness: What level of service—in terms of speed, reliability, and security—must be guaranteed?

The telecommunications services platform map helps explain in business terms the degree of integration the organization needs in its platform, and it helps the designer develop and justify the technical architecture and standards needed to ensure compatibility among individual telecommunications, computing, and information systems.

▸   *Step 3: Identify the Network Design Priorities and Trade-offs.* The network design variables checklist (core framework 3) helps the client and designer set priorities and provide clear guidelines for the implementer. It identifies four main categories of design variable relevant to selecting network building blocks: (1) capability—the technical applications and services the platform provides, (2) flexibility—how easily it can accommodate change and growth, (3) certification—how reliable it is and what level of service it guarantees, and (4) cost—fixed, variable development, and operating costs. There will always be trade-offs among these variables, especially between cost and flexibility.

One frequent complaint by telecommunications managers is that they are unable to justify major investments in telecommunications infrastructures because of the focus on short-term cost savings of business executives who are increasingly concerned with controlling telecommunications costs. The network design variables checklist provides client, designer, and implementer a simple approach to exploring costs in terms of trade-offs—"If you want to cut costs, here's what you have to give up on flexibility. How important is that to you?" or "If you want to ensure rapid disaster recovery (one element of certification) in case of some breakdown, here's the impact on cost. Are we ready to take the risk of not having this because it is too expensive?"

▸ *Step 4: Make the Economic Case.* The **quality profit engineering** framework (core framework 4) identifies where and how the proposed services will contribute to the organization's economic health. It identifies six targets of economic opportunity: (1) profit management and alerting systems, (2) traditional costs, (3) quality premium costs, (4) service premium costs, (5) long-term business infrastructure costs, and (6) revenues. Historically, telecommunications has been viewed in most companies as a cost, mainly part of overhead. When telecommunications meant a telephone and telex utility that was part of the organization's overhead, the economic focus was naturally on cost displacement. Now that telecommunications is both a rapidly growing component of business capital and a central element of business innovation, it demands a far broader financial justification.

Each core framework is described in detail in a separate chapter in Part III. The next sections of this chapter briefly explain each step in the business sequence and outline the relevant core framework, providing real-world illustrations. The main aim of this chapter is to give you the "Big Picture." As you read it, we suggest you focus on the overall decision sequence, how each step leads into the next, and how they ensure a true dialogue among client, designer, and implementer.

| | | | | | |

Step 1: Define the Right Opportunity: The Telecommunications Business Opportunity Checklist (Core Framework 1)

With respect to the opportunities firms have used to exploit telecommunications to significantly enhance their competitive strengths, the patterns are remarkably consistent across industries, size and type of company, and era. Figure 3-2 summarizes these opportunities in the telecommunications business opportunity checklist and provides some specific company examples. The next three sections discusses these opportunities in turn.

Run the Business Better

There are five proven "must do's" for running a business better:

1. *Manage distributed inventories*—speed up all aspects of ordering and distribution by moving to "just-in-time" ordering, manufacturing, marketing, and delivery.

Opportunities	Strategies	Examples
Run the business better	Manage distributed inventories	Levi Strauss used EDI to speed customer delivery, eliminating distribution delays and buffer stock
	Link field staff to head office	Merced County, California welfare agency decentralized application processing, reducing response time from 6 weeks to 1–3 days
	Improve internal communications	Use videoconferencing for electronic meetings; coordinate teams via groupware; use business TV to personalize management and keep staff informed
	Improve decision information	Capture information at point of sale, point of reservation, or point of order to spot trends; screen information and get it to decision makers to match pace of events and operations
	Reengineer processes and simplify organizational complexity	Eliminate steps, delays, documents; reduce processing time from days or weeks to minutes
Gain an edge in existing markets	Provide workstation access in customer's home, office, or on the street	Use telephone as access device for getting customers to correct person quickly
	Differentiate a standard product	Provide customer self-service and up-to-the-second information on transactions
Create market and organizational innovations	Launch preemptive strikes	Change an industry's rules of competition; Federal Express took over small package delivery from airlines and freight companies
	Piggyback services onto an existing platform	British Airways added hotel reservations to existing airline reservations infrastructure; McKesson added insurance claims processing to pharmacy order-entry system

Figure 3-2

The Telecommunications/ Business Opportunity Checklist

2. *Link field staff to head office*—use laptop computers so that personnel can send and receive messages and transactions wherever they are, instead of having to wait until they get back to their office.

3. *Improve internal communications*—speed up information flows and make it easy for people to contact each other, regardless of time and place.

4. *Improve decision information*—provide managers with up-to-date information about what is happening in the marketplace early enough for them to be able to take action.

5.   *Reengineer* **business processes** *and simplify organizational complexity*— use telecommunications to reduce the many steps, documents, and people that clutter up many aspects of organizational life.

Each of these five "must do's" is examined in the following subsections.

Manage Distributed Inventories   Although inventory is listed on a firm's balance sheet as an asset, the world of just-in-time business treats it as a liability.[2] The goal is to carry just the right amount of goods needed to meet expected or scheduled needs and at the same time be able to respond quickly to unanticipated shifts in demand. Typically, firms that have telecommunications-based information, scheduling, and ordering systems to monitor and match demand cut their inventory levels by 15–40 percent.

Levi-Strauss provides a typical example of how streamlining customer-supplier links can reduce inventory levels. Levi's jeans are so famous worldwide—they are a form of currency in many Eastern European countries—that many people have no idea of how close the company came to disaster in the mid-1980s, with sales slipping and costs rising. A management team made a leveraged buyout of the company and made information technology a key element in their new plans. It introduced Levilink, an electronic data interchange (EDI) service for handling all aspects of orders and delivery. Customers can place even small orders as needed, say, every week. Goods are delivered by UPS within two days. Using Levilink, one of Levi-Strauss's customers, Design Inc., a chain of 60 stores, was able to entirely eliminate its warehouses, which act as a buffer supply to deal with shifts in demand and the long lead times for ordering and delivery.

Many car dealers use simple telecommunications services to locate spare parts or vehicles. They can dial up a host computer, either accessing a private leased line or using the public phone system, to tap into a master database; when they find the items they need, they place an order on-line. This service reduces the inventory they must keep and avoids out-of-stock situations, which generally mean lost sales.

High levels of inventory represent an unnecessary expense. Low levels of inventory can lead to stockouts, which create either a breakdown in operations or a loss of sales. Clearly, the only way in which inventory can be matched to demand is by eliminating delay, and telecommunications is the most powerful way to do this. The simplest tool is a fax machine for ordering goods that are then delivered by Federal Express the next day or by UPS in two days; the most complex tools are the integrated systems that connect nearly every link in the

---

[2]The origin of just-in-time, which was invented by Toyota in Japan, was the American supermarket. After its defeat in World War II, when Japan was short of money and materials, a Toyota executive observed how supermarkets kept the shelves full by constantly replenishing them as consumers removed goods. He also saw that this could be applied to manufacturing. More recently, through telecommunications Wal-Mart has eliminated a storage area in the shop for holding its inventory; instead goods arrive by truck just-in-time.

supply chain, with point-of-sale or point-of-manufacturing software automatically updating central databases and with software systems automatically placing orders using EDI links to the suppliers' systems. In turn those systems link to the freight forwarder's, insurer's, bank's, and trucker's systems.

The technology base for systems that aim at streamlining ordering and delivery typically rely on EDI, with the main standards being industry-specific agreements on terminology and types of transaction. Most industries, however, have based their standards on either ANSI's X12 or on EDIFACT. **X12** is the EDI standard that describes how firms can send documents to each other without having to adopt one or the other's formats; **EDIFACT** is the international X12-based standard for electronic trade.[3] X12 is a new business language that does not require a complex technical infrastructure; it can be implemented using a plug-in board on a personal computer and dial-up links either to the supplier's computers or to a node in its telecommunications network. This means that small as well as large firms can exploit EDI to streamline key elements in their management of inventory.

Link Field Staff to Head Office Without telecommunications, field staff such as sales reps, maintenance engineers, and branch office employees often find it difficult to keep up-to-date and in touch with the head office, where most corporate information is stored, most records are kept, and most administrative processes are coordinated. A sales rep, for instance, may need to go into the office every morning to deal with all three of these constraints on his or her ability to get out and meet customers.

With telecommunications, more and more companies are providing their field staff with portable computers that can link to the head office. This helps them handle paperwork; get access to inventory, product, and price information; send and receive electronic mail messages; and respond more quickly to customers' requests and questions instead of having to say, "Oh, sorry. I'll have to wait till I'm back in the office to find out about that for you."

The key need in linking mobile staff to the head office is making telecommunications access cheap and simple. Notebook computers now are powerful and inexpensive and fit into a briefcase. With a modem, their users can simply unplug the standard wire from the back of a phone, plug it into the PC, and then use the public phone system to dial up a remote host computer or electronic mail service. This provides a simple but very limited capability. Users cannot download large amounts of data, cannot get access to their own or their department's files on an office LAN, and cannot run most transaction processing software (to place an order, for example). For these functions they need more complex tools that connect them directly to the corporate network.

Linking field personnel to the head office is a major challenge to the imple-

---

[3] X12 is a U.S. standard. EDIFACT, defined by a committee sponsored by the United Nations, is more widely used in Europe for the obvious reason that there is far more movement of goods across national borders there. In the United States, industries drive EDI; elsewhere, shippers and customs organizations do.

Figure 3–3

**Laptop Add-on
Devices that
Facilitate Networking**

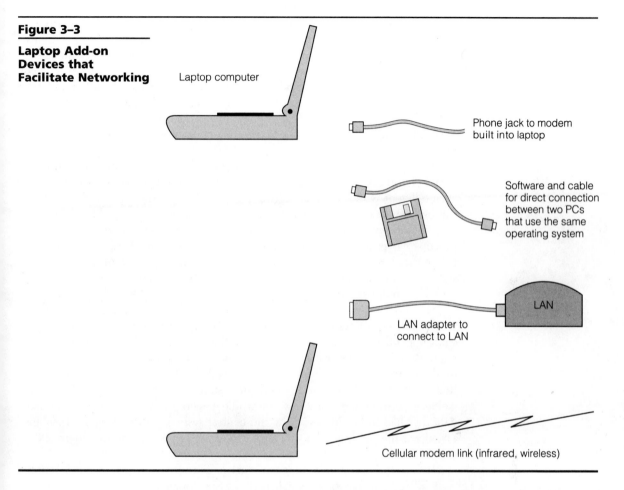

Laptop computer

Phone jack to modem
built into laptop

Software and cable
for direct connection
between two PCs
that use the same
operating system

LAN adapter to
connect to LAN

LAN

Cellular modem link (infrared, wireless)

menter. According to *Network Computing* magazine, "Connecting mobile computers can be expensive, a support nightmare and frustrating to both the user and the network manager. . . . Laptops have grown up in an era where connections were rarely if ever considered in a purchase decision, and indeed where often the entire purchase process was geared against connecting them to corporate networks." (Magidson 1992) Restated, this means that the client wants a cheap, easy to use device, but also one with access to more and more information resources. The designer can offer the first option, and whereas the implementer can offer the second, it comes at a high cost for everyone. To resolve this dilemma and frustration, more and more vendors are offering laptop add-on devices that improve ease of connection, including high-speed modems with error detection and error correction, plug-in adapters for connecting to LANs, fax modems, cellular data transmission, and software and hardware for file transfer. Figure 3-3 illustrates some of these devices.

Distributed computing systems that are accessed from a desktop computer in the office are far easier to implement because they use standard terminals, transmission links, and fixed connections. Although they lack the advantage of complete mobility, they are helping organizations become more decentralized and less bureaucratic. Historically, computer applications are handled on large central computers with limited (if any) telecommunications links to field offices, which helped to create layers of administration and delays in processing. Two government agencies show how distributing computer power through telecommunications can dramatically change this.

The State of Virginia can authorize income tax refunds on the spot from its local field offices. Staff quickly enter a few key figures from your tax return into an intelligent workstation. A software system screens the data, and the workstation both automatically links to the central state computer to check historical records and sends electronic authorization for issuing the refund. The refund is processed overnight on the mainframe computer. It would be far too expensive to use dumb terminals to process the full return on-line. The intelligent workstation thus uses an expert system to estimate from a few figures whether the return is complete and accurate.

Merced County, California, has similarly used distributed workstations, LANs, and expert systems to reduce the time required for new welfare applicants to get their first check. The former reliance on central mainframe processing of the application meant an average delay of six weeks: A new applicant had to fill in a minimum of 15 pages of forms, wait until the central system checked them against the more than 2,500 federal, state, and local government rules on welfare, be interviewed by a welfare officer who generally needed additional data that had to be processed by the central system, and wait again. In many instances, the applicant had to visit several agencies and fill out many extra forms.

Today, the maximum wait is three days, and often it is just one day. Applicants fill in a single form, which is processed immediately through an expert system on a LAN, via a file server. As in the Virginia tax example, it would be prohibitively expensive to process the full application on-line to the host; the distributed system handles the main steps, accessing the host only at the end. Figure 3-4 shows the main features of the Merced County system, which has become a model for state and local government.

Improve Internal Communications Telecommunications is by definition a medium for communication, and more and more uses of it are designed to enhance teamwork, personalize management, and coordinate activities scattered across many locations. The main tools for improving internal communications are:

▶   *Videoconferencing*—two-way telecommunications transmission between two cameras in separate locations that provides electronic "face-to-face" meet-

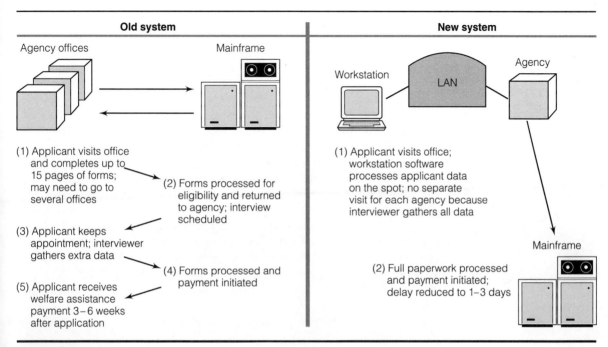

Figure 3–4

The Merced County, California, Welfare Application Process

ings. Until recently, it required an expensive investment, but systems coming onto the market now provide low-cost videoconferencing via desktop PCs.[4]

▸ *Groupware*—a generic term for tools for coordinating work. All are PC-based and include calendaring, workflow management, bulletin boards, schedules, anonymous voting, and the like.[5]

▸ *Electronic Mail*—a simple tool for sending and accessing messages; the two parties neither need to know where the other is located nor need to make direct contact.

---

[4] In 1992 and 1993, PictureTel and Hitachi seemed to be in an almost monthly competition to offer the lowest ever price for a videoconferencing system; their prices dropped in 18 months from around $100,000 to under $20,000. The key technology here has been data compression and a special type of chip, called digital signal processor (DSP), that has the speed to handle multimedia applications on PCs. Data compression speeds up the transfer of the picture, and DSPs speed up its processing and display.

[5] Lotus Notes and Microsoft's Windows for Workgroups are among the market leaders in groupware software. Groupware is a PC application that depends entirely on telecommunications; without it, there is no group to coordinate.

▶   *Business Television*—the broadcast of TV programs that are for reception only by a specified audience within or outside the company.[6] It is growing in use as a way of personalizing senior management, ensuring that employees are kept up to date and well-informed, and providing training.

One problem in the evolution of many of these tools has been the lack of standards, which limits their use. They need special equipment, may or may not be compatible with existing systems, and cannot communicate except with a subpart of the target community—which is contrary to their major purpose. The technical blockages to progress are being surmounted. The **H.261** standard, which defines the protocol for coding and decoding a high-speed digital video signal, is making it practical to link videoconferencing systems.[7] **X.400,** the international standard for electronic messages, including E-mail, fax, and telex, links different electronic mail services. **Lotus Notes,** a PC software package that organizes and coordinates the messages, schedules, and specified workflows of individuals and groups, is becoming a quasi-standard for groupware, but until late 1993 it was available only for MS.DOS and Windows, not for Apple Macintosh or Powerbook computers. Business television can use a wide range of standard transmission links, including T1 and satellites, and, because it is not interactive, it does not pose problems of incompatibility.

More significant than the technical limitations are the social and psychological ones.[8] Communication is among the most personal of activities, and people are naturally somewhat unwilling to shift away from the tools and procedures with which they are most comfortable. Many are reluctant, for instance, to try out videoconferencing, expecting it to be impersonal; groupware software often seems to them to be restrictive and awkward to use, and electronic mail seems impersonal and clumsy. But there is plenty of evidence that most people don't fear using technology; the growth of fax machines demonstrates that. Fax is easy to use because it requires the same actions and corresponding skills as dialing a phone and inserting a document into a photocopier. Still, computers remain far more difficult to use.

---

[6] All a firm needs is either a satellite link that can be rented by the hour or a private link that uses a specific frequency and satellite transponder. Business TV can also be carried over the firm's existing private network at off-peak hours. Kmart, for example, uses its VSAT network, whose main purpose is point of sale, credit card authorization, and voice transmission for its chairman to "meet" with employees daily. The total cost is just 50 cents per store per hour. The network is effectively free when the store is closed.

[7] Another key standard for compressing slower signals is V.21 bis, which is an extension of the V.21 standard for modems. It increases data transfer by a factor of four.

[8] There is a rich literature on the human context of computers and telecommunications. One of its constant themes is that technical innovations too often neglect the human context and that technical professionals are ignorant of and/or uninterested in these issues. Unfortunately, too much of the literature is in academic journals and remains unread by professionals. A landmark book was Shoshana Zuboff's *In the Age of the Smart Machine* (1988); it is tough, but essential, reading.

Improve Decision Information Traditional management information systems have been built on a firm's accounting system, with reports on sales, inventory, profit and loss statements, and the like being generated on paper at the end of the month. This means that the information managers get is often too old to allow prompt action. For instance, in retailing, purchase patterns can shift in days. Supermarkets' stock and pricing decisions may need to change daily because of weather. Airlines need to track flight reservations each day; they make their money on the last five seats on a flight—they are the difference between profit and loss.[9] Manufacturers need to spot quality or production problems just-in-time.

In these situations, information that is 15–45 days old is of little value. More and more leading firms use their telecommunications network capability to capture information at its source—point of sale, reservation, ordering, delivery, and so on—and move it as frequently as is needed to a central computer, which screens and stores the data and makes automated decisions for ordering while simultaneously looking for any trends, anomalies, and news. When managers and planners log onto their workstations to access data, the executive information system draws their attention to such information.

A classic, well-known example of such management alerting systems illustrates the general principles. Frito-Lay was one of the first large firms to provide its delivery staff hand-held computers to handle invoicing and accounting. At the end of the day, the information is uploaded to Frito-Lay's head office computer complex. Managers can now spot trends in just a few days. The executive information system software alerts them to the implications of what happened yesterday so they can act today.

Reengineer Processes and Simplify Organizational Complexity Business **process reengineering** combines fresh thinking about processes with the use of technology. It has become one of the rallying cries of the information services profession, mainly because it offers a way to use telecommunications and computers to do more than just automate the status quo. The basis of reengineering is taking a fresh look at work processes and removing as many documents, people, administrative procedures, and delays as possible. Telecommunications is a central part of this streamlining for the obvious reason that it replaces physical documents with electronic ones that move in millisecond speeds instead of days, allows on-line access to information wherever it is stored, and brings the work to the person.

Examples of effective reengineering include cutting the time to issue a life

---

[9]One travel agent may be informed that no seat is available on a generally fully booked flight at exactly the same time as another agent learns of the availability of the final seat. Guess which computerized reservation system the latter is using—the airline's own or a competitor's? Walter Wriston, the former chairman of Citibank, once recommended that information about money (for example, in electronic foreign exchange trading) is now more important than money itself. Information about airplane seats is more important than airplane seats, too. The reality in both of these situations is the direct result of telecommunications.

insurance policy from 22 days—during which there is typically just 11 minutes of decision making—to four hours. The masses of paper that move through departments and mailrooms are largely unneeded; information is moved electronically. Software screens the forms and applies standard rules of evaluation and pricing. Most applications are routine and are thus processed by a single agent at a workstation by accessing outside information via a WAN and/or a public network and by accessing software and other data via a LAN server. Shearson Lehman (see Minicase 1.2) uses the same approach and technology to a workflow, as does Merced County, California.

Reengineering, streamlining, simplifying, and redesigning workflows are closely aligned to the **total quality management (TQM)** movement that has become a core part of business innovation in the 1990s. In many respects, reengineering is a technology-focused version of TQM, for it is a conscious effort to shift the entire application of information technology away from automation—from technology as the driver and customers and people as the followers and secondary concern, and from efficiency of operations as the business rationale, toward a completely fresh view of work, with technology as the secondary element and customer service and workers as the drivers, and with service and quality as the core of the business case.

Figure 3-5 lists examples of successful reengineering projects. Although telecommunications is just one of the tools for reengineering, it is a key enabler because without telecommunications, there is no means of either cutting out paper and the delays and administration they create or of moving work and documents to a single contact point where they can be processed immediately, instead of step-by-step and department-by-department.

Gain an Edge in Existing Markets

The 1990s are a time of constant competitive stress, for there are no easy markets. Firms must control their costs aggressively. Service and quality are not special add-on features that can be charged for but are the entry fee for being in the game. In this context, telecommunications has been a key element in helping companies find an edge, however small, that gives them a chance of standing out among their many competitors. In more and more industries, telecommunications is a competitive necessity and firms must use it to achieve two goals:

1. *Provide workstation access* in the customer's home, office, or on the street to create a new delivery base.

2. *Differentiate a standard product* through speed of service, ease of access, and information byproducts.

In the next subsection we examine each goal in turn.

Provide Workstation Access in the Customer's Home, Office, or on the Street The term *workstation* covers a wide range of access tools, from the telephone to the personal computer to specialized computer terminals. The crux of telecommunications is to make it easy to do business with a supplier;

Example	Activity	Base	Reengineered	Reference Definitions
Productivity indices Staffing efficiency: Plains Cotton Cooperative Assn. (TELCOT)	Transaction processing	9	450	Base = industry average, thousands of units processed per worker per year, 1991
Staffing levels: Phillips Petroleum Company	Corporate staff	36	12	Corporate staff per 100 employees, 1986–1989
Transaction costs: C. R. England & Sons, Inc.	Invoicing	$5.10	$.15	Cost of sending invoice, 1989–1991
Asset turnover: Toyota Motor Corp.	Work in process	16	215	Asset turnover, industry average annual turnover, 1990
Velocity Progressive Insurance	Claims settlement	31 days	4 hours	Base = industry vs Progressive's Immediate Response service, 1991
Quality Florida Power & Light Co.	Power delivery	7 hours	32 min.	Base = competitor, power outage per customer per year, 1992
Business precision Farm Journal, Inc.	Product variety	1	1,200+	Number of unique editions per issue, 1985–1990
Customer service L. L. Bean, Inc.	Order fulfillment	61%	93%	Base = industry average, percent of orders filled in 24 hours

Figure 3-5

**Examples of
Successful
Reengineering**

Source: *IBM Systems Journal* (1993), 32(1):67.

simply pick up the phone or log on to the computer and get what you need when you need it. Consider the following examples of workstation access:

▸ Electronic Cash Management: **Electronic cash management** systems are software and telecommunications links that process corporate banking transactions electronically. These services include funds transfers, letters of credit, and foreign exchange. The bank provides software and network access; the workstation is generally a standard PC that exploits the available software to add reporting features, spreadsheet analysis, and access to corporate information.

▶   Federal Express: Fedex's legendary network-based services are built on mobile communications, tracking systems, the company's private fleet of airplanes, and its well-trained and proud culture. More recently, Federal Express has given its high-volume customers workstations they can use to initiate their own shipments without the need to phone the company for pickup. The workstations tie directly into Fedex's own systems, simplifying paperwork and administration for the customer.

▶   Dell: Many of the fastest-growing and most profitable retailers have turned the telephone into a workstation. Customers use a catalog to shop by phone, are served by an agent at the other end of the line who has access to inventory data and can complete a transaction in just a few minutes, and get next-day delivery. Although many companies provide this type of access, some, like Dell, complement it with voice messaging systems that route the call to the right person, whether for information, service, or technical assistance. Dell is under ten years old but has already reached annual sales of over $1 billion. It and the other mail- and phone-order companies now have over 20 percent of the personal computer market.[10]

Figure 3-6 provides other examples of using workstation access to provide customers a level of convenience and service that gives them little if any reason to go elsewhere. In these situations, customers increasingly concentrate their purchases with the on-line provider. Federal Express's aim is for customers to choose to use its package delivery services exclusively. A bank's aim with cash management systems is to capture more and more of the customer's business. The terminal becomes the basic contact point in the customer-bank relationship.

Differentiate a Standard Product   Closely related to making the workstation the primary contact point is using telecommunications and information by-products as a differentiator in commodity markets—markets in which the only two forms of differentiation are price and convenience. In competitive markets, price cuts are usually matched by all competitors; discounts, special deals, and promotions have only temporary impacts. Convenience in the form of speed and ease of access can have a more lasting impact, especially if it is based on a telecommunications and computer platform that competitors cannot quickly replicate.

For many standard transactions—ordering an airplane ticket, buying stationery, buying mutual funds, and applying for a car loan—dozens of suppliers

---

[10]Traditionally, face-to-face service is thought of as being more "personal" than phone- or computer-based service. Today, however, there is a growing recognition that the reverse is often true. First, people are very comfortable talking over the phone and may be less comfortable in someone's office or store. Second, the technology Dell uses gets a customer to the correct person very quickly. As a result, Dell provides a high degree of "personal contact" and establishes continuing relationships, from sales to after-sale technical support, without a customer ever making any face-to-face contact with members of the firm.

Figure 3–6	Service	Organization	Previous delivery/access point
Examples of Workstation Access to Services	Telephone application for car loan	USAA	Mail or visit to bank
	Renewal of auto registration	State of New Jersey	Mail or visit to Dept. of Motor Vehicles
	Purchase/sale of securities	Fidelity Investments	Phone call to broker
	On-line news and information	Dow Jones	Visit to library
	Self-diagnosis of medical symptoms	Harvard Community Health Plan	Visit to physician (or ignore symptoms)
	Make own airline reservations	American Airlines	Phone call to travel agency
	Get airline ticket printed at your hotel	American Airlines	Courier from travel agency
	Get message to traveling colleague	Skytel	Cross your fingers

are readily available; many can be quickly contacted by phone. Each of these routine transactions has been differentiated by leading firms through a combination of phone and data communications. CitiTravel, a Citibank subsidiary, has become a major provider of travel services by offering a five percent rebate on airline tickets once they have been used, plus delivery of the tickets to anywhere in the United States via Federal Express. Quill processes stationery orders sent by fax the same day. USAA can process a car loan application over the phone in under five minutes, fax the authorization and the insurance policy to the car dealer, and update police databases to indicate that the driver is covered. Charles Schwab provides customers day and night push-button phone access to make stock purchases and get account information.

American Airlines, for 20 years one of the most creative firms in applying information technology, initiated its telecommunications-dependent frequent flyer program and held a competitive edge for around seven years, until frequent flyer programs became so widespread that most passengers who travel often are enrolled in at least three. One American Airlines innovation created a new AAirpass program. In it, travelers pay a fee that buys them a given number of miles (between 5,000 and 100,000) a year; enrollment ranges from one year to the customer's lifetime. For example, for $102,000, a 45-year-old traveler could get 40,000 miles of travel a year for the rest of his or her life. AAirpass travelers write their own tickets and make their reservations through a special phone number.

The AAirpass program involves complex telecommunications and database linkages for accounting and for recognizing the passenger in a variety of databases. Only a few airlines have the technology platform needed to incorporate these linkages; that is why the program has not been imitated.[11]

In each of these examples, there is no grand "innovation," and there may be no distinctive, sustainable competitive edge. That said, there is an almost certain competitive disadvantage for firms that fall behind the leaders in using telecommunications as the differentiator of service through ease of access and speed of response. In addition, firms such as Quill, CitiTravel, and USAA are also among the price leaders. The streamlining of service to create market differentiation (which falls under the heading of gaining an edge in existing markets) often also streamlines internal processes, which falls under the heading of running the business better.

Create Market and Organizational Innovations

The first two areas of opportunity—run the business better and gain an edge in existing markets—focus on today's activities. The third area of opportunity—market and organizational innovation—is far more radical and explicitly aims at inventing something new. Capitalizing on this opportunity demands very fresh thinking, but in many instances it does not need space-age technology, but only a careful application of existing technology. Telecommunications is a relatively new competitive weapon; the widespread availability of digital communications dates from the mid-1980s. Thus firms have only just begun to explore what can be done to use telecommunications as a force for business innovation. The challenge is to build a dialogue that maximizes the chance to do two things:

▶ *Make Preemptive Strikes,* the rare innovations that change the dynamics of competition and create a sustainable edge for a few leaders (and often a disastrous erosion for the laggards). The nature and duration of the competitive edge heavily depend on (1) whether it is built on simple technology that can be quickly copied or on a comprehensive technology platform that takes years to replicate, and (2) the nature of the business/technology dialogue.[12]

These preemptive strikes do not suddenly and magically appear. It takes around seven years to create the business, organizational, and technical bases needed and for customer response to reach a critical mass. Generally, then, there is a seven-year window of opportunity for the innovator to exploit its lead.[13]

---

[11] One of the authors of this book, who is an immensely satisfied customer of AAirpass, suggested to a major foreign airline that it implement such a program. The marketing staff was eager, and the CEO was even more eager. However, the airline's telecommunications and data management systems lacked just two vital interconnections. The project died.

[12] When every firm in an industry has access to the same technology, and yet a Wal-Mart, American Airlines, Federal Express, BancOne, or USAA consistently makes more effective use of it, there has to be some explanation. The evidence increasingly suggests that differences in management establish the competitive advantage. Those differences begin in the relationship between the senior management team and the leaders and planners in information services.

[13] All the best-known examples of telecommunications as a source of a major competitive edge have included this seven-year window of advantage. Of course, because books and articles discussing innovation often appear about seven years after their implementation, there are many instances of

▸ *Create New Services by "Piggybacking"* and network interconnection. Piggybacking involves using existing telecommunications delivery infrastructures. Once the telecommunications "highway system" is in place, it can be easy to add new services to it; many of these new services are the traditional territory of other firms and even of other industries. An airline can add hotel and car rental reservations to its airline reservation system, and a pharmacy supplier can add electronic processing of insurance claims to its customer ordering network.

Each of these two methods of innovation is discussed in more detail next.

Launch Preemptive Strikes Preemptive strikes use telecommunications to redefine the basics of an industry. There is substantial debate in both academic and business circles as to whether or not the launchers of such strikes gain a sustainable competitive advantage. The argument is that, as with ATMs and airline reservation systems, competitive necessity forces laggards to catch up, and thus every firm ends up with the same market size but a higher cost base.

The rebuttals to this argument are that even if it is true, it takes five to seven years for laggards to catch up, during which period up to half the firms in the industry disappear; that at least one of the losers will have been a previous leader; and that there are plenty of examples of firms that created their own leadership through their preemptive strike. Examples cited as evidence to support the rebuttals include the following:

▸ USAA brought together all customer information to create a level of service unequaled by all except a handful of firms.

▸ Citibank launched into telecommunications-based electronic banking and credit cards that provided it with around $800 million a year in profits and a dominant position in retail banking, even as late as 1992, when most of the rest of its core business was in a shambles. Viewed as a separate company, Citibank's electronic retail banking remains a star, with close to 20 years of sustained success.

▸ Wal-Mart, Dillards, Toys "R" Us, and a few other retailers did not simply carve out a lead through point of sale and quick response; they drove many of their previously strongest competitors out of business.

▸ Federal Express took over the package and small cargo market from the airlines and such well-established carriers as UPS.

In each of these and comparable examples, a distinguishing feature of the com-

---

erosion of the advantage at the very time it is hyped, leading business managers to doubt their original success. Max Hopper, the IT executive of American Airlines whose name is most associated with its successful use of technology, published a controversial article in 1990 claiming that there is no longer any competitive edge to be gained by IT. Many commentators concluded that his real message was, "We've got our value from Sabre and left the others behind. Let them buy pieces of what we have built—and let's move on and still keep our edge by knowing how to use technology as a business resource." (Hopper, 1990) pany is the fusion between the thinking of its business leaders and of its information technology managers, to the extent that it is almost impossible to sort out which of them created the innovation. They both did, through a dialogue that both maximized the chance of the business leaders spotting where telecommunications could make a major competitive difference and linked the information technology managers into the business strategy, so that their contribution would be considered as well. And in each instance, competitors were pushed onto a long defensive effort to catch up. Sears, Kmart, Delta, Bank of America, UPS, Aetna, and many other previously well-positioned firms have had to spend literally billions to close the gap.[14]

Preemptive strikes take a long time to launch. If they depend on small-scale software packages and standard telecommunications links, there is no sustainable edge; but if they require a comprehensive integrated platform, then time, skill, and dialog to design, build, and operate them are required as well.

Piggyback Services onto an Existing Platform   Piggybacking refers to adding a new service—often traditionally supplied by another industry—onto an existing telecommunications capability at low incremental cost. Examples of piggybacking include the following:

▸   British Airways added international hotel reservations to its international airline reservation system at a time when the major U.S. hotel chains lacked an international reservation capability. Travelers do not make their hotel reservations before they have first booked their flight, so a natural by-product of doing so is to ask, "Can I help you with a hotel?" BA was able to command large commissions from hotels and pushed several of them, including Marriott and Hilton, into a catch-up effort that turned out to be a disaster, costing around $125 million before it was abandoned.

▸   McKesson added insurance claims processing to its pharmacist order-entry system, thereby becoming the third largest processor in the country and taking business away from insurers.

Some of the most striking recent examples of innovation through piggybacking come from the public rather than the private sector. State government agencies, for instance, are looking to find ways of providing a new level of customer service by using existing telecommunications delivery bases at minimal incremental cost. As mentioned in Chapter 1, Maryland allows use of ATMs for making welfare benefit payments, and New Jersey enables car owners to renew their auto registration through ATMs.

Piggybacking provides a source of revenue that offers unusually high margins, even at low volumes, because the delivery base is already paid for. Of course, piggybacking is successful only if customers want the service. Hotel

---

[14]Wal-Mart had just a four-year lead over Kmart. That advantage was a major factor in pushing Kmart into a race for survival—and into a $2.2 billion investment in store refurbishment and in information technology.

reservations are a natural cross-selling opportunity for an airline, as is insurance claims processing for a pharmacy goods supplier. But offering a car loan or a chance to buy mutual funds is not a sensible effort at piggybacking for either type or firm—or is it? The customer decides.

Summary of Step 1: Think Opportunity, Not Technology

Chapter 10 goes into greater detail about each of the categories on the telecommunications business opportunity checklist. Each of the examples given here has been illustrative; each points toward particular choices of telecommunications and computers, and in many cases each is enabled by specific technical building blocks. But the technology is not the opportunity. By focusing on the categories in the checklist, the dialog among the client, designer, and implementer is centered on the client and is carried out in the client's language. At the same time, because the categories have been derived from proven examples of telecommunications, the dialog positions the designer to move forward to Step 2—to identify the business criteria for the telecommunications platform, shifting the language toward issues of connectivity, integration, architecture, and standards. Step 1 also positions the implementer, who can draw on a range of proven blueprints and recall that Company XYZ used this technology to meet this type of opportunity and that most companies are using these standards to meet that opportunity.

| | | | | | | | |

Step 2: Identify the Business Capability of the Telecommunications Services Platform Map (Core Framework 2)

The business opportunity checklist does not directly address technology; instead, it zeroes in on where telecommunications can offer a major chance either to run the business better, gain an edge in existing markets, or change the rules of competition. Before the technical design of the network can be defined, the obvious question to answer is: "What are the business requirements for the technical platform that delivers the service(s) identified as targets of opportunity?" The second core framework of *Networks in Action*—the telecommunications services platform map (see Figure 3-7)—has proven to be useful for translating business to technology along three dimensions of business functionality (that is, the nature of business services that can be delivered by the technical platform). Each dimension—reach, range, and responsiveness—runs in a different direction on the map and varies from simple to complex. The greater the complexity along dimensions, the greater the need for integration of business and technology.

Next we examine each of the dimensions of business functionality in turn.

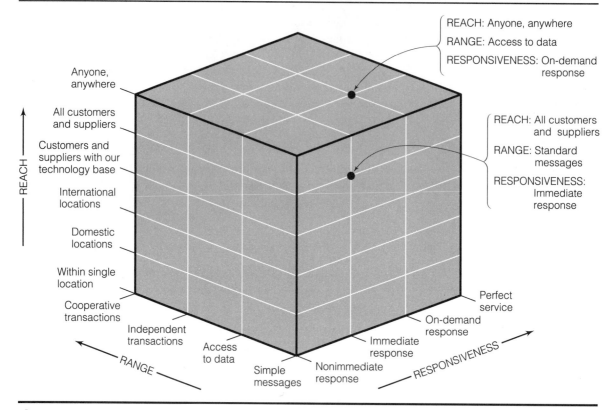

Figure 3–7

The Telecommunications Services Platform Map

Reach

The dimension of **reach** concerns whom we need to connect our network and business processes to, and where:

- *Within a Single Location,* such as a head office department: This situation requires only local telecommunications facilities that need not be compatible with other departments' facilities.

- *Across a Firm's Domestic Locations:* This degree of reach requires compatibility between the various local and wide area facilities. Shearson Lehman, Merced County, and New York State's Empirenet have this degree of reach, which is essential when the main aim is to use telecommunications to run the business better.

- *Across International Locations:* International and telecommunications is a morass of regulatory, economic, political, and technical uncertainties and constraints. Minicase 2.4, "International Telecommunications: The View from Volkswagen," listed many of these problems and also made it clear that they must be resolved by any company that needs to coordinate its worldwide operations. THEi's network is international, but it faces far fewer problems than Volkswagen because its main nodes are in countries with both responsive PTTs and strong national telecommunications infrastructures.

▶ *To Customers and Suppliers with the Same Technology Base* as the firm's: This degree of reach makes it essential that the firm adopt an architecture that is in the business mainstream. Companies such as Levi-Strauss, and many retailers and manufacturers, have made electronic links to customers and/or suppliers a cornerstone of their business. The architecture used must adopt standards that ensure interoperability of telecommunications for interoperation of business functions.

▶ *To all Customers and Suppliers,* regardless of technology base: This degree of reach requires a level of integration that today is close to impossible to guarantee across all services. Emerging standards for EDI and electronic messages are, however, rapidly improving the situation; in more and more industries, electronic customer-supplier links are becoming the norm, not the exception.

▶ *To Anyone, Anywhere:* This degree of reach is the ideal. Although it remains impractical today, it is not impractical to prepare for. The phone link is one base for this, as is cable TV. The needed technology is becoming available, though the investment cost will be huge.

Range

Range involves the information and transactions that we must be able to share directly and automatically across business functions and processes:

▶ *Simple Messages:* These require the simplest form of computer operating systems and telecommunications transmission. Laptop computers with a modem can send and receive electronic mail messages through a dial-up connection to the public phone network. The California telecommuters can do this (see Minicase 1.3).

▶ *Access to Data Stores:* In this degree of range, a single workstation or business application can access data from different databases. This requires an operating system on an intelligent workstation that is either compatible with the operating system of the host computer that stores the data, or routers, smart hubs, and gateways that can ensure that the receiving and sending devices do not just connect to each other but also format the bit stream into meaningful data. The California telecommuter's laptop computer, for instance, cannot directly access mainframe database. THEi's point-of-sale terminals can send and receive data only from specific types of computer systems. Many companies with IBM PCs need software or hardware that emulates IBM 3270 dumb terminals for them to be able to access databases on an IBM mainframe. (**Emulation** is a software and/or hardware technique that converts the PC's protocols into those used by the relevant device.)

▶ *Independent Transactions:* In this degree of range, workstations can make transactions, but each transaction is independent of any other, with no cross-linking or knowledge of each other. For example, when a travel agent makes a plane reservation, then a car rental reservation, and then a hotel reservation, each transaction is entirely separate and specific to an operating system. Thus, an Apple Macintosh cannot substitute for an IBM PS/2 for ATM transactions and at the same time process a transaction on an UNIX processor. Intelligent networking devices and software are needed to ensure that a single terminal can carry out all of them.

▶ *Cooperative Transactions:* In this degree of range, services are interdependent. If the travel agent in the previous example wants to change the date of the traveler's departure, then the airline, car rental, and hotel applications each knows about the others and makes the adjustment accordingly. The two developments that make this new level of service practical, though technically difficult to implement, are client/server computing and cooperative processing. Both these similar innovations divide the processing required to deliver a service among a range of distributed systems; the "client" workstation transfers control to "servers" as needed; whether the "servers" are mainframe host computers, departmental machines that store needed data and additional software, or other workstations.

Responsiveness

Responsiveness refers to the level of service—in terms of speed, reliability, and security—that we must guarantee:

▶ *Nonimmediate Response:* In this case, services and information need not be on-line and can be processed at the end of the day, week, or month. This both minimizes telecommunications requirements and simplifies computer needs. The State of California telecommuters' payroll checks are processed off-line, and, provided they arrive on payday, it does not matter if they are processed at 9 A.M. or 3 P.M.

▶ *Immediate Response:* Transactions are processed on-line, in a few seconds, but only within specified time periods, such as office hours. Merced County's on-line system operates only during the day; in the evening, the mainframe computer processes transactions and updates the databases.

▶ *On-demand Response:* In this degree of responsiveness, a transaction or service is handled immediately 24 hours a day. Merced County's police force needs to access state and national databases around the clock. If you are ever stopped for speeding, you will find that the officer does not get out of the car immediately; he or she first accesses databases via mobile communications in the police cruiser to find out if your car is reported as stolen; and the data must be available at all times. Customers who do not expect their bank branch office to be open at 8 P.M. on Sunday expect its ATM to be "up" at that time.

▶ *Perfect Service:* In this degree of responsiveness, a firm adds resources for backup, "hot restart" facilities, security, and reliability, so that if there is any problem such as a telecommunications switch being out of service or a line being cut, processing still continues. There is a fine line (and an expensive one) between guaranteed on-demand service and perfect service. An ATM network may be down half an hour a week, or routinely taken down for maintenance between 2 A.M. and 4 A.M. on Sunday; that can be an occasional inconvenience. But if the Federal Aviation Agency's Air Traffic Control network is down for only 15 minutes, it is a disaster.

The Implications of Reach-Range-Responsiveness Interactions

The reach, range, and responsiveness of a firm's platform determine the variety and quality of telecommunications-dependent services it can provide and hence, in an increasingly telecommunications-dependent world, its "business degrees of freedom"—the range of business strategies it can consider.

Developments in local area and wide area telecommunications transmission have made it increasingly easier to increase the reach of an organization's platform. Increasing range has been far slower and remains complex because of the lack of open standards, the complexity of existing computer operating systems, investment in existing software, and the as yet largely unproven tools for large-scale client/server applications in a multivendor computer environment.

Federal Express's network has about as much reach as any business in the world, with mobile communications reach to trucks, satellites to most countries around the world, and phones to its business and processing centers. THEi's network reaches to stores in all the countries in which it operates but not to its suppliers or customers. Shearson Lehman's is the classic corporate network that reaches across all domestic locations.

Very few companies have extensive range in their platforms. Consider two insurance companies, the first of which is merely average in its use of telecommunications (and will remain anonymous), and USAA, which is outstanding. Company X has separate processing bases for its three main consumer products: home, life, and auto insurance. As a result, customers must deal with three different units, and the firm does not know which of its customers who have car insurance also have home owner coverage. Even though its entire business strategy rests on cross-selling to meet a customer's total insurance needs, it cannot do this because it lacks range for marketing, service, billing, and pricing. USAA, in contrast, knows exactly which products a customer has, and with a single phone call a customer can contact an agent who can address any issue and handle any product. The agent accesses both databases and image processing files from a single workstation.

The level of responsiveness of the telecommunications services platform varies widely across firms, even those within the same industry. Most securities brokers are open for trading only when their offices are open, but Charles Schwab is open all night via touch-tone phone; buy and sell orders can be placed any time. Most banks require customers to get forms for a mortgage application from their office, fill them out, and wait up to three months while the application is processed; several banks offer approval in 15 minutes. This on-line service is location-independent, for you can phone in from anywhere, and it represents the new baseline for service in more and more industries. Dell, Lands End, and other telephone-based retailers get all their revenues without any physical stores. Most of them are "open" only during office hours, however.

Only when the public phone network is down or a fiber optic has been cut is the difference between guaranteed on-line service and perfect service apparent. When a fire destroyed an Illinois Bell switching center, travel agency business dropped by 95 percent, and floral orders completed by phone ceased. Several insurance companies had to send their staff home for weeks, but Metropolitan Life Insurance was back up in under an hour. Its telecommunications designer and implementer deserve congratulations, as does its senior management, for recognizing that the extra level of responsiveness provided through backup facilities and redundant circuits was not an "expense" but a key business need.

Step 3: Identify the Network Design Priorities and Trade-offs (Core Framework 3)

Step 3 in the telecommunications/business decision sequence begins either the technical design of the platform or the incorporation of the service into an existing platform. There are literally thousands of potential technical choices that vary in capability, features, and costs. Picking among them should occur only when client, designer, and implementer are sure about the relative priorities and trade-offs. Figure 3-8 shows the network design variables worksheet, which has four main categories, each of which breaks down into subcategories:

- ▶ *Capability:* This is the technical applications and services the facility provides. The subcategories are variety of applications, degree of integration of the services, and the range of volumes anticipated.

- ▶ *Flexibility.* This defines how easily the platform can accommodate growth and change; includes speed of change in adding capacity, ease of increasing reach, ease of increasing range, and ease of increasing responsiveness.

- ▶ *Certification:* This category defines how reliable the platform is and what guarantees of service it provides; includes operating reliability, ease of access, security, and response time.

- ▶ *Cost:* This category defines fixed and variable development and operating costs; includes fixed costs of equipment and software, development costs, operations costs, maintenance, support, and education.

The main trade-offs that all designers must address are between flexibility and cost, capability and flexibility, and certification and cost. The skilled designer will always look for standards that permit maximum flexibility without adding cost, that meet today's specific needs while ensuring that the platform can evolve to exploit tomorrow's business opportunities and technology, and that

Figure 3–8	Design Variables	Priority	Specific Needs
Network Design Variables Worksheet	Capability		
	Variety of applications	_____	_____
	Degree of integration	_____	_____
	Range of volume anticipated	_____	_____
	Flexibility		
	Speed of change in adding capacity	_____	_____
	Ease of increasing reach	_____	_____
	Ease of increasing range	_____	_____
	Ease of increasing responsiveness	_____	_____
	Certification		
	Operating reliability	_____	_____
	Ease of access	_____	_____
	Security	_____	_____
	Response time	_____	_____
	Costs		
	Fixed costs of equipment and software	_____	_____
	Development costs	_____	_____
	Operations costs	_____	_____
	Maintenance	_____	_____
	Support	_____	_____
	Education	_____	_____

ensure the lowest practical operating costs without putting the firm at risk. This is not an easy task, but it is one that can have a major impact on the business's success or failure.

One of the main points that underlies the network design variables framework is that there is no such thing as "the best design," any more than there is such a thing as "the best car." Just as auto enthusiasts rate cars in terms of speed, power, safety, reliability, and repair record (and as *Consumer Reports* rates them on value for the money), technical specialists judge network options in terms of speed, technical elegance, and sophistication. The network design variables framework enables designers and clients to focus more clearly on what "the best" and "value for the money" really mean.

| | | | | | | |

Step 4: Make the Economic Case: Quality Profit Engineering (Core Framework 4)

The final step in the decision sequence is to make the business case in terms that are meaningful to business managers. The quality profit engineering framework, which identifies exactly how the proposed investment contributes to improving the firm's profit and cost structures, includes the following elements:

- ▶ *Profit Management and Alerting Systems:* These systems capture information at what is here termed "point of event"—point of sale, reservation, ordering, delivery, and so on—so that planners and managers have timely data on what is happening in the marketplace.

- ▶ *Traditional Costs:* These costs are the standard components of a firm's activities, most obviously labor costs, materials, real estate, administration, and expenditures on information technology.

- ▶ *Quality Premium Costs:* These are the costs a firm must pay to ensure quality, which has become a basic requirement for all firms and not something they can charge extra for.

- ▶ *Service Premium Costs:* These costs are equivalent to the premium a firm must pay to ensure higher and higher levels of service.

- ▶ *Long-term Business Infrastructure Costs:* These costs include research and development (R&D), education, market and product development, and investments in the telecommunications platform. They cannot be justified on the basis of cost displacement. They are a planned profit loss; R&D, for instance, cuts this year's operating profits in order to preserve profits five to 15 years later.

- ▶ *Revenue Improvements:* This element is not the same as revenue growth, but refers instead to opportunities to add revenues without eroding profit margins. This is a major problem for firms in all industries, as deregulation, global competition, new technology, and overcapacity push margins per unit of sales down and down. Piggybacking new sources of revenue onto an existing platform can generate high margin products because there is no added cost of delivery.

Let's consider each element of quality profit engineering in turn.

Profit Management and Alerting Systems

A classic example of profit management and alerting systems is "yield management" systems used by airlines. *Yield management* is the term for the net profit on a given flight. The logic is to adjust prices so that when the plane takes off, it has no empty seats, and no full-fare passengers have been turned away because too many discount seats were sold earlier. The leading airlines monitor sales for a flight a year ahead, fine-tuning prices sometimes several times a day. The key to being able to do this is the capturing of reservations data, which the yield management systems track and analyze as needed. Yield management is increasingly being used by hotels, car rental companies, and other businesses that have a "perishable" product—once a flight takes off, empty seats cannot be filled.

The retailing equivalent of yield management is the use of point-of-sale data to alert managers to trends, problems, and discrepancies. The same principle applies: capture information at point of event—where the activities that determine profits occur—and move it quickly to the managers who need it.

Traditional Costs

The three main elements of traditional costs—those that have always been basic to companies' operations—are labor, inventory, and real estate. In the 1970s and 1980s there were many promises that computers would increase productivity and thereby improve labor costs as a fraction of revenues; that did not happen. In general, computers have had little impact on white-collar productivity, though they have reduced manufacturing labor costs.[15] Adding telecommunications to computers has had a significant impact on all three elements of traditional cost through such applications as the streamlining and reengineering of work, location-independence of back-office facilities, and electronic data interchange.

Quality Premium Costs

Quality is not something companies can charge an extra for, except in very special contexts. In retailing, consumer electronics, computers, furnishings, financial services, and many other industries, the customer has a wide range of alternatives and expects a good price and first-rate quality. In the 1970s, "discount" really meant "crummy"; now, the top discount stores offer excellent products. In manufacturing, total quality management has become one of the central priorities simply because it has to be if the firm is to survive.

Telecommunications contributes to improvements in quality mainly by helping streamline activities. It is a core element of CAD/CAM (computer-aided design/computer-aided manufacturing), customer-supplier linkages, and the creation of and access to databases that provide timely information needed to ensure quality.

Service Premium Costs

Telecommunications and *service* are almost interchangeable terms. Many of the applications listed in the telecommunications business opportunity checklist create new levels of customer service through ease of access and convenience. Examples include workstations for electronic ordering and cash management, telephone services for information and ordering, 15-minute mortgage approval, electronic data interchange to speed up ordering and delivery, and many others.

---

[15] Whether or not you personally believe this or even care about it, reducing costs is *the* concern of almost every senior business executive. Just about every company today is looking to cut costs, and most do not see computers as doing so. As one CEO told an author of this book, "IT people are in the business of asking for money. They do not feel they have to produce value for money. When productivity doesn't go up, they blame the business people, not themselves." The governor of a large state asserted to that author that "To me, computers are bureaucracy. How can I lead a fight against bureaucracy through computers!"

Service is as much a basic requirement now as quality, even for public sector organizations. There are only three practical ways to ensure service; charge extra for it and provide customers with attention that they do not get except by paying for it, pay additional people to provide service, and use technology to ensure service without adding cost and staff; the third is the only practical option today. As with quality, service is not an add-on to be charged for in most industries, and companies are struggling to cut staff and costs. Using information technology does not guarantee service, but there is no other major option.

Long-term Business Infrastructure Costs

Telecommunications is a growing fraction of large organizations' capital investment. Building a large-scale telecommunications platform does not pay off quickly, but then the costs of a new office building are not recovered immediately, and investment in R&D does not generate profits in its first year. Justifying the telecommunications infrastructure must be based on a convincing long-term analysis of the firm's opportunities and necessities.

The three business dimensions of reach, range, and responsiveness provide a framework here for assessing a business's degrees of freedom—that is, the range of practical strategies a company can choose and the ones it cannot select if it lacks capability in its platform. For example, if the industry is moving toward EDI as the basis for choosing suppliers, reach is essential. If a firm is committed to cross-linking products and services, it must have the range to do so.

Revenue Improvements

Several examples of piggybacking to improve revenues were given earlier. In piggybacking, the firm reuses information or adds a new service to its existing telecommunications delivery base. That base is largely a fixed cost, except when a public data network and public voice network is used. British Airways can add hotel reservations to its platform for very little cost; by contrast, hotel chains must invest heavily to build the required platform infrastructure. McKesson already has terminals in pharmacies across the country, so it can easily add insurance claims processing to them. More and more organizations are recognizing that telecommunications opens up new sources of revenue in this way.

Summary: Seeing the Steps as a Process

At the end of these four steps, client, designer, and implementer have worked together to ensure that business opportunity and technical choices have been meshed together. The next step is to begin the implementation sequence, which will involve detailed technical analysis, budgeting, project design, vendor proposals, analysis of outsourcing options, software development, and so on. It will be followed by the operations sequence, which includes installation and testing, network management, disaster recovery, and many other functions and procedures. Part IV of *Networks in Action* reviews implementation and operations.

| | | | | | |

An Illustration of the Decision Sequence: Choosing a PC for Business Usage

To help make the decision sequence both clear and concrete, this section presents as a case illustration the choice of a personal computer for use by Ellen Jackson, a salesperson in a large consumer goods company. Ellen's immediate needs are for word processing to help her catch up on paperwork and for a spreadsheet software package to handle sales forecasts, keep track of clients, and manage expense records. She has looked at a variety of personal computers costing between $800 and $4,000. Several of these had color screens, some were portable, and a few included special features such as sound, pens that can write directly onto the screen, and video facilities. All of them met Ellen's basic needs, but she quickly realized that without a telecommunications capability, none of them could access data or communicate with other personal computers.

Ellen proposed to the head of the sales department that she put together a small team to review options. She stressed the need to talk with telecommunications experts. The group assigned to make a recommendation is led by Ellen, who represents the sales department as the client. The designer, Rich Piccardo, is the corporate telecommunications planner, and the implementer is a group of technical staff from corporate telecommunications.

Step 1: Define the Business Opportunity

The team began its analysis of options by reviewing the potential payoffs from using personal computers. Ellen commented that before talking with Rich and his colleagues she had not thought about the role of telecommunications: "I was thinking of using a personal computer to save me time on administrivia. I haven't considered what new types of analysis I could do if I can get yesterday's point-of-sale data or track my customers' purchases. Is it worth it, though? It sounds great to be able to pull in information from everywhere, but what difference will that really make to how I operate? I'm worried that we could end up with a system that is so complex we can't use it and so expensive we can't afford it."

Another member of the sales team (a yuppie from marketing) pulled out some photocopies of articles from computer magazines. He pointed to some major successes firms have had by providing their sales force with laptop computers. Hewlett-Packard's sales teams report an average 40 percent increase in face-to-face time with customers. They are able to download information to their home instead of going into the office, and they can quickly get a complete picture of customer purchases, sales trends, and updates on products, promotions, and prices. "How many times do we feel embarrassed in front of a customer when we can't answer a simple question about an order, product availability, or price?" Frito-Lay's sales staff get a complete sales report downloaded to their personal computers every morning. "I don't know last week's sales till well into next month; Frito-Lay can spot a trend in a few days."

He argued that a major value of personal computers will come from being able to spend less time in the office. He mentioned that AT&T's New Jersey branch has taken away its sales reps' offices and required them to operate from their home; their office becomes just about anyplace they can put their name-plate on the door. "I don't need a PC in this building anywhere near as much as I do when I'm on the road. I can usually get information here by walking around; I can't get it in Toledo at 9 P.M."

The team agreed that it needed to step back and look at the competitive payoff that should guide their decision. Because the sales department's business goals are to increase market share and to add new accounts, the team agreed that their main priorities must be to:

1. *Improve relationships* with leading retail chains and stores. The company is regarded as behind its major competitors in how it works with buyers in such accounts. One sales manager commented that "we tend to be just order takers, instead of really working with customers to look at their sales planning and marketing. Many of them do not see us as a strategic sup-plier."

2. *Aggressively exploit promotional campaigns* developed by corporate mar-keting. "We need to look ahead more. If we know in July that marketing has a new television and a price rebate ad campaign scheduled for No-vember, we should factor this into all our sales efforts. Sometimes we work in a vacuum."

3. *Increase special bid options* for large customers. "This involves analyzing the overall customer history and relationship. How important is this cus-tomer today, and where can we capture future business? How profitable is the relationship? What will be the impact of cutting prices to increase volume?"

4. *Improve the quality of sales presentations* to new customers. "Buyers and marketing committees see 200 sales presentations a month. You have at best five minutes to get your message across, so you had better be well prepared and have top-rate visuals and supporting material."

5. *Increase the percent of the sales force's time devoted to selling* and reduce time on administration and reporting. "A day a month spent on paperwork is a day a month not spent on selling and meeting the customer."

Ellen asked, "If these are our priorities, where will personal computers and telecommunications make a real contribution?" The members of the team re-viewed the telecommunications business opportunity checklist and noted that all the priorities related either to running the business better or gaining an edge. With respect to running the business better, the checklist highlighted several opportunities they had not previously considered:

▶ *Manage Distributed Inventories:* If the personal computer is linked to up-to-date inventory and sales data, sales reps would be able to spot fast- and slow-moving items and target customers with special deals accordingly. They would also avoid out-of-stock situations more easily.

- ▶ *Link Field Staff to Head Office:* This obvious opportunity relates directly to the priority of increasing time spent on selling. Do the sales staff really need offices? How much extra time could they gain by accessing customer and sales information from home over the phone early in the morning?

- ▶ *Improve Internal Communications:* This opportunity relates to the priority to "aggressively exploit promotional campaigns developed by corporate marketing," but there are many broader opportunities, including electronic mail, access to databases, and so on.

- ▶ *Improve Decision Information:* If the sales reps have access to information, they can also provide information. Karin Jaworski, director of field sales, does not see herself as a client for the system. Ellen argues that perhaps she should; the more she can keep in touch with what is happening in her sales staff's accounts, the quicker she can spot trends and problems.

- ▶ *Reengineer and Simplify Organizational Complexity:* The attention of the sales team is focused on those factors that are most relevant to sales' current activities and priorities. They recognize that as a result they may have blocked out many innovative opportunities. For example, Ellen spoke of using a personal computer to help handle administrivia—forms, reports, expense accounts, and the like. Why not look for ways of removing them entirely?

The team spotted several possibilities to streamline work. One was to submit all orders from customers on the spot, either in the customer's office or in the sales rep's car or hotel, thereby eliminating paperwork and, more importantly, the delay between verbal confirmation of the order and its reception at the head office; another was to use electronic data interchange for all paper-based activities a sales rep must process, including submitting expense reports.

One member of the team asked, "What about hotel receipts and airline tickets?" A response was, "Why do you need the paper receipts? How about using a scanner that captures an image of the receipt in the personal computer, then send a preformatted electronic expense account message by electronic mail?"

"But the IRS insists on paper documentation. . . . "

"Are you sure?"

"No, but we've always required that receipts be attached to the expense account form."

"Ah—maybe that's your problem."

Next, the team turned to gaining an edge, the focus of most of the current priorities, including "improve relationships," "exploit promotional campaigns," "increase special bid options," and "improve the quality of sales presentations." Again, the team saw that there are many broader opportunities.

- ▶ *Provide Workstation Access in the Customer's Office:* Most key buyers probably already use PCs. It is technically very easy to link the customer to the supplier through a public electronic mail system. Customers could also enter their own orders, including late at night or on weekends.

▶   *Differentiate a Standard Product:* This is a priority for sales. Fast response and the ability to put together special bids and synchronize sales and promotional campaigns are examples. The business logic is that the sales process and responsiveness to the customer are the best forms of differentiation.

Finally, the team turned to finding sources of market innovation, but the terms of reference for the sales team's project were too narrow to make it likely that someone would come up with a breakthrough idea. Ellen's comment was, "Let's walk before we run. We lack so many basic tools that we don't need to look at transforming the business; we also don't need the political hassle. So why don't we choose something fairly simple that makes a real contribution to the business and learn from it? Maybe we'll spot some breakthrough opportunity as a result of using it."

The group reviewed the wide range of telecommunications options. In the end, the members concluded that they should not try to be too ambitious or take too many risks. They agreed that reengineering offered many exciting possibilities, but that the sales department was inexperienced in its use of personal computers and data and that clients wanted to get benefits quickly. Ellen concluded that "We will build a fairly simple system first. Within a year, we can be sure that we will want to extend it. Let's learn before we make a big leap." Privately she commented that she felt a little frustrated by the lack of imagination of the telecommunications design and implementation team: "They want to help, but they are not very interested in what we do. I'd like to know more about telecommunications so that I can take a more active lead."

The team's recommendation was to design and implement a system that exploited the opportunity to link field staff to the head office, improve internal communications, and differentiate the firm through improved service and responsiveness to customers, especially in answering questions and making presentations and proposals.

Step 2: Identify the Business Criteria of the Telecommunications Platform

The team then moved on to discuss the scope of the telecommunications platform needed to turn these opportunities into reality. Initially, before the review of telecommunications opportunities, the sales team's interest in personal computers was for stand-alone word processing and spreadsheet applications with no access to data stored on other computer systems. Each user would operate as he or she wished, with no effort to share messages and information or build a departmental information base. In that case, the platform would have little reach and range. Reach to other locations within the firm would be provided through simple dial-up facilities for electronic mail. Range would be limited to simple messages. Responsiveness would need to be immediate; the user of the personal computer needs to have the message processed at once and on-line.

The team agreed that far more business functionality was required. Ellen commented, "A portable fax machine can do most of this." The question was, Just how much more functionality? Rich, the designer, laid out marketing's plans for point-of-sale data capture and finance's development of new reporting systems. He also identified a range of outside information services provided by marketing research firms, advertising agencies, and industry associations. He suggested that sales create a shared departmental data resource on products, promotions, pricing, expenditures, sales, and profits. Rich commented that, "It's clear to me that you have to decide between meeting immediate needs quickly and cheaply—and getting immediate payoff—and planning ahead for a more integrated solution." Ellen's comment was, "If that's the solution, we'd better be sure we define the problem."

At this stage in the decision sequence, the implementer's main role was to highlight key technical issues that may affect the client's decision. Joanna Feldman, a member of the team from corporate telecommunications and the implementer, focused on the volumes of information involved in the various options. She pointed out that the technical problems involved in simply moving a bit stream of digital data through the network are matters of detail, but that combining information from different database management systems is a major challenge. Joanna was very concerned that sales understand the difference between "using a bunch of standard PCs and running a departmental local area network linked to the corporate WAN and databases. I can help you on the LAN level and show you how to tap into specific systems; I can't give you an off-the-shelf, plug-in-and-go capability for accessing everything anytime."

Accessing each of these systems and services independently is an entirely different technical issue from integrating the information from them; the cost of software and network facilities may be very high, and each system codes and structures data differently. Rich emphasized that sales would have far more opportunities to access "heterogeneous" databases if it adopted the **SQL** (structured query language) standard—an English-like grammar and syntax for accessing information without knowing how it is organized or where it is stored and the most widely accepted base for PC systems to interface to **DBMS** (database management system).

At this stage, the designer has three main issues to address:

1. *The extent to which this application can be treated as independent* of any others, versus as part of an integrated facility

2. *The linkages across the telecommunications and computing components,* taking into account opportunities to exploit shared resources

3. *The technical standards* needed to accomplish the first two goals and to maximize later options for linking, integrating, and upgrading applications and facilities

The sales team's decision to create a departmental system that is focused on internal communications provided two extremes of platform choice. The first was to select a local area network that met the specific needs of sales and to ignore issues of linking to the point-of-sale system or to marketing and finance systems. Reach is here limited to the first level, to within a single location. Needs to obtain information from any of the firm's major information systems could be met later by adding special-purpose gateways and software or could simply be handled off-line via a tape dump of the data provided weekly. The other extreme was to view the sales system as one element in a company-wide integrated information management capability. In that case the reach would extend across the firm's locations and systems.

Rich summarized the crucial network design issues in terms of the telecommunications business platform as follows:

▸   *Reach:* "This is to be a departmental system only. That means very limited reach and no need to be concerned about extending the network out to other departments and locations. However, we all agree that access to data from a number of internal sources is a key long-term requirement. The main problem is making sure that we accommodate future extensions of reach to other parts of the company. The standards we base the system on must address this."

▸   *Range:* "The system will not need to access remote transaction processing services; it only needs to access data. That greatly simplifies the design issues. The range we need is only for messages and access to databases."

▸   *Responsiveness:* "We do not need 24-hour service and can afford occasional short out-of-service periods. Response time must be very fast; analysts don't want to spend minutes waiting for data."

Figure 3-9 maps the system onto the platform.

Rich's initial design, which he will review in much more detail with the implementer, assumes that the system will be based on a departmental mid-sized computer with a database management system that uses the SQL interface standard. This will store the extracted information from marketing's systems and, later, from the point-of-sale system. That data will be downloaded overnight, using the telecommunications standards marketing has adopted in its wide area network. Sales reps will access the departmental data through a local area network. The decision about which electronic mail system to adopt will be made after consultation with marketing and finance, because it does not make sense to have different ones.

Rich commented that this may not be the problem it was in the 1980s, when just about every electronic mail system was incompatible. The X.400 standard is being implemented by all major providers, which allows them to exchange messages with any other X.400-based service. It should be possible to link to most of the many e-mail systems in place across the country. He strongly recommends that the sales system architecture be based on SQL and X.400: "I am not trying to say you should choose any specific package—only that if we agree on these standards, you will still have plenty of choice of packages and prices, but we gain the opportunity to integrate your systems with others that have adopted the same architecture."

The issue that the team discussed in most detail was whether the reach of the platform must extend across the firm. It concluded that there was no immediate requirement. The designer pointed out that if the sales department were willing to accept overnight, off-line transfer of data from the mainframe DBMS,

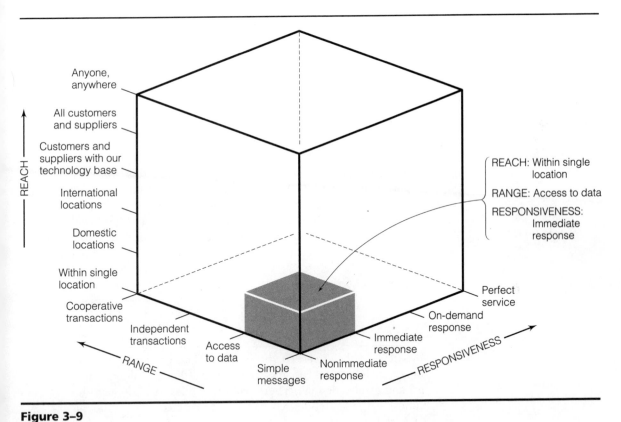

REACH

Anyone,
anywhere

All customers
and suppliers

Customers and
suppliers with our
technology base

International
locations

Domestic
locations

Within single
location

REACH: Within single
location

RANGE: Access to data

RESPONSIVENESS:
Immediate
response

Cooperative
transactions

Independent
transactions

Access
to data

Simple
messages

Nonimmediate
response

Immediate
response

On-demand
response

Perfect
service

RANGE

RESPONSIVENESS

Figure 3–9

**The
Telecommunications
Services Platform
Map for Ellen
Jackson's Sales
Department**

the cost will be far lower than direct access at any hour of the day: "You have immediate response to queries, but the data will be up to 24 hours old. To give you access to up-to-the-minute data, we need an extra level of range that adds significant complexity, given the variety of data processing systems we must link to." In addition, although high-speed transfer of information across the local area network is easy to ensure, on-line transfer over the wide area backbone network is far slower because of the overhead involved in error-checking, protocol conversion, and routing. This overhead averages 100 milliseconds a transaction using an X.25 public network. "Compare that with basic Ethernet at around 2 mbps, other local area networks at 10 mbps or more, with 100 mbps an option through fiber. LANs are built for very fast data transfer, and wide area networks for efficient management of the network resource and flexibility of applications."

Several members of the team were not persuaded. As one pointed out, "We identified as a priority opportunity differentiating our products through outstanding service and responsiveness. We lose much of that if we can't directly tap into any relevant database anywhere in the company." The consensus, however, was to adopt the simpler platform option.

Step 3: Identify the Network Design Priorities for Making Trade-offs

The implementer has a wide range of issues to address in moving from the broad initial design implied by the dimensions of the platform to a detailed technical blueprint. This involves many trade-offs. A firm's policy decisions also guide the implementers' actions. For instance, because this firm views security as a priority, it has adopted standards for encrypting transmission signals and procedures for LAN management, auditing, and accounting. These decisions are part of corporate IT policy, and the sales department must follow them. This may mean that several software and local area network options that are attractive in terms of cost or features must be ruled out. If security standards were not part of the architecture, then the team would have wider options here.

The implementer's summary of the technical requirements for each category of network design variable is presented in the following lists:

Capability

▸   *Variety of Applications:* The basic applications will be fairly standard office technology, including word processing, spreadsheets, calendar management systems, and electronic mail. There will be a widespread demand for graphics capabilities for presentations, as well as for database management and data access tools. These will be added as individuals choose.

Neither the implementer nor designer see any need to accommodate complex number-crunching systems for statistical analysis, although these are being developed in marketing and finance for market research and forecasting tools. The sales needs are mainly for simple analysis and display of data.

▸   *Degree of Integration:* The software tools need to be integrated. The value of the individual applications is increased if data from, say, a SQL inquiry into a database can link into a spreadsheet package for analysis and to desktop publishing for producing high-quality graphs and reports.

The key element in determining the practical degree of integration is the operating system of the personal computer chosen for the sales system, rather than the local area network. Rich felt that Microsoft Windows should be adopted, together with its Microsoft Office software: "This gives you a word processing capability that links directly to spreadsheets. Other packages involve file conversion, and there's always some minor problems of compatibility."

▸   *Range of Volume Anticipated:* Joanna, the implementer, assumed that there will be rapid growth in the number of users, the frequency with which they use the system, and the volumes of data they access. The network must be able to accommodate that growth. At the same time, she does not want to install expensive additional capacity. This is a classic telecommunications problem—meeting today's needs at low cost (with the risk of degraded services as volumes grow) versus carrying expensive unused capacity.

Flexibility

▶   *Speed of Change in Adding Capacity:* Joanna decided to address the issue of handling traffic growth by choosing a local area network capability that allows easy addition of both users and LANs: "I like the Ethernet standard for its simplicity and the fact that you can quickly add subnets with bridges between them. I want to aim for a building block approach in which we add resources in well-defined and affordable units."

▶   *Ease of Increasing Reach:* Joanna was mainly concerned here with extending the local area network to any number of wide area networks. This is not at all an easy problem to resolve. The initial design provides a link from the departmental mid-sized computer into the firm's two main backbone networks to download data every evening.

A key feature of the design is that the users will access data from the departmental computer through the LAN. The choice of computer operating systems for that computer and for the personal computers, the choice of vendors, and the choice of telecommunications standards optimizes this design.

▶   *Ease of Increasing Range:* The design does not provide extensive range. All accesses to separate databases create equally separate data resources, on the personal computers or on the departmental computer. That seems acceptable for meeting both longer-term and immediate needs: "We can provide special software for extracting and combining data from different mainframe resources, and we do not see any need for sales staff to be linking into remote transaction processing or point-of-sale systems. The goal is to get data down to the departmental machine and then make it easy to access and massage locally."

▶   *Ease of Increasing Responsiveness:* This is a potential area of weakness in the proposed design, in that local area networks vary widely in their ability to handle different types of traffic: "Ethernet is not the best option for handling large volumes of data, nor is X.25, which we are likely to use later for the wide area network element. I can envisage some major traffic jams on the LAN a few years from now. That's something to watch for, but by adopting proven standards that are in the mainstream of business usage of PCs and LANs, we should be able to migrate the initial system as new developments emerge."

Certification

▶   *Operating Reliability:* Reliability is crucial here, and Joanna refused to consider any complex combination of advanced telecommunications, operating systems, and database management systems: "We may be losing some degree of flexibility, especially in terms of adding range, and I can see that if we choose an IBM token-ring network instead of Ethernet, we would be better able to plan for integrating data resources and interconnecting the LAN to the WAN more easily. But a token-ring environment with heavy data traffic and power PCs—"Rambomobiles"—adds up to something more complex than the data processing department's mainframe data center just five years ago. I want to go for the safest, most proven LAN base."

▸   *Ease of Access:* This is also crucial here. Users must be able to get onto the system with the minimum set-up time and cost and without having to learn complex procedures. For this reason, all the team agree that the system must have a graphical user interface.

▸   *Security:* This is not a design variable to be traded off against other variables such as cost. Unlike in many companies, this firm has set corporate policies for security, including encryption, data administration, and network administration.

▸   *Response Time:* Joanna has defined some target measures of performance following discussions with the clients about the types of work they do, the variety and volumes of information they initially expect to access, and the type of analysis they anticipate.

Costs

▸   *Fixed Costs of Equipment and Software:* These costs include the personal computers, software, local area network, bridges, printers, modems, and other items.

▸   *Development Costs:* The main costs here are for defining and building the departmental databases.

▸   *Operations Costs:* These costs include staff for data administration and network management. Joanna anticipated that at least four full-time staff will be needed in the sales department: "That's a major hidden cost for decentralized computing and communications. Expertise in these areas is not cheap."

▸   *Maintenance:* The system will need to be upgraded, modified, and enhanced on an ongoing basis. Again, additional staff will be needed.

▸   *Support:* This is mainly in the form of internal consulting expertise concerning the use of the database tools, the desktop publishing software, and electronic mail: "I can look after the network side, but I do not know much about database or sales."

▸   *Education:* Many organizations neglect this and provide only limited training in how to operate the personal computer. They do not commit resources to ensuring that users are comfortable with using the tools in their own work and meshing it with their personal modes of analysis and operation. The sales team views it as critical, especially because they remember several major computer systems that were introduced with very limited training and that subsequently quickly fell into disuse. One recalled a new phone system in the head office: "You can do anything with it—call-forwarding, remote call-forwarding, fast-dial, you name it. It is so complicated to use that only a few people have bothered to go through those unreadable manuals"

Joanna sees several problems, mainly concerning support and education. She suggested that Rich review another design option that is entirely different from this one in order to be sure that the various trade-offs go in the right direction: "We could go for a token-ring link to an IBM AS/400 mid-sized machine within an SNA environment. That is far more expensive than the Ethernet choice, with far larger costs for development and operation. On the other hand, it provides much more powerful and flexible linkage to the corporate databases in marketing and finance. It provides access to a massive central database management system, and above all it reduces the number of local support staff needed for telecommunications trouble-shooting and operations, database, software, maintenance, and so on. We also get far more opportunity to add reach and range and reduce organizational risk. But it will cost more to develop and run. It's your choice."

Choice is the issue. To have good choices, you must know the basis for making trade-offs.

Step 4: Make the Economic Case

By far the most difficult aspect of telecommunications is business justification for "value-added" applications that provide "soft" benefits. This is certainly true in this case. Telecommunications is unlikely to reduce direct costs, and it aims at improving effectiveness rather than measurable efficiency. The sales department's priorities include "improve relationships with leading retail chains," "improve the quality of sales presentations to new customers," and "increase the percent of the sales force's time devoted to selling"; what are these worth in hard economic terms?

The sales department's system needs to be justified in detail, once a clear design has been chosen and a work program defined. The starting point is to identify just where it can contribute to these components of profits; analysis of costs and payoffs comes later. The team's first assessments of the profit engineering opportunity are as follows:

▸ *Profit Management and Alerting Systems:* The system is intended to provide earlier and more relevant information to sales, which should translate into direct improvements in response to trends and to identification and resolution of problems. The project team needs to identify operationally how and where these will occur and what they mean for managing profit.

▸ *Traditional Costs:* There are unlikely to be any direct cost savings in labor, administration, or other traditional costs, unless the system cuts out staff who mainly collect and collate data.

▸ *Quality Premium Costs:* No likely contribution.

▸ *Service Premium Costs:* This is a major aim for the system and is reflected in several of the stated priorities. The goal is to make service and relationship management a differentiator and to move sales reps away from being just "order takers."

▸ *Long-term Business Infrastructure Costs:* The picture here is unclear. As Ellen commented, "I have to believe that we will benefit immensely in the long-run from making PCs a core part of how we work. I can't, though, quantify the size of the benefits. To some extent, this is an act of faith."

▶ *Revenue Improvements:* This is another priority goal for sales, especially in terms of obtaining more revenue from existing major accounts.

The value of the preceding process is that planners gain a clearer picture of their business, technical, and economic requirements than they would obtain simply by trying to answer the question "What PC do we need for sales?" The discussion presented here contains only the preliminary assessments of benefits—finalizing the details concerning costs, technical specifications, and schedules requires further planning and in-depth analysis. By moving through the decision sequence, the team has established priorities for the next steps, which move from making decisions to implementing them.

Summary

The telecommunications business decision sequence builds the dialogue among client, designer, and implementer. It moves from the business opportunity to the criteria for technical design and to the basic for economic justification of the platform. The decision sequence has four steps, each of which applies one of the four core frameworks developed in this book.

The first step, choosing the right opportunity, uses Core Framework 1, the telecommunications business opportunity checklist. This classifies uses of telecommunications under three main headings: run the business better, gain an edge in existing markets, and create market and organizational innovations. Run the business better is divided into manage distributed inventories, link field staff to head office, improve internal communications, improve decision information, and reengineer processes and simplify organizational complexity. Gain an edge in existing markets is divided into provide workstation access in the customer's home, office, or on the street, and differentiate a standard product. Create market and organizational innovation is composed of preemptive strikes and piggybacking.

Step 2 in the decision sequence identifies the business functionality needed in the telecommunications platform in order to turn the opportunities into reality. The three dimensions of the platform are its reach, range, and responsiveness. Reach refers to the individuals and organizations that can link directly to the firm, range indicates the extent to which information is directly and automatically shared across business processes and services, and responsiveness defines the level of service provided, ranging from nonimmediate response through guaranteed 24-hour immediate service on demand.

Step 3 identifies the priorities for and trade-offs among network design variables (Core Framework 3). The variables are capability, flexibility, certification, and cost. Step 4 uses Core Framework 4, the quality profit engineering worksheet, to assess where and how the proposed telecommunications investment contributes to the organization's economic health in terms of six factors: its profit management and alerting systems, traditional costs, quality premium costs, service premium costs, long-term business infrastructure costs, and revenue improvements.

The chapter ends with a case illustration that applies the frameworks to a typical business situation.

Review Questions

1. What are the key features of the approach (described in this chapter) that serves as the bridge between business and technology?

2. List the elements of the telecommunications business opportunity checklist.

3. Give a simple example of how telecommunications can be used to better manage distributed inventories.

4. Briefly explain how electronic data interchange (EDI) can be used to streamline ordering and delivery processes.

5. What are some of the difficulties involved in linking portable computers (laptops) to corporate networks to obtain needed information?

6. List and briefly describe some of the telecommunications tools that are currently used to enhance teamwork, personalize management, and coordinate activities spread across many locations.

7. Briefly describe the business reengineering concept and how it differs from the traditional use of information technology to improve efficiency and make people more productive.

8. Identify and briefly describe two ways in which telecommunications has been used to gain a competitive edge in existing markets.

9. What is meant by using telecommunications to launch a "preemptive strike" to create market and organizational innovation? Explain and give an example.

10. How can "piggybacking" be used to create market and organizational innovation? Explain and give an example.

11. Describe the method that is used to translate the identified business targets of opportunity into the technology needed to support them.

12. Name and briefly define the three dimensions of business functionality associated with the telecommunications services platform.

13. Which dimension of the telecommunications services platform enables cross-selling across a company's product lines to meet a customer's total needs? Explain.

14. Which dimension of the telecommunications services platform is crucial to providing efficient service and rapid turnaround of orders and billing? Explain.

15. List the four main categories of network design variables and give a brief definition of each.

16. List the ways in which an investment in telecommunications can contribute to improving a firm's profit and cost structure, and state what this framework is called.

17. Briefly describe how telecommunications is used to manage and improve profitability by capturing and distributing appropriate information related to a sale or customer transaction.

18. Give an example of how telecommunications has significantly affected one of the traditional operational cost categories, labor, inventory, and real estate.

19. Why is the use of information technology the only practical option for providing the levels of service quality needed to compete in today's global business environment?

20. Give an example of how adding services or applications to the telecommunications network platform increases a company's net revenues (gross revenue minus direct cost of generating the revenue).

21. Briefly explain how salespeople in a consumer products firm can improve their performance through using a telecommunications network to obtain yesterday's point-of-sale data for their customers?

22. What are the implications for the telecommunications services platform of networking PCs to share messages and information and to access external information sources to build a departmental information base?

23. What are the main issues that a designer faces in linking a departmental LAN to multiple databases of different types through a corporate WAN?

24. Using information from the chapter's case illustration, briefly describe how the proposed system will improve profit management and reduce cost through long-term infrastructure improvements.

Assignments and Questions for Discussion

1. Traditional management information systems have been built on a firm's accounting system, with financial reports produced monthly. Explain why this is not adequate for today's management decisions, and describe how telecommunications can be used to improve management decision making and a company's competitiveness.

2. "In our company, we deal with each business application requirement individually and on its own merits, and we create a network solution to support each business need. If this means developing a number of separate, independent networks, that's OK." Explain why you agree or disagree with this statement.

3. There generally are a large number of potential technical choices that meet a given set of business requirements. Describe the framework used for guiding the choice of a particular technical solu-tion, and discuss the impact on this framework of overall company policies regarding risk taking, security, preferred vendors, and so on.

4. Using information from the chapter's case illustration, outline a presentation to the company's senior management that is designed to obtain approval and financial backing for the project; use the quality profit engineering framework.

5. Again, using information from the chapter's case illustration, specifically show the linkage between the goals and priorities of the sales team and the elements of the telecommunications business opportunity checklist. (Hint: Create a matrix with goals and priorities as the rows, and the elements of the opportunity checklist as the columns; alternatively, setup two parallel columns, one for goals and priorities and one for checklist items, and use arrows to connect the individual goals and priorities with appropriate checklist items.)

Minicase 3-1

Conrail, Part A

Turning Losses into Profits

Conrail, a new corporation created by an act of Congress in 1976, was formed from six bankrupt Northeastern railroads. In its first five years—in spite of getting nearly $3.5 billion in federal grants—Conrail lost over $1.5 billion. Its main problems were a morass of labor agreements and government regulations that kept costs and prices high and flexibility low. The situation was the reverse for Conrail's competitors in the trucking industry.

In the 1980s, Conrail turned itself around. Congress ended restrictions on the railroad industry's right to set competitive prices. For over 70 years railroads were not allowed to negotiate with shippers and had to quote a fixed set of prices based on shipment size, type, priority, and length. Now they could market aggressively, and shippers were no longer locked into rigid timetables. Conrail was also allowed by another act of Congress to divest uneconomic routes. It cut its track mileage by almost 30 percent in five years and its work force by almost 60 percent. In 1980, labor costs amounted to 60 percent of revenues; in 1986 they were under 45 percent.

Historically, the railroad industry was dominated by clerks. Tonnage clerks calculated weights of shipments and cars; car-length clerks calculated the length of trains; blocking clerks assembled rail cars bound to a specific single destination; car-checkers walked the length of the train writing

down car numbers. If any of the checkers overlooked a car, it remained where it was, sometimes for weeks. The documents associated with each train, car, and shipment were voluminous. The overall Conrail system was inflexible, expensive, overstaffed, inefficient—and uncompetitive.

Telecommunications was a central element in Conrail's turnaround, particularly fiber optics, optical scanners, and satellites. The scanners are versions of the barcode readers at a supermarket checkout counter that read, instead of the product code for a can of beans, the identifier of the rail car that may be transporting that can of beans.

Managing a railroad is managing movement. By being able to track cars with telecommunications and move information to Conrail's central headquarters, almost all of the clerks' functions were eliminated, with far fewer errors and delays. By knowing exactly where cars are and how close to being on time a train is, Conrail can use central scheduling and planning systems to maximize its use of track. In an average week, around 3,000 Conrail freight trains pass along more than 13,000 miles of track.

Instead of clerks collecting information, the information in effect collects itself. Sensors in freight yards, terminals, and signal points and along the track record data and move it at 9.6 kbps along almost 500 leased lines to regional data centers in 13 locations. The data is retransmitted over 42 high-speed, 56-kbps leased lines to central processing facilities at two locations in Pennsylvania.

In mid-1988, the head of information systems at Conrail commented, "We're totally automated; we don't have paper anymore. If the CRT's [workstations] were to go dark, we'd have to go back to route maps, colored pins, and telephones." To protect against this, Conrail uses redundant terrestrial circuits and satellite links, and the central data center has three large mainframe computers. One is active, running software applications; another backs up data from the first machine twice a minute; and the third machine is on "hot standby."

Here are just a few of Conrail's main telecommunications-based systems and services:

1. TeleTrak, a telephone tracking system that monitors 65,000 railcars on-line and answers around 1,200 customer queries a day.

2. Electronic funds transfer and electronic data interchange to eliminate paper and delays in billing and payments. EDI is one of the key emerging uses of telecommunications in intercompany transactions; it removes paper and reduces delays, errors, and time in ordering goods, invoicing, payments, and scheduling delivery.

3. Video terminals for negotiating contracts and for coordinating arrangements with other railroads, so that customers have a single point of contact.

4. CONSISS, a comprehensive database for sales information support that allows sales staff to know exactly where cars and trains will be and when, and that also includes detailed customer information.

5. The Transportation Management System (TMS). This contains all train schedules and adjustments, pickups, setoffs, and transfers. TMS links to the Locomotive Distribution System (LDS), which tracks and coordinates Conrail's over 2,000 locomotives, to both the Transportation Reporting Inventory Management System (TRIMS), which generates forecasts of traffic and timetables 8, 16, and 24 hours ahead, and to the Computer-Assisted Train Dispatching System (CATD).

In 1987, Conrail was sold for $1.65 billion in stock shares. Bailing it out had required billions of dollars in government cash, but by 1990 Conrail was making annual profits of close to $300 million after taxes. Revenues were over $3 billion in 1991, when Conrail boosted its dividend from 40 to 45 cents a share and arranged to buy back up to $100 million in stock. Its budget for telecommunications and information systems is over $100 million a year.

Questions for Discussion

1. Could the improvements Conrail made in its costs and operations have been practical without telecommunications? What other approaches could it have taken?

2. Many major banks and airlines are facing the same cost problems as Conrail, in terms of administration, complexity of operations, overstaffing, and so on. They, too, are exploiting telecommunications whenever they can. What lessons do you see for a company in either of these two industries?

3. In the 1980s many railroads created subsidiaries in telecommunications that specialized in fiber optics. Why do you think that occurred, and why have most of them sold off the operations?

4. In 1991, Conrail decided to centralize its 14 customer-service centers in a new 90,000-square-foot national center near Pittsburgh. This center incorporates "the latest technology in telecommunications and fiber optics" in processing customer orders. This technology mainly provides massive increases in the volumes of telecommunications traffic that can be handled cost-efficiently. What new opportunities does this open up for Conrail?

5. Using the quality profit engineering framework, identify the contribution of Conrail's network to improving its cost and profit structures. How significant was this contribution?

Minicase 3-2

Conrail, Part B

Cutting Costs Using Telecommunications

The recession that began in 1990 affected the railroad industry, just as it affected the industry's customers. In the first half of 1991, traffic dropped 7 percent compared with 1990, and the profits of the top six railroads fell by 87 percent. Conrail held up well, with just a 1 percent drop. Comparable figures for other sectors of the transportation industry were an 8 percent drop for trucking and a 19 percent drop for air freight couriers. Equipment transportation profits for the top five firms fell from $91 million to a net loss.

Throughout the 1980s Conrail had focused on aggressive cost-cutting and ascribed its adequate 1991 performance to tight cost control and vigorous cost reductions. Conrail's 1991 second-quarter earnings were up by 5 percent the first quarter, and third-quarter revenues and earnings were roughly the same as for 1990, even though the railroad industry as a whole saw yet another drop of close to 90 percent in profits and 8 percent in revenues.

In May 1991, Conrail hired a new head of information services, who commented that "the railroad industry has been through tough times, and Conrail can't get any leaner than it is now." Instead of cost-cutting, Conrail's priority would now be revenue growth through improved customer service. He quoted as an example of the railroad's plans a meeting he had with a car manufacturer that uses Conrail to ship cars to its dealers. The two companies plan to link their systems so that dealer showrooms can find out at any time where their vehicles (and revenues) are. Conrail plans to (1) add to its railcars radio frequency tags (to complement barcoded ones), (2) create new decision support systems for scheduling based on advanced airline industry techniques, and (3) build new customer information systems.

Other railroads, of course, have not stood still. CSX has followed the same general path as Conrail, but far more belatedly. One stock analyst, who sees CSX as well positioned for the mid-1990s, commented that "it was sheer chaos in the '80s under the new CSX name [which replaced Chesapeake and Seaboard] with an organization that confused people. Nobody knew where their cars were, nobody knew who they worked for, and everyone was demoralized." CSX had the worst safety record in the industry. It cut staff from 76,000 to 36,000 in 1991, with an ultimate target of 28,000 to 30,000. It sold off many of its nonrailroad businesses, including resorts, energy, and fiber-optics telecommunications operations. It improved its safety record to second place in the industry. Revenues in 1991 were around $8 billion, versus under $5 billion for its main geographic rival, Norfolk Southern, which restructured five years before CSX.

The railroad industry had to focus on cost cutting in the 1980s just to survive. Its new watchword is customer service, with particular attention to customer retention and additional revenues from established customers. Burlington Northern (BN) won an industry Golden Freight Car Award in late 1991 for its LYNX telecommunications software system, which enables customers to trace shipments on almost every major carrier using a personal computer. Of the five award winners, BN was the only one recognized for information technology; other winners were recognized for a new device that allows grain to be sifted and transferred from rail cars to trucks and for the rebuilding of a 16-mile track to service intracompany transfers of steel. The judges for the award commented that the major railroads did not appear to be as well focused on marketing as in previous years.

BN's LYNX accesses the central computer systems of 14 railroads, including Conrail. Several of these also use LYNX in their own tracing programs. LYNX was developed as an outcome of efforts in the late 1980s to create a generic electronic data interchange software package for shippers. In 1987, shippers working through the National Industrial Transportation League had made a formal request that the carriers create such software and a task force had been set up. Its progress was halting, largely because individual railroads had widely varying communications standards and protocols and because industry standards for EDI were not yet complete or stable. The task force did not feel that generic software could be written to handle these problems and expressed concerns about carriers having to change their own computer software systems.

BN decided to move ahead on its own and built a PC-based system that can trace either cars or fleets and produces customized reports. BN introduced LYNX in late 1989 and added 1,200 customers in just over a year. A new module, introduced in 1991, includes a Time Dependent Transmission feature, which allows customers to send queries at any time of day or night from an unattended PC, with inquiries scheduled up to six days in advance.

A BN manager commented that "LYNX is the first step toward the electronic equivalent of seamless transportation. If the North American rail network is to offer a truly competitive alternative to the highway, it must not only remove those self-imposed artificial barriers to interline shipping, but it must also produce the kinds of benefits that make it easier for the customer to choose rail in the first place. BN LYNX has provided Burlington Northern with an important advantage when attracting new business, as well as maintaining a competitive edge and growing existing business. It has improved customer service through information technology and has enhanced the image of the rail industry through the coordination and cooperation of many railroads."

Questions for Discussion

1. What opportunities do you see for Conrail to exploit its existing telecommunications capability to improve customer service? What will be the likely benefits?

2. What new telecommunications capabilities could and should Conrail add in order to enhance customer service?

3. BN's LYNX system connects to 14 carriers' largely incompatible networks and databases. How do you think the development team achieved this?

4. The railroad industry has for many years been a leader in electronic data interchange. Assuming EDI standards are stable and completely defined, how can individual carriers implement them without having to redesign and rebuild their systems?

Minicase 3-3

Rosenbluth Travel

The "Global Virtual" Company

Rosenbluth Travel, headquartered in Philadelphia, provides services to travel agents around the world through the Rosenbluth International Alliance (RIA). Rosenbluth's growth between 1980 and 1990 is shown below:

	Gross sales (millions)	Number of staff	Number of offices	Sales per employee (thousands)	Sales per office (millions)
1980	$ 40	280	8	$143	$ 5.0
1985	300	900	26	333	11.5
1990	1,300	2,500	480	520	2.7

In just ten years, Rosenbluth achieved a 3,250 percent increase in gross sales. The dramatic growth in sales per employee and just as dramatic drop in sales per office are entirely attributable to telecommunications. Rosenbluth has created a "virtual" global organization by electronically tying to it travel agents across the world. These agents are in effect part of Rosenbluth; they share a technology base that none of them—including Rosenbluth—could have afforded to build alone. This base allows worldwide access to travel services, reservation systems, accounting and administrative tools, information, travel management reports, and payment mechanisms, through RIA's Global Distribution Network (GDN). The consortium gets both advantages of scale in purchasing and discounts from major airlines and hotels that, again, would

not be available to each of them separately. Because the members of RIA share customer records, communications, and software, the alliance has been able to invent new joint services, such as managing travel arrangements for international trips and coordinating groups of people coming together from many parts of the globe.

In mid-1991, RIA operated in 37 countries, with 34 members/partners who have over 1,100 offices. Their total sales amounted to $5 billion. Founded in 1988, RIA was Rosenbluth's response to the major structural changes in the travel industry created by airline deregulation, the growth of the corporate travel market and its sophistication in travel management, the economies of scale and technology that threatened smaller travel agencies and fueled consolidation, the globalization of the entire travel industry, and increasing dependence on information technology, especially reservation systems. Rosenbluth saw three major options for itself in the mid-1980s: acquire other firms or start up operations in selected key markets worldwide, join a "multiple" travel agency chain, or create a global strategic business alliance. It rejected the first two options as being a weak response to the opportunities and challenges and as unlikely to create real growth and profits.

Rosenbluth's criteria for seeking members of RIA include a strong service record and financial base and a strong commitment to technology. In effect, RIA is its technology. That is why Rosenbluth describes itself as a "global virtual organization." RIA's strategy and plans are established by the entire membership, with one vote per member regardless of size. There are RIA standing committees for the hotel business, industry relations (a key issue in the often wary interaction between airlines and travel agents), global back-office technology (management information systems, reports), front-office technology (access to reservation systems), and global contracts and payments.

Rosenbluth defines one of its main products as "leadership." One article describing RIA is entitled "Ahead of the Pack Through Vision and Hustle."

Hal Rosenbluth, the president of the 100-year-old firm, has as distinctive a focus on service as on technology. In mid-1992 he published a book, *The Customer Comes Second: And Other Secrets of Exceptional Service.* He contends that when a company puts its people first and makes an obligation to ensure that their work is fulfilling and enjoyable, "wonderful things result." Negativism, errors, frustration, and turnover drop, loyalty and productivity increase, which in turn improves service: "Ultimately, satisfied workers produce satisfied clients, who learn that by being second they come out ahead." One review of Rosenbluth's book summarized his business methods as "superb people skills" and "jaw-dropping technical systems."

Technical Background

The RIA Global Distribution Network and computer software systems are very sophisticated and will not be described here. This network is a worldwide 24-hour-a-day platform for business; when it is down, the business is down. It must be able to link to many travel providers' networks, including airline and hotel reservation systems, and it must be able to handle a thousand transactions a second and make sure that members do not get busy signals or slow response times. The network building blocks Rosenbluth used are standard and widely available; putting the standard pieces together is the designer's and implementer's challenge.

Questions for Discussion

1. Where is RIA "located"? What do you think Rosenbluth's managers mean by calling it a "virtual" corporation? (Information technology specialists often contrast virtual with physical in talking about computers; a personal computer's physical memory, for instance, may be 640 kilobytes in size, but tricks of software may extend the virtual memory to 2 megabytes or more.)

2. What are the incentives for small or medium-sized travel agents to join RIA? How many of these incentives depend on RIA's telecommunications capabilities?

3. Use the telecommunications/business opportunity checklist to identify any business opportunities Rosenbluth has that a firm without a comparable telecommunications platform lacks.

4. What difference do you think Hal Rosenbluth's focus on putting customers second and employees first has made to Rosenbluth Travel's growth? Say he had been a traditional manager and the firm a traditional travel agency, marked by little training, high turnover, and fairly low levels of skills; would the virtual corporation be just as much a success? Is the technology the key here? The people? A combination?

Minicase 3-4

Whirlpool Corporation

Fitting the Management Pieces Together

When Whirlpool opened its new customer service center in mid-1992, the trade press lauded it as being at "the cutting edge." It uses advanced telecommunications and information systems tools, including ISDN, LAN-based image processing and expert systems, and high-performance workstations. The press did not discuss how close the project came to being a disaster and how much its leader learned about the skills, coordination, and communication essential to a venture that combines business, telecommunications, and information systems.

Walt Coleman, the vice president of operations, who in 1989 was brought in to head the project, spoke candidly to *CIO Magazine* in mid-1992 about what he as a nontechnical executive had learned about handling a technical operation. The goal was to respond to customer surveys that indicated that a key need for Whirlpool was to be more accessible and for its agents to be more responsive. The first step would be to implement a one-call service system—a hotline to diagnose customer problems on the spot and either provide immediate assistance over the phone or despatch a service technician.

Whirlpool makes a million service calls and handles three million phone calls each year. Whirlpool consolidated 26 district offices into two consumer-assistance centers. The scale and complexity of the technology needed to bring together one-call telecommunications, accurate diagnosis, and reliable and responsive answers and actions were substantial:

1. The telecommunications component involves a 1-800 call into an AT&T T1 network. An area-code router sends it to either the Michigan or Tennessee center. Redundant T1 networks provided by AT&T and MCI link the two.

2. Each site has an AT&T Definity 2 PBX. The lines are ISDN Primary Rate Interface. The Definity uses a look ahead feature to route calls to the first available agent. Software in the Definity PBX passes the caller's number to an IBM 3090 mainframe host, which matches it with a customer profile.

3. The data is passed to an agent, who uses an IBM PS/2 personal computer. The PS/2s are linked in work groups of 30 by IBM 16-Mbps token-ring LANs connected to a 16-Mbps token-ring LAN backbone. The PS/2s also use IBM's Extended Edition software, which allows both workstation-to-workstation and workstation-to-host communication.

4. Each LAN supports an image server that houses four CD-ROM drives with around 150,000 pages of scanned product and service data covering 20 years of products. The total paper documents would stretch for 26 miles.

147

5. An expert system on the LAN provides a series of questions for the agent to ask the customer over the phone. It steps through a dialogue that ends either by recommending action the customer can take or scheduling a service call.

Whirlpool estimates that the expert system may reduce service calls by 10 percent; at $40 a call, the elimination of 100,000 calls represents substantial savings.

Walt Coleman had no idea of just how complex the project was. The firm's Information Systems (IS) group was poorly positioned to handle it; for years it had been, according to Coleman, a "background function, involved in payroll, batch reporting, and so forth. It was to be seen and not heard." Rich Koeller, who took over as head of IS on the very same day as Coleman took over the customer service center project and who was to become a key ally, says of the old IS department, "There was no way we could do a massive project on time with our resources." Koeller is widely respected in IS circles for his character, candor, and capability; his job was to stir up IS and bring it into a new service-focused, business-driven environment.

Coleman decided to hire IBM to evaluate the project and outline the technical requirements and costs. Based on IBM's report, he decided to hire the company to design and build the system. The project team consisted of 75 IBM staff and 15 Whirlpool employees. He assigned one of his own people, Bruce Dacre, as project manager. Dacre understood the business well and had some knowledge of information technology. Because IS was being reorganized, he decided not to look for a "champion" in IT to provide technical advice and guidance, but instead to move ahead as quickly as possible using the available resources and contacts.

He now views this as a major mistake on his part. He relied on IBM for technical leadership, and because the systems integrator made the technology sound relatively simple, he was lulled into believing that the emerging technical problems would solve themselves. He did not ask enough questions or provide strong direction. In this very common situation, the project team wandered off course. It did its best, which in this context meant providing the best technical design. Coleman says, "They were going for technical elegance. They were doing what they thought was right, and we were thinking they knew what we wanted. But they got off on tangents, and we did not provide the discipline to stay on course."

Without clear guidance, there is no basis for design trade-offs. For example, is 30-second response time acceptable? Ten seconds? Five seconds? How much will it cost to go from 30 to 10 or ten to five? Is it worth the cost? Koeller points out that it requires an experienced implementer to understand that the cost of moving from 30 seconds to 10 is relatively low; it can generally be handled by selecting and fine-tuning hardware. But going from ten seconds to five is much more difficult and is more likely to involve fine-tuning software; this can tie up an unjustifiable amount of effort and may not be worth it.

IBM's team had to work out what technology was enough and which technical features were essential. Dacre did his best to control the project but lacked technical knowledge. Coleman accepts that Dacre was in over his head and Dacre did his best to shield Coleman from having to address technical issues so that Coleman could concentrate on the business side of the project. When a key delivery date was missed, Coleman asked for help from Koeller in IS. Koeller's judgment was that the original decision to use an IBM AS/400 mid-sized computer that was not integrated into Whirlpool's corporate platform was a mistake; the system needed a much more powerful mainframe machine. Koeller assigned one of his own experienced project leaders, Jan Massey, who worked in close contact with Dacre. Dacre now concentrated on the business side, and the IBM technical team reported to Massey. Coleman comments that "she should have been on board from day one. We need IT leadership that can understand the business case. Conversely, we need business people who can venture into technology to get some ideas. Both need to cross the line to communicate better."

Naturally, senior management was not pleased

by the unexpected news that a project that had been reported as being on schedule was now in major trouble and would cost far more than budgeted. Getting back on track was difficult and politically sensitive. "The wheels had come off and we had to get our confidence back." Massey was the key person here; combining her strengths with those of Dacre were vital to making things work.

The new team completed the project in June 1992, meeting the revised schedule. Coleman summarizes the lessons he learned:

1. "I'm thinking more like [IS] and they're thinking more like me. I have a lot more knowledge and understanding of technology than I ever thought I wanted."

2. Awareness of the technical issues helps assess the business case and the trade-offs between the costs and benefits of technical enhancements.

3. Dialogue leads to deeper levels of collaboration; instead of just supporting Coleman's goals, IS helps the business group establish goals.

4. Sensitivity and collaboration cannot be developed in a vacuum: "Forget about running seminars on mutual understanding. Instead, start with a clear mission and build a team."

5. Familiarity breeds familiarity; sharing successes and failures helped the multi-unit team coalesce.

Questions for Discussion

1. What was the business need that led Whirlpool to develop a state-of-the-art customer service center, and where would this need fit in the telecommunications business opportunity checklist?

2. Walt Coleman, VP of operations at Whirlpool, brought in one of his own people, Bruce Dacre, to act as project manager on the service center project. Was this a good move? Explain.

3. Using information from the minicase, discuss the importance of clear guidance from the users as the basis for design trade-offs. Also, comment on whether there was sufficient user guidance to develop an appropriate design for the Whirlpool service center.

4. Discuss the role of dialogue among user, designer, and implementer in the case by addressing the following:
   a. its importance in meeting the goals of the project
   b. how well the project was actually carried out
   c. ways in which the project could have been improved.

5. What lessons were learned by management about both the role of teams and the mix of skills needed on the team for this type of project?

Two

Network Building Blocks

4

Terminals and Workstations

Objectives

- ▸ Learn the roles and functions of terminals in a network
- ▸ Become familiar with various types of terminals and their particular functions and features
- ▸ Understand the implications of the increasing level of intelligence in terminals and workstations

Chapter Overview

Chapter Overview

Terminals and workstations are the most visible of the telecommunications network building blocks, and for the user of a telecommunications service they are the most important. This chapter describes the roles of terminals in the network, their main features, and the criteria for selecting them. There is a growing variety of choices of terminals, with the two major developments being the exponential increase in low-cost processing power, which adds intelligence to the terminal, and the use of wireless communications to add new levels of mobility.

The word *terminal* is generally associated with computer terminals, but telephones, barcode readers, and personal computers are all types of terminals. In addition, large mainframe computers can act as terminals in a network, especially when they share functions with personal computers/workstations in a client/server application.

After this chapter reviews the main features of terminals and how to choose among them, it then moves on to discuss how terminals interface to a network. Here again, there is a wide range of choice of devices, many of which are highly specialized.

The Roles of Terminals in a Network

A telecommunications terminal is an input/output device that acts as the interface between a user and the telecommunications network resource. Digital terminals are often referred to generically as **data termination equipment (DTE)**. DTE is contrasted with **data circuit-terminating equipment (DCE)**, which is any device whose function is to move and manipulate transmission signals to and from a terminal. The very basis of data communications is to connect a DTE to a DCE through an interface, which generally requires some type of adapter, such as a plug, socket, or slot-in board that connects to a cable or that broadcasts a radio signal.

There are many different types of terminals, some of which are depicted in Figure 4-1. The minicases presented so far in *Networks in Action* have included such terminals as in-store point-of-sale registers (THEi), personal computers with fax boards, high-performance image-processing intelligent workstations, video display devices (Shearson Lehman), mobile phones (Hungary Telestroika), and radio sensors, workstations, and host computers (Conrail). Each of these devices has very different features and types of interface, but to the network they are essentially equivalent; they are addresses in the network directory, and the network manages the flow of traffic to and from terminals. Figure 4-2 lists the terminals used in the networks discussed in the minicases in Chapters 1 and 2.

Figure 4–1

Examples of Terminals

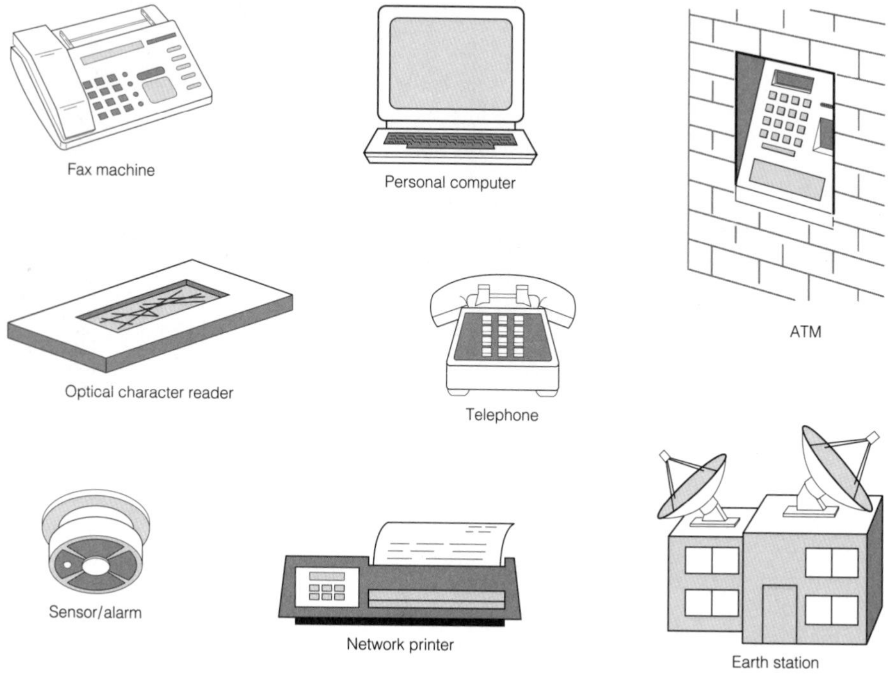

Fax machine

Personal computer

ATM

Optical character reader

Telephone

Sensor/alarm

Network printer

Earth station

Terminals are the reason for having a telecommunications network. Users of the service do not see the network any more than they see the electrical utility when they turn on a VCR or hair dryer. For them, the ATM, phone, or credit card reader *is* the network. Thus the goal of the telecommunications client must be to define the nature of the service, and the goal of the designer is to ensure that the terminal is easy to use and the service easy to access. The implementer must make sure that the terminal gets the service reliably, that the needed interfaces are in place, and that the network manages the flow of information efficiently, securely, and cost-effectively. The client should need to know nothing about how this is done.

A terminal is an address in a network. The network architecture defines the types of terminal that can connect to one, some, or all of its services. In and of itself, a terminal is just a device. Consider the tragic fate of a sailor in a transatlantic sailing race in November 1992. All the competitors had on board a device that signals their location to a satellite. If no signal is received for a given period or if an emergency signal is sent, the Coast Guard is alerted. The sailor disappeared and is presumed lost at sea. Although his onboard device had signaled his location and had sent an emergency message, he had not registered its calling number, so its messages were ignored by the network, for which the sailor did not exist. The device, although a type of terminal equipment, was not a terminal so far as the network was concerned.

Minicase	Terminals	Transmission Links	Transmission Methods	Nodes and Switches	Standards and Architecture
1-1 THEi: Reinventing a Company	Store point-of-sale registers	High-speed cables owned by telecom service providers, plus transatlantic satellites	Packet switching	Public data network switches	X.25
1-2 Shearson Lehman: Streamlining Workflows	Personal computers with fax boards; high-performance image processing video display terminals	Local area network coaxial cables; wide area private network leased lines	LAN collision detection; WAN circuit switching	Cluster controllers; front end processors; LAN-WAN gateways	IBM SNA (WAN); token ring (LANs); proprietary Wang Open/Image
1-3 California State Government: Telecommuting	Personal computers; telephones; host computers; satellite earth stations	Dial-up phone lines; analog and digital satellite microwave; asynchronous	Public phone	Public voice network switches; host computer front end processors	Public switched voice network standards
1-4 "Telestroika:" Hungary's National Telecommunications Policy	Mobile cellular phones; telephones	Line-of-sight microwave radio	Circuit switching	Cellular nodes; digital phone switches	German mobile phone standards
2-1 Terra International: Reaching the Boonies Through VSAT	Point-of-sale registers; host computers; VSAT earth stations	Satellite microwave	Packet switching	Multiplexers; cluster controllers; building switches	Aloha; X.25
2-2 Dun and Bradstreet: Extending the WAN to the LAN	Personal computers; file servers	Leased cable lines & LANs	Mixed	Front end processors; multiplexers	X.25; LAN protocols
2-3 Saving New York State Millions: Building a Massive Network	Computers; workstations; phones; PCs	Wide area network cables	Packet switching	Multiplexers; switches; PBX, front-end processors	WAN protocols and architectures
2-4 International Telecommunications: The View from Volkswagen	Computers; workstations; phones; PCs	Cables; satellites; public data networks	Mixed	Large switches	X.25

Figure 4–2

Terminals and Other Network Building Blocks in Minicases in Chapters 1 and 2

| | | | | | |

Key Terminal Characteristics

Regardless of their individual features, terminals can be classified in terms of generic characteristics and common functions that define their role as input/output devices accessing the network. These characteristics and functions include:

▸ *Input Mode:* How the user generates the information that the terminal processes and passes on to the network. On a PC, this may be a mouse or a keyboard; for a telephone it may be a combination of the mouthpiece into which the person speaks and the push buttons that can be used to select options from a voice messaging menu.

▸ *Presentation Mode* of the terminal: how it presents information to its users. Displays of characters, pictures, and numbers on either a personal computer screen or paper produced by a printer are the most common options.

▸ *Level of Intelligence:* the extent to which it can carry out processing, including error checking and message formatting, versus merely receiving and sending a bit stream.

▸ *Data Rate/Speed of Operation,* measured in bits per second.

▸ *Communications Mode* (asynchronous or synchronous): whether data is sent as a synchronized stream of bits or bit by bit with no synchronization.

▸ *Terminal Control Mechanism:* a stand-alone configuration in which the terminal has its own connection directly to the network, versus sharing DCE connections with other terminals.

▸ *Configuration in the Network,* ranging from being connected to other terminals in the network through a direct transmission link to sharing a transmission link with other terminals.

▸ *Degree of mobility* of the terminal, from wired to wireless.

Figure 4-3 presents the range of capabilities of each category, with the simplest/cheapest/most limited options placed at the left of the capabilities continuum. And just as a bike is more limited than a car, any of the options presented may still be the most efficient and effective tool for a given person and a given need. Many of the terminals that appear in Figure 4-3 were chosen to balance cost with capability.

Next, we examine each of the various terminal characteristics/functions in turn.

Input Mode

Getting information into a terminal is generally accomplished in one of the following ways: by using a point-and-click mouse or typing in characters from a keyboard (a personal computer), by speaking and pushing buttons (a telephone), by scanning a barcoded label (point-of-sale register), or by scanning a document either for fax transmission (facsimile) or for digital transmission and storage (image processing). The differences among them mainly relate to the

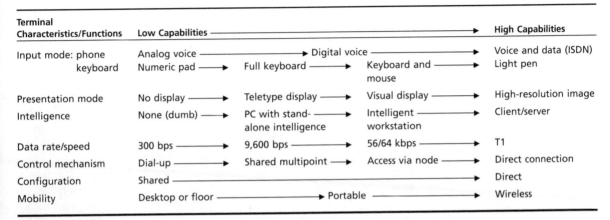

Terminal Characteristics/Functions	Low Capabilities ➝			High Capabilities
Input mode: phone	Analog voice ➝ Digital voice ➝			Voice and data (ISDN)
keyboard	Numeric pad ➝	Full keyboard ➝	Keyboard and ➝ mouse	Light pen
Presentation mode	No display ➝	Teletype display ➝	Visual display ➝	High-resolution image
Intelligence	None (dumb) ➝	PC with stand- ➝ alone intelligence	Intelligent ➝ workstation	Client/server
Data rate/speed	300 bps ➝	9,600 bps ➝	56/64 kbps ➝	T1
Control mechanism	Dial-up ➝	Shared multipoint ➝	Access via node ➝	Direct connection
Configuration	Shared ➝			Direct
Mobility	Desktop or floor ➝ Portable ➝			Wireless

Figure 4-3

Terminal Characteristics and Capabilities Ranges

role of the human user: When the dialogue is interactive, with a human entering data and the terminal responding, the mouse, keyboard, and phone are the natural modes of input. When the purpose of the terminal is to capture data without any human analysis or intervention, the terminal is more likely to use barcode scanning. Image processing generally requires some human confirmation and classification, but fax is a straightforward activity that requires only inserting the paper and dialing the recipient's number.

Conrail's freight car tracking sensors (Minicases 3-1 and 3-2) use barcoding as the input data and radio scanners as the input mode. By contrast, Rosenbluth Travel's agents (Minicase 3-3) all interact with the reservation systems through a keyboard and/or mouse.

Presentation Mode

There are two main modes of terminal presentation: video display terminal and teleprinter. The **video display terminal (VDT)** is the typical video screen found with most personal computers and remote terminals connected to mainframe computers. Display terminals come in several varieties—scroll mode, page mode, forms mode, and graphics mode—each of which has different hardware requirements and telecommunications requirements, with more processing required as you move from a scroll mode device to a graphics terminal. A simple **scroll mode** VDT shows text one line at a time, similar to a hard-copy printer. When a new line of text is displayed, the top line scrolls off the screen and disappears. A **page mode** VDT displays an entire page of text or data on the screen at one time, and the full page is stored in the terminal prior to transmission through the network. A **forms mode** VDT shows a blank form covering the entire screen. The user manipulates a pointer on the screen to a position on the form where information is to be entered. The user then types text or data into the appropriate blank space on the form. Finally, a **graphics mode** VDT is designed to display high-resolution graphics images (engineering drawings, pictures, and so on).

Most common text-oriented VDTs translate digital bit streams into characters (text and numbers) and then move these characters one by one to specific positions on the video screen. The graphics mode terminal, on the other hand, uses bit mapping to provide a high-resolution display that can be quickly updated. **Bit mapping** assigns the digital bit stream representing images directly to specific locations on the screen. This process, without the intermediate step of translation to characters or symbols, results in a higher-resolution image, but it also requires more memory in the terminal to attain this result.

Bit mapping also increases transmission demands. Each bit is mapped to a pixel (picture element) on the screen, so that, for instance, an ellipse appears as a smooth curve with no jagged edges and a person's signature is accurately reproduced. The more pixels a screen displays, the finer the resolution of the graphic image—and the more bits required to create and update it. Some typical VDTs and the number of pixels in each are given in the following list:

VDT Type	Pixels per Column	Pixels per Row
Laptop	320	200
Standard desktop	640	350
with color screen	640	480
High-resolution graphics terminal	1,024	1,024

To store the high-resolution image, about three million bytes of memory are required.

The second principal mode of terminal presentation, the **teleprinter,** is the traditional hard-copy printing device located in a place remote from the computer generating the printed output. The basic unit contains the printer mechanism and a set of electronic controls, including the communications interface. It may also include a keyboard for an operator to enter information to be sent to the distant computer. The teleprinter is a terminal; it can be addressed by the network, so that, for example, a travel agent may make a plane reservation for a customer and specify that the ticket be printed on the agency's teleprinter or on one located in a building at a Memphis airport. The latter teleprinter is significant because it is at Federal Express's air hub, so that the agency can Fedex the ticket to the customer. A printer attached directly to a personal computer is not a teleprinter; the PC is the terminal in the network and the printer is a peripheral device to the PC.

The choice between VDT and teleprinter modes of presentation significantly affects the time it takes to display output information. Clearly, a VDT operating in page, forms, or graphics mode has a higher **response rate**—the time elapsed between transmission and receipt of input and output—than the teleprinter mode of presentation. Because teleprinters generally are electromechanical devices—machines in which electricity drives moving parts—they are slower than microelectronic-based ones such as VDTs.

Intelligence

Terminal intelligence refers to the ability to (1) process or alter the information that is entered into the terminal for transmission through the network or (2) manipulate the information that is received by the terminal for display to the user. With the development and evolution of microprocessor technology, the level of intelligence of VDTs has increased dramatically in recent years. The basic dumb VDT in widespread use several years ago contained no processing or information-altering capability; it was designed for compatibility with a specific host computer, and it either projected on the screen information that was received from the host mainframe or relayed keyboard input to the host computer for processing. In contrast, the modern VDT contains such a high level of intelligence that it is almost indistinguishable from a true PC workstation. These modern VDTs are programmable, have sizable memories, and are capable of protocol and file format conversions.

The increased power and cost-effectiveness of PC workstations continue to drive the transformation from centralized to distributed computing environments, in which data is stored in different remote locations and processing takes place at multiple locations tied together through a telecommunications network. This means that natural workflows among groups, both within and across organizations, will be supported more frequently by peer-to-peer processing relationships managed through telecommunications networks. (**Peer-to-peer** refers to communications between equal, or peer, terminals, as opposed to the original terminal to host arrangement in which dumb terminals were connected to a mainframe computer in a slave/host configuration. In a **slave/host** system, each terminal must link to the host to contact another terminal. Slave/host communications may still have advantages for very high transaction processing systems, in terms of managing and updating a single network directory and reducing communications complexity, but this configuration is increasingly becoming an anachronism.)

Consequently, in a distributed computing environment, PCs, minicomputers, and mainframe computers act as input/output devices accessing a network, and hence they perform terminal functions as well as processing functions. Furthermore, given the relatively small cost differential between intelligent terminals and PCs, it is often more cost-effective to use a PC in place of a simple terminal if there is even a modest stand-alone processing requirement. Thus, for instance, more and more people who travel frequently on business use a portable computer in stand-alone mode for word processing and in communications mode for electronic mail.

The intelligence/capability level of a terminal determines its flexibility and versatility in communicating with other types of terminal devices operating on other network systems. The more intelligence that is in a terminal, the greater its ability to perform conversions, transformations, and translations of protocols and data structures in order to communicate with different makes and types of terminals and other devices connected through the network. For example, have you ever experienced the problem of having created a portion of a document on your IBM or IBM-compatible PC and attempted to hand this over to a friend for completion on an Apple Macintosh machine? This situation often involves an agonizing conversion process requiring special software and much reworking to correct margin problems and so on. This gives you some idea of the difficulties involved in communicating in real-time through a network with different types and makes of terminals and other devices; it also demonstrates one important value of intelligence in terminals.

The level of intelligence in a terminal can also affect the response rate of the network. A highly intelligent terminal (for example, a PC workstation) is capable of local processing of data or transmission of data to another node in the network for processing. Thus the choice between local and remote processing can be made to minimize network response time, given both the computing resources available at various locations on the network and the transmission link capacity of the network itself. If high-capacity transmission links are available and they are not congested with network traffic, response time can often be reduced by the use of these facilities to distribute data to various points on the network, where the actual processing can be done most efficiently (in terms of work load versus capacity available at each location). If, on the other hand, transmission link capacity is limited or congested with network traffic, response time can be reduced if more processing takes place locally, where the data resides. Intelligent terminals permit these choices in order to minimize response time, and, as a result, to increase the business functionality of the network.

A typical example of exploiting local intelligence occurs in ATMs. Dumb terminals cannot format messages; the host does this and sends the full display through the network. Sending "Your balance as of 8 a.m. November 10 was $564.56" requires about 500 bits; sending "080011100056456" and having the ATM reformat the bit stream takes only 120 bits.

Data Rate/Speed

The **data rate** of a terminal—its actual speed of operation—ranges from approximately 75 bps for a teleprinter to 64,000 bps for an ISDN terminal. To illustrate this vast difference in speed, consider a teleprinter that operates at a rate of ten characters per second (approximately 80 bps, assuming eight bits per character); this teleprinter can process (print) 600 characters in one minute. In contrast, the ISDN terminal operating at 64 kbps will process the same 600 characters in 1/13 of a second.

The data rate/speed of the terminal is directly related to its responsiveness; that is, the faster the terminal operating speed, the lower the response time of the terminal. However, this translates into a lower response time for the overall network as seen by the user only if the terminal is the network's **responsiveness bottleneck**—the slowest device or link in the network that reduces the throughput of other elements. If the transmission links or switching components in the network have very limited capacity and are highly congested with traffic, then the overall network response time is limited by the speed of the transmission links, not the terminal. Consequently, using a terminal with a higher data rate will have no effect on overall network response time. You sometimes observe this at an ATM, when the system seems very slow; after you hit the key to select an option, you must wait three times as long as usual. In this case, some aspect of switching or transmission is the bottleneck; the ATM's terminal speed is the same as before, and substituting one that has twice the data rate does not remove the bottleneck. Similarly, sending data at 64 kbps to a printer that operates at 80 bps does not speed up printing. On the other hand, if there is an abundance of high-capacity network transmission links and switching components, increasing the data rate of the terminals can increase overall network responsiveness.

Communications
Mode

A terminal may operate in either asynchronous or synchronous mode, depending on the type of network system being used. As discussed in Chapter 2, asynchronous transmission involves sending one character at a time with start and stop bits included to signal the receiving terminal when a character begins and ends. Also, error control is limited to a single parity bit check (odd or even) accompanying each character. Synchronous transmission sends multiple characters along with control and error detection information in **frames**—discrete units of bits transmitted as part of a message or transaction—that are synchronized in time so that transmitter and receiver are working together and know precisely when a frame of information begins and ends. A message may be composed of multiple frames, each of which includes additional information that frames the message data. To the user of a network, the message is the main unit of information; to the network implementer, traffic is defined by frames, packets, blocks, and bits.

Synchronous transmission is much more efficient for transmitting large amounts of information because control and error detection information are included just once with each frame, rather than having to be sent separately with each character, as is the case with asynchronous transmission. Thus synchronous mode terminals operating at the same data rate exhibit greater responsiveness than do their asynchronous counterparts because with synchronous transmission there are fewer overhead bits (control and error detection information) accompanying a given number of data bits; consequently, the entire bit stream (data plus overhead) can be sent in less time using the synchronous mode of transmission when the transmission rate is the same.

Asynchronous transmission has the advantage of simplicity. It does not require expensive and complex switching equipment to ensure precise synchronization of timing. It is well suited to message-based communications (such as electronic mail) in which short delays do not interrupt the user. Thus a brief hiccup in transmission of a public E-mail service—when a message on your personal computer screen stops being displayed for a fraction of a second—is at most a minor irritation.

Control Mechanism

Terminals can operate independently in stand-alone mode, or they can be connected together in clusters—a group of terminals connected to the network through a **cluster controller,** which is the device that manages their communications operations, including handling both congestion and competition to access the transmission link. A series of stand-alone terminals, each connected directly to the network, generally exhibits a faster response time than a group of

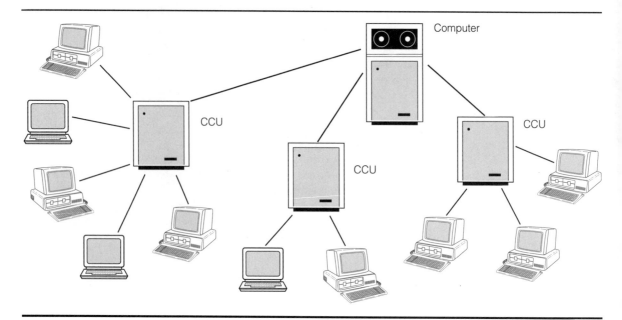

Figure 4–4

Configuration of Cluster Control Units

terminals linked together in a cluster with a single interface to the network; however, the stand-alone configuration is generally more expensive. In the stand-alone configuration, each terminal must contain its own control logic and interface to the network; in contrast, in the cluster configuration the cluster control unit (CCU) includes the control logic and provides a single interface to the network for all of the connected terminals. Figure 4-4 shows how terminals and cluster control units are configured.

The advantage of the cluster configuration is lower cost. Each terminal requires less intelligence because the control and interface functions are concentrated in the cluster control unit; therefore, these functions need not be replicated in each individual terminal. The disadvantage of the cluster configuration is a possibly slower response time. If the terminals in the cluster are active (have a high usage level) and if there is sufficient transmission link and switching capacity such that these components are not responsiveness bottlenecks in the network, then the response time of terminals operating in the cluster configuration is likely to be slower than those operating in stand-alone mode. This slower response time results because the terminals in the cluster are sharing access to the same component, the cluster control unit, that contains the control and network interface logic, and traffic congestion can delay access to this component; every device between the terminal and the service it accesses adds overhead because either software or hardware must inspect the incoming and outgoing traffic and carry out the relevant processes.

Figure 4–5

Front-end Processor

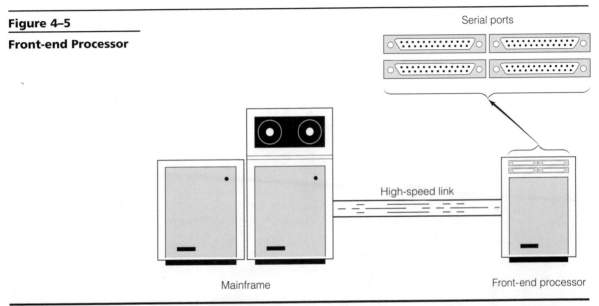

Serial ports

High-speed link

Mainframe

Front-end processor

Front-end processors coordinate the flow of traffic between a fast mainframe computer and slower devices, converting data speeds to make efficient use of the mainframe's processing power. Connections for serial transmission are accomplished via serial ports.

At speeds of 64 kbps, a bit arrives every 15 microseconds, so that a one millisecond delay, fast by human standards, slows down throughput dramatically. This is not a problem for stand-alone terminals, because each has its own control and network interface logic; consequently, network access is not delayed and response time is shorter. It is also not a problem in the many types of transactions in which the terminal is not fully active. For instance, when you use an ATM, you spend several seconds thinking and hitting the keys to indicate that you want to withdraw $50. The ATM is busy only around 2 percent of the time, so that the use of cluster control units to link the ATMs in a lobby is a highly efficient way of balancing network throughput with equipment costs.

A device known as a **front-end processor (FEP)** performs a terminal control function by acting as both a communication network interface for a host mainframe or minicomputer and as a controller for remote terminals accessing the host system (Figure 4-5). The FEP is typically a stand-alone minicomputer or microcomputer (but it can be an internal board in a multipurpose computer) that handles the routine communications tasks for a host computer. These tasks include data formatting, character or message assembly or disassembly, code conversion, message switching, polling of remote terminals, error checking, protocol support and conversion, automatic answering and outward calling, and compilation of network operating statistics for management.

Front-end processors are an essential requirement in networked mainframe transaction processing systems that have a large number of terminals connected to them, such as ATMs, airline reservation systems, and order-entry systems; without them, the mainframe will have added overhead and work load in managing both network traffic flows and computer transactions. Cluster control units and FEPs create the same type of hierarchy as that of the public telephone network. Cluster controllers correspond to the local central office, whereas FEPs are the "traffic cops" at the primary centers. Managing a telecommunications network means managing complexity. Simple local area networks often do not need a hierarchy of devices, but complex firmwide networks invariably do. Cluster controllers and FEPs help manage complexity and reduce cost at the same time.

Configuration

As discussed previously, terminals can be controlled individually in stand-alone mode or in groups through a cluster control unit. Terminals can also be connected to each other through a network in either a point-to-point configuration or in a multidrop configuration.

The point-to-point configuration requires each terminal to have a direct transmission link to the network. In contrast, the **multidrop** configuration has several terminals sharing the same communications transmission line linking them to the network. The multidrop configuration is less expensive than point-to-point because it economizes on the number of transmission lines; however, response time is generally slower due to the need to use a terminal polling and select system to manage contention among terminals for access to the shared line.

Polling and select mechanisms slow network response time because only one terminal on a multidrop line can transmit at any given time. In addition, the polling and select system requires that communication always occurs directly between a terminal and the host computer; one terminal cannot communicate directly with another terminal (Figure 4-6). This use of an intermediary (the host computer) to relay information also slows network response time for communication between remote terminals. Such inefficiency in network response time may be offset by reductions in line costs. Until the early 1990s, transmission links were so expensive that the economics of large-scale transaction processing often favored multidrop lines (and still do in many countries).

Mobility

The mobility and portability of today's modern terminals are important characteristics that can extend the reach of a network. Portable wireless telephones permit continuous voice communications for individuals on the move who do not have access to the public telephone network's wired facilities. The next generation of digital mobile equipment will make portable, over-the-air data transmission as easy as voice communication is today.[1] The laptop and other portable personal computers with a built-in communications interface extend

---

[1] Although this is an easy prediction to make, there are many regulatory and technical issues to resolve in the meantime. For example, the FCC must decide if and how the scarce frequency ranges now allocated to such services will be reassigned. In addition, the almost total lack of standardization in wireless technology makes it the equivalent of early PCs and LANs in that explosive and uncontrolled innovation is occurring. Apple's introduction of its Digital Personal

Figure 4–6

Roll-call Polling

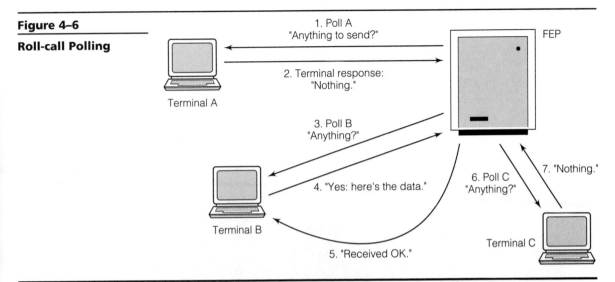

In roll-call polling, each terminal is polled in turn, generally by a front-end processor or a terminal control unit. In fast-select polling, the system does not wait for a terminal with nothing to send to reply, thereby reducing wait times.

the reach of the organization's electronic mail and data transfer capability to virtually anywhere in the world where there is access to a public or private network interface. This interface to the network can be either through a wireless over-the-air system or through a wall-jack connection to a public telephone system.

A recent term in the vocabulary of information technology is **personal communications system (PCS)**—a device that provides its owner with all the services he or she gets from owning a phone at home, from using cellular mobile phones, and from having a laptop computer with a modem. A user can communicate to anywhere from anywhere in a number of possible ways. Apple offers its Newton Personal Digital Assistant; Motorola has offered an ambitious **low earth orbit satellite (LEOS)** service called Iridium (see Figure 4-7), which as of the end of 1992 had not attracted enough investors for its very practical $3.4 billion project to go ahead.[2] Low earth orbit satellites are cheaper to launch, smaller, and easier to operate than the high orbit geosynchronous satellites that

Assistant in mid-1993 may stimulate its adoption as a de facto standard, or it may fuel competition among proprietary systems.

[2] Several foreign governments are reported to have agreed to invest in Iridium, including Indonesia and Brazil. Their interest is obvious: the chance to leapfrog existing technology and obtain a new telecommunications infrastructure with virtually no extra capital expenditure. In the United States, other technologies and providers are seen as alternatives to Iridium, and there has been relatively little enthusiasm for jumping in early with money.

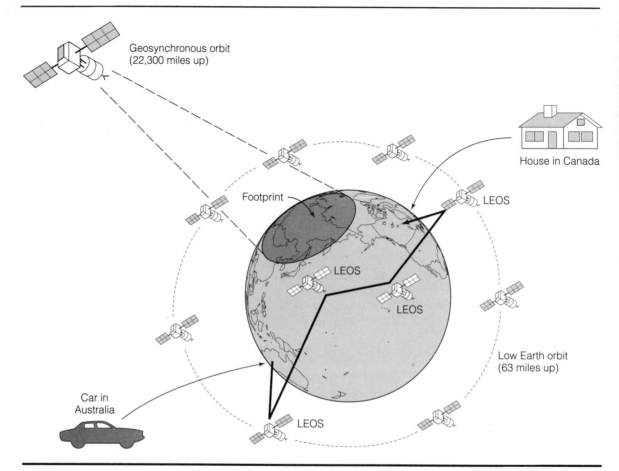

Geosynchronous orbit
(22,300 miles up)

House in Canada

Footprint

LEOS

LEOS

LEOS

Low Earth orbit
(63 miles up)

Car in
Australia

LEOS

Figure 4–7

Motorola's Iridium, a Low Earth Orbit Satellite (LEOS) System

Motorola plans to encircle the earth with 60–80 LEOS. A phone call from a car in Australia is picked up by the nearest LEOS and routed across the world to a LEOS that downlinks to a house in Canada.

are the core of international satellite services.[3] In late 1993 AT&T bought the leading—and as yet unprofitable—national cellular phone company, McCaw, for a cost of over $17 billion. Also in 1993, MCI announced a project to create a joint venture among other wireless communications providers to establish a national wireless infrastructure. The 1980s was the era of breakthroughs in terrestrial fiber-optics transmission links; the mid-1990s may well be the era of

---

[3]The cost to launch a geosynchronous satellite is around $200 million, of which one-third is for the satellite, one-third for the rocket, and the final third for insurance. China, among others, is offering cut-rate launches in competition with both NASA and the European-government–backed ARIANE consortium.

breakthroughs in wireless infrastructures, making telecommunications as mobile as the people who use it.

The growth in mobile communications has accelerated in the early 1990s, but it is too early to identify trends and make firm estimates of their impacts. Recent growth rates for cellular phones suggest that they will reach a market of ten million in about half the time it took VCRs to reach that level; cellular phones will likely become as commonplace as VCRs are now.

Types of Terminals and Workstations

The features of terminals just described permit a wide range of specific devices. The five main categories are telephones, video display terminals, optical recognition and scanning devices, teleprinters, and mini-computers and mainframe computers. The following sections summarize the main trends in each category.

Telephones

Since the deregulation of telecommunications terminal equipment, the number and kinds of telephone sets available have proliferated, including a wide variety of shapes and sizes, cordless phones, and completely portable cellular telephone units. The telephone instrument converts voice sounds to electrical signals (vice versa on the receiving end of the call). It signals the telephone company's central office that a call is coming through and follows this signal with transmission of the destination address (telephone number sequence) using the rotary dial or touch-tone keypad system. (The keypad is also being used more frequently for limited data entry, such as providing an account number to a bank computer to obtain the balance in your account, or to make simple choices among options: "If you wish to speak to a real person, press 3.")

Clearly, the telephone terminal has had a major effect in extending the reach of networks through wireless mobile technology—that is, the cellular telephone. A salesperson in the field (whether on foot or in a car) can continue to communicate with customers and the company, both enhancing the productivity of the salesperson and better serving the customer, because communication is not limited to times when the sales rep is near a wired telephone outlet.

The modern telephone set also has a number of features that lower response time. Speed dialing capability permits faster response time in reaching customers, and incoming number recognition displays accelerate recognition of customer identity, which in turn allows faster access to data and computer records showing the customer's purchase record and buying preferences. Imagine the time saved, and hence the improvement in responsiveness, in dealing with a customer complaint or information query if the customer's purchase records and personal information are displayed on a VDT as soon as the phone rings in the office. An agent can answer the call with "Hello, Mr. Rodriguez. I expect you're calling about renewing your policy with us. I have all the information you need."

In the late 1980s, many companies expected the personal computer to become the key access device for telecommunications-based services. That has not turned out to be the case in the consumer market because most home PCs have no modem and are used in stand-alone mode only.[4] Connecting a PC to an information service, home banking, electronic shopping, on-line news, or electronic mail ties up the household telephone line, sometimes for hours, or requires paying for an additional line. Moreover, dial-up lines are slow, and users quickly become impatient. As a result, none of the many efforts to create critical mass in these services has to date succeeded, and some of the failures have been very expensive. Only Prodigy has stayed the course, and its revenues and numbers of subscribers are still below the break-even point.

Because of the limitations of today's PCs and improvements in basic telephone capabilities, the phone handset is still the main access device for telecommunications-based services. Figure 4-8 summarizes the new features of the phone set—a terminal with transmission capabilities. Many of these, such as **automatic number identification (ANI)**—a controversial feature that displays the caller's phone number on the receiver's handset—are part of the evolution of ISDN. Some commentators see this feature as a threat to privacy and a deterrent to users of emergency phone lines.

When—or if—ISDN is fully deployed nationally, it will make the functions of personal computers and telephone handsets essentially the same. The ISDN handset is a digital terminal that receives and sends data, including voice, at 64 Kbps. The handsets will be far more expensive initially than the much simpler analog voice devices, which are even given away as free gifts for renewing magazine subscriptions, and it is still unclear how quickly ISDN will be installed, what consumer reaction will be, and whether it will be made obsolete by newer forms of transmission, including cable TV.

Video Display Terminals

The video display terminal is the most common terminal in use today. At one end of the spectrum is a dumb terminal that simply consists of a video display screen, a keyboard, and only enough intelligence to control both the display of characters on the screen and the buffering and feeding of characters to the transmission line. The IBM 3270 series is representative of the basic dumb terminals that operate in page or forms mode. These terminals were pervasive in the 1970s and typified the slave/host centralized computing and data communications model of that era. Large numbers of 3270 terminals operating in synchronous mode are connected to a remotely located host mainframe through a front-end processor (FEP). Some of these units are grouped together and linked to the network through cluster control units, whereas others operate in

---

[4] Most published estimates indicate that 10–20 percent of PCs have a modem. The pundits predict that this percentage will double by the mid-1990s, but as yet there has been no telecommunications use of PCs that is the equivalent of ATM services in banking and that generates momentum for radical growth.

Feature	Service
Figure 4–8		
New Telephone Features		
ANI (automatic number identification)	Handset display indicates phone number of calling party
Audioconferencing	Simultaneous three-way, four-way, or eight-way conversations
900-number services	Caller-paid voice communications
700-number services	Routes calls to customer anywhere, not to a phone at a fixed location
ACD (automatic call distribution)	Routes incoming calls to relevant person (sales rep, claims processor, etc.)
Voice messaging	Spoken E-mail
Telset	Combination of phone and PC, with add-on options (fax, copier) available
ISDN handsets	Digital voice plus data for communications

stand-alone mode. These terminals are used to enter data, activate applications programs, or query databases on the host computer. The 3270 is a clumsy device, or rather the software that processes telecommunications transactions through it makes it clumsy to use. It is still in widespread operation in large companies, mainly because the cost of rewriting large-scale transaction systems is too high to be a practical option.[5]

Modern electronic cash registers used in retail operations are VDTs connected to a cash drawer. These point-of-sale (POS) terminals are linked either to a local PC or to a minicomputer via a LAN, or to a central mainframe computer over a wide area terrestrial and/or satellite network. Transaction sales data is consolidated and stored in the computer, and this data is often linked directly to inventory and ordering systems for more efficient management of store resources. Recall in Minicase 1-1 that THEi's POS terminals were used in each of the retail stores to send sales and customer information to the London headquarters mainframe computer over a public data network. Headquarters uses the consolidated data from the various stores to analyze national trends, financial performance, and so on, which enables management to provide better support to local store managers for management of inventory and stock ordering.

At the other end of the VDT spectrum is a powerful desktop workstation that has the full computer power of the giant mainframes of just five years ago.

---

[5] As recently as December 1992, for instance, Novell, a company that did not exist when the 3270 was designed in the early 1970s, announced a new feature in its industry-leading NetWare network operating system that enables DOS-based PCs to emulate a 3270. Novell's business is LANs, but if a PC on a LAN is to connect to a mainframe system, it must use the old set of protocols and act as a dumb terminal. Information systems managers often describe any large-scale transaction processing software that cannot easily be replaced as a "legacy system." The term suggests something left to you by a rich relative, but it really means "stuff that's expensive to run and that we can't get rid of—yet."

The workstation unit includes the basic video display screen and keyboard units, but the display unit is designed for high-resolution graphics images such as engineering blueprints and three-dimensional representations used in computer aided design (CAD). The intelligence level of this type of terminal is extremely high, for it also functions as a powerful stand-alone computer. Consequently, these terminals can emulate a simple input/output device accessing a remote host computer, or they can carry out complex computation and data manipulation functions when operating in a distributed processing mode.

In a fully distributed environment, the various workstations on the network are connected in a peer-to-peer (rather than slave/host) relationship. This means that the data processing, control, and storage functions are dispersed throughout the network among the various workstations, and the determination of which workstation performs what function depends on the particular application and the load on the system at the time. There is currently a strong trend toward distributed processing and peer-to-peer relationships among PC workstations. The incredible computing power and relatively low cost (per unit of computing power), together with the flexibility inherent in a distributed processing environment, makes this form of networking extremely attractive.

Many of these high-end workstations use the UNIX operating system, which is well suited to applications that are computation-intensive, such as engineering. Sun Microsystems, Apollo, and IBM (through its RS/6000 machine) are the market leaders here. Because UNIX is largely incompatible with other major workstation operating systems such as MS.DOS, Windows, and OS/2, compatibility poses many challenges to the network designer. For instance, almost all UNIX applications use a telecommunications protocol called **TCP/IP**—a fairly simple, low-overhead protocol designed to interconnect a wide variety of computer equipment. Adding UNIX machines to an IBM network based on MS.DOS (or to one using Apple's operating system and its local area network protocol, Appletalk) is daunting, though not impossible. The intelligence in the terminal is defined by its operating system, not by the network, and thus the operating system must be adapted to exploit that intelligence and allow data to flow between workstations that use different operating systems. Data flow is accomplished by protocol converters, smart hubs, and gateways, and even then it is rare that they can share applications.

For this reason, many companies have a firm policy on what types of PC and intelligent workstation its departments may use. The most common choice today is MS.DOS and Windows; IBM's OS/2 is becoming a main choice for transaction-based systems in large organizations, and Microsoft's NT is expected to be a challenge to it.[6] Although Apple's Macintosh and Powerbook PCs are far easier to

---

[6] It may be several years before it is clear which operating system will dominate the marketplace. Preliminary reports from beta-test sites in early 1993 criticized NT's lack of enterprise capabilities. Microsoft's roots are in stand-alone personal computing, not wide area telecommunications and large-scale client/server systems. OS/2 2.0, initially dismissed as a failure, sold two million copies in its first 18 months, mainly because of its suitability for handling desktop functions in complex transaction processing systems, reflecting IBM's roots in enterprise computing and

use than DOS and Windows, they have not been a first choice in most large organizations because they are more difficult to incorporate into a companywide telecommunications platform. Although Apple's telecommunications standards have been limited and highly proprietary,[7] it has made improved connectivity a priority for the mid-1990s, in part by creating a joint venture with IBM.[8] The battle for the desktop is moving from stand-alone operating systems to the telecommunications capabilities of those operating systems.

Optical Recognition and Scanning Devices

The range of applications for optical recognition and scanning devices is growing rapidly as the quality and reliability of this type of equipment improves and its cost decreases.[9] One of the most common devices of this type is the **optical character reader (OCR)**, which consists of a photoelectric cell that senses light and dark and converts patterns of light to a character code such as ASCII. A similar device found in many retail outlets is the barcode reader, which is generally connected to a point-of-sale terminal such as an electronic cash register and is used to scan a small strip that contains a series of parallel lines that represent the Universal Product Code scheme. This ten-digit code identifies a particular manufacturer and its specific products.

A facsimile (fax) machine, which has become almost ubiquitous in today's office environment, is also an optical scanning device that senses patterns of light and dark and maps these images directly onto the output medium, either paper or a display screen. Fax differs from OCR terminals in that it does not translate the scanned image into characters, but rather reproduces the scanned image directly onto the output medium. Fax technology has been available for decades but has only come into widespread use in the past few years. Its recent

---

communications. Windows' many inadequacies and lack of robustness—its sudden software crashes—have not slowed its momentum, and it was still selling 100,000 copies a month in 1993.

[7] Apple deliberately chose proprietary, not open, standards for all its Macintosh products, which has resulted in a proprietary telecommunications protocol (its AppleTalk) for LANs. The company believed the Mac was so much better than IBM's computers that users would surely switch from Big Blue. Instead, users needed devices and protocols that connected Apple and IBM computers. The lack of these protocols greatly slowed down Apple's penetration of the business market. In contrast, it has been highly successful in the education market, in which ease of use dominates connectivity to mainframes as a priority.

[8] Taligent, the name of the company formed by the joint venture, has focused on multimedia as its priority. The gamble is that Apple's track record in developing easy-to-use graphical user interfaces and IBM's understanding of how to handle large-scale and complex software and telecommunications would help them leapfrog their joint rival, Microsoft. Initial progress reports were largely favorable.

[9] The development costs of new hardware and software can be massive. A new operating system like NT can include up to two million lines of program code and cost $25–50 million, and a new computer chip costs hundreds of millions of dollars. But because reproducing software on a floppy disk or making chips in volume costs just a few dollars, the prices of information technology products drop 20–30 percent a year. Image processing systems are not yet widely enough deployed for prices to drop as quickly as prices of high-volume systems such as PCs. That said, Canon's stand-alone system, which provides the equivalent of a $1 million machine in 1988, sold for under $7,000 in 1993, and scanners for PCs sell for a few hundred dollars.

explosive growth is attributable to a number of factors: improved quality and reliability, lower cost (both machine cost and telephone network usage charges), a critical mass of accessible FAX users, the pathetic performance of postal systems around the world, and the growing business requirement for faster response times in message communications.

Recall that in Minicase 3-1 Conrail used a form of barcode reader system to keep track of the various railroad cars in use and in storage. Sensors were located in freight yards, terminals, and at signal points along the track to record the data coded on each car and to transmit the information over leased lines to a central computer for analysis and storage. More and more forms of OCR use mobile devices and radio frequency scanners, reflecting the growing importance of wireless communications.

Optical scanning devices are now frequently used in a mobile context to provide greater control and efficiency in the distribution and movement of the various components of a product or service. For example, Federal Express has introduced an instantaneous package locator service that relies on portable, hand-held scanning devices that read and temporarily store coded identification information included on each package. At each key point in the delivery cycle (pickup, sorting, shipping to intermediate transfer points and to final destination city, and so on), the operator handling the package scans the coded identifier information and then transmits the scanned data to the central computer database in Memphis, Tennessee, via Fedex's worldwide network. Thus any Federal Express office in the world can query the Memphis database to get a current update on the location of any package in the system.[10]

Optical scanning devices also permit a wider range of message types and image information (including pictures and handwriting) to be easily digitized and transmitted in a form compatible with the standards of a network. UPS leapfrogged Federal Express by introducing a mobile scanner; customers sign for receipt on the terminal screen using a special pen.

Response time is significantly improved through the use of optical scanning as an input device. Scanning is much faster than using a keyboard to type characters or using a "mouse" to reproduce graphics images by hand. The decrease in response time is clearly evident to most consumers that frequent retail outlets, such as supermarkets, that have introduced scanning devices at checkout counters. In addition to improved response time, operator errors in entering the data are reduced, and more sophisticated inventory and stocking analysis is made possible through the more efficient entry of data.

Teleprinters

Teleprinters are hard-copy printing devices located at a site remote from the computer to which it is connected via the network. A teleprinter unit can range

---

[10] Federal Express's television commercials increasingly emphasize this feature of its service, rather than product or price. Telecommunications is *the* differentiator here, and Federal Express's use of mobile communications was part of inventing a new industry. Today, more and more firms are exploiting it: Progressive Insurance has mobile car-accident claims adjusters (see Minicase 11-4), and Chrysler has mobile warehouse checkers (see Minicase 6-2).

from a high-speed line printer with draft-quality text to a slower, letter-quality printer capable of high-quality text and graphics used in a wide variety of applications (from remote word processing activities to engineering design work). The teleprinter unit contains a set of control electronics, including a communications interface to the network, and a keyboard used to send commands to the remote computer. The teleprinter is used primarily for printing the results of time-shared application programs and for remote printing of documents and billing information. **Time sharing** is the processing of multiple transactions from multiple terminals using a powerful host computer. Airline reservations and ATMs are classic examples.

Even though the teleprinter has been replaced in many applications by the faster and more versatile video display terminal, it is still used extensively in message applications such as telex and for operational control of data center computers. Whenever a hard-copy log or audit trail is required or a customer needs a copy of a document, the teleprinter is the obvious terminal for this purpose.

Minicomputers and Mainframe Computers

It may seem strange to consider a minicomputer or a mainframe computer as a terminal device. But with the strong trend toward distributed processing and peer-to-peer relationship models of information systems, these computers can function as input/output devices or can operate as user interface nodes in a network in a role similar to that of a PC workstation in a LAN. The mainframe responds to database queries and may be programmed to act on behalf of a user in a totally or semi-automated inventory or electronic data interchange (EDI) system. In a fully automated EDI system, purchase orders for products or services, invoices for payment, shipment orders sent to a warehouse, and inventory restocking orders can be handled completely through computer-to-computer communications. In this case, the minicomputers or mainframe computers issuing the orders and authorizing payments are programmed to act as surrogates for the human users (people from the various departments of the organizations involved in the transactions). In effect, the computers are acting on behalf of the users as input/output devices accessing the network, and hence they perform the role of terminals.

The line that separates an intelligent terminal from a true computer workstation has blurred considerably. In today's information systems world, a terminal is more than just an input/output device; it is more broadly construed as any device configured as a node in a network that acts as an interface between the network and user input/output requirements.

Although currently a large installed base of dumb video display terminals and traditional teleprinter terminals is used for sending hard-copy messages (telex), such terminals have clearly been overtaken by more recent technology and are no longer cost-effective for most applications. The personal computer is the terminal of today and tomorrow, for it possesses ever-greater intelligence and the ability to access and process many types of information. Furthermore, tomorrow's terminal is likely to be even more mobile and capable of handling an even wider range of information types; it is likely to be a powerful workstation operating in wireless mode with access to voice, video, and text information on a worldwide basis. It is also likely that this terminal will be available well before the end of this century, for there are few technological blockages; the main constraints are regulatory restrictions on the use of the radio spectrum, financing, and pricing.

Terminal Selection Criteria

This chapter has thus far examined the various functions of terminals and has briefly described the most commonly used types of terminals. This information can be distilled down to a small number of specific criteria for choosing a particular terminal device. It is the designer's role to translate the user's requirements into a set of functions and features; the implementer then selects the specific equipment, using client preferences to select among trade-offs in cost, features, and performance.

The selection process can be handled effectively only after Steps 1–3 of the telecommunications/business decision sequence have been carried out. Step 1 defines the business opportunity and thus provides criteria for assessing functional needs that must be met; Step 2 identifies the reach, range, and responsiveness of the platform, which sharpens the focus on the functionality that must be provided by the terminal in its wider context of computer operating systems, network linkages, and level of service; and Step 3 identifies key design trade-offs, including cost versus performance and flexibility versus specific functionality.

The selection of terminals should include the following considerations:

- General-purpose versus special-purpose features
- Intelligence and processing capability
- Type and form of output
- User capabilities and limitations
- Compatibility with the reach, range, and responsiveness of the telecommunication services platform
- Cost

Let's consider each of these selection criteria in turn.

General-purpose Versus Special-purpose Features

A key issue in choosing the terminal that will be a user's access device for services and information is whether the terminal will be used primarily to support a particular application, or whether it is expected to provide general-purpose access to the network. If use will be limited to special applications, it is often more cost-effective to select terminals that contain only the features and capabilities needed for these applications, rather than paying for extra, unneeded capability and complexity. For example, how many people use the 14-day, seven-event programming capability of their VCR? Most of us just push in a videocassette and press the Play button, occasionally using Fast Forward, Pause, and Rewind.

The hand-held scanner used by Federal Express employees is a special purpose terminal whose capabilities are limited to the specific functions required: reading the package identifier code, temporarily storing the data, and transmitting the data over the network to the central computers in Memphis. This terminal is easy to use and maintain, and it is cost-effective in meeting the special network access requirements of the user. Similarly, all the function keys on an airline check-in terminal have a specific meaning and function, cutting down the agent's time and effort. (Function keys, which are found on the top row of most personal computers, can be programmed to replace a number of commands needed to carry out a frequently used activity. For instance, the F5 key in MS.WORD, a leading word-processing package, means "go to a page number," and F1 activates the Help facility.) Function keys on airline reservation terminals are hardware-defined, and each major airline uses a different terminal design, with special function keys and display formats matched to the special-purpose software in use. Thus, a Sabre terminal (American Airlines) is very different from a BABS terminal (British Airways), even though they each fulfill the same function.

By contrast, the typical PC is a general-purpose terminal, and function keys, display formats, and the like are defined by each specific application, including word processing, spreadsheets, and electronic mail software. Thus a general-purpose PC may sometimes need to emulate a special-purpose terminal, such as an IBM 3270; this is often the case for banks that are introducing IBM PS/2 machines (which run under the OS/2 operating system) for their back office and branch operations. The PS/2s will include a software or hardware capability to emulate the special-purpose 3270 to access transaction processing systems that may be as much as 10–20 years old! Similarly, Digital Equipment Corporation's (DEC) widely used All-in-One electronic mail software assumes a special-purpose DEC workstation that has Gold and Blue function keys. Using All-in-One on an IBM PC or an Apple Macintosh requires that the DEC keyboard be emulated.

General-purpose terminals are replacing special-purpose terminals because users have general-purpose needs, because more and more software is available for them, and because the wide base of users makes them less expensive.[11]

---

[11] In the 1980s, mainframe computer software cost $100,000 and up—way up—and development costs had to be recovered from a small user base. Whereas early PC software packages cost between $2,000 and $5,000, now $500 is often seen as expensive, and general-purpose operating systems today can run on millions of PCs that use general-purpose hardware. Note that the cost of developing those operating systems is just as expensive as developing either an operating system for many mainframe computers or a special-purpose or general-purpose microchip. This is why innovations in technology are so quickly applied in consumer electronics, and why it is difficult for even an IBM to afford innovations on mainframes.

Thus, for instance, by replacing its special-purpose dumb terminals with IBM PS/2s in its Apollo reservation system, United Airlines automatically got access to spreadsheet, accounting, and word-processing software, which it then added to the reservation terminal to offer travel agencies a complete back-office system.[12] A rule of thumb is that if a terminal is to be used almost exclusively for one type or related set of applications in which speed of processing is critical—ATM cash withdrawals or hotel and airline check-in, for instance—a special-purpose terminal is likely the best option. Otherwise, a general-purpose PC or other type of VDT workstation is likely to offer more value and effectiveness.

Intelligence and Processing Capability

The next criterion for selecting a terminal is how much processing capability/intelligence is needed in the terminal, both now and in the foreseeable future. An inexpensive dumb VDT terminal, such as the IBM 3270 line or equivalent, is more than adequate to enter on screen forms data that will be immediately transmitted to a remote mainframe computer for storage or processing. On the other hand, if the terminal will also be used for some stand-alone processing applications, such as developing budgets and performing financial analysis, then a more intelligent terminal or full-fledged PC is required.

The terminal's required capability/intelligence level is also determined by the particular network/computing model—distributed processing, centralized computing, or hybrid—used by the organization. The inexpensive and reliable dumb VDT is ideal for the centralized computing model in which the relationship between terminal and remote computer is slave to host. However, a terminal with stand-alone processing capability, such as a PC workstation, is needed in the distributed processing model. In this environment, the relationship among terminals/computers in the network is peer to peer, and data processing takes place at a number of different network locations.

Terminal intelligence is cheap and gets cheaper by the year. Each new generation of computer chip offers a 5–40-fold increase in price-performance.[13] Although the main uses of this cost-effective innovation have been to upgrade stand-alone personal computers, providing faster speeds, more memory, and more powerful (but also memory-hungry) operating systems, it is also fueling distributed processing by making it easier and easier to dedicate low-cost hardware to network functions, to spread intelligence through the network, and to provide first-rate, easy-to-use displays and graphical user interfaces. It is now cheaper to manufacture a PC than to make a dumb terminal, except for very-special-purpose applications, because of economies of scale in production and marketing.

---

[12] These systems provide travel agents facilities for billing customers and management, accounting, and even marketing information. The reservation software is largely mainframe-based, but these add-on systems reside in the workstation.

[13] Intel's Pentium chip, introduced in mid-1993, has four times the power of its 486 chip, which dominated the PCs of early 1993. When computer intelligence is cheap, this is the way to provide it in terminals and in network equipment. Bridges and routers are really PC chips that run network functions instead of word processing and spreadsheet packages.

Type and Form of Output

The required type and form of the terminal output places constraints on terminal features and capabilities. For example, a teleprinter form of terminal is essential if printed hard copy is required. A video display output is sufficient if the application requires only real-time interaction, without the need for hard-copy records. In other applications, magnetic tape or disk may be the preferred form of output, and this requires a terminal device with the requisite features.

All this extra capability adds cost. A $1,000 PC quickly becomes a small data processing center costing up to $5,000 when a printer, auxiliary disk, and disk backup device are added. The need for high-quality printing has escalated the costs. A few years ago, a dot matrix printer provided acceptable output; now, laser printers offer better resolution and speed, plus a wide variety of print fonts, but at an extra (though still low) cost. Dot matrix printers generate characters as tiny patterns of dots by using dot hammers that move across the page. If you magnify a black-and-white newspaper photograph, you will see dots; the more dots per square inch, the better the quality of the newspaper photo. Laser printers paint high-resolution dots instead of using dot hammers.

User Capabilities and Limitations

It makes no sense to install powerful PC workstation terminals in an organizational environment in which users are only capable of simple fill-in-the-blank tasks. The type of terminal selected must fit the capabilities, style, and culture of the work force that will operate them. However, if the use required of the terminal exceeds the capabilities of the users, then a significant training and education effort is required to bring their knowledge and comfort level up to the level needed to accomplish the required task. Many organizations have simply ignored this issue, installed a sophisticated information technology system and network, and found that the system was not used effectively (or at all), and their investment was wasted. Training, education, and user support are essential components of the effectiveness of a telecommunications network platform.

There are few well-established principles for developing organizational support mechanisms. Many companies use help desks, which are not actually desks but phone numbers used to call a technical specialist who can diagnose most problems over the phone. The advantage of help desks is that the company's central information services organization can cover a large user base with just a small staff, and it can also concentrate its expertise instead of having to build knowledge of individual hardware, software, and telecommunications in many locations.

This support mechanism is generally effective in handling PC software and hardware problems. In more and more cases, however, on-the-spot support is needed to handle departmental uses of complex workstations and the link between terminals and local area networks. LAN administration is a new and rapidly growing role and career path within the telecommunications profession that requires a mix of solid technical skills and a strong focus on service.

The simplest type of terminal to learn to use is a special-purpose device, which not only has a limited range of functions—and things that can go wrong—but is also designed to make the operator as fast and efficient as possible. Learning to use standard personal computers requires learning enough about the operating system to be able to handle routine utilities like backing up disks, installing new software, and managing files. In addition, users must learn the details of the software applications they use. All this can be complex, even for simpler operating systems like DOS. Even with graphical user interfaces like Windows and Apple's System 7, it can be very difficult for even an experienced user to develop in-depth expertise and the ability to diagnose and resolve problems.[14] Help desks are a key support resource here. High-end workstations are even more difficult to learn because they generally rely on local area networks and shared printers and disks. Even when users are knowledgeable about a software application, they are almost certain to be ignorant concerning the telecommunications requirements and impacts of the application; this makes a departmental LAN and PC support specialist a vital resource.

Clearly, choosing a terminal is not a simple matter of "What software do you want to run?"

Compatibility with the Reach, Range, and Responsiveness of the Network Platform

An existing telecommunications network or any planned network architecture has an explicit or implicit degree of reach, range, and responsiveness. For example, the network may be designed to serve six particular points in the continental United States (reach); process electronic mail messages, handle data transfer applications from different types of computers, and carry voice traffic among the six locations (range); and also operate with a terminal response time of two seconds or less for data queries or electronic message traffic (responsiveness). In order to satisfy these business functionality requirements, the terminal chosen should have characteristics and performance levels consistent with reach, range, and responsiveness measures for the overall network design. In some cases, that network design will require that the terminal itself contain the intelligence required for protocol conversion and data structure transformation (range), and the transmission speed (responsiveness) necessary to achieve the desired range and responsiveness levels for the network. In this instance, the terminal is the key component responsible for attaining the desired performance levels for two of the network's three business function dimensions— range and responsiveness; reach is determined by the transmission links.

In other cases, the range capability of the network is built into or limited by the network's other building blocks (transmission links, switching devices, and transmission methods), and the responsiveness bottleneck is caused by the limited capacity of the transmission link. In this situation, the choice of terminal is not critical in achieving the range and responsiveness requirements of the network; consequently, these characteristics are not relevant in selecting the terminal.

Note that a client cannot be expected to understand the complex decisions and technical choices involved in ensuring that needed services are provided

---

[14]The more "user friendly" a system is, the less users need to know about it and thus, of course, the less they know when something goes wrong. The aim is to hide the system's complexity from users.

reliably, conveniently, and cheaply. Equally, an implementer cannot be expected to guess at the variety and scope of business uses and applications that a client needs now and will need in the future. Each of these parties looks at telecommunications from a different and sometimes conflicting perspective. For clients and the users to whom they direct the business service, the terminal is the "system." They may take strong personal positions on MS.DOS versus Apple Macintosh machines, or on Windows versus UNIX, for these are personal tools, and they may have a personal commitment to them.

Implementers look at what lies behind the terminal, which may affect their recommendations or even requirements. The network is not "personal," and quite often an implementer will place departmental or corporate needs ahead of individual ones. This can create conflict. It is an unfortunate fact that the PC manufacturers most committed to making PCs easy to use do not in general place the same emphasis on making them easy to connect to services beyond those on a local area network. Thus, for example, when *PC Magazine* (July 1992) held a "shootout" in which a panel of computer users and information systems professionals tested a range of portable computers, the clear winner was the Apple Powerbook. One of the finest pieces of industrial design in the computer field, it is easy to use, powerful, and perhaps the simplest machine to learn.

Several of the judges who rated the Powerbook as the winner added that it could not be used in their organization and that they would not recommend that it be adopted. Apple's telecommunications are not yet in the mainstream of corporate platform standards; this is the main reason why the clumsier MS.DOS operating system is a required or de facto standard in so many large firms.

Here we have the worst possible situation for dialog. Recall the case illustration in Chapter 3 that addressed the sales department's choice of a personal computer. Suppose that Ellen Jackson had already purchased a Powerbook and enthusiastically argues that this is the machine for the department, and that Joanna Feldman, the telecommunications implementer, then tells her "client" that this excellent machine is unacceptable: "Our corporate standard is DOS and Windows. The Apple is incompatible with our database management systems, and if we install Appletalk as the sales LAN, the department will be an island of automation, isolated from the rest of the company." Ellen's likely—and reasonable—response might be, "You're telling me that Corporate Telecoms won't accommodate our needs, even when we've chosen the best available laptop and one with which Apple had a billion dollars in sales in its first year. This is service?"

The case illustration in Chapter 3 was, of necessity, a simplification to show the telecommunications decision sequence in action. By now you should be able to see the value of the frameworks that make up the decision sequence, especially the services platform map, in bridging Ellen's and Joanna's perspective. Rich Piccardo, the designer, is the key translator here. Ellen's position on the Powerbook may be modified if she understands the sales department's business needs for reach and range, and Joanna may be better able to explain that she is not trying to impose rigid technical standards for standards' sake. The sales team may be able to explore options, including the Powerbook, without getting hung up about DOS versus Apple.

Cost

The final selection criterion for a terminal is cost, which is an important consideration in any selection decision. However, cost must be weighed against the other factors just described. It is important to understand each of a client's preferences among the many selection factors, in as much detail as possible, in order to establish appropriate trade-offs for deciding on a specific terminal. It is also important to consider full life-cycle costs (that is, purchase price, maintenance, and depreciation based on both the physical and technical life of the equipment), and not to focus only on the purchase price of the terminal. In addition, the most expensive cost of all—support—must be included right at the start. The Gartner Group estimates that it costs between $8,000 and $15,000 to support a $5,000 PC in a networked environment.[15] Part Three of this book addresses the issue of telecommunications costs in greater detail. The main point here is that the cost of a terminal is a factor to be traded-off against other costs, such as transmission links, support, DCE, and other factors such as flexibility, reliability, and so on.

| | | | | | |

Terminal Interfaces to Networks

The terminal/workstation is the most visible of the network building blocks because it serves as the input/output device that links the user to the network. It is similar to the steering wheel and gas and brake pedals in an automobile. The interface between the terminal and the network is not a trivial issue; there must be a high degree of cooperation between the terminal device and the network if a reliable communications channel is to be established.

First, the physical connection between terminal and network must be specified through a series of electrical parameters, and the connection usually entails more than one wire, each of which has a specific function. Second, the coding of information for transmission and data flow rules and procedures must be specified. Finally, the timing of the information flow between the terminal and the network equipment must be consistent and coordinated. If these matters are not managed properly, the information that is sent will not be the information that is received.

Next we consider three aspects of a terminal's interface to the network.

---

[15] This estimate has been confirmed by many studies of the real costs of personal computers and LANs. A useful rule of thumb is that the price of any information technology hardware or software is only 20 percent of the total cost; support, education, maintenance, and security make up most of the remaining 80 percent.

Telephone Interfaces   The telephone set (terminal) is physically connected to a wall outlet with a pair of wires attached to a standard **RJ-11 jack connector.** The wall outlet is in turn connected to a copper twisted pair cable that runs to the telephone company's local central office, where it feeds into the telephone company's network.

The telephone terminal has a relatively simple set of coding and data-flow rules and procedures. First, the telephone set codes information by converting sound waves into electrical analog signals that represent the voice sound pattern. These analog signals are sent through the network, and a telephone terminal at the destination point decodes and converts the signal back to a reproduction of the original sound waves. However, before voice information is coded and sent over the network, a call setup procedure must be invoked to ensure that the message is sent to the correct receiving terminal and that there is someone or something (a telephone answering device) present to receive the call.

When the telephone receiver is lifted from its cradle, an electric current is sent over the line from the telephone central office to the destination location to signal dial tone (see Figure 4-9); this signifies that the line is ready for a call to be placed. The network address of the receiving party is signaled through a series of tones or pulses activated in the dialing process. The telephone central office uses the network address (telephone number of the receiving party) to establish a circuit path through the network between the two telephone terminals. Once the circuit path is established, the central office sends a ringing signal to the receiving party, indicating that there is an incoming call. When the telephone receiver is picked up, the ringing signal stops, and the conversation can commence. When the call is completed and the parties hang up, the transmission link facilities dedicated to the call are released and made available for other calls.

Clearly, the rules and procedures for the interface between the telephone terminal and the network are predetermined and contain logical steps necessary to ensure a successful transfer of voice information between two parties. The steps are executed serially in sequence: produce ready signal (dial tone); dial destination number; set up circuit path; ring; answer and complete connection; exchange information; hang up; disconnect circuit path. The analog public telephone system has been standardized over the years so that even though there is a wide range of telephone types available from numerous vendors, they all use the same coding and signaling methods, and all have a common physical interface. However, with the movement toward digital transmission and switching equipment, digital telephone terminals are now becoming more prevalent in the office environment. Thus a digital coding-scheme standard has been adopted for terminals that converts the voice sound waves into a digital bit stream that is then sent over the telephone network.

Data Communications Interfaces   The interface between data communications terminals and the network is more complicated because data communications has not achieved the same degree of standardization as the public telephone network and because several different methods are used to transmit data across a network. In general, data communi-

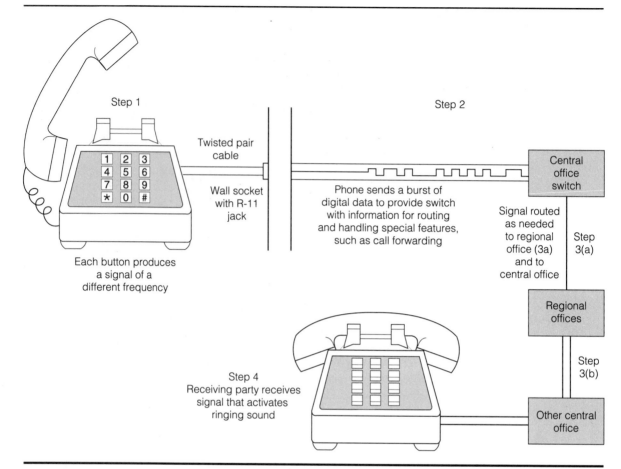

Figure 4–9

Linking to the Phone Company

cations terminals, designated as data termination equipment (DTE), interface with the network through an intermediary device known as data communications equipment (DCE). Examples of DCEs include modems, channel service unit/data service units (CSU/DSUs), and packet assembler/deassemblers (PADs). A DCE intermediary device is needed because data terminals have limited transmission capability and generate a simple digital signal that must be modified both for transmission over long distance and to accommodate different types of networks (for example, an analog telephone network versus digital private data networks).

The typical direct link between a terminal and a nearby computer, or between a computer and local printer, is a cable connecting the two devices. Information is sent between them using **parallel transmission**, which means that all bits in a given grouping (a byte) are sent simultaneously, each over its own separate wire. The receiving end (computer or printer) processes the byte intact. A typical byte sent using parallel transmission will consist of eight, 16, or 32 bits. The advantage of parallel transmission is speed: eight, 16, or 32 bits are sent in the same time it would take to transmit one bit using **serial transmission,** which sends one bit at a time. The disadvantages associated with parallel transmission methods are limited distance capability and difficulty in control and synchronization of the transmission. Most parallel transmission links are limited to 15 feet or less.

The typical link between DTE and DCE, acting as the interface to the network, is a cable with 25 separate wires, called interchange circuits, and a 25-pin connector at each end. The data bits are sent using the serial transmission method; each bit in a byte is sent one at a time in sequence (like the old Morse Code telegraph signal). The most commonly used serial interface standard of this type is called RS-232-C (see Figure 2-21). Each of the 25 wires (interchange circuits) connected to a separate pin in the connector has a specific function and a standard electrical specification (voltage range) for transmission of either data or control information. The types of functions assigned to the various interchange circuits and pins include data signals, control signals, timing signals, and electrical ground. For example, there is a data terminal ready pin, a request to send pin, a transmit data pin, a receive data pin, and so on.

The RS-232-C standard is widely used, but it is limited to a transmission rate of 20 kbps and a distance of 15 meters. Other standards have been developed to get around these limitations and to provide a few additional features not available in RS-232-C. These are the RS-400 family of standards: RS-449/422/423.

The most common DCE providing an interface between a data terminal and a network is the **modem,** a device that converts digital signals to analog signals and vice versa. A modem detects a binary 0 or 1 signal sent from the terminal and modulates a self-generated analog carrier wave such that it represents the pattern of ones and zeros present in the original signal. The analog carrier wave is transmitted through the network and received by a companion modem at the other end of the link. The destination modem demodulates the carrier and retrieves the binary zeros and ones, which it then forwards to the destination terminal in the form of a serial bit stream.

Modems are used extensively for transmitting data over the analog public telephone network, and they are generally classified by speed (number of bits per second) and type of transmission method (asynchronous/synchronous and full/half duplex). Typical modem speeds range from 300 to 9,600 bps on normal dial-up connections; higher speeds can be attained using specially conditioned or high-capacity leased lines. Modems using asynchronous transmission methods incorporate start and stop control bits into each block of data sent to inform the receiving terminal when a given character begins and ends. Figure 4-10 lists key modem standards and developments.

Synchronous transmission methods require special clock circuitry that synchronizes the transmitting and receiving modems so that both know that blocks of data will be sent during certain time slots and that control information will be

Figure 4–10	Standard	Description
Modem Standards and Developments	V.32	CCITT standard; full duplex transmission at 9.6 kbps over dial-up lines
	V.32bis	Improves transmission speeds to 14.4 kbps
	V.fast	Maximum throughput on analog lines; speeds expected to reach 24.0 kbps
	V.42	Standard for error correction in modems
	V.42bis	CCITT standard for data compression; approximate fourfold improvement of any modem

sent in other time slots. Modems using synchronous methods are more complex and costly, but they are much more efficient for sending large blocks of data because start and stop timing bits are not required to define each character separately. Modems using full duplex transmission methods can transmit and receive simultaneously using two different frequencies. Half-duplex modems must perform an additional line turnaround function: shifting from transmit to receive mode and vice versa. This reduces the maximum transmission speed available.

In networks that exclusively use digital transmission links and switching equipment, modems are not needed for digital to analog conversion. However, there is still a requirement for a device to act as an interface between the terminal and the network. In end-to-end digital networks, this device is generally a combined **channel service unit/data service unit (CSU/DSU)**, a device that performs synchronization and clocking functions, digital signal regeneration and reshaping, line conditioning, and some testing activities.

Another important type of terminal-to-network interface device used in special packet-switching types of networks is a **packet assembler/deassembler (PAD)**—a device that converts a message into packets (assembler) for transmission and reassembles the message (deassembler). In a packet-switched network, the PAD frames data and destination address information into blocks called packets. These packets are sent through the network from node to node; the destination address of the packet is read as it arrives at each node, and the packet is then forwarded on a path through the network toward its final destination. Upon arrival at a PAD at the distant end, the destination address and error detection bits are stripped from the packet, and the data is passed in serial form to the host computer or other destination terminal.

PADs can be owned and maintained by the network provider of public packet switching services and can be accessed through dial-up or dedicated leased lines. Alternatively, the user can own the PAD in order to save on access line costs if a large number of terminals are being used.

Virtual Terminal Service

A somewhat different type of terminal interface to the network is a **virtual terminal service**, a software package that is an abstract representation of a real terminal, including all of its basic functions and characteristics such as display mode, character code, data rate, and so on. The virtual terminal service interface is designed to solve the problem created when there are many different types and makes of terminals remotely accessing multiple host computers made by different vendors. Each type of host computer generally supports only a small number of terminal types, each type with its own specific display format and other characteristics. In order for a given type of terminal to gain remote access across a network to a host computer manufactured by a different vendor, a special software package is needed on the host to communicate with the distant terminal.

For example, suppose there are four different terminal types and four different host computer types, and each type of terminal must communicate with all four types of host computer. Thus each host computer must have four separate software packages to support each of the terminal types, which means that 16 separate software packages are required to support the full range of terminal-to-host combinations. With the virtual terminal system, each host computer in the network is equipped with a single software package that emulates the functions of a generic terminal. First, the virtual terminal service software translates the terminal parameters and screen format information of the actual terminal accessing the host into the parameters and format of a generic virtual terminal. Then, a second translation takes place: Format and terminal parameter information are converted into a form compatible with a terminal type supported by that particular host computer. Therefore, instead of 16 separate software packages, the virtual terminal service requires only four separate software packages—a single virtual package for each host computer in the network.

Finally, in an environment in which remote terminals access one or more host computers, the cluster control unit (CCU) and front-end processor (FEP) manage groups of terminals and act as their interface with the network. A **network interface unit (NIU)** for a local area network is really a form of FEP. It is a microprocessor device that performs the communications functions necessary to link a PC workstation to the network. It accepts and buffers data from the terminal (PC), addresses and transmits packets, scans the network for packets with a matching address and reads them into the terminals, and transmits data from the terminal to the network at the proper data rate of the LAN. Like the CCU, the FEP is an efficient way of handling network interface functions by placing them all together in a stand-alone computer or in a separate internal or external circuit-board card connected to a PC workstation in a LAN configuration. By consolidating these network access functions into a specialized device that is optimally designed to perform these functions, it frees up the mainframe computer or PC workstation to run applications and handle other data processing functions.

Summary

Terminals are the most visible element in a network, from the users' perspective. The more technical term for a terminal is data termination equipment (DTE). Terminals include a wide range of devices, including telephones, dumb terminals that have no processing intelligence, personal computers, intelligent workstations, and minicomputers and mainframe computers. Terminals vary in characteristics, including their mode of input (keyboard, mouse, scanning, or spoken voice), presentation (printed or screen display), level of intelligence, data rate/speed, communications features (asynchronous or synchronous transmission), how they are controlled (by cluster control units and front-end processors), configuration (multidrop or point-to-point), and their degree of mobility. Wireless technology is opening up many new opportunities for fully mobile communication.

The main types of terminal in use today are telephones and video display terminals, which are generally personal computers with communications capabilities. In addition, optical scanners (used in point-of-sale applications) capture barcoded data, and networked teleprinters produce hard-copy output. Minicomputers and mainframe computers are terminals that share communications and transactions with workstations in a distributed processing environment.

The main criteria for selecting terminals include the need for general-purpose rather than special-purpose features; the standard personal computer is increasingly the general-purpose choice. Large-scale transaction processing environments, such as ATMs and airline reservations networks, frequently use special-purpose terminals to speed up operations and simplify training and use. Other criteria are the terminal's level of intelligence (a central need in distributed client/server applications), the type of output required, cost, and the capabilities needed by users that will operate the terminal. In addition, the terminal selected must be compatible with the requirements for reach, range, and responsiveness in the telecommunications services platform.

Terminals interface to a network through a variety of devices, including modems, cables, adapters, packet assemblers/deassemblers (PAD), and channel service units/data service units (CSU/DSU). In addition, software interfaces can be used to create a virtual terminal service.

Review Questions

1. Name five different types of terminal devices and briefly describe how they are used.

2. Name three different types of video display terminal devices, and briefly describe the types of applications appropriate for each.

3. Why would anyone ever use a dumb terminal when intelligent ones are available?

4. What is the relationship between terminal intelligence and distributed processing or distributed computing?

5. Under what conditions does the data rate at which a terminal operates determine network response time to the user?

6. Are all terminals equipped to operate in both synchronous and asynchronous mode, and which mode yields a faster response rate? Explain your answer.

7. What role does a front-end processor (FEP) perform in a network?

8. A multidrop terminal configuration is economical because the same transmission link is shared by several terminals; therefore, why not use the multidrop arrangement for all terminal connections to a network?

9. There is a strong trend toward increased use of mobile and wireless terminal devices. What are the advantages and disadvantages of these types of terminals, to the user and to the network manager?

10. Briefly describe how a fax machine works and how it differs from an optical character reader.

11. Explain the advantages of using hand-held scanning devices, such as those used by Federal Express, in package delivery operations.

12. What is a teleprinter? Describe some of its uses.

13. What are the criteria for choosing a terminal? Briefly explain the trade-offs among them.

14. List the major steps required to set up a voice telephone call through the public network.

15. Why is the interface between a data communications terminal and a network more complicated than the interface for voice?

16. Name three types of data circuit-terminating equipment and the conditions under which they are used.

17. Briefly explain what a modem is and how it operates.

18. What is a virtual terminal, and why use it instead of the real thing?

19. How do terminals interface with a local area network, and how does this interface differ from access methods used to connect groups of remote terminals to a wide area network?

Assignments and Questions for Discussion

1. The head of the marketing department remarked that his people don't need anything more than a simple, dumb terminal to carry out their work; anything more would be a waste. You have been asked by the director of telecommunications to outline the appropriate factors and trade-offs to consider in making such a decision. Produce such an outline.

2. Some people are contending that the increased level of intelligence built into terminals is changing the way information systems and networks are constructed and used. Evaluate and discuss this contention.

3. Federal Express undoubtedly spent a considerable sum of money to design and produce the specialized hand-held terminal it uses to scan packages and send information to the headquarter's computer database that is used to track packages. Wouldn't it have been more cost-effective simply to purchase and issue off-the-shelf laptop computers with a modem and portable scanner? Then, employees would have the additional benefits of a full-service computer for word processing, spreadsheets, and electronic mail applications. Evaluate and discuss.

4. What are some of the important issues for telecommunications managers related to the rapid proliferation of mobile and wireless terminals for both data and voice applications?

Minicase 4-1

The Internal Revenue Service

Filing Tax Returns by Touch-tone Phone

The IRS operates one of the most complex information factories in the world. Around March 1 of each year, millions of pieces of paper flow into its regional centers from taxpayers expecting fast processing and fast payment of refunds. On April 15, procrastinators race to meet the deadline for filing. Phone lines are jammed as people call in to get answers to their questions. Changes in tax laws mean changes in extraordinarily complex software systems. Checks that are attached to 1040s must be processed. Congress has reduced the IRS's funding at a time when its workloads have increased, and citizens' dislike of the Service has increased while compliance has decreased.

The IRS has a bad image. People see the IRS as the worst of bureaucracies and a threat to civil liberties, dread the thought of an audit, and worry that this or that might not be tax-deductible. The IRS does care, though it may not show. It understands taxpayers' frustrations and has continually looked for ways of improving its services, mainly by seeking ways of removing paper as the medium of communication with taxpayers. It introduced electronic filing of tax returns in 1990, thereby offering faster refunds. In 1992 it created a new form of citizen service—customer service for the public sector—through touch-tone filing of tax returns with voice response. This two-year trial in Ohio, called TeleFile, was aimed at the more than one million citizens who file the simplest available form, 1040-EZ. They often need advice but cannot afford tax advisers.

A database server in the Cincinnati office of the IRS holds the citizens' social security numbers, addresses, and passwords for using TeleFile. Citizens access the system by calling an 800 number and using their touch-tone phone to answer voice-response questions and commands. The system computes how much the taxpayer owes or is owed and then lets the caller file a tax return electronically. The target is to halve the time it takes to receive a refund. Once the data is entered it is available in digital form, and IRS mainframe computers can immediately begin processing, with no data entry required.

Each computer consists of an audio-response processor and a database server. They contain two Intel 30386 chips, which are widely used in mid-range personal computers. Two machines are used to ensure backup; if one computer fails, the other is not affected. They are linked by an Ethernet local area network that uses a special-purpose LAN peer-to-peer operating system. The servers link to a parallel processing computer that is made by Sequent Computer Systems and runs under the UNIX operating system.

The computers are connected to four T1 circuits provided by Sprint as part of the Federal Telecommunications System 2000 network.

Questions for Discussion

1. Identify the various types of terminals/workstations used in the IRS's TeleFile system, and briefly describe the role played by each in the IRS network.

2. Would you anticipate any taxpayer resistance to the TeleFile concept specifically because it is *not* based on submitting a paper document? Explain.

3. How could the TeleFile system be extended to accept more complex filings of regular 1040 forms and accompanying schedules? Discuss the strengths and weaknesses of such an extension.

Minicase 4-2

Info/California

Citizen Service at a Kiosk

In the 1980s there were many efforts to use electronic kiosks to market goods and provide information. Such kiosks were fairly commonplace in hotel lobbies, for instance, where you could find out about restaurants or local attractions, and at airports, where you could while away your time by looking at offers for insurance. You entered your name and address and a representative would contact you later, or you could order flowers, jewelry, and gifts to be delivered to family or friends.

These systems were slow and clumsy, the graphics were poor, and they were stand-alone systems; most of them lacked telecommunications links, so they could not make transactions. In 1985 there were an estimated 15,000 machines in use, but only 1,000 handled transactions. They generated around $50 million of sales—around $4,000 a machine, far less than their costs. In addition, usage fell off very quickly after the initial fad had passed. The State of Hawaii, for instance, installed kiosks in 1990 to provide information on government services. For the first six months, an average of 33 people used them per day; usage dropped to 25 a day after that, and the project was put on hold.

Info/California may turn out to be another disappointment, but there are many signs that this new-generation, multimedia kiosk is a breakthrough in citizen service. It is designed to put the state's bureaucracy at your fingertips. The pilot system, launched in 1991, provides answers to the questions most frequently asked of state agencies. It has over 90 packages, including getting job listings, applying for a fishing license, or ordering a copy of your birth certificate. A new California driver's license with a magnetic stripe, like that on a credit card, provides the security for authorizing transactions, but anyone can get the free information. After the 1992 Los Angeles riots, one kiosk helped citizens find community-assistance programs.

The system is explicitly intended to cut through red tape and end the frustration of trying to find out which agency handles assistance for your problem, or how to apply for a particular type of aid program, or how to get the application forms for admission to a state college, and so on. The director of the state's Health and Welfare Data Center summarizes its aims: "Right now if you are a citizen you have to worry about 'Do I want to talk to city government or county government or state government, and is it EDD or the building department or motor vehicles, and is it the building on Main or the building on Oak?' With Info/California, we're talking to people like student aid, Fish and Game, licensing for health professionals, hairdressers—we're talking literally to every program in state and local government. . . . State workers spend a large amount of their time doing repetitive things, so that when you want to talk to a live body, you're busy stamping forms. Info/California should help government staff take the appropriate role, which is dealing with people."

The kiosks are built on new multimedia technology that provides full-motion video. *Business Week* reports on someone who while browsing in San Diego's Parkway Plaza Mall spots the seven-foot structure wedged between a T-shirt cart and a lottery ticket booth: " 'Touch my screen,' implores the video monitor. . . . Intrigued, the bearded 34-year-old places his index finger on the glass. Suddenly, Governor Pete Wilson appears and talks about the many services Info/California offers. . . . Webb selects data on area beaches. 'Cool,' he says as he tears off a printout."

Info/California is meant to be a little more than cool. Even before the 1992 earthquakes and the riot that followed the Rodney King trial, California faced a massive and escalating budget deficit. Agency hiring had been frozen, and all units of government were strapped for cash. The kiosks reduce the workload on an already overstretched system and add a new level of service. They provide a new revenue base from transaction fees while cutting transaction costs, and they also help people strapped for time get answers to queries faster than by phone. The multimedia features make it practical to get away from the word-based menu of options found on an ATM. A TV-quality image of a person can appear on the screen and guide you through your options. There is no need to type, only to touch boxes on the screen.

The multimedia feature addresses a problem found in an earlier California pilot system to renew driver's licenses at a bank ATM: Close to a third of the population could not follow the menu, either because of language problems or functional illiteracy. With Info/California, the dialogue is spoken and can be in a variety of languages.

Many private and public-sector organizations are piloting kiosks, most of which still lack transaction capability. Stop and Shop, the New England supermarket chain, has developed a deli-counter kiosk for workers to check products and prices. They can memorize only a limited amount of information, but the workstation can handle masses of items. The designers of the system see information-based kiosks as radically reducing the need for $6–8 per hour staff.

Kiosks are now widely used; *Business Week* reports that about 60,000 have been installed. Tire-kickers can get details on Cadillacs at auto shows and in some dealers' showrooms. At Expo '92 in Seville, Spain, multimedia kiosks provided directions and accepted reservations. Multimedia technology is also being added to existing bank ATM locations, although because well over 80 percent of them handle straightforward cash withdrawals, there seems as yet little reason to add extra investment.

In the first eight months of the 15 Info/California kiosks, usage was 50 percent higher than forecast, with an average of 50 people using each kiosk each day. Typically, each person spends five minutes at the kiosk.

Questions for Discussion

1. Where does the Info/California information kiosk fall on the spectrum of dumb terminal to intelligent workstation? Explain.

2. Describe the kiosk's benefits to users and to the State of California. Is it a win/win situation for both sides, or is it merely a clever gimmick that will fade with time? Explain.

3. What are the implications of using the new multimedia information format in the kiosks?

4. Discuss the implications of connecting the kiosks to a network that would enable communication among the kiosks, and between the kiosks and the state's mainframe computers. Would you recommend networking the kiosks? Why or why not?

Minicase 4-3

Health Link

Point of Sale in the Doctor's Office

Insurance claims administration costs over $8 billion a year in the United States. One solution to this problem is Health Link, an electronic eligibility and claims-processing system that has cut costs for its users by about half and delays on getting payments to doctors by two-thirds. Health Link is a form of value-added network provided by Health Information Technologies of New Jersey. It is used by providers and clients of such major insurers as Prudential, Travelers, Aetna, and several Blue Cross/Blue Shield health-management organizations. Health Link is very much like a credit-card authorization and payment system; doctors can use the Health Link member's magnetic card to check member eligibility and then submit the claim for service electronically. If a patient's health plan does not cover the office visit, Health Link automatically dials the insurer's customer service unit for immediate help.

About two-thirds of the insurance claims submitted to Aetna are "clean" and require no human intervention; Health Link replaces human intermediaries in such instances. A fault-tolerant computer with backup ensures fast response time and close to perfect availability for the 75,000 transactions a month. Doctors' offices and insurers had identified this as a priority and as a reason that other electronic claims-submission systems had failed to reach critical mass.

Insurers also needed a third-party network and

processing system because their own software could not easily be adapted to handle remote submission of transactions. Health Link is simple to access through an 800 number for national access, and local access occurs via BT Tymnet's packet-switched X.25 value-added network. Special-purpose terminals provide customizable software that asks doctors the questions needed to complete the submission of a claim. Having a single terminal that can access a large number of major insurers simplifies their offices' operations.

Health Information Technologies was founded in 1985 with the intention of exploiting "smart cards," credit-card-sized devices with a computer chip built into them. They are in effect the smallest truly personal computer with storage and processing power. Many commentators believed that smart cards were a key technology for health care because they could store patient data, update records, and process payments. The cards cost $8 to $15, and although there have been isolated successes, they have not taken off as expected. Health Information Technologies refocused its efforts on using the well-established magnetic cards, which require access to a central computer that stores the needed data for authorization and payment. The company expected to be mainly a regional processor, as are most credit-card processors. Once major insurers like Aetna signed contracts to deploy the point-of-sale terminals, Health Link became a national player.

Questions for Discussion

1. Where does the primary intelligence for the network reside in the Health Link system: the terminal, the network itself, or the central computers processing the claims? Explain.

2. Explain how the Health Link system is able to reduce costs for its users by half and reduce payment delays to doctors by two-thirds.

3. Why didn't the large insurance companies use their own claims-processing systems instead of paying to use a third-party system like Health Link?

4. Describe a scenario in which it would make sense to transform Health Link from a system accessed by magnetic-stripe cards to one that uses true "smart cards." What implications would this have for the network architecture and for the costs and benefits of the system?

Minicase 4-4

The National Cargo Bureau

PCs at Sea

Personal computers are not "high tech" any more than using telephones or airplanes is high tech. The technology of all three is complex and always changing, but the ability to use any of them rests more on common sense, creativity, and specific need than on technology per se.

The opportunity to make creative use of personal computers and phone lines is illustrated by an individual who spearheaded innovation in a firm that in 1987 had just one PC in its New York head office. It was not just the only PC; it was the only computer.

The National Cargo Bureau is a private nonprofit marine survey with around 100 employees in 17 locations. It surveys cargo traffic unloading and loading and provides reports and calculations shipowners use to charge cargo owners for shipment. Bulk cargo, such as iron ore or wheat, is difficult to weigh directly, so the Bureau's 60 surveyors provide essential data. They look at the weight of the ship before and after loading, how deep it is in the water, and how much fuel, water, and ballast it has used. This has traditionally been done by using hand-held calculators; secretaries typed the figures into standardized survey reports that were sent to New York. An outside contractor was paid $30,000 a year to examine customer invoices to work out how much the Bureau's customers were owed. The reports were three months or more behind events.

The Bureau's surveyors are former ship captains who have no computer knowledge or experience. Sam Sammons, who spearheaded its move to transform its work processes through PCs, had bought a home computer in 1985. In 1987, he showed the Bureau's senior managers how this very limited machine could input data and generate reports. They were impressed and gave him the go-ahead to introduce personal computers in four offices. He bought four standard machines and added software packages for word processing, spreadsheets, and simple database management. Modems and an off-the-shelf communications program made it easy to send activity reports to the head office over the dial-up phone system.

Sammons's main concerns were human, not technical. As *Computer Buying World* commented in April 1992, "initially, employees welcomed the equipment as if it were a stack of pink slips." Computers represented bureaucracy, potential layoffs, fears of looking stupid, and simply the unknown. They could easily disrupt a small, effective organization whose surveyors, retired ship captains, understood ships, cargo, and the sea but not the world of "management" and computers.

Sammons relied on a mixture of show and tell, gentle persuasion, and positive reinforcement. He was an insider and did not dismiss wary resistance as technophobia. He sent computer games to the offices where the personal computers were installed, hoping to tempt people to try them out. He focused on how the PCs would save time on secretarial chores, eliminating work that they all saw as drudgery, such as constantly typing the same report with only minor variations in the numbers. The productivity benefits soon became very visible: Up to several hours were saved while on board ship, survey reports were produced in one-third the time, and managers had access to financial reports in minutes instead of months.

In 1989, the pilot PC system was extended to all offices. Sammons's growing concern was lack of support from the retailers. The Bureau did not have any in-house technical staff, nor would it make sense to have them in a company of 100 employees scattered over 17 offices across the country. In addition, the Bureau's purchases were far too small to get volume discounts or for vendors to assume a dedicated sales rep. The retailers lacked in-depth technical knowledge, and their technical support staff always seemed busy.

Sammons contacted several direct companies, which are basically mail-order suppliers except that they compete on support as much as on price. Sammons chose Zeos almost entirely because of that company's technical knowledge and attitude: "If they have a problem, they work with you until it's solved." He kept ahead of the Bureau's PC users by trying out new software, which he then introduced to the 17 offices. For instance, he bought a database management system that enabled him to develop a vessel-tracking system, something badly lacking in the organization.

By mid-1992, the PCs were fully institutionalized. Sammons had added a heavy-duty 486 PC for the head office, where there was a need for number-crunching and generation of many financial reports, surveys, and analyses. The offices got laser jet printers; special software provides a high-resolution print that has enhanced the quality of the Bureau's published surveys. All the PCs have modems, and Sammons plans to add a LAN in the New York office, which will link the PCs and printers used there, provide a file server, and enable other offices to upload and download data to the server. Some surveyors are using laptops and notebook machines; a few are trying out Hewlett-Packard palmtop machines that are calculator size with small screens and keyboards but that can run Lotus 1-2-3 and have a modem included.

Sammons's strategy has rested entirely on using direct sales channels, commenting that "If you've done your research, buying direct is easy and economical, and you get good quality and good support."

Questions for Discussion

1. Discuss some of the coordination and administrative issues associated with managing the National Cargo Bureau's information system with PCs distributed over 17 locations and connected on a dial-up basis using modems.

2. What are the technical and administrative support implications of installing a LAN in the New York office: Does this move make sense for the company at this time?

3. Discuss the impact on the LAN-based network system of each of the following:
   a. increased use of portable terminals, such as laptop and notebook computers
   b. the addition of wireless data terminals to the mix of workstations accessing the network

5

Transmission Links

Objectives

- Understand the role of transmission links in a network
- Become familiar with the various transmission media
- Understand the role of nodes and switches in a network
- Establish a transmission-link–based network classification scheme

Chapter Overview

I II II I I   # Chapter Overview

Transmission links connect terminals and other devices in a network. This chapter first discusses the various transmission media and their performance characteristics, moves to a discussion of methods for using the transmission medium efficiently, and then describes analog and digital transmission link systems. The chapter then introduces the concept of network nodes and switching and concludes with a classification scheme for telecommunications networks.

Most of the digital telecommunications revolution in recent years has been driven by dramatic improvements in the cost and capabilities of transmission media. Transmission links determine the speed with which information can be sent and thus the type of traffic that is cost-effective. Historically, transmission links were mainly copper cable, radio equipment, and expensive satellites. This chapter describes the new capabilities of fiber optics and other media that offer speeds thousands of times faster than those possible even ten years ago.

I II II I I   # Transmission Principles

Any communication requires a sender, a receiver, a message or information to be sent, and a medium or channel to carry the message from sender to receiver. A **transmission link** is the medium or channel path that carries a message from sender to receiver. More formally, a transmission link is defined as the physical path between terminal devices that carries a coded electronic signal from origin to destination. The transmission link is also often referred to as a circuit (two-way path) or a channel (one-way path); it consists of the physical medium through which a signal travels and includes any intermediate hub points, or nodes, that act as a switch or relay point used to direct the signal on its way to its final destination point.

An obvious and frequently used analogy for a transmission link is a road or highway. A highway consists of the physical pathway, whether dirt, gravel, paved concrete, or asphalt, and the junction points (nodes/switches) where roads intersect. The content of the transmission is independent of the specific medium used to carry the information signal; using the roadway analogy, this means that a given car will travel from point A to point B regardless of whether the road is dirt, gravel, asphalt, or concrete. However, the road surface may affect to some extent the speed traveled and the driving technique used. In telecommunications, any type of medium is capable of carrying the full range of information types—voice, video, or data—in either analog or digital signals; and just like the roadway, the particular characteristics of various media make some better suited than others for certain types of applications.

Transmission links can be simple point-to-point connections, complex multipoint network configurations, or broadcast systems that transmit through the air

Figure 5–1

A Typical Complex Multipoint Network

Fiber cable

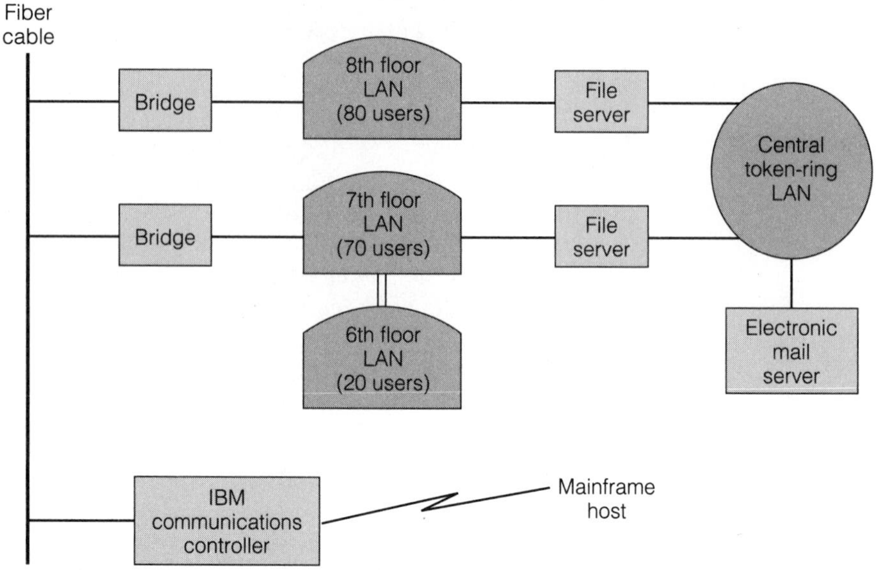

Bank of America's new network, which links several sites and about 180 users, is based on a combination of fiber-optic and coaxial cable transmission links in a complex multipoint configuration. Implementation of the Novell-based system improved interoffice, interbranch, and external communications.

multidirectional signals that are available to any compatible receiver within range of the signal. Figure 5-1 depicts a typical complex multipoint network. Transmission links combined with terminals constitute the physical components of a telecommunications network, connecting individuals to other parts of the company and to other organizations outside the company.

Transmission Media

A **transmission medium** is any substance through which a propagated electronic signal travels from origin to destination; a transmission link can use any of a wide range of media to support an electronic transmission. Within this range of available media, each medium has its own characteristics that make it better suited for certain types of transmission applications than for others. Transmission media are classified as either guided or unguided. **Guided media** are those with an enclosed path, such as a wire or cable, and the transmitter and receiver terminals are connected directly to the medium. **Unguided media** have no enclosed path and generally consist of air, water, or empty space. Signals are usually broadcast (radiated in more than one direction) through unguided media, and transmission and reception are through antennae.

Both guided and unguided media can carry all forms of information: voice, video, and data. For example, although Apollo II's lunar lander *Eagle* sent data and high-resolution video information from the moon to earth through the unguided medium of empty space, these same types of signals are also sent routinely through a guided medium such as a fiber-optic or coaxial cable. The next sections describe various types of guided media, the types of applications for which each is well suited, and performance characteristics of the particular medium.

Guided Media

Twisted Wire Pair A twisted pair, which is the most prevalent guided medium in use today and accounts for a very high percentage of the outside wiring associated with the world's public telephone networks, consists of two insulated copper wires twisted into a spiral pattern. The twisting is used to prevent interference when the wires act as antennae and inadvertently pick up electromagnetic waves in the air. The biggest application of the twisted pair is in the **local loop** portion of the local telephone company's network, in which a twisted wire pair connects a residence or office building to the telephone company's local central switching office. A large number of these wire pairs are bound together in thick cable sheaths; individual pairs are consolidated from adjacent office buildings and residence locations to efficiently link them to the central office. Twisted wire pairs are also used extensively in local area networks to interconnect the PC workstations and file server units on the network. Here, the cable is **shielded twisted pair**, which has protective insulation that reduces attenuation and interference, making it possible to transmit high data rates over short distances, typically up 100 meters. The standard phone line uses **unshielded twisted pair**, which is both the cheapest guided medium in use and the one with the lowest data rate.

A single twisted pair has a bandwidth of approximately 250 Khz; this translates into the capacity to carry approximately 60 analog voice channels over several kilometers without amplification. In digital mode, a single twisted pair can operate at a data rate of 1–4 mbps, but a signal regeneration repeater is needed every 2–3 kilometers. It is possible to obtain higher digital data rates (up to 10 mbps) using a twisted pair for shorter distances under controlled conditions.[1] Figure 5-2 summarizes twisted pair capabilities and developments.

Coaxial Cable A coaxial cable is constructed with two conductors—one an outer cylinder of conducting material acting as a shield, and the other is an inner conducting wire separated from the outer conductor by insulation (see Figure 5-3). This type of construction permits a wider range of transmission

---

[1] Distance, not technology, affects transmission most. LANs are not inherently more efficient than WANs, but they can achieve higher data rates for the same reason that you can shine a flashlight on a book to read it under the bed clothes but Paul Revere had to use a chain of fires as repeaters to signal that the British were coming. Your torch is a LAN, Revere was a WAN.

Figure 5–2	**Device**	**Capabilities and Developments**
Twisted Pair Capabilities and Developments	Standard analog	3.5 Khz (3,500 cycles per second) only; "voice grade"
	Enhanced analog	500 Khz up to 3.5 miles using special digital signal processing chip in customer's home
	ADSL (asymmetric digital subscriber line)	Provides T1 speeds by packing extra data onto the carrier signal; with data compression, is the base for video dial tone; enables full-motion video over standard phone line; signals fade and need regeneration equipment in customer's home
	BALUN (BALanced UNbalanced)	Small device that connects twisted pair to coaxial cable; reduces cost of LAN installation because twisted pair is already in place
	Shielded twisted pair	The more twists, the less electromagnetic and radio interference; data-grade twisted pair that is shielded is used in many Ethernet LANs instead of coaxial cable

frequencies, and hence, more bandwidth, which means more information-carrying capacity. Coaxial cable is used extensively in cable TV network systems because a single cable has the capacity to simultaneously carry 40–60 television channels; it is also used to interconnect computer workstations in local area networks. Coaxial cable has a bandwidth of 300–400 MHz, with 1 gigabit attainable. (Note that this bandwidth is in megahertz, not kilohertz, as for a twisted wire pair.) This is sufficient capacity to carry approximately 10,000 simultaneous telephone calls using analog transmission methods. Coax can operate at a data rate of up to 500 mbps for digital transmission and is much less susceptible to interference from nearby cables because of its special shielding features. A key characteristic of coax is its electrical resistance (measured in ohms). Resistance lowers impedance, which is the resistance of the cable to the data signal. Coax resistance is typically 50–100 ohms, versus up to 200 ohms for twisted pair. An analog transmission signal carried on coaxial cable needs to be amplified every few kilometers, and signal regeneration repeaters are needed every kilometer for digital signals. The main limitation of coaxial cable is its thickness.

Optical Fiber An optical fiber consists of a thin glass core fiber strand (which looks like 1-lb-test monofilament fishing line) surrounded by **cladding**—a different type of glass that is reflective and thus directs stray light rays back to the core fiber strand—and usually covered by an outer protective shield of plastic. The total diameter of the core fiber strand is less than that of a human hair. Individual fibers are bundled together in groups to form a cable configuration. A light source (light emitting diode or laser) is the carrier that transmits information along the glass fibers.

Because of their very large information-carrying capacity, optical fiber cables are now used extensively for long-distance transmission links to carry

Figure 5–3

Coaxial Cable

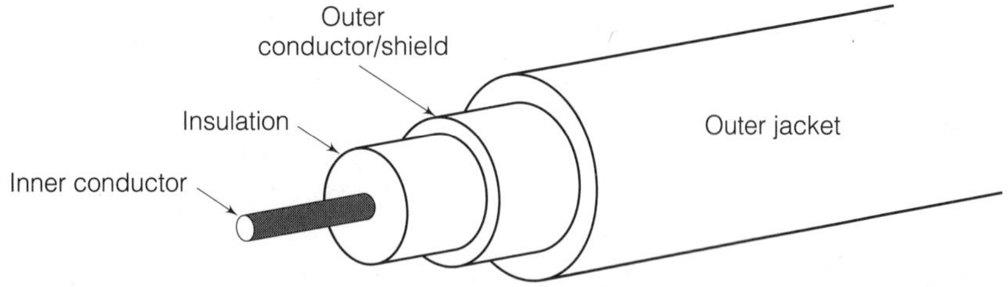

voice and data in the public telephone network. Sprint claims that 100 percent of its long-distance transmission links are fiber-optic cables). Around 95 percent of long-distance traffic is carried over fiber.

Fiber-optic cable is now commonly used both to connect PC workstations in a local area network and to interconnect a number of individual LANs separated by long distances. The high speed of fiber-based LANs is a key to making practical new applications that require extremely high bandwidth, most particularly image processing.[2] Optical fiber is also used for long-distance transmission of broadcast and cable television signals and is slowly replacing the twisted pair as the local loop that interconnects office buildings in large metropolitan areas. The local loop remains predominantly copper cable, mainly because of the high cost of replacing undepreciated facilities.[3] Many industry experts predict that fiber-optic cable will eventually become the local loop to the home, which would open up myriad possibilities for two-way video, high-definition television, and high-speed data communications applications to and from the home. Cable television companies increasingly rely on fiber because it is the best medium for offering subscribers a wide range of program choices. For the RBOCs, an increasingly attractive option is fiber optics to the sidewalk, and from there twisted pair into the houses along the street.[4]

---

[2] Video demands the largest bandwidth, and high-resolution image processing the next largest. So, sending a video of you saying "Hi, Mom! Please send cookies and cash" needs millions of bits a second, but sending a fax with the same message takes just tens of thousands. The electronic mail equivalent is under a thousand bits, and Mom's electronic funds transfer of even millions of dollars requires a few hundred bits.

[3] Regulation of telecommunications varies by state, and rates are set on the basis of complex formulae by which the provider recovers its costs and makes an agreed-on rate of return. Replacing undepreciated plant and equipment means taking a write-off; not doing so may mean losing the chance to improve service and cut operating costs. The more competitive the market, the more likely providers are to write off obsolescent facilities. MCI, Sprint, and AT&T, for example, replaced billions of dollars of analog equipment in their race to exploit new technology, at the very time when Regional Bell Operating Companies were very reluctant to mar their balance sheets and profit-and-loss statements.

[4] Note that in both instances, ownership of rights of way and a large installed base of subscribers make it relatively easy to upgrade transmission media. Cable TV now reaches 60 percent of homes in the United States, and over 95 percent of households have phones. One virtually certain move in the next few years will be telecommunications companies and cable TV firms entering into partnerships.

Considering its astounding performance characteristics, it is not surprising that the use of the fiber-optic cable is becoming prevalent in a wide range of applications. A single fiber strand has a bandwidth of 2 Ghz or more. This translates into a maximum operating data transmission rate of approximately 2 gbps, which represents a significant performance advantage over other guided media. For example, a long-haul telephone fiber line can carry up to 60,000 simultaneous telephone calls; a fiber LAN transmits information at 100 mbps, compared with 16 mbps using a coaxial cable.

In addition, fiber optics exhibit a much lower rate of signal attenuation compared with other guided media; consequently, repeaters/amplifiers are needed only every 50–100 km, compared with coaxial cable and twisted pair's need for repeaters/amplifiers every few kilometers. Furthermore, because the fiber system is optical, it is not subject to interference from ambient electromagnetic radiation and does not create any electromagnetic interference itself; this property makes it highly secure from eavesdropping. The core and cladding of the fiber are made from very pure silica glass with a low index of refraction, which is a measure of how much a ray of light will bend when it enters the glass. The wavelength of the laser beam that sends the light ray through a fiber cable is as small as 50 millionths of a meter, and the cable is only 300 millionths. **Monomode fiber** has a core that is so small that the light has only one possible transmission path; thus no divergence occurs along its path and consequently no slowing down results. **Multimode fiber** has a wide core, which makes it cheaper to produce; light rays propagate slightly, reflecting off the boundary between the core and the cladding.

The main problems associated with fiber are the complexity of connecting and installing it, the cost of replacing existing facilities, and the cost of the fiber itself. Fiber cable costs roughly eight times as much as the shielded twisted pair used for Ethernet LANs. For this reason, there has been a concerted effort by many transmission suppliers and LAN equipment providers to upgrade the performance of twisted pair—to provide an alternative to the extremes of leaving in place cheap but slow twisted pair and upgrading to powerful but expensive fiber.[5] That said, the comment of two leading experts on telecommunications aptly summarizes the role of fiber: "Almost the whole future of telecommunications revolves around the new technology that usurped waveguides."[6]

Unguided Media

Twisted wire pairs, coaxial cable, and fiber-optic cable constitute the principal guided media created to form transmission links for telecommunications networks. Unguided media are nature's natural carriers of electromagnetic waves,

---

[5] Laying cables is an expensive business that relies heavily on skilled labor. Rewiring a building is a major project. In addition, the adapters needed to connect terminals to fiber-optic LANs are more expensive than anticipated, which has slowed down adoption of FDDI and stimulated efforts to increase transmission speed of existing media. Untwisted pair and coaxial cable offer plenty of bandwidth; the trick is to find ways of using as much of it as possible.

[6] Pierce, John R., and A. M. Noll, *Signals: The Science of Telecommunications* (New York: Scientific American Library, 1990).

including air, water, and the vacuum of space. The transmission is generally a multidirectional broadcast signal that uses antennae to send and receive the information. The performance characteristics of unguided media depend on the frequency band used and the method for propagating the transmission signal. Three primary unguided media are in use today for telecommunications: terrestrial microwaves, satellites, and broadcast radio systems.

Terrestrial Microwaves Terrestrial microwave systems use a directional radio broadcast transmission method operating in the 2–40 Ghz super-high-frequency band. This band requires a line-of-sight relationship between transmitting and receiving antennae, and the typical microwave antenna is a parabolic dish attached to a tall tower within sight of the next tower relay station. It is similar to a wireless, high-tech telephone pole system. Terrestrial microwave systems are used for high-capacity, long-distance communications links, particularly in areas where the natural terrain makes it difficult and expensive to construct alternative guided media such as coaxial cable or fiber-optic cable. Microwave systems carry the full range of types of information: telephone calls, data, and video. These systems are also used to interconnect separate LANs located in different buildings within sight of each other. They also act as high-capacity, local loops connecting one office building with another or feeding directly into a long-distance network.

Microwave systems experience relatively low signal attenuation compared with twisted wire pairs or coaxial cable systems; therefore, repeater spacing is farther apart than for these other media, but they must be along a line of sight. However, there is potential performance degradation from interference (other radio systems and stray electromagnetic radiation) and rain. The characteristics of microwave systems make them well suited for high-capacity analog or digital transmissions, either local or long distance. They are very cost-effective in situations in which line-of-sight construction is feasible and frequency band congestion is not extensive. It is generally much easier and less expensive to erect two towers with antennae 50 km apart than it is to install coaxial or fiber-optic cable over that distance, particularly if the terrain is hilly or mountainous and there is no physical infrastructure in place to support the cables.

Figure 5-4 shows an example of how microwaves are used by firms with heavy telecommunications traffic to bypass the local phone company's central office or to avoid having to install cables to link buildings. The supplier, Diginet Inc., offers these companies high-frequency, digital microwave transmission in Chicago. Diginet uses the roof of the Chicago Opera House for this transmission; this is one of the highest points in the city and has an unimpeded line of sight to the main business area. Microwave is used by Diginet to leap over buildings, rivers, busy streets, and parking lots. Microwave is quick to install and to relocate if conditions or demand changes. Diginet's system is manufactured by Motorola, and it can transmit up to 28 64-kbps channels for up to ten miles of line-of-sight visibility.

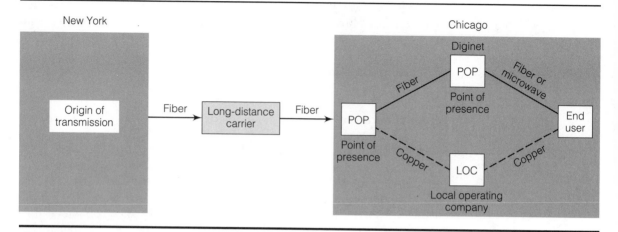

Figure 5–4

How Diginet Inc. Uses Microwave Transmission to Bypass Local Phone Service

Satellites A communications **satellite** is really a microwave relay system in the sky. It is essentially a repeater or amplifier that receives a signal from a ground (earth) station, amplifies the signal, and retransmits it to one or more receiving earth stations on the ground. Most communications satellites are located in geosynchronous orbit 22,300 miles above the earth's equator. They carry long-distance telephone traffic (both domestic and international) as part of the worldwide public telephone network, and they distribute television signals around the world, which are then locally rebroadcast. Some countries have a **direct broadcast satellite service** in which the television signal is received directly from the satellite using a small earth station antenna on the roof of a home or apartment building.

Satellites are also being used more frequently as long-distance transmission links for private business communications networks. They are particularly important in international private networks because many countries do not have adequate terrestrial long-distance transmission facilities, and the cost and time required to construct an adequate terrestrial network within these countries are prohibitive. Fortunately, the cost of installing several earth stations that are strategically placed within a country is relatively modest, and such installations immediately link these locations with the rest of the world via satellite.

Two characteristics of satellite transmission make it particularly well suited to long-distance applications linking a large number of locations: distance insensitivity and the broadcast nature of the transmission signal. A satellite transmission signal is broadcast over a wide geographical area, and it can be received by any properly equipped earth station that is located within the satellite's coverage area and whose antenna is focused on the satellite. Thus hundreds and even thousands of different earth stations in different locations can be instantly part of the same network. Adding another node to a satellite network is simply a matter of installing an earth station at that location. In contrast, adding another node to a nonsatellite terrestrial network using guided media requires the construction of a cable or microwave extension that spans the entire distance between an existing network node and the new location. Extending a terrestrial network 100 kilometers requires twice the length of cable and twice the number of repeaters/amplifiers as it does to extend the network 50 kilometers; with a satellite network, the same equipment is used (adding additional earth stations as needed), regardless of the distance the network is extended. The omnidirectional, broadcast nature of the satellite transmission signal also makes it well suited either for applications in which the same information is destined for a number of locations (such as broadcast television and data distribution from a central location) or for collection from remote sites of data that is processed and stored in a central computer facility.

The data distribution and collection types of applications have been facilitated by the development of the VSAT (very small aperture earth terminal) network concept (see Minicase 2–1). The low cost of VSAT allows firms like Terra International to construct a network consisting of a large number of earth stations that are inexpensive and very compact (with antennae of 1–2 meters in diameter). These small earth terminals are located at remote sites (at retail outlets, on top of oil rigs, with bank teller machines, and so on), and each transmits data in digital form to the satellite. The relayed data streams are received by a central hub station, which routes most of the data via a terrestrial transmission link to a central computer facility for processing. The remaining traffic destined for other remote terminals is sent back up to the satellite and is then beamed down again to be received by the designated remote earth terminals.

This system is very economical because all of the intelligence and complexity (that is, cost) is concentrated in a central hub site, which eliminates the need to duplicate the required intelligence capability at each remote site. Therefore, a given budget can support a much larger number of these simple, reliable, and inexpensive remote earth stations. Although to date VSATs have been primarily used for private data network applications for business, they are an excellent method for extending basic telephone service to rural sites in underdeveloped countries without the need for extensive terrestrial backbone network facilities. Unattended remote terminals with a single telephone outlet could be installed at the local rural post office or general store, and the hub station/network control center could be maintained either in a larger city within the country or even in another country, so long as the location is within the coverage area of the satellite.[7]

---

[7] For this reason, many countries restrict the use of satellite earth stations, fearing that they will be unable to control the flow of information. China is notorious in this regard. In Europe, PTTs have restricted VSAT because they will lose revenues from international traffic. Countries such as

The capital cost of a VSAT network is around $10,000 per site plus (typically) $1–2 million for the shared communications hub; the cost can be reduced by leasing spare capacity from another user or VSAT service provider for around $100 per month per site. Over a five-year period, the operating cost per month, including amortization of the investment, is $300–$500 per site. Users report that the payback period in installing a VSAT network to replace leased or dial-up lines is three to five years.

Satellites operate in the microwave frequency band (primarily at 4 and 6 Ghz and 11 and 14 Ghz). As a result they are subject to interference from terrestrial microwave systems in major metropolitan areas, and rain also affects the signals in the higher part of the satellite frequency band (11/14 Ghz). There is also an insurmountable one-quarter-second delay associated with a round trip to the satellite due to its distance from earth. This delay is noticeable on telephone calls and can create problems in managing data transmissions. There are also potential security problems with satellites because the signal can be received by anyone within the broadcast range of the particular satellite; consequently, some form of signal encryption is needed to obtain secure transmission.

Broadcast Radio Systems Broadcast radio is an omnidirectional, over-the-air transmission technique that uses a lower frequency band than microwave transmissions. Unlike microwave systems, broadcast radio methods do not require either a dish-shaped antenna or line-of-sight alignment. The frequency band from 30 MHz to 1 Ghz covers the FM, VHF, and UHF broadcast bands. The transmission signals in this range of frequencies are not sensitive to rain, but they are subject to interference from other signals and from reflection off buildings and other physical objects. Because of these characteristics, broadcast radio applications are generally restricted to a local (metropolitan area) geographic scope.

Typical applications for broadcast radio methods include entertainment radio and television services, cellular mobile telephone systems, and paging services. There is currently a strong trend toward the development of enhanced mobile and wireless communications technology and applications, including digital cellular systems designed for mobile voice and data communications, wireless local area and personal communications networks connecting individuals within a given building or small geographic area, and mobile satellite services using small transmitting and receiving earth terminals that can be packed into a suitcase or installed on a car or truck.

Figure 5-5 depicts two examples of broadcast systems used to provide innovative services. In Figure 5-5a, a customer selects information and makes an order from a menu of options displayed on a TV screen. The device used, similar to a standard channel selector, sends a radio signal to a local cell site, which beams it up to a satellite. The signal is then sent back down to TV An-

---

Indonesia, by contrast, have welcomed VSAT technology because it opens up huge increases in domestic communication volumes.

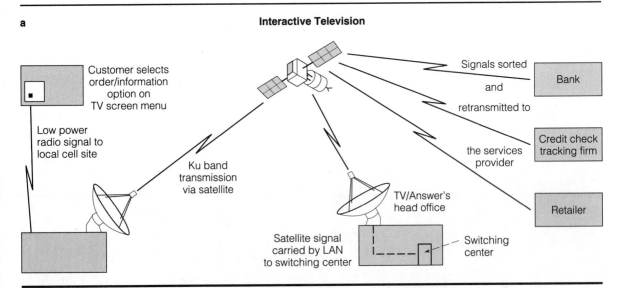

a

Interactive Television

Customer selects order/information option on TV screen menu

Low power radio signal to local cell site

Ku band transmission via satellite

Signals sorted and retransmitted to the services provider

Bank

Credit check tracking firm

Retailer

TV/Answer's head office

Satellite signal carried by LAN to switching center

Switching center

Figure 5–5

Two Broadcast Radio Systems

In TV Answer's interactive TV service (a), customer's orders, chosen from a video menu, are beamed to a satellite, which simultaneously sends signals to TV Answer's head office and to service providers. In Motorola's EMBARC wireless e-mail service (b), EMBARC users can broadcast messages (for example, price quotes or news) to Newstream paging devices, which can then upload the received message to a portable or office PC.

swer's head office, where a LAN routes it to a switching center for handling, accounting, and administration. The satellite also identifies which service providers should receive the signal; all of them are within the satellite's footprint, but obviously only JC Penney should be able to receive orders for JC Penney products.

Figure 5-5b shows Motorola's EMBARC service, which broadcasts messages that can be picked up by portable paging devices and then transferred to a PC. Radio towers pick up the satellite signal and rebroadcast it from terrestrial radio towers, amplifying the signal.

Price/Performance Trends in Media Use

The price/performance characteristics of optical fiber cables and satellites dominate those of other media, and this has led to their widespread use for most applications, except for mobile services, telephone company local loops, and interconnection within a building (which is now moving toward the use of fiber-optic cables). Fiber-optic cable is the most cost-effective medium for high-capacity point-to-point links when the physical terrain permits installation at a

b   **Motorola's EMBARC wireless e-mail service**

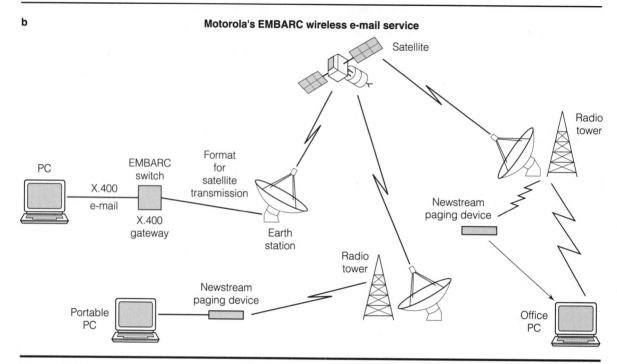

Figure 5–5

(continued)

reasonable cost; satellite is the most cost-effective unguided medium for most broadcast and point-to-multipoint applications covering a broad geographic scope and involving a large number communicating locations. Broadcast radio will continue to be an important unguided medium transmission method used for local mobile/wireless communications, such as cellular telephone systems, paging, and wireless local area networks.

Transmission links are a scarce communications resource whose cost is significant; it is generally the largest portion of the corporate telecommunications budget. This places an important burden on designers and implementers to design and construct networks that use transmission links efficiently, whether by minimizing the link capacity needed to meet a given business communication requirement, or, alternatively, maximizing the utilization of a given transmission link facility.

There are significant economies of scale in transmission link systems because most of the costs are fixed and independent of the capacity of the system. The cost of a given transmission link system consists mostly of (1) labor for installation and maintenance and (2) the cost of the system's basic components, both of which are largely independent of the capacity of the system. For example, it generally costs very little extra to add a few more fiber-optic strands to a cable during construction and doing so could in fact double or triple the capacity of the cable.

The predominance of fixed costs in transmission link systems leads to declining cost per unit of information moved ($/bps) as the volume of information increases up to the maximum capacity of the link. Also, the low incremental cost of constructing higher-capacity systems means that they generally have lower unit costs of information movement than lower-capacity systems. For example, a system with a data rate capacity of 1.544 gbps will generally have a lower cost per bps than a transmission link system with a 64-kbps data rate capacity. This is because the basic construction, installation, and maintenance costs of the two systems using any single medium is approximately the same, even though the larger system has roughly 24 times the capacity of the smaller system.

Thus the unit cost of information transmission can be minimized by either consolidating traffic onto high-capacity systems whenever possible or finding ways to utilize a given transmission capacity more efficiently. These two means of cost reduction are generally accomplished through one of the following methods: multiplexing, signal compression, and multidrop lines and polling. Multiplexing combines smaller, lower-capacity data or voice information streams into one large stream to increase capacity utilization and take full advantage of efficient high-capacity transmission links (similar to an airline feeding passengers from small commuter planes to central hubs where they are combined to fill a fuel-efficient jumbo jet for long-distance transport). **Signal compression** reduces the amount of information that is actually sent across the network such that less transmission capacity is required to send a given voice conversation or block of data (similar to dehydrating bulky materials, transporting it in compressed powder form, and then adding water and reconstituting it at its destination). Multidrop lines and polling share a given transmission link by having terminals take turns transmitting one at a time while the others remain idle. We examine each method in the following sections.

Multiplexing

There are two principal techniques of multiplexing: frequency division multiplexing and time division multiplexing. **Frequency division multiplexing (FDM)** (Figure 5-6a) is a technique that divides the bandwidth of the transmission medium into partitions or slots, each consisting of sufficient bandwidth to carry the required information and have a little left over between each slot—a guard band—to provide a buffer against interference. The separate frequency band slots assigned to each voice conversation or data stream are called channels. For example, suppose we have a range of frequencies from 101 to 120 Hz (a bandwidth of 19 Khz) allocated for voice transmission. We can assign the frequency band from 101 to 105 Khz as the slot for the first voice signal; the frequency band from 106 to 110 Khz is the slot for the second voice signal, with the 1 Khz slot from 105 to 106 Khz left empty to act as the guard band; the third voice signal is assigned to the 111–115 Khz frequency band (110–111 Khz is the guard band); and the fourth voice signal is assigned to the 116–120 Khz frequency band (115–116 Khz is the guard band).

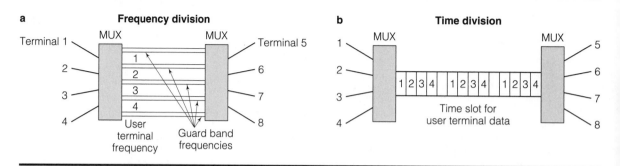

a Frequency division   **b** Time division

Figure 5–6

**Multiplexing:
Frequency Division
and Time Division**

This frequency division multiplexing (FDM) method of stacking separate data streams, telephone conversations, or messages permits more efficient use of a given transmission link by ensuring a high-capacity utilization rate, and consequently, a lower unit cost of information transfer. FDM has been used extensively in the analog public telephone network, and it is also used to multiplex several television signals onto a coaxial cable that carries cable television into the home. With the movement toward digital transmission facilities, FDM is being replaced by a multiplexing technique that is better suited to digital transmission systems: **time division multiplexing (TDM)**.

The synchronous form of TDM is used to consolidate and combine multiple digital information streams when the data rate of the transmission medium is greater than or equal to the sum of the data rates of the individual digital streams being multiplexed. The multiplexer device combines the separate bit streams by interleaving bits from each of the separate source signals into prearranged time slots (Figure 5-6b). For example, the first bit from signal source 1 is assigned to time slot 1, the first bit from signal source 2 is assigned to time slot 2, and so on, until all the bits are arranged on the transmission carrier in a specified time sequence. This technique is similar to a method used to load passengers onto a roller coaster at the local amusement part: the first row of seats in the roller coaster is filled with the first five people from line 1; the second row of seats is filled with the first five people queued in line 2, and so on.

The form of a multiplexed signal is a series of frames, and each frame represents a complete cycle of time slots assigned to the various sources of information being multiplexed (that is, separate voice conversations, messages, or data streams). One or more time slots in each frame is assigned to each source, and the given set of time slots assigned to one source for a given frame remains the same for all subsequent frames. This can be illustrated by returning to the roller coaster example. Imagine that it is local high school day at the amusement park, and officials have decided to ensure order by assigning the first row of roller

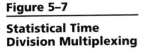

Figure 5–7

Statistical Time Division Multiplexing

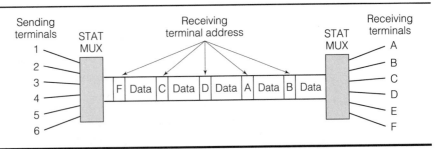

coaster seats to students from Palmetto High School, assigning the second row of seats to students from Killian High School, and so on. These assignments would hold for all roller coaster runs throughout the day at the park, and if there are not enough students from the designated high school to fill their assigned row of seats for any given run of the roller coaster, then those seats are to remain empty.

With FDM, each assigned range of frequencies is a channel; for TDM, a channel is the time slot allocated for each source. Although the synchronous TDM channel is physically different than the FDM channel, they are logically equivalent in the sense that the channel refers to either a constant bandwidth slot (4 Khz for FDM) or a constant data rate slot (9.6 kbps for TDM). Multiplexing information sources with different bandwidths or different data rates is handled by matching the number of time slots assigned per frame to the bandwidth or data rate requirement of the source signal. For example, a 19.2-kbps signal source will be assigned twice the number of time slots per frame as a 9.6-kbps signal source.

The transmitting and receiving multiplexers in a synchronous TDM system must be synchronized so that the demultiplexing process (unbundling the combined signal) will leave the individual source signals intact, with the correct bits in the correct order, for delivery to the destination terminal at the proper data rate. To ensure this result, the receiving multiplexer must know that the bit in time slot 1 of a given frame is the first bit belonging to information source 1, the bit in time slot 2 of the given frame is the first bit from information source 2, and so on.

The other form of TDM, called **statistical TDM** or **asynchronous TDM**, eliminates predetermined time slots assigned to particular devices or information sources (see Figure 5-7). Input from source terminals are buffered in a queue until the multiplexer device scans the input buffers and assigns data to available time slots until the frame is full (no empty time slots are permitted), and then the multiplexed signal is sent through the network. Destination address information is included in each time slot because the slot assignments vary from frame to frame depending on load requirements and availability. The receiving multiplexer reads the destination addresses from the slots of the incoming frames and distributes the associated data to the designated output device. Returning again to the roller coaster example, statistical TDM is equivalent to a regular day at the amusement park. There are no preassigned seats on the roller coaster; seating is first come, first served, and the roller coaster does not begin its run until all seats are filled.

Statistical TDM permits more attached devices than there are time slots because the slots are not reserved, and not all devices are transmitting continuously. This also means that the overall multiplexed transmission data rate will generally be less than the sum of the data rates of the individual information source devices attached to the multiplexer. The statistical TDM approach is designed for efficient use of the transmission link for intermittent and variable communications requirements from multiple sources. Efficiency is achieved through maximum utilization of the transmission capacity; there are no empty slots, and there are more attached devices than there are slots. In addition, statistical TDM generally requires lower-capacity (slower speed) transmission links than are needed for comparable synchronous TDM systems because most of the time not all attached devices are transmitting at once. However, responsiveness suffers when loads increase and more attached devices transmit more frequently; under these circumstances, data to be sent are buffered for longer periods of time waiting for available time slots.

On the other hand, synchronous TDM is ideal for high-volume, continuous-load conditions; each time slot is reserved for an attached device, and the transmission link data rate is greater than or equal to the sum of the data rates of the individual attached devices. This leads to a very efficient and highly responsive use of the transmission link capacity for high-volume, continuous data streams. Given these performance attributes of the two time division multiplexing methods, statistical multiplexers are used most frequently for terminal-to-host computer communications because this involves variable and intermittent load factors. Synchronous multiplexers are used most often in long-distance private and public networks that carry both voice and data traffic. Statistical TDM systems are generally not well suited to this type of traffic because of their longer and more variable response times.

Multiplexing is closely associated with nodes and switches, the topic of a later section in this chapter. Multiplexers handle the flow of FDM or TDM traffic, and their performance largely determines the performance of a large-scale network. FDM has been the reliable service base for analog networks; time division multiplexing is the future direction for large transmission service providers. In 1988, AT&T wrote off all its analog TDM equipment in order to speed up its shift to fully digital transmission. Digital TDM allows the use of pulse code modulation, the core of digital transmission. In **pulse code modulation (PCM)** (Figure 5-8), an analog signal is converted into digital pulses by sampling the frequency of the signal 8,000 times a second, coding the signal as a 7-bit number (using zeros and ones) to represent the frequency (between 0 and 128), and adding an error control bit. The data rate of the line thus needs to be 8 times 8,000 bits per second; 56 kilobits contain the digital voice traffic and 8 kilobits the control data.

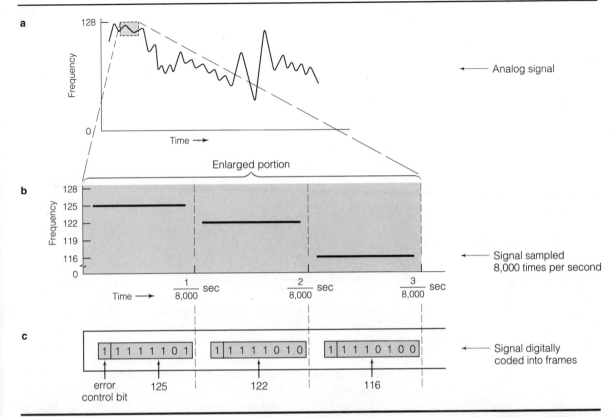

Figure 5–8

Pulse Code Modulation

In pulse code modulation, the frequency of an analog signal (a) is sampled 8,000 times a second (b) and digitally coded into frames (c).

Signal Compression

Signal compression techniques for making more efficient use of transmission link capacity are based on the principle that all forms of information have natural, built-in redundancies. Therefore, data used to convey the underlying information can be compressed by temporarily removing the redundant elements prior to transmission through the network. These redundant elements are then regenerated and reinserted at the receiving end of the transmission, and the data is restored to its original form. For example, in the transmission of video information, the static (stationary) part of an image (say a person sitting in a chair) is redundant; the same image is sent repeatedly when there is no movement to the image. The portions of the image that are changing must be transmitted continuously while the static portion of the image can be temporarily stored; the stored static portion can then be combined with the incoming changed portions of the overall image to generate a current view of the image.

Large-scale signal compression techniques have been applied successfully in business teleconferencing applications. A high-quality digital transmission of a full-motion television signal usually requires a 45-mbps data rate transmission link when compression techniques are not used. However, because of the paucity of motion associated with most business meetings (substitute "dullness"), a 1.544-mbps data rate transmission link using signal compression techniques will provide acceptable quality full-motion video conferences. In fact, signal compression techniques are used to transmit video conferences using a 64-kbps data rate channel, but motion can be somewhat blurred.

Multidrop Lines and Polling

In a typical slave/host computing environment, the terminals connected to a central computer facility transmit only in intermittent bursts. Thus the transmission link connecting terminals to the host computer will be more efficiently utilized if it is shared among a number of terminals. One method of sharing a given link is the multidrop line, a single transmission circuit to which several terminal devices are directly attached. A familiar example of a multidrop line system is the almost extinct "party line" telephone. In a party line telephone system, several homes share a single line to the telephone company central office; only one party on the line can talk at once, and each terminal drop (home) on the line has a slightly different ring to signal the appropriate party that they have an incoming call. A party wishing to call out must pick up the handset and listen to see if the line is clear before dialing. This is an efficient use of a single telephone line, but it is quite inconvenient when more than one party wants to use the phone at the same time.

The principal difficulty with multidrop lines is avoiding contention for the single circuit by two or more terminals attempting to tranist at the same time. This problem is solved by sequential **polling**, in which the host computer polls each terminal in turn to determine whether it has something to transmit. This process is very similar to a police officer directing traffic at a busy intersection. He or she permits only one lane of traffic to cross the intersection at a time while holding all other lanes in place. For transmissions in the other direction—from host to terminal—the messages or data streams are addressed to specific terminals connected to the multidrop line (just like the special ringing signal on the party line). The multidrop form of link sharing increases the utilization rate of transmission capacity and works well for terminals with bursty, intermittent transmission requirements. However, if the terminals on the multidrop line transmit often or send large, continuous blocks of data, waiting time for the other terminals on the line will increase, and hence the responsiveness of the system is reduced.

| | | | | | |

Transmission Link Systems

Transmission links have been discussed so far in this chapter in terms of the medium used to establish the physical connection between the sender and the receiver. The characteristics of any medium help shape the performance properties of the link and establish the link's maximum capacity in terms of overall bandwidth and data rate in bits per second. In practice, the physical link is almost always partitioned or subdivided into units of capacity conforming to the most common end-user application requirements, such as a voice circuit or a video channel.

The basic unit of capacity in the public telephone network is the analog voice circuit, which generally consists of 3 or 4 Khz of bandwidth allocated on some communications medium (twisted wire pairs, coaxial cable, terrestrial microwave, satellite, and so on). Strictly speaking, the term *circuit* refers to a two-way communications path; therefore, two separate 3 or 4 Khz bandwidth allocations are needed for each unidirectional channel (one in each direction) to make up a complete circuit.

Analog transmission link systems consist of standard multiples of basic voice circuit units obtained by multiplexing individual voice circuits together using the frequency division multiplex technique: 12 4-Khz voice channels multiplexed together, is called a **group**; five groups (of 12 channels each) multiplexed together constitute a super group (60 voice channels); and a master group consists of ten super groups multiplexed together (600 voice channels). Different combinations of groups, super groups and master groups may be placed on the same high-capacity medium (such as a fiber-optic cable), for long-distance transmission to another region of the country or the world. Individual voice circuits and multiples of the basic voice circuit capacity can also be used for data transmission; a 4-Khz bandwidth voice circuit is roughly comparable to a 64-kbps digital data stream.

Although the analog transmission link system has been the backbone of the world's telecommunications networks for many years, it is rapidly being overtaken and replaced by an evolving digital transmission link system that supports a fast-growing data communications requirement and is a more efficient means of transmitting voice signals over long distances. The digital system uses the same media as the analog telephone network, but with a growing emphasis on fiber-optic cable systems. In the digital system, the FDM hierarchy (that is, groups, super groups, and master groups) is replaced by a synchronous TDM system. The United States, Canada, and Japan have standardized on one hierarchical system, developed by AT&T, whereas Europe and most of the rest of the world use a slightly different hierarchy:

North American		International	
DS-0	64 kbps	Level 0	64 kbps
DS-1	1.544 mbps	Level 1	2.048 mbps
DS-2	6.312 mbps	Level 2	8.448 mbps
DS-3	44.736 mbps	Level 3	34.368 mbps
DS-4	274.176 mbps	Level 4	139.264 mbps

The basic unit in both systems is the 64-kbps circuit, known as a DS-0 circuit (DS stands for "digital service") in the United States and as a level 0 circuit in

Europe. The DS-1 level consists of 24 64-kbps circuits multiplexed together with an overall transmission data rate of 1.544 mbps; T1 is the tariff for the line that carries DS-1 signals, and T3 similarly carries DS-3. The equivalent "level 1" in Europe is 30 64-kbps circuits with a total transmission data rate of 2.048 mbps. Because of the different standards for Europe and North America, transmission links operating between the two areas, such as a submarine transatlantic fiber-optic cable, must utilize equipment for conversion from one standard hierarchy to the other.[8]

The multiplexed digital transmission link systems can carry both voice and data simultaneously. As far as the transmission system is concerned, a bit is a bit; it is irrelevant whether it represents part of a voice conversation or a number/character. Each higher level in the digital hierarchy is formed by multiplexing together signals from lower levels in the hierarchy. For example, a DS-2 level digital carrier may contain four DS-1 systems, which in turn may consist of either four groups of 24 64-kbps circuits multiplexed together, or alternatively, three groups of 24 64-kbps circuits and a single 1.544 mbps high-speed data circuit.

The transmission link facilities provided by the various long-distance carriers utilize the full range of media: twisted pairs, coaxial cable, terrestrial microwave, and satellite. The bandwidth available on these media are subdivided and configured to correspond to the digital multiplex hierarchy using a T designation in place of the DS. For example, the T1 transmission link system operates at a data rate of 1.544 mbps and is compatible with the DS-1 level of the North American digital multiplex hierarchy. Similarly, the T2 and T3 systems correspond to the DS-2 and DS-3 levels in the multiplex hierarchy.

Use of the T1 system is widespread and growing rapidly in the United States;[9] more organizations are switching to higher-speed transmission links to accommodate the growth in data and voice communication requirements and to increase responsiveness of the networks. As we have discussed, the cost of a transmission link system is driven by fixed cost factors (labor, construction, and materials) that are largely independent of capacity. Therefore, the actual cost of providing a transmission system with T1 capacity (1.544 mbps) is not much different than the cost of providing a transmission system with a 64- or 128-kbps capacity. Although the actual price to the end-user customer does not always

---

[8]This is not always a routine process. For example, on a T1 multiplexer, the bit sequence that in the United States signals "nothing to transmit" means "please disconnect" in France and Mexico. Similarly, although ISDN is an international standard created in Europe with the explicit goal of ensuring interoperability between countries' services, there are nine incompatible switching systems in Europe and none of the ISDN services can link to the others.

[9]In addition, the T1 rate of 1.544 mbps has become the base unit of transmission for emerging public network services, such as SMDS and video dial tone. The next emerging base unit is 45 mbps, the T3 rate; it will be used more and more as multimedia applications, especially video, increase. Today, there are very few of these applications. In March 1993, the announcement by a bank that it will offer videoconferencing contacts with staff from an ATM was newsworthy enough to get front-page coverage in the trade press. For the moment, T1 speeds will dominate business use of telecommunications outside a LAN.

Figure 5–9

Monthly Costs for Dial-up and Two Leased Lines, 1992

Distance	Monthly Cost		
	Dial-up[a]	56 kbps	T1 (1.54 mbps)
10 miles	$ 13	$ 137	$ 3,735
100 miles	211	732	4,465
500 miles	303	1,008	6,632
1,000 miles	316	1,224	9,332
2,000 miles	330	1,759	14,732
5,000 miles	330	1,931	17,432

[a] Monthly cost for 1 hour per day of dial-up service.

mirror the underlying cost structure, the relative cost of different capacities is reflected to some degree in the prices for the various transmission facilities. For example, at current prices offered by the long-distance carriers, the price of a T1 link is roughly the same as the cost of three 64-kbps circuits, even though the T1 system has a capacity equivalent to 24 64-kbps circuits. This bulk capacity pricing structure that heavily favors high-capacity transmission links is certainly fueling the rapid growth in the use of T1 and higher-capacity transmission systems in all types of networks. Figure 5-9 contains typical prices as at the end of 1992 for dial-up services, leased 56 kbps lines, and leased T1 lines.

More recent service offerings by the carriers include a **fractional T1 service**, which provides for the leasing of any fraction of a full T1 data rate transmission capacity, down to a minimum of a single 64-kbps channel. This service enables a user to easily add capacity in increments warranted by particular applications and growth in demand. For example, a limited-motion videoconferencing application may require a 256-kbps data rate, which is one-sixth of a full T1 data rate; this amount of capacity is available through a fractional T1 service offering.[10]

Nodes and Switching

Until now, this chapter has discussed transmission links as if they were simply a direct connection between two terminal end points. In the real world of networks, however, many terminals are interconnected using multiple transmission

---

[10] Fractional T1 services are likely to be supplanted over the next few years by SMDS and frame relay, which offer bandwidth on demand at speeds matching fractional T1, full T1, and T3. The three main U.S. long-distance companies are redefining the market for high-speed data communications, with AT&T focusing on maintaining its 90 percent share of private networks, MCI on packaging a wide range of fast packet-switching services and virtual private networks, and Sprint on aiming to maintain its lead in value-added networks.

Figure 5–10

Examples of Network Configurations

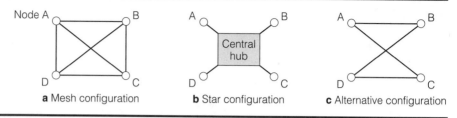

a Mesh configuration   **b** Star configuration   **c** Alternative configuration

In the mesh configuration (a), connecting four nodes requires six links. In the star configuration (b), only four links are required because the central hub directs information flow to the correct node. In an alternative configuration (c), each node contains intelligence for routing and traffic control; if the link between nodes A and B fails, A can still reach B via A → C → D → B.

paths that are themselves connected at hub locations called nodes. A network node can be an interface junction between a terminal work station and the network, or a hub where several separate transmission link systems are joined together, much like a regional airport or an intersection of several freeways. Therefore, a true network consists of multiple nodes interconnected by transmission links.

To illustrate, one type of network configuration has a separate transmission link connecting every pair of nodes; this is illustrated in Figure 5-10a for a network consisting of four nodes. This type of network configuration, or topology, is called a mesh network. Six separate point-to-point transmission links are needed to completely interconnect all four nodes in a pairwise fashion. Another configuration for connecting the four nodes is shown in Figure 5-10b; this is called a star network topology because the four nodes are interconnected through a central hub. The hub acts as a switch, or traffic control center, to direct the information flows to the correct destination nodes; all communication is through the hub switch, and it is clear from the figure that only four separate transmission links are needed to completely interconnect the four nodes of the network.

This example clearly illustrates the primary advantage of the hub or switching concept in network design: It economizes on the need for direct point-to-point transmission links, which are costly to install and maintain. There is an additional advantage to the hub/switch configuration beyond the direct cost savings derived from eliminating separate transmission link systems: It concentrates in one location the intelligence required to route and manage the traffic flow through the network. In contrast, the mesh network requires that routing and traffic flow control intelligence be duplicated at each node in the network; such duplication makes the equipment required at each node necessarily more complex and expensive than if this function were concentrated in one location. On the other hand, the redundancy means that the network is less affected if a central node goes out of service.

A third alternative is shown in Figure 5-10c. This arrangement requires only four transmission links but does not include a central hub or switching center.

Each of the four nodes contains some routing and traffic flow intelligence that permits it to act as a relay to pass on appropriate information to an adjacent node. This configuration contains features common to both the true star and mesh networks: It has fewer direct transmission links (similar to the star network), but it also has intelligent relay capability at each node (similar to the mesh network).

The philosophy underlying the star network topology was adopted by Federal Express in the design of its overnight package delivery service. Rather than dispatching planes on multiple point-to-point routes to deliver packages, each night all flights converge on a single switching hub in Memphis. There, packages are sorted, loaded, and then flown to the final destination city. A large part of the success of Federal Express can be attributed to the efficiencies inherent in the hub/switching star network concept.

Another example of the power and efficiency of the star network design is the rapid growth in VSAT satellite systems previously described. The standard VSAT network uses the star topology, with all switching and routing intelligence concentrated at a single hub/switching station. All traffic on the network flows through the hub station, and from there it is routed to its final destination, either another remote station or a data center co-located at the central hub station itself. Concentrating all of the network routing and control intelligence in the central hub station enables the many remote earth station terminals to be simple and inexpensive, which in turn minimizes the cost of the total network.

The VSAT star network configuration is well suited for point-of-sale data collection applications in retail businesses, for which it is important to obtain quick and accurate reports of sales or transactions trends in order to better manage inventories and purchasing/delivery decisions. VSAT networks are also attractive for applications like credit card verification, electronic funds transfer, and other similar types of real-time data collection or distribution applications involving a large number of remote sites.

The switching/hub function is a key part of an efficient network because it reduces the number of direct transmission links required, and it minimizes duplication of network intelligence software by concentrating it in the hub. Nevertheless, there are important trade-offs in choosing between switching and transmission links in designing a network. The advantage of additional direct transmission links, instead of additional switching capability, is speed and responsiveness. If there are not a sufficient number of direct transmission links, large volumes of traffic will be funneled through relatively few switching hubs and excessive congestion can occur, slowing network response time. Thus the ideal mix of switching nodes and direct transmission links varies from network to network and depends on specific traffic patterns and the capacity of the various network components.

This issue is similar to that faced by the airlines in determining the number of direct (point-to-point) flights versus those with intermediate stops at regional hub/switching cities such as Atlanta, New York, or Chicago. A direct flight is justified if there are a sufficient number of passengers traveling frequently enough between the two cities; otherwise, it is more efficient to provide an indirect flight through a regional hub city. Under the hub and spoke approach, passengers bound for different locations can share the flight to the hub and transfer to other flights at the regional airport switching center.

The key point in this discussion is the importance of evaluating the trade-off between direct transmission links and switching/hub facilities. The resulting network design should strike an appropriate balance between economic efficiency and acceptable response times during peak use periods.

Switching Devices

Switching devices are typically microprocessors or specialized computers located at nodes in a network. They have evolved through a primary emphasis on supporting voice communications requirements; however, modern digital switches are capable of handling both voice and data traffic, and some specialized switching devices are designed exclusively for data.

Switches are arranged in a hierarchical configuration in public networks or large private networks. In the public telephone network, a telephone call placed from a home or office is carried over a twisted wire pair to the telephone company's local central office. At this point, a local call enters a switch and is routed directly to another local central office switch in the area; the destination of the call is outside the local area, it is directed to a local switching device of one of the long-distance carriers. Once the call enters the network of the long-distance carrier, it may be routed through several long-distance switching centers before entering the switch of the local central office in the destination city and then on to a local wire pair leading to the residence or office of the called party.

A public network is designed to be cost-efficient and provide high-quality service. This requires balancing the low response time and high cost of using more direct transmission links against the lower cost and higher response time associated with the use of more hub/switching centers.

Most medium to large organizations have their own switching device called a PBX (private branch exchange) or PABX (private automatic branch exchange). A PBX is essentially a scaled-down version of a telephone company's central office switch. It is designed primarily to switch voice traffic, but it is also capable of switching moderate- to slow-speed data traffic. Organizations purchase and operate their own switches because doing so generally costs less than leasing additional direct lines between an office building complex and the local telephone company. Figure 5-11 shows the main functions of PBXs.

The rationale for buying a PBX is based on the following logic: If the organization's office building has a number of individual offices, each with its own telephone set, and does not have its own switch, then a separate twisted wire pair local loop is needed to connect each office's telephone set to the telephone company's local central office. The cost of these local loops begins to add up, even for a modest-sized building. It is also a very inefficient way to design a network if a significant proportion of the calls are between offices in the same building; such calls will simply travel to the local telephone company's switch and then be routed right back to another office in the same building in which the call originated.

Figure 5–11

PBX Functions

Call Forwarding
Individuals can automatically send incoming calls to another line when they are away.

Call Back When Free
When another user's phone is busy, the PBX automatically lets the caller know when the receiver is free by ringing again.

Call Waiting
A tone lets a user know if a call is received while in the midst of another conversation.

Hunting
If no one answers at one number, the call automatically jumps to another line.

Voice Messaging
This service uses a flexible answering machine shared by PBX users that is capable of receiving, storing, and replaying messages to individuals or a group.

Least Cost Routing
Users are directed to lowest-priced telephone lines. (Some delay can result.)

Programmable Stations
When employees move offices, the system can be reprogrammed so they can retain their original number. Hunting and other services can be maintained as work groups change, without changing physical connections.

Account Management
This service involves reporting of calls by station and destination to assist in cost control and client billing. More sophisticated systems allow the use of account codes to assign costs to projects.

Network Management Statistics
This service provides reports on availability and reliability for the switch and for individual stations. This helps in maintaining the least-cost routing algorithms and planning expansion. Sophisticated systems tie information on blocked lines to customer profiles for account management.

Data Connection and Switching
Lines can be switched for data transmission. Maximum transmission rates vary between 4 and 16 kilobits. Communication to the switch is usually digital (it must be for higher rates). Modems are then needed for external communications only.

External Communications
This device provides support for multiple phone company services, including tie lines and multiplexed digital trunks, usually the T1 standard.

Source: International Center for Information Technologies

It is often more efficient to invest in a private switch (PBX) located in the organization's building and to connect each office telephone directly to the PBX. Then, calls placed to other offices in the same building are switched through the PBX without any need for a local loop to the telephone company. As a result, the number of local loops between the building and the telephone company's central office can be reduced to that number needed to accommodate calls going out of or coming into the building. This number is determined with the aid of special traffic engineering studies conducted to ascertain how many circuits are needed between the PBX and the telephone company's central office to achieve a specified service quality level (that is, the probability of obtaining a busy signal due to circuit unavailability for any given call attempt).

Some nodal/switching devices are designed to handle data traffic exclusively. Such devices can be stand-alone minicomputers or microcomputers programmed to operate as a nodal relay or switch, or they can be specially designed microprocessors that interpret destination address and routing information to ensure that the data is sent to the correct location. These microprocessors also incorporate a buffer memory to temporarily hold blocks of data in a queue waiting to be forwarded to another node in the network.

Servers

Another device that is used as a central hub or node in a LAN is a **network server**, typically a microcomputer or a minicomputer that acts as the central routing and control node of the network. A network server is the central storage repository for files, application software programs used by other devices on the network, and electronic mail and messaging services. It also controls access to the various files, manages file transfers, and ensures data integrity.

Bridges, Routers, and Gateways

LAN technology itself restricts the geographical scope of a given LAN to a true local environment; however, there is a growing need to interconnect two or more LANs to form a wide area network for accessing remote databases, exchanging files, and coordinating an electronic mail system for the entire enterprise. Linking separate, autonomous LAN subnetworks through long-distance transmission links is called internetworking, and three principal node/switching devices were developed for this purpose: bridges, routers, and gateways.

The bridge, the simplest of these internetworking node/switching devices, is designed to connect LANs of the same type that use identical protocols, standards, and transmission methods. A bridge simply reads the addresses on the incoming blocks of data and forwards blocks destined for terminal stations on another LAN to an appropriate node in that particular LAN, without modifying the format or content of the information. This is analogous to leaving the local road system of a given city in an automobile and entering an interstate highway destined for another city in the United States. The rules of the road and signs are uniform throughout the country, and driving from a local street to a distant city is simply a matter of choosing the correct route and locating the final destination in the distant city. The bridge acts as the transfer point in the local network, similar to the junctioin between the local road and the interstate highway system.

The router is another internetwork nodal/switching device designed to link two different types of LANs. The two LANs may differ in terms of addressing schemes, data block sizes, network interface standards, and so on. Therefore, the router must have a level of intelligence sufficient to convert the addressing and control information conventions and protocols of one LAN system to those of another LAN system to which it is connected. To illustrate, consider taking the new underground tunnel from France into England. The road signs and addressing systems in the two countries differ, as do the languages in which they are displayed. Furthermore, cars drive on the left side of the road in England

and on the right side in France. These conversions and translations are made as you cross over the subterranean border; at that point (analogous to a router), you immediately shift your driving to the other side of the road and start translating the other language to interpret road and street signs on the way to your destination.

A router is appropriate for connecting subnetworks that differ in operational details but have a similar overall system architecture and use compatible formats and data structures at the user application software level. **A gateway** is the internetwork nodal/switching device that is used to interconnect totally dissimilar subnetworks, such as a LAN of IBM PCs and an Apple Macintosh LAN. A gateway is the most complex of the internetworking devices because, in addition to performing the same translation functions as a router, it must also convert and reformat the underlying data and application program structure. In the travel analogy, the gateway is the equivalent of a border crossing where you not only change countries but also the mode of travel, such as from land to sea. In such a case the car must be completely unloaded at the border, and everything in it must be repacked in a different manner on board a ship headed for the desired destination.

Routers and related equipment such as smart hubs (a hub is equivalent to a LAN node) are the fastest-growing part of the telecommunications equipment market, mainly because of the increasing use of LANs and the equally increasing need to interconnect LAN to LAN and LAN to WAN.

The Roles of Client, Designer, and Implementer with Respect to Transmission Links

Transmission link systems determine the tangible structure of a network. In spite of this visible presence, the client does not really need to know anything about the transmission links used to form the network. The client's role is to establish guidelines for the trade-offs among performance, reliability, and cost. The designer's role is to choose a transmission link system design that both provides the appropriate business functionality (range, reach, and responsiveness) necessary to support the strategic and competitive goals of the company and conforms to the client's performance, reliability, and cost guidelines. This clearly must include an evaluation of the performance characteristics of alternative media and the use of techniques (such as multiplexing and polling) that are designed to ensure efficient use of the media. In fact, it is likely that a designer would configure the network to include several transmission link systems that make use of different media: a fiber-optic or coaxial cable system for some parts of the network and satellite or terrestrial microwave systems for other parts. The role of the implementer is to evaluate and choose the specific equipment for the transmission links and to construct, install, and maintain them. The implementer chooses the specific transmission link systems, but they must conform to the overall design features and performance standards determined by the designer.

Summary

Transmission links are either guided or unguided. Guided links are enclosed in a wire or cable, whereas unguided links are a wireless signal that is usually broadcast. The main guided medium in use today is twisted pair cable; this is the standard telephone wire that connects your phone to the wall outlet. All communications signals attenuate—fade and lose clarity—over distance. Unshielded twisted pair has a more limited speed than shielded twisted pair and can operate over a smaller distance.

Fiber optics are the most powerful transmission medium, with a bandwidth of 2 Ghz or more. It offers speeds on LANs today of up to 100 mbps, compared to 16 mbps using coaxial cable. Its signal has less attenuation over distance, which reduces the number of repeater devices that must be placed at intervals along the link to regenerate and amplify signals.

Unguided media include terrestrial line-of-sight microwaves, satellites, and broadcast radio. The broadcast nature of these media make them particularly well-suited for mobile communications, such as cellular data and voice communications, and for communications in which a large number of terminals receive the same information, such as retail stores receiving price data.

Transmission links are an expensive resource, generally the largest single component of the telecommunications budget. High-speed links are shared among many simultaneous users through multiplexing and polling. Multiplexing shares the link on the basis of either frequency or time division. Frequency division multiplexing divides the signal into smaller channels that are used by individual terminals; time division multiplexing allocates capacity in time slots. Polling is a process of checking which terminals are ready to send and receive and then allocating the link to each of them in turn.

Digital transmission links typically provide channels of 64 kbps up to 274 mbps. These levels of service are provided by long-distance telecommunications and wide area network services. U.S. and international levels of service are slightly different. U.S. providers offer DS-0 (64 kbps) up to DS-4 (274 mbps). International providers also offer a 64 kbps (level 0) service; level 4 service provides 139 mbps.

Nodes and switches interconnect transmission links. They reduce the number of lines needed by organizing the links in the configuration that is most cost-effective for the given applications, rather than providing an impossibly expensive point-to-point link between every pair of terminals on the network. Switching devices that route traffic through the network and manage its flow include PBX, bridges, routers, gateways, and smart hubs.

Review Questions

1. List five different substances that are commonly used as transmission media.

2. Is satellite transmission a form of guided media? Explain.

3. Are unguided media capable of carrying all forms of information (voice, video, data, images, and so on)? Explain.

4. Compare twisted copper wire pairs, coaxial cable, and fiber-optic cable in terms of transmission capacity and the conditions under which each is most appropriate and most likely to be found in a network.

5. What does "bypass" mean in a telecommunications context, and how has the availability of fiber-optic cable facilitated this activity?

6. Compare terrestrial microwaves, satellites, and broadcast radio in terms of the types of applications that are most appropriate for each and why.

7. Current trends indicate that satellites and fiber-optic cable are dominating the use of other media. Why is this happening?

8. Explain the concept of multiplexing and its impact on the efficiency of transmission facilities.

9. Briefly explain the difference between frequency division multiplexing (FDM) and time division multiplexing (TDM), and describe appropriate uses for each.

10. Briefly describe the difference between synchronous time division multiplexing and asynchronous or statistical time division multiplexing, and give appropriate uses for each.

11. What is the basic principle behind signal compression techniques? Give an example of the use of each technique.

12. What is the main difficulty with a multidrop line or circuit, and how is this problem solved?

13. Explain the differences among a 4-Khz circuit, a DS-0 circuit, and a level-0 circuit.

14. What are the differences among a T1, DS-1, and level-1 circuit?

15. What is the primary advantage of using a hub or switch to create a star network configuration?

16. What is the currently popular satellite system that uses the star configuration with a central hub/switching center? Briefly describe how it works and why it is enjoying increasing popularity.

17. What are disadvantages of the hub/switching concept that limit the applicability of pure star networks?

18. Why do so many companies have their own switching devices in the form of private branch exchanges (PBX) instead of relying on the local telephone company to switch the calls; isn't this a wasteful duplication of equipment?

19. How is a local area network server like a true switch, and how is it not?

20. Briefly explain the differences among a bridge, a router, and a gateway.

21. What do clients, designers, and implementers need to know about transmission links, and what are their respective responsibilities?

Assignments and Questions for Discussion

1. If the content of any message is independent of the transmission medium used to carry the information, then what difference does it make whether we use air, water, fiber-optic cable, or some other substance for the transmission links in the network?

2. Use the analogy of an airline transportation system to illustrate the role of transmission links in a telecommunications network. Be sure to identify the airline version of nodes and switches, and discuss the trade-offs relative to direct connections.

3. A new chief financial officer has just been appointed in your company, and you, as a member of the telecommunications department, are to report to her. She knows little about telecommunications technology, and you have been asked to prepare an overview briefing on some methods to more efficiently utilize network transmission link facilities in the company's private network. List the main points you would address.

4. Discuss the implications for transmission links in the design of corporate networks of the following trends:
   a. the widespread availability of high-capacity transmission facilities at lower prices
   b. the growing demand to access corporate networks with mobile and wireless terminals

Minicase 5-1

European Payments System Services

The Role of Switches in Improving Network Performance

European Payment System Services (EPSS) is owned by several of Europe's major credit card providers: Eurocard, Eurocheck, and MasterCard International. It routes authorization data among merchants, ATMs, and the banks that issue the cards and handle the accounts. It routes Master-Card transactions over private transatlantic links to its U.S. BankNet network.

Until 1992, the EPSS network was a traditional hybrid network—a mix of public X.25 data network links and private lines that route data between IBM Series 1 minicomputers in 27 countries; the computers act as communications front-end processors. The Series 1 is a workhorse; developed in the early 1980s, it is reliable, limited in power, but well-suited to the bursty, short messages involved in credit card operations and to accessing large databases held on powerful IBM mainframes.

However, the Series 1 machines are slow and cannot efficiently handle today's volumes. It takes up to 12 seconds to authorize a transaction, which is unacceptable in, say, a large discount store at 7 P.M. during the week before Christmas. EPSS has used a phased timetable to replace the Series 1 with integrated switches from Netrix Corporation. "Integrated" here means that the switch handles both circuit switching and packet switching automatically at the hardware level, which is far faster than through software. These #1-SSS switches are the base for a new mesh backbone network, with five main nodes, one each in Belgium, France, and Germany and two in the UK. The mesh network links each of the switches to the others, ensuring a high level of backup and reliability.

The #1-SSS switches improve response times by 30–40 percent. Uptime is over 99.98 percent, versus around 90 percent for the old network. According to EPSS's network operations manager, "Response time and availability are important competitive issues" for EPSS.

EPSS took over a year to select a vendor. It chose Netrix for a number of reasons: The #1=ISS supports a wider range of speeds than its competitors and includes powerful network management facilities, with detailed diagnostics and reports on network operations and errors. Its design supports the defined but not yet fully implemented OSI network management standards. It is easy to operate—"a technician's dream and an imbecile's delight"—and allows a phased migration; EPSS began with X.25 packet switching and is moving selectively to circuit switching for high-volume applications. Netrix also had an edge by having support staff in 19 countries, as compared with 15 for its closest competitor.

Questions for Discussion

1. Discuss the weaknesses in EPSS's traditional hybrid network that led the organization to consider other alternatives.

2. What are the advantages to EPPS of using an integrated switch that handles both circuit switching and packet switching? Are there any disadvantages?

3. Didn't EPPS waste a lot of time by taking more than a year in selecting the switch vendor for the new network? After all, how much difference can the choice of a vendor make? Explain and discuss.

Minicase 5-2

Rockwell International

Building a Global Platform

Rockwell International has 28 divisions within its five lines of business in manufacturing electronic, aerospace, automotive, and graphics products. Much of its work is application-specific and customized; it takes orders from Europe and fills them in the United States because it is too expensive to replicate high-tech manufacturing operations on both continents. Nor is it easy to replicate skills; Rockwell employs around 1 percent of all scientists and engineers in the United States.

Up to the mid-1980s, Rockwell had very limited international telecommunications links. Its many computers used stand-alone applications and were incompatible; they were selected by the various business units to best meet their specific needs. In the early 1980s, engineers at the fabrication plant in Newport, California, would fly to Europe, 15 hours and at least eight time zones away, to pick up computer tapes of design plans. These tapes would be processed on the California CAD (computer-aided design) system, and the engineers would take another trip to Europe to reload the CAD work on the European computer.

In the late 1980s, more and more of Rockwell's key customers, especially those in the automotive industry, demanded electronic links between themselves and the manufacturer. Rather than add these links piecemeal, Rockwell decided to create a pan-European-to-U.S. platform, called EURONET. The business drivers for this were the recognized needs to speed up access to information, to handle product specifications and changes more rapidly and with more control, and for leverage for the advanced engineering and scientific computation that is Rockwell's distinctive business edge.

The key feature of the network is that it is designed to handle data traffic that is very text- and graphics-intensive, unpredictable in volume, and transmitted across great distances. This traffic includes CAD files, engineering designs, product configurations, documentation, drawings, and commercial transactions, such as order entry, bills of material, and electronic data interchange. The director of information systems technology at Rockwell commented that such traffic is very different from "nice little neat financial transactions."

Small transactions of that type could easily be handled by an X.25 packet-switched network; Rockwell needed very high-speed, low-overhead transmission. The backbone network links the main European data center in London to Dallas and Seal Beach, California. The Seal Beach location has three giant supercomputers, the fastest commercial machines on the market, plus six top-of-the-line IBM mainframes; Dallas has six large mainframes, and London has comparable capabilities. Each of the data centers has a large switch that connects it to the others through 71 T1 digital circuits; each of these operates at 1.544 mbps. The main network architecture is IBM's SNA. In addition, the network supports DECnet and TCP/IP, the most widely used standard for transmitting data from computers that use the UNIX operating system. Engineers can access any processing resource across Rockwell's geography. A major beneficiary of the new ability to transfer engineering data virtually instantly has been Rockwell's printing business; it was able to get a new color liner product to market far ahead of expectations—and of the competition.

Smaller and "nice little neat" transactions cost nothing because they require little of the network's capacity. Electronic mail, smaller project designs, and orders can be economically handled, but the cost of creating a network just for these uses would have been excessive. Rockwell has used the platform to help set up a pan-European mail system, migrate U.S. supplier databases to London, and integrate the London office with all Rockwell's U.S. offices. New application software releases are distributed through EURONET.

The network is being continuously upgraded. It is too slow! Even with T1 circuits, there are bottlenecks in transmission of massive, high-resolution graphics files and large-scale databases. The main needs are for bandwidth on demand, which points toward a shift from a private network to a virtual private capability. But it will take time to evolve in a multivendor international environment. Security is an issue, too, because TCP/IP lacks sophisticated security features. It is intended to facilitate ease of access across heterogeneous networks. Access and control conflict, by definition.

Rockwell's director of IS credits customers for the innovation. The demand for electronic data interchange stimulated a rethinking of Rockwell's business operations, and that in turn created the technical design: "The customers defined the requirements for the network, and all the different pieces of Rockwell got together to make it work. It was just like the Harvard Business School says it should be. It was a strictly business-driven approach. We try to do only the things where we're responding to a business need. We're not building networks just because it would be nice to build them."

Questions for Discussion

1. What were they key business drivers that led to the development of Rockwell's global network, and what implications did they have for the types of transmission links selected for EURONET?

2. Why did the integrated platform concept make sense for Rockwell, instead of two separate networks: a high-capacity net for graphics-intensive CAD files and a low-capacity X.25 net for e-mail, order entry, and electronic data interchange applications?

3. Even though EURONET was considered a high-capacity network when it was built, it is already too slow for some applications, including transmission of high-resolution graphics and large database files. Is the solution to simply increase the capacity of the private transmission links? Discuss.

4. Describe the range, reach, and responsiveness levels of the EURONET platform. Are they adequate for Rockwell's needs? Discuss.

Minicase 5-3

Analog Devices

Evolving an International Network

Analog Devices's experience in developing its European network is typical of many large firms. It initially operated a number of separate networks that handled data processing transactions; as each new software system was installed, data communications links were added on an as-needed basis. Different business units used different accounting and sales systems. Each country processed its own transactions, including order entry and finance. This largely ad-hoc approach has been standard for most companies into the early 1990s and reflected both the relative unimportance of telecommunications and the emphasis on decentralized decision making.

What is also standard is Analog Devices's shift to replacing the costly and inefficient multiple stand-alone systems with an integrated network. The company has sales of over half a billion dollars, half of them coming from Europe, and it saw an opportunity to save millions of dollars by using both private lines and X.25 public data networks to route order and financial data to a central minicomputer in Zug, Switzerland. It wanted to improve its internal efficiency, upgrade its software, and reduce unnecessary processing and administrative costs, and the network was they key to achieving this. One manager commented in late 1991 that "our network is driven by our desire to become much more efficient. I don't think we could do what we're doing and achieve reductions in expense levels without the network."

Analog Devices phased the network in carefully. First, it used simple dial-up X.25 lines that provided each European office with access to Zug, where it had installed a new centralized finance and order entry system. Data could be routed to offices in Santa Clara, California, over private leased lines provided by BT (British Telecom) North America's Tymnet international value-added network. BT had acquired Tymnet, a well-established X.25 VAN, specifically for the purpose of capturing the business of international firms, who needed "one-stop shopping," the ability to implement international links without having to deal with a multitude of vendors and PTTs, a major problem for decades.

The dial-up links were then migrated to dedicated 9.6-kbps Tymnet lines. This took eight months, which reflects the immense difference in speed and quality of service of many international PTTs with whom Tymnet has to deal. The first link was installed in the Netherlands, which had privatized its PTT to better respond to business needs for communications. The last link was installed in Italy, a country with notoriously poor telecommunications; the Italian PTT has entered into a multibillion-dollar contract with AT&T to transform its entire telecommunications infrastructure.

The main advantage of the dedicated links is cost. Analog Devices pays a flat rate of just under $2,000 a month for each office. (With the dial-up links, it paid for each transaction.) Just under a year later, the company cut over—a term that means operations were quickly transferred, without being taken out of service—to a high-speed 64-kbps private line that linked its British network hub to the Zug host. The UK hub links into the firm's private network connecting the United States and Asia; the links between the United States, Zug, and the UK hubs constitute the backbone network. The 64-kbps line was needed to handle backbone traffic, which comes from many offices, whereas the 9.6-kbps links handled traffic to and from a single office. Sales and product managers now are able to access data directly, across the firm's entire geography, particularly from the main host computers in its Massachusetts headquarters. The most difficult problem has been reliability. The Tymnet link in Switzerland was down for two full days in late 1991, and the firm was considering choosing another value-added carrier or installing its own private net.

The experience of Analog Devices is typical with respect to the way a fairly simple but practical network can be developed. The starting point, obviously, is to choose the business opportunity—in this case, to improve internal coordination and streamline and simplify administrative processes. There was no decision to coordinate complex engineering operatiions (compare this with Rockwell International in Minicase 5-2). Analog Device's business choice helped clarify the choice of network architecture, in this case X.25.

The development was cautious. Within the X.25 architecture, Analog Devices moved from 9.6-kbps dial-up (the simplest option) to dedicated links and then to a private 64-kbps capability.

Although this adds capacity and cuts costs, it increases risks and dependence on the telecommunications provider. BT Tymnet's European problems are common ones; Analog Devices may decide to move to a private network to add reliability, even if that does not directly cut costs.

Many U.S. companies rely on VANs for international links, and many telecommunications providers have acquired foreign VANs to position themselves for a rapidly growing market: BT (United Kingdom) bought Tymnet at a huge premium over its market value, AT&T bought the British Istel, and MCI became the main shareholder in Infonet, originally created by a group of European PTTs.

Questions for Discussion

1. What are the key business drivers that led to the development of Analog Devices's European network, and how did they affect the types of transmission links selected?

2. Explain the trade-offs between dial-up links provided by international VANs and dedicated private lines for carrying Analog Devices's traffic. Also, discuss the trade-off implications of moving to higher-speed, dedicated, transmission links (that is, 9.6 kbps to 64 kbps) in an international environment.

3. Why did it take Analog Devices eight months to migrate from dial-up links to a dedicated, private-line backbone network. Explain and discuss.

4. What suggestions would you offer to the management of Analog Devices on how to deal with unreliable transmission link facilities in some countries in Europe?

Minicase 5-4

Reliance National

Bridges and Routers for "Mission-critical" Applications

Reliance National, founded in 1987, provides casualty insurance. Its annual revenues are close to $1.5 billion. It views several of its processing systems as "mission-critical"—as vital to the firm's operations, reputation, and competitive positioning. In developing these systems it took care to separate backbone LANs from departmental LAN applications, to make sure it could isolate problems and respond to them quickly without putting other systems at risk. The core mission-critical system was written in the most widely used business programming language, COBOL, and installed on a Novell NetWare LAN. Volumes grew rapidly, and the COBOL system could not handle them efficiently. Reliance purchased a substitute insurance processing system that ran on a mid-range IBM AS/400 computer, and it kept departmental word processing and spreadsheet applications on departmental LANs. The AS/400 manages the corporate database, accounting transactions, and insurance services.

The logic of separating departmental and mission-critical applications led to a profusion of bridges. Each of the 18 departments' LANs needs access to the AS/400. An interdepartmental NetWare backbone LAN required a bridge from each of them, as did the AS/400. To ensure redundancy, each bridge had a backup bridge, so that there were two IBM bridges and two Novell bridges per LAN, each a potential point of network failure.

The telecommunications manager needed a simpler and more reliable solution. He examined a range of suppliers of multiprotocol bridge/routers that could halve the number of internet devices. Most of the products were expensive and offered more features than Reliance needed. He selected Proteon's P4100+, which allows two departmental LANs to share a common internet unit. It has three token ring slots, so it can attach two departmental LANs to the backbone LAN. Two P4100+s can connect two LANs to both the IBM and Novell backbones; one links the two LANs to the AS/400 network, and the other connects them to the Novell one. Previously, it took eight devices to do the same thing.

One valuable byproduct of installing the P4100+ devices is that Reliance now has unified network management because the bridge/routers use the Simple Network Management Protocol.

Questions for Discussion

1. How did Reliance National configure its network system to ensure that the "mission-critical" processing applications were protected and that problems could be identified and resolved quickly?

2. Evaluate the use of bridges in the original network configuration. How were they used, and was this use appropriate to meet the business need? Explain.

3. Explain how substituting multiprotocol routers for bridges in the Reliance National network greatly simplified the internet configuration, and discuss the overall benefits of such a change.

6

Transmission Methods

Objectives

▸ Understand the role of transmission methods in network operation

▸ Learn the function of communications protocols

▸ Understand the characteristics and appropriate use of circuit- and packet-switching techniques

▸ Examine in detail some key transmission methods

Chapter Overview

| | | | | | |

Chapter Overview

This chapter describes the additional functions that are required to transform the physical structure of a telecommunications network into an efficient and reliable means for transporting information. Terminals, switches, and transmission links constitute the physical network, but their operations must be coordinated through transmission methods. Transmission methods generally fall into one of three broad categories: methods to establish and terminate network connections, rules for the orderly transfer of data across the network, and procedures to manage and control transmission link operations. The rules and procedures for performing these three functions are called protocols, and the chapter contains some examples of widely used protocol systems.

The two methods used to establish and manage connections across the network are circuit switching and packet switching. The chapter reviews these and the new fast packet-switching methods that permit the integrated digital transmission of voice, data, and video information at very high speeds across the network. It ends with an analysis of selected protocols that are both widely used and representative.

| | | | | | |

The Role of Network Transmission Methods

The two network building blocks discussed so far—terminals and transmission links—represent the physical shell of the network. Transmission methods are additional functions that are needed to bring the network's physical structure to life and to ensure that a reliable bit stream or analog signal representing some form of information reaches the correct destination. Three functions are necessary to the operation of any telecommunications network: connection establishment, data transfer rules, and link control.

Connection establishment involves making a connection between two terminals on the network and releasing this connection upon termination of the session. This process requires four steps: (1) locating the destination device through a network addressing scheme, (2) establishing a path through the network that links the two terminals, (3) signaling the receiving terminal to prepare to exchange information, and (4) releasing or tearing down the connection upon completion of the communication session. Data transfer is the movement of information from sender to receiver using precisely defined procedures for formatting the data and synchronizing transmission; data link control procedures ensure that the data is correctly sent and received and that any errors are recognized and resolved.

The transmission path that is established may be a physical connection that is dedicated to the communicating parties for the duration of the session; alternatively, it simply may be a virtual path through the network. In a **virtual path**,

Figure 6–1

A Virtual Path

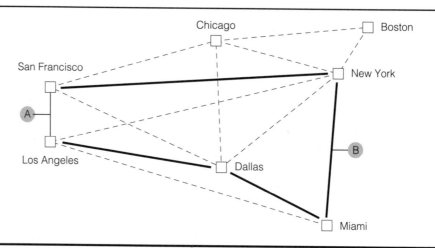

A virtual path is used only for the duration of transmission of a specific message. When the network shown here is congested, path A from Los Angeles to San Francisco may be much less efficient than virtual path B: Los Angeles to Dallas to Miami to New York to San Francisco.

the route along which information flows between two terminals is established through the network, but it is not dedicated exclusively to these two terminals and may be shared in other communication sessions connecting other terminals (see Figure 6-1). A dedicated path connection is used for voice traffic and some types of data communications; a virtual path connection is used primarily for data transmissions when the information is assembled into packets and sent as discrete blocks.

Each device, or entity, on the network that either sends or receives information must have a unique address in order to ensure that the circuit path that is established does in fact connect the two terminals that requested the communications session. One important problem that the connection establishment method must solve is how to connect two parties attached to different subnetworks that use different addressing schemes. Internetworking connection/switching devices such as bridges, routers, and gateways are designed to handle this problem. A network is in effect defined by its directory of addressable units. Until the emerging X.500 standard is fully implemented in practical products, it will be impossible for networks to interoperate directly without special equipment and/or software.

Once the connection is established and a path or virtual path defined, procedures and methods need to be established for how data will be transferred over the network. For example, are both terminals permitted to transmit an information stream at the same time, or is only one-at-a-time transmission allowed? A procedure is needed to synchronize the terminals involved in data transmission so that the receiving terminal knows which parts of the incoming bit stream represent control information and which parts are data. Also, a method is needed to ensure that the incoming information stream is assembled in the correct order, for all parts of a message or data stream do not necessarily arrive in the order in which they are sent under some transmission schemes.

Finally, once the data transfer procedures have been established, methods are needed to ensure that the information arrives intact and without error. Flow control methods are designed to adjust the speed of the transmission flow so that a fast transmitting device does not overwhelm a slower receiving device. Error control methods are designed to detect errors in transmission and to request the sending terminal to retransmit the block or segment in which an error was found.

All of these functions must be performed in some manner that is understood by all the communicating entities if reliable and effective communication is to take place across a network. It is not enough to simply construct transmission links and switching nodes that connect a group of terminal devices and then to assume that somehow messages and data will arrive at the appropriate destination still bearing a resemblance to what was sent. Sets of specific rules or procedures for performing one or more of these network functions are called protocols and are the subject of the next section.

Communications Protocols

The simplest definition of a protocol is the set of rules or conventions by which two machines talk to each other; perhaps more accurately, a protocol is a strictly defined and precisely coordinated set of rules and conventions. Protocols are procedures, and there are protocols for just about every function of a telecommunications network, starting, for example, with the rules for sending a signal from the phone receiver to the central office. The protocols approved as standards by the CCITT, for instance, include: X.51, which defines a method for multiplexing data across the international interface between two synchronous data networks; X.75, which specifies terminal and transit call control procedures and data transfer on international circuits between packet-switched networks; X.25, the dominant packet-switching protocol; X.400, which defines procedures for routing electronic messages, such as electronic mail, across different networks (and hence across directories); and X.500, the emerging protocol for handling network addressing across networks.

Individual vendors have their own protocols. Apple Computer uses its own AppleTalk protocol for PC communications within a local area network. Digital Equipment defined its Digital Data Communications Message Protocol (DDCMP) as part of its DECnet architecture. IBM similarly defined its own Binary Synchronous Communications Protocol (BSC), and Sperry Univac and Burroughs (now combined to form Unisys) defined their Universal Data Link Control and Burroughs Data Link Control protocols. IBM added its SDLC (Synchronous Data Link Control), and the International Standards Organization adopted HDLC (High-level Data Link Control). All these protocols carry out essentially the same function.

The concepts of a protocol and a standard are closely related. A standard is generally defined as the *rules* for interfacing two services or pieces of equipment; a protocol is the *procedure* for achieving the interface. In practice, the two terms are often used interchangeably. A network designer rarely needs to understand the protocol, only the standard; he or she can assume that the engineering has been correctly defined and carried out. An implementer needs to know far more about specific protocols, especially when they have been implemented slightly differently by equipment providers.

Protocols refer to individual functions within a network's operations or to operations across networks. An architecture, the subject of Chapter 7, is a hierarchy of protocols and standards that define the full capabilities and characteristics of a network. The architecture for a public data network, for instance, generally includes the X.25 packet-switching standard, the X.21 interface for full-duplex data transmission, the HDLC protocol, and many others. (The "X." is one of the conventions used by CCITT in categorizing standards; the "X." series relates to data communications, and the "V." series relates to telephony.)

IBM's and Digital Equipment's communications architectures are built on a hierarchy of protocols that are very different from the protocols of packet switching. Local area networks differ widely in the protocols they use, with Ethernet and token ring the main ones. Each of these networks increasingly incorporates standard protocols and/or adds interfaces that allow intercommunication across architectures. Standards-setting organizations frequently try to resolve differences among protocols by adopting one as a standard or by defining a common interface between them.

The entire subject of protocols, standards, and architectures can be bewildering. As telecommunications professionals often wryly state, "Standards are wonderful—there are so many of them." There are even more protocols, thousands and thousands of them.[1] Think of protocols as procedures—"This is how to XYZ"—of standards as interfaces that assume specific protocols, and of architectures as a comprehensive hierarchy of standards and protocols.

Connection Establishment Protocols

A set of specific rules for performing the required tasks of establishing a connection for a communications session and terminating it at the conclusion of the session is called a connection establishment protocol. In a nonelectronic, person-to-person communication, the connection establishment protocol consists of the following steps: (1) Look up the address of the person with whom you wish to converse, (2) find the address on a map to determine the most direct route to follow to reach the destination location, (3) travel to the destina-

---

[1] The reason for so many protocols is twofold: the need for absolute precision in procedures, right down to the individual bit level, and the escalating pace of technical innovation in areas in which either existing protocols are inadequate or the new application is unique. For example, the expanding range of digital wireless technologies and uses is breaking new ground and requires new protocols and standards.

tion location following the designated route, and (4) upon completion of the conversation, return home along the designated route.

In a telecommunications network, the connection establishment protocol is in the form of computer software that includes the electronic equivalent of the same steps. The software interprets the destination address, establishes an efficient path through the network for the duration of the session, and releases the facilities (transmission circuits and switching components) constituting the path upon completion of the session. The connection establishment protocol may be a connection-oriented service or a connectionless service.

A connection-oriented service designates a specific path through the network for the duration of the session; the public telephone network is one example. In **connectionless service**, used in some packet-switching systems, a communication session between two terminals does not establish a specific path through the network to carry all of the session traffic. This is analogous to mail delivery by the post office in which a complete set of communications—equivalent to a communications session—may involve individual pieces of correspondence sent in separate envelopes addressed to the same location. Each envelope may take a different route to the destination address, and the envelopes may not arrive in the order in which they were mailed. Although a connectionless service protocol does not establish a specific path through the network, it includes logic in each switching node in the network that interprets addresses and routes information to adjacent nodes closer to the ultimate destination, such that traffic congestion in the network is minimized.

Data Transfer Protocols

Following connection establishment, a data transfer protocol must specify the rules for the actual exchange of information across the network. In a nonelectronic, person-to-person context, an example of a data transfer protocol might be the following set of rules governing a conversation: (1) Only one person speaks at a time (even though both parties are technically capable of speaking simultaneously); and (2) Both pauses in speech and prefatory remarks such as "And now I am going to describe the facts of the case to you" tell the listener what type of information will follow; it will be either substantive information (facts or opinion) or control information relating to beginning and ending the conversation, asking a question, requesting clarification or reiteration of a point, or changing the subject matter.

In a manner similar to this example, a data transfer protocol is needed in a telecommunications network to specify whether simultaneous transmission in both directions is permitted (full duplex), or whether only one terminal may transmit at a time and, in essence, share the same transmission circuit (half duplex). In addition, the data transfer protocol also must include a procedure for synchronizing the transmitting and receiving terminals so that the receiving terminal is able to distinguish control information from data and can determine when a particular message or block of data begins and ends.

There are protocols designed to implement both asynchronous and synchronous data transfer methods. The asynchronous method of transmission requires start and stop bit signals that are included with each character to tell the receiving terminal that the next seven bits represent a character. This solves the synchronization problem, but it is very inefficient for transmitting long messages or large blocks of data because timing information is included with every character. In contrast, the synchronous method of transmission incorporates start/stop timing information at the beginning/end of a block of data, which may include 1,000 or more characters. Synchronous transmission methods have much lower overhead associated with them, and therefore, they are more efficient for handling large traffic streams or sizable blocks of data.

Duplex/half duplex and synchronous/asynchronous transmission were described in Chapters 4 and 5 in relation to terminals and transmission links. These basic transmission processes become parts of a protocol only when there is a fully defined set of rules that indicate, for example, the specific timing information for synchronous transmission or the exact start and stop bits for asynchronous transmission. Thus many protocols use synchronous transmission techniques, but in itself synchronous transmission is not a protocol. Even though IBM's SDLC, for instance, is byte-oriented—message information is organized and processed a full byte at a time—and HDLC is bit-oriented, they are both synchronous protocols.

Link Control Protocols

In addition to specific procedures for handling connection establishment data transfer, a link control protocol is needed to manage the flow of information across the network and to detect and correct errors. In the nonelectronic, human communication example, visual and direct verbal feedback are used to manage the flow control problem. Thus, if the speaker begins talking rapidly, and the listener begins to look overwhelmed or actually requests the speaker to slow down, then the speaker adjusts her or his talking speed based on these feedback signals. Error detection and correction are handled in a similar manner. If the listener receives a garbled message or simply does not understand what the speaker is saying, the listener simply interrupts and requests that the previous statement be repeated, and the conversation continues.

A telecommunications network flow control protocol generally includes some form of buffer capacity to temporarily store a limited number of digital blocks of information, which gives the receiving device time to process the preceding blocks. The flow control mechanism also typically includes some form of acknowledgment system to indicate that specific blocks have been received. Some versions of this method send a receipt acknowledgment to the sending device after each block is received, whereas others acknowledge receipt only after a specified number of sequential blocks have arrived.

Error Control Protocols

Error control is becoming increasingly important in modern telecommunications networks because businesses and individuals have a growing reliance on the information being sent. For example, errors in electronic funds transfer data are clearly unacceptable. Therefore, greater emphasis is being placed on efficient and reliable error detection and correction methods. In addition, error control methods represent a trade-off opportunity with other network design variables that can result in lower network cost and greater flexibility in the types of transmission links used. For example, without a reliable and efficient error control protocol, transmission links in the network would have to be engineered to deliver higher signal-to-noise ratios to ensure error-free transmission of data. This typically requires more costly transmission link components (for example, greater bandwidth in the link, more radiated power in the signal transmission, or greater sensitivity in the receiver device). An efficient error control protocol combined with a noisier, less costly transmission link system can yield the same level of data transmission accuracy as a more expensive, low-noise transmission system with no error control protocol.

An error control protocol performs two functions: error detection and error correction. Error detection schemes for asynchronous transmission methods involve the introduction of some form of planned redundancy of bits in the data stream. A typical asynchronous error detection protocol will use a coding scheme that introduces a redundant bit at the end of each character block, such that the total number of 1s in the binary character code sums to either an odd or an even number. In such a system, the receiving device is programmed to test whether the sum of the 1s in each character block received is consistent with the protocol rule (either odd or even). If it checks properly, the block is accepted and an acknowledgment is returned to the receiver. If it does not sum properly, an error message is sent to the receiver, and the sending device retransmits the character.

Error detection protocols for synchronous transmission methods are more efficient (fewer overhead bits are used) and more powerful (more errors are detected) than the simple bit redundancy methods used in asynchronous transmission systems. These protocols use some form of mathematical calculation performed on the bits in each block of data being transmitted, such that the result obtained is unique with respect to the bit pattern being sent. The resulting number is inserted in a control portion of the data block being sent. The receiving device performs the same calculation on the arriving block of data and then compares the result with the number calculated by the transmitting device.

The most common calculation is **cyclic redundancy checking (CRC)**, which uses a variety of often-complex mathematical formulae that typically add the remainder from a division operation to the transmitted message as a check sequence. CRC techniques allow only about three bits per hundred million to be incorrectly transmitted, a bit error rate of 3×10^{-8}.

The Layered Protocol System Concept

Clearly, the system of protocols necessary for efficient network operation is complex. The complexity increases as technology changes and networks grow in size and sophistication. For example, before the proliferation of the personal computer and local networks of personal computers, telecommunications essentially consisted of both the public telephone network developed and managed by AT&T in the United States and teleprocessing computer networks linking remote dumb terminals primarily to IBM mainframe computers. As a result, protocols for connection establishment, data transfer, and link control were developed for well-defined, homogeneous types of networks whose components were supplied by, or engineered to the specifications of, a single vendor such as AT&T or IBM. International committees defined standards as needed. The CCITT dominated telephony standards because its membership was composed of the monopoly PTTs. The technology did not change rapidly up through the 1970s, so the lengthy process of drafting and defining a standard was acceptable to the monopolists, and users had no other real option.[2] The variety of data communications protocols was limited and dominated by a few providers, most obviously IBM.

As use of minicomputers and PCs grew rapidly and these devices were connected in local network configurations, a demand emerged both to interconnect a number of local networks over large distances and to connect the local networks to large mainframe computer clusters. In addition, many of the local network configurations were using equipment and protocols developed by a variety of different vendors, which made interconnection even more difficult. New types of traffic, such as electronic mail, computer-to-computer file transfers, and video, required new protocols, most of which were vendor-specific. The speed of technical change often made the slow, formal standards-setting process irrelevant. Users defined de facto standards by their equipment purchases and network design decisions. For instance, TCP/IP, one of the most widely used communications protocols for UNIX-based applications, was not explicitly designed as a standard; it was far more the product of graduate students at the University of California, Berkeley developing and extending something that worked and companies picking it up and using it because it worked.[3]

In order to deal with this complexity, heterogeneity, and rapid change, the telecommunications profession has developed what amounts to a new discipline for definition of standards and design of networks. The specification of network architectures partitions the various transmission method protocols into layered, self-contained modules. Each modular layer contains only those protocols designed to address a specific, interrelated set of transmission method functions. The best-known of these, the Open Systems Interconnection reference model,

---

[2] ISDN is an example of lengthy definition and slow rollout: It took well over a decade to agree on the standard and another decade to implement it by monopoly PTTs. By contrast, in today's highly competitive personal computer and LAN industry, time is a critical resource. Vendors will not wait for committees to weave their way through a maze of debates and paperwork until they reach a consensus.

[3] Of course, this pragmatism and ad hoc zeal is the reason that TCP/IP lacks many features that were of little interest to the hackers. The main gap is security. University researchers wanted to ensure ease of access and open transfer of information; businesses, by contrast, want control of access and information.

was briefly reviewed in Chapter 2. The OSI model has seven layers; within each layer are many protocols, each of which is independent of any of the protocols in a layer above or below it.

Telecommunications evolves to more and more interoperability both as individual protocols are defined and implemented to handle a specific need and as protocols are combined to create standards and architectures. For example, OSI (see Figure 2-22) includes an HDLC link control protocol (Layer 2), the X.25 packet-switching protocol (Layer 4), the file transfer and access method protocol (Layer 5), the X.400 protocol (Layer 6), as well as many others that are either fully implemented or in the process of being defined and/or implemented. Interested parties will often propose that a specific protocol be included in OSI or adopted as a standard by one of the major standard-setting organizations.

The layered/modular approach to dealing with technical complexity and rapid change is analogous to the modular design of stereo components for home entertainment systems. Most modern stereo systems incorporate all of the broadcast receiving and amplification functions in one piece of equipment: the stereo receiver, which consists of an AM/FM tuner and amplifier. The sound-input functions provided by an audio cassette player or compact disk player are included in separate pieces of equipment. Finally, all of the procedures and operations related to final sound output are contained in the speaker component subsystem. All of these separate modular components are interconnected through standardized electrical interfaces; thus a compact disk player made by Sony can be connected to a receiver/amplifier made by Kenwood, which in turn can be connected to a set of Bose speakers. These individual modular components are self-contained, and each is interchangeable with another vendor's unit designed to perform the same function, so long as the electrical interface for the input/output of the new unit is compatible with accepted standards.

A typical modular stereo component system is essentially a three-layer design. The first layer is the prerecorded source-input component consisting of either a compact disk player, an audiocassette player, or an AM/FM radio tuner. The output from this first layer is the input to the amplification layer, which is typically packaged together with the AM/FM tuner into a single piece of equipment, the stereo receiver. Finally, the output of the amplification layer is the input to the broadcast sound layer consisting of a set of speakers. Note that each layer of the stereo system is independent and performs its own set of related, well-defined functions. Layer independence enables interchangeability of equipment to accommodate products of different design made by different vendors (allowing, for example, substitution of a Sony compact disk player for a Panasonic audiocassette player as the sound-input source). This also means that if, say, the engineering design of the compact disk player is changed, there is no need to modify any of the other system components that are performing functions assigned to other layers (such as amplification or final sound output), so long as the output from the modified CD player design adheres to the electrical interface standard for input to the amplification layer. (In contrast, with a fully integrated stereo system embodied in a single unit, a change in the specific design of one subsystem component will likely require modification to other subsystems within the integrated unit, necessitating a costly redesign of the entire system.)

Now that we've examined the stereo system analogy, consider a hypothetical, modular, three-layer protocol system for a telecommunications network. Layer A (data link layer) contains all protocols related to the data link control functions, such as flow control and error control. Layer B (data transfer layer) includes all protocols supporting actual data transfer operations, such as whether transmissions are simultaneously carried in both directions (full duplex) or are carried only in one direction at a time (half duplex). Layer C (connection establishment layer) contains all of the protocols dealing with the connection establishment and termination function, such as the specific procedure for both allocating transmission facilities that constitute a path through the network between two telephones and releasing these facilities upon completion of the conversation.

Theoretically, a network could use any number of layers of protocol. A three-layered example is used here for explanatory purposes; the seven-layered OSI model is commonly used.

The benefits of using layered protocol structures in the design of telecommunications networks are similar to the benefits achieved with the modular design of stereo systems. Independence of the layers ensures that design changes or modifications to a given layer do not affect other layers. Thus entire systems need not be modified or redesigned to accommodate changes in one function or component of the system. Layered protocol systems also facilitate the use and interconnection of equipment from different vendors, so long as common interface standards are supported.

Three key elements make up a modular, layered protocol system: generic services provided to adjacent layers, interfaces between layers, and the protocols contained in each layer. **Generic services** are the functions that a lower layer provides to the adjacent higher layer. In the stereo analogy, Layer 1 (the sound source-input layer), delivers electrical signals representing musical sound to Layer 2 (the amplification layer). The specifications of the electrical signal (voltage, current flow) and of the physical connection (types of connectors, functions of specific wires, and so on) necessary to connect the two adjacent layers constitute the interface between Layers 1 and 2 of the stereo system. The actual procedures and methods used to carry out the required operations within a given layer constitute the protocol. Using the stereo analogy again, both the procedures invoked to retrieve the sound representations stored on an audiocassette tape and converting them into an appropriate electrical signal to be amplified constitute the set of Layer 1 functions (protocols) for a stereo system.

In discussing layered protocol systems, we have introduced the concepts of generic network services that one layer provides to another and of interfaces that specify the link between adjacent layers. The complete layered protocol system, called a communications architecture, is the main topic of Chapter 7.

A specific example of a widely used protocol designed to handle the data link layer functions for synchronous networks is **high-level data link control**

(HDLC), the protocol used in X.25 networks. HDLC operates on the sending side of the link by accepting user data from a higher protocol layer and packaging it for delivery across the network to the receiving device. On the receiving side of the link, HDLC also accepts incoming data from the network, and in turn, delivers it to the next higher protocol layer.

HDLC's principal function is to provide flow, error, and other transmission link control functions. The protocol first operates by formatting the data received from the user into a frame with designated locations for data and for various types of control information. A typical HDLC information frame includes synchronization information, general control information (indicating the type of frame and its function), flow control codes, user data, and error detection information.

Special control frames initialize and terminate the connection, acknowledge receipt of frames, and indicate errors. A number of data link layer protocols perform essentially the same function for different types of networks. These protocols include SDLC (synchronous data link control), which is a part of the IBM Systems Network Architecture (SNA) system; **link access protocol-balanced(LAP-B)**, which performs the data link control functions in the X.25 packet-switching protocol standard; and **logical link control (LLC)**, which is used in some local area network configurations.

SDLC is IBM's bit-oriented synchronous link control protocol that for years was totally proprietary, making it close to impossible to link SNA and X.25 networks. The translation between SDLC and HDLC was handled first by protocol converters and is now a routine feature of internetworking equipment.[4] LAP-B is a link control protocol within HDLC that simplifies the design of data terminal equipment (DTE) implementation. Logical link control originated as a protocol for connecting an SNA network to an X.25 network. It is very unlikely that you, whether as client, designer, or implementer, will ever have to worry about LAP-B or LLC, but equipment designers and PTTs do; protocols must be defined right down to the smallest detail to ensure that every single bit and electrical signal is correctly interpreted and processed.

Although protocols such as HDLC, LAP-B, and LLC are at the second layer of the OSI model (the data link layer), Layer 1, the physical layer, is the domain of electrical engineers; it defines plugs, wires, and signals. Data link protocols transmit the physical blocks of data that make up the message and needed control information; together with Layer 1 functions, they ensure the basic operations of telecommunications: the generation and movement of signals. At Layer 1, communicating devices send and receive a stream of digital bits. A frame incorporates these bits, and the protocol defines how to interpret them

---

[4] Protocol conversion at the link control level is generally fairly easy to provide because all the main protocols in use structure messages in frames, in which control information surrounds the data packet. The conversion is mostly a translation process. Higher-level protocols are more complex to handle because they do not have a common set of principles for formatting and structuring. When the structure must be converted, there is more work to do and more chances of exceptions or errors.

and what to do next. The X.25 frame is typical. It has each of the following components:

1. Beginning flag: a special bit pattern, 01111110
2. Address: identifies the receiving terminal(s)
3. Control: the type of frame plus other information
4. Message: the packet of data that is part of the overall message
5. Frame check sequence: this is the result from the cycle redundancy check calculation
6. Ending flag: 01111110

Connection Establishment Methods

Connection establishment is a Layer 3 function that routes data between sender and receiver. The implementer needs to know each connection establishment method in detail because many choices of network design rest on them. The main three alternatives are circuit switching, packet switching, and the more recent fast packet switching.

Circuit Switching

The most common connection establishment method now in use is circuit switching, which is the base for public and private voice telephone networks worldwide. Circuit switching is also used for a large amount of data traffic because a substantial amount of data traffic is carried over the public telephone network in almost every country in the world. It has also been the core of IBM's SNA and related telecommunications systems and products.

Circuit switching uses a dedicated path through the network to connect communicating devices for the duration of a session. Three steps are involved in circuit switching: establishing the circuit, transferring the information, and disconnecting the circuit. First, the transmitting device sends a request signal to establish a connection with a particular receiving device on the network. The first switching node receiving the request examines internally programmed routing algorithms and the availability of various paths and then allocates a dedicated channel to the next switching node. This process is repeated until a complete, dedicated path connecting the transmitting and receiving devices is established. At the final switching node, the request signal is sent to the receiving device to determine whether it is available to receive information; if it is, the connection is accepted and the conversation or exchange of data begins.

The information transfer segment of the process typically operates in full duplex mode, with simultaneous transmissions in both directions. In addition, information travels at the same speed/data rate throughout a circuit switched connection; therefore, both the transmitting and receiving terminal devices must also operate at that data rate. The circuit disconnect part of the process occurs when either party terminates the session. At that point, a signal is sent along the dedicated transmission path to release the transmission link (and any switching nodes along the path) and to return them to ready and available status.

Because a circuit-switched path is dedicated for the duration of a session, circuit switching is an inefficient transmission method if the established circuit is not fully utilized for the duration of the session. If there are long pauses in speech or long gaps between actual data transfer bursts, the dedicated transmission link and switching facilities that make up the circuit are both idle and unavailable to serve other users. There is also a time delay involved in setting up the circuit path at the beginning of the session, but no subsequent delays occur once the circuit is established; the actual speed of transmission is constant except for any propagation delay properties associated with the particular transmission medium being used (for example, the approximate one-quarter-second delay inherent in satellite transmission systems).

Circuit-switching applications are widespread. Among them is the public telephone network, which provides worldwide voice connections and an increasing amount of dial-up data service. The private branch exchange (PBX) is a circuit-switching device used extensively in business and government organizations both for intra-organizational telephone switching and to obtain cost-effective links to the public telephone network for access outside the company. Wide area private voice networks generally consist of several PBXs linked together through either dedicated leased transmission links or a **software-defined network (SDN)** service, a wide area virtual private network provided by long-distance carriers.[5] Customer organizations lease a certain transmission path capability between designated locations that is made available on demand by the carrier. However, specific transmission and switching facilities are not dedicated to the customer full-time; instead, the specific transmission/switching resources required for any communications session are allocated from a pool of available capacity on the network at the time service is needed, and these facilities are released upon termination of the session. The data-only version of a PBX, a data switch, is also a circuit-switching mechanism used both to interconnect terminals within a local area and to provide efficient interconnection of these terminals to distant locations through either dedicated leased lines, an SDN service, or access to a public data network.

Packet Switching

The circuit-switched transmission method was designed for use with voice traffic in the public telephone network. Although public circuit-switched networks are used to carry data traffic, they are not efficient for this purpose. The intermittent, bursty nature of most data traffic results in inefficient use of transmission capacity dedicated for the duration of a session; this is especially true for

---

[5] Software-defined networks add immense intelligence to switches, which historically were mainly hardware-driven. Because its foundations were in engineering, the telecommunications industry was very slow to build software development capabilities. SDNs require complex software and hardware to control the allocation of resources to meet guarantees of capacity.

interactive sessions in which a person sits at a terminal and spends time typing or pointing and thinking and looking, with only occasional communication traffic to be transmitted or received. Also, the requirement in circuit switching that all terminal devices along the dedicated path must transmit and receive at the same data rate makes it difficult to interconnect a wide range of terminals and computers that are transmitting/receiving at different speeds. In the early 1970s, a new transmission method, packet switching, was developed to make more efficient use of network facilities in the transmission of data traffic.[6]

With packet switching, a bit stream representing a message or block of data is subdivided into groupings called packets. A typical packet size is 128 characters (bytes) of information. The packet contains user information (for example, message or data) and control information representing the destination address, sequence number of the packet, and error detection/correction codes. Packets are formed either at the user's terminal or at the packet-switching node linking the user terminal to the network. Such a node contains programmed intelligence, including routing algorithms representing paths through the network to various destinations and congestion status information on various paths in the network. The node receives the packet in a holding buffer, reads the destination address, and queues the packet for transmission to an adjacent node that is on a path to the final destination. This process is repeated at each intermediate packet-switching node until the packet is delivered to the node connected directly to the receiving terminal. At this point, the control information is stripped away and, if necessary, the message or data block is reassembled in the original order and delivered to the destination terminal device.

Packet-switching networks can accommodate terminal devices that operate at different data transmission rates because the switching node interconnecting the terminal to the network can be programmed to transmit and receive at the speed of the attached terminal device. The switch uses its packet storage buffer capacity to adjust the transmission rate of packets entering or leaving the network.

The packet network routes packets through the network using one of two alternative methods: the datagram method or the virtual circuit method. In the **datagram routing method,** the first switching node determines a route for each packet based on current network congestion status. Thus packets representing various parts of the same message or block of data may travel on different paths through the network and arrive at different times and out of sequence. However, the switching node at the final destination queues the incoming packets and uses the sequence number attached to each packet to reassemble the message or data block in the original order before delivery to the destination terminal.

---

[6]Packet switching resulted in the 1970s from developments sponsored by the U.S. Department of Defense, which funded the ARPANET network that linked researchers at laboratories and universities across the United States (and subsequently across other countries). Traffic on the ARPANET was intermittent and did not require real-time processing. Packet switching was the natural path for innovation for connecting together thousands of occasional users.

In the **virtual circuit routing method**, a single path through the network connecting the communicating terminals is established at the beginning of a session, again based on current congestion conditions; all packets travel the same path for the duration of the session. However, packets sent from other devices involved in other communications sessions also share the same transmission facilities because they are not dedicated to a given pair of terminals for the duration of the session, as is the case in circuit-switching networks. In the virtual circuit method, packets arrive in the order in which they are sent because they stay in sequence while following the same path through the network.

The differences between the datagram and virtual circuit methods can be illustrated in the following transportation analogy: Consider a large tour group of 150 people assembled at the airport in Miami and headed to Los Angeles. One approach an airline representative could use is to book three subgroups of 50 people on three consecutive flights from Miami to Los Angeles through the airline's hub in Dallas (the virtual circuit method). Another approach is to fly one group to Los Angeles through Dallas, a second group from Miami to Los Angeles via Chicago, and the third group from Miami to Los Angeles with stops in Houston and Denver (the datagram method).

Although organizations with large data transmission requirements can construct their own private, packet-switched networks, a number of public time-sharing packet-switching network services are available in most countries around the world. These networks are public data networks (PDNs) and value-added networks (VANs). PDNs are the PTT's national public infrastructure. VANs are offered by many PTTs and by others, depending on the degree of liberalization and deregulation of telecommunications in a country. U.S. VANs that provide packet-switched data services include Telenet, Tymnet, IBM Information Network, Compuserve Information Services, and many others. VANs permit the use of terminal devices that operate at various transmission speeds, and they accept a range of different protocols. Therefore, the VAN service can form the basis of a corporation's backbone data network connecting a full range of terminal devices in diverse locations, and the company either pays a flat (nonusage-sensitive) fee for access or pays on the basis of terminal connect time or number of packets sent.

The well-known X.25 international protocol standard is a packet-switching interface standard for connecting a terminal device host system and a packet-switched network. The X.25 standard has three levels of protocols operating in a layered structure. It consists of a physical layer protocol, a transmission link layer protocol, and a packet layer protocol. The packet layer protocol converts a raw data stream sent from a terminal into a frame structure consisting of blocks of data and address information. The address information includes a destination address and a virtual circuit identifier (because X.25 uses the virtual circuit method). The frame is then passed to the transmission link layer, where control information is appended to the frame to form a complete packet.

The transmission link layer protocol for X.25 is designed to ensure reliable and accurate transmission of information across the network. The protocol it uses is LAP-B, which specifies the procedure for initiating the session by having both terminals agree on which of the optional rules and procedures for data transfer will be used for the session. LAP-B also includes an error detection and correction scheme. Finally, LAP-B provides a disconnect procedure to end the session.

The complete packet containing the user data, address, and control information is then passed to the physical layer. The physical layer protocol prescribes the method of physically connecting data terminal equipment (DTE) to a DCE (data circuit-terminating equipment). The X.25 physical layer protocol is itself another standard, **X.21**, which specifies the physical connection of terminals to the X.25 network. It is equivalent to the RS-232-C, 25-pin connector used in the United States. (Recall that in the RS-232-C specification, each of the 25 pins in the connector has a designated function with precise electrical parameters.) This three-layered handoff process is reversed when incoming data is being received from the network.

Recall that in Minicase 1-1, public data networks and value-added networks were used by THEi to set up three separate data networks: one for the United States, one for the UK, and one for 13 countries in Europe. These public data networks use the X.25 set of protocol standards to implement the packet-switching service, which permits the various THEi stores to send to and receive data from corporate headquarters, and for corporate management to have up-to-the-minute information on buying trends in particular regions, which makes for more efficient ordering and inventory management. The principal advantage of the X.25 protocol system for THEi is its worldwide availability and its reliability. The X.25 standard has a proven track record in a wide range of operating environments, and its error detection and correction scheme, embedded in the transmission link layer, is very effective in delivering error-free data to its destination.[7]

Packet- and Circuit-switching Methods Compared

Circuit switching requires a certain amount of delay in setting up a connection. A request signal is sent through the network both to establish the call path and to determine whether the receiving device is available for the requested session. This setup delay is experienced at each node in the network as the path is established. However, the return acknowledgment signal experiences no delay because the dedicated path is already established. The conversation, message, or data stream experiences no delay (other than that associated with the transmission medium itself) as it travels along the dedicated path through the network for the duration of the session. Communicating terminal devices must operate at the same data rate and use compatible protocols because the dedi-

---

[7] It is too early to discern how quickly X.25 will be displaced by ISDN and fast packet-switching technologies, most obviously frame relay and SMDS. Opinion in the industry is mixed. The current consensus seems to be the X.25 will still be in widespread use in the year 2000, mainly because it meets most needs, capital costs have been amortized over many years, operations have been fine-tuned, and costs are acceptable. However, if multimedia applications take off, X.25 will surely decline in usage.

cated circuit path is a transparent connection that merely carries the signal along the network path and lacks capability to change speeds or protocols.

Circuit switching is ideally suited to continuous streams of traffic such as file transfers and transmission of batch data, which make efficient use of the dedicated transmission and switching facilities forming the circuit path. Circuit-switching methods are well suited for voice and video transmission because both involve streams of information.

The virtual circuit version of the packet-switching method experiences similar delays due to circuit switching in the setup phase of the transmission. To establish a virtual circuit, a call request packet is sent through the network, and a delay is experienced at each node as the packet is queued and a routing algorithm is invoked at the switching node. A call acceptance packet sent back to the originating terminal along the established virtual circuit path also experiences queuing delays at each node. Once a virtual circuit has been established, the message or data packets also experience queuing delays at each node, with the length of delay dependent on traffic on the network. The datagram version of the packet-switching method experiences no setup delay, and therefore it can be faster than either circuit switching or virtual circuit packet switching for short messages or blocks of data. However, for longer messages or blocks of data, queuing delays will be experienced at each node. Also, because all related packets do not follow the same path through the network, they are likely to arrive at the destination node at different times and out of sequence. Therefore, an additional time delay is incurred at the destination node for collecting all related packets and placing them in proper sequence for delivery to the destination terminal.

Both the virtual circuit and datagram versions of packet switching permit the exchange of data between two terminal devices operating at different data rates. The switching node queuing process buffers the data arriving at one speed and permits retransmission at another speed to accommodate the receiving device. Intelligent packet-switching nodes can also perform protocol conversion to enable devices using different protocol systems to communicate.

Packet-switching methods are very efficient for intermittent, bursty types of data traffic, such as remote terminal data entry and transactions that process applications, because the circuit path facilities are shared by other users and because packets from various sessions are interleaved as they move through the network. Therefore, the idle periods between bursts for a given pair of terminals engaged in a communications session do not result in inefficient use of network facilities, as would be the case in a circuit-switched network.

Fast Packet-switching Methods

Packet switching is a reliable and efficient method for transmitting bursty forms of data traffic. However, the growth in the number of high-speed LANs and the need to interconnect large numbers of them across long distances have created a need for a fast packet-switching technique. Regular packet-switching methods have significant speed limitations because at each node in the network every packet is subject to error and flow control checks before being forwarded to the next node. These checks are required because in the early days of packet switching the underlying network transmission facilities were prone to noise, which caused frequent errors in the data stream; consequently, flow and error control were needed at each point-to-point segment on the network to ensure accurate transmission of data. In contrast, today's modern high-speed transmission facilities operate virtually error-free, and it is thus sufficient for end-user terminal devices to perform flow and error control on an end-to-end basis. Thus a packet can be relayed through a fast packet-switching node with essentially no delay, except for queuing delays due to traffic congestion. This can create a tremendous speed advantage over standard packet switching.

Before the advent of fast packet-switching methods, the way to avoid delays at packet-switching nodes was to increase the transmission link speed (capacity) and reduce the number of nodes in the network. But this was a very costly solution that involved the use of many point-to-point leased lines for a wide area backbone network connecting a significant number of locations. A fast packet-switching hub can substitute for a number of point-to-point leased line connections with no loss in responsiveness or throughput speed because of the substantially reduced throughput delays of the fast packet-switching hub. This arrangement significantly reduces the cost of the backbone network connecting a number of LANs without any comparable performance degradation.

Frame relay and **cell relay** are specific techniques for implementing fast packet switching. Frame relay systems encapsulate LAN packets of variable length and transmit them across the backbone network at speeds ranging from 56 kbps to 1.5 mbps. Cell relay uses a fixed-length frame that encapsulates an existing LAN packet without altering its format for transmission across the backbone network. If the LAN packet is larger than the fixed cell length, the packet is partitioned into more than one cell and then reassembled at the destination switching node. Cell relay systems operate at speeds ranging from 1.5 to 30 mbps, a range that is suitable for public fast packet service offerings that multiplex together packets from a number of different user sessions. With one such public network service, **SMDS (switched multimegabit data service)**, it is possible for an organization to provide high-speed interconnection among a number of LANs by simply providing a local interface connection from a LAN bridge or router to a carrier's local point of presence that offers SMDS.[8]

Although X.25 has been the international protocol standard of choice for data transmission for many years, the requirement to transmit more data at higher speeds has created a demand for fast packet switching. It appears that frame relay will be the successor to X.25. In addition to faster and more cost-effective data transmission, fast packet switching (particularly the cell relay version) can transmit voice and video information in packet form. This makes possible a true broadband integrated services digital network (BISDN), which

---

[8] In the 1970s, fast wide area networks handled traffic from slow terminals; in the late 1980s, the WAN became the bottleneck as LAN speeds improved exponentially and WAN speeds improved only incrementally. The importance of fast packet switching, the base for SMDS and asynchronous transfer mode transmission, is that it removes the bottleneck and moves fast LAN data over fast WAN links.

means that all types of information (data, voice, video, and image) can be carried together as one high-speed digital bit stream. Chapter 7 contains a more detailed discussion of ISDN and BISDN.

| | | | | | | |

Key Protocols and Transmission Methods

This chapter concludes with a review of selected transmission methods and protocols, focusing on those that will be major options and to some extent on conflicting options for the design of networks through the mid-1990s. So many protocols are in use, emerging, and under discussion that it is impossible to cover more than a few of them. The ones chosen are both representative and widely used, and comparing them gives a generalizable picture of most transmission methods and protocols and highlights important topics from the implementer's perspective that are not covered in detail. These topics include how congestion management is handled, how paths are chosen to route messages through local area networks and across networks, and how security is handled.

The protocols chosen for review here are used in many companies, especially when interoperability is a priority. They are:

▸ TCP/IP: This relatively simple protocol is the option of choice in most UNIX environments and, increasingly, for multivendor networks, including those that use SNA.

▸ APPN: This is IBM's peer-to-peer extension of SNA that fills a major gap in its telecommunications capabilities, one that has helped both advance TCP/IP and lead firms to move away from SNA toward other multiprotocol options.

▸ DECnet Phase IV: Digital Equipment's proprietary equivalent to and competition for SNA.

▸ AppleTalk: Apple Computer's LAN protocol.

▸ IPX: Novell's proprietary LAN protocol that is the core of its very successful NetWare network operating system.

The first three of these protocols are for wide area and the last two for local area networks. The discussions of each of them below address LAN-WAN interoperability. Because frame relay (along with ATM) is the fast emerging base for the next generation of network technology, it will be the primary focus of the analysis in assessing the interoperability among the protocols.

TCP/IP

The Transmission Control Protocol (TCP) was designed for use across the first world's first packet-switched network, ARPANET, which was sponsored by ARPA, the agency within the U.S. Department of Defense that funds civilian research. By making ARPANET available for use by universities and other organizations, ARPA hoped to and did develop a new telecommunications technology. First proposed in 1973, TCP is a connection-oriented protocol, for end-to-end transport over unreliable networks; for that reason, it included error-checking algorithms and a timer to detect lost or unacknowledged packets.

The growth of LANs in universities in the early 1980s led to a need to extend TCP to interconnect them to ARPANET and other LANs. The Internet Protocol (IP) is an elementary datagram-oriented protocol; if a network node is congested, the protocol simply throws away incoming packets. TCP's timer mechanism detects that they have been lost and signals back through the network to the sender.

The company that developed the original ARPANET switches was contracted to design new internet routers (called "gateways" at the time); it needed to define new protocols for exchanging routing information between devices, including the Interior Gateway Protocol (IGP) and Exterior Gateway Protocol (EGP). These protocols were the base for several protocols that have become standards for internet equipment, including Routing Internet Protocol (RIP), Open Shortest Path First (OSPF), and Intermediate System-Intermediate System (IS-IS). Within the TCP and IP protocols, academics developed many extensions and improved implementations. They added, for instance, the Simple Mail Transfer Protocol (SMTP).

ARPANET evolved to become the Internet, with first hundreds and then thousands of user institutions. When the Department of Defense created the X.25 Defense Data Network (DDN) in the mid-1980s, it became mandatory for nonresearch departments to use the network. Computer vendors such as IBM and Digital Equipment then needed to create reliable (that is, nonstudent, non-ad hoc, and not undocumented) versions of TCP/IP. UNIX had emerged as the operating system of choice in the academic community, with the University of California, Berkeley playing a leading role. UC Berkeley decided to bundle UNIX with TCP/IP, making it the de facto standard. TCP/IP is now supported across most transmission methods and protocols, including Ethernet and token ring LANs, X.25, X.21, and the major emerging protocols, FDDI, ATM, and frame relay. New companies like Sun Microsystems (UNIX workstations) and Cisco (routers) focused on the UNIX/TCP/IP market. Cisco did not even have a sales force but sold by word of mouth through the Internet.

The history of TCP/IP shows two things: that standards do not necessarily emerge through a formal standards-setting process with official standards-setting agencies and that it is not necessarily the "best" standard that wins in the marketplace. TCP/IP's strengths are its wide adoption and thus the wide supply of products and services that use it. (Over 300 companies provide TCP/IP products.) It is also topology-independent; TCP/IP networks can be star, mesh, ring, point-to-point, or any combination. It is vendor-independent and has over 20 years of development and operation. It can operate both over slow 1.2-kbps links and over 100-mbps ones, such as FDDI.

TCP/IP's weaknesses are that its address space is very small, limiting the number of terminals that can be part of the network. It has limited file transfer, remote log on, and other management features needed in a complex commercial network both to provide additional services and simplify users' operations. TCP/IP handles routing through a variety of options, including a distant vector algorithm in its Routing Information Protocol (RIP). **A distant vector algorithm** chooses paths through the network on the basis of **cost and hop**—the lowest-cost path with the fewest number of nodes (hops).

More and more commercial networks use TCP/IP as the backbone protocol. That means that traffic is moved via the TCP/IP protocol and other protocols converted to it (often referred to as "mapping from one protocol to the other"), including DECnet traffic or X.25. The dominance of TCP/IP in the UNIX world has resulted in standards being defined and adopted by router vendors who carry TCP/IP traffic over frame relay.

APPN

APPN (Advanced Peer-to-Peer Network) is a 1992 addition to IBM's flagship SNA architecture, which is described in more detail in Chapter 7. SNA, by far the most widely implemented communications architecture in large organizations' platforms, has evolved over a 30-year period and provides a capability for WANs that handle heavy transaction volumes. It is based on the 3270 dumb terminal, which is a classic slave/host terminal-to-mainframe configuration. Today, SNA looks clumsy, but it was a superb design for getting the maximum possible performance out of the slow-speed circuits that were the main practical source of transmission until the late 1980s.

The main limitation of SNA has always been its hierarchical star configuration. Although the central directory control allows for thousands of terminals to be coordinated in an SNA network, it is poorly suited to peer-to-peer, terminal-to-terminal traffic. In addition, if a link or node fails, the session dies even though there may be plenty of alternate routes through the network. The user must log back on to the network and be allocated a new path.

SNA is rich in features and complexity. APPN is IBM's response to the need for peer-to-peer features. Although implementing APPN involved many revisions to and extensions of SNA, the core change was the routing protocol. When a new communications session begins, a path is automatically selected to the first node, and each node selects the next link. The path is fixed throughout the session. Routing tables at each node simplify selection of the path, keeping track of previous sessions and the destination location and routing path. Previously, SNA did not allow personal computers and LAN internet devices to be included as addressable units; APPN now includes them. This will facilitate interoperability among SNA, TCP/IP, and LANs, but problems remain. SNA is connection-oriented, and internet is connectionless. SNA's roots in circuit switching make it unreliable in handling highly bursty traffic, especially a flood of unsequenced packets and/or long delays in transmission.

One feature of SNA that facilitates developing SNA-to-frame relay routers and routers that handle other protocols is that the architecture does not specify a particular Layer 2 link control protocol, but instead assumes that a reliable connection-oriented transport mechanism exists beneath the Layer 3 transport layer. (The layers here are from the OSI model, but the lower layers of SNA directly correspond to OSI's). Router vendors can thus include multiple protocols for SNA internetworking, including token ring, Ethernet, FDDI, and SMDS. All the leading router vendors have already announced SNA-frame relay products. IBM has also implemented a frame relay interface to its FEPs for token ring LANs.

It can be a lifetime work to learn all the details of SNA and keep up with it, with internetworking to SNA, with developments to IBM's other telecommunications architectures, and with IBM support for OSI, FDDI, ATM, and other innovations. Even skilled telecommunications implementers generally must draw on specialized technical experts in the computer as well as the communications field. A point to consider here is a simple but critically important one: APPN will not be available before 1994, and the router vendors' products do not yet exist. APPN solves some but not all problems of internetworking and peer-to-peer communications. Designers in this situation need to balance carefully the trade-off between network capability and flexibility; they also need to talk closely with implementers to assess not just a particular standard, product, or architecture, but its implications for interoperability, integration, and evolution of the platform. For instance, a small but often vital aspect of APPN is that it is backward compatible with the 3270 terminal and major software systems that use it. **Backward compatible** means that existing products can still run under the new protocol, even though they may not be able to exploit its features; this is analogous to emulation. APPN/3270 compatibility means that firms need not retool and rebuild their existing network and software to get the benefits of APPN. If they did, the proportion of users likely to adopt it would be closer to zero than even to 50 percent.

DECnet Phase IV

For almost a decade, Digital Equipment's DECnet was the main rival to SNA. It was also a proprietary architecture, but it differed from IBM's in that it was peer-to-peer, giving it an edge in many markets until the maturation of TCP/IP, which DEC initially and fairly publicly did not support. In the mid-1980s, DEC moved from maintaining a **closed architecture**—one in which interfaces to other architectures are either not permitted or impractical—to adding interfaces to SNA. DECnet Phase IV is a proprietary peer-to-peer protocol whose main features are that it extends the number of total addresses to 64,000 from a previous limit of 255 subsystems, called areas, and that it distributes the routing process to router devices, making it a fully distributed routing protocol, a major step forward in internetworking. DECnet Phase IV offers a range of applications services that is at the opposite end of TCP/IP's, including Distributed Name Service, Distributed Time Service, File Service, Queuing Service, Remote Procedure Call, and Management Architecture. It includes in addition a two-level routing concept. A DECnet network is divided into up to 63 areas, each of which can contain up to 1,023 addresses. Routing within an area is termed Level 1 routing, and routing across areas is Level 2 routing. This reduces overhead traffic between routers. Instead of having to update all directories and routing tables when there is an addition or change to the network or when a link or node fails, updates are made only to the relevant area.

The key limitation of DECnet Phase IV is that it is a proprietary protocol.

Router vendors support DECnet but most computer manufacturers do not, including IBM, Hewlett Packard, and Apple. Phase V will support OSI, but OSI is losing out to TCP/IP. Digital Equipment is implicitly relying on router vendors to ensure interoperability for DECnet and is simultaneously focusing its efforts on making DECnet efficient through proprietary protocols while moving toward OSI. This gamble, which enables DEC to fine-tune its systems rather than adopt inefficient general-purpose protocol standards, may or may not pay off.

AppleTalk

As mentioned in Chapter 4, Apple Computer's Macintosh and Powerbook machines are among the easiest to learn and use, and Apple created the move from typing and command-driven systems to GUIs and the use of a mouse. Yet, these machines have not been adopted by most large companies as the PC base for their platform. AppleTalk, Apple's LAN protocol derived from a 1960s protocol developed by Xerox, is part of the reason: It is solidly centered on LANs, with almost no facilities for LAN-WAN internetworking; it is intended to link Mac users on a LAN simply and cheaply; and it lacks sophisticated routing algorithms, address management, and congestion management.

These facts are not necessarily problems in a context of individual LANs, but they are major blockages to integrating Apple LANs into a business platform. For example, AppleTalk routers are updated with information about links, nodes, and terminal addresses through a time mechanism. Every router then receives the complete routing database, which could be as many as 16 million entries, though in practice it will be no more than 300. This happens every ten seconds, making the overhead on the network the main traffic. Because AppleTalk is built to handle individual LANs, multiple LANs may have duplicate addresses, making internetworking a muddle.

Within the AppleTalk protocol, the Zone Information Protocol (ZIP) broadcasts to all routers updates to tables that map network resources to two different types of name: the zone name (function of the device plus where it is located) and the network address number. The map is needed because although the GUI shows the zone name to the user, the network needs the network number. When ZIP broadcasts the message, it starts a timer that indicates the period of time before which it should receive an acknowledgment that the other routers have received the broadcast. If, as often happens in large AppleTalk networks, it does not receive the reply in time, it broadcasts again and again and again, if need be. ZIP storms can bring down AppleTalk networks and can also swamp a WAN to which they interlink with the electronic equivalent of junk mail.

All this may sound like a criticism of AppleTalk. It is certainly fair to say that Apple neglected the issue of internetworking and the importance of efficient LAN communications protocols for many years. Yet, TCP/IP is an inefficient and unsophisticated set of protocols in many ways; the difference is that TCP/IP is strong in ways in which AppleTalk is weak: LAN, LAN-WAN internetworking, and routing. Both are relatively ad hoc protocols that meet a given need; as the need changes, the ad hoc features often create limitations. Even with TCP/IP, there are growing concerns among large "power" network users that its basic design will not meet their needs beyond a few more years. AppleTalk assumes the existence of small networks with limited and local traffic, but that was the LAN of yesterday.

IPX

The final protocol briefly reviewed here was also derived from a Xerox protocol, as was AppleTalk. IPX supports Novell Corporation's NetWare network operating system, which dominates the LAN market. NetWare provides a wide variety of network services, including print sharing, file sharing, directory services, and network management. The IPX protocol sets up communications between client and server terminals on the LAN, including PCs and servers. It is a low-level, topology-independent protocol. The largest diameter through which the protocol can route is 16 hops. Novell is planning to increase this to up to 256 hops, extending internetworking capabilities. As with AppleTalk, IPX generates substantial overhead through broadcasting of updates to routing tables and does not provide strong WAN features. Novell is upgrading IPX but will find it difficult to replace because it is embedded in so much of its software. Some of the improvements include the use of the Service Advertising Filter NLM (Network Loadable Module—that is, plug in and go). In the pre-1992 versions of NetWare, servers broadcast a message every minute, indicating the services it could provide to the network; this created a situation comparable to ZIP storms. Novell has also improved the IPX routing algorithms. The distant vector algorithm it previously used could take up to 15 minutes to redefine the network topology if a link failed in a large (100-node) network. **A link state algorithm**, which selects paths on the basis of many factors (including reliability of a link, delay, and maximum data rate), rather than just on cost and number of hops, takes under two seconds.

Internetworking the Various Protocols

Multiprotocol routers are the primary tool for internetworking these protocols. Bridges can handle this for networks that use the same protocols, but they have many problems handling ZIP-type storms. Several protocols widely in use cannot be routed, including IBM's Netbios and DEC's LAT. Router vendors naturally target their products to support the protocols most widely used; they also target them to get to market as quickly as possible when a new protocol such as frame relay is defined and looks like a winner.

This section on protocols may seem very technical to you if you are focused on the client's perspective, and far too lacking in technical detail if you take the implementer's view. Still, it provides some insights into the nature of protocols and some of the practical problems they involve; it also provides a transition to Chapter 7, which reviews standards and architectures.

Summary

Transmission methods have three main functions: establishing the connection between two terminals, providing the rules for them to transfer information to and from each other, and managing the communication link to ensure that the data is accurately transmitted and received and any errors detected and corrected.

Communications protocols are the precisely

defined sets of rules for carrying out these procedures. Many of these protocols are standards defined by such organizations as CCITT, whose X.25 protocol is the best-known international standard; de facto protocols adopted by such a wide range of vendors and users that they have the same force as an officially defined one; or proprietary protocols provided by a single vendor.

Protocols for establishing a connection may be either connection-oriented, in which a fixed path is set up for the duration of the communications session (as with a phone call), or connectionless, in which individual units of data flow through the network using the next available and/or most efficient route. Packet switching is connectionless, and IBM's SNA is connection-oriented. Data transfer protocols specify asynchronous or synchronous transmission. Link control protocols include such error-detection techniques as cyclic redundancy checks. The HDLC protocol is used by X.25; the corresponding SNA protocol is SDLC.

Protocols are layered in a hierarchy in which the lowest layer handles electrical signals, the next one handles the data link between two adjacent devices, and so on up through routing of messages, establishment of a communications session, and presentation and conversion of information. IBM's SNA and the OSI model are both layered in this way, although the exact layers do not correspond directly. The layering ensures that each individual layer is totally separate from the one above and below it and can assume that other layers have correctly carried out their relevant functions.

There are thousands of protocols. Apart from X.25, OSI, and SNA, major developments in architectures and standards include fast packet switching, FDDI, extensions to SNA, and many LAN-based innovations. Roughly defined, a protocol is a specific set of procedures and rules; a protocol may become a standard, and a standard may include many protocols. An architecture is the master blueprint of standards (and hence of protocols) that governs the entire operation of the specific network plus its ability to interoperate with networks built on other protocols.

Review Questions

1. Once the physical components of the network structure (terminals and transmission links) are in place, what additional functions must be performed for the successful operation of any network?

2. What are the necessary steps for making a connection between two terminals on a network?

3. What is a virtual path or virtual circuit through a network?

4. What is the connection between the concept of "internetworking" and the addresses of devices on a network?

5. What is the difference between a connection-oriented and a connectionless service protocol?

6. Give a simple, human (nonelectronic) communication example of a data transfer protocol.

7. In a transmission across a network, how does the system know when a particular message or block of data begins and ends, and whether a given group of bits are in fact data control information?

8. What is the purpose of a link control protocol?

9. Evaluate the following statement: If transmission systems just had signal-to-noise ratios that were just high enough, there would be no need for error control protocols.

10. What is a connectionless protocol?

11. What are the three key elements of a modular, layered protocol system?

12. Briefly describe the purpose of the "High-level Data Link Control" (HDLC) protocol and how it functions.

13. Briefly describe the concept of circuit switching and the major steps involved in call setup and takedown.

14. Under what conditions is circuit switching considered an inefficient form of connection establishment?

15. Briefly describe packet-switched transmission.

16. Explain the major differences between the datagram and virtual circuit methods of packet switching.

17. What is the most widely used international protocol standard for packet switching, and what does the standard actually specify?

18. Make a list of three applications that are most appropriate for a circuit-switching network, and three applications that are well suited to a packet-switching system. In the margin, list key characteristics for each of the applications that make it appropriate for the method chosen.

19. What is fast packet switching, and how does it differ from regular packet switching?

20. What are the major differences between frame relay and cell relay?

21. What is "switched multimegabit service" (SMDS)? What type of transmission method is the basis for it?

22. Briefly describe the key features of the TCP/IP protocol system, and explain why it has achieved such widespread use in spite of the fact that it did not receive early formal approval by international standards bodies.

23. What was the driving force behind IBM's creation of the Advanced Peer-to-Peer Networking (APPN) protocol, and how does it differ from previous IBM network protocol systems?

24. What is the common weakness shared by Apple's AppleTalk and Novell's IPX protocol systems?

Assignments and Questions for Discussion

1. Describe a complete transmission method system for conducting oral communications among four friends. The system should include a human version of the following:

a. a method to establish and terminate a conversation link;

b. rules for the orderly exchange of information;

c. procedures to manage and control the conversation link.

2. You have been asked by your supervisor to prepare a brief memo describing the concept of a layered protocol system and indicating conditions when such a protocol system is appropriate. Prepare such a memo.

3. A bank is in the process of designing a private data communications network to handle its worldwide operations requirements. Extensive analysis has shown that the vast majority of the network traffic will be relatively short messages and transfers of funds that require only short bursts of information. Furthermore, the bank is very concerned about reliability and error-free operation. You have been retained as a consultant to produce a report to help them choose between a circuit-switching or packet-switching system for the network, and the report is expected to provide a justification for the choice. What information would you include in such a report?

4. In a recent conversation, several of your colleagues said that the new fast packet-switching technologies are just so much hype from the vendor community, which is trying to sell more equipment and services to beleaguered end-users. Your colleagues said that the speed advantage is oversold, and that for most users, fast packet technology is like owning a Ferrari to drive to the grocery store. Do you agree or disagree with this statement? Evaluate and discuss.

Minicase 6-1

University of Miami

Implementing SONET

SONET (Synchronous Optical Network) would have been dismissed as fantasy in the mid-1970s. At that time, the designers of ISDN, mainly European PTTs, thought that 64 kbps represented the practical transmission target for the public networks of the 1990s, but fiber optics make it possible to reach gigabit-per-second speeds. SONET will compete with frame relay and SMDS (switched multimegabit data services) for the high bandwidth services of the near future.

Medicine is an immediate target of opportunity for providers of massive bandwidth. In particular, hospitals and medical centers see major benefits in teleradiology, the transmission and interpretation of high-resolution X-rays and other medical images (see Minicase 8-2). The University of Miami announced in June 1992 that it had become the first U.S. university to implement a SONET link between its cancer research center and its medical sciences facility. SONET, a private link provided by AT&T, operates at 155 mbps. The school's director of telecommunications emphasizes that the school's decision to use SONET instead of the originally planned four T1 lines reflects tomorrow's needs rather than just today's: "We knew that this application [exchanging X-rays and magnetic resonance images] was coming and that the existing data network would not support it. So we had to go with a design that would."

The main issues in implementing high bandwidth transmission are less the transmission link than the switches needed to synchronize transmission. The University of Miami is using an AT&T Network Systems DDM-2000 Optical Carrier multiplexer at each end of the link. Voice traffic is routed onto the 155-mbps link via Definity PBXs. Instead of four T1s to carry voice traffic, there are eight. Each T1 operates at 1.54 mbps, barely a fraction of the total capacity. A 45-mbps channel is used for imaging. The rest of the bandwidth will be used for point-to-point videoconferencing, which needs another T1 equivalent per conference link.

Questions for Discussion

1. Explain why it is not feasible to transmit high-resolution X-ray and other medical images over ISDN links (or even over T1 circuits).

2. The University of Miami SONET system is operating at 155 mbps. Isn't this much more capacity than is needed for the imaging applications identified? Explain.

3. Why is there so much interest in a fiber-optic network for high-resolution imaging? Connecting a couple of buildings in a medical complex with a fiber link isn't such a big deal (certainly not worth a Minicase). Evaluate and discuss.

Minicase 6-2

Chrysler

Halving Ordering and Shipping Times Through Wireless Communications

By installing wireless terminal networks in its four national parts centers in 1991, Chrysler Corporation halved its order processing and shipping times from seven to three and one-half days. Hand-held radio-frequency–based terminals (RF) record orders and shipments, cutting out masses of paperwork. Previously, clerks keyed in dealer orders that were sent via Chrysler's SNA network to mainframe computers that generated paper order forms, which were hand-delivered to loaders at shipping docks, who then filled the order and recorded relevant information that had to be keyed in and processed.

Now, shipping staff enter the information themselves into the hand-held terminal. The data is sent to and from the mainframes via a wireless network. Chrysler has 280 RF terminals, 22 base stations, and 16 network controllers in use. The hand-held computers emulate an IBM 3278 terminal, so that to the mainframe they appear to be just another device on the network.

The designers of the system expected negative reactions from many shipping staff, some of whom do not speak English well and some of whom have college degrees, because they were being asked to take over clerical functions in addition to their shipping activities. In practice, the workers welcomed the innovation. They had more control over their work, and they were in charge of the system instead of administration and documents being in charge of them. A secondary benefit was that Chrysler was able to move mainframe transaction processing to real-time from batch. The hand-held terminals send radio signals over the 450–470 MHz bandwidth radio spectrum assigned to Chrysler to be picked up by the base stations in the four distribution centers. Communication controllers are hardwired to the stations and receive and route the data to the SNA networks. A Chrysler employee punned that "RF is definitely the wave of the future."

Questions for Discussion

1. Draw a simple diagram showing the various nodes and components in the hand-held network system and how they are connected.

2. Explain how the use of the hand-held terminals was able to reduce processing time by one-half. (Hint: Describe how it changed the way actual work is performed.)

3. The new system required workers with minimal education to assume clerical duties in addition to their regular shipping activities; this must have led to negative reactions from some of the workers. Explain and discuss.

Minicase 6-3

Adaptive Corporation

Getting Ready for Asynchronous Transfer Mode Transmission

ATM (asynchronous transfer mode) is one of the three most promising technological innovations that will drive telecommunications in the 1990s and challenge vendors and users to make it work; the other two are SONET (see Minicase 6-2) and frame relay (see Minicase 6-4). All the major high-speed technologies used today divide bandwidth among a number of users; this can degrade network performance significantly as more and more devices compete for their share.

ATM takes the 0-1, on-off nature of digital communications to a new level. It pulses digital voice, data, image, and video through the network using all the available bandwidth; packet size is fixed, simplifying network protocols and reducing overhead. It offers gigabit speeds, for a fiber-optics link can move the data as fast as it is pulsed; but pulsing it requires ultra-fast switches. Adaptive Corporation's ATM LAN switch and workstation adapter cards are the first products that implement the new transmission mode, which is expected to be a cornerstone of public networks and make the old distinction between LAN and WAN increasingly less meaningful.

Bear, Stearns and Company, a leading New York security broker, began beta-testing Adaptive's products in late 1992. It plans to bypass the state-of-the-art LAN transmission mode, FDDI (fiber distributed data interface), just as it became stable, standardized, and practical. Several other organizations, including the U.S. Air Force and Texas Instruments, are making the same leap of faith. Bear, Stearns hopes that it can use ATM instead of the multiple routers and bridges that are widely used for LAN internetworking and that inevitably degrade performance by the very fact that the software and hardware operate in milliseconds and microseconds, whereas the data bit stream moves at nanosecond and picosecond speeds.

Proponents of ATM argue that however useful and convenient bridges, routers, and hubs may be, they are a band-aid. They reflect a reactive "interconnect-as-you-go" approach to internetworking. ATM makes it practical to produce a clean network design and get away from the multiplicity of devices needed to move a bit from LAN to LAN. Public networks that use ATM, as most major U.S. carriers are planning to do, will make LAN-WAN-LAN interconnection seamless.

The main barrier to adopting ATM is the installed base of Ethernet, FDDI, token ring, and other LANs. So, too, is price; Adaptive's products require an investment of around $9,000 per connection, versus $2,000 for FDDI. Adaptive offers the same 100-mbps transmission speed as FDDI potentially provides. Bear, Stearns's managing director of communications estimates that only 20 percent of his users need 100 mbps, mainly for imaging, yet his company dropped FDDI as ATM became practical. The main reason is simple: scalability. All the firms adopting ATM see it as the technology that will move them up to gigabit speeds without the constant reactive upgrading and adding of new equipment that is the norm in today's environment. Just as frame relay is attractive for private network users because it preserves their existing X.25 architectures while increasing throughput by a factor of ten, ATM will help firms design their LAN internet today in the knowledge—or belief—that they can maintain the integrity of the architecture through the next several decades.

Questions for Discussion

1. Explain how ATM technology differs from other high-speed transmission methods and why it will eventually render the old distinction between LAN and WAN meaningless.

2. Why is Bear, Stearns adopting an ATM system instead of the more established FDDI standard that is just as fast (100 mbps) and significantly cheaper ($2,000 per connection vs. $9,000 per connection)?

3. Does ATM spell the end for LANs as we know them? Discuss.

Minicase 6-4

Covia

Linking Private and Public Frame-relay Networks

Covia is the technology subsidiary of United Airlines, with minority ownership by six European airlines, including British Airways and KLM. It has been a major innovator in telecommunications and leads the industry technologically, although its main rival in computerized reservation systems, American Airlines, dominates it in terms of business leadership.

In early 1992 Covia announced a new "first," a test link between its private frame-relay network and Pacific Bell's public one. Up to then, frame-relay traffic has been transmitted only within, not across, networks. Success in linking public and private high-speed networks will give telecommunications designers more flexibility in network configuration and will reduce leased-line costs. The frame-relay links replace 56-kbps dedicated lines. Observers see fast packet switching as the new wave for large users, first through frame relay and later through asynchronous transfer mode lines (also known as cell relay). This change will greatly reduce the need for dedicated facilities and create a new generation of virtual private networks.

But transmission links are not the focus of the trial; interoperability of multivendor switches is. The Frame Relay Forum, a group of users and vendors, was in the process of finalizing the specification for NNI (network to network interface), a standard for switches. Covia tested to determine whether the StrataCom frame-relay-outfitted integrated packet exchange (IPX) multiplexers it uses in its own network were interoperable with Northern Telecom's DMS supernode link peripheral processor that Pacific Bell put in place for the trial.

Covia sent data between its IBM host mainframe computer and a token ring LAN in Denver and its offices at the San Francisco airport. The frame-relay link between Denver and San Francisco operates at 1.544 mbps, T1 rate. Racal-Datacom data service units (DSU) are the interface between the Covia IPX and Pac Bell DMS Supernode LPP. Data travels across the Pac Bell part of the network that links San Francisco and Oakland, California, at 56 kbps. Figure 6-2 shows the hybrid network.

Questions for Discussion

1. What are the implications for corporate telecommunications departments of successfully linking public and private frame-relay systems?

2. What are the key technical issues associated with linking public and private frame-relay networks? Discuss.

3. If the Covia trial is successful, does this success signal the end for X.25 packet-switching networks? Explain.

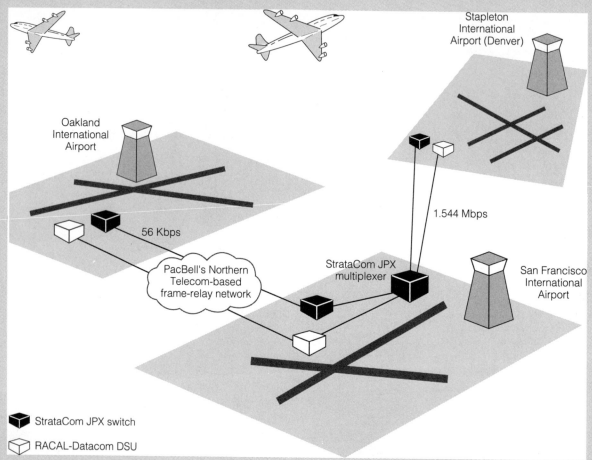

StrataCom JPX switch

RACAL-Datacom DSU

Figure 6–2

Covia's Hybrid Network that Links Its Private Frame-relay Network to Pacific Bell's Public Network

7

Standards and Architecture

Objectives

- ▶ Learn how standards are determined and their impact on integration
- ▶ Understand the nature and implications of open systems
- ▶ Review the components of a telecommunications architecture
- ▶ Learn the basics of the seven-layered open systems interconnection (OSI) model
- ▶ Understand the criteria for choosing an architecture

Chapter Overview

| | | | | | | |

Chapter Overview

The evolution of telecommunications has historically been driven by the development of multiple, vendor-specific, proprietary networks, each designed for a particular application or to solve a particular problem. In general, these independent networks have not been able to interconnect and share information. As a result, it is difficult for some banks to immediately determine the status of your checking, savings, and mortgage accounts, for they are maintained on different computer systems that are part of different networks, and each must be accessed separately, often from different terminals. This proliferation of incompatible networks is the result of a lack of widely accepted standards for telecommunications protocols and for interfaces that link telecommunications equipment.

Chapter 6 showed how different are the protocols used by IBM, DEC, Apple, and Novell, how some protocols are connection-oriented and others connectionless, and how these and many other companies built their networking capabilities on specific assumptions and specific core protocols. These companies' services were incompatible for a variety of reasons: the vendors' wish to protect their installed base, difficulties in making the technology work—the companies breaking new ground when they developed their protocols—and lack of established standards in most areas of data communications. Only in basic voice telephony were standards well established, and then only in individual countries.

Rapidly changing business needs have created a strong and growing requirement for integration of separate networks and for open network systems that can provide the appropriate information in the desired form to anyone, anywhere. The old situation of "horses for courses" no longer applies; the phrase refers to picking the best individual system that met a specific need. Now, companies increasingly choose the best system within a stated set of standards—the best UNIX-based workstation, the best IBM-compatible PC, the best Ethernet LAN, and so forth.

When telecommunication was in its infancy, companies sought vendors that could deliver specific capabilities. Now, all leading vendors provide the same functionality within their market niche—if they do not, they go out of business very quickly. Interoperability and integration, not individual features, are the drivers of modern communications. Progress to that end is the main topic of this chapter.

The chapter begins with a discussion of the role and importance of standards and open systems in today's business environment, and then proceeds to describe how standards are determined. Next comes a discussion of the concept of a telecommunications architecture as the design blueprint for creating and developing the network over time in response to changing technology, uses, volumes, and geographic locations. The concept of an architecture is examined at two levels: the communications protocol structure, which defines how the devices on the network work together to transfer information, and the network configuration structure, which defines how sending and receiving devices identify and locate each other. The seven-layer Open Systems Interconnection (OSI) model illustrates the protocol structure, and IBM's Systems Network Architecture (SNA) illustrates the configuration structure. Although there are many other protocol configuration structures, these two are both widely adopted and demonstrate the generic design principles of an architecture.

The chapter concludes with a discussion of the key factors that should determine the choice of a specific telecommunications architecture: the organization's requirements for reach, range, and responsiveness in the telecommunications platform.

Standards

The concept of standards is central to the development and evolution of telecommunications networks. A standard is a specification that describes the performance requirements and interface parameters for a piece of equipment, a network, or a software program. A standard can be narrow or broad, specific or vague, complex or simple, or formal or informal.

A widely accepted standard that is properly defined becomes invisible to the user, but it enables products developed by different vendors, or multiple products developed by the same vendor, to function together. For example, every household phone outlet in the United States is standardized both in terms of the size and shape of the phone jack that it will accept and with respect to the parameters for the electric current available on demand. Consequently, in preparing to purchase a Mickey Mouse phone, a handset that offers speed dialing, or a handset that has such features as redial, no thought is given—or needs to be given—to whether or not the jack will fit the outlet at home or at the office. Everyone relies on the national standard, and it is taken for granted that all manufacturers have engineered their equipment to be compatible with that standard.

However, matters are different elsewhere. British and U.S. phone wires look very much the same, but trying to force one into the other country's wall outlet will not work. Dialing from one country to the next works because the necessary protocol converters, voltage converters, and related equipment have been installed. The telephone systems around the world are very different, but standards and interfaces remove the differences.

The goal of a corporate information technology platform is to remove such differences and make full, shared use of the technology infrastructure, including the following: exploiting economies of scale, combining information from separate applications in order to create new services and products, avoiding redundant and duplicate facilities, using the same delivery base for a growing range of services, facilitating cross-organizational information flows, and linking a firm's system with those of customers or suppliers. To realize these shared-use opportunities, a high degree of compatibility among an organization's computers, software, and telecommunications networks is essential.

Establishing a single, monolithic system is not necessary, but ensuring that the technical components share common standards is vital. Just as the world's phone systems are not a single network but a set of interoperable networks, companies want to make their transaction processing systems, local area networks, wide area networks, and international networks seamless—that is, the interfaces are invisible and the interconnections inapparent. This goal can be accomplished only through standards.

Standards can be formally developed and sanctioned by one or more standards-setting organizations, or they can be informally determined by a dominant product in the marketplace, the specifications of which become the de facto standard. For the most part, telecommunications standards are voluntary and unenforceable; compliance is based on the desire to ensure broad marketability of a given product, which in turn depends on its compatibility with other equipment in the network. The buying power of major government agencies frequently can drive vendors to implement standards quickly; we noted in Chapter 6 that when the Department of Defense mandated its agencies to use the Defense Data Network, leading vendors had to adopt TCP/IP if they were not to lose business with those agencies. TCP/IP became a market-led standard, as were MS.DOS and Windows.

Once a standard has been widely accepted by the market, compliance by all manufacturers is virtually guaranteed out of basic self-interest. As discussed in Chapter 6, leading vendors have rushed to announce APPN to frame-relay internetworking routers because they know that this is a priority for large organizations that have SNA networks. Those same vendors have been in less of a rush to offer AppleTalk to SNA routers because those companies have not made Apple Macintosh PCs the base for access to their networks. It is fashionable among many telecommunications professionals to view SNA as a relic of the 1970s, but any company that does not offer SNA-interconnection locks itself out of a major part of the networking market. In turn, IBM fell close to locking itself out of important markets through its lack of OSI and TCP/IP-compliant products.[1]

It appears likely that user- and market-driven standards will increasingly replace committee-based standards-setting procedures, mainly because the rapidity of change in the technology and marketplace makes speed rather than

---

[1] Once IBM decided to drop its deliberate posture of ignoring OSI and TCP/IP, it quickly changed the situation. In 1990 and 1991, it announced literally hundreds of OSI products. However, many of these products enable firms whose systems are built on IBM architectures and operating systems to distribute more functions and thus reduce their reliance on IBM, using IBM's own products to do so. For instance, many companies use UNIX workstations from companies such as Sun to connect to IBM mainframes by carrying SNA traffic via TCP/IP. They then selectively remove systems from the mainframe and downsize it.

consensus the main driving force.[2] That said, a standard is not a standard until it is defined and published as such. It needs a forum for definition, publication, adoption, and certification. Without the CCITT, for instance, it is extremely unlikely that the world's many providers of electronic mail systems on LANs and WANs would have worked together to create the X.400 standard. At a time when the Ethernet and token ring protocols for LANs were rivals in terms of philosophy and vendor, it needed the forum of the IEEE to negotiate a standard that stabilized individual definitions and thus made it possible to interconnect them.

Thus standards-setting organizations are the equivalent of the United Nations and the GATT talks in a less-than-perfect world. In general, when any of these standards-setting organizations issues a definition of a standard, it will become one. However, that does not mean that users will choose it rather than some de facto standard or that users and/or vendors will not generate additional and competing standards through the free play of market forces.

A standard can be almost anything at any layer of the OSI model. The easiest standards to define and get accepted are at the lower layers—plugs and sockets—yet even there a plethora of offerings exists. The highest layers are as difficult to standardize as is the format of magazines. Magazines typically use standard print fonts—Times Roman, Helvetica, and the like—and standard mailing labels, barcodes, and spelling, but not standard layouts and style. Print fonts are like Layer 1 physical signals and layouts like Layer 6 presentation services; style is the equivalent of Layer 7.

Figure 7-1 lists major telecommunications standards and the organizations that were responsible for their definition and that remain the forum for updating and/or extending them. To this list could be added the many industry groups that attempt to influence or coordinate the direction of the standards-setting process; examples are the Open Systems Foundation, the Committee for Open Systems, and the Internet Engineering Task Force, all of which are informal alliances of users and vendors.[3] The organizations shown in Figure 7-1 and discussed in the next section are official standards-setting bodies; though their decisions on standards do not have the weight of law and they cannot ensure the success of a standard, they carry substantial weight.

---

[2] FDDI products, for instance, appeared on the market even before the standard was fully defined; the same situation occurred with switches for fast packet switching.

[3] These organizations are generally funded by large user and/or provider companies that each contribute a fixed sum and assign one of their staff as a representative. Their success has been mixed. OSF, for instance, took a lead in designing a Distributed Management Environment (DME) that used components from several OSF member organizations. The first products were delivered on time, but internal disagreements, including a threat from Hewlett Packard to pull out of OSF, has put the project in jeopardy. Cooperation is the core of standards-setting, and it often requires that competitors collaborate.

Standard	Organization that defines/approves/oversees the standard	Nature of the standard
X.25	CCITT = Consultative Committee on International Telephony & Telegraphy (Geneva, Switzerland): standards-setting committee of the International Telecommunications Union, composed of 150 PTTs and private telecommunications agencies	The most widely used standard for packet-switched networks and the base for almost every public data network
X.400	CCITT	Defines formats for transferring electronic messages across networks, including electronic mail
X.500	CCITT	Global electronic directory, analogous to the printed phone book
V.32 and V.32 bis	CCITT	High-speed modem transmission
ISDN	CCITT	Blueprint for interfaces and operating parameters for PTT and phone companies' digital services
802.1 to 802.9	IEEE = Institute of Electrical and Electronics Engineers (New York); professional society whose work covers many fields	Set of standards for local area networks; IEEE 802.3 defines Ethernet standards; 802.4 & 802.5 define token ring
ISO 7498 (OSI)	ISO = International Organization for Standardization (Geneva, Switzerland); composed of each member country's standards organizations; member of CCITT but operates independently (though cooperatively with it). OSI's most important recent standard, ISO 9000, certifies companies' quality-setting programs in manufacturing and is becoming a key element in trade among and with European nations	The OSI reference model. Subsets of OSI include FTAM (file transfer and access management; ISO 8571) and CMIP (Common Management Information Protocol; ISO 9596)
EDIFACT	United Nations SITPRO (Standardization of International Trade Procedures), with ISO supervising the standards process	The core standard for international trade EDI; voluntary committees develop industry-specific subsets of EDIFACT
X12	ANSI = American National Standards Institute (New York); composed of close to 1,000 U.S. organizations; coordinates, rather than defines, standards; ANSI is the U.S. representative to ISO	EDI standard for U.S. domestic trade
FDDI	ANSI	100-mbps fiber-optic LAN standard; Fiber Distributed Data Interface; ANSI is coordinating other nonfiber 100-mbps standards, including CDDI (Copper DDI) and fast Ethernet
SONET	Bellcore (Bell Communications Research), formed by the seven RBOCs after the divestiture of AT&T to coordinate the industry's research and development	The blueprint for ultrafast fiber-optic transmission at 53 mbps to 2.4 gbps and above (Synchronous Optical Network)
RS-232	EIA = Electrical Industries Association; accredited by ANSI, members are manufacturers of electrical and telecommunications equipment	The basic standard for connecting cables

Figure 7-1

Key Telecommunications Standards and Standard Setters

| | | | | | |

*International
Organization for
Standardization
(ISO)*

Standards-setting Organizations

ISO is a voluntary, nontreaty, international organization whose members are the official standards bodies of the participating countries; it also includes nonvoting members from observer organizations. The U.S. member of the ISO is the American National Standards Institute (ANSI). The purposes of ISO are to promote the development of standards and related activities that will facilitate international exchange and commerce and to facilitate cooperation in scientific, technological, and economic activity. An important aspect of ISO standards development is the Open Systems Interconnection (OSI) architecture model described later in this chapter.

ISO, headquartered in Geneva, Switzerland, uses a seven-step process to develop standards, requiring a minimum of 14 months. Negative votes received on the initial balloting lead to revisions to the draft standard in an attempt to develop widespread support for the final version. Draft standards are published in what are often very long and wordy documents; the 1982 draft of OSI is over 80 pages long. The review process can be very slow; although study of OSI began in 1977, a first draft was not published until December 1980. Comments were incorporated into a revised draft that was approved in August 1981. The ISO Secretariat then convened a technical advisory group that prepared a revised text, which was issued in April 1982 and adopted by the full voting membership.

In the mid-1980s many commentators saw OSI as the only model for telecommunications for the rest of the century.[4] Many also saw OSI as a single entity, as the architecture. ISO defined OSI as a basic reference model, not as an entity, and OSI may never be completely implemented, particularly at the higher layers. The 1982 draft omits many requirements for fully integrated open systems that were not apparent at the time; the model has had to be revised to address network management, for example.[5] What ISO provided was both a framework that has helped vendors and users create an evolutionary strategy for moving toward open systems and a continuing forum for defining standards within OSI. For instance, in 1984 CCITT published the specifications for X.400, the OSI standard for messaging. (The standard was revised and

---

[4] One 1992 article even announced that the United States's entire competitive future was at risk because American companies had not implemented OSI, whereas European companies had already done so. The author claimed that by the end of 1992, all seven layers of the OSI model would have been fully implemented, right down to the LAN protocol stack. Such claims are nonsense; the field of telecommunications standards-setting and implementation is full of uncertainties and complexities that lead to a whole industry of hype, claims, and wild forecasts and that too often lead to a backlash. That has happened with ISDN and OSI, and it may happen with frame relay, FDDI, and ATM, too. The best way to avoid being victimized by such hype is to track what real companies have done, how well their standards worked, and whether their experience points to a new standard that is likely to create a critical mass of users and suppliers.

[5] Rather than wait for OSI, many users of open systems have adopted the Simple Network Management Protocol (SNMP).

extended in 1988.) Similarly, CCITT defined X.500 (network directories), ISO defined FTAM (file access and transfer), and many other organizations have proposed that standards be adopted by ISO as OSI standards. Even IBM, whose SNA was the dominant alternative to OSI, proposed that its DIA/DCA protocol (document interchange architecture/document content architecture) become the OSI standard for electronic document internetworking.

ISO has a strong European flavor, and its work covers many fields of standardization, not just telecommunications. As part of the European Community's Open Market 1992 initiative, ISO issues standards on just about everything: contracts, weights and measures, and even food labels. In telecommunications, OSI is and will remain the central contribution of ISO. European companies and vendors have, not surprisingly, seen ISO as the central driver of telecommunications. In the United States it carries weight, but it is just one of many parties shaping open systems. It is also very laborious and slow: In late 1989 the ISO/IEC (International Electrotechnical Commission) Joint Technical Committee 1 had 16 subcommittees and 75 working groups responsible for over 700 projects, 270 published standards, 125 draft international standards, 150 draft proposals, and 175 working drafts.

Consultative Committee on International Telegraphy and Telephony (CCITT)

CCITT is the telecommunications standards-setting committee of the International Telecommunications Union (ITU). The ITU is a United Nations Treaty Organization, and its members are governmental entities. The charter of the CCITT is to develop necessary standards for telecommunications operations and techniques to attain end-to-end compatibility of services linking various countries. The ITU, which was founded in 1865 and has over 150 member countries, is the forum for PTTs, and its focus is on the standards PTTs need for international operations. In general, CCITT's standards are quickly and widely adopted, though implementing them may take many years. Thus even though X.400 was defined in 1984, X.400 products did not reach the market until 1991–1992.

The CCITT works in four-year cycles, using individual study groups that concentrate on a particular area within the field of telecommunications. Traditionally, the CCITT has focused on telephony and data transmission issues; the ISO has traditionally focused on computer communications and distributed processing issues. As the line between computing and telecommunications continues to blur, the work of these two standards bodies increasingly overlaps, which has led to increased cooperation between them to avoid the proliferation of competing standards. Thus, X.400, X.500, and X.25 are all CCITT standards that have been made part of OSI.

CCITT's equivalent of OSI is ISDN, which was defined in the late 1970s. Implementation has been very slow; France Telecom took the lead, introducing ISDN in Brittany in 1985. Initial interest in the United States was mild and then almost disappeared until the early 1990s. Other CCITT standards have had far greater impacts. These generally have either the prefix V (for telephony) or X (for data communications). CCITT also issues recommendations in other areas of telecommunications. Its D.1 text, for instance, defines principles for PTTs and users to adopt in pricing and regulating the use of international leased circuits.

The ITU's most important role in the coming years will be in allocating the radio spectrum for international telecommunications. It will also play a major role in GATT negotiations,[6] and in European and Asian policy on open networks and international telecommunications policies. CCITT will continue to focus on ISDN and OSI-related issues and on emerging mobile communications developments.

The American National Standards Institute (ANSI)

ANSI is a nonprofit association of standards-making and standards-using organizations in the United States. Its members consist of professional societies, trade associations, government and regulatory bodies, industrial companies, and consumer organizations. ANSI coordinates voluntary standards in the United States, and it is the U.S. representative to the ISO.

ANSI does not actually develop standards themselves; it merely publishes and disseminates them. Other groups are accredited by ANSI to develop standards for the institute. ANSI operates in a manner similar to the CCITT: Committees focusing on specific standards areas develop through a consensus process a draft standard that is then published by ANSI for public review and comment. If there is widespread support for the proposed standard, it is adopted as a U.S. national standard.

ANSI thus plays a coordinatory and fairly passive role in the telecommunications standards-setting process. It does not drive standards so much as help those who are already driving them. For example, the ANSI X12 standard for EDI emerged from businesses' need to create a standard so they can work together. The ISO, CCITT, and other vendor-dominated or PTT-dominated organizations are often marked by conflicts, with individual companies or countries trying to get their own choice (or product) adopted as a standard.

X12 is one of the key standards from a telecommunication client's viewpoint. (Note that it is correctly "X12", not "X.12", though it is often shown as the latter.) In the computer field, ANSI has played a strong role in the standardization of computer languages and database management systems. It is far more responsive to business users than is either OSI or CCITT, and it provides them with a forum to work together. For instance, it has industry-specific groups working on proposals for X12-based standards in their sector. An ANSI ASC (Accrediting Standards Committee) X12 developed EDI standards for insurance and played a major role in developing the ANSI X12 835 standard for electronic insurance claims processing. FDDI is another key ANSI standard, and ANSI is also providing the base for coordinating frame-relay standards.

---

[6] Lack of awareness of telecommunications among most government figures, economists, and trade representatives may explain why an issue so basic to U.S. competitiveness in global markets was neglected in the Uruguay round of GATT talks. U.S. firms can take their business strategies abroad, but in many countries they will be restricted in their use of, say, private networks or VSAT, greatly limiting their effectiveness.

In general, an ANSI standard is a little like the Good Housekeeping Institute's seal of approval. An ANSI-compliant product has been proven to follow the ANSI standard through exhaustive **conformance testing**, the process by which laboratories, PTTs, and other agencies certify that a product meets a standard. In the United States, the National Institute of Standards and Technology handles conformance testing for federal government procurement; the Corporation for Open Standards (COS) provides seals of approval for OSI standards compliance.

Institute of Electrical and Electronics Engineers (IEEE)

The IEEE, a major contributor to ANSI standards, is an independent association that develops standards in many areas. Its main impact on telecommunications has been in an area entirely ignored by CCITT and ISO: local area networks. Its IEEE 802.X series was the first to bring some order to the LAN field, which was dominated by competing technologies and vendors, with Xerox's Ethernet and IBM's token ring LAN operating on entirely different technical principles. IEEE did not endorse any single type of LAN but decided instead to stabilize LANs as a set of related standards and let the combination of the standard and the market pace developments.

Vendor/User Standards Organizations

As the importance of standards in ensuring compatibility among multivendor networks has grown, vendor and user organizations have emerged to ensure that these standards are accepted in their respective communities of interest. Of particular importance is the Manufacturing Automation Protocol/Technical and Office Protocols (MAP/TOP) group, which emerged from two separate user organizations created to ensure both that its members comply with accepted standards and that vendor equipment is manufactured to meet those standards.

MAP's purpose is to define LAN protocols for terminals, data processing resources, and robots within a manufacturing environment. MAP operates either by finding and adopting existing international standards or proposing and supporting the development of new standards that serve its needs. MAP has widespread, though waning, support among manufacturing corporations.

TOP's purpose is to ensure that appropriate protocol standards both exist and are met for nonmanufacturing, office automation networks. MAP/TOP is now a combined user organization working toward the development of, and ensuring compliance with, standards that serve the interest of its members.

The Corporation for Open Systems (COS) is a joint venture among a large number of major suppliers of computer and data communications equipment. Its purpose is to accelerate the introduction of compatible, multivendor products and services that comply with international standards and lead to open network systems throughout the world. In particular, COS is involved with the development of test methods and certification procedures to ensure that a given vendor's products and services meet international standards and will in fact work with another vendor's certified equipment. The Open Software Foundation has had a significant effect on UNIX developments.

The growth in support for these standards organizations is testament to the crucial role played by standards in the evolution of telecommunications networks. Multivendor, mix-and-match networks that are modified frequently to meet the changing needs of the global business environment are today's reality. Without widespread acceptance of standards that ensure compatibility and interoperability of different components and services, it is not possible to have telecommunications network platforms with the extensive reach, range, and responsiveness needed to meet the global business challenges of the 1990s.

Open Systems, Integration, and Standards

Standards are the key to integration of different networks and equipment manufactured by different vendors, and thus positioning a telecommunications platform requires choosing standards. This often involves either selecting specific vendors and their proprietary standards or choosing open system standards. An open system is a vendor-independent specification that ensures compatibility among different manufacturers' equipment.

For example, ribbons for impact computer printers are based on proprietary standards that are specific to a particular vendor's printer. Although all serve the same function, a ribbon for one type of printer will not work on another type of printer. In contrast, photocopier paper adheres to an open system standard—8 1/2-by 11-inch dimensions—and thus is compatible with any vendor's copier designed to meet the open system standard for paper size.

In practice, vendor-specific standards are frequently more "open" than true open ones, when they are both fully defined, stable, and published in a form that allows vendors to incorporate them into their products. Many of IBM's so-called proprietary standards fall into this category of de facto open standard. The IBM personal computer's DOS operating system, for example, began as a proprietary system and quickly became an industrywide, de facto, open standard. An "IBM PC-compatible" industry has evolved on the basis of numerous vendors' design and manufacture of hardware and software to meet the DOS operating system standard. In contrast, many formally open standards have fuzzy definitions; as a result, the implementation of those standards will vary from vendor to vendor, which can lead to incompatibility of equipment that actually has been designed to meet open systems specifications.

The challenge for the telecommunications designer is to integrate the current multivendor, multitechnology telecommunications base without discarding the pieces and starting over again. The first step toward this integration is to adopt a set of standards that is currently fully implemented. Some of these standards are almost certain to be at least partly proprietary with respect to one or more vendors. Evolution toward open system standards will occur as such standards become more fully defined. The need to balance today's operational needs with the longer-term movement toward open systems and integration means that the criteria for selecting strategic vendors must go well beyond a consideration of which ones have the "best" technology. The following factors should also be considered:

▸ The ability to help rationalize today's installed telecommunications resources

▸ Proven support for today's mainstream proprietary and nonproprietary systems

▸ Proven capabilities in implementing new standards

▸ Breadth of capability in all areas relevant to the telecommunications platform, including computing and information management

▸ The abilities to upgrade facilities smoothly and quickly and to perform efficiently as processing volume and transactions traffic increase

In addition to the selection criteria for strategic vendors, the following three facts are largely ignored or overlooked in most discussions of standards, integration, and open systems:

▸ Open and proprietary *definitions* of standards are different from open and proprietary *implementation* of standards.

▸ The software development costs involved in moving along the path to integration and open systems are far greater than all but the largest vendors are able to bear.

▸ To reach a critical mass in the marketplace and be considered a primary supplier to a large organization's information technology base, a vendor must be able to offer IBM compatibility and/or interoperability.

A fuzzy open system standard definition, or multiple versions of an open system standard, can lead to incompatibility or lack of interoperability among products designed to meet the open standard. Thus although the UNIX operating system has been touted as a true open system standard with complete portability across computing platforms, several different versions of UNIX have emerged, and not all of them are compatible. In contrast, a proprietary standard definition can become open in implementation if the specifications are made widely available on an unrestricted basis. This has been the case with the IBM PC design, which has become a de facto open standard in implementation, and numerous vendors are manufacturing IBM PC clones.

IBM is moving toward interoperability and open systems with all of its new developments. However, given the industry's history of proprietary system standards, IBM's market dominance, and the glacial movement of formal open system standards bodies, the practical path to integration is through IBM-compatibility as a de facto standard. This does not mean that users must purchase IBM's products, but merely that the equipment be compatible with IBM system standards. Both Digital Equipment and Apple have recognized this need to be IBM-compatible. In the mid-1980s, DECnet switched from a completely proprietary system to providing linkages to IBM's Systems Network Architecture standard. Similarly, recent versions of the Apple Macintosh computer provide file and format conversion capability that makes the "MAC" interoperable with DOS programs and IBM-compatible computers.

These developments suggest that the practical path toward integration and open system standards is through increasing interoperability and compatibility with IBM system standards, which are themselves converging on formal open systems standards. The reason that the choice of standards is so important is that standards define the telecommunications architecture, which in effect is a firm's technical strategy. Note that it is now fully practical to adopt an IBM architecture without having a single piece of IBM equipment. The mainframe can be, say, an Amdahl machine because Amdahl's entire strategy rests on being fully compatible with IBM's major operating systems; the workstations can be from any one of hundreds of companies that provide DOS and Windows-based systems, and Northern Telecom's switches can carry SNA traffic over an X.25 backbone.

| | | | | | | |

Architecture

The telecommunications architecture is the technical strategy of the organization; it is the design blueprint for developing the telecommunications network platform over time and across technologies, uses, volumes, and geographic locations. The telecommunications blueprint is similar to a city plan, which consists of a mixture of fixed main routes, zoning regulations and ordinances, and procedures for extending and modifying existing buildings and expanding or adding roads. The higher layers of the city plan specify terms and standards for buildings; the lower layers specify the standards for fixtures and building subsystems, such as electrical equipment, heating and air conditioning systems, and so on. It does not describe the details of buildings, though it may impose standards of size, construction, and safety.

The telecommunications equivalent of this city plan specifies at the lowest levels of the architecture the protocols for interconnecting equipment devices; at the highest levels it specifies the standards (and even specific software) that ensure that the devices can share transaction services and data. Some architectures are selected for reasons specific to the firm or its industry, just as individual cities choose their own ordinances, but more and more of them conform to generic standards for reasons of cost, availability of equipment, and the like. Just as the details of a city plan can involve thousands of individual specifications, so can a telecommunications blueprint, but again analogous to a city plan, an understanding of the overall principles is often all a designer must know or demonstrate; most firms' architectures can be summarized in a paragraph. The city plan analogy is most useful as a reminder that the architecture is a blueprint, not an artifact, that provides guidelines and requirements for a continually evolving mix of developments, devices, and applications.

The alternative to a telecommunications architecture is a series of piecemeal, network applications implemented by individual business units on a case-by-case basis. The individual application approach has value in enabling decentralized business units to take responsibility for telecommunications decisions and to match closely the specific network technology to the characteristics of the application. However, the architecture approach leads to a coordinated telecommunications platform that meets a company's business requirements for range, reach, and responsiveness without unnecessary duplication and redundancy (and the excess cost and management attention associated with them).

The architecture concept imposes common, organizationwide technical standards for choosing equipment, software, and telecommunications services. These standards should have the same force within the organization as its accounting rules. An effective architecture will balance central coordination of the telecommunications infrastructure with decentralized use: The more flexible and adaptive the architecture is, the wider the variety of shared and interlinked resources it enables.

The technical standards chosen depend on vendors' available products and systems. Though choosing a standard is not the same as choosing a vendor, the choice of architectural standards and primary vendors will remain highly interdependent until truly open systems standards are proven and comprehensive.

The telecommunications architecture concept can be viewed as defining two structures that together determine the reach, range and responsiveness of the platform. The first is the protocol structure, which accomplishes all the communications tasks and functions necessary to transfer information among end-users reliably and accurately; this determines the effectiveness of the platform: The second is the configuration structure, which defines the permissible configurations and interconnections between these components; this determines the efficiency of the platform—the costs and overhead incurred as the protocol structures make communication practical.

The next sections of this chapter review the two types of structure.

Telecommunications Architecture: Protocol Structures

Recall from Chapter 6 that the layered/modular approach to a communications architecture is an attempt to reduce design complexity and to minimize the impact on the entire system of modifying individual protocols. Each layer in the architecture should perform a well-defined function, and layer boundaries should be chosen to minimize the information flow across the interfaces. Finally, the number of layers should be large enough to keep distinct functions separate while remaining small enough to be manageable.

Both proprietary and open protocol structures exist. We will examine the open systems interconnection (OSI) model architecture developed by the ISO standards body, and later we will compare it to IBM's proprietary protocol structures. We focus on the OSI model for two reasons: (1) It is a worldwide standard that is supported by most vendor and user organizations to some degree, and (2) its seven-layer structure can be taken as a generic representation of the protocol structure telecommunications architecture concept, with all required communications functions and tasks addressed by the model.

The purpose of the standards developed for the OSI architecture model is to provide a common basis for interconnecting dissimilar open systems. The model does not address the internal workings of the individual systems themselves but is concerned only with ensuring the cooperative exchange of information among systems. An open system consists of applications packages, an operating system, and other system software, such as a database management system. To be an open system it must also include communications software that links it to other systems across a communications network. Today, virtually all vendors are providing communications software that conforms to the OSI model standards.

The OSI model is a seven-layered generic structure that defines communications tasks and functions to be carried out by specific protocols, but it does not specify the particular protocols to be used. Strictly speaking, then, it is not a complete architecture. However, the ISO group has separately produced and published specific standards for all of the layers to provide a complete open systems architecture. Figure 7-2 illustrates the seven layers of the OSI model architecture.

A fully open system contains protocol and interface software for each of the seven layers, and each layer in the model provides interface services to the adjacent higher layer in the structure. The model is designed so that each layer communicates with its peer counterpart in another open system through the protocols for that given layer. However, the actual physical transfer of information is between adjacent layers of one system until it reaches the bottom physical layer; then it is transferred across the network to the corresponding physical layer of the receiving system, and finally it is sequentially handed off to the next higher level in the receiving system, where appropriate processing takes place based on instructions from the peer layer in the transmitting system. The handoffs to adjacent higher layers in the receiving system continue until the application layer is reached and the data is processed by the end-user.

The process is similar to the workings of a postal system. When person X writes a letter to person Y, X doesn't deliver the letter directly to Y. Instead, she places it in an envelope, addresses it to person Y, and then gives it to a letter carrier, who in turn hands it off to others in the local district post office for sorting and loading onto a truck or an airplane. The letter is then *physically* carried to the destination post office, where it is sorted and handed to person Y's

Figure 7–2

The OSI Model

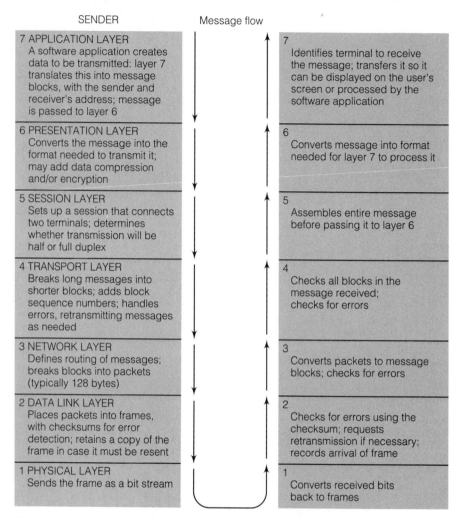

SENDER   Message flow

7 APPLICATION LAYER
A software application creates data to be transmitted: layer 7 translates this into message blocks, with the sender and receiver's address; message is passed to layer 6

7 Identifies terminal to receive the message; transfers it so it can be displayed on the user's screen or processed by the software application

6 PRESENTATION LAYER
Converts the message into the format needed to transmit it; may add data compression and/or encryption

6 Converts message into format needed for layer 7 to process it

5 SESSION LAYER
Sets up a session that connects two terminals; determines whether transmission will be half or full duplex

5 Assembles entire message before passing it to layer 6

4 TRANSPORT LAYER
Breaks long messages into shorter blocks; adds block sequence numbers; handles errors, retransmitting messages as needed

4 Checks all blocks in the message received; checks for errors

3 NETWORK LAYER
Defines routing of messages; breaks blocks into packets (typically 128 bytes)

3 Converts packets to message blocks; checks for errors

2 DATA LINK LAYER
Places packets into frames, with checksums for error detection; retains a copy of the frame in case it must be resent

2 Checks for errors using the checksum; requests retransmission if necessary; records arrival of frame

1 PHYSICAL LAYER
Sends the frame as a bit stream

1 Converts received bits back to frames

Note: bits are sent only at layer 1; the sequence is 7-6-5-4-3-2-1-1-2-3-4-5-6-7.

letter carrier, who in turn delivers it directly to person Y. Person Y then removes the letter from the envelope and "processes" the information.

In this analogy, the postal system is a four-layer architecture. The highest layer, the read/write layer, is designed for communication on a peer-to-peer basis—person X to person Y. However, the actual movement of the data (the letter) is through adjacent layers of the same system: from writer (Layer 4) to her letter carrier (Layer 3); from her letter carrier (Layer 3) to the local district post office complex (Layer 2); and from the local district post office complex (Layer 2) to a truck or an airplane (Layer 1). The letter physically moves from one system to the other only through the Layer 1 transportation system. When it reaches the destination post office, it is passed upward through adjacent layers in that system, with appropriate processing at each layer, until it reaches the reader at Layer 4.

The protocol procedures at each layer are designed for peer-to-peer interaction, but the communication is not direct; it takes place only through the adjacent-layer handoff process within each system. The letter writer and reader may agree on conventions (protocols) for language, grammar, and so on, but they do not communicate directly with each other. Similarly, both post offices agree on procedures (protocols) for pickup, delivery, addressing, and routing of the letters, but the actual exchange of information is through the adjacent-layer handoff process within a district post office complex.

Now that we have described the structure of the layered architecture model and how it functions, we turn to an examination of each of the seven layers in the OSI model.

Physical Layer

The physical layer specifies the physical interface between the open system's transmission/receiving device and the transmission medium (that is, it specifies the RS-232 serial interface standard). This layer also includes specifications for electrical standards and signaling parameters to make and break the link connection, as well as the rules for moving bits on and off the transmission medium. Note that this layer is the only one in which data actually moves between separate systems.

Data Link Layer

The purpose of the data link layer is to transform a crude transmission path into an error-free link. This layer structures the data into frames and addresses the issues of error detection and correction and flow control. As discussed in Chapter 6, the HDLC protocol, adopted as an OSI data link layer standard, was developed to address these issues.

In LAN systems, the data link layer is divided into two sublayers: a media access control layer (MAC), which specifies rules for access to the LAN transmission medium, and a logical link control layer (LLC), which performs the error correction/detection and flow control functions on the LAN.

Network Layer

The network layer is primarily concerned with network addressing, establishing and terminating connections across the network, and the routing of information through the network. Also, when a message is received from the transport layer, the network layer is responsible for partitioning it into blocks or packets that are appropriately sized for transmission. At the receiving end of the transmission, the network layer reassembles the packetized data received from the data link layer into the original message format before passing it up to the transport layer for delivery to the end-user. If a network consists of only a single transmission link connecting two end points, there is no need for the network layer because all the necessary tasks can be performed by the data link layer.

Figure 7-3 illustrates X.25 network switching, a network consisting of both end-users and intermediate nodes that act as switching relay points in the network. In this case, the intermediate relay nodes use the lower three layers of the

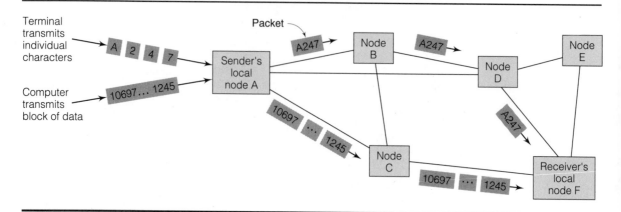

Figure 7–3

X.25 Network Switching

Node A creates packets. Node B acknowledges receipt of packets from node A, node D acknowledges receipt from node B, and node F both acknowledges receipt from node D and disassembles the packet. A similar receipt-of-message acknowledgment process occurs between nodes A, C, and F.

OSI model, and each relay node has two sets of data-link and physical-link layers supporting a common network layer. The network layer in the relay node switches and routes the data while the separate data link and physical link layers perform their assigned functions for their respective point-to-point links.

To illustrate this concept further, consider an extension of the post office analogy previously described. If person X in Miami is sending a letter to person Y in New York, the letter will likely be routed through one or more intermediate relay points, such as Atlanta. The physical layer transportation system delivers the letter from Miami to a regional post office in Atlanta (call this office the Southeast physical layer). At this point, the letter is checked to ensure that it was not damaged in transit and is temporarily stored (by the Southeast data link layer). Next, the destination address is read, and a route is selected for the next leg of the journey (common network layer function). The letter is then transferred to the section of the Atlanta post office complex responsible for Northeast corridor mail traffic (the Northeast data link layer) to prepare it for the Atlanta-to-New York trip. Finally, it is loaded on a flight to New York (by the Northeast physical layer) for ultimate delivery to the end-user/recipient.

Transport Layer

The purpose of the transport layer is to ensure total end-to-end reliability of the information exchange between separate systems. It performs the following functions: (1) It ensures that entire messages arrive at the desired destination error-free and in the proper sequence; (2) it controls the choice of network service options available (that is, leased line transmission versus packet switching, if both are available on the network); (3) it provides the requested quality of service (that is, meets acceptable error rate and transmission delay limits, satisfies security criteria, and so on); and (4) it makes efficient use of network transmission facilities through the use of multiplexing techniques (when appropriate).

Some of these functions appear to be performed by the data link layer of the model; the difference is that data link layer protocols are designed to ensure the integrity of data flows between individual links, whereas the transport layer provides end-to-end message and data integrity. Specifically, Layer 4 is concerned with the arrival of entire, intact messages at the destination node, whereas Layer 2 ensures that individual packets and blocks of data arrive intact at the next node on the path to the final destination. This can imply some redundancy if the underlying transmission network consists of highly reliable individual links; in that case, an extensive Layer 4 protocol may not be needed. On the other hand, if some of the underlying segments of the network are not highly reliable (that is, if a node or individual transmission link may fail), both Layer 2 and Layer 4 protocols help to ensure end-to-end reliability.

Session Layer

The session layer is responsible for establishing and terminating a connection between end-user terminals or application programs and for managing the exchange of information according to an agreed-on set of rules and procedures. The session layer establishes whether the transmission is half duplex (two-way alternate) or full duplex (simultaneous two-way) and when a given entity transmits and receives. If the link connection is broken during the session, the session layer protocol is responsible for reestablishing the session connection.

The network layer both establishes a chain of individual link connections that forms a path across the network and routes data over this path. It is the session layer protocol that for the duration of the session establishes the connection between communicating terminals or application programs residing on separate computer systems.

Presentation Layer

The presentation layer is primarily responsible for formatting data. It transforms data from the format of the sending terminal or application program to a format compatible with the lower layers of the mode, transforms data received from lower layers of the model into the format of the receiving terminal or application program, and performs code conversion, data compression, and data encryption when needed.

Application Layer

The application layer is responsible for coordinating communication between applications that reside in different systems linked across a network. Because application programs can differ in so many ways, the application layer is used to standardize generic elements common to different classes of applications.

To date, standards have been developed for the following application activities: (1) log on and password identification, (2) file transfer, access, and management (FTAM), (3) electronic mail and messaging exchange (CCITT X.400 standard), (4) the transfer and editing of documents across multivendor systems, and (5) virtual terminal services used to facilitate the efficient reformatting of data for compatibility with different terminal standards.

That specific standards have been developed for all seven layers of the OSI model does not guarantee that communication will, in fact, take place between two open systems adopting these standards. Implementation of a given standard is subject to interpretation, and one manufacturer may interpret that standard differently than another manufacturer. Consequently, adhering to common standards in an open systems, interconnection environment is necessary, but not sufficient, to ensure communication between separate systems. Extensive testing of individual components is needed to ensure true compatibility and reliable exchange of information between separate systems.

SNA: A Proprietary Protocol Structure Architecture

The OSI model has been generally accepted as the architecture framework for the current and future design of new open systems. However, there exists a large embedded base of information systems linked through vendor-specific, proprietary architectures. In particular, IBM's proprietary **System Network Architecture (SNA)** has been the architecture of choice for the vast majority of large computer system networks.

In 1974 IBM announced the creation of SNA to help its customers integrate new IBM products into their existing IBM system networks. The problem that led to the creation of SNA was that different protocols and network access methods were available on different IBM computing platforms; thus it was difficult to link different IBM computer systems so that proprietary applications developed for particular systems could be accessed and used by other systems.

Until 1980, SNA remained a closed, proprietary architecture linking only IBM systems. By 1980, a growing number of customers pressured IBM to incorporate the X.25 international standard for an open packet-switched network into the SNA architecture. This effectively made SNA an open architecture because non-IBM systems could exchange information with IBM systems through the open system X.25 protocols. With time, SNA has become even more "open," and it now supports interconnection with systems based on the OSI architecture.

SNA is a seven-layer, modular, protocol structure architecture similar to the OSI model, but the individual layers do not correspond completely. Figure 7-4 compares the OSI and SNA models. Both architectures perform the same set of communications functions, but the allocation of these functions among the layers is different in the two models.

Layers 1, 4, 5, and 6 are comparable in both architectures; the same set of communications functions is performed in the same layer in both models. However, the manner in which each function is carried out (specific protocols) in SNA and OSI is not necessarily the same. The combined data link and path control layers in SNA are equivalent to the combined data link and network layers in OSI, but the functions are divided among the individual layers some-

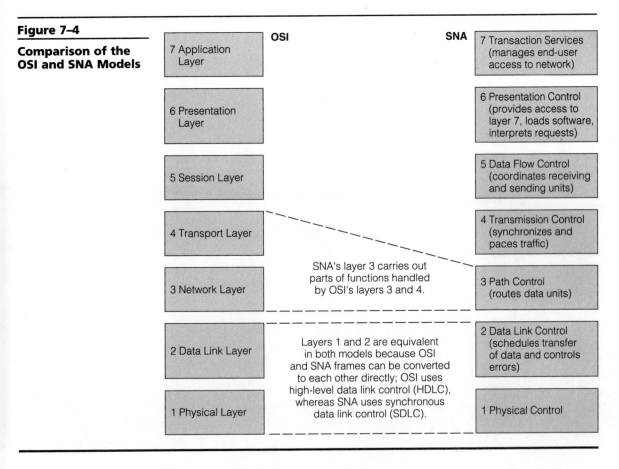

Figure 7–4

Comparison of the OSI and SNA Models

OSI		SNA
7 Application Layer		7 Transaction Services (manages end-user access to network)
6 Presentation Layer		6 Presentation Control (provides access to layer 7, loads software, interprets requests)
5 Session Layer		5 Data Flow Control (coordinates receiving and sending units)
4 Transport Layer		4 Transmission Control (synchronizes and paces traffic)
3 Network Layer	SNA's layer 3 carries out parts of functions handled by OSI's layers 3 and 4.	3 Path Control (routes data units)
2 Data Link Layer	Layers 1 and 2 are equivalent in both models because OSI and SNA frames can be converted to each other directly; OSI uses high-level data link control (HDLC), whereas SNA uses synchronous data link control (SDLC).	2 Data Link Control (schedules transfer of data and controls errors)
1 Physical Layer		1 Physical Control

what differently in the two architectures. For instance, some of the functions included in the data link layer of the OSI model are incorporated in the path control layer of SNA. Finally, the transactions services layer of SNA, generally comparable to OSI's application layer, provides application-oriented services for such functions as document interchange, data distribution and management, and file management.

Internetworking

We have discussed how the OSI and SNA architectures enable communication among different systems across a network. In addition, there is a rapidly growing business need for different systems to communicate and exchange information across multiple networks because it is often inefficient and impractical to merge multiple networks into one large network.

Figure 7–5

Bridge functions

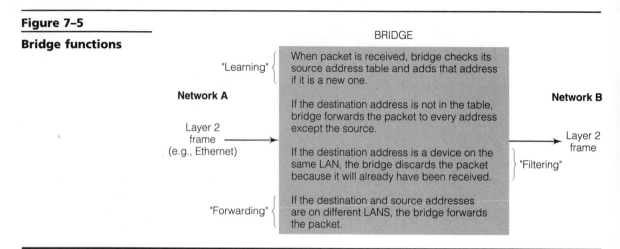

Bridges join networks at Layer 2 of the OSI model; they connect similar networks that use the same protocol.

The mass migration to distributed processing and client/server computing, driven by the efficiency and power of the personal computer workstation, has led to a proliferation of LANs as the core computing and data communications resource in today's business environment. As more and more processed and stored data is off-loaded from mainframes or minicomputers to LANs, there is a growing need to interconnect individual LANs and interconnect LANs with mainframe and minicomputer host systems. The union of interconnected separate networks, in which each retains its own identity, is called an **internet;** the individual networks making up the internet are called **subnetworks.**

Three generic devices connect subnetworks in the formation of an internet: bridges, routers, and gateways. We next consider each device in turn.

Bridges

Bridges are used for interconnecting the same types of subnetworks that use identical protocol structures. In effect, a bridge increases the number of addressable nodes on a LAN and makes two remote LANs that use the same protocol operate as if they were a single LAN.

Bridge protocol structure contains only Layers 1 and 2 of the OSI protocol because these are sufficient to transfer packets between subnetworks with identical protocol structures. A bridge (Figure 7-5) contains two physical layers, one for each of the connected subnetworks, and a single Layer 2 protocol system. The physical layers transmit and receive bits on their respective subnetworks.

The Layer 2 protocol reads the address header on incoming packets and determines which of the two connected subnetworks contains the destination address. If the destination address is on the originating subnetwork, the bridge ignores it; if the addressee is located on the connected subnetwork, the bridge retransmits the packet onto the connected subnetwork.

Why bother with using a bridge to connect identical types of networks? Why

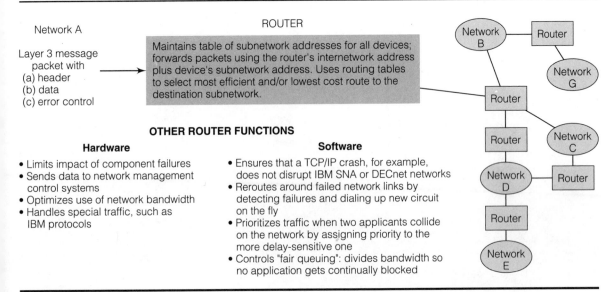

Figure 7–6

Router functions

Routers connect dissimilar LANs at OSI Layer 3 (the network layer).

not create one large network instead? The reason for maintaining two separate subnetworks is really a matter of reliability and performance efficiency. If in an internet system a fault occurs in a subnetwork, the impact can be limited to the users on that subnetwork; users of other subnetworks are unaffected. If the system operated as one large network, all users would be affected by the fault.

With respect to performance on a LAN, response time and system throughput are generally a function of the number of users on the network and the distance between communicating devices. Performance can often be improved if users with similar requirements and close geographic proximity are grouped into subnetworks connected by a bridge to handle any inter-LAN traffic.

Routers

A router (Figure 7-6) is used to connect dissimilar types of subnetworks, but only those with protocol structures that are compatible with the OSI model architecture. A router is treated as an addressable entity by each of the subnetworks that it connects, and it operates at Layer 3 of the OSI model. (Layer 2 has no routing capability; hence a bridge can transfer packets directly to another LAN, but does not have the intelligence to select *which* LAN and which route.)

Because a router links different types of subnetworks, it is more complex than a bridge and must deal with several additional issues, including: (1) different addressing structures on the subnetworks, (2) different packet size limits for the subnetworks, (3) different hardware and software interfaces for the various subnetworks, and (4) differences in reliability among the subnetworks.

A router works in the following manner: If a message that originates on subnetwork A is sent to a user residing on subnetwork B, the network layer

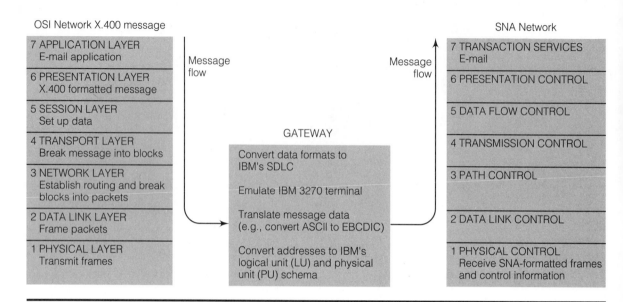

OSI Network X.400 message

| 7 APPLICATION LAYER
E-mail application |
| 6 PRESENTATION LAYER
X.400 formatted message |
| 5 SESSION LAYER
Set up data |
| 4 TRANSPORT LAYER
Break message into blocks |
| 3 NETWORK LAYER
Establish routing and break
blocks into packets |
| 2 DATA LINK LAYER
Frame packets |
| 1 PHYSICAL LAYER
Transmit frames |

Message flow

GATEWAY

Convert data formats to
IBM's SDLC

Emulate IBM 3270 terminal

Translate message data
(e.g., convert ASCII to EBCDIC)

Convert addresses to IBM's
logical unit (LU) and physical
unit (PU) schema

Message flow

SNA Network

| 7 TRANSACTION SERVICES
E-mail |
| 6 PRESENTATION CONTROL |
| 5 DATA FLOW CONTROL |
| 4 TRANSMISSION CONTROL |
| 3 PATH CONTROL |
| 2 DATA LINK CONTROL |
| 1 PHYSICAL CONTROL
Receive SNA-formatted frames
and control information |

Figure 7–7

Gateways

Gateways translate protocols between entirely dissimilar networks

protocol in the originating system recognizes that the addressee resides on another subnetwork and sends the message to a router that connects the two appropriate subnetworks. The message is received by the subnetwork A physical layer of the router and is then processed up the protocol stack to the network layer. At this point, the destination address is read, and the message is passed down through the router's Layer 2 and 1 protocols for subnetwork B and then on to final delivery to the destination workstation on subnetwork B.

Gateways

A gateway (Figure 7-7) enables the exchange of information between very dissimilar subnetworks. It is generally used to link a subnetwork based on a proprietary architecture with subnetwork that conforms to the OSI architecture. Because a gateway links two totally incompatible architectures, all seven layers of the OSI model and the complete set of protocols for the proprietary architecture are incorporated in the gateway device. Because the conversion takes place at the application layer, the gateway must contain both the OSI version and the proprietary version of any application that is linked across the two subnetworks.

The system originating the transmission in the OSI subnetwork sends a message via an application program based on the OSI X.400 protocol, which is standard for electronic mail. The ultimate destination is a user on an SNA proprietary network connected through a gateway. The message is routed to the gateway node, where it is received by the OSI physical layer and passed up the protocol stack to the application layer. The application layer software in the gateway removes all X.400 protocol information from the message frame and substitutes the corresponding SNA application protocol information. The message then moves down the SNA protocol stack to the SNA physical layer, where it is sent across the SNA subnetwork to its final destination.

Gateways are much more complex than routers because a complete protocol transformation is required for all layers of both the OSI and proprietary architectures. The extensive processing required can cause internet bottlenecks and slow response times; this may create the need for additional gateways to relieve the congestion, which in turn requires modification to host or server system software in the subnetworks to include the added gateways in routing decisions.

Bridges, routers, and gateways are all a combination of microprocessor hardware and software. The simpler the internetworking function, the more it can be handled by hardware (bridges); the more complex the function, the greater the reliance on software (gateways).

Telecommunications Architecture: Configuration Structures

The configuration structure architecture is the blueprint that specifies standards for the permissible configuration of entities on the network and the manner in which they are connected and interact.

For example, a company's configuration structure architecture might specify that when a new work group is formed, the new workstations will be connected to a file server in a LAN configuration. The blueprint might specify that the LAN will conform to either the IEEE 802.3 Ethernet or the IEEE 802.5 token ring configuration standard. If the workstations on the LAN need access to a database located on a remote mainframe computer, the architecture might specify the use of an X.25 packet switching standard to link the LAN to the mainframe's front-end processor.

Although the goal of the OSI model architecture is to specify standards for protocol and configuration structure functions, the development to date has concentrated on protocol issues. In contrast, the proprietary architectures of the major vendors have well-developed specifications for configuration structure architecture features, such as the components of the network, the function and operating role of each component, and the configuration and management of the network components. Because IBM's SNA is the predominant architecture in use today and has become a de facto open system standard supported by most equipment vendors, we will use it to illustrate the features required for a complete configuration structure architecture.

SNA was developed as a hierarchical configuration with control of the network housed in mainframe computers because all processing took place in the mainframe and access was through clusters of remote dumb terminals. The basic entities in SNA networks are **network addressable units (NAU)**—software components of a network node that support communication between end-users (terminals or application programs residing on computers) and between end-users and other entities that provide network services (control and management activities). Each NAU has an address on the network and is classified as either a physical unit, a logical unit, or a control point.

SNA divides hardware into **physical unit (PU)** classifications. However, PUs are actually software modules that are used to manage the hardware resources at a given node and to allocate the use of those resources to particular communications tasks.

Logical units (LU) are software representations of end-users (either terminals or application programs). Every end-user is represented to the network by a logical unit that enables it to communicate with other end-users. Because application programs are defined as end-users, there can be more than one logical unit at a network node. **Control points (CP)** are software entities that both direct link activation and deactivation and provide network management and control services. LUs at the network nodes request sessions to gain access to network resources and services. These sessions can be terminal-to-terminal, program-to-program, or terminal-to-program. Sessions are also classified as interactive, batch, or printer sessions.

In earlier versions of SNA, all sessions were controlled by host mainframe computers and their communications controller units because terminals had little intelligence and generally interacted with host application programs. The overall network was partitioned into subarea networks and further partitioned into domains within each subarea. All sessions and devices within a given domain were managed by a control point software package residing in the designated host computer for that domain. Sessions requiring information flows across domain boundaries needed cooperation among the control points of the respective host computers in each domain to manage the sessions.

With the growth in intelligent workstations, distributed processing, and client-server computing through LANs, the number of terminal-to-terminal sessions has increased dramatically. This increase placed a strain on the rigid, hierarchical control mechanism built into the SNA structure because each terminal-to-terminal session had to be managed and controlled through a host computer. To accommodate this change in the computing environment, SNA was modified to include a feature called **advanced peer-to-peer networking (APPN)**, which allows peer nodes to establish and manage sessions without the need for a designated host to provide centralized support and control, as described in Chapter 6.

SNA also provides for the interconnection of separate SNA networks by establishing gateway nodes and associated control point software. In addition, SNA has included open system interface specifications to accommodate non-SNA devices and to interconnect non-SNA networks. These standards (implemented through software) enable SNA nodes to communicate across an X.25 packet-switching network or permit non-SNA devices to communicate with an

SNA node through an X.25 interface. It is also possible for non-SNA terminals to connect to an SNA network.

The SNA structure has demonstrated that a complete configuration structure architecture must define and classify both the set of devices that can be connected to the network and the manner in which they are connected and controlled (for example, hierarchical or peer-to-peer). The configuration structure architecture must also establish standards for growth and expansion of the network by specifying how new nodes and subnetwork segments can be added to support additional business activities without inhibiting or degrading the performance of the existing network.

Specifically, a configuration structure architecture should specify which types and makes of mainframes/minicomputers, workstations, terminals, controllers, and front-end processors are permitted on the network and in what configuration. It should also specify standards for the various types of connections among these devices, including: LAN to mainframe, LAN to LAN, mainframe terminals to LAN, subnetwork to subnetwork, terminals to mainframe, and terminals/workstation to outside public networks.

In addition to meeting current communication and connectivity requirements, a configuration structure architecture blueprint must also provide answers to the major "what if" questions facing the company: What if the marketing department buys ten new PCs and wishes to connect them to the network? What if division X is sold and its LAN and data center must be removed from the network? What if we just purchased company Y with its 200-node AppleTalk network that needs communicate with our IBM mainframe and some of our Ethernet LANs?

These and other similar questions and issues involving current and future telecommunications needs must be thought through in order to choose an appropriate architecture for the organization. The process of choosing an architecture is our next topic.

Choosing an Architecture

Ideally, a specific architecture blueprint should be developed through a choice of standards (and hence of strategic vendors) to provide the business functionality needed to support the strategy of the company. From the blueprint, a telecommunications platform is constructed that provides a specific level of business functionality measured in terms of range, reach, and responsiveness. In practice, it is difficult to specify the exact levels of range, reach, and responsiveness required, and this makes the choice of an architecture challenging at best.

For example, SNA provides reasonable range, but its reach is limited. SNA reach can be extended, however, through its connectivity interface to X.25

packet-switching nodes, which essentially provide a worldwide extension of the network. Responsiveness is limited with X.25's cumbersome error detection and correction scheme that is invoked at every node in the network, but this can be improved considerably with migration to fast packet switching through the use of frame- or cell-relay techniques.

It should be apparent at this point that developing an architecture is currently more art than science. It involves both specifying a blueprint to guide future development and planning and traversing a migration path from the old to the new while continuing to maintain a link to the embedded technology base. There is no simple formula for doing all this, but we will consider six key business policies that affect the company's need for range, reach, and responsiveness, and thus should heavily influence any architecture choice.

Policy 1: Choice of Architecture Should Not Block Business Initiatives

The telecommunications platform base must never block a practical and important business initiative. So long as a network can be built or leased separately to meet a particular business need, then the choice of technology does not affect the rest of the firm's activities. However, it is a problem to maintain multiple networks with different skill needs, operating requirements, equipment, and software; it results in added cost and complexity. Moreover, when likely new business initiatives involve linking the company's separate networks (such as materials management and financial systems) with their own and other firm's payments and EDI systems, all of these separate network systems must be interconnected to provide additional reach within and across firms.

At the same time, multivendor network systems must be able to interlink and connect with the network systems of other firms in such a way that different technologies, such as image-processing systems and database management systems, are able to share information. This requires additional range.

Finally, the additional reach implicit in interconnected systems, and the extended range required to automatically share information across systems based on different technologies, must not adversely affect the timely receipt of information. Thus adequate responsiveness capability must be designed into the telecommunications platform so that the appropriate information is delivered to the desired location when it is needed.

For some applications, it is acceptable to have the appropriate information assembled and stored anytime before it is needed as input either for analysis or for a key decision. However, for an increasing number of applications the assembly and storage of information before it is needed results in needless storage cost and stale information. As more and more business operations go on-line, the demand for just-in-time (JIT) information grows. JIT information eliminates information inventory (storage) cost and ensures that the latest data is included in the analysis or decision. This requires a higher level of responsiveness in the telecommunications platform to avoid blocking key business initiatives that depend on JIT information.

A well-thought-out telecommunications architecture will have anticipated the need for additional reach, range, and responsiveness to support new linkages and their associated information needs, and such needs will have been factored into the choice of standards and vendors. Consequently, a telecommunications platform based on a sound architecture enables business growth rather than inhibiting it.

Policy 2: Choice of Architecture Should Allow Response to Competitors

If its competition uses telecommunications as the base for a successful initiative, a business must not be automatically locked out of countering or imitating it. Leading firms in many industries have been unable to respond to IT-based initiatives by competitors during the past decade. Delta, Johnson & Johnson, and Bank of America were left at a competitive disadvantage by American Airlines, American Hospital Supply, and Citibank, respectively. It took those firms over a decade to close the gap.

The ability to respond to a competitor's moves depends both on having information technology skills in place and on the technical architecture of the two companies. An initiative based on either an off-the-shelf or a custom software package is less likely to yield a sustainable competitive advantage than an application based on an information technology platform that few (or no) companies can easily match. There is an immense difference between developing a competitive application and developing a competitive platform.

American Airlines's AAdvantage frequent flyer program has succeeded mainly because it is based on the Sabre reservation system platform, which enables American to cross-link passenger data from Sabre to AAdvantage and to its yield-control and hub-planning systems. Its competitors cannot match that cross-linking capability; they have the same individual applications as American, but not a platform that integrates them and allows them to add application after application with automatic cross-application linkages.

The separate application approach yields the same reach as the computer reservation system to which it is linked, but it does not have the range inherent in platform-based, cross-linked applications. Responsiveness is also greater through American's reservation system because reservations and multiple queries are processed much faster with automatic cross-application linkages than under a series of separate applications. The increased range and responsiveness built into the Sabre platform created a competitive advantage for American Airlines that imposed a competitive lockout on most of the airline industry.

Policy 3: Choice of Architecture Should Facilitate Electronic Alliances

Increasingly, standards for exchange of business information among companies are becoming as important as the technical standards and architectures upon which telecommunications platforms are built. Many industry associations have defined specific standards for document formats, procedures, and electronic processing, including the U.S. grocery industry's uniform communication standard (UCS), the international banking community's SWIFT standard, and comparable standards in transportation, warehousing, automotive, and aerospace. The international EDIFACT and closely related U.S. X12 standards are the basis for many of these business standards for electronic data interchange.

Such interchange standards are crucial requirements for intercompany uses of information technology, such as point-of-sale data exchange; EDI for purchase orders, invoices, and funds transfer; and customer/supplier links. As more businesses move their core operations on-line and create electronic links to other firms, they will choose electronic partners (and be chosen by them) on the basis of telecommunications reach, range, and responsiveness, as well as on the basis of information interchange standards.

In the 1990s, on-line operations will inevitably be extended across organizational boundaries, with telecommunications technology as the enabler. This extension of operations poses new challenges for telecommunications designers and business executives. What range of business interchange capabilities must they plan for? Must a bank's platform be designed to ensure transactions and information sharing with petrochemical companies and automakers concerning point-of-sale data and electronic payments? Should a parts supplier assume that its engineering department must be able to interchange technical specifications with vendors and customers? Must the network be able to provide intercompany links to all parts of the world? What level of responsiveness must be designed into the system to ensure that appropriate information is available just-in-time for key decisions?

The only satisfactory answer to all of these questions is a telecommunications platform with appropriate range, reach, and responsiveness. A series of individual applications will never be sufficiently flexible to support a rapidly changing business environment that relies increasingly on information exchange across organizational boundaries to support on-line operations.

Policy 4: Choice of Architecture Should Allow Flexibility

In today's business environment, it can be safely assumed that mergers, divestments, relocations, and reorganizations will occur with greater frequency and become more commonplace. Ideally, a telecommunications platform should include the flexibility to change with a minimum of disruption and dislocation whenever the organization changes. Companies that acquire another firm or propose a joint venture are very likely to have information technology applications that differ significantly in terms of hardware, software, and information formats and structures. The ease with which an acquiring or joint venturing firm can mesh external systems into its own technology base can be a major determinant in the decision to make a deal.

The almost certain future importance of information technology in assessing the workability of acquisitions, restructuring, and joint ventures suggests that a telecommunications architecture must embrace standards that support mainstream technologies in an open systems context. In addition, a telecommunications platform should include sufficient range, reach, and responsiveness to accommodate organizational expansions, contractions, and restructurings without interruption to ongoing business activities.

The airline industry provides an example of how important architectures can be in mergers. When British Airways acquired British Caledonian, it was able to mesh the entire operations of both carriers, including reservation and baggage-handling systems, over a weekend. By contrast, the mergers of Eastern Airlines and People Express and of Republic and Northwest Airlines were chaos, and neither of the new companies really recovered. BA has for many years explicitly viewed its architecture as a corporate asset and a competitive resource.

Policy 5: Choice of Architecture Should Block Third-Party Intrusions

One of information technology's distinctive features is its ability to erode and eliminate boundaries between industries.

For example, electronics funds transfer at the point of sale blurs the distinction between retailing and banking. Publix, a Florida supermarket chain, installed an on-line debit card machine at its checkout stands. This service, together with its own ATM network, has made Publix a leading supplier of banking services at the expense of the major banks in the state. Similarly, airline reservation systems now include many nonairline services, such as hotel, rental-car, and theater-ticket reservations.

A firm with a telecommunications platform that has substantial range, reach, and responsiveness will naturally look for opportunities to add traffic and new applications. If an outside firm's platform is superior to the platforms of firms already in the industry, it may be able to provide superior electronic services at lower cost and capture a significant share of the market in a very short time because the new services are merely extensions of the existing technology platform. Therefore, in addition to supporting the firm's positive thrusts described in policies 1–4, the telecommunications platform must also have sufficient range, reach, and responsiveness to block third-party intrusions and sustain market share by discouraging unmatched electronic entry into the industry.

Policy 6: Choice of Architecture Should Provide Sufficient International Capability

All of the preceding policies are applicable in an international context. It is vital for many firms to have a strategy for competing in a global marketplace that is supported by a worldwide telecommunications platform. Unfortunately, very few firms have such an architecture, and very few vendors can fully support it.

International telecommunications is a political, regulatory, and technical minefield. Nevertheless, no firm will be able to operate efficiently in a global environment in the 1990s unless it has a global telecommunications platform that includes (1) international reach, which means adopting international telecommunications standards; (2) sufficient range to support information sharing across all business functions in all locations; and (3) adequate responsiveness to permit real-time interactions among widely dispersed operational entities.

Although these six policies highlight the need for extensive range, reach, and responsiveness in a telecommunications platform, they clearly do not provide sufficient guidance to unequivocally choose a particular architecture. The summary message concerning these policies is to choose a standards blueprint that, on the high side, will enable the company to exploit current and future opportunities arising out of the major structural trends in the global business environment, and that on the low side will not inhibit any reasonably likely business initiative. This outlook also implies choosing strategic vendors who have staying power, who are committed to providing extensive reach, range, and responsiveness in their products, and who move with the mainstream of the industry toward integration and the implementation of key standards.

As with any extensive planning effort, the real value is engaging in the process required to create an architecture, not in the specifics of the architecture itself.

Summary

Standards are at the heart of modern telecommunications, with the main forces in the industry shaping trends away from proprietary systems toward open systems. The main barriers to open standards have been the rapid pace of technical change, plus vendors' wish and/or need to develop product- or technology-specific protocols. Business needs now include integration and interoperability, making standards the new priority.

Standards-setting organizations include ISO, the sponsor of the OSI reference model; CCITT, whose main focus is telephony and data communications; ANSI; and a number of vendor–user alliances. In addition, marketplace forces often lead to de facto standards that are never defined as such. Some of these standards were initially proprietary and were adopted by enough vendors and users to turn them into open standards.

IBM's standards—particularly its SNA, token ring, and DOS, which are the base of many large-scale transaction-oriented corporate networks—have been a powerful force in the industry; the OSI movement is the other major driver of the overall direction of standards. These comparable layered architectures are the basis for many organizations' protocol structure architecture, which is the blueprint for the standards and protocols that bind together the many internal computing, data, and communications resources. The seven layers in the OSI model are physical, data link, network, transport, session, presentation, and application. Internetworking is the emerging priority for large organizations that need extensive reach, range, and responsiveness. Bridges, routers, and gateways are the main tools for linking LANs to LANs and LANs to WANs.

The configuration structure architecture for a telecommunications platform extends the designer's focus from the internal linkages in the network and across networks to the configuration of the many entities that use the platform. This is not a major concern of the OSI model, but the business-oriented architectures provided by leading computer vendors provide many facilities and services both for configuring the network and for making distributed processing and client/server computer applications practical through the design of the telecommunications platform.

Choosing an architecture is a business decision, even though the architecture is defined by technical standards and protocols and is implemented through choices of specific products and services. There are a number of policy guidelines for a designer, including the following: ensuring that no business initiative is blocked because of lack of reach, range, and responsiveness in the platform; allowing response to a competitor's move; ensuring that the firm can enter such electronic alliances as customer-supplier EDI relationships; allowing sufficient flexibility to make mergers and acquisitions practical, for meshing operations increasingly means integrating computing and telecommunications resources; warding off third-party intrusion on the firm's central business activities; and positioning the firm to operate efficiently and effectively in a global business environment.

Review Questions

1. Why are standards important in telecommunications?

2. True or false: Telecommunications standards are enforceable by law. What are this statement's implications?

3. What is the ISO, what is its purpose, and who are its members?

4. What is the CCITT, and how does it work?

5. What is ANSI, how does it work, and who are its members?

6. What type of organizations are MAP/TOP and COS, and what purpose do they serve?

7. What is an open system, and why is it important?

8. Explain the following statement: Open and proprietary definitions of standards are different from open and proprietary implementation of standards.

9. What practical path to integration uses IBM compatibility as a de facto standard? Explain.

10. What is a telecommunications architecture, and why is it important to have one?

11. Describe the difference between protocol and configuration architectures.

12. What is the OSI seven-layer model, and what is its primary purpose?

13. Make a list of the seven layers of the OSI model, and briefly describe the functions performed by each layer.

14. Explain the following statement: Each layer of the OSI model operates on a peer-to-peer basis with its counterpart in another system.

15. What is IBM's System Network Architecture (SNA), and how does it compare with the OSI seven-layer model?

16. Briefly describe the roles of bridges, routers, and gateways in internetworking and the appropriate uses of each.

17. Instead of using a bridge to connect to identical types of networks, why not simply create one large network?

18. IBM's SNA was created as a hierarchical system, but the new version of SNA includes a feature (APPN) that permits peer-to-peer networking. Explain the implications of the new version.

19. Are configuration architecture elements included in the OSI seven-layer model specification? Explain.

20. How should a company go about choosing a particular architecture?

Assignments and Questions for Discussion

1. Suppose that you call your local bank and ask for the current status of all your accounts, which include checking, savings, and two separate brokerage accounts. The bank's response is, "We'll have to get back to you in a couple of days; it will take time to gather the information." What is the most likely reason for the delay, and is this situation likely to change significantly in the near future?

2. Discuss the advantages of a shared-use telecommunications platform relative to separate networks created to support individual applications, and explain whether a single monolithic system is necessary to implement the shared-use platform.

3. Describe the role played by standards in the integration of different networks and equipment manufactured by different vendors, and evaluate the choice between open system standards and vendor proprietary standards in achieving integration.

4. Very few people would consider building a house or an office building without a set of plans representing an overall architecture. In contrast, very few organizations have a true telecommunications architecture plan for the design blueprint for the evolving network platform. How do you explain this apparent discrepancy in rational behavior?

5. As the manager of telecommunications planning and strategy for the Wakefield Widget Co., you have been asked to prepare a short memo describing the key business factors in choosing a telecommunications architecture for the firm. Write the memo, explaining why each factor is important.

Minicase 7-1

Dana Corporation

An Award-winning Manufacturing Network

Dana Corporation's Lima, Ohio, facility won an award in 1991 for leadership in computer-integrated manufacturing (CIM). Its integrated factory network has helped reduce equipment setup time by up to 1,000 percent and inventory by 360 percent, and it has virtually eliminated the use of paper on the shop floor. The network is shown on page 301.

The terminals and workstations include:

1. Personal computers in finance and engineering (IBM PS/2s) and "industrially hardened" old IBM PCs on the shop floor

2. IBM RS/6000 workstations used to develop numerical control programs

3. IBM 9370 midsize computers and a large IBM mainframe, the 3090

4. A PC linked to a cell controller that runs program logic controllers (PLCs) that run shop-floor machines, such as industrial robots. The PLCs are not directly part of the network but link to the cell controller, which is

5. A file server on a large personal computer

The transmission links consist of:

1. A local area network (LAN) that uses coaxial cable

2. Wide area network (WAN) links to Toledo and Marion, Ohio; these links use a mixture of copper cable and optical fiber

3. A satellite link for other U.S. and international locations; this link uses microwave radio transmission

The transmission methods are:

1. The local area network: token ring technology (this is contrasted with the most widely used LAN transmission method called CSMA/CD, used by Ethernet)

2. The wide area network: IBM's proprietary circuit-switched SDLC (synchronous data link control)

3. The satellite: packet switching

The nodes and switches are not shown in the diagram. An IBM switch links the token ring LAN to the 9.6-kbps WAN. The WAN includes many intermediate switches that route traffic to and from other cities.

The network standards and architectures are:

1. The local area network: the token ring standard IEEE 802.5

2. The wide area network: IBM's SNA architecture

Figure 7-8

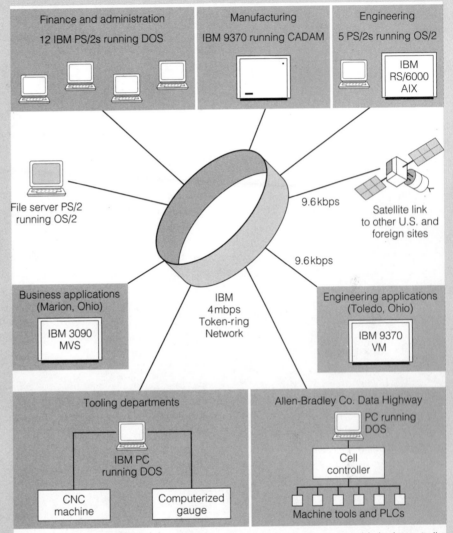

CNC = Computerized numerical controlled   PLC = Programmable logic controller

Dana's Spicer Universal Joint Division in Lima has integrated business, engineering, and shop floor activities using an IBM 4mbps Token-Ring Network.

Questions for Discussion

1. To the uninitiated, computer integrated manufacturing (CIM) simply consists of electronic robots connected together on an assembly line. However, Dana Corporation's integrated factory contains much more. Describe these other elements.

2. Three different transmission methods linking several different types of terminals and workstations are used in the CIM network. What problems does this create, and how are such problems managed?

3. The Dana Corporation network is a single-vendor system; all its systems are provided by IBM. Discuss some of the key issues involved in shifting to a multivendor environment using equipment manufactured by different companies.

Minicase 7-2

Department of Commerce

The Federal Information Processing Standard (FIPS)

In September 1992, all federal agencies were required to adopt the ANSI X12 and EDIFACT standards for electronic data interchange. Many agencies had implemented proprietary EDI formats that were industry- or application-specific. This caused many problems for businesses that exchanged data with several agencies, for they had to implement a wide range of standards in dealing with different agencies for the same basic interorganizational transactions. The new FIPS mandate has already blocked development of still more proprietary EDI systems; agencies have until 1997 to migrate to the new standard. In late 1992, the statuses of several leading agencies were as follows:

1. Using proprietary systems: Healthcare Financing Agency (HCFA), which handles Medicare and Medicaid

2. Using proprietary systems, but migrating to X12: Environmental Protection Agency, Internal Revenue Service, Customs Service

3. Already using X12: Department of Defense, General Services Administration, Department of Veterans Affairs

The initiative was made by the National Institute of Standards and Technology (NIST), an advisory group to government that has no formal authority. The EDI manager of Baxter Healthcare Corp., which is the largest single supplier of medical products to the Department of Veterans Affairs, welcomed the mandate but identified some practical problems for NIST. X12 is an umbrella EDI standard; an X12 purchase order defines mandatory fields that must be included, but there are also many optional fields in which trading partners determine the exact information to be provided. These "trade sets" are very industry- or transaction-specific. The EDIFACT standard (which is closely related to X12) has many trade sets for insurance, freight forwarding, payments, and so on.

The Baxter EDI manager commented that "If we have many purchase orders coming into us from various government agencies, even though they're all using the X12 standard, they could each be sending us different information. We'll still need to do a first-time map with each government agency before it's truly standardized."

A number of federal agencies initially opposed the mandate. The Healthcare Financing Agency (HCFA), for instance, had been moving in a very different direction from X12 and needed to protect its heavy investment in existing systems. One consultant summarized the views of the non-X12 agencies: "There's not adequate time or money to make the switch from their current systems to X12. It's for this reason that I don't think we'll see rapid implementation at the government level."

In early 1992. HCFA decided to adopt X12 835, a subset of X12 developed for electronic data interchange and electronic payments in the health care industry. Previously, the industry had an estimated 400 data formats in use. The change of mind appears to result from direct pressure from the Office of Management and Budget (OMB). Hospital managers and insurers had become concerned that HCFA's push to impose its own standard would be expensive for them and would add new layers of bureaucracy and administration. OMB forced HCFA to coordinate its EDI plans with ANSI, which in turn agreed to modify X12 835 to meet the agencies' needs.

HCFA is a major player in health care; it pays for about half the total annual U.S. medical bill of around $800 billion. Almost 24 percent of this massive drain on the economy—the United States pays twice per capita for health care than either Germany or Japan, yet lags badly behind them in most relative measures of results—goes to administrative processing; that is nearly $200 billion a year, or over $3,000 per person. HCFA spends over $150 million a year on check processing alone, an amount that can be cut by at least 50 percent through electronic funds transfers. It costs between 3 cents and 10 cents to process a medical payment electronically; in contrast, the stamp needed to mail a check costs 29 cents. Replacing a lost check is estimated to cost HCFA $47, and the health care industry handles 500 million checks a year. The average accounts receivable period is 54 days, but this period is shortened by at least seven days by using EDI.

The EDI translator software for X12 835 costs from $700 to $30,000, but the telecommunications software may add another $100,000. HCFA plans to issue electronic payments three days after it receives the EDI instructions, in order to keep the interest on the float that it gains from paper checks. Small hospitals and clinics face financial problems in moving to EDI; large health care organizations will surely gain.

The consensus of commentators on government EDI in general and on HCFA in particular is that the NIST initiative ensures that standards-based EDI is the wave of the fairly immediate future.

Questions for Discussion

1. The EDI manager for Baxter Healthcare Corp. stated that simply adopting the ANSI X12 standard is not sufficient to eliminate the problems associated with multiple proprietary standards that currently exist. Explain and discuss.

2. The Healthcare Financing Agency opposed a common standard for EDI that would improve efficiency and reduce the cost of interagency transactions. Why? Explain and discuss.

3. Explain how the use of a common EDI standard to improve the efficiency of HCFA's administrative processing could significantly improve the United States's competitiveness in world markets.

Minicase 7-3

IATA (International Air Transport Association)

Agreeing on Shared Standards

Almost all of the world's airlines belong to IATA, whose main responsibilities are to establish policies and regulations concerning travel agency rules, security procedures, information and accounting codes and formats, and many other areas of airline cooperation. One of IATA's most recent actions has been to recommend computer and telecommunications standards. The airline industry was the world leader in telecommunications in the 1970s, but it fell badly behind in the 1980s, when individual airlines went their own way. A few, such as American Airlines, United Airlines, and British Airways, recognized the opportunity provided by information technology and moved ahead aggressively. Others, like Aer Lingus, created cooperative arrangements to buy and sell software to other small to medium-sized carriers. Many major airlines, including Air France and Lufthansa, fell well behind the leaders.

As a result, by the early 1990s the industry as a whole was a patchwork of systems and capabilities. One major problem was that the industry had adopted its own proprietary telecommunications protocol called Airline Link Control (ALC), which uses a six-bit protocol for coding data, unlike the standard eight bit one; among other limitations, this means that ALC messages are in uppercase only. The use of a protocol outside the mainstream of telecommunications developments meant the industry could neither exploit such widely available standards as X.25 nor use the many public networks and software and hardware products that exploit X.25. American Airlines began to implement it own move to X.25 in 1990 and expects to take until 1994 to complete it.

Airline reservation systems were a small market in the 1970s and early 1980s. Only a few carriers used them, and there was no incentive for the two mainframe providers of computerized reservation systems (CRS) hardware and software, IBM and Sperry (which later became part of Unisys), to make heavy investments in products that had just a few customers. Only when deregulation fueled the expansion of CRS usage and competitive exploitation of information technology did that change. Growth was accelerated by new alliances (such as Covia, led by United and British Airways, with its main rival, Amadeus, led by Air France and Lufthansa) and by extension of CRS into the hotel and car-rental markets.

Given the importance of industry-wide information sharing for reservations, scheduling, interairline payments and operations, the IATA Architecture Strategy Group developed a recommendation that IATA approve the Airline OSI Profile guidelines. If adopted by a vote of the membership, these guidelines would establish the industry as the first to make a wholesale migration to OSI (only the U.S. government has announced its long-term commitment to OSI). The guidelines were announced in May 1992 at IATA's annual conference.

Reaction was very mixed. Advocates of OSI argued strongly that carriers that do not move to open systems will soon find it increasingly difficult to communicate and share information with other carriers or to be able to purchase software from them and from third-party suppliers. The Architecture Strategy Group selected OSI rather than TCP/IP as the base for open systems because it is better designed to support any-to-any links. OSI would provide standard application program interfaces (APIs), which would reduce programmer training and development costs. APIs are the software codes that transfer control from one program to another.

Modularity of code—being able to write codes as small building blocks that can then be hooked together through APIs—would be immensely helpful in upgrading airline systems and in simplifying migration to such new technologies as frame relay, thus helping free the industry from its ALC history. Industrywide adoption of OSI would also make it much easier to connect additional parts of the industry and the industry to outside partners and suppliers. Currently, for example, interairline payments are made through cumbersome IATA-approved systems that take weeks to complete many transactions and have heavy administrative costs. OSI would allow airlines to link directly to each other and to banks on-line.

The opposition to the guidelines was muted but quite widespread. One problem is that the industry is going broke. The airlines have lost billions of dollars annually throughout the 1990s. Indeed, since Orville Wright first took flight in 1905, the compounded profit of the entire airline industry is zero. Since deregulation in 1978, U.S. carriers have earned under 0.3 percent return on their investment. The Architecture Strategy Group could hardly have chosen a worse time to announce a costly proposal, for many airlines simply do not have the funds to replace existing communications equipment and software. The benefits of OSI are at least a decade away. Many parts of OSI are still not fully defined and finalized as standards that can be implemented in real products. TCP/IP is here now and it works. The position of a supporter of OSI is that 90 percent of TCP/IP implementations are on LANs, with bridges, routers, and smart hubs interconnecting TCP/IP nets. OSI applies to LANs and WANs. (While true, this position is rather misleading. The world's largest network, the Internet, is based on TCP/IP.)

In parallel with and reinforcing IATA's Airline OSI Profile recommendations, another consortium has announced recommendations for the Aeronautical Telecommunications Network (ATN), an OSI/X.25 architecture designed to facilitate communication between aircraft and ground-based airline and air traffic control systems. Its members include only two airlines, United and American. The other players are the Federal Aviation Authority; airline network service suppliers, of which ARINC and SITA are the most important, and computer vendors. United expects to take three years to implement ATN.

Proponents of the OSI moves accepted at the IATA conference contend that there is no way of justifying them in the short-term. One attendee pointed out that without some evidence of a positive rate of return on investment, it will be extremely difficult to sell OSI to airline companies' top management.

Questions for Discussion

1. Describe some of the difficulties the airline industry faces in operating with a patchwork of systems and capabilities, including their own ALC proprietary protocol.

2. Discuss the trade-offs for IATA of adopting the OSI architecture versus standardizing on the well-established TCP/IP protocol system.

3. The IATA Architecture Strategy Group recommended approval of a set of Airline OSI Profile guidelines, which would establish the industry as the first to make a wholesale migration to OSI. Reaction was very mixed and opposition was widespread. Why would the airline industry oppose a move to enable its members to communicate more effectively among themselves and with others outside the association? Discuss.

Minicase 7-4

Compression Labs and PictureTel

Standards—Difficulties Moving from Idea to Actuality

In 1991, CCITT ratified the first of a set of standards for interoperability among different vendors' videoconferencing equipment. The international standard is called H.261, and it addresses compatibility between the coder/decoders that digitize and compress audio and video so that it can be transmitted over public networks.

By early 1992, progress in implementing the standard was badly lagging behind expectations. The two leading U.S. vendors, Compression Labs Inc. (CLI) and PictureTel Corporation, had committed to deliver within six months products that complied with H.261. Ambiguities in the definition of the standard had blocked implementation. The two companies have 87 percent of the U.S. and 62 percent of the world market and are fierce rivals. Implementing the standard meant testing the mutual interoperability of all firms' equipment. The tests showed that most vendors can connect with only some of their competitors' products. Only two companies were able to interconnect their systems reliably and completely; these companies hold only 2 percent of the market in the United States.

Communications Week commented on the problem of asking the two leading competitors to cooperate. It quoted a spokesman from PictureTel as saying that "We have extended invitations to [CLI] but they have not accepted invitations to in-teroperate with us." CLI responded that it had not received any such invitation, but "would not participate in one-on-one tests anyway." *Communications Week* added: "But many analysts have said CLI and PictureTel are dragging their feet with regard to standards because they want to ride the wave of their proprietary systems." The view of one analyst is that "the vendors' proprietary algorithms [mathematical formulae for data compression] continue to be better than H.261" and that "Most experts agree."

CCITT expected to resolve the problems of ambiguity before mid-1992 by reviewing and clarifying the standards called H.221 and H.242, which determine how video will be multiplexed, transmitted, and decoded. That said, the consensus in the industry is that true interoperability will be a problem for many years to come. One vendor commented that "There's some vagueness in those standards and it leaves a lot open for interpretation." Despite these problems, analysts and vendors agree that H.261 is specific enough to move the field forward in terms of interoperability.

H.261 is just one relatively minor standard among thousands. It illustrates the general pattern of standards-setting and the difference between defining a standard and implementing it in real products.

Questions for Discussion

1. From the end-user's perspective, discuss the implications of having complete interoperability among different vendors' videoconferencing equipment.

2. Why hasn't the emergence of the international videoconferencing standard, H.261, resulted in complete interoperability among the equipment of different vendors, and what are the implications?

3. Discuss and evaluate the following statement from the perspective of senior management at PictureTel Corporation: "Universal standards are open architectures for videoconferencing systems and are good for everyone—it's a win-win situation for users and vendors in the long run."

8

Types of Networks

Objectives

- ▸ Review how networks are classified and what the classifications imply for design and implementation
- ▸ Learn the key characteristics of LANs, WANs, and public networks
- ▸ Assess trends in technology

Chapter Overview

| | | | | | | |

Chapter Overview

In the previous four chapters we have described the essential technical building blocks of a telecommunications network. These building blocks can be combined in many different ways to form an often bewildering array of network types, which range from high-speed local area networks connecting powerful desktop workstations to global satellite networks carrying CNN news broadcasts around the world. This chapter pulls together the material in the preceding chapters by focusing on the designer's and the client's central concern: What type of network do we need?

To make sense out of the morass of types of networks, it is helpful to have a classification scheme that aids the designer in choosing the appropriate type of network to meet a given *business* need. The first section of this chapter develops a five-dimensional classification scheme for networks that takes into account the historical evolution of network development. The next three sections are devoted to a more in-depth examination of three of the major types of networks: LANs, WANs, and public switched networks. Finally, the chapter ends with a brief look at trends in network development.

A designer should look at types of networks from a broad perspective and should focus on issues related to the choice of an overall telecommunications architecture. Such a perspective necessitates grappling with a variety of issues: How much of our telecommunications requirement should be outsourced?[1] Should our enterprise's backbone network be an internet connecting multiple local area subnetworks, or a centralized, hierarchical, wide area network? How much of our requirement can best be met using public switched network services? How many separate networks do we need given our diverse service and application requirements?

| | | | | | | |

Classification of Networks

As telecommunications network technology has evolved and the regulation of the telecommunications industry has changed over time, different types of networks have emerged. Historically, networks have been classified by some characteristic of the underlying technology, the type of service provided, or the geographic coverage of the network. More recently, two other classification

---

[1] Outsourcing, discussed in Chapter 17, is the contracting with an outside firm to operate facilities and/or services that are currently managed in-house. A more useful term is "multisourcing," of which outsourcing is just one option. Many firms keep, say, network management in-house, outsource network operations, and enter joint ventures for network and data center disaster recovery.

Figure 8–1	Dimensions	Examples
Dimensions for Classifying Networks	Type of information transmitted	Voice; data, special-purpose (fax, image); multimedia (voice + data + text + image)
	Type of underlying technology	Cable; microwave; satellite; (separate or interoperating)
	Type of ownership	Public; private; semipublic, and semiprivate VANs
	Geographic scope	LANs; MANs; WANs; international
	Type of topology	Star; bus; ring; mesh; hybrid LAN; backbone WAN

dimensions have emerged: the type of network ownership (public or private) and the physical layout or topology of the network.

Networks are often grouped into more than one of the five categories, for classification occurs informally through industry convention, and no standard or official classification scheme exists. For example, an "X.25 network" is shorthand for a wide area network (geography) that uses a packet-switching method (underlying technology) based on the X.25 protocol system. A token ring network is a local area network (geography) configured in a ring layout (topology) using the token passing method for access to the transmission medium (underlying technology). Figure 8-1 summarizes the five dimensions that best classify a network capability, and each of them is discussed in the following sections.

Type of Information Transmitted

The oldest electronic networks were classified by the type of terminal device used to access the network—telephone, telegraph, or telex terminal—which was closely related to the type of information being sent over the network, such as a voice or text message. Until relatively recently, networks were generally dedicated to one particular type or form of information, either voice, data, message (telex), or video, which led to the classification of networks by the predominant type of information carried on the network.

Until only a few years ago, the public telephone system was almost exclusively a voice network, and companies operating their own private networks tended to have separate networks for voice and data applications. Networks carrying video information were also separate from voice and data networks, primarily because of the high capacity requirements of video transmission.

In recent years, there has been a significant increase in the use of the public telephone network for dial-up data applications such as fax, e-mail, and file transfer. The increase in demand for nonvoice applications, together with vastly improved network technology (that is, high-capacity, digital transmission) has led to greater integration of different types of information on the same network, which generally reduces costs and improves network management because there are fewer networks to operate. As a result, the current trend is away from classifying networks by type of information, although remnants of this classification method will probably persist for some time due to habit and established conventions. Most of those simple conventions—voice versus data networks, in particular—are being made obsolete by the rapid emergence of <u>multimedia</u> applications in which voice, data, image, and text are all part of a single bit stream.

Type of Underlying Technology

Networks have also been developed to exploit the capabilities and characteristics of particular technologies or transmission media, and thus networks have been classified on the basis of the technology used. Networks can be <u>classified</u> <u>according to switching method used</u>; thus there are <u>packet-switched</u>, <u>circuit-switched</u>, and **tandem-switched networks**. (Tandem-switched refers to processors that are <u>directly connected together</u> in a transaction environment.) There are also <u>fiber-optic, satellite</u>, and <u>microwave networks</u>, which are classified on the basis of the dominant transmission technology or medium. Currently, networks are less often classified by type of technology because of the increasing use of <u>mixed technology</u> and <u>mixed media</u> within a single network system; with increasing frequency a particular technology is no longer the distinguishing feature of a network.

Type of Ownership

Deregulation of the telecommunications industry, both in the United States and abroad, fueled a trend toward the increased use of private networks by large organizations; classification of networks as either <u>private</u> (<u>owned and managed</u> by private companies) or <u>public</u> (<u>owned and managed by authorized common carrier companies</u>) is one result. In a <u>true private</u> network, a firm typically owns its own <u>PBX</u> and <u>data-switching equipment</u>, as well as <u>modems</u>, <u>multiplexers</u>, and <u>other circuit interface devices</u>. However, a firm generally leases <u>long-distance transmission</u> facilities and local access lines from a common carrier company. <u>Public networks</u> include the long-distance networks of <u>AT&T, MCI,</u> Sprint, and others; the <u>local distribution networks</u> of the former Regional Bell Operating Companies and other local carriers; and specialized <u>value-added networks</u> (VANs) provided by <u>private companies</u>.

A <u>VAN</u> is a semipublic network that <u>provides some additional feature</u> or <u>service beyond the basic transmission of a data stream.</u> It is semipublic, rather than <u>public</u>, because user-firms often need some qualifications to use the service, and individuals are not encouraged or may not be allowed to use the VAN. Value-added features of networks include <u>format</u>, <u>protocol</u>, or <u>speed</u> conversion; <u>store and forward services</u> designed to deliver messages at a later time; and <u>EDI services</u> or particular <u>encryption</u> and security features to protect information against unauthorized access. Examples of VANs are the SWIFT international banking network and the following public packet-switched data networks: Telenet, Tymnet, General Electric Information Services (GEIS), and the IBM Information Network. VANs are typically accessed through local dial-up nodes, toll-free 800 numbers, or dedicated leased lines connecting the customer to the nearest VAN access node. Only banks can use SWIFT, another indication of the reason VANs are not fully public networks.

The distinction between public and private networks began to blur when the major long-distance carriers introduced a new service concept called the <u>virtual private network</u> (VPN) or <u>software-defined network</u> (SDN). A SDN/

VAN

Public

Private

Feature	BT-TGN	X-25 Public Network	Private Network
Roundrip character echo delay	Under 500 msecs	2–5 secs	1–2 secs
X.25 supported?	Yes	Yes	Yes
SNA supported?	Yes	No	Yes
Protocol conversion	All major protocols	None	Vendor-dependent
Public dial-up	Global	Local only	Via gateway
Network management	Fully supported end to end	Only up to the gateway	Own support staff
Cost	Constant or variable	Variable	Capital investment plus monthly lease
Configuration flexibility	High	Inflexible	Moderate
LAN interconnectivity	Via frame relay	None	With capital investment

Figure 8-2

The World's Largest Value-added Network and Typical Public and Private Networks

VPN has the look and feel of a private network, but it retains the flexibility and the carrier's ultimate responsibility for performance that are inherent in a public network service. In a SDN/VPN, the carrier guarantees the user access on demand to a specific network capability (that is, transmission speed/capacity, performance parameters, and points of access) based on access to a pool of available transmission and switching facilities.

Unlike a pure private network, specific facilities are not dedicated to a given user in a SDN/VPN. This arrangement provides for more efficient use of the carrier's transmission and switching facilities while giving the user unblocked access to needed capacity on demand. In addition, the user is able to reconfigure the network (add or subtract nodes or capacity) through a software modification to the carrier's SDN/VPN control program.

As a result, if a company acquires a new division through a merger or loses a division through a divestment, a SDN/VPN can be instantly reconfigured to accommodate the change. Such changes would be much more difficult, costly, and time-consuming to implement for a true private network with dedicated facilities, and the company could be stuck with equipment that it no longer needs. Finally, with a true private network it is generally the company's responsibility to manage problems on the network: with a SDN/VPN, the carrier is responsible for performance problems and for maintaining network facilities.

Classification of networks by ownership is still quite prevalent, but the hybrid VPN/SDN blurs the distinction somewhat. Indeed, a VPN/SDN may be considered a semiprivate network and the VAN a semipublic network. Figure 8-2 compares BT's TGN (The Global Network, the largest VAN to date) with a

VPN/SDN = semi-private

VAN = semi-public

typical public network and a typical private network. The VAN offers a combination of features (and resultant flexibility) that neither of the other networks provides, but the cost per unit of traffic is higher than either a public or private facility.

Geographic Scope

The proliferation of private networks and virtual private networks has created another dimension for classifying networks based on their geographic scope. Local networks have emerged to connect personal computer workstations in nearby offices either to share software and expensive printer facilities or to interconnect office telephones to more efficiently handle voice communication requirements. These networks are labeled local area networks (LANs) because at most they have a reach that extends to nearby buildings in a campus environment. Even though a local, intra-office, voice and data network connected through a PBX switch technically qualifies as a local area network, the term *LAN* has become synonymous with a particular type of high-speed data network linking computer workstations in a client/server configuration. Today, the main geographic distinction between types of network is LAN versus WAN; to some extent, this is a political as well as a technical differentiation, with proponents of LANs championing decentralization and freedom from bureaucracy and advocates of WANs valuing coordination and freedom from chaos. LANs will be discussed in detail in the next section.

Other private networks whose reach extends beyond a local concentration of buildings are considered wide area networks (WANs). Some of these are regional in scope, and others have extensive reach that spans continents and literally circles the globe. Unlike most LANs, which are limited to data communication applications, WANs are long-distance networks that transmit the full range of types of information. A WAN is made up of leased circuits provided by common carriers and switching facilities provided either by the user or a carrier. Generally, the user supplies the circuit interface equipment, such as multiplexers and modems.

The transmission capacity of WANs typically ranges from 9,600 bps to 1.544 mbps (T1), although a few WANs now have segments that operate at 45-mbps speeds (T3). A digital bit rate of 32 kbps is adequate for voice transmission, and video conferencing is now conducted very satisfactorily at 64 kbps (full duplex). Digital compression techniques have become so good that full-motion, broadcast-quality video requires less than 3 mbps of transmission capacity. This is a remarkable achievement because just a few years ago the digital transmission of full-motion video required a channel capacity of 45 mbps.

A new geographical scope classification is emerging through the standards work of the IEEE to extend the LAN concept beyond the corporate campus to include complete metropolitan areas. The metropolitan area network (MAN) is a high-speed digital network with reach throughout a metropolitan area that links LANs and host computer subnetworks with fiber-optic and coaxial cable transmission lines. These high-speed network facilities are designed to carry voice, data, image, and video information. Many of the backbone fiber-optic links for the newer MANs are provided by private companies that are installing

Bypass co. [handwritten margin note]

their own cables and are competing directly with the local telephone company. The "bypass" companies offer high-speed, local transport service and local access to long-distance network nodes, typically at lower rates than the local phone company charges for comparable facilities. The Regional Bell Operating Companies, whose revenues will shrink as a result of growth in bypass companies, are moving to exploit MANs.

Type of Topology

define top [handwritten margin note]

In certain instances, networks are classified according to network topology—the physical configuration of the various nodes and the manner in which they are linked. Topology is often the basis for classifying LANs and satellite networks in particular.

star [handwritten margin note]

In a star network topology, all terminal nodes connect directly to a central node, usually a host computer, file server, or a PBX. This topology facilitates central control of the network and economizes on the need for network intelligence in remote nodes because the central hub manages all network activity. On the other hand, the central hub node can become overloaded during peak traffic periods or can fail altogether, which is the vulnerability of this topology.

Some LAN architectures use the star topology; all workstations connect directly to the file server hub. The most popular VSAT satellite architecture is also a star configuration, with all remote earth terminals linked directly to the hub station. This is economical for large VSAT networks because all switching and routing intelligence is concentrated in the hub station, and thus the remote nodes are relatively simple and inexpensive.

mesh [handwritten margin note]

In its pure form, a mesh network topology directly connects all nodes in the network. This is ideal from a performance standpoint because a direct point-to-point connection delivers the lowest response time and minimizes the chance for transmission errors resulting from transiting intermediate nodes. It is similar to a direct airline flight versus a flight with an intermediate stop at a hub city: The direct flight is faster and there are fewer chances for something to go wrong (such as inadvertently removing your luggage during the stop) or for an accident on the extra landing and take-off.

mesh modified [handwritten margin note]

The pure mesh topology becomes prohibitively expensive, however, when the number of nodes becomes large; consequently, most actual mesh networks directly link all the major nodes but connect minor nodes to only a few locations, at most. Most traditional satellite networks are examples of a mesh topology: Each earth station contains the equipment necessary to establish a direct circuit path through the satellite (acting as a relay) to every other station on the network. *How is this diff than the star?* [handwritten note]

hierarchical [handwritten margin note]

A hierarchical network topology is similar to a layered organizational chart. End-user nodes are linked directly to lower-level intermediate nodes, which in turn are linked to higher-level intermediate nodes, and so on. The number of layers is determined by the overall size of the network, the trade-off between a higher-performance level attainable through a direct point-to-point connection, and the greater overall efficiency of limiting direct connections to only those nodal pairs with high traffic volumes. The public switched telephone network is a hierarchical topology. It contains several layers of switching centers that direct through the network large blocks of traffic that represent a combined stream made up of smaller individual traffic streams that are consolidated at switching centers found at lower levels in the hierarchy.

A bus network topology is represented by a single length of a guided transmission medium, such as coaxial or fiber-optic cable. The terminal devices are directly attached to the fiber or coaxial cable bus. The bus concept provides very high transmission speeds, and each terminal device on the network shares the same transmission link. There is no single point of control on the bus system, as with the star topology; consequently, the causes of failures and faults are difficult to locate. On the other hand, a single point of failure does not generally affect the overall functioning of the network.

The bus topology is usually found only in LANs or in networks with very limited reach because the high-speed operation of the bus severely limits the distance over which the network can function effectively. However, with recent improvements in transmission technology, the distance limits for effective operation of a bus network have been extended. Recent trials have demonstrated that a 100-mbps transmission rate can be attained using unshielded twisted pair wire over distances up to 100 meters without the use of repeaters.

A ring network topology is somewhat similar to the bus network in that it consists of terminals directly connected to some type of guided transmission medium. However, unlike the bus, the ends of the length of cable are connected together to form a circle or ring. In the ring topology, each terminal checks the address of signals passing by on the shared ring. If the address of the given terminal matches that of the incoming signal, the message is retrieved; otherwise, the terminal regenerates the signal and sends it on around the ring. This regeneration feature results in less signal attenuation than occurs on a bus network, which lacks this regeneration capability.

The successful operation of a ring network requires all of the connected terminals to function correctly in reading and regenerating data streams. If a connected terminal malfunctions, the network will crash (unless there is a special mechanism for bypassing the failed terminal) because the passing signal cannot be regenerated. One type of special mechanism is the use of two parallel rings that transmit data in opposite directions; then, if one terminal fails, its node is essentially bypassed. Data moving in one direction that is blocked by the failed node can reach its destination terminal by traveling around the other ring in the opposite direction.

In practice, almost all networks other than small LANs are a mix of topologies. For example, a hierarchical network may have a ring or star segment attached, or it may include a mesh configuration for its major nodes. Also, in most modern LANs, the bus or ring configuration is physically contained in a wiring hub chassis. Twisted pair wiring is used to connect the various workstations and the server to the wiring hub. Within the hub chassis, a ring or bus card is used to link the various workstations through the appropriate topology. This same wiring hub also has slots for bridge and router cards used to interconnect separate LANs. The physical concentration of the LAN medium and internetworking devices in a common hub facility enhances control and management of the network and makes changes to the network much easier to implement.

The remainder of the chapter is devoted to a more detailed description of LANs, WANs, and public switched networks, and a brief look at future trends.

Local Area Networks

As described previously, the basic LAN concept is the real-time interconnection of end-user devices in a local environment for the purpose of sharing information, files, software, and hardware peripheral devices. In principle, the LAN concept can range from PCs linked together to share software, common databases, and printers at one end of the spectrum, to the interconnection of telephones within an office building through a PBX at the other end of the spectrum. Although technically a LAN, the intra-office, PBX-based voice network is not included in the common usage of the term.

The primary impetus for development of the modern LAN is attributable to three causal factors: (1) the availability of powerful and inexpensive personal computers; (2) the lack of responsiveness of management information systems (MIS) departments in meeting the applications needs of company work groups; and (3) the need to share software, files, and hardware peripheral devices among the work-group members. A key feature of the modern LAN concept is that all end-user devices on the network are given equal status; there is no slave/host or hierarchical arrangement of nodes. This approach takes advantage of the distributed processing power available through low-cost PCs to meet the common information-sharing requirements of local work groups.

LAN growth is exploding and is taking over as the computing/communications platform of choice for a large number of firms, including a significant number of the world's largest corporations. The emphasis today is on managing LAN growth and the challenge of linking multiple LANs to form an enterprise-wide internet that is the company's backbone information system.

Because LANs began as a form of grass-roots guerrilla tactic to bypass the monolithic, unresponsive MIS department, there were literally no standards or accepted industry methods of connecting PCs. Consequently, many different LAN architectures and protocols emerged, and only recently has there been a consolidation around a small handful of dominant LAN architectures and methods based on accepted industry standards.[2]

---

[2] In the mid-1980s, Ethernet dominated business uses of LANs, with token ring second. In the 1990s, sales of token ring networks grew very rapidly, at about 30 percent a year. Ethernet is simple, cheap, and easy to install and was once well-suited to departmental uses of personal computers; it is less effective and efficient for today's more complex client/server applications.

LAN Technology

There are three broad categories of LAN technology: circuit-switched, broadband, and baseband. Circuit-switched local area networks, such as PBX and data switch systems, are technically local area networks, but they are not usually included in the common usage of the term *LAN*. Circuit-switched local area networks are based on a star topology, with the PBX or data switch acting as the central hub. The end-user devises are all connected directly to the switch, which can link any combination of terminals on the network to establish a session. Modern digital PBX networks can accommodate both voice and data traffic, but the transmission speed is generally limited to 64 kbps.

A broadband LAN uses analog transmission techniques, and the total capacity of the cable medium forming the network is subdivided into separate circuits or channels through multiplexing. Separate subnetworks can be created on the individual multiplexed transmission channels, and all forms of information (voice, data, and video) can be accommodated on a broadband LAN.

The most prevalent type of LAN technology in use today is the **baseband LAN**, which consists of computer workstations connected directly to a shared transmission medium, such as a twisted wire pair, coaxial cable, or fiber-optic cable, configured in either a star, bus, or ring topology.

One of the workstations on a baseband LAN is generally designated a dedicated server and acts as the repository for network file storage, application software, and the network operating system; a single LAN may have more than one server attached. There are general purpose servers, which manage shared databases, files, and application programs for the network, and there are specialized servers dedicated to a particular function; such servers include disk servers, file servers, fax servers, communications servers, and so on.

LAN transmission occurs in digital form, and the entire capacity of the medium is allocated to each signal on the network. Thus there is potential for congestion and interference among separate transmission signals vying for the same shared medium capacity. To avoid this problem, the transmission speeds must be very high, and there must be some means of managing access to the shared medium. This situation is similar to the problem of managing access to a single-lane road that is shared by a number of homes on the street; if the cars on the street move slowly and several cars attempt to enter the street at the same time, there will clearly be heavy traffic congestion, and possibly a collision.

Recall from Chapter 7 that a complete network architecture requires both a protocol and a configuration structure. The OSI model provides an open protocol system for linking two separate computer systems, but it does not include protocol standards for linking multiple computers in a LAN configuration. A configuration structure system is needed as well to specify standards for the components of the network, the naming and addressing of network devices, the provision of network services, and methods for interconnecting these components in a particular LAN topology.

Although a number of complete LAN architectures are in use today, the industry is dominated by two in particular: Ethernet and token ring. The others are variations on the key elements of these two architectures.

Ethernet LANs

[Handwritten margin notes: IEEE 802.3 CSMA/CD Layers 1 & 2]

The Ethernet architecture is an outgrowth of research done by Xerox Corp. to develop a simple and reliable system to network terminals. The Ethernet concept evolved into an international standard known as IEEE 802.3: CSMA/CD baseband bus. The Ethernet architecture consists of protocol standards that fulfill the functions of Layers 1 (physical) and 2 (data link) of the OSI model. A network Layer 3 protocol is not needed because all messages are broadcast to all nodes on the shared medium, and there is no need to establish specific connections between users on the network. Layers 4 –7 are independent of the particular LAN specification and therefore are not part of the LAN architecture itself.

[Handwritten margin notes: Layer #1 – RS-232-C]

Layer 1 of the Ethernet system specifies the physical interface between the end-user device and the LAN medium. This is equivalent to the standard RS-232-C serial interface protocol. The Layer 2 data link protocol in a baseband LAN is actually divided into two sublayers: a logical link control (LLC) sublayer and a **medium access control (MAC)** sublayer. The LLC sublayer fulfills the same functions as the HDLC data link protocol: It encapsulates the message or block of data into a standard frame specification and attaches an address for a particular user, which can be an end-user or an application program. The LLC also includes a mechanism for error detection and correction and manages flow control by buffering incoming and outgoing data streams.

[Handwritten margin note: Layer #2]

The unique aspect of the Ethernet architecture is the specification of the MAC sublayer protocol, which receives the frame from the LLC sublayer and appends a terminal station address; it then manages access to the network to avoid congestion and interference among transmissions vying for the shared medium. Recall that there is no direct user-to-user connection specified on a LAN, as there is with circuit-switched network systems, because LANs use a shared medium and operate in a broadcast transmission mode.

[Handwritten margin note: no direct user to user connection on LAN]

In the Ethernet specification, access to the shared medium is managed through a contention protocol system called **carrier sense multiple access with collision detection (CSMA/CD).** A contention system relies on each terminal device on the network listening to the network to determine whether the bandwidth is available or occupied with a transmission. If it is available, the terminal transmits its message, and control of the medium by the transmitting terminal is retained until the message is complete.

[Handwritten margin note: Contention Protocol: CSMA/CD]

The CSMA/CD system is a specific type of contention system used in the Ethernet architecture. In addition to listening before transmitting (the CSMA portion of the protocol), there is also a collision detection and retransmit feature (the CD part of the system). Two terminals may sense that the broadcast medium is available and may transmit simultaneously, which results in a collision. In listening mode, the terminals will detect the collision, and each waits a random amount of time before retransmitting the message. If a second collision should occur, the retransmission sequence is repeated.

The CSMA/CD system is effective when traffic on the network is light, but during heavy traffic periods a large number of collisions and retransmissions can occur. This can slow network response time significantly, which means that

CSMA/CD is generally inappropriate for applications that require a uniform response time.

The physical, LLC, and MAC protocols of the Ethernet level 1 architecture are implemented through a network interface unit (NIU) device, which is either an external hardware device or a microprocessor card inserted into an expansion slot in the PC workstation and connected directly to the LAN medium through an RS-232-C type interface (physical layer function). For outgoing messages, the NIU receives the data from the computer terminal and packages it into frames containing an end-user address and error detection bits (LLC sublayer functions). A workstation address is then appended, and the packets are temporarily stored in a buffer until the network medium is available. The packets are then sent onto the network at the appropriate speed (MAC sublayer function).

For incoming messages, the MAC sublayer in the NIU scans packets on the medium, seizes those with a matching station address, and passes the frame to the LLC sublayer. The LLC protocol checks the packet for errors, strips away the address and control information, and sends the message or data block at the proper speed to the destination end-user (person or application program) over the workstation's internal bus system.

The Ethernet configuration structure architecture specifies standards for both the configuration of the network nodes (topology) and the transmission medium itself. The topology is a bus configuration, and the medium is either twisted pair wire or coaxial cable operating at 10 mbps. However, in most modern Ethernet LANs, the physical topology is not the same as the logical topology. The physical layout of an Ethernet network is generally a star configuration in which the individual workstations are connected directly to a hub located in a wiring closet. The hub structure contains workstation ports that are connected directly to the backbone Ethernet bus. This configuration eliminates the difficulty of attaching workstations directly to the bus in awkward locations such as the ceilings and walls of office buildings; it also makes it easier to isolate the location of network faults because all workstation connections to the bus can be tested at the hub.

In addition to the formal Ethernet standards spelled out in the IEEE 802.3 specification, a complete configuration structure architecture should include specifications for the role and configuration of servers on the network, a network management and control method, and a plan for accommodating network growth requirements. The LAN architecture should address the following questions: What is the maximum number of workstations assigned to a single server on the network? What conditions dictate the use of specialized servers on the network, such as the presence of print servers, fax servers, and possibly communications servers? What network operating system features are needed to control and manage the basic functions of the network and accommodate growth? What conditions justify the grouping of workstations into separate subnetworks?

Token Ring LANs

The other principal LAN architecture is the token ring system, which is based on a concept developed by IBM that has evolved into an international standard (IEEE 802.5: Token Passing Ring). The token ring protocol structure architecture is similar to the Ethernet protocol system for the physical and LLC layers, but it uses a completely different approach to the MAC sublayer functions. The token ring LAN uses a token passing access protocol instead of the Ethernet contention system.

The token passing approach to media access control is the electronic equivalent of an old method used to prevent school children from talking all at once and disrupting the classroom. With this approach, the teacher designates some object (an eraser or colored pencil) as the speaking token, and permission to speak resides with the token. The child who possesses the token speaks until he or she is finished and then passes the token to the next child, who either speaks or passes the token on to another child. This system eliminates simultaneous conversations and interference in the classroom, but it can be cumbersome and can sometimes interfere with spontaneity and normal human dialogue.

The electronic token passing system works in a similar manner, but at such high speeds (4 or 16 mbps) that the handoff of the electronic token is not generally a hindrance to the electronic exchange of information across a LAN. The electronic token is a particular sequence of bits in a packet or frame of information. When the packet containing the token arrives at a station on the ring that wishes to transmit a message, the station seizes the token packet, changes the bit pattern of the token to read "not available," and appends its message and an address header before sending it on around the ring. When the packet arrives at the destination workstation, the encapsulated message is read and processed internally. The workstation then attaches an acknowledgment message and sends the packet onto the ring bound for the originating node. When the packet arrives at the origination point, the workstation reads the acknowledgment, changes the token bit pattern to "available" status, and retransmits the token onto the ring to be seized by another node that has information to transmit.

The token passing system generally exhibits lower variability in response times than does the Ethernet contention method because there are no collisions and retransmissions to slow things down when traffic is heavy. The response time on a token ring network is primarily a function of the number of stations on the network and the average message size, and it does not vary with traffic peaks nearly as much as the CSMA/CD system does.

The physical, LLC, and MAC protocols of the token ring protocol structure architecture are incorporated in a NIU device similar to that of the Ethernet system. This NIU device is generally a card placed in an expansion slot of the network workstation, and the protocols are executed the same way as for Ethernet LANs.

The token ring configuration structure architecture specifies a ring configuration for the topology, and the transmission medium is either unshielded or shielded twisted pair operating at 4 mbps or 16 mbps, respectively. In most actual token ring networks, the physical network configuration resembles a star configuration, with each workstation connected to a central hub unit located in a wiring closet. All of the workstation ports in the hub are connected directly to the ring medium, which facilitates connecting or disconnecting workstations from the LAN and isolating the location of network faults without the need to examine ring connections throughout a building. In addition to the formal token ring standards spelled out in the IEEE 802.5 specification, a complete configuration structure architecture should include specifications for the role and configuration of servers on the network and for how a network operating system manages and controls the network, as well as plan for accommodating network growth requirements.

Fiber Distributed Data Interface

The fiber distributed data interface (FDDI) standard has been developed to provide very high-speed LANs and backbone links between LANs. The medium access method is token passing, and the topology is two rings transmitting in opposite directions at a speed of 100 mbps. One ring is the primary medium and the other is a backup. In case of a single point failure in the ring or in a workstation, the network can continue to function by bypassing the fault and using the portions of both the primary and backup rings that are still operational. With two counterrotating rings, a closed loop can be maintained around a single failed node.

The FDDI standard is similar in most respects to the IEEE 802.5 Token Passing Ring specification but includes some significant enhancements. The counterrotating pair of rings is a significant improvement in reliability and fault tolerance over the single ring specification in the 802.5 system. FDDI also includes a more extensive priority system than 802.5; workstations designated as high priority are permitted to access the ring for longer periods of time than are low-priority stations. The maximum distances allowed for FDDI are significantly greater than those for the 802.5 system: An FDDI ring can be up to 200 kilometers in length, which enables either the creation of a single LAN spanning several miles or a backbone ring interconnecting several individual LANs located in the same section of a city.

The high-speed FDDI standard also enables the use of multimedia applications in a LAN environment. Even with advanced digital compression techniques, the integration of voice, data, image, and video information into multimedia applications requires fast computer processing speeds, large storage capability, and very high bandwidth transmission capacity. The 100-mbps transmission speed of FDDI solves the transmission speed problem and makes networked multimedia applications feasible.[3]

---

[3] In the meantime, efforts have begun to create a 100-mbps Ethernet standard, which would provide 10–50 times more speed than the original Ethernet implementations. Increasingly, users are looking for standards that do not require throwing away old equipment, applications, and expertise.

*LAN Applications
and System
Performance
Trade-offs*

Support for the office work group and a desire to link work groups across an enterprise have been the driving forces behind the mass migration to LANs and the client/server computing concept. LANs are ideally suited for the following types of applications: managing local databases and processing database queries; supporting work group office automation activity, such as shared document preparation, e-mail, and fax distribution; and reducing cost through sharing software and expensive hardware peripheral devises such as printers. Linking LAN systems across the enterprise, and linking LANs with systems in vendor and customer organizations through internets, extends the reach of the work group support platform to meet the changing needs of an organization.

The local PBX network links an office for voice communication and provides circuit-switched data links at speeds up to 64 kbps. This setup is sufficient for electronic mail and for limited file transfer applications, but it does not provide access to the server-based features inherent in a baseband LAN: software and printer sharing and large-scale database management and storage. Furthermore, the LAN operates at significantly higher transmission speeds, which means that more information can be moved faster (greater responsiveness).

In comparing the performance features of the various types of baseband LANs, Ethernet is the simplest and generally the least expensive system, but it is subject to high variability and a lower mean level of responsiveness during peak traffic periods. On the other hand, the token ring system exhibits less response variability and generally better average responsiveness under heavy traffic loads; however, it is more complex and generally more expensive than a comparable Ethernet system. Also, because each station on a token ring LAN regenerates the signal, the failure of a station or a break in the ring can cause the entire network to crash unless special recovery capabilities (such as a second ring transmitting in the opposite direction) have been incorporated into the design. The Ethernet system is not subject to this problem because the information packets are broadcast once on the shared bus and there is no relay mechanism.

Clearly, the FDDI architecture provides responsiveness under all traffic load conditions that is superior to token ring and Ethernet systems. In addition, a FDDI LAN can link workstations that are spread over a much wider area without the use of repeaters, and the dual ring configuration protects the system from a ring or workstation fault. Furthermore, the 100-mbps transmission rate of the FDDI LAN is sufficient to accommodate video and multimedia applications. The availability of desktop video and audioconferencing and LAN-based delivery of multimedia information has the potential to radically change the way information systems support the work environment. Although FDDI was designed specifically for fiber optics, it has been successfully implemented on twisted pair (see Minicase 14-1), which preserves existing investments in cabling while opening up opportunities to exploit FDDI-based innovations.

Wide Area Networks

[handwritten: define] Wide area networks (WANs) are private or semiprivate networks whose reach extends beyond a metropolitan area. A WAN can be regional, national, or international in scope, and it may either carry only one type of information or be designed for the integrated transmission of voice, data, and video information. A WAN may be restricted to one technology or medium, such as a satellite network or a fiber-optic cable network, or it may utilize a mix of technologies and media configured in a star, mesh, hierarchical, or hybrid topology.

WANs are growing in importance and capability as modern businesses move more of their operations on-line and as increasingly sophisticated applications require greater reach, range, and responsiveness in a company's telecommunications platform. The trend in WANs is toward high-capacity, digital backbone links that interconnect LAN subnetworks and provide long-haul carriage of integrated voice, data, and video information streams.

WAN Technology and Services

WANs have evolved from completely private networks composed of dedicated circuits and central office switching capacity leased from local and long-distance carriers, to semiprivate software-defined networks. The private networks interconnect a company's PBX systems, host computer front-end processors, or LANs using leased facilities that are dedicated exclusively to the company's use. The company has complete management responsibility for its private network, and long-term lease arrangements make it difficult to reconfigure the network to accommodate organizational or geographic changes.

[handwritten: Private]

Some private backbone WANs consist of only one technology or medium. For example, a company may construct a fiber-optic cable WAN or a VSAT satellite WAN to link various data centers or retail outlets. A currently typical private WAN system consists of leased T1 circuits operating at a transmission speed of 1.544 mbps (see Minicases 2-3, 2-4, and 5-1). To accommodate growth, some WANs are constructed from leases of fractional T1 capacity. The company's network is configured with T1 links between its major nodes but pays the carrier on the basis of the fraction of the T1 capacity that is actually used (in multiples of 56 or 64 kbps). Eventually, as traffic grows over time, a full T1 circuit capacity may be utilized.

With a growing trend toward high-capacity backbone WANs, fast packet transmission and switching methods such as frame relay and cell relay (described in Chapter 6) are being incorporated into private and semiprivate networks. This is important because the overall capacity of a network is determined by the lowest-capacity segment. Consequently, if high-speed switching nodes are not available, the specified throughput can be maintained only with additional high-speed, point-to-point circuits configured in a mesh layout. This can be a very costly solution for large, multinode WANs. Fast packet equipment and services are currently available for private networks through frame relay and SMDS (cell relay) technology and services.

Fast packet-switching technology enables the transmission of integrated streams of voice, video, data, and image information over a single backbone WAN. The integrated WAN eliminates the need for multiple, single-purpose networks, which in turn results in lower costs and significantly greater range for a company's telecommunications platform.

In the late 1980s, the truly private WAN gave way to the software defined or virtual private network (SDN/VPN). Instead of having particular circuit and switching facilities dedicated to a particular company's use on a full-time basis, the SDN/VPN concept guarantees only that the required circuit and switching capacity specified in the lease will be made available on demand. This capacity is allocated on a priority basis to SDN/VPN networks from a pool of available facilities owned and managed by the carrier.

A SDN/VPN WAN is superior to a dedicated private network in many respects. It is a much more efficient use of carrier facilities because idle time of dedicated, leased circuit capacity is eliminated, and the common pool of facilities can be allocated when and where they are needed. Furthermore, network configurations are flexible and easily modified to accommodate changing requirements through a simple software program modification. Finally, a company has overall control of the network, but the carrier retains responsibility for the day-to-day operations and maintenance of the underlying network facilities.

Another form of semiprivate network is **wide area telecommunications service (WATS)**, a bulk-rate, switched network service priced at a discount over regular long-distance service and provided by long-distance carriers primarily as a voice service. There are two versions: inbound and outbound service. The inbound service is better known as toll-free 800 service because all calls routed over the inbound WATS system to the subscribing company use an 800 prefix. The inbound WATS, whose coverage area includes the United States and many foreign countries, has become a significant marketing asset for many companies, particularly as a means for customers to communicate easily with the company and for telemarketing applications. Inbound WATS is priced in two parts: a fixed monthly charge plus a price per minute of use. The per-minute usage charge is lower for higher volumes of calls.

Outbound WATS, available as a separate service offering, enables a company's employees to dial out under a special discount rate schedule. Outbound WATS is offered for different mileage bands that extend out from the subscribing company's home base. The price charged for outbound WATS includes a fixed monthly fee per mileage band plus a charge per minute of use, with lower rates for higher-volume usage.

WATS uses the carrier's switched network facilities in the same manner as any dial-up call. However, special value-added features are available, such as automatic routing of incoming calls to different locations (depending on time of day) and automatic prompts asking the caller to select from a menu of options. WATS also can be used for dial-up message and data transmissions (such as fax), but its use for data is limited to the capacity of a single voice circuit.

In classifying WATS on a continuum between a pure public network and a pure private network, WATS is closer to a public network service than is a SDN/

Offerings	AT&T	MCI	Sprint	WilTel
SONET (synchronous optical network) -readiness	Available now as needed	Bulk of network SONET-ready by 1994	Full network SONET-ready in 1994–1995	Under development
ATM (asynchronous transfer mode) implementation	"Aggressive deployment" planned	Underway in 1993	Wide availability as of late 1993	Under development
Frame relay availability	Current	Current	Current	Current
X.25 availability	Current	None	Current	None
ISDN availability	Primary rate offered	Basic and primary rate offered	Primary rate and X.25 transport of ISDN data	None
SMDS availability	As needed by users	Current	Available as of late 1993	Not planned as general service
Availability of other services	Switched 56 kbps	Switched 56 kbps	None	Switched T1

Figure 8-3

Competitive Offerings of the Leading U.S. Long-distance Providers as of Mid-1993

VPN but contains many private network, value-added features, such as special call routing and call accounting. Nevertheless, it does not have the configuration flexibility and the transmission capacity options to extend range and responsiveness for nonvoice applications that are available in a SDN/VPN. Figure 8-3 shows the competitive offerings of the four leading long-distance providers as of mid-1993.

Public Switched Networks

Public switched networks are the oldest form of networks and for many years were the only source of network services. Public switched network transmission and switching facilities are owned and maintained by authorized carriers who offer telecommunications services to the public at tariffed prices. Until the breakup of the Bell System in the United States in 1984 and the introduction of limited competition in other countries, each country generally had only one public switched network that was owned and managed by a monopoly organization. In the early 1900s, the telephone became the dominant means of electronic communication, and public switched networks have been designed to support public telephone service since that time. The use of the public switched network for data communications has been incidental until the recent development of integrated services public switched networks and separate public data networks.

Public v. Private [handwritten margin note]

The choice of public versus private networks to meet organization communications needs has historically been cyclical and has depended to a large extent on the available range of services and the relative prices of the two alternatives. The decision has also depended on the extent to which an organization wants either to operate and manage its own telecommunications activities, or to outsource them—the familiar "make or buy" decision. Generally, larger companies demand more telecommunications capability in the form of capacity, reliability, flexibility, and management control than is available through a public switched network service that, by definition, must serve a large cross section of the population. Therefore, customized private network solutions are generally preferred by companies with large and diverse communications requirements.

However, the trend in public switched networks toward flexible, high-capacity digital systems that integrate various types of information enables networks to meet many corporate communications demands in a cost-effective manner. Furthermore, carriers are now cleverly brokering excess capacity on their public switched networks in the form of virtual private networks. Public network services appear even more attractive in the current environment of corporate downsizing, cost cutting, and emphasis on the short-term "bottom line." In this climate, a company's decision to eliminate its telecommunications department with its high-priced equipment can be a shot of adrenaline to next quarter's income statement and balance sheet.

Public Switched Network Technology and Services

Early – analog/voice [handwritten margin note]

The largest public switched network in most countries is the public telephone network, which is generally controlled by one entity. In the United States, the public network is segmented into local and long-distance service components. The former Regional Bell Operating Companies and a handful of other telephone companies provide local exchange service in nonoverlapping areas; the long-distance portion is supplied by AT&T, MCI, Sprint, and several other carriers. The local and long-distance networks are linked to form one large interconnected national network. Reach is extended internationally by connecting the various national public networks through satellite and fiber-optic cable transmission links.

For many years the public switched telephone network used analog transmission and switching methods to accommodate the predominance of voice traffic. In recent years the network has rapidly converted to digital switching and transmission systems, which has become a more efficient method for carrying both voice and data traffic.

ISDN B & D channels (Basic [handwritten margin note]

In keeping with this trend, standards have been developed for a universal, public switched, integrated services digital network (ISDN), which is being groomed as the replacement for the world's analog telephone networks. The ISDN standard specifies two types of digital channels: a bearer channel (B channel) carrying 64 kbps of information and a delta channel (D channel) carrying 16 kbps of signaling and control information. There are currently two defined ISDN service offerings. Basic access service consists of two 64-kbps bearer channels and one 16-kbps delta channel for signaling. A shorthand for the basic access service is 2B+D. Basic access service will be delivered to the *twisted pair*

terminal device and therefore will be available in homes and offices using twisted pair wiring. The other service offering is called **primary access ISDN**, which consists of 23 64-kbps bearer channels and 1 16-kbps delta channel, or 23B+D. The 23 bearer channels have a transmission capacity almost equivalent to a T1 circuit (24 64-kbps channels).

Primary 23B+D

ISDN uses a circuit-switched transmission method and is designed to carry digital bit streams that represent any type of information—voice, video, data, and image. Also, the total bearer capacity available in either the basic or primary access service can be subdivided in any manner. For example, the two 64-kbps channels in the basic access service can be combined into a single 128-kbps channel for video conferencing or divided into four 32-kbps voice channels. Because ISDN is being developed as a worldwide standard, it includes an open system interface that is fully specified, which facilitates the construction of multivendor ISDN systems.

Unfortunately, technological changes are so rapid in telecommunications, and the formal standards process so slow, that ISDN may be obsolete before any large-scale implementation. Already, specifications are being developed for a broadband integrated services digital network (BISDN) that will support transmission speeds greater than the primary rate. The growing requirement for video, image, and multimedia transmission across networks is feeding the appetite for higher-capacity public switched networks.

BISDN

As this requirement grows, it is accompanied by advances in technology that enable the creation of very-high-capacity switched networks capable of rapidly moving large quantities of information in any form to anyone, anywhere. In its full implementation, a worldwide BISDN will be a public switched telecommunications platform with levels of range, reach, and responsiveness that can hardly be imagined today. It will support multigigabit transmission rates capable of worldwide retrieval and sharing of information from multiple sources in multimedia format, including on-demand, real-time distribution of educational and entertainment programming to the home or office. The U.S. Internet, built on TCP/IP as a standard, T1 as the backbone transmission, and T3 as its next increment, shows that this is not an idle dream. Researchers and scientists worldwide use this (so far, free) network to handle simple e-mail and transfers of massive databases. Al Gore (then a U.S. senator) made support for the Internet a personal interest several years ago, pushing for its funding and also running up against opposition from the chairman of AT&T over private versus public funding of the Internet.[4]

---

[4]This issue is likely to grow in importance over the coming years. In almost every other country, economic and political forces have led to the assumption that its government will sponsor the national telecommunications infrastructure through its PTT. In the United States, modern data communications has developed since the divestiture of AT&T, so that there is no such national network. The Internet, which is the largest network in the world, offers the basis for it. Should the government continue its existing role? Or should it rely instead on the competitive forces that have been so effective in long-distance communications?

fiber

SONET
ATM

A BISDN network will require transmission-link speeds greater than 500 mbps and high-capacity switching equipment capable of switching many 150-mbps streams of information. These speeds will require both fiber-optic cable links to the desktop and the home and a new generation of fast packet and optical-switching equipment, which is currently in development. The standards being developed for BISDN include an ANSI specification for a synchronous optical network (SONET) and a standard for an asynchronous transfer mode (ATM) switch. SONET, essentially a specification for a time-division multiplexing method over an optical transmission medium, includes a hierarchy of transmission rates that are multiples of a synchronous, 50-mbps, base-rate signal. The standard also includes a frame structure, an error detection and correction scheme, and a channel identification and control mechanism.

High-speed synchronous traffic can be switched relatively easily through circuit-switching equipment without the need for large storage buffers because the bit streams are regular in their timing, and the circuit-switched path is configured only once at the beginning of the session. Switching asynchronous traffic at high speed is much more difficult. Circuit switching is not well suited to asynchronous traffic because the intermittent bursts of data create large amounts of idle time for the dedicated circuit path, which is an inefficient use of the facilities. Packet switching is a more efficient method for handling asynchronous traffic streams because packets from different sources can be interleaved and can share the same network path. A high-speed ATM switching standard is being developed for efficiently handling asynchronous traffic in BISDN networks. The ATM switch is a very fast packet-switching device connecting optical-fiber links operating at transmission rates of 150 mbps and up.[5]

Value-added networks are also growing in popularity and evolving with technological change in the industry. A few years ago the VANs' data rates of 2,400 and 9,600 bps were more than adequate for most applications; throughput rates of 64 kbps are now common using the X.25 packet-switching protocol. The trend is toward even faster transmission speeds both to accommodate greater amounts of data transfer and to include other types of information being sent in digital form, such as voice, video, and image. After all, why restrict the service to data, when a bit is a bit regardless of whether it represents voice, data, or image? Infonet, the technological leader in international VANs, offers a "worldwide LAN" through its Infolan service. Leading VANs are currently offering frame-relay service, which is a fast packet-switching protocol that uses variable length packets and achieves throughput rates up to 1.544 mbps; it is the likely successor to the X.25 protocol. Local telephone carriers and others are conducting trials of switched multimegabit data services (SMDS), the cell-relay, fast-packet switching method that uses fixed-length packets and operates at speeds substantially

---

[5] Early ATM switches, introduced in 1993, supported only T1 rates. It will be several years before the full power of ATM is harnessed through new generations of switches.

higher than 1.544 mbps. The specification for the SMDS cell is very close to that of the ATM cell defined in the BISDN standard, which means that SMDS is likely to become part of the emerging BISDN concept.

The movement of VANs into fast packet technology (frame relay)—which frees them from the data-only restriction—and the entry of the traditional local and long-distance telephone carriers into the BISDN version of fast packet technology (SMDS)—which gets them into the provision of value-added, nonvoice services—suggests a major market realignment. The traditional, data-only VAN market segment is disappearing rapidly, and it is being replaced by an application-neutral, high-speed, digital, public network service that is available in somewhat different form from the traditional VANs and the former telephone carriers.

Trends in Networks   There are currently two clear trends in the evolution of types of networks. The first is the growing dominance of distributed processing and client/server computing models that are supported by LANs and internets of multiple LANs connected through backbone WANs. In the LAN environment, the concentration of more functionality in the intelligent wiring hub—the single point at which individual workstations are connected to the shared medium (bus or ring) and also the location at which bridging and routing to other LANs takes place through links to the backbone WAN—has become the focus of attention.

The likely next step is the further integration of high-speed, digital switching devices, such as the ATM cell-relay switch, into the wiring hub to provide more efficient interconnection of workstation devices across multiple LANs. If, in addition, LAN servers are also included in the hub chassis for convenience, economy, and network control, then in essence the hub becomes the LAN itself. In a sense, this makes the LAN hub the client/server equivalent of the data center computer room.

The second clear trend is the development of the BISDN. The worldwide availability of a very-high-speed, public switched network with open system interfaces is the manifestation of the universal information utility—full range, reach, and responsiveness on-demand in the home, office, or on the move. Future private networks will likely consist of customized access to segments of the BISDN on a virtual or software-defined basis to interconnect LANs and specialty networks.

On the one hand, it appears as if we have almost reached the promised land of BISDN implementation and the universal information utility; on the other hand, the telecommunications industry battlefield is strewn with empty promises of imminent interoperability, open systems, and universal standards. As a result, guarded optimism and appropriately hedged bets regarding the coming of BISDN is probably the appropriate strategic stance for the wise telecommunications designer.

Summary

The five main distinctions among types of networks are (1) the type of information they can transmit, with voice and data the main distinction, but one that is becoming increasingly obsolescent as massive bandwidth networks provide multimedia applications; (2) the type of technology, with packet-switched and circuit-switched services the main options for both LANs and WANs; (3) type of ownership: private versus public; (4) geographic scope: LAN, MAN, and WAN; and (5) topology: star, ring, mesh, hybrid, and bus.

LAN technology has been the main area of innovation in telecommunications in recent years. The two main types of LAN are Ethernet and token ring. All LANs use digital transmission. FDDI is emerging as the base for the next generation of LANs, exploiting fiber optics although not limited to it.

WAN technology is increasingly interdependent with LANs. High-speed backbone networks are needed to internetwork high-speed LANs. The public networks are increasingly exploiting new protocols and transmission media, particularly fast packet switching. Users now have options of public data networks, VANs, ISDN, frame relay, cell relay, and SMDS. SONET and BISDN are already in early use and extend the options.

Review Questions

1. Is the current classification system for networks adequate to truly capture similarities and differences that are important for comparisons and choices among different network systems? Discuss.

2. Explain the following statement: Over time, networks originally dedicated to transmission of one form of information (voice, data, and so on) are now carrying multiple types of information.

3. Why is the current trend moving away from classifying networks on the basis of the underlying technology (satellite, fiber optics, packet switching, and so on)?

4. True or False: In a private network, the company generally owns and operates *all* equipment constituting the network. Explain.

5. Give a brief description of a value-added network (VAN), and explain its unique features or services.

6. Briefly describe the concept of a virtual private network or software-defined network, and explain how it differs from a true private network.

7. Explain the differences among LANs, MANs, and WANs from a geographic perspective (locations, coverage, distance, and so on).

8. Because a pure mesh network maintains a direct, point-to-point connection among all nodes in the network, why isn't this the preferred and most prevalent network topology?

9. Briefly explain how a hierarchical network works. What are the advantages that result from all the apparent complexity in routing and switching?

10. The bus and ring topologies are usually associated with a shared medium transmission concept. Explain what this means, including how these two topologies differ in their method of sharing a given transmission medium.

11. What key factors are primarily responsible for the emergence of the LAN concept?

12. What are the important features that distinguish a LAN from other types of data communications networks?

13. What is a server, and what role does it play in a LAN?

14. Briefly explain the concept of a contention protocol system, and describe how carrier sense multiple access with collision detection (CSMA/CD) works.

15. What role does a network interface unit (NIU) device play in the Ethernet LAN system?

16. The actual physical layout of most Ethernet networks is a star configuration with the individual workstations connected directly to a hub in the wiring closet. Explain this apparent contradiction with the bus system specification for Ethernet.

17. Describe the token passing approach to accessing the shared medium in the token ring LAN architecture.

18. Briefly describe the fiber distributed data interface (FDDI) standard, and explain why it is needed when a token ring specification already exists.

19. List four applications that are ideally suited to a client/server computing environment and are implemented through a LAN architecture.

20. What are the important cost and performance trade-offs relevant to choosing an appropriate LAN configuration (either Ethernet, token ring, or FDDI)?

21. Describe the concept of a "fractional T1" WAN, and explain how the "fractional" part works.

22. What is the importance of fast packet-switching technology (frame relay and cell relay) to WAN performance and efficiency?

23. Explain the recent resurgence in the use of the public switched network to meet corporate communications requirements.

24. Explain the Integrated Services Digital Network (ISDN) concept and include a brief description of the 2B+D circuit design.

25. What is the broadband integrated services digital network (BISDN), and why even consider it when the use of basic ISDN is not yet widespread?

26. Briefly explain the role of the Synchronous Optical Network (SONET) standard and the asynchronous transfer mode (ATM) standard in the evolution of broadband ISDN.

Assignments and Questions for Discussion

1. LANs and interconnected LANs are fast becoming the dominant type of network configuration used to meet corporate data communications requirements. What caused this shift away from traditional terminal-to-mainframe data networks, and what are some of the implications of this shift for the future of corporate communications?

2. As the newly hired director of telecommunications for a public accounting and consulting firm (about 200 professionals and 75 support staff), you are responsible for evaluating and deciding on the appropriate local area network configuration for the firm. Before undertaking a detailed analysis of the firm's requirements, you have been asked to prepare a brief memorandum discussing the key features of the Ethernet and token ring LAN technologies and to highlight the strengths and weaknesses of each with respect to what is likely to be important to the firm. Prepare such a memo.

3. It seems as if we are entering an era in which all required network functionality necessary to support wide area corporate telecommunications requirements can be obtained through carriers and service providers in the form of virtual or software-defined networks, ISDN, and BISDN. Why not dismantle the corporate telecommunications department and simply order service from one or more of the carriers? Explain and discuss.

4. What are the implications of VANs implementing fast packet technology and of traditional local and long-distance carriers planning the introduction of BISDN services. Discuss in terms of the impact on traditional market segments and specialization of the respective service provider groups.

5. The promise of BISDN is the worldwide availability of a very-high-speed, public switched network with open system interfaces that can provide full range, reach, and responsiveness in a cost-effective manner. Drawing on the history of telecommunications network evolution, what are some of the important and formidable obstacles to the realization of this promise?

Minicase 8-1

Redstone Arsenal

Building a Complex of LANs

Figure 8-4 shows the complex of LANs used at the U.S. Army Missile Command's Redstone Arsenal. The arsenal deployed all Patriot missiles in Saudi Arabia during the Gulf War.

The complex of LANs is a campus network linking the arsenal's 9,000 users. It has a three-tier architecture, which is glued together through the TCP/IP (Transmission Control Protocol/Internet Protocol) standard. The outer tier consists of 9,000 workstations that are a mix of dumb terminals, personal computers, and powerful workstations manufactured by Sun Microsystems. These workstations are directly linked into 16 Ethernet LANs. The arsenal has over 250 departmental computers, each linked to a departmental LAN. A router connects the Ethernet LAN to a token ring backbone LAN. The routers use T1 links, operating at 1.544 mbps, or FDDI links, which transmit at 100 mbps. Gateways link the backbone LAN to other locations, including Saudi Arabia. The links on these wide area networks transmit data at 56 kbps.

There are 25 miles of fiber cable in—or rather under—the arsenal. Interconnecting the LANs requires many devices, including bridges for departments that have several smaller Ethernet networks. Routers connect the Ethernet and token-ring backbone LANs, and gateways connect the LAN and wide area networks. The gateways also convert traffic that uses the TCP/IP standard to and from the proprietary protocols used by the IBM mainframe-centered network.

Note how this complex provides departments with an efficient local capability through the Ethernet LANs, provides a campuswide facility through the routers and backbone token ring LAN, and accomplishes interorganizational connections via the gateways and wide area network links. The complex took 11 years to develop.

Questions for Discussion

1. Discuss the advantages of the three-tiered LAN architecture at Redstone, compared with the alternative of linking all users directly to the departmental computers through dumb terminals and cluster controllers (with direct, high-speed links among the various departmental computers).

2. Why have two different LAN architectures (Ethernet and token ring) been included in the same campus network? Discuss the implications of this choice for the use of internetworking devices (bridges, routers, and gateways).

3. Discuss the role of the gateway internetworking devices in the Redstone network. Focus on the functions they perform and the reasons they are the appropriate devices.

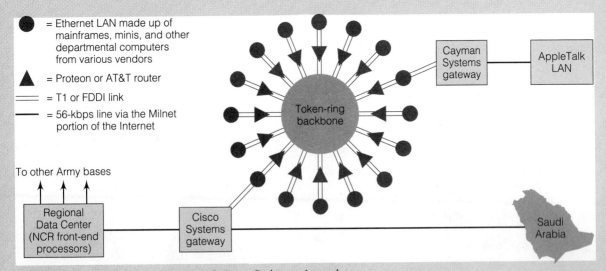

Figure 8–4

Source: U.S. Army, Redstone Arsenal

Redstone Arsenal's
Token Ring Campus
Network

Minicase 8-2

Health One Corporation

Supernetworking

Supernetworking refers to massive bandwidth platforms that provide individual applications with extremely high-speed telecommunications. This is very different from sharing broadband transmission links among many users. Health care is a natural target of opportunity for supernetworking in such areas as radiology, physician video-conferencing, multimedia patient reports, cardiology, and other applications involving rapid transmission of high-resolution images.

Health One, Minnesota's largest health-care provider, hopes that supernetworking will not only help it improve service and cut costs but keep it in business. This not-for-profit chain of nursing homes, hospitals, and other health-care operations in Minnesota and Wisconsin faces a very competitive environment. Half the Minnesota population is enrolled in health management organizations (HMOs), and there are three other large hospital chains. Hospitals everywhere are in a cost crisis, as the federal and state governments are cutting Medicaid, Medicare, and welfare support, insurers are tightly monitoring and controlling rates, and the nation's health-care system is increasingly seen as economic madness. Only 2 percent of hospitals make a profit. At the start of 1991, Health One's total liabilities were three times its net worth. It has closed and consolidated many operations and laid off staff. It is merging with LifeSpan Inc., a competitor of about the same size as itself.

Even as it continues to cut costs it must also maintain quality and remain competitive. It wants to consolidate high-technology, high-expertise activities such as heart surgery in just a few medical centers, and it wants to make specialists in one facility quickly available to other facilities. This is where supernetworking offers an opportunity. Together with 13 other Minnesota companies, Health One took part in a 1992 trial of US West's COMPASS (COMmunications Program for Advanced Switch Services), network switched DS1 (T1) links operating at 1.544 mbps. The switched service delivers this bandwidth on demand. The main innovations in COMPASS are the switches and data compression technology that make multimedia video practical. VideoTelecom, one of the equipment providers working with US West, is testing its VTC units that use the new H.261 standard for data compression algorithms. These algorithms reduce the bandwidth needed for full-motion video to 384 kbps. The VTC units also provide a range of bandwidth categories, including primary rate ISDN. The network also uses AT&T class 5 central office digital switches and ISDN switches, with a network bridge managing multipoint video-conferencing; this is the first time a bridge has been used in a local exchange carrier's network. The bridge allows a conference connection to be set up in the same way as a normal phone call, with no intermediaries, delays, or scheduling.

Some of the applications Health One is piloting are the following:

1. Radiologists who are on call for off-hour emergencies will receive medical images on graphics workstations that will decompress and display them. Emergency-room physicians will no longer need to wait while the image is delivered to the radiologist's home or for the radiologist to come into the hospital.

2. Doctors in any one of the organization's facilities will have their patients' images interpreted by the most suitable expert. These specialists will be able to carry out interpretations for many more patients, too.

3. Practitioners in different locations will hold on-line conferences without having to travel to meet. The network bridge and switched service will let them make connections on demand.

4. Radiologists' analyses of images will be relayed instantly via voice and/or electronic messaging instead of being either transcribed and faxed or (for emergencies) phoned in.

The amount of bandwidth needed for these applications depends on many details. For example, the radiologists' screens may need a resolution of 1,024 by 1,024 pixels (dots) per square inch. Health One hopes that 512×512 will be acceptable but recognizes that what is considered high resolution to a telecommunications manager may be totally inadequate for a radiologist. A trial project showed what was required and also highlighted technical issues of screen refresh rates and performance of compression algorithms.

In late 1992, Health One began to test the COMPASS videoconferencing capabilities. As it centralizes operations in order to cut costs, it needs to maintain close communication between key headquarters managers and local ones. The system will become the main organizational linkage for doctors in rural hospitals, for training, and perhaps later for transferring video documents such as patient records.

Major areas of concern that the trial project explored include cost, physicians' receptivity to the new technology, and patients' comfort with a very new approach to consulting and radiology. If the results of the trial are considered positive, Health One's own future looks more positive, too.

Questions for Discussion

1. Explain the difference between "supernetworking" and shared broadband networking, such as LANs, and explain why supernetworking is better suited to Health One's set of applications.

2. What is the real innovation in the proposed network system that has Health One and US West's executives excited, and why is this innovation important?

3. How is Health One's supernetworking system likely to solve its financial crisis? Explain and discuss.

Minicase 8-3

The Travelers

Integrating Wireless Local Area Networks

Typically, local area networks are physically linked together by cable. One consequence of this is that adding a personal computer to the LAN or moving someone's desk can require additional cabling, and installing a new LAN is a substantial construction project involving drilling holes in walls and paying electrical contractors a lot of money.

The Travelers Corporation has cut the time required to install a LAN from between 30 and 150 days to a few hours through its use of Infralan, a product that is compatible with IBM's token ring local area network. As its name suggests, Infralan uses infrared spectrum transmission—light instead of cable—for transmission. Cabling a workstation can cost up to $100; with Infralan, the cost drops to $10.

Travelers' staff often work out of leased field offices and move around frequently. Many of the offices were not cabled; in 1989, 100 of 140 planned LAN installations were going into buildings in which it would take up to five months to lay cables—and this only after negotiating with the owners of the buildings for wiring space and finding qualified electricians to pull (literally), terminate, and test the wiring. Travelers had looked at wireless LAN systems as early as 1986, but the available systems were not compatible with the IEEE 802.5 standard used by IBM token ring LANs, the core of Travelers's systems. They also cost up to $30,000 per node; Infralan costs only $3,000 per node.

Questions for Discussion

1. Why is the wireless LAN concept appropriate for Travelers? List and briefly discuss the reasons.

2. What is it about the Infralan system that finally caused Travelers to decide to proceed after looking at wireless LANs for over four years?

3. Will wireless LANs dominate the LAN market in five years and relegate wired LANs to historical relics? Explain your reasoning.

Minicase 8-4

Byer California

Using a Frame-relay Value-added Network

Speed and flexibility are two critical elements for any garment manufacturer, for fashions change rapidly, not just in terms of overall tastes, but in such details as color, buttons, necklines, and materials. Fashion changes sometimes require shifts in production virtually daily.

Byer California is one of the largest garment manufacturers in the San Francisco area, with around 1,000 employees. It has seven main garment showrooms, where sales staff take orders for its lines of women's clothing, in such far-flung locations as Atlanta, New York, Dallas, and Charlotte, North Carolina.

Byer California has developed its telecommunications capabilities to meet its business needs by moving with the mainstream of technological innovation and by keeping to the practical leading edge and thus avoiding betting its future on unproven vendors. Given that Levi-Strauss is just down the road, Byer California cannot afford to drop behind; even a five-day delay between order and production can have retail buyers going elsewhere.

Until 1987 Byer relied on overnight couriers, fax, and phones to move orders and reports between its manufacturing facilities and the showrooms. It then added a simple asynchronous dial-in service provided by BT North America, so that sales staff could tap into its host computer to check on the status of shipments. This was a cheap and simple system that worked well, but it was inflexible and cumbersome to operate. Changing a password or phone number had to be done from the head office. Communications to headquarters could be initiated only from the remote sites, not the other way around. The computer system hardware and software was outdated and required substantial technical expertise to change or update it. The system controlled the people, not the other way round. An IT manager at Byer said, "We didn't want our people worrying about [the system]. We wanted them selling garments."

In the late 1980s, Byer's San Francisco headquarters updated its technology base and moved to an Ethernet LAN that connected a range of UNIX minicomputers, Apple Macs, IBM PCs, and other workstations. It added an X.25 connection for the dial-in showrooms, but the LAN and all the applications on the minicomputers and personal computers were inaccessible to the salespeople.

In 1991, Byer implemented XLINK, a service of BT North America that provided a port-based X.25 capability; port-based means that the Ethernet LAN could connect directly to BT Tymnet's frame-relay VAN through a router box, in effect pushing the LAN into the showrooms. Thus two-way communication could occur, instead of just dial-in; there is no difference now between having a terminal 50 feet or 3,000 miles away.

XLINK is a bundled service; Byer pays a flat fee, with no itemization or usage charges. This is very much like a private network. XLINK provides electronic mail as well as X.25 connectivity, which has cut Byer's phone bills by over $100,000 a year. Whenever they had any problem with the old system, frustrated salespeople would stop trying to use the dial-in facility and phone someone in San Francisco for answers. Those people would have to take time away from their primary activity, which is shipping garments as quickly as they are made, to track down the needed information about production and shipping. In addition to cutting phone call services at headquarter by 99 percent, Byer saves on fax and courier; reports can be sent at night to remote printers instead.

The connection between the LAN and XLINK was originally through a 9.6-kbps leased line, which soon became overloaded by the new data communications traffic. In 1991, Byer went back to its VAN supplier, BT North America, for a solution. BT was preparing to announce its new ExpressLANE 56-kbps frame-relay service. Byer had become a beta-test site for this, before it went operational in September 1991, by using the network service before it was "released" onto the market. The idea is that Byer would help BT track down any remaining errors and problems of installation and use and provide feedback. Byer understood that there would be "crashes" and bugs, but they would gain the advantage of moving through the learning curve ahead of other organizations.

ExpressLANE offered Byer "plug-and-play." The firm's technical support manager commented in early 1992 that "It's transparent connectivity without having to own a great deal of technology. . . . I can take my Ethernet LAN and just plug it into the network so that we can move data to our showroom." ExpressLANE provided all the network interconnection facilities, including dedicated port access, software, bridge/routers, and 56-kbps access lines. The purpose was to be just a utility, on the same basis as the electric utility.

The difference between 9.6 kbps and 56 kbps was more than just one of transmission speed; it allowed Byer to add a new application of immense importance to its business operations: color images sent quickly to the showrooms. Byer's business rests on getting enough orders for a garment before it commits to production. A Levi-Strauss can make literally 10 million pairs of its 501 jeans and put them into inventory, but Byer would go broke trying to make to stock instead of manufacturing to order. If it brings out a new line at the first of the month and then decides it wants to add an additional design a week later, it takes five days to get samples made up and shipped to showrooms. That leaves 18 days to take orders for the new design before planning next month's production, which is not enough.

Now Byer makes up just one sample, makes a color slide, and transmits the slide to every showroom. Facsimile machines do not provide adequate clarity of color and resolution of image. The X.25 network can handle transmission, but frame relay is 10–15 times faster. It used to take 20–25 minutes to transmit a single image, which takes 4–5 megabytes of data to code (about 40 megabits); now it takes around three minutes. The process of creating, sending, and receiving the image remains complex and takes around an hour. The photographer's color slide is scanned into a multimedia workstation. Next it is cropped, cleaned up, and annotated on an Apple Mac, which accesses the workstation through the Ethernet LAN. It is then transmitted via frame relay.

When it reaches the showroom, the image is printed on a Macintosh color printer. The output is an 8½-by-11-inch color photo as good as the original slide. Byer can now send around 200 color images a month, which adds up to a significant advantage for a relatively small firm in a very competitive business. "Our salespeople can be taking orders for a new garment the day we add it to our line. Another company that has to send the actual garment will be at least five days behind. We can be booking orders during that time and if we can get the buyers to commit their budgets to us, they won't be spending it on someone else."

Frame relay did not exist as a product before 1991, and in 1992 only a few companies were even testing it. Most long-distance providers announced their own frame-relay service around mid-1992, with beta testing occurring in late 1992.

Questions for Discussion

1. Byer California once relied exclusively on overnight couriers, fax, and telephones to transmit orders and reports between its factories and showrooms. What caused them to update their technology base and install a new telecommunications network system?

2. What role did BT North America's XLINK service play in both expanding the capability of the Byer network and in reducing communications costs?

3. What role did the ExpressLANE 56-kbps frame-relay service play in the evolution of the Byer network and its ability to support the company's evolving needs with additional services?

9

Networks in Action: Examples of Four Companies

Objectives

- ▶ Review four case examples of corporate networks that focus on their technical design
- ▶ Learn about how leading firms go about developing their platforms and incorporating new technological building blocks
- ▶ Use the cases to review and test your understanding of the material in Part Two

Chapter Overview

Chapter Overview

This final chapter in Part Two examines the network designs and strategies of four leading companies: Texas Instruments, the Royal Bank of Canada, British Airways North America, and Digital Equipment. Each case illustrates effective approaches to ensuring that telecommunications addresses overall business and organizational priorities and exploits existing and new technologies, especially such devices as intelligent routers and high-speed transmission links.

At the end of the chapter are assignments that can help you check your understanding of the important terms and central concepts used in preceding chapters.

Texas Instruments: A Global Capability

Texas Instruments, whose annual revenues are close to $10 billion, makes semiconductors, computer equipment, missile guidance systems, and other industrial electronic products. It is also widely recognized as a world leader in using telecommunications, winning several awards from trade associations and publications. *Network World*, which picked Texas Instruments (TI) as winner of its 1991 User Excellence awards, cited the "relentless pursuit of perfection" as the main reason for TI's success ("User Excellence" 1991). Its corporate telecommunications staff of around 170 people raised network availability from 99.05 percent in 1987 to 99.64 percent in late 1991. TI pushes the practical state of the art aggressively and claims to be ahead of most comparable users in technical innovation by some 12 to 18 months; it works closely with its vendors to make sure it both alerts them to its thinking and priorities and is able to be an alpha- or beta-test site for their new products. Its network spans 30 countries and links over 60,000 workstations to each other and to 23 mainframe computers, with response times of a second. The topology of the network is shown in Figure 9-1.

TI's success is not a matter of instant implementation. Like other firms, it has had to struggle with issues of coordination, integration, LAN-WAN incompatibilities, uncooperative PTTs, challenges of providing end-user support, and so on. The telecommunications unit explicitly focuses on service quality as its business driver, with rigorous metrics for tracking it and continuous commitment to improving performance in relation to those metrics. For example, statistics on network availability, errors, and response time are captured in a stored database and analyzed monthly. These performance figures are discussed with TI's vendors, whom it expects to offer a service well above their published guarantees.

TI's telecommunications capability evolved over a 20-year period beginning in the 1970s, when the company developed its own private global network. This

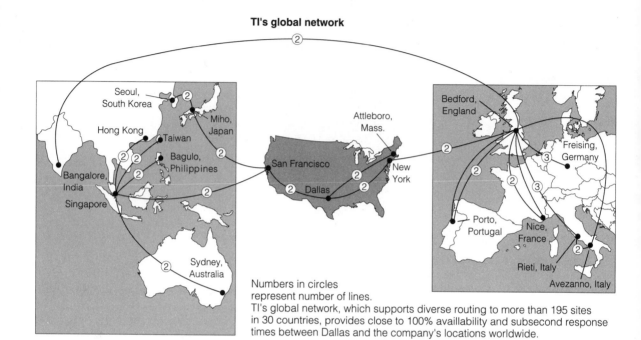

TI's global network

Numbers in circles
represent number of lines.
TI's global network, which supports diverse routing to more than 195 sites
in 30 countries, provides close to 100% availlability and subsecond response
times between Dallas and the company's locations worldwide.

Figure 9–1

**Texas Instruments
Network Topology**

development required many ad hoc and proprietary solutions because PTTs
were not positioned to provide either technology or service. Only a small subset
of the PTTs would lease data communications circuits to companies. TI built a
packet-switched network with a proprietary protocol for transmission over satellite links; it leased voice-grade analog lines, the only choice available.

This network was gradually migrated to a backbone network of 85 T1 circuits built on IBM's Systems Network Architecture, which connects to local
area networks in 25 countries. TI's manufacturing applications rely heavily on
large mainframe computers in Dallas that require workstations on LANs to
emulate the standard IBM 3270 dumb terminal. TI's goal is to ensure "seam-
lessness" across its entire geography, including being able to set up and run
factories, warehouses, and design facilities anywhere in the world, with direct
linkages to the Dallas mainframes, LANs, and workstations worldwide. By
1986, TI had probably the lowest unit costs for communication of any multinational firm. An electronic mail message could be sent anywhere in the world for
under four cents, and a database inquiry cost less than three cents. At that time,
TI's network interconnected 13,000 terminals and conducted about a million
transactions a day, with a maximum response time of six seconds. The compara-

ble figures for late 1991 were 60,000 workstations, eight million transactions a day, and a subsecond response time.

Texas Instruments's backbone global network was optimized to provide wide reach, low cost, and high reliability. The U.S. hubs are in Dallas, San Francisco, and New York; the three regional international hubs are in Miho, Japan; Singapore; and Bedford, England. These hubs are linked by pairs of T1 circuits operating at 64 kbps via satellite and fiber. Each regional hub has twin data centers, and transmission links are routed "diversely" using different physical paths to ensure protection against disasters, such as a fiber cut. (For the same reason, each hub has two data centers.) The hubs use IDNX multiplexers provided by Network Equipment Technologies. Within a region, sites link to the hubs via fractional T1, running at up to 64 kbps. The network supports SNA traffic, voice, and inter-LAN traffic. PBXs at T1 locations connect to a T1 multiplexer that routes voice calls through the backbone and also handles offnet calls—calls using long distance and local carriers' switched services.

Like most large decentralized firms, TI did not recognize the implications of rapid and unmanaged growth of local area networks in time to develop a strategy for ensuring integration and coordination. User departments had full freedom to build their own LAN capabilities using any vendor, product, and architecture they chose. By 1990, this uncoordinated growth had resulted in 125 LANs that used a wide variety of completely incompatible protocols, including Digital Equipment's DECnet, TCP/IP, Ethernet, Novell's IPX (Internetwork Package Exchange), and Apple Computer's AppleTalk.

As user departments began to want to interconnect these LANs, mainly to exchange files and electronic mail messages, they looked to the central corporate group for help. Initially, the network simply used point-to-point links to connect the LANs, bypassing the backbone; there were over 150 such links, a very expensive and "chaotic" solution that was replaced by a coordinated approach that succeeded mainly because TI's new "Network Interoperability Forum" (NIF) emphasized support to end-users and not central bureaucracy and control. This approach provided plenty of incentives for departments to give up much of the responsibility for day-to-day operations of their LANs. The manager for TI's communications strategy understood that "the majority of users don't want to be in the business of developing networks and will walk away from it once they know there is a corporate group that will provide quality service and cares about their needs."

Such caring and quality service rests on close interaction and coordination. The Network Interoperability Forum meets once a month and includes network support staff and end-users from every business unit. Through the NIF, TI set up a companywide requirement that all LAN operating systems must have a hardware interface to IEEE 802.3 Ethernet LANs and must be able to transmit data to a remote LAN using either TCP/IP or OSI protocols instead of its own native protocol. TI gradually adopted Cisco Systems Inc.'s line of routers, as well as concentrators from Ungermann-Bass, which became de facto standards, as did Ethernet, which is now used by 95 percent of the company's units.

TI divided the existing backbone network into two data channels; one channel is for IBM SNA traffic that is mainly to and from the Dallas mainframe computers, and the other is for LAN-to-LAN traffic using TCP/IP. The local area networks feed traffic onto the backbone through either a bridge (between local LAN segments) or a router (across LAN domains). The bridges and routers connect to the backbone through an attachment to the IDNX multiplexer.

Network management was an early priority for TI. It developed its own proprietary method of testing all its circuits on a continuous basis, logging performance data in its SAVAIL (Systems Availability) database. Specially developed software tests each SNA circuit every ten minutes, 24 hours a day. One test measures the time for a 200-kilobyte database to move end-to-end through the network and back from the Dallas host; a second test transmits the same file to terminal controllers on the network. These two figures are monitored in real-time at the Dallas network control center. Whenever performance falls outside acceptable parameters, the system generates a trouble ticket that alerts the network staff, who immediately get to work to fix the problem. Trouble tickets and their resolution are reviewed at a daily telephone conference between Dallas and Bedford.

TI meets monthly with all its carriers, which include AT&T, MCI, WilTel, local Regional Bell Operating Companies, and international value-added services. Together they review every trouble ticket for the past month to track any trends that indicate potential problem areas. The carriers trust the accuracy of the figures, and there are no arguments or self-justification. (TI's figures are more accurate and detailed than the carriers' own network statistics in most instances.) TI informs each carrier of its relative performance. Currently, Japan's NTT (Nippon Telegraph and Telephone Corporation) is the benchmark leader, with just 12 minutes of downtime for the entire year out of over three million minutes of transmission; this is over 99.999 percent availability.

Given the many and much-publicized problems that U.S. carriers have had in network failures in recent years, TI has had to take action to ensure improvement. In 1991 it set up a special quality program with AT&T as a result of a fiber cut that left it without communications to Europe for over seven hours, even though AT&T had supposedly provided physically diverse Dallas-to-London T1 routes. TI now regularly reviews circuit routes with AT&T; in turn, AT&T brings in TI's T1 experts at its own planning meetings for engineering new circuit paths. Senior AT&T network management is automatically informed if any T1 problem has not been resolved in two hours.

With its equipment suppliers, TI carries out exhaustive tests designed to find the point at which equipment fails. It looks to establish long-term relationships with suppliers and keeps them well-informed about its plans. Instead of keeping information secret for use as a weapon in negotiations, TI views sharing of information as vital. TI has a long history of cooperative alliances and joint ventures across its operations, including both its 1992 development and marketing agreement with MCI to co-market TI's EDI capabilities, which had won it the Yankee Group's EDI User of the Year award, and its many cooperative ventures with other firms in advanced computer chip development, a major area of strength for TI.

The earliest major success in TI's use of telecommunications was electronic mail. It developed its own in-house system, still in use after close to 20 years, with around a million messages a day flowing across the network. The system is host-based, with X.400 gateways used to transmit e-mail messages to locations outside the company. End-to-end delivery time is under three seconds.

Today, TI is widely seen as one of the world leaders in electronic data interchange. It began early and sustained its innovation, and it consistently wins awards for being in the top five to ten in EDI use. It links close to 2,000 trading partners in 20 countries, making over 11,000 transactions a month. It subscribes to 16 value-added networks and uses almost every established EDI standard, with EDIFACT and X12 the main ones and X.400 a key transport standard for its EDI Translator and Gateway software. It handles over 50 types of documents and can directly route them into its core business applications, including purchasing, accounting, and shipping. Close to half its purchase orders from customers are processed without a human intermediary. It has a strong consulting, training, and support group for EDI and offers EDI systems integration services.

The foundations underlying the 20-year evolution of TI's network platform have been consistent and explicit: a global window, cross-process integration and coordination, and disciplined technical innovation. The business driver has been the vital need for global processes for a global organization. According to a senior TI information systems manager, "The network is absolutely vital for TI to operate on a global basis." Such a comment has become a truism in many firms; TI, however, defined it as the priority that shaped its entire information technology plans and organization in the late 1960s. It relied on strong central planning, reflected in the design of the backbone and the use of mainframes not so much as "computers" but as libraries and shared information storage, access, and coordination points. It did not rely on the typical piece-by-piece, case-by-case, bottom-up approach to international telecommunications. The network had to serve the organizational priority of managing TI as a single entity and a "global factory." John White, the head of TI's Information Systems group, commented in 1989 that TI sees the next generation of computerization as "a global networking of machines and applications in nonlinear fashion, driven by the information-browsing users rather than being merely process-driven." White, who has headed TI's IS group since 1976, built his strategy around strong central coordination plus strong local support. The focus on service quality is at the heart of this strategy, with metrics as the key feedback mechanism and measure of progress.

Unlike almost every other company, TI thus began with a viewpoint that all telecommunications and services should be integrated. Its early development of electronic mail was intended to help project teams of designers, engineers, and manufacturing staff work closely together across three continents. The backbone network allowed TI to consolidate its financial data almost in real-time, again across three continents. Its equally early commitment to EDI was part of a business drive for process integration, not an administrative move to speed up the flow of documents. EDI is seen as part of the intelligent global factory of the future, for it links every core business process.

TI is a leading manufacturer in many areas of IT. It is at the forefront of research, development, and production of high-speed chips used in voice recognition applications, high-performance intelligent workstations, high-bandwidth multimedia applications, and programmable processors. Thus it is a natural innovator in its own use of technology. White recommends that information systems and telecommunications organizations build close links with leading suppliers and track their innovations continuously. Information officers, in his view, need a technical awareness as well as business alertness, so that they know what is possible and where the leaders are moving. White's group takes pride in their technical expertise and is encouraged to push the practical state of the art.

One of the obvious risks in being a leader is that TI has often had to take action before standards are at all clear. This was the case in electronic mail, EDI, the almost out-of-control explosion of LANs, and TI's original development of the backbone network, and TI could easily have become spread too thin. In mid-1992, the Information Technology Group, the business unit that sells software and equipment to outside customers and that White also heads, sold off its division that provided commercial UNIX-based multi-user systems, which had performed poorly in the marketplace.

However, TI has gained many advantages of lead time over its competitors, has been able to exploit emerging technologies and applications quickly by not waiting until standards are there for all to use, and has been able to optimize its systems to provide the highest performance, often at the lowest cost. TI has the skills and experience to exploit new technology, and its general approach is to move fast and early, work closely and openly with vendors, reduce risk through exhaustive testing and monitoring, and adapt existing systems as standards and equipment for interoperability emerge.

| | | | | | | ## The Royal Bank of Canada: Internetworking

The Royal Bank of Canada (RBU) is one of the world's leading banks in the traditional use of computers and telecommunications—that is, in the reliance on very large mainframes for transaction processing, on huge databases for customer service and marketing, and on a very large backbone network capability for coordinating operations across a 3,000-by-2,000-mile area. In the 1980s, RBC was among the top five banks worldwide in almost every aspect of large-scale retail electronic banking.

RBC was a relatively recent newcomer in the use of personal computers. Initially, its highly centralized corporate information systems group placed strong controls on the use of PCs while it developed policies for acquisition and operation. IS was seen as bureaucratic and unresponsive and far more interested in "building Cadillacs" than in meeting business needs. The criticisms were largely justified, though the IS group pointed to the vital importance of its telecommunications architecture and to integrated data resources as one of the keys to RBC's success in Canada. The reliability of its operations, its transaction processing capabilities, and security were expensive to develop and took years of effort, but the result was years of competitive payoff, as RBC's return on assets grew faster than its competitors' from the mid-1980s on and as it earned wide respect for the quality of its IT activities. Many corporate customers reported that the Royal Bank's network delivery system was a key factor in their deciding to increase the share of their business they gave to the bank.

By the early 1990s, personal computers were widely dispersed throughout the company, and corporate IS played a stronger supportive role in helping business units get value from them. PCs were viewed as part of the corporate platform, and heavy investments were made in branch automation and the teller's multifunction workstation. The Royal Bank now had two very powerful IT bases; the new problem was that each type of base—PCs and LANs, versus mainframes and large-scale transaction processing and databases—was entirely different. Spreadsheets, word processing, data analysis, and electronic mail were best-suited to PCs; transactions and data access were best fitted to mainframes; and bank personnel needed to run both types of applications.

IS's strategy for providing for this was to create a new internetworking capability that would ensure automatic access to any application and data. Starting in 1990, RBC began to migrate its 1,600 branches away from network controllers that connected to the backbone SNA network, one of the largest in the world, and onto LANs. The strategy demanded careful planning and phasing over two to three years. The internetwork had to support the cores NA transaction processing base, which demands extremely high levels of reliability, speed, and security, with automated network management tools central to IS's efficiency and effectiveness. It was essential that the LANs were connected in a uniform network that permitted continuation of centralized network management through IBM's Netview. LAN network management software does not provide for end-to-end management across the LAN-WAN-LAN environment that constitutes the "network."

Additional requirements were that the internetwork accommodate all the existing LAN operating systems environments and protocols in use, including Novell NetWare and IBM LAN Server. Most LANs in the bank use token ring, but Ethernet connection is required to interconnect the Tandem host computers that work as front-ends to the IBM mainframes to ensure nonstop on-line processing. All of the existing LAN environment is coax cable-based, but the internetwork must accommodate fiber optics and be flexible enough to allow graceful migration to FDDI and emerging standards and protocols.

These are extremely demanding requirements that involved imaginative planning and careful implementation. The Royal Bank's team decided to create a series of test environments before bringing the production systems on-line. It could not risk outages of even a few hours. In 1991, a tiny error in a new funds-transfer system caused major errors in processing corporate customers' payments; the outcome was that the chairman and president had to cancel all planned activities for two days and spend the time phoning customers to apologize. This error occurred despite IS's exhaustive testing and systems audit procedures. The internetwork must not put the bank at similar risk again.

A natural test site for the internetwork was the Systems and Technology Division's own office building, which has 17 floors, all of which have at least one token-ring LAN operating at 4 mbps. These LANs are connected to a backbone fiber-optic link that runs the height of the building and operates at 16 mbps. The link is built on four wiring closets on three different floors; each is equipped with a multistation access unit (MAU) that has a ring in and ring out port. The ring in fiber cable on one MAU is attached to the ring out on another; this builds the 17-floor link. Devices connected to the MAUs to connect it to the bank's backbone network include IBM 3745 front end processors, SNADS[1] gateways for access to the bank's host-based electronic mail services, cluster controllers, asynchronous gateways for dial-up and dial-in communications, and 3270 terminal gateways. Figure 9-2 shows RBC's internetwork.

Each floor's LANs are connected to the building backbone by intelligent routers that can concentrate up to three LANs, reducing connection costs and also keeping traffic that flows between the three LANs off the backbone and thus cutting transmission loads. IBM's token ring bridging device, a simple alternative, would connect only one LAN to the backbone, but users on each floor run two different LAN operating systems: IBM's OS/2 LAN Server and Novell's NetWare. The IBM server, which handles the powerful PS/2 personal computers that are the base for branch and administrative workstations, makes host connections through a 3745 front-end processor or 3174 cluster controller. The NetWare users must make a connection through a 3270 gateway. The two different modes of connection allow the very different LANs to access the same services. For example, electronic-mail users may access LAN-based and host-based electronic mail through Microsoft's PC e-mail, which handles addressing and routing. The bank has committed to purchasing 15,000 licenses for PC e-mail.

The fiber-optic campus backbone uses monomode cable and it can be upgraded to 100 mbps FDDI without being rewired; only closet equipment need be replaced. Currently, the backbone runs at just 16 mbps, as mentioned previously. Each routing device includes a token ring LAN adapter for connection to the backbone plus a T1 adapter for connecting to a T3 multiplexer, which in turn connects to a planned fiber campus backbone that will link four buildings in Toronto, including the one previously discussed. Fiber cable has already been laid, and a private T1 network already in operation will become a campus 16-mbps token ring network operating at around eight times the T1 speed (16 mbps versus 1.544 mbps) at no extra transmission costs. The multiplexer will

---

[1] SNADS stands for SNA distribution services, one of the elements of SNA that has extended its capabilities from a hierarchical star topology to distributed and client/server configurations.

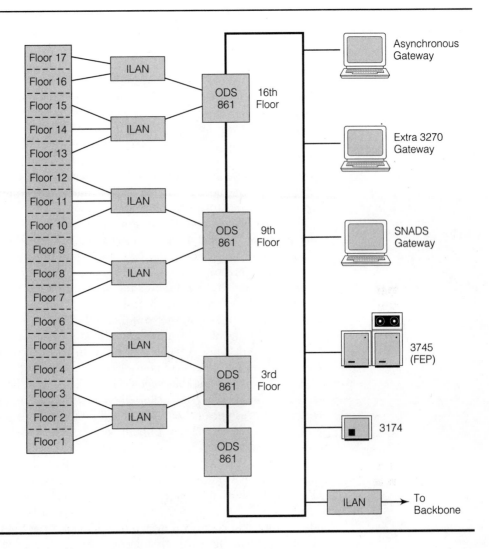

Figure 9–2

The Royal Bank of Canada's Internetwork

then be removed and each router equipped with 16-mbps token ring interface cards. The campus network will have a gateway to the Royal Bank's massive processing systems, which form a complex called OPC (Ontario Processing Center). Internetwork management will be centered here, with all diagnostic equipment monitoring the entire network, right down to individual workstations. IBM's LAN Manager[2] station gathers diagnostic information from all token ring devices. The design of token ring LANs keeps such information within the LAN; it is quite literally a local network. The intelligent routers, a product called ILAN that is supplied by Cross Com, can forward this information from ring to ring, moving it to the OPC. The ILAN devices can be managed as part of this overall token ring environment, so that the network management unit can immediately locate a broken bridge, faulty adapter, or out-of-service access unit. The OPC center has a complete view of the network, not just parts of it.

The new campus ring network will also link the bank's treasury system with the nonstop Tandem host computers, replacing a private T1 link. Both these facilities use Ethernet LANs. The ILAN routers are equipped with Ethernet adapters as well as token ring adapters.

The cost savings from the internetwork will be immense. Today, remote users are linked to mainframes through slow 4.8-kbps lines provided by Bell Canada. The new service removes literally hundreds of telephone lines and increases transmission speed by a factor of hundreds (or even a thousand). The use of MAUs and ILANs removes the need for cluster control units. Network management is brought entirely in-house because the phone company is no longer involved in any internetwork link.

Technical professionals often talk with admiration of an "elegant" solution. The RBC internetwork is highly elegant, with a simple design, a clear phasing plan, and clever exploitation of new telecommunications hardware (such as intelligent routers) without adding technical risk. The Royal Bank has moved incrementally from a totally IBM mainframe and wide area network environment to one that balances personal computers and local area networks with mainframes, in the process getting the best of both options. The overall architecture remains IBM-centered; SNA is the network architecture; IBM's flagship MVS operating system and flagship DB2 database management system are the core of the bank's transaction processing environment; the main LANs are IBM token ring; and OS/2 is the key personal computer operating system. One criticism of the central Systems and Technology group was that they were locked into IBM mainframes and mainframe thinking. However, the backbone carries SNA traffic over an X.25 network by exploiting Northern Telecom intelligent switches, the OS/2 personal computers are NCR 486 hardware, token ring coexists with Ethernet, and Tandem hosts are the front end to IBM mainframes.

The Royal Bank's information services managers have constantly and loudly emphasized the importance of architecture and an integrated delivery platform. Perhaps at times the corporate unit has been slow to encourage business units' autonomy, but there can be no doubt that the architecture is the source of the elegance. That elegance has given the Royal Bank of Canada a continuing business edge.

---

[2] IBM's main network management system, Netview, manages the enterprise network; LAN Manager manages a single LAN.

| | || | |

British Airways North America: High-speed Access to the Public Network

Our third example of networks in action illustrates how an effective design can both minimize the risk of failure and get the best value for the money in choosing suppliers of transmission services. In 1987, British Airways (BA) moved its U.S. headquarters from Manhattan to Queens, New York. Its telecommunications facilities had become badly overloaded, with 15-minute waits in many instances to process a reservation. The network was inefficient, expensive, and woefully unreliable; much of the problem came from the facts that BA's main processing systems and databases were in London and that it had to rely on a single local telecommunications provider's central office through a single cable access point in New York. Any failure was a disaster. Although BA could provide redundant routes for the long-distance (interexchange) carrier's links, it still had to depend on one route for what is usually termed the "last mile," the local exchange carrier's (LEC) link. The LECs offered cost-effective bulk access rates for large users, in multiples of T1 circuits. Slower-rate circuits (9.6 kbps, 56 kbps, or 384 kbps) could be multiplexed onto the T1s, but if the link failed they could not be protected. In addition, there were no facilities for routing traffic to multiple destinations from the LEC's POP (point of presence).

British Airways designed and implemented a network that uses T1 multiplexers and what is called automatic protection logic switching (APLS) between redundant T1 lines with diverse physical routing. This approach drastically cuts reliance on LECs and risks of failure in an environment in which the telecommunications network is as much the business as are airports and airplanes. (BA is the most profitable airline in the world and a formidable competitor; it makes well over half its profits on its transatlantic routes, making its New York hub a critical element in operations.) New York City is at the center of a backbone network that links it to London, where the mainframe computers are located, to Kennedy Airport in New York, where it owns and operates its own airport terminal, and to Miami, Chicago, and Toronto. When any link is down, so too is the airline. Reservations, boarding, catering, maintenance, spare parts, aircraft weight and balance, baggage handling, and crew scheduling depend on the network and computer systems being available in real-time. BA aims at a 98.5 percent availability of information and communication services.

Redundancy and diverse routes required an alternate exchange carrier. BA chose LOCATE, a firm that offered a custom microwave link from Queens to an AT&T POP, bypassing the terrestrial LEC cables. This service uses two radios operating at a frequency of 18 gigahertz. The sending and receiving dishes are two meters in diameter. Two multiplexers provide four T1 circuits for each radio. Because there is no direct line of sight between BA's Queens headquarters and the AT&T POP, a repeater is required. T1 cables leased from New York Telephone provide redundancy.

When one of the T1s fails, voice calls can be switched fairly quickly, with at most a few minutes loss of service. For data communications, loss of seconds is unacceptable. T1 links need to be automatically switched as soon as the bit pattern indicates a problem. BA's North American unit uses three types of digital interchange circuits: fractional T1 for the Transatlantic backbone; 56-kbps DDS (digital data services) on the North American backbone; and subrate DDS of 2.4, 4.8, and 9.6 kbps for tail circuits. All of these circuits terminate in BA's New York headquarters.

An automatic protection switch electronically (instead of manually) shifts this diverse traffic onto the alternative routes. AT&T's Verilink 551V equipment converts the T1 message frame—the bit stream at the beginning of the transmitted unit—to a special format that is inspected by the receiving 551V switch to detect errors. If the bit error rate exceeds a user-specified threshold or if the multiplexer signals that there is no data to send, transmission is switched to the alternative routing.

British Airways NA decided to use this technique to protect its most crucial circuits: a single 384-kbps transatlantic link, a domestic 56-kbps link, and 20 subrate tail circuits. BA's time division multiplexers had to be specially adapted to convert message frames to the format that AT&T's equipment could recognize and interpret. BA had to work at the most detailed level of hardware to accomplish this. For example, T1 coding substitutes a patter of 1s for every eight 0s sent consecutively; this is because T1 design requires a density of 12.5 percent for 1 bits to ensure that transmission is efficient and synchronized. T1 transmission links operate at over 1.5 mbps; a bit arrives at a piece of hardware at microsecond speed, but software runs at millisecond units. The "1s density requirement" is a trick to bypass software processing that requires "stuffing" channels with 1s, which can cause problems of synchronization or can eat up expensive bandwidth. Implementers must consider extremely complex details. An article describing BA's system includes as one of its more readable passages the following: "In order to use FT1 channelization with AM1 line coding, AT&T has gotten around the problem of excessive 0's by specifying the use of alternate (instead of contiguous) DSOs in its technical publication 54019A Addendum 1. . . . Although this method uses twice as much bandwidth as a B8ZS line for a given amount of throughput, it only applies to the portion of the T1 frame allocated to FT1 channels."

This situation clearly has a well-defined and simply stated business need— "Make sure that if the local carrier's link goes down, the message gets rerouted"—that demands very specialized implementer skills. The client, of course, never sees the details, but it knows that the financial payback from savings on equipment takes just 18 months. Operating costs will also be cut significantly because failures in the local access link to AT&T's central site do not require restoration of connection, and BA can take advantage of rates for bulk purchase while retaining customer control.

| | | | | | | |

Digital Equipment: Productivity Driven by Networking

The fourth and final case example in this chapter concerns Digital Equipment Corporation (DEC), the second largest U.S. computer company and one with a long-established reputation for innovation in hardware design. Its 1992 introduction of its Alpha series machine was a step up in performance of mid-range computers. In the 1980s, DEC posed a consistent challenge to IBM through its workhorse VAX series computers and the easy connectivity it offered through its DECnet telecommunications architecture and use of Ethernet. Its history of software development, implementation of standards, and quality of marketing has not matched that of its engineering, and DEC has struggled in the 1990s to maintain its position. Many of its problems have been those of the computer industry in general, which has been marked by overcapacity, rapidly falling margins, and reduced purchasing by customers.

For close to 20 years, DEC has viewed telecommunications as a key element in organizational coordination, and it explicitly regards its network as one of the most important determiners of productivity. Its chairman and founder, Ken Olsen, has often said that the ability to bring together a worldwide team and coordinate its activity through telecommunications will be one of the main differentiators in time-based global competition. In an article (Brown 1991) on DEC's adaptation to the "white water of change," a senior DEC manager wrote: "Digital Equipment Corporation faces some of the most difficult challenges the company has faced in its history. People are dealing daily with complexities never experienced before. . . . The network at Digital is the unifier and enabler that enhances these management models and business results." This perspective helps explain why the company operates the world's largest peer-to-peer private data communications network, which links over 85,000 computer hosts. Its seven-digit dial-up private voice network links every DEC location worldwide, providing easy communication among over 100,000 employees.

The peer-to-peer design of the network reflects its focus on linking individuals to each other, versus Texas Instruments's focus on linking business processes and the Royal Bank of Canada's on linking workstations to services. The basic design of DEC's platform is employee-centered: It is decentralized, peer-to-peer, and bottom-up, with multiple links and simple architectures. The DEC community requires five types of access: (1) from home, (2) when traveling, (3) between DEC buildings, (4) to outside non-DEC systems and databases, and (5) by customers and suppliers.

Meeting all these types of communication needs cost-efficiently requires a flexible network design. For instance, access from the home or when traveling requires dial-up service through low-cost modems; such access would be expensive and inefficient for global communication. International peer-to-peer links makes the ubiquitous X.25 standard a natural base; X.25 is far less suitable for computer-to-computer links. The wide variety of traffic, which ranges from e-mail to transactions, requires a mix of low- and high-speed lines; those in use today range from 9.6 kbps to 384 kbps.

Initially, the network was built around DEC's proprietary DECnet protocols, X.25, DECnet bridges and routers, and statistical multiplexers. The U.S. network, which is separate from the international one but connects directly to it, has 200 nodes and uses X.25. It is a meshed topology, with 9.6-kbps and 19.2-kbps circuits. These circuits are well-suited to message-based traffic such as electronic mail, but they cannot handle transfer of large-scale databases. The private network connects via gateway nodes to public X.25 networks, extending reach easily and inexpensively. The reliance on Ethernet and avoidance of multitechnology LAN chaos, the problem Texas Instruments had to address, have enabled a complex network to grow out of simple building blocks. Bridges that support 56-kbps to T1 rates can be used cheaply to interconnect LANs with low overhead on the network. Statistical multiplexers running at 9.6–56 kbps make efficient use of circuits.

This core design of the network, based on terminal-to-terminal communication for simple messages and transactions, needed extension to handle the new era of distributed computing. It was based originally on a very proprietary architecture; statistical multiplexers, for instance, do not conform to established major network standards. The DEC architecture is internally focused, which is why it is being extended between 1991 and 1996. The goals are to migrate terminal traffic to an OSI platform, implement DEC's open Enterprise Management Architecture (EMA), and reduce terminal traffic by 70–75 percent through continued migration to distributed applications. Much of this has already been implemented in the United States and parts of Europe. The core network architecture was flexible in terms of extending reach, but it was inflexible and inefficient in extending range and in allocating resources and optimizing performance. Because it is terminal-based, it does not exploit opportunities to distribute intelligence across applications and services. The new design focuses on a distributed user interface with access to centrally located transaction processing and data stores. Distribution may be implemented in all three components of access, processing, and storage.

The limited speeds of the core network neither exploit today's transmission capabilities nor support key protocols. Accordingly, the network is being upgraded with high-speed multiprotocol routers that replace bridges and is interconnected by T1 and fractional T1 links.

The design assumes a migration in the mid-1990s to T3 and SONET. The multiprotocol routers, which support such protocols as OSI, TCP/IP, DECnet, and X.25, allow a smooth migration to a fully open environment. The backbone is based on the OSI-compliant Digital Network Architecture (DNA) Phase V. DEC expects to implement the OSI virtual terminal protocol ("VTM" or virtual terminal management) defined by OSI in 1992, which will facilitate terminal-to-host communications. Network management in the highly distributed environment will be handled through CMIP (Common Management Information Protocol), which conforms to the OSI model for distributed management and collects a range of network information in standard metrics, including configuration, accounting, fault management, security, and performance. EMA also supports SNMP, the network management protocol used in TCP/IP networks.

EMA allows an enterprise network to be divided into several cooperative management domains, which both enables telecommunications organizations to be set up in various regions or countries as administrative domains and supports DEC's philosophy of decentralized decisions. The corporate network offers a guaranteed level of service at low cost. Domain managers may decide to extend that service—and pay for it themselves—within the overall EMA architecture.

The major area of vulnerability for DEC here is the difficulty of implementing all the relevant OSI standards. EMA is ambitious, and DEC has had to announce many delays in releasing OSI-based products. DEC's commitment to be the broadest range of OSI and related open standards spread its development capabilities very thin in the early 1990s. The old strengths of the core network were its simple peer-to-peer terminal-based architecture, which is obsolescent in today's world of distributed intelligence. DEC is taking three to five years to move toward an EMA-based (and hence OSI-compliant) open architecture, in realistic recognition that this is not a single, straightforward step but a phased progression. Multiprotocol routers and T1s are easy to implement, and such implementation is a natural first step. Distributing intelligence in applications requires new software; implementing DNA Phase V and (VTM) requires complete definitions of standards plus their effective implementation in software and hardware.

The payoff should be large, for DEC has already gained so much in terms of productivity from its commitment to telecommunications as a core organizational resource. For example, the routine for developing a new product is for DEC to form a team that cuts across multiple functions and is located across the world. External suppliers and customers are often members of the team, which is more than just an e-mail bulletin board; the team has total responsibility and accountability for design, development, manufacturing, marketing, selling, and delivery of the product. Other organizations have not been able to make DEC's early shift to team-based, collaborative structures because without a high-reach platform, doing so is almost totally impractical.

In sales and marketing, DEC offers customers free access to its Electronic Store, an on-line service that provides information on every DEC product and allows orders to be made directly into its order-processing system. The network has reduced the planning cycle for materials management—from the creation of the production plan to placing orders—from 120–150 days to under 30. It now takes less than an hour to consolidate DEC's worldwide financial data at the end of each month, compared with up to eight weeks before. The network has replaced 97 general ledgers, 30 accounts payable and 30 accounts receivable systems, and over 100 charts of account with a single financial information system.

DEC's corporate manager of communications states that the network has helped reduce any "not invented here" syndrome and has encouraged and reinforced a corporate culture of "free and open communication between people

	Growth of Network, Employees, and Revenue					
	1986	1987	1988	1989	1990	1991
Revenues	$7.6B	$9.4B	$11.5B	$12.7B	$12.9B	$13.9B
No. of employees	95,000	111,000	122,000	126,000	124,000	121,000
Computers on network						
DECnet	14,600	22,300	32,638	42,500	57,000	65,113
TCP/IP	—	900	2,500	4,200	8,500	15,245
Appletalk	—	—	—	—	80	1,351
T1 links	111	215	306	610	874	906

	Growth of E-mail, Videotext, and Computer Conferencing					
	1986	1987	1988	1989	1990	1991
E-mail accounts	61,000	78,000	90,997	105,315	111,839	110,079
Videotext infobases	22	103	144	233	224	296
Computer conferences	738	940	1,171	1,450	1,705	1,848

Figure 9-3

Growth in Digital Equipment's Business and Use of Telecommunications

and the ubiquitous access to data and information." He quotes the DEC adage that "the network is the system" and cites productivity gains directly attributable to the network, including an increase in inventory turnover from two to 4.6 times a year; savings of over $1 billion annually; ability to produce the annual report a full month earlier; and an average halving of cycle times in manufacturing and time to market. The growth in DEC's business and in its use of networking is shown in Figure 9-3. The percentage of employees with e-mail accounts was 64 percent in 1986 and 91 percent in 1991. Computer conferences per thousand employees were 7.8 and 15.3 respectively, in those two years, and videotext "infobases" per thousand employees were 0.23 and 2.45. Revenues per employee, a key measure of productivity, were $80,000 in 1986 and $115,000 in 1991.

| | | | | | | |

Some General Lessons

These four case examples discuss companies with very different business priorities, cultures, and processes. Each firm has tailored the specifics of its network architecture and operation to its individual needs. There are, though, several common patterns:

▶ A focus on a guiding architectural blueprint for the platform

▶ A view of telecommunications as a major business and organizational enabler, not just an add-on utility

▶ A growing emphasis on ensuring interoperability

▶ An awareness of the challenges of adapting existing services and technology to new needs and opportunities

▶ A reliance on routers, gateways, and related devices to handle multiple protocols

▶ A growing view of the network as LAN plus WAN, with "seamless" links between all components

▶ A realization that there are no quick fixes; "integration" and open networks demand time, effort, and constant work

Summary

This chapter presents four examples of the best of today's practices in the use of telecommunications. Texas Instruments has been a leader for 20 years through its focus on telecommunications as a resource for coordinating worldwide operations, by ensuring technical leadership, and by providing first-rate service and coordination. TI's backbone network is SNA-based and has been adapted to handle LAN traffic across international locations.

Texas Instruments emphasizes close links and long-term relationships with its suppliers. It monitors their performance very closely and shares information with them. It also aims at technical leadership, which often means that it must make major moves before standards have been defined. It thus developed its own proprietary electronic mail system, which is still in use after 20 years.

The core standards for TI's corporate architecture include SNA, Ethernet (used by 95 percent of its business units), TCP/IP for carrying LAN traffic over the international backbone, and X12 and EDIFACT for EDI. The backbone is T1-based, with nodes in Dallas, San Francisco, and New York and in Japan, Singapore, and the United Kingdom.

The Royal Bank of Canada case describes the bank's extension of its backbone processing network to handle LANs efficiently and to internetwork them. Its corporate architecture is based on IBM's SNA, token ring LANs, and OS/2 workstations, but it uses many non-IBM suppliers that support the IBM architecture. Its internetworking capability accommodates all major LAN protocols, including Ethernet. It is built on a fiber-optic campus backbone with intelligent routers. Although LANs currently run at 16 mbps using token ring, the configuration allows direct migration to FDDI without having to replace existing cable.

British Airways North America designed a new capability to ensure that its reservation and airline management systems, centered in London, were not vulnerable to problems in the local exchange carrier's networks and switches. Although it could provide redundant routes for its long-distance traffic, BA previously had to depend on a single provider's central office. BA designed and implemented a T1 network that includes multiplexers and network software for automatic rerouting of traffic when a link is down. It contracted with a firm that offers a custom microwave link to bypass the local exchange carrier.

Digital Equipment is the number two firm in the industry, after IBM. Telecommunications has been an explicit core element in its efforts to create a worldwide collaborative culture. DEC's network is peer-to-peer and is designed to make it easy for its employees to connect to it and to each other from their homes, and on the road. The network, which links over 85,000 host computers, is built around DEC's proprietary architectures that will be evolved in an OSI-compliant environment over a three- to five-year period via DEC's Enterprise Management Architecture blueprint. The major difficulty DEC faces is the high costs of implementing all major OSI standards and upgrading its own proprietary architectures.

The productivity gains DEC ascribes to its network investments are substantial. Its staff members often say that "the network is the system."

Some common lessons from these four very different organizations and different approaches to the use of telecommunications are a focus on an architectural blueprint and a view of telecommunications as a core business and organizational enabler that ensures interoperability while maintaining existing facilities and service.

Three

Linking Business and Technology

10

Choosing the Right Opportunity

Objectives

- Learn a systematic way of identifying competitive and organizational opportunities to use telecommunications
- Review practical examples from a wide range of industries

Chapter Overview

| | | | | | | # Chapter Overview

Part Two of *Networks in Action* focused on the technological building blocks of a network. Part Three shifts the focus back to the telecommunications decision sequence, and this chapter addresses the key starting point—choosing the right business opportunity.

Chapter 3, which introduced the telecommunications decision sequence, included a case illustration of a simple "need" for personal computer capability in a sales department. In that case, exploring a broad opportunity uncovered a variety of additional opportunities and the issues and trade-offs relevant to design and implementation. As Chapter 3 showed, a PC is not in itself an opportunity; it is just a tool. Without telecommunications it is a limited tool that lacks facilities for accessing data resources and for sharing information, messages, or reports.

At first sight, the case illustration in Chapter 3 is a simple exercise in technical assessment of personal computers and costs, but very quickly many additional questions jump out, each of which has business, economic, and/or technical implications. To take a situation at its initial face value is to lose an opportunity; a constant danger for a designer is to prejudge the "problem" and hence predetermine the "solution." Choosing the right opportunity and reserving judgment is the essential first step and the topic of this chapter, which introduces Part Three, Linking Business and Technology.

The telecommunications business opportunity checklist (see Table 3-2) reflects the common patterns underlying successful use of telecommunications in any industry. Its three opportunities—run the business better, gain an edge in existing markets, and create market and organizational innovation—are applicable across industries and can be realized through a wide range of telecommunications technologies. Any single technical building block can be applied across the entire range of business opportunities. It is important to realize that whenever a specific technology or application is under consideration, it is easy to overlook the opportunity to extend the technology beyond what is generally considered its "natural" use.

For instance, 900 telephone numbers, referred to in the telecommunications industry as "caller paid voice" (CPV), are most widely associated with pornographic rip-offs at $X per minute. But in 1992, Northeastern University in Boston began a pilot with MCI to let students handle many routine interactions with the university through a 900-number service bureau at under $1 a minute; services include registration, queries on courses, and the like. The university reduces administrative costs, the bureau exploits its low-cost telephone lines and skills in handling queries, and the student gains a new level of convenience and service at very little cost. The opportunity here is to improve ease of access to services and information.

In another instance, the long-distance carrier Sprint has exploited 800 numbers, which are mainly associated with businesses, to create a new service for consumers. Personal 800 numbers allow family members to call home without having to use a call box, make a collect call, or run up bills on their own phones. Sprint identified students and the elderly as the greatest beneficiaries of such a service. The business opportunity here is to differentiate a standard product. In both these instances, a service that is generally seen as specific to a segment of the business or consumer market can easily be extended to many other contexts.

One of the main barriers to effective exploitation of telecommunications is the false assumption that the technology defines the opportunity; instead, defining the business opportunity clarifies the requirements for the technology. Another barrier is associating a technology with a specific type of application, as in the examples of 800 and 900 numbers. Thus an ATM is not limited to financial transactions only; it is—or can be turned into—a general-purpose tool for accessing electronic services, including making electronic welfare benefit payments, renewing automobile registration, and registering for university classes.

The tool is not the opportunity. The message here is: First think opportunity, then technology. Accordingly, we turn next to the first business opportunity in the checklist—running the business better.

Running the Business Better

There are five main proven strategies for exploiting telecommunications to make significant improvements in an organization's efficiency, coordination, responsiveness, and quality of operations:

1. Manage distributed inventories
2. Link field staff to head office
3. Improve internal communications
4. Improve decision information
5. Reengineer processes and simplify organizational complexity

In the following sections we will consider each strategy in turn.

Managing Distributed Inventories

Consider the retailer that has a 100-day supply of Panasonic television sets in stock: It had better get ready to mark them down, for Panasonic markets over a thousand entirely different TV products in the United States, and the average life cycle is just 90 days. This retailer's inventory is certainly not a real business asset. The dramatically shortening cycles for just-in-time manufacturing, quick response in retailing, and next-day delivery in distribution reinforce the trend away from stockpiling inventory as an asset and the trend toward cutting it as a liability and a cost.

Whenever a business unit has high inventories, almost surely a telecommunications opportunity exists. Anything that is carried on a firm's balance sheet as a current asset but cannot be turned directly into money today is an inventory.

Accounts receivable are an inventory, as are work in progress, raw materials, and finished goods.

Consider two business opportunities that address managing a widely distributed inventory: office supplies and cash:

▸ A major office supply firm has added to selected clients' PCs an ordering capability for paper, toners for copiers, envelopes, and other standard office supplies. The firm recognized that secretaries make many managers' travel reservations and also order office materials. Both of these tasks are time-sensitive; the flight needs to be booked and confirmed today, as does the order for office supplies. The firm offers the same deals as do airlines—frequent purchaser bonuses, special offers, and corporate discounts—and believes it has a head start over its main rivals of at least two years.

▸ A large multinational distributor has justified the entire cost of its international network on the basis of the savings from electronic cash management alone; all the many other applications the network made possible come free. The business logic is to handle all cash management as if it were centralized while in fact keeping it decentralized. Each of the company's business units is responsible for its own financial operations, including billing, collections, deposits, lines of bank credit, and so on.

The new system links each of the units electronically. Three times a day the central computer "sweeps" the system. Spare cash in a business unit can be "offered" to the center for investment or for foreign exchange trading, and additional cash needs may be "requested." One unit may need to draw down on its line of credit at the local bank, where the interest rate is X%; another unit may have excess cash in a country where the interest rate on short-term deposits is lower; still another unit may need Japanese yen and have surplus German Deutschmarks.

The system first checks all offers and requests and then calculates the optimal set of transactions. It may meet the first unit's need for cash by transferring funds directly from its central cash management system so that the unit need not use its local line of credit. It may similarly move the second unit's excess cash to the central system for overnight investment, or it may decide that the local rates offer the best deal. It moves funds automatically to and from units where needed, and it can send electronic mail messages or faxes to its banks, instructing them to make foreign exchange transactions. The net improvement in cash balances and the reduction of interest and short-term debt were substantial; the debt alone was reduced by over 40 percent.

The technology base for managing distributed inventories depends on the specific application. The office supplies firm is using a public network for the obvious reason that it wants to get as much reach as possible; access to the service is through a standard personal computer and modem on a secretary's desk. The system uses the standard public phone system; customers dial an 800 number, which connects to a private high-speed leased T1 line. The firm already operates an on-line purchasing service for large buyers that uses 800 numbers and is being extended to include EDI linkages so that customers can tie directly into its computer systems. The bank with which the firm is piloting the EDI service buys over $120 million of stationery a year; the bank expects to reduce its inventory by at least 25 percent.

Linking Field Staff to Head Office

The most rapidly growing part of the personal computer industry in the 1990s has been laptop computers (or, as they are more commonly termed now, notebook computers). The movement of PCs from the desktop to the briefcase has taken about a decade, and laptops and notebooks now match the power of desktops. In early 1992, for instance, Dell offered by mail a four-pound notebook computer with a color screen that used the (then) top-of-the-line Intel 486 chip. By mid-1993, the price had dropped from $4,000 to well under $1,000. In 1991, Apple introduced its elegant Powerbook, replacing its cumbersome portable that was closer to a "luggable." Apple sold over $1 billion worth of Powerbooks in the first 18 months. A laptop with all the capabilities of a standard desktop can be bought for about 30 percent more than a desktop; such a purchase creates a new degree of convenience and portability for a wide range of software and storage options.

Progress in the portability of computers has been paralleled by progress in telephones, fax machines, scanning devices, and even printers; several companies now offer an electronic briefcase that contains a personal computer with a modem, a printer, and a cellular phone. Now sales reps can operate a personal computer and a cellular phone in their cars by plugging into the cigarette lighter. Telecommunications thus adds to the laptop's portability additional access to information and services.

This combination of technologies expands a business opportunity that few firms exploited in the 1980s. Computers had historically been considered a head office function, and when personal computers first appeared, the "obvious" place to put them was the head office. The software that moved the PC from being a toy for the hobbyist was the spreadsheet, and its obvious use was in finance, most particularly for budgeting and financial analysis; indeed, the very idea of a spreadsheet is derived from the accountant's rows and columns.

One principle of telecommunications as a business opportunity is to "put the technology at the firm's point of event." *Point of event* is a term we introduce in this book to refer to the exact location of the basic transactions that create a firm's reputation, customer image, and processing and administration. For instance, in the life insurance business, the point of event is generally the customer's home, where the agent does the selling and the customer provides information and signs the application form. For a sales rep in Frito-Lay, the point of event is the store where he or she delivers goods and takes orders; for an airline, it is the travel agent's reservation terminal, where the agent makes a traveler's booking; for a retailer, it is the point of sale; for the IRS, it is the mail box where taxpayers post their tax returns.

Putting the technology at the point of event is an almost universal rule of sound business. Almost invariably, a key point of event is in the field, not at the head office. If, for example, an insurance agent has a laptop, he or she can connect it to the customer's phone, dial up the head office computer to get rates and payment terms, use the personal computer's software to check that the form is complete and correct, and then send the application electronically so that processing can begin immediately; the physical application can follow at leisure because all it requires is the actual signature of the client. The alternative—to take or send the paper form to the branch office and wait for it to work its way through the mail room and data entry operations—adds days, errors, photocopies, and storage space. One insurance firm, Mutual Benefit, whose streamlining of insurance policy issuing is discussed later in this chapter, found that typically there is only 17 minutes of decision making in the 26 days it takes the average insurer to handle this process.

Hewlett-Packard estimates that its sales force gained almost 40 percent more face-to-face time with customers through the use of laptops in this way. Sales reps need not go to and from the office to get information and bring in sales orders; they can go directly to see clients at 8:30 A.M. instead of driving to the office. Frito-Lay similarly calculated that its truck drivers save half a day each week on paperwork by being able to print out invoices on the spot and to confirm available inventory and prices through data that is downloaded overnight to a hand-held computer. AT&T no longer provides offices for its sales reps in New Jersey; they operate out of their homes and go to the office only as needed, sliding a nameplate into a slot on the door of a spare room. The extreme example of linking the field to the head office is telecommuting, described in Minicase 1-3.

Not all field locations involve mobile sales reps, obviously. Many field locations are customer contact points or branch offices that are some distance away from the central units that process key transactions and manage—or sometimes guard as their own bureaucratic fiefdom—key information. The state of Virginia was able to reduce both the costs of its state income tax commission's budget by 19 percent and the time it takes to issue tax refunds, cutting it to a maximum of a week. Some 800 internal and 1,000 external users across that state, including other agencies, are linked to a complex central system that has 1,200 software programs and 50 databases with over 20 billion bytes of data. The point of event is generally either the office to which the tax forms are sent or a state employee receiving a query from a taxpayer. The state's estimated savings are $50 million a year. Here the technology base is a desktop, not a laptop, but the business logic is the same: Use telecommunications to extend the reach of a necessarily centralized processing system out to the many field offices.

The telecommunications base used to link the field to the head office or other locations obviously depends on the exact nature of the field work. In general, a notebook with a modem is the simplest set of tools; dial-up phone lines can then be used from just about anywhere. The limitations are security and inability to send more than simple, short messages.

Many field-head office links use mobile communications. Federal Express's trucks are connected by radio to the company's computer centers; Ryder has put small VSAT dishes on its trucks, and UPS has installed a nationwide digital cellular network as part of its effort to catch up with Federal Express. Companies such as Frito-Lay use data communications lines to handle downloading and uploading of transactions. When its truck drivers return to the depot at the end of the day, they plug their hand-held computers into a special interface, and the data is sent over a high-speed, high-quality digital line.

Improving Internal Communications

Telecommunications and a business's internal communications obviously should go together. Thus whenever a company's top management or human resource group talks about "improving communication," the telecommunications designer has an opportunity to make a significant contribution to the organization. It is very difficult to communicate in many organizations. Top management is far away in the head office, and fairly junior staff in the corporate finance department know more about what is happening than senior managers out in the hinterland, where the rumor mill prevails and the head office is considered to be "them," not "part of us." The paper and administration of paper that dominate most organizations make it difficult for people to find out what's going on.

In this context, a designer can bring to the client's attention a wide range of proven tools for making it easier to communicate. The five main tools are electronic mail, fax, voice mail, groupware, and videoconferencing.

Electronic Mail Electronic mail has been in use in many organizations for over a decade, but as of 1993 less than a quarter of large companies have made it an integral part of their management processes. Those that have done so consider it indispensable. There are an estimated 10 million users of e-mail in the United States; the growth rate has been steady but not as rapid as predicted. In 1980 there were under half a million users, and in 1984 a million. Once e-mail reaches critical mass in a company, it generally grows explosively because it has become a key communication tool instead of something useful to only a few.

Consider the following typical uses of e-mail:

▶   Hughes Aircraft uses it to coordinate its many aerospace projects, including linking to other companies involved in a project. Like many large manufacturing firms, it operates projects across the globe, and timeliness is an increasingly crucial competitive issue. Engineers are very relaxed about using what may seem an impersonal medium; it is convenient and allows them to send a message at 6 P.M. on the West Coast to Europe, where it is almost dawn, confident that the message will be picked up and replied to. Project managers can use bulletin boards and mail lists to improve ease of coordination, and the system allows spreadsheets and reports to be attached to messages.

▶   Deere and Company's sales branches have ended what they call "the nightmare of ringing telephones and call backs" through e-mail. Proposals and price data are requested and supplied as needed, thereby reducing turnaround for issuing sales-related documents and bids from a week to a day.

▶   In 1986 the World Bank began a "workstations for all" program to address the need to build critical mass for its e-mail service. The system initially had only a few users, but over the next six years its easy availability made it an indispensable part of the bank's operations, to the extent that the Jakarta office is now called the "Washington, D.C. night shift"; there is now an around-the-clock work cycle between the two locations. The bank's work is very document-intensive, but electronic mail reduces many delays for economic analysis, project planning, and program implementation.

Fax   The second widely used tool for simplifying communication is facsimile. The current availability and low cost of fax machines make it easy to forget how recent has been the explosion of fax and how old the technology is. Until the late 1980s, fax was confined to large organizations and was very slow, and telex was the most common mode of interoffice and intercompany communication. The main driving force for fax was the development of low-cost machines by the Japanese. By focusing on one of the four types of international fax standards and exploiting microcomputer technology for scanning, storage, and transmission, the new machines have transformed the market. Personal computers can be turned into fax machines by adding a single board, and modems routinely offer fax receiving and sending. Now fax machines are everywhere—in homes, shopping malls, airports, and hotels.

Although fax is more limited than electronic mail in terms of storage, quality of transmission, and options for manipulating the information it contains, it dominates e-mail in everyday life. In the Far East, where distances are immense, 60 percent of telecommunications is by fax. The combination of analog phone and analog fax offers small trading firms a convenient, relatively cheap mode of communication across huge distances. The geography of the United States and Europe offers advantages in the use of private networks and digital data communications, but for many years to come analog fax will meet the needs of many companies, especially those that need international communications only intermittently.

Voice Mail   Voice mail is the digital equivalent of answering machines and is very much like electronic mail. Because people are very used to both phones and answering machines, many find voice mail far more "natural" than e-mail. In basic systems, messages are stored and forwarded, just like e-mail; advanced systems include features for notification of receipt, redirection, message-sending priority, time stamps, and interworking of different vendors' voice mail systems.

The main users of voice mail are people who call into the office when they are traveling; consultants and sales staff are obvious examples. They need not carry a computer with them and can quickly scan their calls from any pay phone. It is very common now for firms to add voice mail to their internal phone system, which may be why the main vendors are PBX manufacturers and local Bell companies. The market for voice mail is still fairly small but is growing fast. In 1989 it totaled just under $200 million and reached $700 million in 1992.

Figure 10–1

Using Telecommunications to Help Business Teams Overcome the Obstacles of Distance and Time

		Team members' location:	
		Same	Different
Time	Same	Same time, same location Tools: face-to-face meetings; blackboards	Same time, different location Tools: videoconferencing; audioconferencing
	Different	Different time, same location Tools: bulletin boards; memos	Different time, different location Tools: groupware; e-mail; electronic bulletin boards; Lotus Notes

Source: Based on Johansen, et al., 1993.

Groupware is a catchall term for software designed to help business teams work together across locations. Given the growing emphasis of teamwork within organizations, groupware is an obvious extension of electronic mail, fax, and voice mail. Figure 10-1 provides one useful way of placing groupware in the wider context of teams collaboration and coordination.

Voice mail and e-mail apply whenever teams work in different places at different times. Videoconferencing (discussed below) falls in the same time/different place square. Telecommunications, by definition, does not facilitate work done in the same time and place or at the same place at different times.

Thus different time/different place becomes the primary target for using telecommunications to improve teamwork. Software such as Lotus Notes or Microsoft's Workgroup for Windows aims at helping groups handle joint writing and revisions, hold conferences on-line, structure conversations, organize and file messages, manage forms, and share data files.[1] The systems run on a personal computer, with a master node acting as a server for the group; a directory defines the group members.

Groupware is a new style of software that is built around the designers' conception of how communications should be organized and decisions made. It provides many features that attempt to improve teamwork. By contrast, electronic mail is passive; it is the telecommunications equivalent of memos and letters. Groupware tends to be more proactive by including facilities specifically

---

[1] An emerging debate in many organizations that have used groupware products concerns whether such products increase coordination and communication or create chaos and lack of control by making it too easy to set up work groups and to bypass official channels, by making it difficult for central units to keep track of decisions and events, and so on. For the telecommunications and information services manager, an additional concern is loss of control over network management and operations. Workgroup for Windows is explicitly intended to allow anyone to communicate with anyone, with no intervening IS "bureaucracy." Groupware looks like a repeat of the "power to the people" PC and LAN movement, which will likely be followed by "all power to everyone, but can we please coordinate all this multitechnology, multitraffic, multirisk muddle."

designed to enhance teamwork, such as anonymous voting on issues, reminders, calendar management, and the like.

Groupware can help make collaboration a competitive advantage. The pharmaceutical industry provides a typical example. The industry depends on R & D; it spends a large fraction of its resources on research and must develop new drugs to ensure that its products and revenues are not eroded by generics, expiration of patents, and the ferocious competition that marks the industry. Computers have become an increasingly important element in R & D, as have databases that contain scientific data on drug trials, reported side-effects, patents, and so on.

Mirrason (a pseudonym for a pharmaceutical firm) is using telecommunications to redefine the nature of teamwork in R & D. Historically it has concentrated research at its head office, but as it has acquired other companies in a number of countries, more and more research groups are working in isolation from each other. Each subsidiary has its own distinctive strengths in terms of specialization of training and focus of research.

Mirrason has created a network to link its worldwide research groups. Initially the aim was to access the many databases Mirrason routinely creates and updates. More recently, though, Mirrason is redefining the research team. Scientists in Italy now "work" in Canada by using a combination of very powerful workstations, high-speed telecommunications links, and database management systems. Mirrason's research teams in Japan now link directly to their colleagues in other countries.

Mirrason's view of R & D is that it requires the very best scientists "armed to the teeth with any weapons that speed up their progress and enhance their creativity." It sees research as a network of experts who need to talk to each other across the world. Its scientists report far more communication and far better use of specialized skills and knowledge.

What Mirrason is providing is the vehicle for a new form of team that coordinates its activities regardless of location or time of day. As its head of R & D commented, "I suspect we are reinventing research. I have no data to support this hunch but I see a new breadth in our scientists' thinking. If I am right, we will see a new source of competitive edge—communicating as the base for researching."

Videoconferencing   Videoconferencing and business television (the first involves two-way and the second one-way transmission) are technologies whose times always seem about to come. For two decades now, many companies have built conferencing rooms to bring together for meetings people at separate locations. Although a typical room costs between $50,000 and $250,000 to equip, the cost of transmission, which is usually via satellite, has dropped rapidly. A one-hour coast-to-coast videoconference that cost $750 in 1990 cost under $250 in 1991. PictureTel's LAN-based system costs around $40,000. Hitachi has a $13,000 device, and codecs (data compression coders/decoders) are under $10,000. Data compression is the key to making videoconferencing cost-efficient and practical over standard phone lines and local area networks. Video-conferencing has even moved to the desktop.

The main barrier to wide use of video is simply unfamiliarity. People tend to be very cautious about risking loss of personal contact; they worry about how they will look on camera, and they are afraid the meeting will be awkward. They also think of videoconferencing as a substitute for travel, and that is indeed how it has traditionally been justified. In practice, most of these fears are completely unfounded, and within 15 minutes people are usually fully at ease. That said, such valid concerns are a reminder that communication must be natural, re-laxed, and open for it truly to be communication. High-tech substitutes for face-to-face meeting, complete with body language and body space, have often been overtouted.

Thus, videoconferencing per se will not lead directly to better communica-tion. That said, there are many proven examples of its effective use, including the following:

▶   Project meetings are routine in aerospace and construction firms, in which specialists such as engineers need to keep in touch, review progress, evaluate design changes, and so on almost continuously for projects that can take ten years and involve thousands of people.

▶   The University of Maine's interactive television education program has over 4,000 students and has brought 85 percent of the population of this largely rural state together in an electronic classroom.

▶   The regular and routine use of **business television** helps such CEOs as Fred Smith (Federal Express), Jo Antonini (Kmart), and Ken Olsen (Digital Equipment) to personalize their leadership to the company, "meet" employees, make sure important news is disseminated internally and not through the press, and so on. The CEO of Domino's Pizza believes that video is the way to keep people fully informed and well-trained as well as to make sure the head office is small, flexible, and responsive. Antonini of Kmart credits it as a major element in his $2 billion program to renovate this retailers' dingy stores and rejuvenate the company. He meets regularly with employees across Kmart's 2,200 stores, answering their questions and seeking their opinions. "The satellite is fantastic. I would never believe it. You walk into a store and it's like they've known you for 15 years."

A final example of the use of videoconferencing shows how it can become a part of the firm's overall business processes. The launch time for a new car model used to be around 84 months. Ford cut this to 60 months, but its Japanese competitors have kept ahead of U.S. firms; one of them is rumored to have cut its launch time to 18 months. For Ford, every single week is on the "critical path"; a week's delay adds a week to the launch time. Ford of Europe now uses videoconferencing to bring together ad hoc teams for trouble-shooting in the last stages of a launch prior to full production, when a car is tested literally to destruction. Any breakdown that occurs is a "job stopper"; work stops while engineers, designers, and production experts are called together to solve the problem. The typical job stopper for Ford of Europe lasted four days; installing a videoconferencing link between Ford's UK and German locations reduced job stoppers to an average of half a day. Instead of flying to the German test facility, the relevant people drove or walked to the videoconferencing room.

The two main technical issues involved in using videoconferencing are high-speed transmission—usually through satellite downlinks, data compression, and coder/decoders (codecs)—and, as is so often the case, the lack of established standards. All the manufacturers of video equipment use proprietary systems. The H.261 standards defined by CCITT have been implemented by leading vendors, but there have been problems of interoperability, and the leading manufacturers have not rushed to eliminate them; they want to maintain their proprietary advantage as long as they can. Minicase 7.4 reviewed the wary maneuvers by the leaders in the industry, PictureTel and Compression Labs, to address H.261 testing.

The Company as a Communication System The following example shows how the various tools discussed in the preceding subsections can be put together. It involves VeriFone, a company whose structure, culture, and entire business processes are in essence its telecommunications. VeriFone manufactures the machines sales clerks use to swipe your credit card to get credit approval, which is obtained when the machine dials up the relevant mainframe host computer. The company also provides customized local area communications and application software. Founded in 1981, VeriFone had grown within ten years to 12,000 employees, with sales of $150 million and 30 offices across the world.

William Melton, the founder of VeriFone, had previously built a firm in Hawaii that developed software for real estate applications. "We were totally unsuccessful. I spent ten years learning how not to do business." He could not have been too unsuccessful, for he obtained the $3 million needed to set up VeriFone by selling his software company. He brought to the new venture two central ideas for creating an "intelligent organism." The first idea was to send himself and his people out to where the action is: "Real intelligence—the information you need to understand and solve business problems—is on the frontiers. It's with the customers, on the manufacturing floors, with the component vendors." The second idea came from his experience in software development: "Programs run better when you have tight, fast loops. I was sure a company would run better if its information loops were short and tight."

A VeriFone employee implemented these two central ideas by building a network based on Digital Equipment computers and telecommunications (to avoid a mess of incompatibilities). At any one moment, about 35 percent of the company's staff are logged onto the network; managers receive an average of 60 electronic mail messages a day. Half the employees have terminals linked to DEC VAX computers, located as far away from VeriFone's California headquarters as China and India; the other half have PCs, originally Dell MS.DOS IBM clones but now comparable DEC machines.

There are no secretaries. Melton comments that "a lot of executives pay lip service to the idea of an open-door policy. But they still have a secretary guarding the door and screening mail and calls. Anybody in the company can reach me anytime through electronic mail and they know I'll read that message." Melton's CEO (Melton is the chairman) refused to open a software development office in India, where telecommunications links are poor, until a dedicated link had been installed. He announced the opening of that office by electronic mail, from Bangalore, India.

This attitude is not a gimmick. Electronic mail is the organizational structure in action and is at the heart of every employee's interactions with the firm. There is an official ban on internal paper transactions; hiring, capital budgets, travel arrangements, and so on are handled through electronic forms. Getting approval to hire someone anywhere in the world takes about two hours of to-and-fro E-mail. Employees have access to just about any information: "We allow so much access to our financials that about a quarter of the company is registered with the SEC as insiders."

Although VeriFone's sales have grown about 60 percent a year for a full decade, turnover of employees is high, about 18 percent a year: "There's a tremendous amount of intensity here. People expect answers, and expect them pretty damned soon. We're a real-time company, and that's a difficult environment to maintain. When 35 percent of the company is interacting on the system at any given time, you have to recognize that that's where things are happening. To a large degree, that's where we do business."

Improving Decision Information

It is a truism that managers need information, but too often they get it in a form that cannot help them make decisions. Traditional management information systems (MIS) are notorious for their volume of paper, the bureaucracy they represent, and the uselessness of what they provide. They are a product of the computer era, not the telecommunications age. Because the main applications of large mainframe computers through the early 1980s were accounting-centered and database management technology was still expensive and immature, the base for MIS was last month's accounts. Sales reports, exception reports, comparisons against budgets, profit analyses, and the like lagged behind the events that generated them; thus managers could use them mainly to maintain control, but not for proactive decision making.

Telecommunications changes that. Earlier in this chapter, we discussed "point of event," which is where information for decision making should be captured. The term *executive information systems* is often applied to those uses of information technology that provide managers the up-to-date and timely information they need to spot trends and anticipate problems early enough to be able to respond constructively. We prefer the term *management alerting systems*, but the label used is far less important than the actual business opportunity, which involves giving business leaders the information they need to lead instead of react.

Frito-Lay provides an outstanding example of ways to use telecommunications to capture and move information to help run an entire business better; its management alerting systems are a by-product of linking the field to the head office. Each of Frito-Lay's 10,000 salespeople uses a hand-held computer to handle all their transactions. When they go into a store, they call up the customer's account from the computer; the information is updated daily by Frito-Lay's central mainframes in Dallas, which downloads the information, together with price changes, product promotions, and any other news. Instead of filling out order forms, the salesperson enters the data into the hand-held computer and then goes outside to the truck and connects the computer to a small printer; the sales ticket is printed immediately. At the end of the day, all the orders are uploaded via telecommunications to Frito-Lay's main offices.

Frito-Lay's management has precise and immediate information on what is selling in which regions of the country. Customers get accurate billing and delivery information; they can check it on the spot, and any errors can be corrected at once. The sales force saves three to five hours a week on paperwork, and thus the reps have more time to look for new accounts and to provide better service to existing ones. If Frito-Lay decides to change the price of a product in a specific region of the country, it makes the change in the central computer system's database and the information is downloaded to the relevant salespeople only. Top management receives an accurate picture of the entire business. Product and sales managers can spot trends instead of being informed about them weeks later, and thus a fresh supply of products that are in demand can be moved to retail outlets continuously.

A simple way to identify managers' true information needs is to ask them the following question: "If, when you come into the office each morning, you could switch on your PC and get just one screen of information automatically, what would be on that screen?" The idea here is to identify their "comfort" information—the data that enables them to quickly see that things are well under control, or just as quickly alerts them to a potential problem. When most managers are asked a much broader question—"What information do you want?"—they tend either to ask "What have you got?" or choose elaborate reports that cover just about every possibility. By focusing their attention on just one screen of information, you can generally get a good idea of the data they really need and will use.

In most instances, that data desired will relate to operational figures, not just to accounting matters. Here are a few examples:

▸ *Retailing data:* indicators of any region, city or major store in which, during the past three days, sales are out of line with previous trends

▸ *Stockbrokerage data:* a graph of total cash in customers' accounts; a downward trend of the graph indicates that there are likely to be margin calls, and if it is up, the question is why aren't the members of the sales force moving to get this cash invested?

▸ *Manufacturing data:* yesterday's stoppages in production and quality indicators

The next step is to locate the relevant point of event that captures—or should capture—the information. In many instances it is already part of the firm's

IT base. The retailing data is captured at point of sale, the stockbroking information is automatically generated daily by the accounting system, and the manufacturing information is recorded on operators' logs. The role of telecommunications is then to move this data quickly and accurately, so that it can be "massaged" and delivered to the managers' personal computers.

Reengineering Processes and Simplifying Organizational Complexity

Companies' desires to automate and streamline their business functions, and to decrease their internal bureaucracies as well, are behind such popular terms as *business process reengineering* and *business process redesign*. Telecommunications simplifies and streamlines. It is central to reengineering by the very fact that it creates location-independence and time-independence, which add up to freedom from the tyranny of documents and delays. Here are some examples of the ways in which telecommunications can substitute simplicity for the complexity that dominates most large organizations:

▶   *Insurance:* One of the earliest and best-known examples of business process reengineering comes from Mutual Benefit, which cut the average time to issue a life insurance policy from 24 days to four hours. The old process was dominated by documents that were handed off among close to 30 separate departments. Examination of the workflow revealed that only 17 minutes of decision making were needed to process the typical application.

Streamlining the process began by anchoring the work to a single individual and moving documents electronically to him or her. Software replaces people in screening the forms and carrying out routine checks and calculations. Only exceptions that require special analysis must be routed through a longer workflow. The average application is checked, priced, and approved in under two hours.

▶   *Manufacturing:* Ford Motor Company made an early contribution to the field of business process redesign by eliminating invoices. For several years, all its purchasing was handled on-line through electronic data interchange links to suppliers. Using EDI to send and receive invoices was expected to reduce staff from 500 to 400. Ford personnel noticed that Mazda had under ten staff handling supplier payments and that no invoice was required. Ford's obvious question was, Are invoices necessary? The answer was no; Ford's on-line purchasing system already provided an electronic audit trail. When goods are received, the shipment can be checked against the original order and terms and payment authorized immediately. This cut Ford's staff from 500 to 100.

▶   *Securities:* A Norwegian bank reduced the staff needed to handle the physical movement of stock certificates by 98 percent by regulation that allowed "dematerialization" and "immobilization." Dematerialization means that when a stock is bought or sold, the paper certificate need not be transferred; an electronic notification and database record is adequate. Immobilization means that the original certificate can be kept in one place. Many countries still require that the certificate be a paper document and that it be transferred whenever the security is sold, which accounts for why it takes up to 30 days for settlement of trades. The G-17 committee of industrialized countries is aiming for five days.

The Norwegian system has reduced settlement to under a day. This example shows the general impact of document-dominated workflows on organizations' costs, levels of administrative staffing, and lead times. Telecommunications is an obvious force for streamlining.

In each of the previous examples, the main issue was not technology but rethinking the ways things "have always been done." The insurance company's workflows had evolved over many decades to the degree that they had been taken as a given. Invoices were just as much an embedded feature of Ford's operations, and stock certificates were the established base for securities trading.

In the 1970s, when computers were first widely applied for all core transaction processing systems, the tendency was to automate workflows, rather than to streamline them. Computers generated many documents; Levi-Strauss estimated in 1990 that 70 percent of the output of one computer system is rekeyed into another system. Computers created a new form of bureaucracy; indeed, they made such bureaucracy practical, generating the flood of paper and administrative complexity that marks far too many simple transactions.

It is easy to see why reengineering became the information technology equivalent of the total quality management movement in the early 1990s. It did not depend on new technology but instead offered the chance to get significant, concrete benefits by backing off from technology, by taking an entirely fresh view of workflows, and then by targeting technology specifically to streamline and simplify.

How New Technology Can Transform a Problem into an Opportunity

Once in a while a new technology offers the chance to transform a business's core activity. EDI can transform the mass of paper, administration, bureaucracy, and minutia involved in handling purchasing documents into the chance to streamline and simplify operations. Image processing can transform unglamorous aspects of customer service. Telecommunications is most effective when it is deployed to improve a basic aspect of operations. It is in these directions that "innovative" and "advanced" technology should be focused.

Mercy Hospital in Springfield, Massachusetts, is a small hospital that has gone well beyond any Fortune 100 firm in using voice recognition to streamline its operations. Over 20 doctors use a system built on a combination of a local area network and a special-purpose computer to apply voice-recognition technology to report making.

Each of Mercy Hospital's physicians sees 30 to 50 patients a day. They must compile a detailed report for each one, partly to protect against the epidemic of malpractice suits. The voice-recognition system prepares a standard one-page report in two minutes, versus four to six minutes for a handwritten one. The physician speaks into a standard telephone handset and can include "trigger phrases," the same sort of verbal shorthand managers use in instructing their secretaries to produce a longer sentence or paragraph in a letter. The report appears on a computer screen as the doctor speaks.

The technical sophistication required for voice recognition is immense. Consider the following sentence: "I would like you right now to write a letter to Mr. Wright." Recognizing this sentence (as a massive IBM prototype in its laboratory in Nice, France can do) is not an issue of interpreting sound but of determining meaning. Complex hardware and software are needed to add semantic analysis and ascertain whether "write" or "right" is the most likely word in the context of the rest of the sentence. Deciphering the differences between a Boston, Dallas, Atlanta, or Chicago accent compounds the problem. Mercy Hospital's system is not as sophisticated as the IBM prototype, and the doctor may need to spell out some words; the voice-to-text conversion is displayed on the screen.

Some of the physicians once dictated their reports into a tape recorder. Transcription, review, and revision took up to five days and cost the hospital close to $100,000 a year. The voice-recognition system saves another $50,000 to $100,000 by automatically inserting into the report the codes that are required by insurers to determine how to reimburse medical expenses. The doctors each save $4,000 on malpractice insurance. The head of the hospital's ambulatory care unit commented that "once in every 30,000 cases—twice a year in our case—a lawyer will send for a transcript of a report. Many cases are lost where the care was good but the documentation was terrible. The reports won't stop malpractice suits, but our model is that it should cut them by approximately 45 percent."

Although the voice-recognition system that Mercy uses is based on personal computers, it is telecommunications, provided through a local area network, that makes the system practical and cost-effective organizationally.

Gaining an Edge in Existing Markets

The first category of telecommunications opportunities, run the business better, mainly focuses on the organization's own activities, though improved customer service is often a by-product of it. The second category, gain an edge in existing markets, explicitly focuses on the customer. In many ways, this is the area in which telecommunications has had (and will continue to have) the most radical impact on entire industries, not because it creates radical innovations, but because it does not.

This may seem to be a contradiction, but it is easy to illustrate through three well-established examples of how telecommunications has changed the dynamics of competition in an industry: the automated teller machine in banking, reservations systems in the airlines, and point of sale in retailing. None of these "innovations" does anything "new"; they merely create location-independence and time-independence in routine and everyday transactions, which in fact transforms them.

Routine and everyday transactions are the core drivers of any business; they establish the core determinants of service, quality, customer satisfaction, and brand or product loyalty. Most of our lives are built around commodities and commodity transactions, ranging from check cashing, to booking air flights, restaurant meals, or hotels, to dropping into the supermarket, to using a credit card. When any company is able to transform some aspect of service and convenience in such commodity activities and gain substantial customer approval, it redefines the industry. Because convenience and service rest on reducing constraints of time and place, telecommunications has consistently and continuously transformed the basics of industry after industry in this way. It in this sense that it is radical in its effects while being unglamorous and not overtly "innovative."

One consequence of this is that identifying the main targets of opportunity for using telecommunications to gain an edge in an existing market depends on understanding the basic operations of the organization and looking at them from the customer's perspective. Generally, technical specialists in the computer and telecommunications field assume that their "high-tech" knowledge and services should be matched to equally innovative applications. As a result, they are most attuned to looking for and listening to glamorous and exciting new options. That inclination has given us a range of failures: consumer videotext, self-service kiosks, home banking, and many other ideas that seemed logically sound and for which a market should exist, but things in which customers were not interested.

By contrast, the focus in applying telecommunications to a firm's basic services has given us not just ATMs and airline reservation systems, but the use of 800 numbers and UPS to create a rapidly growing catalog shopping industry, led by firms like Lands End; electronic filing of federal income tax returns; express courier services, in which, according to an executive of DHL, number two in the international industry behind Federal Express, "information technology is the defining factor between those companies that are going to make it and those that aren't"; NASDAQ, the world's first electronic securities trading market; rapid check-in via cellular radio at Avis; and *USA Today*, a national daily newspaper that can compete with a major city's daily paper.

All of these applications address the basics of a business. A leading firm may not always establish a sustained competitive advantage by using telecommunications, but it does guarantee that many of its competitors will be placed at a disadvantage such that, typically, 50 percent of firms in an industry whose core processes are transformed by telecommunications will disappear within a decade. Generally, several of these firms will have been leaders; they often fail because they do not continually reexamine the assumptions that created their success and hence overlook the implications of telecommunications. This happened to many airlines, retailers, and banks. Time literally passed them by.

There are two consistent strategies in the use of telecommunications to transform basics of competition: (1) *provide workstation access* in the customer's home or office or on the street, and (2) *differentiate a standard product* through information and/or convenience.

There is no proven formula for finding the key that unlocks customers' pocketbooks. The targeted service must be something customers naturally and fairly frequently need; people can be expected to mesh the use of telecommunications (via phones or personal computers) into their everyday activities if it is natural and easy to do so, but special situations generally demand special response.

For a service to be a success, convenience and fast response must be of premium value, for they largely determine who prefers to shop by phone and catalog versus who wants to visit the store and see the goods. Convenience and fast response also strongly influence what types of transactions corporate customers demand to be able to make electronically (funds transfers), may be willing to consider (purchase orders, insurance claims processing) and are unlikely to want (application for large and complex long-term loans).

An electronic service must offer a self-justifying benefit; if you have to explain why the service benefits someone, the probability is that he or she won't want it. The justification for ATMs is immediately clear: "You can get money any time, day or night, in almost any city. You can get cash at the supermarket or the airport." The justification for home banking is that "you can manage your finances better and make payment electronically instead of by check." That may be true, but it is not self-justifying; what does "better" mean" Why will I benefit from making payments via a personal computer instead of using my checkbook?

The service must also be located at a natural point of event. It is natural to make a hotel reservation via a travel agent, to use your home fax machine to order pizza, or to place an order for spare parts via electronic mail. It may not be so natural to make your own airplane reservations via personal computer, to apply for life insurance by fax, or to order clothes via e-mail.

Whenever you assess an opportunity to "improve" a service via telecommunications, make sure you have convincing answers to the following questions:

▸   Why is this service something customers want and will use? What need does it fill?

▸   Is the service truly easy to use and straightforward?

▸   What makes the service convenient and easy to access?

▸   Is the convenience an important enough feature to persuade customers to change their habits?

▸   In one sentence, what is the benefit of the service?

▸   Why is the service so much better than existing equivalents?

▸   Can the service be accessed without new effort and learning?

▸   Will customers feel comfortable in accessing the service at the point of event?

Next we examine some strategies for gaining an edge in existing markets.

Putting a
Workstation in the
Customer's Office

Telecommunications needs an access point. For simple transactions and access to people, the standard telephone provides access, and more and more companies exploit 800 and 900 numbers, ANI (automatic number identification), and voice mail for this purpose. These services have many limitations. They are analog and will remain so until ISDN and its successors are widely deployed; thus messages cannot be routinely stored and digitally processed, except for simple push-button menus of options. These services cannot exploit software and database management systems. They also tend to take time, tying up a phone line and a service agent, and they generate frustration if the phone system gets overloaded and customers are put on hold or constantly get a busy signal.

Workstations, including personal computers, can add a new dimension to service, if they offer a natural point of event. Workstations provide all the advantages of the phone equivalent, with few of the disadvantages, if they are easy to install, learn, and use. They can operate at all hours of the day (unlike phone calls answered by service agents) and can keep transactions short, if they are simple to operate and exploit high-speed transmission. Users will not get a busy signal if the network is appropriately configured.

The "ifs" are important but mainly are of concern to the implementer rather than the designer. The client and designer need to address the point of event, customer benefits, and targeted service; the opportunity then is to make the workstation the customer's access point. Frequently, the opportunity is obvious. For electronic banking, cash management via personal computers is now a core service; the point of event is the corporate treasurer's office; the targeted benefits are fast, flexible management of all aspects of the firm's cash, including funds transfers, "netting," drawing down lines of credit, foreign exchange, and access to information on balances. In manufacturing, customer-supplier links for purchasing are a natural point of event. EDI allows the customer to place orders, access pricing and availability data, and track shipments at any time of day.

The competitive advantage derived from putting a workstation in a customer's office to handle core services rests on "occupancy"—on displacing other companies as the key supplier. The airline reservation systems wars of the 1980s centered around gaining occupancy. United Airlines and American Airlines had close to 80 percent of U.S. travel agents' business for their Apollo and Sabre systems. Delta, Eastern, and Northwest were not only left with a small market, but the low-volume agents as well. No travel agency wanted two sets of equipment, two sets of software, two communications systems, and two sets of training programs. The battles were vicious, with several of United's and American's competitors offering agencies large sums of money to break their contracts and promising to cover any resulting legal costs.

The benefits customers accrue from electronic service can be immense, and it is often the smaller customers who gain. An example is Guy Marine, a small farm-equipment supplier in St. Mary's County, Maryland. Perry Guy and his brother run a company that sell tractors, "bushhogs," and ploughs in a very rural community. He has no college degree, and his business is the epitome of the best of American small business: hands-on management, personal attention to customers, and no frills or unnecessary overhead. One of his main suppliers provided him with a low-cost personal computer-based system for ordering and inventory management. Guy initially found it difficult to use the system; the manuals did not make sense, and the training programs assumed too much knowledge of computers. He asked for more help and soon became a complete convert.

From Guy's viewpoint, the system makes his entire business more responsive and efficient. It saves paper work by providing him with accurate and timely records. He can locate special-purpose equipment more quickly, reduce his inventory, and meet customer needs in a fraction of the time it used to take. Everyone gains; both the supplier and Guy Marine reduce paperwork, reduce lead times and inventory, and provide a far higher level of customer service.

A large firm can put a complete back office and management information capability in the customer's personal computer. Guy Marine gets a quality of accounting information it could not otherwise afford. Transactions that previously consumed time, administrative staff, and paper are now greatly simplified. The direct contribution to Guy Marine's cash flow may have made the difference between survival and bankruptcy in the recession of the early 1990s.

Putting a Workstation in the Customer's Home

Putting a workstation in the customer's office is now a basic requirement of business in manufacturing, distribution, securities trading, banking, airlines, and retailing. To date, no one has managed to put in the home a workstation that fulfills all the principles listed earlier: establishing or meeting a real need; featuring straightforward procedures; and providing comfort in selecting options, convenience and fast response, self-justifying benefits, incentives to displace good enough existing alternatives, reduced marginal economics of effort, and a natural point of event.

The natural point of event for most of us is the telephone, not a personal computer. PCs remain difficult to use for most people in a telecommunications environment; merely advertising a product as "user-friendly" does not make it so. To date, despite many efforts to introduce telecommunications-based PC products to the consumer market, there have been no major successes and many costly failures. Only a few companies have built a critical mass. Prodigy, backed by a well-funded consortium, advertises aggressively and offers special deals, which helped it build a market of about half a million customers by mid-1992. That figure was well behind its projections and was barely at the break-even point. Previously, other consortia offering a similar mix of information services, promotions, shopping offers, games, and educational programs had failed, including Trintex and a NYNEX gateway service into more than 100 information providers. These large ventures were well-planned; they all bombed.

The only proven success in the consumer videotext market is widely cited as evidence of a potential mass customer base: France's Minitel system. It was part of the French government's efforts to dramatically upgrade the nation's telecommunications and computing infrastructures. In order to stimulate the market, the French government produced Minitel terminals at very low cost and provided them free; their basic use was for directory inquiries. The terminals connected to France's Transpac packet-switched network, the most advanced public data network in the world, providing a low-cost access point to a growing range of services. Groups of businesses could create their own bulletin boards and share information resources; similarly, entrepreneurs could create their own information products and offer them through Minitel. Use of Minitel grew very rapidly from its inception; by 1992 it had over three million users, many of which were households, not businesses. Several U.S. companies, including Baby Bells, licensed Minitel from France Telecom, hoping to copy its success.

Success, however, is relative. First, Minitel is heavily subsidized; it lost over $800 million in 1991. Second, it succeeded largely because there was no alternative service—France Telecom had a monopoly on all telecommunications and information services. This is not the case in the United States, where there are many providers, including Compuserve, Prodigy, Dow Jones News Services, and many public electronic mail services. The very fragmentation of the American market has made it difficult for any player to build a critical mass.

The main problem is apparent in the list of principles for selecting opportunities previously listed. So far, no provider has either created a self-justifying benefit or identified an application that meets regular and routine needs. Prodigy and Compuserve offer a wide range of services, but none of them stands out as a "must have" service.

There are hints concerning the source of success in the future: the combination of introduction of fiber optics into the home, multimedia, and entertainment. The world population's appetite for video, which seems insatiable, is mainly met today by videocassettes, satellites, and cable. By contrast, the limited bandwidth provided by the telephone system greatly constrains information providers. For example, Prodigy's full-color, personal-computer–based videotext is extremely slow, and screen resolution is poor. Massive bandwidth will remove that barrier to ease of use.

Multimedia—which refers to the complete integration of the personal computer (and, by implication, across a network) of image, voice, video, data, music, and text—will add many features to and improve the quality of presentation of today's limited offerings. A typical multimedia application is education; a student can access an encyclopedia that, say, displays text on Mozart, flashes pictures of Salzburg, and includes sound of a chamber orchestra playing "Eine Kleine Nachtmusik." Similarly, a teacher can prepare a lecture that includes photographs, short videos, sound tracks, and simulations, all selected interactively during class time.

Entertainment will benefit from and exploit the combination of fiber, multimedia, and direct broadcast satellite (DBSs), which are "third-generation super sats" that offer greater bandwidth, increased power, and compression of digital signals. DBS will accomplish many of the goals of HDTV (high definition TV). Hughes Communications Inc. launched its national DBS system in late 1992. The president of Hubbard Broadcasting Inc., another company planning to enter the market, estimated that it will cost only $2 of investment per home to reach every household in the United States and 90 percent of households in Canada. Receivers, including antennas that are two feet in diameter, are expected to cost up to $700 initially. DBS will offer 100 channels of broadcasting on a single satellite signal.

Obviously, fiber and DBS open up many potential applications to reach consumers in their homes. The technology itself is not the opportunity, however, nor will it be fully in place soon. The target date for a national fiber-optics network envisaged in Senate bill 1200 is the year 2015. Today, the basic terminal in the customer's home remains the telephone, with the television set an ancillary used for home shopping in conjunction with the phone. Telephone companies, advertisers, retailers, publishers, sports teams, educators, and entertainment providers are all aggressively looking at ways of using telecommunications to reach the home. The announcement by Telecommunications USA in early 1993 that it would offer subscribers 500 channels of cable television before the end of the year signals the beginning of a new era of telecommunications for consumer services. Those channels can carry many types of traffic with no additional capital investment by Telecommunications USA, which can in effect franchise the bandwidth to, say, long-distance carriers (MCI was rumored to be discussing a major joint venture), electronic-game providers, information services firms, and many others.

These trends suggest that the personal computer of the late 1990s may include the television set and telephone handset. The TV will access telecommunications services interactively through on-screen menus, and the phone will do the same through voice messaging. It is too early to tell when the new generation of "personal digital assistants"—Apple Computer's term for hand-held palm-sized computers with wireless digital communications—will be in widespread use, but the question is when, not if, because wireless communications is expanding dramatically in terms of volume of use, technology, and competition. The new opportunity here will be to put a workstation where the customer is, extending location-independence and reach.

Differentiating a Product via Telecommunications

Automated teller machines do nothing that a teller does not do; they merely cash checks and accept deposits. And they even turn the customer into a data entry clerk; you stand outside typing in your password and following a menu of instructions, and you validate the transaction and wait to get your money. Viewed in these terms, ATMs sound like an unattractive customer option. Of course, they have proven to be the opposite; they have provided new convenience and transformed service. They extend the bank into the street, office complex, airport, shopping mall, and supermarket, in effect creating an electronic bank branch. They provide easy access to the bank's basic services while reducing the bank's operating costs per unit of revenue.

ATMs are a classic example of how telecommunications can be a significant differentiator of commodity services. Consider some examples:

▶   Charles Schwab, the discount broker, highlights in its television ads how clients can place orders by touch-tone phone from an airport, even late at night.

The commodity element of the actual security commission is downplayed because it is a source of differentiation only against full-commission firms, not other discount brokers.

▸ American Home Shield (AHS) provides service-warranty contracts for plumbing and electrical systems. It is rapidly shifting its entire service to telephones, creating the possibility of reinventing its products. AHS generates 200,000 new contracts a year, in part by using a new feature of advanced telecommunications networks—automatic number identification (ANI)—to transform its operations. ANI is conceptually simple (and highly controversial, in that some observers see it as an intrusion on personal privacy). When a homeowner or a real estate broker calls in to order a service, the AHS agent knows exactly where he or she is phoning from; the caller's records are pulled up on the agent's computer screen at once through an automatic message to AHS's main computer.

Calls are routed to the relevant department because the ANI facility can recognize the type of customer. One major advantage of the system involves handling major problems and upset customers. According to one AHS manager, "This is a key part of the business. The customer who is calling is already stressed out, so the quicker we can solve their problem, the better. With ANI we get the history of the customer immediately. Our goal is to make the process the easiest for the customer. We want to exceed their expectations."

Is the product in these examples simply the service warranty or is it the service response?

There is a clear distinction between using telecommunications to differentiate a product and using it to support one; support does not directly add value. For example, it is standard now for manufacturers to provide customer service 800 numbers. Only a few of these numbers become a part of the product in the customer's mind, mainly because of the speed of response, quality of staff, and the information they provide. In the highly competitive personal computer market, Dell has established as a differentiator its combination of phone service for ordering, fast fulfillment of orders, and immediate service plus replacement of hardware. This rests on telephones, but it is a comprehensive system of service that eliminates many traditional retail chains. When personal computers first began to attract consumer interest, retailers like Computerland flourished. New chains sprung up everywhere. Now, many of them are empty and the companies are struggling or bankrupt. Meanwhile, Dell has sales of over a billion dollars.

In the spring of 1992, CNN reported on IBM's decision to enter the low-cost personal computer market as a marketer, not just as a manufacturer. The entire report focused on the use of telephones to offer help desks, access product information and expertise, and reach out to customers through direct selling and telemarketing. CNN identified "thousands of operators" as the marketers and sellers, highlighted Dell as IBM's competitor and model for emulation, dismissed traditional retailers as no longer in command, and saw IBM's initiative as a major market move. Historically, IBM's initiatives have been product-based, and the differentiation has been its brand name, technology, or service and support. Now, CNN sees as newsworthy IBM's strategy in making telephones the differentiator in a "high-tech" market.

More generally, the search for differentiation through telecommunications focuses on information, rather than on access, partly because access is easy to imitate, whereas information is difficult to create, structure, and manage, making it a longer-lasting source of advantage. Many banking services are little more than a packaging of data accessed via a personal computer. Electronic cash management systems, for instance, allow customer self-service for standard inquiries and funds transfers. ATMs do the same for basic transactions, but the differentiation, which for the leaders provided several years of competitive advantage and a longer residual edge from occupancy, comes from providing (1) self-service and immediate completion of the customer's request; (2) on-line response, instead of having to call, wait, write, and wait again for an acknowledgment of completion of the request; and (3) accurate and complete information relevant to the customer's needs or concerns.

Improving Ease of Access

The crucial element in exploiting telecommunications to gain an edge in existing markets is ensuring ease of access. Doing so mainly involves extending the firm's reach, as discussed in Chapter 3 and described in more detail in Chapter 11.

Today's 800 numbers provide a simple and increasingly cost-effective way of providing ease of access. They are best suited to transactions that do not involve complex information and analysis, for such transactions require some form of data communications. The combination of telephone access and customer service databases and workstations is increasingly transforming peoples' expectations and raising the cost of not meeting those expectations. We expect to get through on the first four rings—not to get a busy signal; be put on hold with music and repetitious messages about how "all our agents are busy at this time"; face complex voice mail menus that amount to a test of memory, IQ, and nimbleness of finger; or find that a question about price, availability, delivery, or features cannot be answered.

ISDN is part of the emerging combination of access and information. ANI (or "caller ID," as it is more often called) makes it practical for the agent to reply "Hello, Mr. Jansen" and to bring up on the computer screen the customer's records: "Are you calling to renew your policy? I see it expires at the end of this month." Image processing goes one step further and ensures that relevant documents as well as computer data can be accessed.

In passing, it is worth noting that "caller ID" offers extra advantages to the company. Domino's Pizza had $5.1 million worth of bad orders in 1991. In the 300 of its 4,800 stores that use caller ID, the number of such bad orders dropped by more than 90 percent in 1992. Michigan Bell, which introduced the feature to a million households in March 1992, offers a free per-call blocking feature for customers who see it as an intrusion on their privacy, but Domino's will not deliver orders to callers who block transmission of their identification.

Providing a New Level of Government Responsiveness

Many of the most imaginative uses of telecommunications to improve convenience and ease of access are found in state and local government, organizations that are not well known for being easy to deal with. The best agencies are working hard to provide a level of citizen service that matches the best of customer service in the private sector. For example, Merced County, California, introduced a welfare assistance system in mid-1992 that shortens claims processing from over three weeks to as little as a day (see Chapter 3). MAGIC (Merced Automated Global Information Control) is a LAN-based system that uses client/server technology and expert systems to replace mainframe processing and reduces operating costs by $1.5 million. The old system required a welfare recipient to come into a branch office or the main office and fill out 12–14 pages of forms and to return in three weeks for a meeting with a welfare representative. Three weeks after that, the representative would determine the applicant's eligibility and manually calculate the client's budget. The applicant had to apply separately for each type of assistance.

The project director for MAGIC commented that "if a client is in a crisis situation and we tell them it will be three weeks before we can see them and another three weeks before they see their first check, the applicant may sink further into that crisis. . . . Now we can catch them before they have a crisis." The new system uses a one-page form that is entered into the mainframe computer via a dumb terminal. A clerk then sets up an initial meeting with a welfare representative within three days. The night before the meeting, the mainframe downloads the application forms to a Hewlett-Packard minicomputer in the relevant branch office; the users of MAGIC's 150 workstations access the data through an Ethernet LAN. The host computer has a large database that tracks applications and stores data on clients, decisions, and laws. All types of welfare requests are handled in a single meeting.

Making It Easy to Handle Transactions

Electronic data interchange has become one of the fastest growing applications of telecommunications for reducing costs and speeding up transactions (this falls under the category of running the business better). It can also transform service, which places it in this category of gaining an edge.

Levi Strauss's LeviLink is an impressive example of a firm using EDI to both differentiate its products through its service and dramatically improve all aspects of convenience, speed, access, and responsiveness. Levi Strauss is a $3 billion company serving 17,000 retailers who operate over 200,000 stores. In the 1980s, Levi Strauss had become a mediocre performer in terms of growth and profits, but a management buyout changed the situation. The new team moved aggressively to reposition the firm, seeing telecommunications and computers as a major source of opportunity.

Levi Strauss had calculated that 70 percent of all of its and its customers' business data (for orders, invoicing, and so on) was manually entered into a computer, and the output from the computer (such as reports) was then again manually input into another computer. In effect, EDI lets the telecommunications network do the work. The Levilink system eliminates all the steps involved in inputting information; data is sent electronically between computers and from workstations into computers. Levilink provides electronic purchase orders, barcoded carton tags, electronic package slips and invoices, and automatic stock and point-of-purchase data.

One of Levilink's users is Design, Inc., a chain of 60 stores. The system reduced Design's stock replenishment time from 14 days to three days and turnaround for delivery from nine days to three days. In addition, it enabled individual stores to make small orders whenever they wish and to get the goods delivered by UPS, instead of having them sent to Design's headquarters, where they accumulated until the end of the month. As a result of Levilink, Design Inc. was able to eliminate its regional warehouses.

EDI is rapidly becoming one of the major uses of telecommunications across industries worldwide. Hong Kong and Singapore have made it a national priority; each of them has created a government-sponsored EDI value-added network that handles all aspects of trade documents and transactions. The goal—to reduce the time needed to deal with any of these from one or two days to 15 minutes—has helped Singapore move from number 10 to number 1 in the world in terms of volume of goods shipped through its port. The Port of Rotterdam clears goods electronically in 15 minutes through EDI; this effort is an explicit part of the Netherlands's plans to retain its position as the logistics center of Europe.

Why do business with a company that is difficult to deal with? Telecommunications has become the single most powerful force in ensuring ease of access and convenience. Even if it does not establish a competitive advantage, at the very least it avoids a competitive disadvantage for firms in industry after industry.

Creating Sources of Market Innovation

The third category of telecommunications business opportunity is the most aggressive: the search for breakthrough services that go beyond the status quo and that have the chance of creating a major new source of revenue and profits.

If innovation were easy, everyone would be an innovator. Few are. We can expect many firms to stay within the mainstream of telecommunications usage, sensibly keeping up with trends and moving quickly to exploit any provider opportunity, so that even if they do not gain a competitive advantage and become one of the leaders, they are not pushed into a competitive disadvantage by lagging behind. Some firms will trail the pack, risk erosion of their market and their earnings, and, in any situation in which telecommunications redefines the base level of service across the industry, become a likely target for acquisition as the owners of a comprehensive delivery base exploit telecommunications and fuel consolidation of the industry. This has been the pattern in the distribution, banking, retailing, and airline industries.

A few firms in this context will be continual innovators, with their occasional failures being offset by their overall frequency of successes and occasional breakthroughs. Consistent U.S. leaders in this regard are Federal Express, American Airlines, Banc One, USAA, and Frito-Lay.

It is far more difficult to generalize about the use of telecommunications to create real innovation than it is to assess successes in using it to run the business better and gain an edge in existing markets, because by definition effective innovation is something unusual and a departure from the norm. We should not expect such initiatives to be everyday occurrences. Figure 10-2 selects a few recent examples of business innovation through telecommunications, mostly from medium-sized organizations. Because telecommunications involved massive capital costs in the 1970s and early 1980s, before local area networks and value-added networks created new options, most of the striking competitive successes came from Fortune 100 giants.

The examples shown in Figure 10-2 show that innovation is a function of creativity, not money. In each instance, the technology was available to all companies in the industry, but only a few connected it to the business idea that made it a source of innovation. There are two main ways to create market innovation:

1. *Launch a preemptive strike*, a major business innovation made possible through telecommunications specifically and that other competitors cannot quickly match or for which they cannot find a substitute

2. Exploit ownership of an electronic delivery base to *piggyback new services* onto it

We will consider each strategy in the sections that follow.

Launching Preemptive Strikes

Preemptive strikes, the most exciting uses of telecommunications, are initiatives that may take years to launch but that quickly redefine the basics of competition and give the leader a springboard for exploiting its competitive advantage of lead time. Legendary examples of preemptive strikes include those by Federal Express, which took away from the airlines the entire small-package cargo market in the United States; Merrill Lynch, which took $80 billion of deposits away from the banking industry through its Cash Management Account; American Hospital Supply, whose success in using customer-supplier links in effect created the information-technology-and-competitive-advantage movement of the early 1980s; American Airlines, which preempted the competition with its IT-based AAdvantage frequent flyer program, its Sabre reservation system for yield management, and its generation of revenue from selling its systems amounting to over 10 percent of its sales and about 25 percent of its profits; and USAA, whose use of image processing transformed the level of customer service in insurance.

Figure 10–2	Organization	Innovation
Recent Examples of Business Innovation via Telecommunications	Makati Medical Center (Philippines) and Stanford University (California)	Teleradiology across the Pacific; Stanford provides on-line, interactive diagnosis with Makati physicians
	Travelers Insurance CareNet	20,000 PCs access and share health-care data for seven million enrollees; data gathered for each stage of insurance, from sale to treatment to payment; data used by corporate customers to manage costs (dial-up access)
	Harvard Community Health Plan	On-line self-diagnosis; patient keys in symptoms to get advice at any time of day; screen provides list of treatment options or alerts physician's office, on-line, to schedule appointment
	University of Missouri	50,000 students, $900 million budget network provides 45–60 hours a week of classes, saving four-hour round trip to remote campuses; classes retransmitted for state employees; faculty and administration hold multipoint conferences across all computers; network used for regular business meetings
	USS *Iowa*	One of five ships included in the Navy's Paperless Ship Project; uses image processing and LANs to reduce and manage 40 tons of paper-based information needed for daily operations, gaining an 8:1 productivity increase
	Otter Tall Power (Minnesota)	Electrical generator with $180 million in 1992 revenues; manages its grid on-line, including fault monitoring; downloads customer usage data nightly via automated meter reader system. No rate increases for ten years, with reductions in power outages every year since 1962.
	H&R Block	Electronic tax filing, using its Compuserve subsidiary; captured 20 percent of this new market, with fees of over $150 million
	Home Shopping Club	Sells goods "impulse purchases" 24 hours a day over two cable television networks; 200,000 calls a day; caller information fed instantly into terminals and fed back at once to studios, production control rooms, and offices, so show's producers can make split-second decisions about products and programming. Sales close to $1 billion a year; this growth built in seven years.

There is significant skepticism among many observers about the likelihood of comparable new preemptive strikes through telecommunications and computing, partly because firms are now far more alerted both to the technology and to what competitors are doing with it. In addition, these sceptics argue that open systems make it far easier for companies to acquire technology and are making technology more and more a commodity.

That may be true, but there are plenty of signs that only a small number of firms are able to exploit the commodity, just as McDonalds has exploited the commodity of hamburgers better than any other fast-food company. When all the leading firms in an industry have access to the same technology, the competitive edge comes from management. The feature that most marks the firms that have set the pace for the competitive use of telecommunications is the quality of dialogue between the senior business leadership and the information services planners and implementers. Indeed, it is almost impossible to distinguish between these two groups' contributions.

Preemptive strikes still occur. Two of the more recent ones have been in the telecommunications industry itself: MCI's Friends and Family and WilTel's frame-relay service. Friends and Family is a service in which people who phone each other regularly get extra discounts for their calls to each other. In 1991, five million customers had signed up, and by 1992 the customer base was ten million. Friends and Family has taken at least four percent of the consumer market away from AT&T and has repositioned MCI from a successful but small competitor to a powerful $10 billion firm. In 1988, AT&T had a 73 percent market share; now it is down to 61 percent. Since 1988, MCI's revenues have doubled, while AT&T's have barely kept up with inflation.

Friends and Family was the breakthrough product, for MCI's gains had been incremental and its margins thin. Friends and Family exploited an apparently trivial difference between MCI's and AT&T's billing systems. AT&T had subcontracted many aspects of on-line processing of customer accounting and recording of data to RBOCs, consolidating the information monthly to produce the bill. It could not at the time of the call capture information about the telephone number receiving it; MCI could, and thus it could recognize that the call was between two people signed up for Friends and Family.

AT&T lacked the intelligence in the switch to create a comparable product, and its main response was a barrage of aggressive television commercials built around the themes of "your friends are your own business" and "I'm being badgered by people intruding on my privacy trying to get me to join up in this Mickey Mouse fad." Now, ten million customers later. . . . Who would have thought that in a high-tech business in which all the firms provide the same basic product, an accounting and billing system could break open the marketplace? Someone in MCI thought so, and that's what makes a preemptive strike possible.

Someone in WilTel also spotted how the trivial could be made into a totally new thrust in the industry. A subsidiary of Williams, a petroleum company, WilTel was the outcome of a task force's review of what to do with unused oil pipes that spanned thousands of miles. One possibility was to lease rights of way to companies that wanted to install fiber-optic links; many railroads had spotted that opportunity years before. But the task force came up with a different idea. Fiber-optic cable must be buried two to four feet underground (which is why a farmer with a backhoe can inadvertently bring to a halt the entire reservations system of a major airline), and this is expensive and labor-intensive. Why not place the fiber in the oil pipes instead? A "pig," the device that forces oil through the pipes, can be used to install the fiber.

In 1992 WilTel was the leader in frame-relay services. It did not have any existing phone or data communications infrastructure in place and could thus technologically leapfrog AT&T, MCI, and Sprint. It may or may not be able to hold its lead, but this example shows how often a major opportunity to use telecommunications is recognized by a firm outside the industry. Just as WilTel's lack of an infrastructure for yesterday's business context helped it position for and create tomorrow's business context, in Fidelity Investments's and USAA's announcement in 1992 that they will provide full-service banking, their lack of physical bank branches was not a liability, and the banks' ownership of branches was not an asset.

The fusion of business and technology—and of culture and human resources— is well demonstrated by one of the most far-reaching preemptive moves made in recent years: USAA's use of telecommunications and image processing to transform just about every aspect of customer service. The innovation reflected a 20-year venture in which the CEO, General Robert McDermott, led the management process for information technology that enabled his information services executives to manage it; McDermott helped establish the business drivers for IT within the broader business context.

USAA is one of the most efficient companies in the insurance industry and is consistently rated among the best in service. Its underwriting expense ratio—the cost of underwriting policies divided by the total premiums—is 9 percent, versus 13 percent for the next best in the industry. USAA has shown that big does not have to mean bad and bureaucratic. It is the largest single user of incoming WATS lines (800 numbers) in the world, with 150,000 calls received a day. It is the largest IBM database site, handling 6.2 million transactions a day. It receives close to 100,000 pieces of mail each day. It operates the largest business building in the world, 7.8 million square feet, bigger than the Pentagon. Its computer disk drives cover an acre of space. It is the fourth-largest travel agent for cruises, the fifth-largest MasterCard provider. . . . The list goes on.

USAA is best known for its $100 million investment in image processing that began in 1981 and included a joint venture with IBM that created Imageplus, the first large-scale image processing software system. USAA eliminated 99 percent of all documents in its business units using Imageplus. Previously, it maintained close to 300,000 boxes of files that were stored for an average of seven years. Finding an archived document took from two hours to three days. At any one point in time, 50 percent of documents were unavailable: either lost, in transit, or in use elsewhere. USAA employed 35 college students to go around at night looking for mislaid documents; it typically took two weeks to find them.

Now, documents do not get lost, and any stored document can be retrieved in seconds. Incoming mail is scanned, indexed, and then destroyed. (Scanning and indexing take around 11 seconds.) Outgoing mail is imaged and matched against the incoming mail that prompted the response. Work is routed to people; high-priority documents go to the top of a supervisor's or claims processor's action list; work does not get buried in an in-tray. Supervisors can see where there are backlogs and what their staff are working on. When customers phone in, a service agent can pull up on the screen every single document and record about the customer, including application forms, handwritten letters, and photographs. This cross-linking of information transforms service because there are no functional or department handoffs in which an agent must say, "I'm sorry, but I don't have your X form" or "That's handled by our Y department."

The Automated Insurance Environment project that led to the invention of the image processing system began well before any such technology existed or could even be predicted. It was driven by a clear top-management business vision and commitment and a long-term focus on implementing that vision. The vision was "the single company image" and a single customer point of contact; USAA should be faster, better, and easier to deal with than living next door to an insurance salesperson. The services across products should be so integrated, so simple, and so easy to obtain and use that people would feel they were losing something of value by going elsewhere for any one of them. USAA should develop its products to meet its members' expanding and changing lifetime needs for insurance and financing. The basis of USAA's relationships should be mutual trust and respect, values that reflected USAA's origins as a company providing insurance to military personnel, who moved location frequently and thus often had problems getting and changing policies. One USAA executive explained that "we operate on the basis that an officer's word is his bond and the officer puts his trust in us to provide quality products and services." This may explain why, when USAA Federal Savings Bank first sent out applications to insurance policy holders for MasterCard credit cards, 52 percent applied, versus the industry average of 10 percent.

The "single image" vision set clear business criteria for the use of telecommunications, computers, and information. Ease of access was essential to service; hence the reliance on first-rate phone service. Don Lasher, president of USAA Information Services, states that "we live or die by the phone" (personal communication). Between 1985 and 1990, USAA cut its cost per call minute from $1.50 to 75 cents; the length of the average phone call went up, from 4.6 minutes to 5.3. However, through the use of information that cross-links customer product data and policy data, there are fewer needs to call customers back and to leave messages. Lasher estimates that answering a customer's question or request typically required four or five calls to make contact; "Now, it's zero." The image processing system streamlines work (and helps show why image technology is widely seen as a key to business process redesign). In the 1970s, producing a policy took 57 steps and several weeks; now, one is produced in a single step and is mailed within two days. Increasingly, though, there is no need for the mailing. Lasher comments, "How many of us look at our auto policy or really understand what it says when we do? We tell our members they don't need the printed policy; they can phone us and they know they will get excellent and fast responses."

The keys to USAA's success are top management vision that is clearly stated and clearly committed to; a long-term focus; an emphasis on platform and technical architecture as the base for the strategy; constant dialogue and partnership among the business units and information services at all levels; and continued sharpening of the vision. During its 20-year course USAA made mistakes in organization, in choices of technology and of suitable decentralization of information-technology decisions, and in planning. It could afford these mistakes as part of the inevitable risks of innovation because its vision provided an anchor. General McDermott's personal and active involvement with IT and his belief in the strategic importance of technology guaranteed the drive: "Our success and even survival will be closely tied to how well we can compete in the marketplace on technical grounds." The focus on the strategic technical architecture as the base for integration ensured that all the processing and information systems would contribute to the single company image.

When the history is written of the shift of telecommunications from being just phones, cables, and overhead to its becoming a key business resource, the five preemptive strikes that are likely to be highlighted as most pivotal are: (1) Citibank's launch of ATMs and credit cards as the key to implementing a vision of service through a "telephone/mail" relationship; (2) American Hospital Supply's recognition that putting a workstation in the customer's office could create an entirely new theory of business, logistics, and service; (3) American Airlines's breaking away from the rest of the industry in seeing reservation systems as the key to distribution, and its telecommunications- and database-dependent frequent flyer program as the key to marketing and customer loyalty; (4) Wal-Mart, Toys "R" Us, Dayton Hudson, Dillards, and a few other retailers recognizing that point of sale and fast information movement through telecommunications was the key to merchandising and store management; and (5) USAA's creation of the era of image. In every case, a business vision drove and even created the technology.

Creating New Services by Piggybacking

One of the most effective ways to create market and product innovation through telecommunications is to exploit an established delivery base and "piggyback" new services onto it. This strategy should be the priority of any company that has either successfully used telecommunications to put a workstation in the customer's office, established a strong point of event for service, or attracted customers to carry out transactions electronically, for the company can then add services to this base. If such piggybacking takes away business from other companies and they do not have a comparable electronic delivery base, a company has a chance to establish a sustainable competitive edge, the dream of every telecommunications designer and an item on the wish list of many senior executives.

The key principle of piggybacking is placing technology at point of event as is evident in the following examples:

▶   British Airways, as mentioned earlier, piggybacked international hotel reservations onto its system because this is a natural opportunity to ask "Do you need a hotel room?" and (2) the major hotel chains did not have a reservation system (and would have to spend well over \$100 million in an unsuccessful effort to catch up.)[2]

▶ McKesson, which had captured a large part of the U.S. pharmacy market for ordering goods electronically, piggybacked insurance claims processing onto its Economost system. Again, it owned the point of event and had the platform in place to add electronic claims processing at a time when no insurance firm could match their effort. It quickly became the third-largest processor in the United States.

An innovative approach to exploiting an existing electronic delivery system is that of the Public Broadcasting System (PBS). The same infrastructure that delivers "Masterpiece Theater" to television sets can deliver any type of information that can be piggybacked onto the same signal. PBS needs new revenues and is positioning itself to use a portion of its broadcast signal—the "vertical blanking signal," the black bar you see if your TV picture starts to roll—to transmit credit card and check cashing information for stores. Local PBS stations will pull down that information, encoded to ensure security, from the satellites that carry PBS's signals and then will broadcast it through the vertical blanking signal to clients. Retailers and other customers for the PBS innovation will capture the broadcast information through a special system provided by PBS's partner in this joint venture. Today, clients of the new service must make a long-distance call to check almost every credit card transaction; the new service will be faster and cheaper.

PBS recognized that the unused vertical blanking signal can carry many types of information. It is a "data pipe" into homes and businesses across the country. PBS's innovation may or may not be successful. Firms like McDonalds and Wendy's were very positive in their evaluation of the pilot system, but PBS has yet to decide on pricing and marketing and has yet to prove it can manage the service, and the level of demand is not yet apparent. The innovation looks like a sound business idea, but customers will decide whether such a telecommunications-based initiative will take off, not the provider. PBS is intruding on the traditional turf of banks and credit card providers, just as McKesson used its network to intrude on the insurance industry's turf.

---

[2]The failed project was CONFIRM, a joint venture among leading hotel chains, car rental firms, and American Airlines, with the aim of exploiting American's legendary, but technologically cumbersome, Sabre system. A few months before it was due to go live, CONFIRM was found to contain major "bugs" and was abandoned. Heads rolled, faces were red, and lawyers were hired in abundance. What strategy would you recommend to a hotel chain that has lost control over its own distribution base to British Airways? Lack of a telecommunications platform created a major business blockage and provided BA with a sustainable advantage.

Summary

Choosing the right opportunity is the starting point for the telecommunication decision sequence, and there are now plenty of examples and lessons to draw on. The three opportunities in the telecommunications business opportunity checklist provide a framework for applying telecommunications as a business resource, rather than as "new" technology. The checklist highlights the key importance of making sure a proposed innovation meets real needs, is simple to access and use, offers convenience, provides real and easily explained benefits, does not involve new learning and effort, and is accessed from a natural point of event.

The three opportunities in the checklist address the internal operations of the organization, under the headings of running the business better, gaining an edge in existing markets, and creating sources of market innovation.

Under running the business better, the proven strategies are to manage distributed inventories, link field staff to head office, improve internal communications, improve decision information, and re-engineer and simplify processes. To gain an edge in existing markets, the main strategies center around providing workstation access to services from the customer's office or home and differentiating a standard product through ease and speed of access and information. Telecommunications here can often provide a competitive edge in commodity markets by providing a new source of differentiation.

Creating market innovation through telecommunications comes from the relatively infrequent preemptive strikes that change the rules of competition and from exploiting ownership of an existing telecommunications platform to add to it services that involve limited extra investment and that can capture business services that have traditionally been provided by firms in another industry. This "piggybacking" process is one of the main strategies for firms with a strong platform and a problem for firms that lack a good platform.

Review Questions

1. Explain the notion of "prejudging the problem and hence predetermining the solution" in the context of telecommunications, and illustrate with an example from your work or school environment.

2. Explain how telecommunications can be used to reduce undesirable inventories, and illustrate with an example from your work or school environment.

3. Explain why "point of event" capture and processing of information is crucial to improved business performance.

4. Does using information technology to link field staff to the head office require extending the range, reach, or responsiveness of the telecommunications platform? Explain.

5. Why is the use of fax machines so widespread and growing when superior technologies such as e-mail (with the ability to send spreadsheet or document attachments) and voice mail are readily available?

6. Explain how "groupware" enhances teamwork in organizations, and describe the work conditions under which it is most effective.

7. Describe three ways in which video-conferencing and/or business television have been used to improve an organization's internal communication, and explain the type of business result that is expected from each.

8. Explain how VeriFone's corporate network and business processes embody the following ideas:
   a. Real intelligence is on the frontiers (with customers, on the shop floor, and with component vendors)
   b. A company runs better if its information loops are short and tight.

9. Explain why traditional MIS generated reports are useful primarily for control purposes but are inadequate for proactive decision making.

10. How does a "management alerting system" overcome the weaknesses of the traditional MIS reporting system? Use information from the Frito-Lay example in the text to support your reasoning.

11. Explain why telecommunications is an essential core of business process reengineering and design.

12. Explain how the physician's voice recognition system in Mercy Hospital constitutes business process reengineering (by comparing the old and new processes), and describe how this new process saves time and money.

13. Telecommunications has been very effective in helping companies gain an edge in existing markets by establishing closer links with customers. Explain the underlying basis for this type of competitive edge gained from the use of automated teller machines, airline reservation systems, and point-of-sale retailing.

14. Describe the competitive advantage obtained by putting a workstation in the customer's office, and identify the conditions necessary for this advantage to be realized.

15. The French Minitel videotext system has been successful, whereas other videotext experiments have failed. Explain its success in terms of the eight principles for obtaining service improvement via telecommunications, and highlight the key factors that separate Minitel from other failed systems.

16. A bank's automated teller machines do nothing that a human teller can't do. In fact, they turn the customer into a mere data entry clerk. Yet, they have been extremely successful and have changed the way customers interface with the bank. Explain the success of ATMs in terms of product differentiation using telecommunications.

17. Explain the distinction between using telecommunications to differentiate a product and using it merely to support a product. Also, discuss why differentiation focuses on providing value-added information rather than simply facilitating customer access.

18. Explain how electronic data interchange (EDI) can be used to transform service and provide an edge for competing in existing markets.

19. True breakthrough innovations using telecommunications and information systems to achieve preemptive market strikes are rare. Briefly describe how USAA achieved a preemptive strike with image processing.

20. Describe the use of the Public Broadcasting System's satellite distribution network to link banks to retail establishments, and explain how this may create a sustainable competitive advantage.

Assignments and Questions for Discussion

1. Discuss why there has been such rapid growth in nonreal-time electronic communication (voice mail, fax, e-mail, and so on), and explain why many of the technologies supporting nonreal-time communication were available many years before they attained widespread acceptance and use.

2. In the Mirrason example in the text, telecommunications is being used to redefine the approach to R & D; telecommunications provides a new source of competitive edge—"communicating as the basis for researching." Explain this new approach to research, and discuss the key advantages that make it superior to the old style of R & D and why it can lead to a sustainable competitive edge.

3. Telecommunications and information technology have been used for many years to help companies run their businesses better. Is the concept of reengineering work using information technology simply old wine in new bottles, or is this concept something fundamentally different from what has gone before? Explain and discuss.

4. It does not appear to be sufficient to gain a competitive edge in existing markets by following the general rules of providing workstation access to the customer or differentiating a standard product through information or convenience. Automated teller machines have been a huge success, whereas electronic banking and most videotext systems have not. Discuss what else is needed for success in using telecommunications to gain an edge in existing markets.

5. Consider the following statement: "Using telecommunications and information technology to obtain true market/product innovation is a function of creativity, not of money." Explain why this is likely to be truer in the 1990s than it was in the 1970s and 1980s, and support your reasoning with two examples: one representing a preemptive strike and the other an instance of piggybacking new services on an existing electronic delivery base.

Minicase 10-1

U.S. Defense Department and CIA

Turning Swords into Electronic Ploughshares

By far the heaviest users of satellite telecommunications have been the military and intelligence agencies. *U.S. News and World Report* estimates that the United States spent over $100 billion gathering information through satellites and spy planes since the end of World War II. Now that the Cold War is effectively over, this military technology offers major opportunities for environmental analysis and planning. The same satellite technology that can make detailed maps of subsurface ocean temperatures to help find submarines hidden in "shadow" areas that trap or distort their sound can track warming trends in the North Atlantic and the Mediterranean, combining the Defense Department's historical data with ongoing monitoring. The United States has over 40 years of such data, still classified as top secret.

Civilian satellite technology usually represents the commercial rather than the research state of the art. Military technology is often as much as 20 years ahead. For example, civilian satellites such as LANDSAT can provide images that can spot only general overall features about wooded areas, such as the advance of the tree line over a period of time. Military satellites designed to spot a tank on the ground can pick out the changes in the species of trees as the tree line moves forward.

With the full set of historical images, scientists could work out which oceanic changes are most likely natural regional fluctuations and which ones signal a general warming trend. There is a long-standing debate about the impacts of Himalayan deforestation on the recurrent flooding of Bangladesh. Access to the military's and intelligence agencies' archives would almost surely help clarify this and many other ecological issues.

The atmosphere of secrecy—Warning: Destroy this document before reading it!—blocks the availability of such resources. Government officials, according to *U.S. News and World Report*, refused even to acknowledge the existence of an organization that informed journalists know exists: the National Reconnaissance Office which supervises many satellite data-gathering operations. Government officials fear that releasing some photographs could reveal the capabilities of U.S. satellite technology, as well as information about countries that are U.S. allies.

Even considering all the concerns about national security, not using all satellite technology seems a waste of essential information for environmental analysis and action. When many informed scientists now view the problems of ocean warming and depletion of the world's ozone layer as far more threatening to humankind than nuclear weapons, why not use the technology designed to defend against the H-bomb to protect against the greenhouse effect?

Questions for Discussion

1. Give three examples of how communications-satellite technology designed for military use can be used in nonmilitary applications.

2. Explain why commercial satellite technology lags as much as 20 years behind its military counterpart, and discuss ways in which this gap can be closed.

3. Discuss ways in which archives of data collected from military satellite scanning operations can be made available to commercial ventures without compromising national security.

Minicase 10-2

Fidelity Investments

Telecommunications and Market Innovation

Fidelity Investments, with $137 billion of assets under management, has six million customers. *Forbes* magazine described Fidelity's use of telecommunications and computing in an article entitled "Watch Out, Citicorp." Above the title is the punch line of the article: "While Congress haggles over changing our six-decade-old banking laws and the banking brass talks merger, technology is making the system itself obsolete."

Fidelity is "a bank in all but name." Fidelity offers as much or more in range of services as Citicorp, with lower fees in most cases. It provides checking accounts, Visa cards that can be used at almost any ATM, discount brokerage, and 200 mutual funds (its original business). "Note that it wasn't Congress that permitted an investment firm to become a near bank. It was technology. Fidelity spent over $1 billion in 1989 and another $150 million last year [1991] (13% of revenues) to ensure that its computers and communication networks are every bit as sophisticated as those of the big banks like Citicorp whose customers it wants."

Fidelity has 233 security analysts, support staff, traders, and portfolio managers. It has 850 information technology staff. It operates a high-speed fiber-optic telecommunications network across the country. In August 1991, a hurricane forced many firms in Boston, where Fidelity's head office is located, to shut down at a time when Europe's securities markets were "in turmoil." Calls jammed the switchboard of other firms; at best, callers got through to an answering machine. Fidelity's network automatically rerouted calls to its service centers in Cincinnati, Salt Lake City, and Dallas.

The network handles over 100,000 customer phone calls a day and processes about the same number of electronic transactions. "This is the essence of the business—servicing accounts. Managing money is almost secondary to that. In the end, its customer-handling capabilities may be more important to the future of Fidelity than whether the Fidelity Magellan Fund goes up or down."

Ongoing developments in Fidelity's electronic customer service include allowing clients to track their account status through a phone call (and soon, from a personal computer) using automated number identification to retrieve a customer's account records as soon as he or she enters a password from a touch-tone phone, and computerized voice confirmation of orders that reads back the transaction from a discount broker's personal computer just as the trade is made.

The cost of this strategy is substantial. "To pay for this technology spree, [the CEO and majority shareholder] has cut deeply into his profit margin. His company netted $32 million last year, a tiny 0.027% of average assets under management." A "traditional" competitor netted 0.19 percent.

As *Forbes* concluded, "Selling mutual funds isn't banking, issuing credit cards isn't banking, offering a checking account isn't banking, stocking teller machines isn't banking. But tie them all together with computers and phone lines, and you have something that looks very much like banking."

Questions for Discussion

1. Explain how Fidelity's telecommunications network makes it a "bank" in all but name, rather than simply a securities firm.

2. The Fidelity network has clearly created a business opportunity. Discuss how you would classify this opportunity (running the business better, gaining an edge in existing markets, creating market innovation) and support your reasoning.

3. The cost to implement Fidelity's network strategy is substantial and resulted in reduced profitability in 1991. Discuss and evaluate whether or not it is a worthwhile investment.

4. In early 1993, Fidelity announced that it will offer full-service banking to its customers. Since Fidelity does not have any bank branches and does not intend to build any, how will it be able to offer such services? What advantages and disadvantages will it have in competing with traditional banks? Who do you think will win and why?

Minicase 10-3

JC Penney

Working with Far East Suppliers As If They Were Next Door

Until 1989, the process by which JC Penney worked with its private-label apparel suppliers in the Far East was costly and very slow. When Penney designed a new product, samples were sent by air to the chain's agents in Asia, who then sent them to suppliers, who then "spec'd" the goods to establish a price structure. That information worked its way back to Dallas; several iterations were often needed to complete negotiations. Even with express air freight, the time to reach the Far East was anywhere from a week to ten days.

One option Penney looked at to cut lead time was satellite communications. Penney has been a leader for many years in using business television and videoconferencing in the United States so that its buyers and merchandisers could "meet" electronically. Penney buyers using videoconferencing can see the goods that are being shown by merchandisers on a large TV-like screen and can place orders frequently, without having to wait for samples or to travel to the head office. Buyers hear about new products almost daily and can see them displayed; they can ask for products to be held close to the camera so they can check the stitching, and they have the authority and information to tailor their purchases to the specific nature of their stores and to local buying patterns. Every new Penney buyer gets training in videoconferencing skills

of etiquette and communication; the video-conferencing system is the core of its buying operations in the United States.

Extending this to the Far East was too expensive to consider. Satellite transmission costs several thousand dollars an hour. Instead, Penney used a specially developed still-image video camera that takes a picture and sends it over standard phone lines. The picture has high resolution and is in full color. Buyers in Penney's Dallas office now bring samples into a studio, where a camera records the picture onto a computer floppy disk, which is inserted into a personal computer that then sends it over dial-up telephone lines to agents in Seoul, Singapore, Osaka, Manila, Taipei, and other cities. The pictures are printed as a hard-copy reproduction. A Penney executive described the process: "For example, we might send an image of a sweater that we want to be produced. Along with that, we might send a message to the buying agent saying that we want to have a 3/4-inch sleeve instead of full-length, or have some sort of new treatment on the shoulder, or some different collar treatment. So the supplier will then manipulate the image a bit (on a personal computer), mock up a sample for us, and send the new image back to us so we can make sure they understand what we meant."

There is very little cost involved in repeating

these steps many times within a few days." "We're able to do more visual verification, rather than hope that the words we chose were clear." Suppliers can make sure an item is right before they make and ship it, using the telephone at a cost of about $4 per minute. Previously, there would be at least a 14-day lead time for sending physical samples. The time difference between Dallas and the Far East is about 12 hours; because images can be sent and received automatically, images are waiting for the buyers when they arrive at work.

The cost of the equipment for each location is $30,000. The quality of images is far superior to facsimile but inferior to a conventional photograph. Penney expects the gap to close as telecommunications technology advances. Penney plans to link Europe, the Far East, and the United States; European cities like Milan and Paris are where buyers go to spot the latest trends. The faster they communicate with Penney's Dallas merchandisers, the quicker Penney can respond in an industry in which fashions can change very quickly indeed. Penney has also used CAD (computer-aided design) techniques to speed up the design process. A blouse can be designed right down to the stitches,

for instance, on a CAD workstation, and a still-video image can be created and transmitted anywhere. Images can also be altered at the computer, so the designer can see how a blouse will look in different colors and sizes. The altered images can be immediately retransmitted to the Far East.

Questions for Discussion

1. Satellite videoconferencing is too expensive today to be an option. Assuming it were cost-effective, how could JC Penney use it in its relationship with its Far East suppliers? What would be the benefits to Penney, Penney's buyers, its agents, and the suppliers?

2. How would the agents' roles be changed?

3. Most of JC Penney's competitors operate in the old way, with it not atypical to take six months to go through the process just described. What effects do you think the difference has on relative competitive strategies? What can Penney now do that they once could not? What are the likely business consequences for Penney and for its competitors?

Minicase 10-4

Mexico

The Growth Opportunity of the Telecommunications Industry

One of the hottest stock markets in 1991 and 1992 was Mexico's. The country broke out of a decade-long slump through aggressive government and business action that has generated an economic growth and energy matching the breakaway "dragons" of the Pacific Rim—Hong Kong, Singapore, Korea, and Taiwan.

Telmex (Telefonos de Mexico) was one of the fastest-growing stocks, jumping 237 percent in 1991 even though its phone system remains dreadful. There are only five phone lines for every 100 Mexicans (Canada has 80 for every 100 Canadians; the United States has four *cellular* phones per 100 population), ranking Mexico 83rd in the world. The average waiting period for a phone line is three years, repairs take months, and most pay phones operate for free because repair crews cannot keep up with vandals. Lines are frequently crossed; repair crews traditionally sell their services to the highest bidder. Clearly, the state of the phone system affects the daily lives of Mexicans: "[The village of] Ayolta was so poor it didn't have its own public phone. . . . [Adeleida] had no way of calling

the doctor to stitch up the wound. She had no way of calling us to say she would be late for work because she had to go to the hospital. In emergencies, she had no way of contacting Felix, who worked nights."[3]

Some of the problems are the result of the 1985 Mexico City earthquake that badly damaged the Telmex network and switching centers. Bad management and gross underfunding compounded the difficulties. The country's huge foreign debt prevented access to capital, and Mexican organizations were prevented from importing foreign equipment by the government's policy of national technology development and manufacturing. In 1992, a U.S. telecommunications manager commented that "Telmex didn't get this bad overnight, and it's not going to get better tomorrow."

Why, then, has the price of Telmex's stock shot up, and why do so many experts and investors see it as an exciting company? The confidence is based on two factors: the privatization of Telmex, which was sold to a consortium of Mexicans plus Southwestern Bell and France Telecom in 1990, and the

---

[3]Patrick Oster, *The Mexicans: A Personal Portrait of a People* (New York: Harper & Row, 1989), p. 31. This book is about the business and organizational side of telecommunications. The quote here is a reminder of what we take for granted—that telecommunications is a central element of the quality of our lives, as much as electricity and clean water, and that for most of our fellow human beings on this planet, all three are absent.

explosive growth of the Mexican economy. According to the director general of Banco International, as quoted in the *Wall Street Journal* in early 1992, "Buying Telmex is very much buying Mexico." A Paine Webber report on Telmex in late 1991 described it as "the right company in the right country at the right time."

The new owners installed a record number of lines in 1991 and cut costs by several hundred million dollars. Earnings are growing faster than those of the Regional Bell Operating Companies. Southwestern Bell's 10 percent holding in Telmex cost it $953 million and is now worth almost $3 billion. Telmex installed an 8,000-mile line of fiber optics connecting the nation's main cities. Switches dating from the 1940s have been replaced and literally sent to a museum. Telmex got tough with the two main equipment suppliers who have had a local monopoly through a lack of competition due to the old foreign import restrictions. (These restrictions do not apply now.) Telmex brought in AT&T as the third supplier.

All this may sound as if a very poor Third-World company in a Third-World economy will make progress but won't catch up to the United States. Perhaps. Mexico is still poor. So is Korea. So was Singapore.

Cemex (Cementos Mexicanos) is the largest cement producer in the Western Hemisphere and will be the largest in the world by 1994. It is already the lowest-cost producer in the world. Even with the abysmal Telmex infrastructure, the accompanying bureaucracy and the problems it generates, Cemex has been able to build the world's best telecommunications network in its industry, using satellites initially and then Telmex's fiber links. Cemex is at the forefront of process control, and each day its chief executive gets precise and up-to-date operating data from Cemex's plants. Less than 40 percent of the cement sold in the United States is made by U.S. firms, and only 5 percent of Canada's is made by Canadian companies.

Indeed, buying Telmex is buying Mexico, and buying Mexico is buying Telmex.

Questions for Discussion

1. In addition to poor, outdated equipment and lack of investment capital, discuss some of the other factors that are crucial to Telmex's continued turnaround and development.

2. With Telmex being in such bad shape and providing poor service, how has Cemex been able to build the best telecommunications network in the industry? Also, explain how this network has given Cemex a strong competitive advantage in world markets, and assess the sustainability of the advantage.

3. Advise the CEO of Telmex regarding a strategy for the next steps to further facilitate the global competitiveness of Mexican companies like Cemex.

11

Defining the Telecommunications Platform

Objectives

- ▸ Understand and be able to explain the business needs for integration
- ▸ Learn and apply the telecommunications services platform map to define the business capabilities needed in a telecommunications delivery base

Chapter Overview

| | | | | | | |

Chapter Overview

This chapter describes the step in the telecommunications decision sequence that provides the bridge between the business-centered—and client-driven—search for business opportunities and the designer- and implementer-centered definition of the technical architecture for the platform that is needed to make those opportunities practical. Of all the chapters in *Networks in Action,* this one is most central to the overall client/designer/implementer dialog because the issues it addresses are simultaneously business-focused—what business functionality do we need in our telecommunications resource?—and technology-centered—what architecture and what degree of integration and interoperability of networks do we need?

The chapter reviews the business forces that are driving the need for integration, focusing on the LAN as the emerging technological and organizational center for innovation. It discusses the reach, range, and responsiveness that define the capabilities of the platform and relates them to technical choices of architecture and standards.

| | | | | | | |

The Business Forces Driving the Need for Integration

The telecommunications platform is a business resource that must be able to turn a client's business opportunities into practical applications. This requires a technical architecture, as described in Chapter 8, built on specific standards and choices of products and services. But trying to talk to clients in terms of SNA, TCP/IP, token ring, SMDS and the like is hardly the basis for a dialogue; there must be a bridge between a client's business-focused concerns and an implementer's technical ones.

This chapter uses the second of the core frameworks of *Networks in Action,* the telecommunications services platform map,[1] to provide this bridge. In terms of creating an effective dialogue with business managers, it is perhaps the single most important of our frameworks because it provides a business perspective on the central technical issue for the designer and implementer: the choice of architecture and standards. It is with respect to this issue that managers generally are most bewildered or frustrated when they try to talk with

---

[1] This framework was first presented in Peter G.W. Keen, *Shaping the Future: Business Design Through Information Technology,* which focuses on the role of senior business management in deploying IT. One of the main findings in interviews with executives in close to 30 large firms was that they did not hear any compelling business message about the need for integration of the IT base. This problem was at the root of their own and their information services managers' difficulties in working together.

telecommunications specialists; so much of the jargon of the field relates to architecture, integration, and standards. Given that most managers either do not have a comfortable understanding of those three conceptually simple words or cannot see why they are so important for business choices and technical decisions, it is not surprising that, in most firms, multitechnology chaos has become the norm. Here is what two experienced researchers said after surveying companies' experiences with CAD/CAM (computer-aided design/ manufacturing), a telecommunications-dependent activity that is at the core of modern manufacturing:

> Incompatibility created a kind of Tower of Babel effect; different computer systems used incompatible software mounted on different hardware. Communication was possible only in the most fortunate of circumstances. . . . Even different generations of the same system were sometimes unable to communicate with each other.

These comments remain true today, though matters have improved, largely because they had to. The many incompatibilities that mark service and manufacturing firms have become a growing business problem, so that companies are now recognizing that something must be done. About the time the comments were made, a new term was entering the information services vocabulary: **systems integrators**—companies that provide other companies integrated solutions to business needs, ensuring that software, hardware, and telecommunications work together. Leading systems integrators include Andersen Consulting, EDS, and computer vendors such as IBM and DEC. The revenues of this new industry segment have grown very rapidly for a simple reason: There is so much systems disintegration to deal with.

Business clients generally have very limited awareness of the problems of incompatibility until such problems become apparent—that is, when they create a barrier to business development and/or operations and it is too late to prevent them. Their priority then becomes getting a business need met at the lowest cost and as quickly as possible. To them, "integration," "standards," "architecture," and "interoperability" are technical terms, not business issues that they must address.

In practice, however, they really *are* business issues. The rise of systems integrators is the proof; in many cases they are being paid more to fix up incompatibilities among existing systems and services than is being spent on developing new ones. Some common trends that make it essential for companies to be able to communicate across departments, functions, locations, and business processes include the following:

1. *A move away from the traditional hierarchical organizational pyramid,* with its tightly defined boundaries between business functions. This structure is too inflexible to be able to meet the pressures of constant change and time-based competition that mark the 1990s. The newer forms of organizational design aim at streamlining work, moving toward a team-based mode of operations, improving communication at and across all

levels, and empowering staff by giving them more authority. All of these aims are practical only with first-rate telecommunications and information sharing. There is no point in giving an employee the authority to make a decision if he or she can only tell a frustrated customer, "I'm sorry. I'd love to help, but I don't have the figures you need. They're in the head office." Similarly, firms can encourage communication between the field and the head office as much as they want, but good intentions easily get frustrated if there is no comprehensive electronic mail system, easy-to-access data resources, and well-publicized directory of names and locations.

2.   *Cross-selling of products and "relationship" management.* This trend is especially marked in financial services. A typical bank, for instance, may have over 250 different consumer products, but the average customer uses fewer than four of them. Banks are moving rapidly to target products to customers, provide consolidated statements, and provide a single contact point for service and queries. Insurance companies are similarly aiming at providing a customer with a combination of insurance and savings plans that meet their lifetime needs. Companies like Merrill Lynch and Fidelity Investments provide a mix of banking and security services, including checking accounts; their aim is to capture the customer relationship and tightly cross-link services.

Many firms committed to such cross-selling and integration of products have found that their telecommunications and computer systems lack the essential integration. One major insurance firm, for instance, was unable to implement its intended strategy because it could not easily find out just which of its products a given customer uses; the automobile, home, and life insurance divisions operated entirely separate technology bases that could not be yoked together. Another found its marketing strategy for "seamless" customer service undermined by the fact that it had nearly two dozen 800 numbers—seams—through which customers had to maneuver to find the right person. It is in this sense that the firm's technology platform strongly influences for the next three to five years its "business degrees of freedom"—the practical range of business strategies and innovations the firm can consider.

3.   *Electronic linkages (instead of paper and mail)* to a growing range of business partners. Increasingly, this is becoming a basic condition, not an option, for getting business. By 1995 it will be as difficult for even a small firm to deal with major customers and suppliers without electronic data interchange and on-line information and transaction links as it is now to operate without telephones.

These trends both make telecommunications an increasingly basic business and organizational resource and require a far greater focus on the linkages across services and facilities than has generally been the case in most firms. Anticipating the needs for future linkages, either as a competitive opportunity or a likely competitive necessity, is part of the *business*-centered steps in the decision sequence. Unless clients and designers together examine the business criteria for integration and interoperability, it is very likely that the technical solution implemented will overlook potential needs in these areas.

"The Network from Hell": The Challenges of Integration

Corporate Computing magazine reported on the practical problems of integration, using as a test case a company that wanted to connect four desktop computer systems to a Novell NetWare file server and a Digital Equipment VAX computer that acted as both a file server and a database server. The four systems were based on Macintosh, DOS, Windows, and Unix. The need was—and is typical for most companies today—to share data files, share electronic mail messages, and access corporate databases. The article began with an example concerning a different company, Associated Press, which had easily integrated another DEC VAX system with a network of Sun workstations, used by 2,000 reporters worldwide. The AP integration "worked like an IS manager's dream," but the other situation was the "network from hell." "Many large enterprises don't have a homogenous computing environment like AP's. They have various computer platforms and many different networks, which together create a huge puzzle. Linking the disparate pieces of this puzzle is no dream—it's a nightmare."

Forty-eight different connectivity products were needed to glue together the four desktop capabilities and the servers. Two separate approaches were tested, one centralized and one distributed. The centralized strategy relied on the NetWare file server to coordinate communication; it "speaks" the language of the desktop systems. The distributed one used TCP/IP as the common transport mechanism, in effect translating from, say, the language of AppleTalk used by the Macintosh system into a TCP/IP-related protocol, such as its Network File System for UNIX. Neither method was perfect. For example, the distributed system could not handle e-mail messages larger than 64 kilobytes because TCP/IP relies on the Simple Message Transport Protocol (SMTP), designed over a decade ago, when the slow transmission rates then available did not make it practical even to envisage sending large files via e-mail.

The article summarized the practical state-of-the-art at the end of 1992 as follows:

> There is still no single operating system, software package, or protocol that can connect everything without a lot of jury-rigging. So your connectivity strategy must be founded on your installed base and future needs, and on an assessment of how soon promising technologies will become reality. Whether you centralize or distribute your network, choose many protocols or one, you must consider both your users' immediate needs and your long-range business mission. . . . [T]he kind of

success AP had with its network comes only with careful planning, as many others have found.

Where does this planning start and who is involved in it? The 48-product test the magazine described followed an implementer's plan. Earlier decisions had created incompatible platforms, and technical ingenuity was required to create coherence. The (unnamed) company's networking plans had been application-centered, not platform-based. There was no architecture.

The term *enterprise network* is often used to describe a companywide communications platform. In reality, the "network" is a set of networks that meet the three main needs of integration: shared messaging, shared files, and shared access to corporate databases. The platform needs many standards, internetworking equipment such as routers and gateways, and special-purpose software if it is to be a "network from heaven." It will be built on a core standard or set of standards that determine the overall architecture. IBM's SNA remains the dominant choice here. *Corporate Computing*'s research unit reported that 28 percent of the overall network load in typical large firms is SNA-based, versus just 3–4 percent for TCP/IP, OSI, DECnet, and IPX (Novell). However, most companies now mix protocols and standards. Seventy percent of Fortune 1,000 companies use SNA, 58 percent TCP/IP, 32 percent IPX, 46 percent Decnet, and 20 percent OSI. There is to date no single, universal architecture for the enterprise platform.

In this context, "planning" must span the entire decision sequence, from business need and opportunity right down to choice of protocols and operating systems.

Reach, Range, and Responsiveness

The telecommunications services platform map was developed specifically to help business managers, especially senior executives, both understand the issues in business terms and contribute directly to planning and decision making. Recall that it defines three dimensions of *business* capability:

1. *Reach*: How far inside and outside the organization can we directly link people and organizations to our telecommunications resource?

2. *Range*: What information can we directly and automatically share across business processes?

3. *Responsiveness*: What level of service can we guarantee?

By identifying the required levels of reach, range, and responsiveness, a designer establishes the business criteria for selecting an architecture and standards. A designer and an implementer can then work together, without having to bother clients with telecommunications jargon; the client has no need to know. Yet, in effect it is the client who has designed the integrated architecture by defining the platform requirements.

In the sections that follow we will address the issues of reach, range, and responsiveness in some detail.

Reach

Levels of Reach Reach, the most purely telecommunications-centered of the three dimensions of the platform, refers to the extent of connectivity and interconnection of communications facilities. In discussing reach we consider the following six levels:

1. *Within a single location,* such as a head office department, branch, or single building. The main technical network building block here is obviously the local area network. Links between adjacent buildings or among a closely spaced group of buildings, such as an office complex, can be handled by bridges and routers. The SONET network implemented by the University of Miami (see Minicase 6-1), in which the university sought to link its cancer research center and medical sciences building, is an example of this level of reach.

2. *Across a firm's domestic locations.* This level of reach requires some combination of LAN-to-LAN connections and LAN-to-WAN connections. Routers have become the main tool for linking LANs, with smart hubs rapidly emerging as the next major innovation. Frequently, the standards used in implementing a firm's backbone WAN are the most binding constraint here. For example, as recently as 1992 there were few proven tools for linking LANs to SNA-based WANs. The proliferation of new vendors in the LAN and LAN-WAN interconnect market points to both the opportunity and the difficulty. Even with tools such as bridges, routers, gateways, and smart hubs, increasing reach is not a simple and automatic exercise in a multiprotocol LAN environment. The new manual published by IBM in November 1991, *IBM/APPLE Enterprise Networking Guide for SNA Products,* contains over 300 pages of nonintroductory, in-depth specifications—and this manual is for one of the longest established and most basic interconnection needs.

3. *Across international locations.* Extending reach across international boundaries is a totally different issue from extending domestic reach, regardless of the countries involved. The differences are not so much matters of technology as of regulation, service, cost, and quality. As yet, efforts to provide "one-stop shopping" for international telecommunications have largely failed; instead of being able to work with a single supplier, companies must negotiate with most or even all the national PTTs individually. Costs vary widely, and in many cases they are several times those of U.S. suppliers. Many restrictions on the use of particular types of network exist in many countries, with the public data network, based on X.25, the only option.

    In many situations, the choice of private leased lines versus public data networks versus value-added networks for international operations comes down to the option that provides the most reach at reasonable cost and with high reliability. In several of the minicases presented in earlier chapters, reach was restricted by PTT capabilities. Volkswagen (see Minicase 2-4), which faced constant problems in restructuring its international corporate backbone network, had to use satellites to link its Czech Republic plants to Germany, and Spain's PTT, Telefonica, provided only low-speed circuits.

As reach extends from local area networks to domestic and to international networks, the available transmission speeds are drastically reduced. LANs operate in millions of bits a second, ranging from a typical 2 mbps for standard Ethernet LANs to 100 mbps for FDDI LANs. In the United States, speeds of backbone networks vary from 56 kbps up to 1.54 mbps (T1) and 45 mbps (T3). SONET is in early but real implementation, with speeds of up to 2.4 gbps. International PDNs offer 9.6 kbps (typical of older X.25 networks), and ISDN provides 64 kbps. In this context, a firm like Rockwell (see Minicase 5-2) that needs global reach plus a high degree of range relies on private leased T1 and T3 equivalent links.

Chapter 14 describes the international telecommunications industry in detail. International reach is becoming a priority for firms that need global communications for global business operations, service, and competition, and more and more PTTs and VANs are providing services to meet the demand. Privatization and deregulation are greatly improving quality and cost, and alliances among PTTs and between PTTs and U.S. services providers are growing. That said, international telecommunications is a political, technical, regulatory, and economic maze. Volkswagen's experience, described in Minicase 2-4, is typical rather than an exception.

4. *Customers and suppliers with the same technology base as the firm's.* Here, the main issues are common standards, both technical and business-based, with the most relevant ones being the general standards for EDI (such as X12 and EDIFACT) and for electronic messaging (such as X.400), and industry-specific EDI standards such as VICS in retailing or SWIFT in international banking. Links to other companies are most likely to include a VAN or VANs by the very fact that these networks are designed for shared services; they differ from PDNs in that they are configured to meet special-purpose needs rather than general-purpose ones. For instance, Infonet offers X12 and EDIFACT-based global VAN services, which simplify customer connection and reduce the investment in equipment required to interface to public networks.

5. *All customers and suppliers.* Obviously, if it is difficult to link a firm's internal networks, each of which uses different protocols, vendors, and computer operating systems, then the same limitations apply to linking to other firms. Telecommunications designers must track very carefully and continuously the technical architectures of their competitors, their customers, and industry leaders. There is now a business risk inherent in being outside the industry mainstream.

A major bank lost a $400 million client in early 1990 because it could not implement a widely established standard for electronic payments. The client, a supplier to automotive manufacturers, had eliminated all the paperwork associated with orders and deliveries and wanted to move on to doing the same for payments. It needed to connect its supply system platform to the bank's, but it was forced to drop the bank when it became clear that the connection was not feasible.

British retailers abandoned working with five leading banks to implement EFTPOS (electronic funds transfer at point of sale) because their own POS standards and systems were too far ahead of the banks'. The retailers instead looked to value-added networks to handle their needs. Once electronic linkages for EDI and payments become the norm in an industry, customers look to suppliers who can meet such standards of service and drop those that cannot.

6.   *Anyone, anywhere.* The ideal, as yet unattainable, but not inconceivable. At some point within the next ten years, we can expect to see the telecommunications equivalent of an electrical utility across the United States and within leading countries such as France, Holland, the United Kingdom,[2] Japan, and Singapore. It may be another decade before such a utility is fully deployed and stable. Many new standards must be defined, especially for the multimedia transmission via fiber optics into the home, which most commentators see as a key to creating a mass consumer market for telecommunications-based products, including new forms of entertainment, education, home shopping, financial services, and the like.

The reach of a firm's telecommunications platform increasingly determines *basics* of service and *basics* of organization and coordination. The 1980s were marked by the shift away from large, central computer facilities that could be accessed through dumb terminals only to, first, stand-alone personal computers and then to local area networks as an extension of personal computer capability. Now, the focus in the telecommunications industry is almost entirely on interoperability. Terms such as *Internet, multiprotocol,* and *any-to-any connectivity* have entered the technical specialist's vocabulary, and the firms whose stock prices shoot up and down are increasingly those competing in the interoperability market. The field is very volatile, with new developments occurring every few months. Network management across networks is the emerging, but as yet very immature, area of innovation in both equipment and software.

TCP/IP has become the single most effective standard in providing reach to businesses. Its limitations—security, the number of users it can support in a single network, and inefficient use of transmission methods—are offset for

---

[2]The areas in which Europe leads are cellular technology and ISDN. In both instances, much of the impetus reflects geography. Countries smaller than many U.S. states have their own technology and standards, and linking, say, Germany to Ireland may involve five or more incompatible systems. Thus, the Germans are leading efforts to create a pan-European standard for mobile communications, and ISDN was intended to create pan-European communications of all types.

many companies by its two main advantages: It works, and it is topology-independent. An increasingly pragmatic approach to internetworking is to use routers and smart hubs to interconnect LANs, with TCP/IP the main transport mechanism from LAN to WAN. Bell Atlantic's $2 billion plan for a new customer-service infrastructure adopts UNIX workstations within an SNA environment via TCP/IP. Many companies encapsulate SNA traffic as TCP/IP packets; to **encapsulate** is not to convert the protocol but simply to treat the SNA frame—control data plus user data—as a straightforward TCP/IP communication and to add extra control data. This is inefficient because the TCP/IP overhead is added to the SNA overhead. Today's low-cost, high-speed network services make this an effective option, however inefficient it is, for it allows an extension of reach without having to convert equipment or software.

Connection Is Not Communication It is important to understand what reach does not provide and why we distinguish between reach and range. Reach refers to connectivity and the ability to move bits within and across networks so that they can be received and their format recognized and used by the receiver. A bridge, for example, simply allows two LANs that use the same protocol to operate as if they are a single LAN; Ethernet to Ethernet bridges are commonplace. A multiprotocol bridge/router will link the most widely used LAN protocols, such as token ring, Ethernet, Banyan Vines, Novell's IPX, and TCP/IP, by converting the traffic from the original protocol to that used in the receiving network. X.400 ensures that an electronic mail message in MCI Mail's addressing and transmission format can be received by a mail service like Bitnet, the worldwide academic and research network. X12 goes somewhat further; it defines the format of the contents of a transaction message—"This first field is the customer account number, the second is the type of transaction," and so on. The standard enables the receiving system to recognize the format.

In all these instances, the process does not interpret the content of the messages; it moves them via a protocol or message format that enables the receiving service to use the incoming bits. The range dimension of the telecommunications services platform addresses the extent to which the *content* of the message can be used across business services. The difference is directly analogous to making an international telephone call. The technical features of the world's phone systems vary widely, including voltages, dialing conventions and codes, interfaces, type of switches, and transmission protocols. They are interconnected through the equivalent of bridges, routers, and gateways so that from the United States you simply dial "011" to access international service, then the country code, then the area code and number. Thus, to call Bela Horizonte in Brazil, you just dial 011 55 31 and the local number. This is anyone/anywhere reach in action. But how fluent is your Portuguese? When the person at the other end picks up the phone, reach is not enough unless you both speak a common language; reach ensures that your words are heard but not that they are understood.

In most firms, there is plenty of reach for users of personal computers. For example, they can share word processor files, sending a report over a local area network to someone in the same department or through a wide area network or over an electronic mail system to someone located elsewhere. However, the received file may be unusable without additional conversion and processing. An MS.Word file created on an Apple Macintosh is incompatible with the MS.Word software on a DOS machine. A WordPerfect file created on, say, a Toshiba 1200XE personal computer cannot be directly processed by MS.Word running on another 1200.XE. Even when the file is converted, the format of the report remains different, and most special formatting characters, such as bold, underlining, and boxes around figures, are lost or misinterpreted. This is not a problem of reach because the file arrived without difficulty.

Many telecommunications specialists too easily equate connectivity with communication. Connectivity is the essential prerequisite for, but is not the guarantor of, communication, as in the example of phoning Brazil. In the 1980s many information-systems professionals complained of the over-narrow perspective of their telecommunications peers on connectivity; to be fair, the technical problems of just ensuring connection were so large that telecommunications designers and implementers had little attention to devote to broader issues. The IS community, for example, saw point of sale as a complex application involving message formats, linkages across databases and transaction processing systems, and interfaces between different operating systems. For the traditional telecommunications implementer, POS was simply a matter of linking the firm's transmission systems—of moving the bits. Point of sale really requires integrating a wide range of information systems, operating systems, and computing resources, in addition to the far simpler need to connect transmission facilities.

Reach is the core of telecommunications as a business resource and connectivity the bedrock. Range is increasingly the constraint, opportunity, and challenge. Many of the architectures described in Part III focused on reach. The most obvious example is X.25. For a firm like THEi (see Minicase 1-1), X.25 provides the simplest base for defining an international telecommunications platform for relatively simple and low-volume traffic. The company's reach can easily be extended to international customers and suppliers for most standard types of message traffic, including EDI.

By contrast, SNA provides excellent international reach within a firm and can handle a far broader range of transaction traffic than a standard X.25 network, which is best suited to bursty traffic. Linking non-SNA and SNA-based transaction-centered platforms is not at all easy. For a private network that must handle complex, transaction-based traffic, SNA may be a better choice than X.25 for a firm—in terms of efficiency of operations but at the cost of less reach across the world outside the firm. For this reason, all major international VANs offer X.25 capabilities.

Range

Before we discuss the levels of range, we summarize the historical trends in the computer field that have contributed to major and continuing problems in cross-linking information and applications. Until the mid-1980s, a company's use of information technology was driven by its information systems organization; the telecommunications function was mainly concerned with operations of telephones, telex, and cables. The tradition of application development in the IS field led to several sets of programs for, say, sales accounting, production scheduling, inventory management, and finance, each one built on separate—and, of course, incompatible—hardware and software. The general pattern in large firms has been that IBM's flagship products and architectures dominated financial applications and applications that required massive databases. Digital Equipment Corporation dominated engineering applications, although just as IBM has been displaced by many of the niche vendors that offer alternatives to its centralized systems, DEC has encountered powerful competition from companies such as Sun Microsystems, Apple, Hewlett-Packard, and Novell, who provide either hardware, software, or communications that challenge its historical strengths. Tandem, Amdahl, and Fujitsu offered IBM-compatibility in mainframe computers, challenging IBM's business core.

Many of the leading niche players of the early 1980s were in serious trouble within a few years. For a decade, Wang dominated word processing, but in 1991 it essentially withdrew from its main markets and signed an agreement with IBM to sell IBM's products. It will concentrate on its new niche in image processing, in which its WIIS system has strong customer loyalty. (Its main competitor here is IBM's Imageplus and Filenet.) Wang's troubles were paralleled by other former leaders' troubles. Prime and Data General were profitable minicomputer vendors before that segment of the market came under attack from larger personal computers and workstations and smaller, mid-sized mainframes. Unisys, which was formed from a merger of Sperry and Burroughs, launched an ambitious but ineffectual strategy to displace IBM as the major vendor to large companies and looked powerful for a while, mainly through its early commitment to open systems and OSI. By 1991, Unisys was in serious financial trouble; the promise of open systems is not the same as proven delivery. Cullinet dominated the field of database management systems in the early 1980s but had disappeared as an independent entity, having been acquired and broken up after years of eroding financial and market performance.

These companies largely lost their original edge for a combination of two reasons: failure to invest enough resources in research and development, and failure to move their products with the mainstream of development in standards and integration. Wang is the most obvious example of the latter problem. In the early 1980s, Wang dominated word processing; by the late-1980s, word processing had moved from being handled mainly on special-purpose machines to being a standard part of personal computer usage. Office technology expanded beyond word processing to electronic mail, desktop publishing, local area networks, and PC-to-mainframe links. Even customers *most* satisfied with Wang's systems began to drop it as a strategic supplier. Wang had gambled that it could successfully leverage its position through proprietary telecommunications. Even though its WANGNET was very advanced, exploiting broadband transmission, Wang's deliberate decision not to move with the mainstream in standards and its weak computer hardware cost it heavily.

For designers and implementers, a major lesson to be drawn from the history of the computer industry over the past few years is that in evaluating vendors, the specific features and cost of a product are often less important than the vendor's relative strengths in R & D, sustained product innovation, and product evolution in harmony with mainstream trends in standards. Although the history of the digital telecommunications industry is far shorter than that of computing, dating mainly from the mid-1980s, the same pattern is likely to emerge, if anything at a far faster pace; as will be discussed in Chapter 14, the industry is marked by a bewildering rate of change. Just as Microsoft, Lotus, Dell, Compaq, Novell, and Oracle were all start-up firms in a computer industry that achieved half a billion in sales in under a decade, a telecommunications firm such as Cisco in just seven years reached $300 million in sales in a market—routers—that literally did not exist when it was created. And we can expect many casualties in telecommunications. The winners will certainly be firms that can add to range. In the 1980s, the need was connectivity (reach); today the need is reach with range.

The Business Impacts of Lack of Range The previous summary of trends in the computing industry is of necessity oversimple, but it highlights the extent to which databases and processing systems in large organizations became and still largely remain incompatible. They were developed for individual applications in individual business units, using incompatible hardware, vendor-specific operating systems, and vendor-specific database management systems. Vendors focused their product and R & D on a narrow base. Computing was vendor-driven, not standards-driven.

Whereas it is telecommunications that provides the tools and standards for extending reach, increasing range depends far more on developments in the information systems field. The structured query language (SQL) interface is rapidly becoming a key to making it practical for any single application to access data from a variety of databases; the major providers of database management software are moving to ensure that their products are SQL-compliant. In telecommunications developments, TCP/IP has dramatically improved sharing of data in a UNIX operating system environment.

Such progress is fairly limited, though. The most typical situation in large organizations is that a personal computer/workstation can access most of the firm's databases, but only as separate operations. The user does not realize this; the operating system or menu of options hides the communications activities needed to make connections, and each of the accesses is independent of the others. For example, in most firms a client's name and address, most recent checking account deposit, mortgage balance, and the status of his or her loan application are in separate databases.

Business forces are pushing the trend toward cross-linking such data. The organizational and technical problems are immense, though so too are the advantages to the relatively few firms that have integrated their customer data. Two of the best-known instances are the Royal Bank of Canada and USAA, both of whom explicitly defined a long-term architecture for their IT platform as a key business policy. The Royal Bank gets response rates of over 40 percent, versus the typical 2–4 percent for mailshot promotions and telemarketing, through its Service Reference File (SRF), which contains information on customers' demographics, use of products, and the profitability of the relationship. When the Canadian government changed the rules for personal pension investments, the bank was able to query the SRF that afternoon to identify customers of a given age range, mix of savings and investments, and the like. By evening, it had brought in retired employees, who then called the prospects over the telephone. The Royal Bank took 60 percent of this new market by the weekend.

USAA was the world leader in adding image processing to the range of its platform. Like the Royal Bank of Canada, its response rates in targeted marketing average over 40 percent. There are many stories of its level of service. A recent one was told to one of the authors of this book by a student, who went one Saturday to look at new cars, not intending to buy one. He found exactly the car he wanted but needed to be sure he could get a car loan from USAA, so he phoned a customer service agent. A USAA agent can pull up on a single screen every piece of information about a customer, even including handwritten letters. In this case, the agent asked the student for the manufacturer's identification number for the car and then said, "Please put the phone down." The student was a little surprised and did not know if his request for a loan would be met. But just as he put the phone down, the dealer's fax machine began printing out a loan approval and a new auto insurance policy. The agent had been able to check the student's credit history and driving records and had then issued from the workstation all the data needed to make the loan, issue a policy, and update police records about ownership and insurance.

This is not a gee-whiz business innovation, but it is a very new way of doing business. It is also one that only 5 percent of USAA's competitors can now match or will be able to match within the next five years. Telecommunications reach opened up many business opportunities in the past decade, and a lack of range becomes a competitive barrier.

RBC, USAA, Federal Express, BancOne, Wal-Mart, and Frito-Lay, to name but a few of the companies best known for effective exploitation of IT, all explicitly recognized the business importance of integration along the entire distribution and management chain as a driving force for investment in telecommunications and computing. This is why, even for smaller projects, the *client* must define the business functionality needed in the platform. The designer is the bridge here between the business client, who has never before needed to think of integration in such terms, and of the implementer, who thinks of it fairly constantly, but mainly in terms of standards and products. Telecommunications implementers and the equivalent information systems or database management implementers see only part of the technical picture, and little, if any, of the broader business picture. To the traditional telecommunications specialist, the issue is mainly reach; to the IS technical expert, it is range.

Levels of Range The range of a firm's platform defines what information can be directly and automatically shared across business functions and business processes. In discussing range we consider the following four levels:

1. *Simple messages.* These include electronic mail, spreadsheet files, word processed files, and, increasingly, diagrams and multimedia photographs, images, and even music and video. Many older computer operating systems treat these as entirely independent units, so that the range is essentially at level zero, but this is becoming the exception because of the trend toward such operating systems as Microsoft Windows, Apple System 7, IBM's OS/2, and their newer extensions, such as Microsoft's NT and IBM/Apple's Kuleida project. These operating systems routinely allow spreadsheets to be inserted into reports, photographs to be combined with text, and data to be downloaded into a spreadsheet. Even when the operating system does not directly enable these functions in the way that Windows does, software is available to the information systems developer to facilitate them. Depending on the standards used in the architecture, specific types of personal computer and software may share messages.

    In these cases, what is transmitted is the equivalent of a posted envelope with something inside it. It is sent to a business process, or more technically to a software application accessed by a terminal. An electronic mail message, for instance, can be sent as far as the reach of the network extends. The message is a discrete unit, with no transaction capability; such capability is added through the software application. The message also needs software processing to organize it as part of a database or comparable share information resource. This first level of range amounts to "I can get the information to you—you decide what to do with it."

    X.400, the OSI standard, is positioned to become a key element in extending messaging systems across organizations and architectures. Its main limitations are the complexity of addressing it creates—to locate someone, a user must specify lots of locations separated by "@"—and the complexity of definition of the original 1984 standard and the even more complex 1988 one, which impedes easy implementation of X.400 in products. That said, it is a comprehensive standard that is being widely adopted by vendors and users alike.

2. *Access to separate data stores.* We distinguish here, perhaps somewhat artificially, between simple messages and data stores. A message is data, of course, but it does not involve a complex structure that needs special processing to access and interpret it. A message is sent as a whole block and is not organized so that its elements—subunits—can be accessed. A unit is a memo, file, spreadsheet item, or picture; a subunit would be a phrase, a column of figures, or a face. Users of an electronic mail service can enter a command such as "list messages" or "read message 1," but they cannot enter "find any message that includes last year's sales figures." Similarly, they can request that a spreadsheet of financial data be sent over

the network, but not "send the figures from any spreadsheet that shows monthly budget estimates for Department 21."

Providing the ability to access information at the element level, instead of at the message level, requires that the information be organized, catalogued, indexed, and cross-referenced using complex database management systems (DBMS). DBMS have become one of the three major areas of innovation, investment, and breadth of use in the field of information systems, to the extent that the term *computing* is a misnomer. Information management is the priority for most organizations. (The two other areas of priority, described later, are client/server computing and object-oriented software development methods.) The technology and applications of DBMS are areas of specialization in and of themselves, and neither telecommunications designers nor implementers can be expected to know much about them, any more than the DBMS expert can keep up with the details of telecommunications. That said, access to shared information is a driving force in business that demands attention from and communication between telecommunications and database specialists.

The main trends in DBMS are toward distributed relational database management. "Distributed" means that the databases needed to handle a query or transaction are on separate hardware systems; telecommunications is absolutely essential here. "Relational" refers to how the data is organized to allow retrieval at the most detailed level of elements and to ensure the maximum degree of cross-referencing of elements. From the perspective of a telecommunications designer, the main standards are those that permit ease of access; SQL is the most important of these. From the viewpoint of information systems professionals, the key issue is the DBMS itself, with IBM's DB2 and Oracle dominating the field and Sybase and INGRES rising fast.

Database management systems are heavily dependent on other aspects of software and hardware, especially operating systems and the design—or often lack of real design—of existing transaction processing systems. A common problem is the lack of data management disciplines: consistent definition of terms, accurate and reliable rules for updating or changing items, and redundancy (that is, the same item being defined and input to many different systems).

Accessing heterogenous distributed databases generally requires the use of APIs (application program interfaces) via gateways that contain additional intelligence beyond that provided by routers, plus servers specifically designed to handle DBMS functions.

3.   *Independent transactions.* Here, a single workstation can initiate a variety of transactions, regardless of the hardware, software, and DBMS environment in which they operate. As with access to separate data stores, the transactions are independent of each other. Historically, different applications required different workstations or dumb terminals, and any single type of transaction might also require different workstations, depending

on the department or organization involved. For example, in 1983 one major oil firm had 17 different terminals in its corporate treasury department for internal financial transactions, access to several banks' funds transfer systems, and access to Reuters, Telerate, and other suppliers of financial information. In late 1990, one of the top manufacturers in the United States had 89 major applications running on 132 different telecommunications and computer vendor's equipment and software, with 43 different networks and 64 different types of workstation.

Telecommunications alone cannot solve this problem, but it is a major part of any solution. Open systems are the key. Communications gateways—powerful processing systems in their own right—are beginning to link previously incompatible systems that require interconnection at the transaction level of such incompatible environments as IBM's SNA, Apple's LAN-based proprietary communications, and UNIX-based services. Because many transaction processing systems now in use were built on IBM's SNA and even earlier teleprocessing systems, IBM's 3270 terminal protocols remain in use, though not on those terminals; instead, software emulates their functions and protocols. Minicase 6-2 provides an example; when Chrysler adopted hand-held computers to link to a remote mainframe via wireless communications, they emulated IBM 3278 terminals (of the 3270 series).

4. *Cooperative transactions.* One of the major trends in innovation in the field of information technology is the client/server model and the closely-related cooperative processing. These innovations transform the concept of an "application" from a software and/or information service that resides in a specific location (either on the LAN or on a remote "host" computer) and that is independent of other transactions made from the workstation, to a software unit that interacts with other software units. Instead, the computing and information resource is viewed almost as a set of lego blocks that create services together. The workstation is a client that links to a server, as needed. For instance, a client workstation may initiate a customer service inquiry to find a customer's name and address, identify the most recent checking account deposit, and list the mortgage balance and the status of his or her loan application. The workstation—the client—handles input-checking, then *automatically and directly* links to the server that holds the checking account data (almost certainly a mainframe machine), and then links to the mortgage database and to the status file on loan applications. The key phrase here is "automatically and directly"; this is a single application, not a set of separate steps. "Remote program calls" (RPC) make the connections as needed; "application program interfaces" (API) provide the software equivalent of telecommunications protocols.

The difference between cooperative transactions and independent transactions can be quickly illustrated by considering a travel agent arranging a customer's business trip. A platform that is at level 3 (independent transactions) along the range dimension handles this as three separate transactions; the first arranges the flight, and the others make hotel and car rental reservations. To the travel agent, this may seem as if these reservations are made through a single application, but in practice three independent communication links are required. In addition, none of the transactions has any knowledge of the others. If the client phones the agent to change the day he or she travels, each of the other transactions must be reprocessed, with the agent making the decision: "I've changed the airline reservation. Now, I need to rebook the hotel."

With cooperative transactions (level 4 of range), the agent can be taken out of the loop. The application itself transfers control to each of the relevant processes, which may be on processors anywhere in the world. The airline, hotel, and car reservation subsystems each know about the other and act interdependently and cooperatively. Covia, the subsidiary of United Airlines and a group of foreign carriers (see Minicase 6-4), has been a leader in cooperative processing, with exactly this example as one of its targets. Cooperative processing pushes the practical state of the art in both telecommunications, software, and database management; that is why it offers a chance of a significant competitive advantage for the firm that can implement it when others are limited in either reach or range.

The technical basis for this new fusion of computing and telecommunications is complex and still emerging. The client/server model has emerged from rapid innovations in personal computers and local area networks. Cooperative processing is a related innovation stimulated more by developments in wide area networking (hence, Covia's early tests on linking public and private frame-relay networks). The logic is the same: to create an entirely new generation of services that exploit the various strengths of all the new telecommunications and computing tools acting in harmony.

Object-oriented Thinking The third major area of innovation in information technology complements cooperative processing and the client/server model. This is object-oriented programming, a term that is too narrow to fully describe a transformation of virtually every aspect of how software is designed and developed; it is really object-oriented *thinking*. The logic is that anything can be an "object": a database, piece of music, transaction processing routine, or picture. Objects can operate on or be accessed by other objects. Each object is entirely self-contained; it includes knowledge about the characteristics it inherits from the broader class of objects to which it belongs. For instance, the object "car" inherits features of the class "vehicle." Each object packages data and related procedures; it is this "encapsulation" that makes an object self-contained.

The combination of encapsulation and inheritance is the core of object-oriented programming systems (OOPS). Application development via OOPS focuses first on defining the relevant objects that correspond to business activities and entities, then identifying their "responsibilities" and "collaborators." Procedures and processes involve "messages" that specify the relevant objects. In Smalltalk, one of the two most well-established object-oriented development languages (the other main programming language is C++), "Car43 MoveTo location12" is a typical and self-explanatory message. Changes to the object designated MoveTo do not affect any other object. The same is true for changes to the object Car43; other objects are unaffected. No object need know about how another object operates. Objects are thus entirely self-contained and *reusable*. Complex applications can be assembled using libraries of objects. Generic objects such as "Draw" can operate on a range of objects such as "circle," "square," or "time-series." The link between object-orientation and client/ server and cooperative processing is, of course, that objects can be on separate processors.

Object-oriented programming has evolved over the past 20 years and was the basis for the Apple Macintosh's use of icons, which in fact indicate objects. When you click on an icon for "Open Mail" and then move the mouse to a new window on the screen or to a memo, you are activating a message comparable to "Car43 MoveTo location12."

There are growing indications that object-oriented methods are the long-sought breakthrough in software development, which has barely improved in quality, cost, and productivity in the past 30 years. Such methods will be essential to client/server computing and cooperative processing in that applications are distributed across the network on multiple computers and must be able to work together interactively. The systems must be developed as independent, reusable, and self-contained building blocks for the methods to be easy to design, maintain, and develop. There can be no doubt that the new mainstream of software engineering will be based on object-oriented thinking.[3]

If this discussion of range leaves you a little bewildered, do not be dismayed. This is a bewildering time in the information systems and telecommunications fields, with many promising lines of innovation—and with uncertainty and risk as well. The coming together of these two historically separate disciplines and professions opens up entirely new opportunities. The very idea of a telecommunications platform that can be a central determinant of a firm's basic business options is one of these opportunities. Reach defines *who* and *where* the firm can target those options, and range defines *what* is quite literally the scope of those options.

Responsiveness

The third dimension of the platform, responsiveness, relates to quality and reliability of service. Managing a complex on-line utility 24 hours a day is very expensive. Ensuring security, backup, guaranteed response times, and levels of availability of at least 99 percent adds complexity and cost. In many instances, such as ATMs or airline reservation systems, there is no business choice but to provide the highest practical level of responsiveness, for even short and infre-

---

[3] See James Martin, *Principles of Object-Oriented Analysis and Design* for a comprehensive and very readable summary of what he terms OO. Martin highlights the implications of OO for reach as well as range as follows: "In conventional distributed systems, data can be accessed in ad hoc ways. There may be little or no assurance of who is doing what to the data via the network. . . . To build distributed systems open to masses of users, without encapsulation, is dangerous. Encapsulation protects the integrity of data."

quent failures in network performance translate to lost business and damaged customer confidence. In other instances, however, there is a broader range of options. The key implication for the telecommunications business/technology dialogue is that the choice of level of responsiveness constrains the quality of service that can be offered and the quality of products that can be added to the platform. If a firm does not explicitly decide to invest in additional responsiveness, it may run severe business risks or add unnecessary costs. The decision to invest or not to invest should not be a decision by default or omission.

In many of the minicases presented in this book it is easy to identify the effects of breakdowns in responsiveness: problems of downtime, security, accuracy, and so on. A major problem in ensuring responsiveness is that the network is only as responsive as its weakest link. For example, although over the past decade many companies have built up substantial experience in managing wide area backbone networks, now they must do the same for local area networks and for LAN-WAN interoperation. If the WAN is up 99.99999 percent of the time but the LAN only 95 percent, the availability to the user is just 95 percent. Similarly, if the LAN transmits data at, say, 200,000 packets a second but the router can handle only 120,000 a second or the network switches add millisecond delays, the user still gets degraded performance.

The major implication of the technical aspects of network design and total quality operations described later in Part Three is that implementers must wrestle with these issues, some of which are close to intractable. Keep in mind that what this chapter is addressing is how to clarify the *business* criteria prior to design. Once that is done, designers and implementers can tackle the technical specifics.

Levels of Responsiveness The four levels of responsiveness are the following:

1.   *Nonimmediate response.* At this level of responsiveness, information and services need not be on-line and can be processed at the end of the day, a week, or even a month. This level minimizes the telecommunications requirements and simplifies the computing needs. Typical examples are end-of-day transmission of sales or production data from individual branches, stores, or plants to the head office. In the reverse direction, the head office sends price changes, inventory summaries, and so forth to the distributed locations. If the network is down for an hour or so, the information can be transmitted later.

THEi provides this most basic level of responsiveness (see Minicase 1-1). Each store's point-of-sale terminals capture transaction data, which is "squirted" at the end of the day directly to the London mainframe computer using public data network links. A network failure of an hour would not damage THEi's efficiency or effectiveness. For Health Link (see Minicase 4-3), however, when the network is down the business is in effect closed. The doctors that use this insurance-authorization and claims-processing system must be able to access it whenever a patient comes in during office hours.

2. *Immediate response.* In this level, the transaction is processed on-line, within a few seconds, but only within specified time periods, such as office hours. At peak times, response may be slow or connections difficult to make. Most ATM networks provide this level of responsiveness; services are unavailable during scheduled periods for maintenance, and occasionally machines are inoperative (sometimes because they simply run out of cash to dispense). Downtimes can be irritating to customers, of course, and can even be a disaster, as anyone who has been stranded cashless at an airport at night because the ATM there is "down" understands.

    Health Link offers immediate response within a 12-hour time window (8 A.M. to 8 P.M. Eastern time). This is one reason why it uses a value-added network service rather than its own leased lines; there is no sense in paying for capacity that is unused for half the day. Conversely, of course, international operations require 24-hour immediate response; Rockwell and Volkswagen lease private lines to ensure this level of responsiveness cost-efficiently.

3. *On-demand service.* This level avoids the ATM's downtime problems, except in dire emergencies, such as when a fire put a Bell Illinois switching center out of service for weeks in 1988. Service is provided 24 hours a day. Airline reservation systems must operate at this level because they are accessed worldwide; a travel agent in London must be able to book a domestic U.S. flight at 8 A.M., even though it is 3 A.M. in New York.

    One of the most far-reaching innovations in telecommunications has been the emergence of massive bandwidth on-demand. Until early 1992, a company that wanted to transmit large files on an ad hoc basis had no choice but to lease T1 circuits. Consider the following example: (*Communications Week*, May 11, 1992):

    > Dr. James Leas wants ISDN service to his home outside Denver. . . . Hospitals from all over the United States that don't have staff with his expertise send him images of x-rays, CAT scans or other patient medical records for consultation and analysis. Leas wants to have the images sent to his home at nights and over the weekend so he "can always provide coverage." Receiving an image at 9.6 kbps, which he currently can do over ordinary phone lines, takes 5–60 minutes. The 56 kbps service he uses at his hospital cuts that by a factor of more than four. NYNEX's SMDS service cuts it to around two seconds, using T1 and DS-3 lines (1.544 mbps and 45 mbps).

    It is important to note that neither Dr. Leas nor his hospital could afford a leased T1 or DS-3 line. (DS indicates the level of service and T the tariff.)

    LANs have provided the capacity for on-demand high-bandwidth services; SMDS and ATM (cell relay) are creating the same for MANs and WANs.

4. *Perfect service.* For this level of responsiveness, a firm adds resources for backup, "hot" restarting, rerouting (if lines are down), encryption and other security features, and diagnostic network management tools—and a very high level of professional expertise in network management as well. This level may also include duplicate facilities. For instance, Caisses Populaire of Quebec rents two data centers, ten miles apart, in order to safeguard against disasters. Caisses Populaire's processing systems handle 200 transactions a second at peak times. Backing up disk files averages 30 minutes, during which time 360,000 transactions could be lost. The 56-kbps link between the centers was used by many applications, so on occasion it took 90 minutes for backup data to arrive. Installing a 45-mbps fiber link at a cost of $200,000 cut the lag time during backup from 30 minutes to 17 milliseconds.

Conceptually, most firms would like to provide guaranteed-on-demand (perfect) service; in practice, relatively few can. Networks go down. According to a 1990 survey by *Benchmarks* magazine, the average business experiences two hours of downtime a week, which adds up to 13 eight-hour days a year. The firms in the survey estimated the cost of this downtime to be between $5,000 and $50,000 an hour, or in excess of $500,000 to $5 million a year. The more a business depends on its network for core services, product delivery, customer confidence, supplier relationships, internal coordination, and efficiency of operations, the greater the direct and indirect costs.

One obvious reason why so many firms have not invested in network operations and network management tools is the cost. In addition, many business managers are unaware of the options and risks and do not see networks as part of their own business responsibility. Approximately 85 percent of the New York City financial institutions that together process $1.9 trillion in electronic transactions every *day* do not have backup and disaster recovery capabilities.

| | | | | | | |

The Customer's Perspective on the Business Platform

One way to highlight the business need for reach, range, and responsiveness is to look at the platform dimensions from the customer's perspective. Figure 11-1 recasts the platform map; reach becomes "ease of access," range becomes "breadth of service," and responsiveness becomes "quality of interaction."

This perspective links back to the discussion in Chapter 10 on choosing the right opportunity. Consider, for example, applying for a mortgage. How long should it take? It should take 15 minutes to get a preliminary approval. How long does it take? It generally takes weeks or even two months. The reason for this is that the application and approval process is dominated by time-

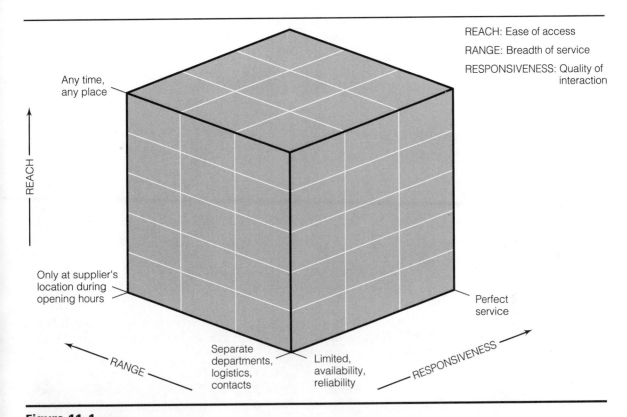

REACH: Ease of access

RANGE: Breadth of service

RESPONSIVENESS: Quality of interaction

Figure 11–1

The Telecommunications Services Platform Map from the Customer's Perspective

dependent and location-dependent processes. Customers must phone the bank to get application forms, which are then mailed to them, or go to the bank to pick the forms up. The completed forms move through departments, mail rooms, and in-trays. Credit records must be checked. The branch sends material to head office, and each customer becomes a "file." If a customer's file gets lost, so does the customer.

Consider the platform dimensions for processing a mortgage application from the service provider's perspective:

▸   *Reach*: Application for a mortgage can occur from the real estate agent's office or the customer's home. The customer's "moment of value,"[4] when

---

[4]See Keen, P.G.W., J. Rollins, R. Puryear, and L. Woodman, *Stealing the Moment of Value: Information Technology Comes of Age*, unpublished manuscript.

the customer receives banking services, usually passes because the bank's operations are dominated by place of event.

▶ *Range*: An agent processes the customer's application via a workstation that completes the forms, accesses relevant records such as credit history, and assesses the application, often through expert systems software. The agent can give a conditional approval and initiate the processing needed for closing the purchase of the house.

▶ *Responsiveness*: Network users do not get a busy signal, and even if there is a hurricane in Boston they still get an answer from an agent.

Now consider the same transaction from the customer's perspective:

▶ *Ease of access:* "I phoned them from our realtor's office. They phoned back and told us that they'd approved the loan, subject to confirmation of the valuation of the property, which they'd get done this week, so we can go ahead and schedule the closing when we want."

▶ *Breadth of service:* "The agent asked us a lot of questions about our finances. I was waiting to hear 'We'll get back to you when we've checked the information' or 'A loan officer will contact you.' Instead, I heard 'That's fine. We're approving the loan but we need you to send us a few records. Don't worry, it won't hold up your closing. The computer is already processing your mortgage.'"

▶ *Quality of interaction:* "I got through on the third ring! When I phoned XYZ last week to book a flight, I got the 'All our agents are busy' message and had to listen to the same song for five minutes before I hung up."

A platform must combine reach, range, and responsiveness. Reach alone is an answering machine: "Thank you for calling XYZ. Please leave your number and one of our agents will get back to you." Range without reach requires you to locate the right person yourself: "Why don't you set up an appointment with Julie. She's the person to cut through the red tape. . . . Oh, you've already phoned her and got her answering machine?" Responsiveness by itself is meaningless: "When I phone, I always get through—always—but never to a person."

The key difference between the customer's and the organization's perspectives on the telecommunications platform is that the customer never sees the complex set of building blocks the service provider must assemble. One purpose of "service" is to hide the business process and all its complexities from customers. Responsiveness hides these by ensuring that when you dial up a service, it is processed as you expected. Reach hides the service's location—the 800 number creates location-independence—and range hides the separation of business functions, applications, or data resources. Lack of reach, range, and responsiveness makes it very clear to the customer—sometimes painfully clear—where the company is, when it operates, and how its functions are organized.

| | | | | | | | | # Moving the Platform from Idea to Implementation

The concept of a telecommunications platform does not imply a monolithic technology base. Bridges, routers, technical- and business-based standards, choice of operating systems, and selection of vendors on the basis of integratability rather than on individual product features are all ways of creating a platform that exploits the best available building blocks. The key is to plan in advance how to ensure that the technical components fit together to provide the required business functionality.

In Part Two of this book we reviewed a wide range of standards and highlighted some that are obvious priorities for any company looking to create a platform with high reach, range, and responsiveness. Although each organization's specific needs vary, the following list of standards both indicates the mainstream for business operations across most industries and highlights gaps and uncertainties in standards:

▸   *Reach*: No single LAN product or standard to date has become dominant in the business marketplace. Ethernet is the preferred choice of many manufacturing companies that do not have the heavy transaction-processing volumes of a bank or airline. For these companies, especially those whose workhorse systems are built on IBM's architectures, token ring is the LAN standard; because it is more complex than Ethernet, it allows more complex applications to be run efficiently. UNIX-based environments generally rely on TCP/IP, which has become the standard of choice for organizations that want to move traffic between UNIX and IBM-based applications.

The importance of range for sharing data in the 1990s has led to a growth in sales of token-ring networks of 30 percent a year, overtaking Ethernet, whose strength is reach.

These standards are by no means the only choices. Banyan VINES has become a robust LAN base for many firms. TBase10 (an Ethernet-related standard) and FDDI are widely used. The two main approaches to choice of LANs are (1) to select the most efficient LAN and use bridges and routers to handle interconnection or (2) to select the LAN that most easily meshes with a firm's existing WAN capabilities. For wide area networks, X.25 is an integral element, either as a basic standard or through protocol converters and gateways. X.25 has been superseded as a standard by frame relay, as a technology by ISDN, and as a service by value-added networks that incorporate more recent protocols and architectures, including LAN-based ones. It will not be superseded as part of the international telecommunications infrastructure simply because many lesser-developed or smaller countries cannot afford to drop it; their existing PDNs rely on it, the quality of circuits requires X.25's node-by-node error checking, and the cost of switches for fast packet switching will be far too high for them to afford until at least the end of the 20th century.

▸   *Range*: Operating systems and DBMS drive the choice of telecommunications here. TCP/IP, SQL, and IBM's SNA/APPN architectures seem to be the most practical options. X12, X.400, and EDIFACT, together with industry-specific EDI standards, are the base for extending an organization's platform to other organizations' platforms.

IBM's DB2 is increasingly the database engine for very large firms that handle very large volumes of transactions. No other DBMS can match its breadth of capabilities. Oracle is the preferred choice of many firms that are not so constrained. DEC's Rdb, Sybase, INGRES, and SQL-compliant systems are other proven options. All these companies' ads emphasize cross-application and cross-vendor interoperability as much as or even more than their products' specific features.

In software development, client/server computing and cooperative processing are not yet fully implemented or stable. They will ensure increases in range, but most applications to date are LAN-based; range demands LAN-WAN-LAN interoperability.

Object-oriented methods are rapidly gaining support, but it will be at least five years before they become the mainstream. Many VANs aim at ensuring range, whereas PDNs mainly offer reach.

▶   *Responsiveness*: Reach and range mainly depend on choice of standards; responsiveness depends much more on choice of equipment and transmission techniques, plus strategies for backup and recovery, a topic discussed in Chapter 16. Implementers, not designers, have the most responsibility and expertise in this area. Choice of network management tools is one of the central (and most complex) responsibilities of implementers.

The telecommunications services platform framework identifies the business criteria for defining the technical architecture. The architecture really *is* the strategy. The clearer clients and designers can define the dimensions of the platform, the more easily implementers and designers can then select standards and equipment.

Dealing with Multitech Chaos

The "network from hell" example highlights a reality of networks in most companies: the existing mess of incompatible systems that prevents the creation of an enterprise platform. It would be easier to start from scratch and select the standards for a platform than to mesh all the many computer and communications systems now in use. Even if open systems were fully implemented, firms would face the difficulty of migrating to them while preserving existing investments. This means that they must phase the development of the platform, as shown in Figure 11-2.

There are three interrelated phases. The long-term goal is complete openness: complete interoperability of all computer and communications components of the information technology resource. Because that is not yet possible, firms need to define what we term here its target architecture. That architecture will be based on several core standards, such as SNA, TCP/IP, and/or selected OSI standards, including X.25 and X.400. It will provide for LAN-WAN interconnection, probably through choice of multiprotocol routers and hubs. It

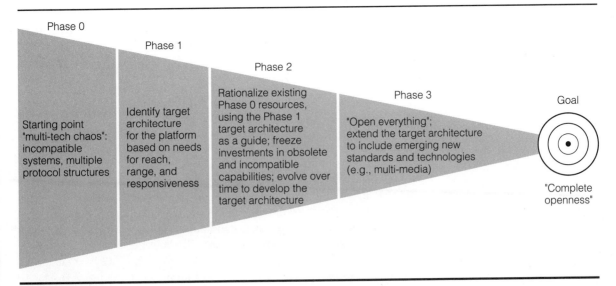

Phase 0

Phase 1

Phase 2

Phase 3

Goal

Starting point "multi-tech chaos": incompatible systems, multiple protocol structures

Identify target architecture for the platform based on needs for reach, range, and responsiveness

Rationalize existing Phase 0 resources, using the Phase 1 target architecture as a guide; freeze investments in obsolete and incompatible capabilities; evolve over time to develop the target architecture

"Open everything"; extend the target architecture to include emerging new standards and technologies (e.g., multi-media)

"Complete openness"

Figure 11–2

Developing the Platform

will either specify a required desktop operating system, such as DOS or Windows, or, again, use internetworking devices to allow some degree of interoperability between PCs and workstations that use different operating systems. The chosen levels of reach, range, and responsiveness will determine where the platform must be based on such business standards as EDIFACT for EDI and on specific database management standards and products.

The target architecture will also address network management, the most challenging aspect of network design and operation in a multivendor environment. For a straightforward TCP/IP-based network, a firm is likely to adopt SNMP (Simple Network Management Protocol); for a straightforward SNA network, it will use IBM's Netview and Systemsview products; and for a straightforward LAN environment, it is likely to adopt Novell's NetWare.

The target architecture may take a long time to reach; it is the blueprint for progress. The main problem many companies face is the multitechnology chaos already in place; the "network from hell" described by *Corporate Computing* is relatively simple compared with many firms' morass of processing systems, database software, and network components. For this reason, the key first phase in moving to the target architecture is to rationalize today's installed base. This can be best handled through a combination of three courses of action:

1. Replacing old hardware and software with systems that conform to the target architecture's standards. This step may involve leaving in place some core systems that are too expensive or too complex to redevelop and accepting some inefficiencies in operation. For instance, companies that

replace IBM mainframe-based systems with UNIX-based client/server systems often retain the central transaction processing software, transmitting the SNA traffic as TCP/IP packets. This course of action adds overhead; the SNA header and control data are transmitted via TCP/IP, which adds its own data.

2. Installing internetworking devices, such as routers, gateways, and protocol converters. This can be very expensive in terms of capital cost and adds delays as traffic moves through the additional network nodes.

3. Requiring all new systems and services to adopt the target architecture standards. This course of action can be politically difficult to justify and make work. Telling the marketing department's managers, for instance, that they may not use Macintosh PCs can be the start of an organizational civil war. Increasingly, skilled designers and implementers focus on stating the communications requirements and not the software and operating systems requirements. For instance, they will allow business units to adopt any desktop system on any LAN they wish, provided that system can connect to the wide area network; the target architecture includes the facilities for doing this.

Over time, the target architecture will be adapted and extended to address new needs and to include new technology. For instance, the architecture should be able to include relevant open systems standards as they become stable and proven. Decisions concerning FDDI, CDDI, Ethernet, 10-base 10, and whatever new developments may occur throughout the 1990s should be made within the framework of the architecture as a master blueprint, instead of as a constant reaction to a multitechnology muddle. For example, the move to frame relay is a natural and incremental step for a firm whose platform includes X.25; X.500 will similarly be an incremental step. By contrast, moving to frame relay from, say, an environment of Ethernet and NetWare-based LANs requires a new target architecture and rationalization process.

There may never be completely open systems in telecommunications because the pace of change is so fast. For example, 1992 saw an explosion of low-cost multimedia PCs well before there was time to debate, define, and implement standards. 1993 saw the proposal for (but not the implementation of) many new wireless systems, plus the introduction of a new generation of PC operating systems and major competitive moves by cable companies, RBOCs, and long-distance service providers. Constant change is the only constant.

That said, the telecommunications platform approach previously described provides a practical framework for moving toward integration and open systems in a phased manner and in a way that always focuses on both business and technical issues.

Ending the Chaos   This chapter ends with some examples taken from a *I/S Analyzer* article on how some companies have successfully moved from chaos to a platform. Litton Industries' Advanced Circuitry Division faced growing problems from its lack of a platform. Staff were unable either to track products as they moved through manufacturing or make accurate forecasts. Each step in the logistical process relied on entirely incompatible systems. Products often piled up on the shop floor, and no one was able to locate them. The manufacturing resource planning (MRP) software was not on-line (zero reach). The computer-aided manufacturing (CAM) system used a stand-alone Ethernet LAN. The CAM and MRP systems could not talk to each other. Some 250,000 transactions a month were batch-processed by an IBM AS/400 mid-size computer that connected PCs via a token-ring LAN. The same data had to be reentered into the CAM system. Three different communications protocols were in use: TCP/IP, IPX (Novell), and SNA.

Achieving the reach, range, and responsiveness needed to make the systems interoperable took just under a year. Barcode scanners were added for data collection into a PC; wand readers scanned the data at each of the 100 steps in the manufacturing process. The information was sent to a centralized UNIX server, which sent data to the AS/400 as needed. The AS/400 became the server for MRP data. To link the IBM and UNIX Sun Microsystems servers, Litton used a special-purpose client/service software package to handles protocol conversions. The package uses the software company's proprietary database management system on its three main subplatforms: UNIX, AS/400, and Novell. The Ethernet networks are bridged and linked to the three interconnected token ring LANs via the servers, which contain software for protocol conversion.

The new platform required adding over 100 new products, which were screened by using a checklist of features that emphasized open systems capabilities. None of these products was a pure OSI product, but each could meet key OSI specifications, and, more importantly, each was designed for an open systems environment. Litton found off-the-shelf products to handle three of the largest areas of interoperability: AS/400, by far the most successful, widespread, and profitable non-PC computer in use today; Novell, by far the leading LAN NOS; and TCP/IP, by far the most widely adopted LAN-to-WAN transport protocol. Had Litton been running a less mainstream set of systems, it might not have been so successful in moving from chaos to platform.

The business payoff from the shift is reportedly growing by the week. Data that once was as much as six weeks old upon receipt is now 6–20 minutes old; supervisors no longer spend 30 minutes to four hours a day tracking physical inventory; forecasts are not made by guesswork, as they were before.

UNUM Corporation took a very different approach than did Litton. UNUM is the leader in disability and long-term care insurance. Whereas Litton used open systems products to interconnect incompatible ones, UNUM took a proprietary approach, focusing on families of products that allow it to move applications from large to small processors. It installed a proprietary software package that links its mainframe core systems to Windows- and OS/2-based PCs and to UNIX workstations. For UNUM, data integration is the key to its business plan of ensuring that it meets its customers' needs to be treated as individuals who have a relationship with UNUM, not as purchasers of a variety of specific products. UNUM has taken a case-by-case, pragmatic approach to interoperability, essentially ignoring the siren call of OSI, which plays no role in UNUM's plans to date.

Bell Atlantic's $2 billion plan to rebuild all its core customer services applications brings together two entirely different sets of tools: IBM's mainframe MVS operating system and the UNIX front-end workstation operating system. The core links between them are provided by Bellcore's massive Operations Systems Computing Architecture (OSCA), which defines a three-layered approach (data, process, user) that specifies how each layer works together. Software agents communicate with each other, allowing, for instance, the SNA backbone network to mesh with the TCP/IP-based front-end systems that use UNIX. Although Bell Atlantic supports OSI initiatives, it sees TCP/IP as the best available working solution to interoperability.

In each of these instances, the role of the implementer was vital in working out imaginative and practical ways to integrate what was often a mess of systems. The relatively few large organizations that have not needed to overcome chaos benefited from having an architecture for integration in place years ago.

Summary

Integration and interoperability are the main forces driving development of information technology. Business increasingly depends on cross-functional communication, the sharing of previously separate information and transaction processing systems, new forms of customer service, and streamlining of organizational processes. Historically, incompatibility has been the norm, not integration, and many organizations today must deal with a "network from hell."

The telecommunications services platform map provides the basis for defining the blueprint for design of an integrated platform that avoids creating problems of incompatibility that block business opportunities and needs. The map has three dimensions of business capability: reach, range, and responsiveness. Reach defines which people, devices, and locations can access a firm's electronic services. It has six levels: (1) within a single location, (2) across a firm's domestic locations, (3) across international locations, (4) to customers and suppliers with a compatible technology base, (5) to all customers and suppliers, and (6) to anyone, anywhere. Range, which determines what information can be shared directly and automatically across business functions and services, has four levels: (1) simple messages, (2) access to separate data stores, (3) independent transactions, and (4) cooperative transactions. Extending reach may not extend range. Reach mainly refers to connectivity and the movement of data, and range refers to interpretation and use of information. An analogy that illustrates the difference between reach and range is a phone call to Brazil: Reach connects you to Brazil, but unless you speak Portuguese or the other person speaks English, there is no communication and thus no range.

The third dimension of the map is responsiveness, which defines quality of service in four levels: (1) nonimmediate response, (2) immediate response, (3) on-demand service, and (4) guaranteed on demand (perfect) service.

From the customer's perspective, reach corresponds to convenience and ease of access, range to breadth of service, and responsiveness to quality of interaction.

Developing a platform in a context of multiple and incompatible systems requires a phased strategy that moves from today's chaos to the target architecture that is the translation of the platform map into specific systems and standards, which can then evolve as new standards and open systems emerge.

Review Questions

1. What is the central technical issue for designers and implementers? Explain why.

2. Explain why it is more important than ever for companies to communicate and share information across departments, functions, locations, and business processes.

3. Briefly explain how a firm's technology platform influences its business degrees of freedom.

4. Explain the following statement: Reach determines the extent of connectivity and interconnection of communication facilities, but it does not address communication per se.

5. Give an example of how reach is extended across a firm's domestic locations in several different cities.

6. Explain the following statement: Extending reach across international locations is a totally different issue from extending reach to domestic locations within the boundaries of a given country.

7. Describe the main issue(s) related to extending reach outside a firm to customers and suppliers.

8. Explain why increasing the range of an information technology platform depends far more on developments in the information systems field (such as database management systems) than in the telecommunications field.

9. Explain the following statement: Today, most personal computer/workstations can access most of a firm's databases using a common interface, such as structured query language (SQL); this is an improvement in the range of the platform, but it is still quite limited by the manner in which access is obtained.

10. Describe the key development trends in database management systems that are having the greatest impact on the range of a telecommunications platform.

11. Explain the difference between independent and cooperative transactions, and their relationship to the range of a telecommunications platform.

12. Briefly describe the concept of object-oriented thinking, and explain how this relates to the range dimension of a telecommunications platform.

13. Explain how the choice of the level of responsiveness for a telecommunications platform constrains the quality of services offered on the network.

14. Explain and discuss the meaning of the following statement: A firm's failure to invest in additional responsiveness capability for a platform may create severe business risks or add unnecessary costs.

15. Classify the appropriate level of responsiveness of the telecommunications platforms for the following services, and briefly justify your choice:
   a. Daily remote printing of the *Wall Street Journal* in various U.S. regional offices
   b. Credit card verification links between retail sales outlets and various financial institutions
   c. Compilation and production of weekly sales figures reports for management at company headquarters
   d. Collection and analysis of global remote sensing and intelligence data for use by the U.S. military and the National Security Council

Assignments and Questions for Discussion

1. Describe the purpose of the telecommunications services platform map, and explain how it explicitly shows the business capability of the telecommunications technology base.

2. Explain and discuss the meaning of the following statement: Reach is the core of telecommunications as a business resource, and range is increasingly the constraint, opportunity, and challenge.

3. Evaluate and discuss the following statement: The history of the computer industry is likely to be repeated by evolving telecommunications industry: Many emerging industry leaders provide key products and services, which then flounder and fail a short time later. The major lesson designers and implementers should learn from this trend is to choose products based strictly on specific features and cost, because the supplier base is so unstable.

4. Explain the relationship among the client/server computing model, cooperative processing, and object-oriented thinking, and discuss the implications of the integration of these three factors for the three dimensions of a telecommunications platform.

Minicase 11-1

Sea-Land

Client/Server Networking for Seaports

Sea-Land, a leading international shipping services provider, operates shipping terminals at major ports, where it manages cargo delivered and picked up by trucks or ships. This is an extremely time-dependent and complex business, with many steps, procedures, agents, and documents involved. The hub of Sea-Land's communications is its data center in Elizabeth, New Jersey, which runs a complex of large IBM 3090 mainframe computers that until recently were accessed through the standard IBM 3270 dumb terminals that have been the norm in large firms' on-line processing for several decades.

The client/server model of distributed computing makes this configuration obsolescent. Sea-Land is phasing in a client/server network that will greatly transform operations. Pushing intelligence out to the ports will greatly simplify access to mainframe data, permit the cost-effective use of powerful decision support and reporting tools, and improve ease of use of workstations. The hardware base is a SynOptics Series 3000 wiring hub that supports a 16-mbps token-ring LAN. The workstations are IBM PS/2 Model 55s that access a SQL database server on a powerful IBM PS/2 Model 95 or a comparable SystemPro server from Compaq. To call the Model 95 and the Model 55 "personal computers" and to view them as roughly equivalent is misleading. The two IBM systems and the Compaq are compatible—Compaq was the first firm to offer IBM "clones" and relies on excellence in en-

gineering as its main edge—but the Model 95 and SystemPro servers outperform many minicomputers in raw processing power. Thus Sea-Land's client/server system has software compatibility across the seaport workstations, the IBM token-ring LAN, the SNA backbone, and the 3090 mainframes.

The hardware and software elements do not in themselves add up to a client/server network. They can be and mainly are used for distributed computing in which the workstations handle straightforward data entry, formatting of messages, and local analysis. A client/server network provides far more interactive communication between local and remote applications and databases.

The old system relied on 3270 terminals that link to the mainframe through IBM's CICS teleprocessing software. For a variety of technical reasons, CICS generates complex screen displays and does not permit local customization because the 3270s have no local intelligence. In Sea-Land's client/server network, the mainframe retains its central role in processing transactions and in managing data resources. Microsoft's LAN Manager is the network operating system for the LAN. Digital Communications Associates' Select Communications Server (CS) enables workstations to transparently access the host mainframes. John Darienzo, manager of Sea-Land's terminal automation systems, told *Network World* in early 1992 that the

437

goal is to bring a miniature data center infrastructure right down to the PC LAN. For users, it is as if the mainframe were in their own PC.

Most seaport employees are computer novices. Sea-Land is using Microsoft's Visual Basic PC-based programming language to create simple and customized user interfaces that ask only the questions users need to answer in accessing the host mainframe and that hide the irrelevant data that is displayed on the 3270 screens. The DCA CS software assembles shipping information from a variety of points, without the clerk having to take any action, and automatically routes the transaction to the host for confirmation. Given that a clerk's request to verify whether a ship container can be released or checked into the year in, say, Hong Kong, may involve checking records from the origination point in Rotterdam, accounting information in New Jersey, and yard management data in Hong Kong, the management of client/server communications is not a simple process. Straightforward 3270-to-mainframe computing has the advantage of technical simplicity; the mainframe is the master coordinator. Straightforward distributed computing adds an economic advantage while retaining technical simplicity; the equivalent of Santa's helpers takes the load off Santa (the mainframe), but Santa still handles delivery and answers childrens' letters addressed to the North Pole.

Client/server computing adds collaborative communication; the elves have brains, too. The network takes on all responsibilities for determining how to answer a user's request; this simplifies the human side of the process but adds immense technical complexity, which is one reason why client/server computing is easy to talk about and difficult to make work.

One goal for Sea-Land's client/server network is to greatly enhance reliability. In the old system, if the mainframe was down, so was everything else. Security, software updating, and performance monitoring were also handled by the host. Now, data is captured locally and the local workstation decides whether to continue processing or to stop the operation because, say, the mainframe is not responding. In this situation, the intelligent workstation will immediately cut over to a SQL database server that will act as a repository for local transactions and will automatically update the mainframe later. The DCA Select CS software similarly adds intelligent functions, such as off-loading communications traffic onto other telecommunications links to reduce congestion.

The client/server model is the obvious and accepted framework for the integration of information technologies: local area networks, wide area networks, personal computers, mainframes, and databases. Sea-Land is at the forefront of this innovation, and not until the mid- to late 1990s will firms have the tools and experience to make it the mainstream. As John Darienzo warned in January 1992, "Anyone that thinks they can go client/server today without looking at all the possible implications is crazy. Things are just getting to the point where you can trust the products. Six months ago, many of the pieces we needed and are using today just were not there."

Questions for Discussion

1. Explain the difference between client/server and distributed processing methods of computing, and discuss why this difference is important to the future of Sea-Land's business.

2. How does Sea-Land manage the complexity of their client/server architecture so that novice end-users of the system are not overwhelmed?

3. Compare the reliability of Sea-Land's client/server network with its old terminal-to-mainframe network system.

Minicase 11-2

Towers Perrin

Building a Global Voice Messaging System

Towers Perrin is a large international consulting firm that consists of four operating companies: TPF&C, the world's largest independent consultancy in compensation and benefits; Cresap, a management consulting firm that specializes in organizational innovation; Tillinghast, an actuarial consulting company; and TPF&C Reinsurance, a major U.S. reinsurance intermediary. Towers Perrin has 5,300 staff scattered among 60 offices across the world; it does not have a typical head office, and the top management team is not in one location. Communication is very much "peer-to-peer" and informal; consultants need to talk with colleagues in other locations and travel frequently.

In 1988, the company carried out a study of its communications needs—communications, not telecommunications. Each office handled phones, fax, and so on as it wished; spending decisions in Towers Perrin were made locally, in the absence of any corporate communications strategy. The study concluded that this had to change and that the firm should implement a voice messaging platform worldwide. Steve Hallowell, the director of telecommunications for the firm's internal consulting group, had no line authority; in this situation of extreme decentralization and local budgeting, "we as a corporation had to decide whether we were going to let the network evolve naturally or give it some help."

They gave it some help. "Our study showed that we would benefit from a voice mail system now. It convinced management that a centrally planned network and coordinated implementation plan would assure the greatest acceptance and flexibility. . . . Our goal is a worldwide network with a mailbox for every one of our 5,300 employees. We want our staff to be able to communicate easily with any single employee or group of employees in the world."

Voice mail is essentially spoken electronic mail. Towers Perrin's system is based on Rolm's PhoneMail, which has many advanced features:

1. Outcalling places a call to a subscriber's specified phone or pager number whenever a new message arrives. ANI (automatic number identification) provides the external caller's area code and phone number in the header message.

2. Future Message Delivery lets a subscriber record a message to be delivered up to 364 days from now, at a specified date and time.

3. Urgent Message Delivery adds a top priority indicator to a message so that it gets delivered ahead of others.

The system was not an easy sell. In a client-focused business in which personal contact is considered essential, a number of consultants felt PhoneMail would be too impersonal and would make clients uncomfortable. Hallowell commented that "we've been careful to educate our staff and let them become familiar with the product before introducing it to clients and other outside callers. We also let the local office control how their clients interface with the PhoneMail. We visit each site three months before installation and determine how they want the system to work for them. Some offices have implemented PhoneMail's telephone answering capabilities for all calls. Others have elected to have a secretary or receptionist answer external calls and give callers the option of leaving a PhoneMail message."

The network is designed to be as easy to use as possible. For instance, an employee can send a message to any other employee just by dialing the letters of his or her name; there is no need to find a phone number. Distribution lists allow information sharing and bulletins in a highly decentralized and widespread organization; one call reaches anyone on the list.

One top manager in Towers Perrin commented that he was initially "somewhat leery." Now, "between a portable phone and a phone in my car, I'm in touch wherever and whenever. . . . Because I travel a lot, it's my link to what goes on. We don't use PhoneMail just for messages; we use it for issues. When I get into a taxi, I can solve two or three problems before I get to the next meeting." Once, a client's company called an emergency meeting of the board of directors for the next day. The manager urgently needed information, but when he called Towers Perrin's office, no one was available to help. "He left me a message on PhoneMail. I got into the voice mail of one of our consultants, who I knew could help this client. Our consultant was traveling to the West Coast, but I knew he'd check PhoneMail before arrival. I asked him to call the client anytime before morning. He called the client at 2 A.M. But the client didn't mind; he was able to go into that directors' meeting prepared. And we picked up a $50,000 job." The manager adds that

people in the service business must prove themselves daily: "Anything that gives us an edge helps our clients. So we employ technology to give us that edge and we work at making it as personal as possible."

The choice of Rolm as the supplier of the voice mail system was based mainly on its global presence. Its alliance with IBM and Siemens means that it has operations in over 120 countries.

The next step is to build on the new platform. Hallowell explains that "as a global enterprise seeking global solutions, we must build on the infrastructure we've put in place. We see PhoneMail as a dynamic, rather than static, tool. We anticipate finding additional creative applications for use both internally and with our clients. As both PhoneMail and CBX (computerized branch exchange switches) evolve, we are positioned to take advantage of new developments. Further integration of voice and text, enhanced call processing applications and tighter networking capabilities will all address business needs and help us remain a global leader in our field."

Questions for Discussion

1. Describe the business need for and the potential benefits of Towers Perrin's worldwide voice messaging system. Focus your answer on some of the advanced features of the Rolm PhoneMail system in particular and how Towers Perrin uses them to advantage.

2. Describe the range, reach, and responsiveness of the Towers Perrin voice messaging platform, and discuss whether the level of each is sufficient to meet the company's needs.

3. The next step for Towers Perrin is to build additional capability into the new platform. Identify some additional creative applications that could be added to the platform that would enhance customer service and/or employee productivity.

Minicase 11-3

Maryland State Government

Electronic Benefits Systems

State governments everywhere are under financial siege. One American child in seven now lives on welfare. The costs of health care, unemployment compensation, and welfare are draining already-depleted resources. There is no magic solution to the many problems state governments face, but every small improvement in cost and quality of service helps. Maryland, which is regularly ranked in surveys among the top five states for fiscal performance, has shown how electronic systems can benefit recipients of aid, stores that handle food stamps, and government itself.

Maryland's Independence card, which looks just like a credit card, is a debit card used by 300,000 recipients of state benefits in grocery stores, at bank automated teller machines, and at some state agency locations. It is used to buy food and collect welfare payments. Instead of having to collect and present food stamps, a benefits recipient presents the card in stores that have installed special point-of-sale terminals at checkout counters. Similarly, instead of receiving welfare checks that can be delayed, lost, or stolen and that cost the state 75 cents a month to print and mail, recipients collect their money from an ATM.

When they purchase food, the cashier runs the card through the POS register to verify the recipient's personal identification password and to calculate the balance remaining in his or her account. The figures are stored and updated on a database

in Dallas. (The reason the database is out of state is that although Maryland operates the food stamp program, the federal government provides all the funding and half the total funding for the Aid to Families with Dependent Children program.) Stores no longer need to sort stamps, which saves hundreds of thousands of dollars a year for larger chains; the state saves around $1.2 million a year in administration. In addition, there is much less risk of misuse of benefits. The initiator of the project commented about a CBS television "60 Minutes" investigation that "we aren't giving the POS terminals to prostitutes or drug dealers."

One key principle in the design of the telecommunications platform for this and several other electronic benefits systems supported by state and federal government is to use existing, standard, and simple packet-switched networks to support point-of-sale and automated teller machines. The logic is to avoid expensive development of special capabilities and to be able to roll over new programs quickly. (The Maryland pilot was put in place in two months.) Although the telecommunications part of the electronic benefits transfer (EBT) system is simple to implement, the database management side is far more difficult. A new SQL translator gives all users access to both relational and nonrelational database structures so that they can access recipient and program data easily through a standardized interface.

Ongoing EBT projects include a two-year federal pilot program to pay social security and government pensions electronically in the Houston area. Of over 750 million payments a year the U.S. government disburses, almost 60 percent are made by checks that are sent through the mail.

Questions for Discussion

1. Describe the range, reach, and responsiveness of the Maryland POS/ATM platform, and discuss whether the level of each is sufficient to meet the state's needs.

2. Describe the need and potential benefits of the State of Maryland's electronic benefits transfer systems. Focus your answer on some of the electronic features of the system and the ways the state and the welfare recipients use them to advantage.

3. Discuss the extent to which the POS/ATM platform is adequate for expanded use of the system to pay social security and government pensions electronically in other states. Frame your answer in terms of the range, reach, and responsiveness capabilities of the platform.

Minicase 11-4

Progressive Insurance

Reengineering Instead of Automating

Business process reengineering, which combines fresh thinking about processes with the use of technology, replaces the historical tendency of firms to use technology to automate what they already do. Progressive Corporation, a specialty insurer based in Ohio, is widely regarded as outstanding in its growth, productivity, and profitability. Progressive specializes in selling coverage to drivers who have been rejected by other insurers or whose policies have been cancelled. A five-year, $28 million project—the largest single investment in telecommunications and computing Progressive has ever made—helped the firm make radical changes in a period in which it faced cutthroat competition in the automobile insurance market.

Costs were growing out of control, and regulation in many states was highly unfavorable to the industry. As Progressive's COO commented, "People just hate our product. The way insurance works, it isn't a good value. Other than the consumer, there's no one who really wants fewer accidents. Car mechanics and repair people, lawyers and hospitals all love accidents. The incentives of the system are to encourage more accidents and more grotesque injuries." In 1986 the firm considered itself to be in a "crisis situation" that demanded a total rethinking of how it did business.

The starting point for rethinking was the claims department. Progressive's 175 offices across the country operated in the traditional, paper-dominated way, on a state-by-state basis. The goal was simple and clear: dramatically reduce the time between a claim being filed and the money being paid. At the heart of the project, which was called Pacman, were an IBM mainframe computer and Compaq portable personal computers. Policy underwriting is handled at the workstation. Expert systems provide more accurate cost estimates and fraud detection; the latter is critical for Progressive, given its market niche of high-risk drivers. The most important single feature of the new system is that claims adjusters no longer sit at their desks. *Computerworld* describes the results:

> Midnight, downtown Atlanta. A late-model sports car skids out of control on rain-slicked Peachtree. The teenage driver careens into a utility pole, making a perfect letter V of the hood of his car. . . .
>
> Moments after the police arrive, a van driven by an adjuster from [Progressive] appears. Less than one hour later, the car has been towed to a garage, and the driver has been driven home with a check in his pocket to cover the costs of repairing his smashed but salvageable car, hiring a rental car and even replacing his $75 stone-washed designer jeans which were torn at the knees during the accident.

Adjusters on call 24 hours a day are despatched by radio to the scene of an accident and in most instances close out the claim on the spot. In an ex-

443

tension of this Pacman project, adjusters in Atlanta work out of roving vans equipped with a personal computer and modem that links them to Pacman via cellular telephones and a fax machine. The adjusters can sit in the van and handle claims information, issue payments, call a tow truck, or book a hotel reservation, and they now spend most of their day with customers and new prospects. Progressive was surprised and pleased by their favorable response to the new job and new working hours.

What it did not foresee was "the staggering rippling effects" on the rest of Progressive's 6,000 employees. Agents were reluctant to give up control of their own client base and turn over phone calls immediately to the adjusters. Many staff were jolted by the new "fast-response" pace demanded by the new environment. Progressive spent an additional (and unbudgeted) $1.3 million on training; implementation was slower than expected. The COO commented that "when you're talking about funda-

mental changes in jobs, you have to give people time to get used to the idea."

Questions for Discussion

1. Describe the approach used by Progressive Insurance to reengineer a central aspect of its business. Focus your answer on how telecommunications and information systems were used to achieve the goal of the company's effort.

2. Describe the range, reach, and responsiveness of Progressive's information technology platform, and discuss whether the level of each is sufficient to meet the insurance company's needs.

3. Discuss the major implementation problems encountered by Progressive, and devise a strategy to speed up implementation and minimize the problems.

12

Making the Economic Case

Objectives

- Relate the opportunities created by telecommunications directly to the cost and profit structures of an organization
- Understand and apply the fourth core framework of *Networks in Action*, quality profit engineering

Chapter Overview

| | | | | | | # Chapter Overview

This chapter addresses the fourth step in the telecommunications decision sequence, making the economic case for spending capital on telecommunications rather than on some other business resource. This step is especially difficult when the telecommunications opportunity is not intended to cut operating costs and save money; when savings are intended, the payoff can generally be estimated through cost-benefit analysis. In most instances, though, the opportunities address customer service, communication, responsiveness, or "better" information, and placing a monetary value on these qualities is extremely difficult. The problem is compounded by many business executives' skepticism about the payoff from telecommunications.

Making the economic case is generally the most contentious step in the telecommunications business decision sequence, and it is one where an imaginative and practical proposal often founders. For that reason, skill in both making and presenting the case is a key part of the designer's craft. This chapter presents a quality profit engineering framework that guides clients and designers in using telecommunications to contribute to any of six factors that are crucial to a firm's economic health: (1) profit management and alerting systems; (2) traditional costs, including labor and real estate; (3) quality premium costs—the costs of ensuring maximum practical quality; (4) comparable service premium costs; (5) costs of investments for future infrastructure; and (6) revenue improvements that provide a high margin and low incremental investment.

The quality profit engineering framework, like the business opportunity checklist, does not quantify the payoff from the telecommunications investment, but instead provides a systematic set of categories for identifying high payoff areas. It is in the later formal planning and implementation process that follows the telecommunications decision sequence that detailed forecasts, calculations, and quantification are made.

| | | | | | | # Why Spend Money on Telecommunications?

The arguments presented in *Networks in Action* treat telecommunications as a capital investment in infrastructures that are an important part of the base for an organization's future competitive, economic, and organizational health. For many business managers, though, telecommunications is an expense, just as a monthly phone bill is. When someone offers to double your phone bill to provide you with what he or she calls a "student learning advantage," the benefit is hypothetical and delayed, and the cost is real and immediate. When your phone bill goes up every month, perhaps because of daily chats with a close friend 2,000 miles away, your natural instinct is to find ways of cutting it.

This has been the attitude of many business executives; they see telecommunications as a cost to be controlled. They appreciate its business necessity, just as much as most of us consider our phone a necessity. They want to minimize the cost of that necessity, especially because it has grown far more rapidly in recent years than other business costs. Ten years ago, just before the divestiture of AT&T took effect, most companies had no idea what they spent on telecommunications, which was confined primarily to telephone and telex expenses. These expenses were scattered over many budgets, and companies had little reliable knowledge of their costs. International telephone calls and telexes were charged to overhead budgets and were not consolidated. At that time there were no local area networks beyond simple PC LANs, so as far as most business unit managers were concerned, telecommunications could be left to the administrative group that handled operations; in turn, this unit was concerned mainly with costs and efficiency. Telecommunications services were provided by AT&T and had a very narrow range of customer options. The only large capital expenditure was for PBX equipment to provide an internal switchboard.

When costs of telecommunications began to rise in the 1980s, mainly stimulated by growth in data communications and by an even more rapid increase in the use of phones for marketing, customer service, and the like, the demand from users was, naturally, for cost control. Large organizations were able to exploit private networks, which gave them a substantial discount over dial-up public links; telecommunications managers aggressively exploited the competition among AT&T, MCI, and Sprint to get the best price. Toward the end of the 1980s, business units began to install LANs, and cost was one of the main drivers for this move. The installation of LANs snowballed, with more and more efforts to replace high-cost mainframes and WANs with low-cost workstations, servers, and LANs. Executives began to question whether their firms should try to run their own equivalent of a phone company or outsource it. They questioned, too, why central corporate expenditures in information technology had grown so fast, when there was often no evidence of any real financial payback; the terms "downsizing" and "right sizing" entered the vocabulary of IT.

All these cost-driven forces made it very difficult to justify investments in infrastructures that do not directly and visibly reduce expenditures, especially in a business environment that has become very harsh for almost every industry. The typical answer to the question "Why spend money on telecommunications?" then becomes "Because we have to—and we make sure we spend as little as possible."

A better answer to the question is "Because investing in telecommunications will offer the organization a very good deal indeed." How to create and present that very good deal is the topic of this chapter. These skills are crucial for telecommunications designers, and without them, dialogue with business clients will be very much restricted. Of course, designers and implementers should always exploit opportunities to reduce costs, but they must also be able to think and talk in broader terms about the values of telecommunications. In just about every textbook on telecommunications that views it as a technical resource to be assessed by technical criteria, discussion of the value of telecommunications is absent.

As a result, many telecommunications professionals admit with frustration that they do not know how to argue the case for investments they believe are essential to their company. In viewing telecommunications as a business resource, they need a framework for assessing it according to business and economic criteria. The telecommunications services platform map, presented in Chapters 3 and 11, provides the business criteria. The network design variables worksheet provides the base for specific choices of technical features and trade-offs among them; this core aspect of network design helps keep the technical aspects of design separated from the purely business ones. In *applying* the business decision sequence, the main design parameters of the network must be made clear before it is possible to make the economic case, but in *learning* the decision sequence, we feel it is more helpful to address step four before step three. This helps put step three, network design, in its full business context; this step is framed by, at the front end, the business logic of choosing opportunities and defining the criteria for technical design, and, at the back end, by the economic logic of justifying the chosen design in terms of its contribution to a firm's financial health.

This fourth core framework of *Networks in Action* defines the economic criteria for assessing the investment in telecommunications—either an investment in building or extending the platform, or an investment in a specific application that uses the platform—of which cost and benefits are elements. The focus, however, is not on the cost of telecommunications but on its impact on the costs of doing business, including the costs of ensuring quality and service. Obviously, management then must decide whether these improvements in costs of doing business, plus any revenue-related benefits, justify the cost of the telecommunications investment.

In this way, a designer can often show that telecommunications is not just a good deal, but one of the best deals a firm can get anywhere. One piece of evidence for this position is that the companies best known for service in a number of industries are frequently both leaders in the use of telecommunications as a business resource and low-cost producers with everyday low prices. Examples include Wal-Mart, Dell, Lands' End, Federal Express, and USAA. It is very rare to find that the service leader in an industry is either not making substantial use of telecommunications in ways that the average firm does not or has higher costs of providing its services than its main competitors. If the client and designer can show this and the implementer can make good on the promise, then telecommunications is not an overhead cost but a business resource.

The caveat here is that to effectively make their case, the client and designer must present a convincing economic model—in the language of business, not of technology.

| | || | |

Competitive Advantage Is Not A Convincing Economic Model

In the early to mid-1980s, the most popular tactic to justify major telecommunications investments was to cite the promise of telecommunications as an opportunity to create a competitive edge for a company. This approach generally sought an increase in revenues and market share. Many managers now groan when they hear this line of argument, partly because too often a competitive edge does not emerge. But even if it did, gaining "competitive advantage" is not a convincing economic justification in and of itself. After all, a firm could get a major edge just by halving its prices and giving a free Mercedes to anyone who spends $10,000 on its products. This example is, of course, absurd, but the point it makes is not; many of the claims made for computers and communications—that they provide a source of differentiation or create new strategic systems—ignore the price tag. Very few cases of competitive successes—Merrill Lynch's Cash Management Account, Citibank's rise to prominence in the late 1970s through electronic banking, ATMs, and credit cards—addressed the issue of the profitability of the strategies versus the growth in revenues and market share. Citibank, the exemplar of the 1980s, became a disaster story of the early 1990s, with a loss of $800 million in a single quarter. Merrill Lynch gained huge deposits through CMA but failed to control its cost base. Innovation is not the same as economic gain.

There are no reliable figures on the exact amount of money the best-known exemplars spent to gain their competitive advantages; some of them spent hundreds of millions of dollars. Almost by definition, their tactics were neither cheap nor simple and quick to implement; if they were, they would be quickly imitated by competitors, which would negate the advantage. A number of commentators argue that this is exactly the situation that is created even by successful investments in telecommunications. The leaders gain a clear edge, forcing their rivals to catch up. Catching up may take several years, at the end of which every firm's cost base has increased, but not the size of the market; the customer may gain, but not the industry. The example most often cited is automated teller machines in banking. Led by Citibank, major banks built their own proprietary networks. Within a few years, however, consortia of banks and third parties had built shared networks, and customer pressures pushed for banks to allow other banks' cards to be used on their network. Now there is no competitive edge in owning a proprietary ATM network, and the industry as a whole might have saved a great deal of money by cooperating instead of competing.

Some commentators question whether there ever was a real competitive advantage in the first place, except for a few special situations. As a result of the overselling of claims about IT and competitive advantage, there has also been a growing skepticism in business about the payoff from telecommunications. Study after study by academics and consulting firms shows that productivity has not been improved in financial services or white-collar work as a whole by the massive investments made in telecommunications, office technology, and electronic banking. Given the disastrous condition of the banking industry in the early 1990s, many business executives have closed their ears to the claims. Many of them are looking for ways to outsource networks and computer operations and to drastically cut costs.

In this context, it is clearly vital for telecommunications designers to present a convincing business case that clearly shows where and how the proposed investment will contribute to increasing profitability and efficiency, or reducing staffing levels and expenditures. Without a convincing case, designers will be frustrated by the unwillingness of clients to listen to proposals that offer business opportunities but no "hard" and immediate cost savings. In turn, clients will be frustrated by the inability of technical professionals to demonstrate real business value.

This chapter relates telecommunications to the main *economic* concerns of today's business managers, in their terms. To be listened to, a telecommunications designer must address the following economic issues:

▸ Profit as the "top line" in terms of management priority, not the bottom line of business priority. How can telecommunications help maximize profits, or get the most value from budgets in the case of public-sector organizations? Businesses are being squeezed in terms of margins, competition, and pressures to react quickly to new demands and trends. How can telecommunications managers help spot those trends, avoid waste, and react to problems and opportunities quickly?

▸ Cost structures. Virtually every organization in the United States is worried about costs: staff wages and salaries, health care, overhead, administration, pension obligations, and many other elements of the cost base. How can telecommunications make a significant contribution?

▸ Quality and service. The single new transformation in both business and many aspects of governments is awareness that quality and service are basic requirements, not special favors or something customers are willing to pay for. How can telecommunications help provide firms a service and quality premium without also adding to costs?

▸ Infrastructures. Telecommunications is one of the infrastructures of modern business. How can designers help position a firm to maximize the cost-effectiveness of its overall operations, enable future expansion and innovation, and reduce the cost of essential telecommunications investments?

▸ Sources of new revenues that offer adequate margins. Can a telecommunications platform pay for itself by adding new sources of revenue at low incremental cost?

Previous chapters of *Networks in Action* provide many cases and examples that answer these questions, including the following:

▸ Profit as the "top line." Frito-Lay's management alerting systems and leading retailers' systems driven from point of sale help managers spot a trend in as little as three days; American Airlines's yield management systems are an entirely new approach to pricing and managing inventories that is spreading across other industries. One of the most far-reaching contributions of telecommunications to management is alerting instead of reporting.

▶   Cost structures. Location-independence through telecommunications allows firms to bring in skilled labor from anywhere at a reasonable cost; EDI cuts administration staff by a factor of 2 to 20; providing direct access to information and eliminating administrative intermediaries can remove two layers of management in some instances.

▶   Quality and service. 800 numbers and voice messaging services, EDI, ANI, image processing, and many other contributors to reach add convenience, speed, and service and reduce errors. Databases and processing systems add range and thus add prompt and accurate response to questions and requests. Network management systems and backup and recovery facilities add responsiveness and hence reliability.

▶   Infrastructures (platforms). The exemplar companies in the use of telecommunications have all made technical integration and the creation of a multi-use platform a priority. USAA in financial services and Texas Instruments in manufacturing addressed the following questions: (1) Will an integrated platform ensure that we can meet business priorities more effectively and cost-efficiently than case-by-case applications approach? and (2) Will such a platform substantially reduce our long-term costs, including those in telecommunications? American Airlines's preeminence in its industry can be shown to depend heavily on its integrated set of applications.

▶   Sources of new revenue. British Airways added international hotel reservations to its airline reservation system and McKesson became a leader in insurance claims processing, both through piggybacking; the incremental revenues that accrued did not involve either large capital investments or operating costs. Although the revenues may be relatively small, the margins are high. The best revenues are those that exploit infrastructures that are already in place and paid for, whereas those that decrease margins may increase absolute profits but do not promote long-term economic health.

Clearly, a designer must make a systematic analysis of a firm's specific opportunities and whenever possible should quantify the expected benefits. The point here is that the **quality profit engineering** framework allows a designer to explain telecommunications payoff in a language businesspeople can understand.

Quality profit engineering is not the same as profit maximization, or total quality management, or business process reengineering, although it is fully consistent with their premises. It makes sense because firms now face increasing pressures on margins at a time when they must provide ever-improving levels of quality and service without being able to charge a premium for them. In this context, revenue growth may not lead to profit growth. The term *downsizing* relates to this; it adds up to "We can't afford these revenues—they are hurting our profits."

The next section looks at the economic context that the quality profit engineering framework is designed to address: the "Cruel Economy," the toughest business climate since World War II.

Profit as the Top Line

The economic reality of the 1990s for U.S. and European corporations is increasing margin erosion; for public sector organizations, it is increasing budget pressures. As a result, profit is the top line of management concern. Operating profit margins, not sales revenues, drive almost all companies' concerns today.

Historically, profits were a by-product of revenues. We call profit "the bottom line" for the commonsense reason that it appears at the bottom of a profit-and-loss report. If a firm increased its sales and kept its costs reasonably well under control, profit flowed to the bottom line. Inventory was an asset; if you made something, you could expect to sell it. Revenues drove growth, and business cycles were fairly long. Companies often spent several years to test-market a new product, check the results, update their marketing plans, and launch the full product.

In this economic model, the supplier set the terms of business, using advertising, promotions, and sale prices to move inventory. The economy was sales-driven. Much of business practice was highly oligopolistic, with the U.S. car manufacturers of the 1970s frequently cited as examples of how a small group of large firms with limited foreign competition moved sluggishly to innovate. Large firms were considered to have an advantage just by being large; they could use their size to attain economies of scale in manufacturing and distribution, could afford large expenditures on R & D, and had ready access to capital. But the model organizations of the 1970s and 1980s became the failures of the early 1990s, as flexibility and nimbleness replaced size as the edge in increasingly competitive markets. Established companies were challenged by new entrants, by foreign firms and reinvigorated large companies that were able to offer more innovative products, better quality, and better service, all at a highly competitive price.

The new realities of competition have made profitability the driver, rather than revenues, in a context in which profits must be earned through lower, not higher, prices. The leading firms in the airline, retailing, car rental, supermarket, telecommunications, and banking industries are increasingly shifting toward an economic model that emphasizes *yield management* rather than merely sales. The term was invented by the airlines and refers to the operating margin of a flight. Airlines frequently discount fares—which decreases unit revenue—in order to maximize profits. They can do this *only* because their technology base—telecommunications-centered reservation systems—provides the means to monitor market performance and trends and update prices in real-time. In 1986, when Continental Airlines launched its Maxsaver discount fares, industry prices dropped by over 40 percent, and profits of the top airlines *increased* by 36 percent. This is how to manage profit as the top line. The leaders in yield management, most obviously American Airlines in the United States and British Airways in Europe, monitor every individual flight for a year, fine-tuning prices as traffic patterns become apparent. They use complex mathematical techniques to balance the risk of selling discounted seats too early (thus losing the opportunity to sell them to full-fare passengers) versus having unsold seats left over. The reservation system is the base for anticipating and monitoring what is occurring in the marketplace as it happens. On-line alerting systems allow rapid adjustment instead of late reaction.

The same shift to on-line profit management alerting systems is apparent among retailers. In the 1970s Sears stood out as the dominant force, mainly through its massive purchasing power; central buyers negotiated contracts with suppliers, and decisions about stocking individual stores were made centrally and/or regionally. In the early 1980s, Kmart led its industry through low prices, again with central merchandising and stocking. By the late 1980s, Wal-Mart had used superior networked logistics to overtake Kmart, creating one of the main success stories of modern business. The individual store became the focus of both operations and information. Point-of-sale technology ensured up-to-the-minute data on inventory and purchasing patterns, up-to-the-minute pricing adjustments, and daily information for central buyers, product line managers, and business unit executives.

By the mid-1980s, Wal-Mart had a lower gross margin than Kmart, offered lower prices, and had far higher operating profits; telecommunications and point of sale were key factors in optimizing its information, pricing, merchandising, purchasing, and distribution operations. Through 1991, Wal-Mart's gross margin per unit of revenue was 5 percent lower than Kmart's, but it sold an average of $250 per square foot of store space, versus $190 for Kmart. Retailers with networked logistics are able to have the right goods on the shelf at the right time. In the recession of the early 1990s, profit growth continued for the networked leaders, including Wal-Mart, Toys "R" Us, the Limited, and Dillards, even though revenue growth was flat or two to three times lower than profit growth.

Time-based competition makes time the driving factor in reporting, stocking, distribution, and pricing. It also makes telecommunications crucial in all these areas, because it is the least time-dependent mechanism any business can use. Fax is faster than mail. Point of sale tells managers what goods are selling faster than the accounting system can report the figures. Electronic data interchange matches inventory to purchase patterns as fast as the point-of-sale system can update the relevant databases. Customer-supplier ordering and delivery systems similarly match stocks to sales and/or production, reducing or eliminating inventory and often removing the need for intermediaries in the supply chain.

The technical aspects of profit management systems can be very complex in a system that captures data at the point of event, where profitability is deter-

Figure 12–1

Examples of Effective Use of the Platform for Profit Management Systems

Company	Reach	Range	Responsiveness	Profit Management
Frito-Lay	To stores via hand-held computers	All product and sales data cross-linked	Limited on-line capabilities for uploading; on-demand service for head office's product and brand managers	Alerted automatically to trends, discrepancies, and competitive moves
American Airlines	To over 40% of U.S. travel agents; to all major and domestic airlines' reservation systems	Reservations, operations, schedules, and passenger data cross-linked	Target of perfect service; current availability occasionally erratic	Yield management; on-line tuning of prices and capacity
Wal-Mart	To stores and suppliers, including truckers	Store and product data cross-linked; ordering and delivery processing on-line and automatic	Perfect service	Fine-tuning of entire logistical chain; fast reaction to match inventory with demand

mined, and moves it quickly to the decision makers who need it. Every company known for its profit management alerting systems has a telecommunications platform with comprehensive reach, range, and responsiveness. Figure 12-1 shows examples.

The Economic Realities of the 1990s

Of course, retailing, manufacturing, financial services, and airlines are not exclusively time-dominated; they are increasingly quality- and service-dominated. In retailing, "discount" used to mean "crummy"; now it means superior price, superior inventory, and superior level of service, including after the sale. Firms cannot trade off price and service as before, and very few businesses can neglect quality and service. Obviously, if a business must increase its own costs to provide quality and service but cannot pass on those costs in the form of higher prices, it faces yet another pressure on its margins. This is a major reason why revenues no longer ensure profits. Peter Drucker defined profits as "the cost of staying in business." This remains true in an era when quality and service are also costs of staying in business, and when traditional costs—labor, benefits, real estate, and administration—have become unacceptably high in most public and private-sector organizations.

Profit margins per unit of revenue are not likely to increase in the next decade in any major industry. Firms will have to learn how to manage as if they are a permanent recession. Deregulation, globalization, and overcapacity mark most industries. Individually, these trends erode margins; together, they guarantee a margin crunch.

Deregulation

Based on the experiences of U.S. airlines, securities, and long-distance phone industries, and of European financial services as well, deregulation cuts margins by around 20 percent. We can expect worldwide trends toward deregulation to continue through the 1990s, especially in telecommunications. Of course, there will be pockets of protectionism and reregulation, but powerful forces are driving deregulation, including the revamping of federal banking laws, the business momentum of the European Community's 1992 Single Market, the North American Free Trade Agreement (which will add Mexico to Canada as a key business partner with the United States), and the relaxation of Latin America's tightly controlled markets.

Deregulation is everywhere. Worldwide, the telecommunications industry is rapidly moving from being a virtual monopoly within each country to some degree of competition. Airlines have moved in the same direction, although more slowly, with the U.S. government restricting the percentage of foreign ownership of any U.S. carrier. British Airways tried to be a major owner of United Airlines as part of a management buyout; it later made an unsuccessful move that would have changed forever the nature of airline competition worldwide, when it offered to acquire 40 percent of US Air.[1] The securities and banking industries are still largely regulated, but many loopholes, such as bank holding companies and off-shore tax havens, still exist.

In every industry in which deregulation has occurred, unit prices drop fast. The U.S. long-distance telephone industry is illustrative. The price of a typical call has dropped 40 percent since the AT&T divestiture ended its monopoly in 1984. The local telephone industry remained a regulated regional monopoly in this period, and prices went up, even though the technology of the long-distance providers was the same as that of the local service providers. The difference in prices had nothing to do with technology and everything to do with regulation. In Great Britain, competition in the PBX market reduced prices by 30 percent in two years. Airline seat prices on individual European routes

---

[1] Foreign ownership of U.S. airlines is heavily restricted by both regulation and the desire to curb foreign airlines, which are either national monopolies or restrict U.S. access to key airports. To compete, U.S. carriers need access to capital and to European and Asian landing rights. British Airways owns both of these resources and is aggressively expanding internationally. Look for the battle to continue between BA and its U.S. rivals.

dropped by as much as 60 percent in 1991, once the national carrier's monopoly was relaxed.

The pattern of deregulation cutting margins is clear and unstoppable. In general, the customer gains (although many critics of both airline and telecommunications deregulation argue that the price cuts were offset by a reduction in service). Deregulation obviously puts pressure on any firm, especially when the industry has historically had high costs of administration and overhead. Improving cost structures is an immediate priority when deregulation changes the rules of competition. The U.S. telecommunications industry has cut over 100,000 jobs since the AT&T divestiture; it had no choice. Regardless of one's individual opinion on the political and social issues of deregulation, there can be no question that it puts immediate and sustained pressure on firms to rethink their basics of operations and to recognize that they can no longer pass their costs on to their customers.

Telecommunications has played a major role in many deregulated industries in two ways: (1) by providing opportunities for adding new revenue sources, the fifth component of the quality profit engineering framework, and (2) by imposing immediate pressure to reduce service delivery costs in the face of price erosion.

One of the strategies in the telecommunications business opportunity checklist (Core Framework 1), under the opportunity to create market innovations, is piggybacking, and deregulation facilitates piggybacking. It can be difficult to do if it involves setting up locations, adding a sales force and marketing, but very easy to do if the network already links to target customers. This is one reason why financial service companies have been able to add several banking services to their own business. Deregulation makes it possible, and telecommunications makes it practical, for USAA to use its existing telecommunications platform to operate without the expensive bricks and mortar network of traditional banks.

The second way in which telecommunications is often relevant to firms in a deregulated environment is simply that they are forced to find quick ways of cutting their costs. A protected PTT, a regulated electrical utility, a monopoly RBOC, or a traditional insurance company was once able to benefit from excessive administrative, selling, and service costs, because it can add such costs to its prices; deregulation generally ends that, but not always, in that a regulated industry may have kept prices low through government subsidization or market distortions. For example, AT&T before divestiture had subsidized local phone rates at the expense of long-distance users; competition reduced long-distance rates and initially increased local rates. After divestiture, the RBOCs held a monopoly on the local loop and became notoriously overstaffed and overpriced. The flood of competition that began in mid-1992 forced the RBOCs to move aggressively to cut costs and prices, as the boundaries between local and long distance were blurred and cable TV firms and RBOCs were allowed to offer each others' products. Ironically, the cable companies were becoming the new villains, accused of using their oligopoly position to keep prices unjustifiably high. Both industries have had to move aggressively and rapidly to position themselves for the new competitive environment.

From the perspective of the telecommunications designer in a firm facing the forces of deregulation, the challenge—and opportunity—is to show that telecommunications can make a contribution to quality profits; if he or she does not do so, the price/cost margin crunch makes it very unlikely that senior business mangers will approve expenditures on telecommunications, and it will be a candidate for cost-cutting, not for investment. As later sections of this chapter show, there are plenty of opportunities to make the more positive case.

Globalization

Globalization also cuts margins by up to 30 percent, partly because in effect it is a form of deregulation that breaks the hold of large and often lazy domestic firms, but mainly by raising the base level of service and quality to the level provided by the leading transnational supplier; this might be termed the Sony, Toshiba, and Toyota phenomenon. Companies can no longer charge for service and quality, which are now the entry fee for being players in the game.

A striking example of this is the car industry. The Japanese Lexus is highly engineered and low-priced, with a level of after-sale service that matches its manufacturing quality. American customers are used to getting letters that announce a recall of their car, in which the message essentially is "We messed up again. The government has made us recall your car. Bring it in and we won't charge you. If you don't bring it in, it's your problem." Lexus handled matters very differently. A small number of customers in 1990 reported a minor problem. Toyota, manufacturer of the Lexus, arranged to pick up every car it had sold in the United States to date, drop off a spare, take the owner's car back to Toyota, correct the fault, carry out a tune-up, clean the car inside and out, and then return it. One Toyota employee had to drive 300 miles each way to carry this out.

The Lexus story stands for the new base level of service needed to compete in a marketplace of global excellence. Quality is just as much a new requirement. **Total quality management** (TQM), which has become an established priority for most firms, has moved quality from being an add-on feature to its being an integral element of, first, manufacturing industries, then service industries, and most recently education and government. Although much of the quality movement has been a defensive response to the erosion of U.S. firms' market position by Japanese competitors, many U.S. companies have thrived by maintaining quality in every market. For example, Motorola dominates the cellular phone equipment industry, partly through its early commitment to "six sigma" quality, a term borrowed from statistical quality control, that means error rates of less than one part in a million. U.S. retailing is moving aggressively into Japan and Europe; Toys "R" Us led U.S. companies into Japan in the face of massive restrictions on foreign retailers and is now in the United Kingdom, offering a range of quality products at prices no British firm can match. Lands' End, the catalog clothes seller, has also moved into Great Britain. Of the top ten securities firms in Tokyo in terms of 1991 profits, six were American. A 1993 McKinsey study revealed that the current gap between the United States as the world leader and Germany and Japan in service industry quality and productivity is the same size as the gap by which the United States trailed in 1983.[2]

It is almost certainly not a coincidence that the firms most noted for service and quality are also advanced in their use of telecommunications, ranging from the first-rate telephone service you get from Hewlett-Packard or Dell when you have a query or problem with one of their software or hardware products; to Lands' End and LL Bean shipping goods to you by UPS or Fedex, thus offering through electronic service prices, quality, and range of inventory that only a few mall-based retailers can exceed; to USAA electronically approving a loan and issuing an auto insurance policy by fax to the car dealer's office five minutes after the customer phoned.

Underlying all these examples are two points that are central to quality profits:

1. Quality and service are competitive necessities in the globally driven marketplace.

2. Telecommunications is an essential tool for providing them.

There are three ways to afford quality and service without decreasing margins per unit of revenue. The first is to charge customers for it; that is no longer an option in more and more industries. As Compaq found out in the personal computer market, a Dell or AST will provide a cheaper version of a commodity PC with top-level quality and service, so that "cheaper" goes with "better." USAA's service and prices are the best in its business. Lexus produced one of the first automotive products that was a luxury foreign car at a nonluxury price; it has badly cut into the sales of German automakers, who can no longer charge extra for quality.

The second way to provide quality and service is to add people. This strategy helped get U.S. banking and manufacturing into trouble in the first place. U.S. firms in these and other industries became greatly overstaffed through the 1980s, to the extent that just about every Fortune 1,000 company has been able to reduce middle management by 20 percent with no loss of efficiency. Firms are aggressively trying to control health-care benefits, which have become an increasingly heavier burden for them and for society. Virtually every time a bank merger is announced, the press release mentions reductions in staff by 15–20 percent.

If a firm cannot afford quality and service by charging for it or by adding people, the third option becomes the only option: use information technology—the combination of telecommunications and computers—to add a quality and/or service premium without adding a cost premium. This does not mean that IT

---

[2]Today, U.S. banks are 30–40 percent more productive than the best of Europe, its retailers are close to 50 percent better than Japanese retailers, and its telecommunications service providers are 20–30 percent better. In the 1980s, the United States competed against its global manufacturing rivals mainly from weakness; in the 1990s, it has the chance to compete from its strengths in industries in which telecommunications is the major differentiator.

guarantees quality and service, only that it is difficult to find a solution that does not rest heavily on IT. Later in this chapter, we identify the uses of telecommunications that provide the most reliable base for the solution.

The situation is new for most businesses. "Quality" and "total quality management" will not be found in the indexes to the pop-management books ten years ago, and "telecommunications" is almost entirely missing. Even in leading books on time-based competition, total quality management, and global competition, the topic is rarely mentioned.[3] These topics must be addressed today. Financial services and retailing now face the same pressures as manufacturing faced a decade ago.

Here is the main opportunity for telecommunications in the 1990s. Computer-integrated manufacturing, electronic data interchange, customer-to-supplier links, point of sale, and electronic payments systems provide the chance to change the cost structures of service and quality and thus help rebuild margins.

But the opportunity is expensive, especially for a high-cost labor economy as in the United States. Globalization may open up many business opportunities, obviously, but it certainly cuts margins, regardless of the opportunities. Globalization is also correlated with overcapacity in many industries, including manufacturing, financial services, transportation, and distribution. Much of this overcapacity is the direct result of technology.

Overcapacity

Manufacturers across the world have reduced the labor component of their goods, shortened time to market, and reduced ordering and delivery lead times, much of this through on-line processing and communication links. They are producing goods today with a fraction of the labor and inventory of ten years ago. All this has led to high productivity and high capacity.

The leading airlines have similarly created electronic delivery bases through on-line computerized reservation systems that are each designed to handle 2,000 transactions a second. The problem is that there are too many systems working too well; the total capacity is around 12,000 transactions a second in an industry that currently uses about half that. There is a similar oversupply of transaction capability in the credit card industry, in banks' payments systems and automated teller machines, and in the securities industry. It has become apparent that not only are securities firms in New York City overstaffed, but that the economies of information technology are such that all their back office processing can probably be efficiently handled by one company. In most of the developed countries, there are too many banks and too many branches, the legacy of an era that telecommunications and deregulation have brought to an end. Most commentators agree that in London, New York, Toronto, Amster-

---

[3] A review of over 20 leading books published in the 1990s on these topics showed that not one of them had even a page on the impact of telecommunications. This is an indicator of how little dialogue there has been between business thinkers and practitioners and their equivalents in the telecommunications field.

dam, and other major financial centers, one of the top five banks could disappear and plenty of processing and trading capacity would remain.

How can margins hold up in this era of deregulation, globalization, and overcapacity, plus the commoditization that each often accelerates? Obviously, they cannot hold up, even for pharmaceuticals firms, who for decades have been able to exploit lengthy patents and premium prices for new drugs. They now face the same pressure to cut time to market and the same customer power in pushing prices down as do manufacturing, retailing, and financial services.

Businesses are having to reinvent their cost structures. This is not the same as cutting costs; anyone can do that just by laying off all the firm's employees. The challenge is to keep the business growing while changing the relationship between revenue growth and cost growth and by rethinking profits as the top line, not the bottom line.

Many telecommunications applications contribute to the top line, but managers rarely hear the message couched in economic terms. For instance, electronic data interchange is often discussed in terms of competitive industry trends or of the need for "business process reengineering." EDI may offer no competitive advantage in a given situation, but if it can be shown to be the most effective way to provide service for less cost than through alternative means (most obviously additional staff), it can create a major organizational advantage. The business logic here is that the service premium must be provided anyway; thus margins will be under even more strain next year than they are now. Where, how, and to what degree does EDI both meet the competitive necessity and improve the cost structures of providing first-rate service?

The discussion so far in this chapter has been about business in general, not about telecommunications specifically. A telecommunications designer has a business responsibility first and a technical one second; a telecommunications implementer has a technical responsibility, first and second; a client has a business responsibility, first and second. The rest of this chapter shows how telecommunications can make these two responsibilities essentially identical.

| | | | | | |

Management Alerting Systems: Managing the Top Line

Quality Profit Engineering

Most management information systems (MIS) reflect the origins of data processing in their reliance on historical accounting information and on the revenue- and growth-centered tradition of profit as the "bottom line." Inventories have become a liability, not an asset. The entire logic of just-in-time manufacturing is to avoid carrying expensive inventory; the logic of advanced point of sale is to match ordering to sales patterns as closely as practical. Quick response goes further by reducing and even eliminating the need for intermediate distributors and warehouses. The shift to time-based competition represented by just-in-time inventory implies a need for just-in-time information. In most firms, however, information for management decisions is synchronized not with the flow of the business, but with the accounting system, particularly sales reports. It is not uncommon for a bank that processes millions of dollars in foreign exchange, bonds, and securities not to know its net profit and loss for weeks, as the transactions flow through the settlement system, back into the accounting system, and onto printed statements.

Information systems that rely on historical accounting data are too slow and often too misleading to alert managers to trends, opportunities, and problems in time to take action. The quickly emerging trend in information management is away from traditional MIS and toward alerting systems built on data captured at point of event. The data is routed to the people who need it via powerful backbone networks that are increasingly likely to include VSAT satellite earth stations so that even the smallest business location is part of the electronic information platform.

The exemplar industry here is the airlines, perhaps not coincidentally the first industry in which IT became the core of sales and distribution through computerized reservation systems (CRS) and combined with deregulation to create the stimulus for moving toward an entirely profit-centered view of business. The airlines and airline IT in the 1980s thus became a general precursor of the 1990s, with banks and retailing having much to learn from them. The top airlines use their CRS as the base for yield management, the recognized art form of the industry; yield is the key measure of performance.

Some commentators see airline yield management as specific to the industry because of the perishable nature of its product; an unfilled seat is revenue lost forever. However, the same principles apply in other contexts; slow-moving inventory, out-of-stock situations, or a mismatch between production schedules/product mix and consumer buying patterns quickly penalize profits. Information systems that alert managers to what is happening, fast and simply, are a natural outcome of time-based competition.

Such systems can also be a natural by-product of using telecommunications in core operations, making them a major potential contributor to quality profits. The company whose cash flow is processed on-line—including orders, deliveries, payments, services, design, coordination, and manufacturing—already has in place (and has largely paid for) the information base for management alerting systems. Point-of-event workstations create and capture data as they carry out basic operations. Point of sale captures actual sales within a time frame that can be as short as management desires and the network allows. Point-of-reservation systems for hotels, airlines, tour operators, and any business that is highly seasonal, including advertising, provide early warning signals that can be monitored and responded to within the same conditions of management requirements and network capability. In some cases, daily information is vital, as with airlines and retailers; in others, weekly information may be acceptable. Without a first-rate network, information may be too expensive to capture and too slow to arrive. By contrast, it may well be that the information has already been captured but cannot easily be moved.

One of the advantages of a private network is that at some time in the night it is a free resource; the leased lines are paid for but no one is using them at 3 A.M. Thus they can then be used to send large volumes of data that would be too expensive to transmit over a public, pay-as-you-go network or that would congest a private network during peak business hours.

Obviously, it is vital to make sure that the information sent across a network has value to decision makers and is used by them. The use involved is specific to the nature of the industry and to the specifics of a company. It is generally relatively easy for a client and designer to identify a few key pieces of information that can significantly help to alert managers and enable them to make decisions quickly. For retailers, this data relates to daily sales patterns and inventory/out of stock figures. For corporate treasurers, the key data concerns cash balances, overdrafts, and fund flows, with electronic cash management an essential tool. An airline's profits depend on yield management data and analysis; a manufacturer's profits rest on order, inventory, and scheduling data.

Clients and designers seeking to identify telecommunications-related economic benefits should answer the following questions:

1.   Is there any information that could help our managers add 10 percent to operating profits if they had it when they need it?

2.   Where does that information originate?

3.   Can our proposed platform move the information from its origins to decision makers?

4.   How much would that be worth to the company?

Traditional Costs

When business is thriving, customers are rushing to buy your product, and investors are begging to put money into the firm, cost is a secondary concern to sales. Your company can be inefficient and still be profitable. In the growth years of international banking, executives' offices were measured in acres, flights were first-class, and bonuses and expense accounts were lavish. Banks were greatly overstaffed, especially with mid-level administrative staff. Waste was everywhere. The same was true for many Fortune 1,000 firms that built large headquarters offices, developed elaborate bureaucracies, and spent heavily on fads and fashions—including information technology.

But not now. The 1990s have seen U.S. business make the most aggressive effort on record to cut out waste and improve costs. They have no choice. Nor do state and local governments, hospitals, universities, or libraries. Traditional costs include the obvious elements of labor, materials, and real estate. Telecommunications can contribute to profit engineering in these areas in three ways:

1.   Telecommunications enables firms to build location-independence—to bring work to people, rather than the other way around. Given the decline of education in the United States and consequent problems of labor demographics, this is sure to be one of the driving forces for international business networking. Cities, states, and even countries are recognizing the opportunity here. A common pattern underlies Omaha's becoming the

800-number capital of the United States and Ireland's capturing the back offices of financial services firms in New York for electronic processing. These and the many comparable moves to exploit business networking for economic development reflect: (a) a recognition among business and government leadership of the opportunities location-independence affords, (b) a coherent plan to shift education, especially at the junior college and high school senior levels, toward building skills that can be marketed in relation to these opportunities, (c) close cooperation by local or national telecommunications providers to ensure rapid installation of facilities and service, and (d) a policy and strategy for developing a telecommunications platform that includes everything from intelligent buildings to fiber-optic networks to teleports and satellite hubs.

2.   Telecommunications provides a vehicle by which firms can recreate organizational simplicity. Electronic data interchange and image processing have dramatically streamlined overly complex administrative processes, enabling American Express, for example, to cut its billing costs by 25 percent, USAA to improve productivity per employee by a factor of 6, and Northwest Airlines to increase efficiency in revenue accounting by a factor of 50. IS consultants and managers today place a great deal of emphasis on business process "reengineering," "redesign," and so forth. Two common principles underlie these activities: (a) questioning the basics of the process, including why it should exist in the first place, and (b) meshing process, people, and telecommunications rather than "applying" telecommunications. Observing these principles almost invariably results in streamlining and simplification and, over the longer-term, contributes to reductions in levels of hierarchy and administrative staff. The latter can sometimes be dramatic; purchasing departments that have thought creatively about their basic activities and have eliminated or designated for electronic processing much of their work have lost two levels and 20–80 percent of their people.

3.   A telecommunications unit can improve its own cost disciplines. Most companies' accounting systems and management processes do not accurately capture the nature of telecommunications and computer costs, typically the third largest expense for service firms, after labor and real estate. A reliable rule of thumb is that the apparent price of telecommunications is 20–25 percent of its full actual cost.[4] It costs $4 million over the first five years of use to operate and maintain a system that cost $1 million to develop. To support a $5,000 personal computer costs from $8,000 to $18,000 per year. Chapter 17 reviews telecommunications costs in detail.

---

[4]These figures are reviewed in more detail and justified in Chapter 6 of Peter G. W. Keen, *Shaping the Future: Business Design Through Information Technology*.

The easiest way for telecommunications designers to identify opportunities to improve traditional costs is simply to ask clients what costs they are most concerned about.

Quality and Service Premiums

The simplest way for public and private organizations to cut traditional costs is to reduce quality and cut corners on service. One of the general messages that underlies the wide range of examples of leading companies' strategies is that firms can cut traditional costs and still provide first-rate quality and service.

A central premise of quality profit engineering is that firms have no choice but to ensure ever-higher levels of both quality and service, and that doing so will inevitably erode margins unless service and quality can be provided without adding labor costs, managerial layers, and administrative overhead, which in turn increase the organizational complexity that has been a barrier to quality and service. Service has two key elements: convenience and getting your request or question handled immediately. The reach of a telecommunications platform helps ensure convenience, and its range ensures breadth of service.

Many examples have been given in *Networks in Action* of using telecommunications to differentiate a standard product via service and convenience, to improve ease of access, and to put a workstation in the customer's home and office. The economic benefits of such customer service improvements can be very difficult to quantify, but even when specific figures cannot be reliably projected, the following questions establish if and how telecommunications offers the best deal:

1. Where must we dramatically improve service and/or quality, either as an opportunity or as a competitive necessity?

2. What are the main options in doing so in terms of programs that add staff, increasing prices to cover service improvements, or any other non-telecommunications approach?

3. Are these options practical and cost-efficient?

4. Which if any areas of telecommunications technology can be reliably, quickly, and cheaply implemented to provide visible and significant improvements?

5. Do these technologies offer better and faster payoff than the other options?

6. Is our existing platform (or its opposite: incompatible applications, hardware, and communications) an enabler or a barrier to gaining the benefits?

For almost all companies, this line of analysis will quickly highlight opportunities from 800 numbers, EDI, voice messaging, customer service workstations, electronic mail, or other standard telecommunications services. Telecommunications is so basically about ease of access and communication that this could hardly be otherwise. In the area of quality improvements, telecommunications in and of itself generally plays a lesser role; it is business networking—the combination of access tools, network links, transaction processing systems, and information stores—that provides benefits via CAD/CAM, material resource planning (MRP) systems, and the like. Relatively few firms have the reach and range needed to implement new services to provide a quality premium without adding a substantial cost premium.

Improving Revenues   Improving revenues is not the same as increasing revenues. Telecommunications may be able to help add revenues that do not disproportionately add costs. An emerging trend among industry leaders is to exploit both occupancy at a point of event and the operation of their telecommunications platform to change the economics of innovation. For example, it was cheaper for British Airways to add hotel reservations to its platform than for Marriott and Hilton to build their own infrastructures; similarly, McKesson's ownership of the pharmacist's point of event for ordering and distribution made it easy to add insurance claims processing.

One of the most successful instances of a company using information it already has to create a new stream of revenues is MCI's Friends and Family program, which was built on MCI's own billing system. Its integrated platform had the range to enable MCI to identify both the caller and the receiver of a phone call. Each 1 percent gain in market share is worth about $500 million to MCI, which has made Friends and Family a $2 billion a year product—all from the firm's own billing system. This is a classic example of a preemptive strike, of the value of a high reach/range platform, and of one way a firm can gain revenues from its existing telecommunications and information base.

Getting a real payoff from telecommunications is the single biggest frustration of business executives who deal with it. Quality profit engineering is a framework for thinking creatively and practically about how to do so. The phrase "quality profit engineering" emphasizes that:

1.   Quality profits are not the same as profits; neither are profits the natural bottom-line outcome of revenue growth; telecommunications designers must focus ruthlessly on helping to resolve the dilemma of providing quality and service in an era of eroding margins.

2.   Profits are the main issue; the competitive advantage model for justifying telecommunications has largely missed this point.

3.   This is not an issue of reengineering; the company never did the engineering in the first place.

Summary

Perhaps a designer's most difficult task is to make the economic case for investing in telecommunications. Managers typically are skeptical of the payoff and view telecommunications as a cost to be con-

trolled. Technical specialists who have little understanding of business find it difficult to justify telecommunications in business terms. It is essential to the telecommunications-business dialogue to provide a convincing economic argument. The fourth core framework of *Networks in Action* addresses "quality profit engineering," which has six elements: (1) using telecommunications to help manage profits through management alerting systems; (2) improving traditional costs, most obviously staff and real estate, through such uses of the telecommunications platform as location-independent work and electronic data interchange; (3) providing a service premium to customers, without adding a cost premium to the company, by using a wide range to telecommunications tools, including EDI, voice messaging, image processing, customer relationship databases, network management systems, and access tools; (4) providing a comparable quality premium, with CAD/CAM adding to the proven tools; (5) building the telecommunications platform infrastructure needed to ensure that the business can both meet its future business priorities and needs effectively and reduce the costs of both business and telecommunications; and (6) improving revenues by adding sources of high-margin services to the platform.

Businesses in just about every industry face growing pressures on their operating margins. Deregulation cuts margins by 20–30 percent, with airlines, international financial services, and the long-distance phone industry being only a few examples. Globalization is a form of deregulation that both opens up market entry to new providers and creates new sources of supply for domestic competitors. Globalization has cut margins by up to 30 percent, partly because of the overcapacity that marks most industries.

Globalization also contributes to the new focus on quality and service as part of the cost of staying in a competitive game, instead of being something for which companies can charge a premium. Telecommunications can play a major role here. There are only three options for providing quality and service. The first is to charge extra for it, which is no longer a viable option except in special situations. The second is to add people, but this was what created many of the competitive problems of U.S. business in the 1970s and 1980s in the first place. The third is to apply technology; this option does not guarantee success, but there is plenty of

evidence to suggest that telecommunications can be a key resource in dramatically improving service and quality without damaging margins.

Review Questions

1. Explain why investing in telecommunications networks for the purpose of gaining a competitive advantage for the firm is not sufficient justification.

2. Explain why many business executives see telecommunications mainly as a necessary cost, not as a potential value.

3. Explain and discuss the meaning of the following statement: In the current economic reality of the 1990s, companies need to focus on profit as the "top line" instead of the "bottom line."

4. Briefly describe the concept of "yield management" as it is used in the airline industry, and explain how this concept is applied to profit engineering of telecommunications networks.

5. Explain how Wal-Mart was able to achieve higher operating profits than Kmart, even though it had a lower gross margin and lower prices.

6. Explain the meaning of the following statement: "You can no longer charge for quality and service." Briefly discuss how this view differs from previous perceptions of quality and service.

7. Briefly discuss the impact of deregulation on companies with respect to profit margins, daily operations, and costs.

8. Explain why the phenomenon of globalization has cut profit margins significantly.

9. What is the only viable way to add a quality premium or a service premium without adding a cost premium? Explain.

10. Explain the difference between improving cost structures and cutting costs.

11. What does the shift to time-based competition, represented by just-in-time inventory systems, imply about the need for information? Are historical accounting systems adequate to meet this need? Explain and discuss.

12. Some have argued that airline yield management systems have only limited usefulness in industries that do not have perishable products or services. Describe how the principle of yield management can apply to a nonperishable commodity that experiences frequent out-of-stock conditions.

13. Explain how management alerting systems can be a natural by-product of using telecommunications to support core business and manufacturing operations.

14. Briefly describe the three ways in which telecommunications can contribute to profit engineering by reducing traditional costs (labor, materials, and real estate).

15. Explain the meaning of the following statement: "Improving revenues is not the same as increasing revenues." Describe how telecommunications can add to revenues without disproportionately adding to costs.

Assignments and Questions for Discussion

1. Describe the causes of the breakdown in dialogue between client and designer over the economic justification of investments in telecommunications networks. Then, develop a process for reestablishing that dialogue, and explain how it overcomes the existing difficulties.

2. Develop the concept of an on-line alerting registration system for courses offered by a college or university. Describe the system concept, and explain how a telecommunications system can be used to improve net tuition revenues.

3. Discuss the relationship between telecommunications and a firm's cost structure, and give some examples of specific telecommunications applications that improve the cost structure of the company.

4. As a consultant to a Fortune 500 company that is competing globally, explain to the CEO how telecommunications can be used to provide competitive quality and service levels without increasing costs/prices. Use examples of relevant types of telecommunications applications to illustrate your points.

5. Compare the concept of profit engineering with traditional views of profit and profit analysis, and explain why profit engineering is better suited to meeting the needs of business in the globally competitive world of the 1990s.

Minicase 12-1

American Standard

Measuring Business Value

How to measure the payoff from computers and telecommunications remains one of the major challenges facing the field. There are so many problems: ascertaining how to assess qualitative benefits, choosing suitable payoff metrics, gathering correct and complete cost data, and including all relevant components of an increasingly distributed IT base. One comprehensive and practical program to develop and apply a measurement system was implemented by American Standard, Inc. A four-stage methodology helps IS and business managers identify key business objectives and performance objectives for any IT investment. The four stages are:

1. Manage the business strategy. A strategic business unit (SBU) establishes business goals that break down into more focused objectives that provide the base for specific measures of investment performance. For instance, the goal "Improve time to market" might break down into the objectives "Reduce the time spent on revising design plans" and "Streamline testing time."

2. Manage the investment. This is a baselining exercise in which current expenditures are broken down according to cost components (development, maintenance, and operations). These components are analyzed by category: institutional systems (for example, finance and accounting, core marketing transaction, and reporting systems), factory automation, professional support systems (for example, electronic mail, decision support), and external support services (for example, electronic data interchange, telex gateway). SBU managers then look for opportunities to use existing or new IT investments to meet the focused objectives that were defined in step 1.

3. Manage the portfolio. Both the SBU needs and corporate priorities are combined to define American Standard's IT portfolio, balancing development, maintenance, and operations. The firm has a three-dimensional platform architecture: processors (mainframe/corporate, functional/department, and workstation/personal), business portfolio (institutional, factory automation, professional support, and external support), and standards (application, data, communications).

    Existing applications are rated on a scale from 0 to 100 on functional quality (how well it meet the business's need and how easy it is to use) and technical quality. Business managers rate functional quality, and IS rates technical quality. This provides a base for identifying systems that are performing well and are rated highly by both users and IS, and those that need enhancement (low functional, high technical quality), refurbishment (high functional, low technical) and replacement (low, low). It also helps target maintenance investment to enhancement and refurbishment systems.

4. Manage the benefits. This last step selects specific business measures and performance targets for the investments identified in step 3. There are two different categories of "IT-system value-added": criteria for justifying proposed new investments (justification measures) and criteria for tracking existing investments over time (tracking measures).

Justification measures relate the expected payoffs, time, cost, and capital back to the focused business objectives. Weights are used to generate a cumulative index, which provides a basis for comparing investment opportunities and for identifying trade-offs between payoff and, say, lead time and cost.

Tracking measures similarly define an observable factor that can be tracked and quantified. For example, if a new 800 customer service was justified on the grounds that it supports the goal of "Improve our customer image for after-sales support," tracking measures might be the number of complaint letters, the average time a customer must wait for service, and the number of callers left on hold for more than ten seconds.

The process is complex; its goal is to encourage managers to identify investment performance measures right up front, instead of after the event or by default. It serves as a vehicle for ensuring that measures relate back to goals and that there is a systematic base for assessing the existing as well as the proposed portfolio. The methodology had to be carefully sold to senior SBU managers, and it may take years before it is fully institutionalized. It does face up to the challenge of systematically assessing and tracking business value, a major step forward.

Questions for Discussion

1. Describe the major problems associated with measuring the payoff from computers and telecommunications.

2. For each of the major problem areas identified in question 1, indicate which of the four stages in the methodology used by American Standard addresses that particular problem, and briefly describe how it solves the problem.

3. Explain how American Standard decides which new information technology investments to undertake, and describe how each investment's potential added value is determined.

Minicase 12-2

United Parcel Service

Catching Up to Federal Express

Before the creation of Federal Express in 1973, UPS had close to a monopoly on small package delivery. Only in 1982 did it add overnight delivery to its services. By 1986, Federal Express's profit margin per package was four times that of UPS. Fedex was far leaner in its staff and costs; UPS had almost 170,000 employees handling 2.3 billion packages. Its sales were $8.6 billion. In mid-1992 *Business Week* described UPS in the 1980s as "a dull, brown dinosaur that couldn't match Federal Express in technology and wasn't sure it should." While UPS experimented with every gizmo imaginable, its manual package handling was so efficient that, according to chairman and CEO Kent Nelson, "every time we applied technology, it slowed us down."

Nelson led a move to redesign the basics of UPS's operations and to apply technology to that end, not use technology for its own sake. "We saw that the leader in information management will be the leader in international package distribution—period." Between 1986 and the end of 1991, UPS spent $1.5 billion to transform its operations through telecommunications and distributed computing. It will spend another $3.2 billion between 1992 and 1996. It has already seen benefits, though it will be many years before the full payoff is apparent. Comparisons between UPS's 1986 and 1991 performance are shown in Table 12-1. UPS's 1992 first-half profits were up 25 percent over the corresponding period for 1991.

Table 12–1

UPS Performance, 1986 versus 1991

	Revenues (in billions)	Revenue per parcel	Employees	Parcels (in billions)
1986	$8.6	$3.81	168,200	2.26
1991	$15.0	$5.17	256,000	2.9
Increase	74%	36%	52%	28%

In 1986, UPS operated mainly in the United States; it now has a presence in 181 countries. In 1986, the number of parcels per employee was 13,400; in 1991, it had dropped to 11,300. Revenue per employee had grown from $51,000 to $58,000, a 14 percent increase. These figures partly reflect the recession of the early 1990s. Federal Express had a $250 million loss in a single quarter, after years of compounded growth and profitability; this mainly reflected its failed entry into Europe.

UPS had been a very traditional organization in its use of information technology, whereas Federal Express had been an innovator in such areas as package tracking, mobile communications, scanners, satellite links across its operations, and providing customers with personal computers to handle their own weighing, labeling, accounting, and other functions. In 1986, UPS had about 100 IS staff working on mainframe computers at its New Jersey headquarters; there were only scattered personal computers and no local area networks. Its communication network was a simple and slow point-to-point system that had under 50,000 miles of lines to handle its trans-U.S. operations. Its new UPSnet has 500,000 miles of lines. Its 65,000 van drivers each have hand-held computers that download and upload data to and from the firm's computers, which now include five very large mainframes, 300 mid-range, and 33,000 personal computers. It has 1,500 local area networks. UPSnet carries voice, data, fax, and image traffic. The hand-held computers now capture and transmit customer signatures electronically; the customer writes directly on the screen, using a special pen.

UPS is aiming to match Federal Express in its ability to track parcels from collection to delivery and provide status information to customers immediately and at every stage. It has automated and has linked just about every part of its operations: Yard control includes organizing the loading and unloading of individual trucks; delivery centers now know in advance how many packages they will be receiving and which zip codes they are going to; customs are informed in advance about contents, consignees, and other information for shipments. UPS's fleet of over 400 aircraft use automated systems for scheduling, maintenance, problem alerting, and FAA real-time tracking.

In a few instances, UPS has leapfrogged Federal Express. For example, its Maxiship personal computer system gives customers administrative help and many management reports. It also captures information for UPS *before* the package is picked up. UPS also announced the first coast-to-coast cellular mobile data network that will provide immediate air and ground tracking; van drivers will send information directly from the truck to the central mainframe computers.

Initially, UPS expected that UPSnet would need only a seven-node T1 backbone with 40 district-level switches. Expansion of traffic, adding international locations (UPS now has 1,200 worldwide distribution sites), and relocation of its head office to Atlanta made this network far too small to meet business needs. UPSnet added three DS-3 multiplexers, which handle T3 traffic at up to 45 mbps, and upgraded its X.25 district-level switches. It expects to implement frame relay within its existing architecture. As with any major innovation, some aspects of the three-year implementation went well, but others met problems: UPS developed a new form of high-density barcode that can store up to 100 characters on a square inch; customers refused to adopt it.

UPS has invested heavily in backup and disaster recovery. It has two copies of every piece of network software maintained in two separate locations. For this reason it built a second data center (at a cost of over $100 million) and backup power generators; extra hardware and other equipment are everywhere throughout the network.

A major cost in rolling out the new capabilities was for training. Almost all the organization's staff were unfamiliar with computers. The new systems affect every part of the business, and the new technologies they apply represent a radical change for the information systems and telecommunications staff, who are now required to spend at least 15 days a year on training.

The package market is very competitive now, with the three key factors being speed, reliability, and price. Federal Express continues to move aggressively, and its domestic operations have held up well during the recession. Technology remains a core element of Fedex's strategy. It is focusing heavily on cutting costs per package. The three leaders in the industry, UPS, Airborne, and Federal Express, have created a price war, with airlines also looking to use their spare capacity to capture a larger part of the market.

Questions for Discussion

1. The CEO of UPS said that "'the leader in information management will be the leader in international package distribution—period." Describe how information management affects profitability in the package delivery business, and discuss its impact on quality, service levels, and cost structure.

2. Explain how the new UPS network system exploits the use of point-of-event information capture and analysis opportunities to gain or maintain a competitive edge.

3. Describe the areas in which UPS has leap-frogged Federal Express, and explain how each of these areas contributes to one or more of the following:

a. Profit as the top line

b. Improved quality without a price premium

c. Improved service without a cost premium

d. More favorable cost structures

e. Reduction in traditional costs

Minicase 12-3

Nordstrom

Linking to Suppliers Simply and Cheaply

Nordstrom, the fashion specialty retailer with a legendary reputation for service, operates 64 stores in eight states and has 27,000 suppliers. Nordstrom's highly decentralized store operations mean that a supplier may need to contact ten or more buyers in different states to deal with orders for a single product. Many of the suppliers are small organizations without comprehensive information systems or telecommunications capability.

Nordstrom has found a cheap and simple way for them to link to the retailer's locations: electronic mail. It has opened up its internal e-mail system so that from a personal computer the suppliers can use MCI Mail, a public service offered by MCI Communications Inc., to send any type of message, including shipping information and payment status inquiries. They can send a broadcast message to multiple buyers across the country instead of phoning each one separately. MCI Mail can be accessed from any type of personal computer.

Electronic mail sends messages to people, instead of to phone numbers or fax machines. There are two main types of service: public and private. A private service interconnects only people within a firm or even a department; the reach of the service is usually limited to users of a specific type of computer, local area network, or e-mail software package. Public subscriptions are like the telephone system; you sign up and pay a monthly fee and can access all other subscribers of the service.

MCI Mail messages can be sent in a variety of ways:

1. Standard *electronic mail*. Your message is stored by the system and delivered when the person(s) you sent it to next logs onto MCI Mail. Obviously, this is possible only if the recipient is also an MCI Mail subscriber or a subscriber to a system which, like MCI Mail, uses the X.400 standard to interconnect electronic mail services. X.400, which only recently has been implemented in widespread products, ends the almost complete historical incompatibility among e-mail services.

2. By *fax*. Your message is sent to a fax machine. The service offers options such as time of delivery and number of attempts to send the message if the machine is unavailable.

3. By *courier* service, with four-hour delivery. The message is printed out on paper at an MCI Mail node (the equivalent of a post office) and sent through a package delivery service.

4. By *first class mail*. The message is printed and sent through the mails.

All of these messages may be "broadcast" to groups of recipients. Nordstrom's innovation requires no new investment by the company and no investment for the suppliers beyond a personal computer and modem.

The Nordstrom executive in charge of the project explains that "we're not trying to replace voice mail [in which you phone and leave a recorded message that the intended recipient can access later], the telephone or faxes with e-mail. We're just giving our people and partners another tool to enhance communication between them and to improve customer service." The program is optional, and to date Nordstrom has not invested heavily in electronic data interchange, a more complex form of customer-to-supplier links that communicate directly between computers. "We've read that a lot of retailers are forcing suppliers into EDI, but we're not doing that." Some of Nordstrom's own buyers are worried that they will get "an onslaught of vendor messages and junk E-mail."

Questions for Discussion

1. What are the likely problems—technical, organizational, and economic—that a supplier could face in using the e-mail system? What steps could it take to reduce them?

2. What are the advantages and disadvantages for both buyers and suppliers in using a desktop versus a portable laptop computer for the electronic mail service? What impacts do laptops have in changing either reach, range, or responsiveness?

3. Explain how the Nordstrom e-mail system is likely to affect the profitability of the company. Discuss this from the standpoint of the quality profit engineering framework.

Minicase 12-4

City of Hope National Medical Center

A *Fiber LAN* for Productivity

City of Hope National Medical Center in California has developed a network that is attracting much attention from the press and medical professionals and may represent a wave of the near future. The center specializes in catastrophic illnesses, when saving seconds can mean the difference in a patient living or dying. Its Hopenet 2000 links 51 geographically dispersed departmental local area networks with each other, as well as with the hospital's central mainframe computer. The network is built on nine primary hubs that concentrate traffic feeding in from over 40 secondary hubs in Hope's 1,002-acre facility. The primary hub traffic is routed onto a backbone fiber FDDI local area network that includes a gateway into the world's largest network, the Internet, a worldwide facility for research and education. This configuration gives Hope's researchers access to other research centers and the ability to transmit high-resolution X-rays, CAT scans, other images, and large-scale databases.

This system, built on the FDDI standard, replaces a morass of separate facilities. Previously, everything was isolated; to access the pharmacy system, for instance, required reaching a pharmacy terminal. Internet connections could only be made by slow and limited e-mail dial-up; data files could not be transferred, and departmental systems were not linked. Now, there is a single and unified interface, so that users are not even aware of which physical system they are accessing or how.

All patient records are held on Hope's central mainframe computer, which is connected to the backbone FDDI LAN by a special software gateway that can be accessed by all departmental LANs. Previously, setting up connections between networks meant digging up streets to lay new coaxial cable; the hubs and FDDI backbone eliminate this need. The many individual networks are in effect a unified network now, making network management far easier. A primary workstation console maintains a view of the entire network; if there is any fault on a LAN, an icon flashes on the screen. The user can then click on the icon to see what and where the problem is. The network management system uses SNMP (Simple Network Management Protocol), the standard that is used by both the Ungermann-Bass hubs and Cisco routers that handle internetwork links.

A main aim in developing Hopenet was to create a base for an "infinite" range of applications. At present, Hope is using very little of the massive fiber FDDI ring capabilities. It will add imaging, electronic records, and radiology that will exploit the bandwidth.

Hopenet is a technical innovation, but it is not technology for technology's sake. Hopenet is the solution to the center's information strangulation. With 40 buildings, coordination was a growing problem. It became routine for a patient to be kept waiting for three hours because doctor's orders were waiting on someone's desk. Searches for patients' charts could be frantic, and nurses often spent more time on administration than with patients. The network can entirely change that situation by providing fast and easy access and communication. The only real worry for the network designers is how to ensure privacy and security for patients.

Questions for Discussion

1. The Hopenet network is designed to provide significant benefits to the City of Hope National Medical Center. Describe the nature of these ben-

efits, and explain how the quality profit engineering framework can be used to realize these benefits.

2. Why does "Hopenet 2000" use a high-speed FDDI backbone LAN to connect the various departmental LANs? Is it justified from a "quality profit engineering" standpoint? Explain and discuss.

3. Briefly outline an approach to ensuring privacy and security for patients on the Hopenet 2000 system.

13

Case Study: Landberg Dairies

Objectives

- Apply the telecommunications decision sequence to a real-world case
- Show how the technical topics can be made meaningful and relevant to business managers
- Use the case example as the base for reviewing your understanding of the material covered in Part Three of *Networks in Action*

Chapter Overview

| | | | | | |

Chapter Overview

This chapter, which concludes Part Three, presents a detailed case example of how the telecommunications decision sequence was applied in a large British firm, one that had very limited telecommunications capability and senior management that was very skeptical about its value. The focus of the chapter is on the business logic that led to the recommendations for Landberg's telecommunications platform. It describes trends in Landberg's competitive environment that have eroded its position in terms of both market growth and profitability. It examines a business review that focused on the customer's perspective, the industry, and the company itself and identifies a variety of business opportunities for Landberg to improve its competitive and economic health. It defines the telecommunications priorities the business opportunities imply and that together constitute a new platform for the company.

The main aims of the chapter are to show how the style of thinking and analysis presented in the preceding chapters of *Networks in Action* can be applied to a real-world situation, to illustrate ways to effectively communicate the analysis to business managers, and to show how the telecommunications decision sequence helped convince business executives of the value of telecommunications in addressing major problems and opportunities. Landberg has made little use of information technology and almost none of telecommunications; the company was chosen for the study that is the basis for this chapter because it is weak in telecommunications and thus provided a test of the applicability of frameworks in this book, most especially the telecommunications services platform map.[1]

First we examine Landberg's business operations in some detail.

| | | | | | |

Landberg Dairies: Profile of the Business

Landberg is a British company, with $1.8 billion in revenues. Its exact identity is disguised in this chapter, in order to preserve confidentiality; it is not in fact a dairy foods company, but the dynamics of its industry are virtually identical to those of the dairy industry. Landberg is one of the top two firms in its domestic industry; the other is Heller Select. Together, Landberg and Heller have about 60 percent of the total market, but this figure has dropped from 75 percent in the past ten years, as regional "independents" focus on high margin specialty goods, including yogurts, frozen goods, and diet foods.

---

[1] Information and quotes in this chapter are from unpublished research notes of a study conducted in 1991 by the International Center for Information Technologies.

Landberg and Heller operate nationally. Both of them are the result of acquisitions and consolidations of small companies over a 50-year period. At the end of World War II, Great Britain had thousands of small dairies that operated close to their local markets and that concentrated on the basic products of fresh milk and cream. Those small companies became increasingly unprofitable, and most were acquired by one of the two giants. The British market remains highly localized. Milk still comes mainly in glass bottles, and many local retail companies still offer daily deliveries to the home. For that reason, even though it is a national company, Landberg operates 21 dairies, most of which are in the center of large cities, to ensure that its production facilities are close to its local markets.

The market is moving away from standard dairy products. Long-life milk—either a useful or a disgusting product, depending on your viewpoint—constitutes a growing fraction of the industry's sales. Frozen goods, ice cream, yogurts, and dietary goods have grown from under 4 percent of the market to close to 25 percent. At the same time, the prices for basic milk and cream products have not kept pace with inflation, and Landberg's margins continue to drop; the main reasons are that consumer consumption of milk does not increase with income, so that demand is basically flat, and that Great Britain's retail grocery industry is dominated by only five major chains that have immense buying power and thus pressure suppliers for the best price. The grocery chains—the multiples—are also adding their own dairy departments that buy from local suppliers and sell products under their own brands.

In the 1950s to 1970s, Landberg prospered through economies of scale in production. Its operations mainly have fixed costs, with large pasteurizing machines, bottling lines, and trucks that deliver to customers each morning. Orders come into Landberg's 21 dairies between 8 P.M. and midnight; deliveries to the supermarket chains must be made by 7 A.M., so a truck breakdown or a production delay is a disaster. The chains stock almost no inventory; their selling point is freshness, and they regard milk as a "pull in" product that draws customers on the basis of price. British consumers purchase milk in pint or quart bottles; the one-gallon or half-gallon American paper or plastic container is an alien extreme to the UK customer.

Although Landberg's production line takes in raw milk and produces a limited range of products—whole milk, low fat, and skim—its bottling involves a wide variety of different customer private labels for the large chains, its own national and regional brands, and types and size of container. In addition, Landberg must add pricing labels for many of its customers; these vary widely and sometimes must be changed daily.

Landberg's production lines cannot handle the wide variety of specialty goods that are the only growing part of the market and that provide the highest margins. It would be impractical for each dairy to produce and package fruit-flavored yogurts, for example. As a result, specialty goods are produced in ten of the dairies (each of which typically specializes in just a few product lines) and are transshipped to the others; this process is termed "trunking." Each dairy is in effect an independent operation, receiving customers' orders, scheduling production, and placing trunking requests. The head office handles contract negotiations with large customers, but their orders are still made to the dairy that supplies the individual stores.

Landberg is part of a large food company, Waterfield, whose revenues in 1992 were over $25 billion. Waterfield owns many large dairy farms, so Landberg is a very important outlet for its products. Waterfield has been the target of several hostile acquisition efforts over the past few years; the raiders have pointed to its relatively low profits, weak growth, and undervalued assets. The company has placed a premium on improving profits to ward off takeovers. Landberg is by far its weakest performer, so Waterfield has been unwilling to invest new capital in it, cannot sell it off because it is so tightly linked to Waterfield's farm operations, and is demanding that Landberg improve its performance.

Most of the Landberg management team has been in the company for over 20 years. The managing director (the equivalent of CEO) was brought in from Waterfield in 1984 to turn Landberg around. Most of his subordinates came from the dairies; the head of operations is openly disdainful of "numbers experts who have never had to deal with an eight-hour business response time." Most of the managers in Landberg pride themselves in being "hands-on" operators. The manager of a dairy is in effect a feudal baron who is totally in charge of every aspect of production and distribution, and hence of profits.

Landberg's information technology resources are small-scale. Each dairy has its own IBM System/36 minicomputer and handles all aspects of payroll, billing, accounting, and management reporting. The head office has its own minicomputer, which consolidates data from the dairies and also provides corporate financial information. There is no corporate telecommunications network; Landberg uses value-added networks for electronic data interchange, an innovation that the dairy managers dislike but is required by most of the major supermarket chains. At the time of the study described in this chapter, Landberg was experimenting with hand-held computers in its delivery trucks; the computers had no telecommunications capability and were mainly for providing on-the-spot billing and delivery records.

The Research Study

At the time of the research study (1991), Landberg's top management team was increasingly concerned about three main issues: (1) finding new sources of profits, (2) gaining cost control, and (3) improving the executives' ability to plan and forecast trends. The executives were deeply skeptical about the role of information technology in any of these matters, for they had not seen any benefits from either of two expensive IT strategies implemented in the 1980s. These strategies added extra computer hardware that handled clerical and accounting appli-

Figure 13–1	1. Business scan	Customer-centered analysis Interviews with customers

The Landberg Study Business Review		Industry-centered analysis
		   Interviews with industry experts and associations
		   Analyses of trade publications and secondary library sources

		Company-centered analysis
		   Interviews with Landberg personnel
		   Analysis of nonindustry "exemplar" firms

	2. **Business drivers**	Landberg's IT platform needs: "We insist that our IT platform . . . "
	3. **Platform requirements**	Landberg's requirements for reach, range, and responsiveness; high-payoff operational targets and economic/competitive justification; the target technical architecture
	4. **Planning for implementation**	Application priorities and timing; sourcing strategy (in-house, outsourcing, joint ventures, and so on); priority steps and payoffs; timetables, phasing, and responsibilities

cations and complex reporting software, neither of which improved business effectiveness, only administrative efficiency. Management saw IT as a necessary cost of doing business, not as a business resource.

The research study was funded by a leading computer vendor as an explicit test of the validity and applicability of two of the core frameworks presented in *Networks in Action*: the telecommunications services platform map and quality profit engineering. The company that the vendor selected for the study was about as low-tech as could be imagined and had no real dialogue between the business leaders and information services managers. The logic here was to test whether the style of analysis and the specific frameworks are convincing to business managers who are neither knowledgeable about nor biased in favor of IT. The challenge posed to the research team was: "Prove you can make telecommunications make sense to any business manager in commonsense terms."

The study was carried out by one of the authors of *Networks in Action* and a colleague over a three-month period. As you read this chapter, focus on the *style* the researchers used in bridging the business-technology culture gap. The study team carried out a business scan that had customer-centered, industry-centered, and company-centered parts. Figure 13-1 summarizes the review. This was not a "technology" study, but a business study. The team set out to identify Landberg Dairies' business needs and opportunities and then looked at where and how IT could directly and productively contribute to the dairy's business. A key aim throughout the study was to make sure that the report's recommendations on technology were firmly grounded in Landberg Dairies's business mangers' priorities and perceptions. Without that focus, a management team already skeptical about IT would immediately disregard recommendations concerning spending yet more money on IT.

The main sources of information used in the study were the following:

1. *Interviews with Landberg Dairies' senior managers* and with other staff
2. *Interviews with industry experts*
3. *Analyses of trade publications*, industry association publications, and other library material on the production, retailing, and distribution industries
4. *Interviews with customers*
5. *Analysis of relevant firms outside the dairy industry* that represent the best of current practice in the uses of telecommunications and computing that are directly relevant to Landberg Dairies

A wealth of information is available on just about any industry to anyone who has the curiosity to look. A key element in the style of the study (and in the style that underlies the telecommunications decision sequence) is to focus as much as possible on the needs of the business *in business terms* and then to translate what you find into technological terms through the frameworks of the decision sequence. In a way, this is a form of detective work and of listening—of trying to find telecommunications business opportunities.

Interviews are one of the most powerful and productive sources of information. That may seem to be obvious, but too often technical professionals come to meetings to talk, not to listen. They may feel they have the answers and do not need to ask questions. They are eager to jump to the technology, and they find the lack of structure in business irritating in contrast to the systematic nature of their own field.

The entire logic of the telecommunications decision sequence is to start from the client's perceptions and reality and move toward defining technical requirements—all in the client's own language. This is why the research team not only carried out interviews across the organization but used direct quotations in its report. Rather than say dogmatically, for instance, that "Landberg must reduce the number of dairies in response to market trends," we reported that "just about every manager in Landberg feels that the company must reduce the number of dairies. The following are typical comments:"

> "1999? Fewer dairies, routes, products, trucks, people. There is nothing in the industry that offers growth."

> "My own best guess is that we will have seven dairies in 2000. Twenty-one is too many. There's duplication everywhere and trunking is out of control. About half of all product is now trunked and we don't really know the full costs."

> "The standard bottle of milk will still be our mainstay, but it's declining. It's now 55 percent of the UK market and won't fall below 50 percent, but that 5 percent drop will create even more strains on us. We cannot afford to run 21 dairies. We have to find a way of regaining economies of scale."

In addition to interviews in Landberg Dairies, the research team carried out a broad range of library searches. Their main focus was on likely industry trends over the next three to seven years and what they implied for Landberg's competitive options and financial position. The team aimed at identifying the following:

1. *"Inevitabilities"* These are trends that are so clearly discernible that in every scenario they are expected to continue, even if no one can reliably predict their exact pace or timing.

2. *Likelihoods and possibilities* These are opinions among Landberg managers, the trade press, and outside experts that vary in predictions of what may happen.

3. *Major industry telecommunications-related forces* in distribution, EDI, customer-supplier relationships, and other telecommunications-relevant matters.

Landberg's Marketplace

No one inside or outside Landberg Dairies can guarantee growth for the dairy industry. Neither brand management, promotion, product innovation, nor production technology offers any real chance of increasing the market, whose overall maximum size can be reliably predicted from past trends and is driven mainly by demographics. The demographic trends point fairly bleakly toward an even tougher market for the 1990s than in the 1980s. Estimated industry sales volumes in 1991 were down almost 4 percent from their peak in 1987.

The exact mix of products in the 1990s cannot be forecast with certainty, but the growth trends are toward more varied and upmarket goods that are not necessarily suited to Landberg's optimal use of its production facilities. Additional erosion of the percentage of the market supplied by the major dairies may occur; dried and frozen products and independents taking away sales from Landberg is more likely than Landberg being able to expand its revenues and market share.

The dairy business is about as tough a market as any large and established firm in any major industry must face. Apart from erosion of overall volumes, Landberg faces (1) massive retailer power, which leaves it with less and less leverage in controlling its sales and marketing; (2) an almost insoluble problem of balancing scale versus flexibility in production; (3) low operating margins with escalating costs, especially in packaging, freight, distribution, and service in small accounts; (4) niche independents sitting on the sidelines, picking off the plums; and (5) a powerful and well-financed competitor, Heller Select, that will counter any Landberg initiative in products, prices, or promotion.

Neither any of the Landberg managers interviewed nor any outside expert expects Landberg to expand. In this context, Landberg's business priorities obviously must be improving cost structures and finding some way of countering or coopting the retailers' almost brutal control of prices and conditions of sale. Information technology, however imaginatively exploited, cannot solve the problems of Landberg's limited and decreasing market. In other industries, the main focus of IT investment is revenue-centered, and it seeks to find new sources of product and market innovation through computers and communications. Banks are obvious examples of this, with credit cards, cash machines, and electronic cash management being used to create new revenues.

Trying to use telecommunications in this way seems totally inappropriate for Landberg Dairies. Profits cannot come from trying to expand the market and market share. Instead, the goal must be to increase operating profits from existing revenues and to prevent further erosion of revenues. The business opportunities to address mainly relate to running the business better and gaining an edge in existing markets.

One priority for accomplishing this is to move rapidly toward a two-tier structure for handling customer service and relationships. The situation is no longer one of a single market for dairy goods but of one market for the multiples (tier 1), which will be driven by central distribution and very tough demands for service, and of another market for smaller customers (tier 2), which is highly expensive to service and that may need an entire rethinking of the delivery system.

When 70 percent of Landberg Dairies' revenue comes from five customers but 80 percent of its delivery calls are very small (with the milk crate worth more than the bottles that it contains and the vehicles needed to move the crates worth even more than that), then there are two entirely different "downstream" businesses flowing from the same "upstream" production process. In the upstream process, producing to capacity is everything, and Landberg cannot afford to lose volumes downstream, where it makes its profit on the final 5 percent or so of production. At the same time, it cannot afford to be squeezed on price by tier 1 customers and to be squeezed by the cost of logistics and administration in tier 2. There is a growing need to separate the two downstream processes. The role of telecommunications here is to help improve service to tier 1 and to reduce excessive administrative, accounting, and IT costs in tier 2 operations.

One of the strong recommendations of the study was that Landberg Dairies create a two-tier sales and service structure to match the two-tier customer structure. This opens up the business opportunity to provide workstation access in the customer's office and to differentiate a very standard product through ease of access, communication, and information. Because Landberg's managers came up through the business mainly via operations, they think in terms of product, pricing, and promotion as the main elements in marketing. They had not thought about differentiating the product through telecommunications-based service.

So, we see that Landberg Dairies has two almost entirely separate customer bases: the multiples and individual stores. This is really two markets, not one. Next we examine tier 1: the multiples.

Tier 1: The Multiples   The top tier, the multiples, is marked by the following:

1. *A fundamental rethinking of purchasing and supplier relationships*, with electronic data interchange being the most visible strategic driver. Three of Great Britain's five leading multiples have made EDI a major component of their business strategies, with one, the legendary Marks and Spenser, integrating many of its relationships with suppliers to the extent that it demands the right to input data directly into their computer systems, bypassing the company's own staff.

2. *Movement toward central distribution* for more and more products from more and more suppliers. This is contrary to Landberg's historical distribution system, which was built on local dairies serving local stores by making deliveries in its own trucks. The multiples are finding it more efficient to set up a small number of large central depots to which goods are shipped; they then use their own trucks to handle shipment to the stores. Most large U.S. supermarket chains have moved in this direction, for it offers purchasing power for volumes and efficient coordination of inventory. In the UK to date, the multiples have used central distribution for nonperishable goods only. Dairy products, which mostly have a shelf life up to only four days, are less suited to it. A question the study team had to answer convincingly for Landberg's managers was, "Is central distribution an inevitability, a strong likelihood, or just a possibility?"

3. *Continuing developments and a search for new opportunities in basic areas of retailing strategy* relevant to dairy products, including pricing, location, and mix of brands and labels. Many of the multiples have installed their own in-store bakeries using frozen dough, for instance, and are looking to complement these with their own cakes, fresh yogurt, and any product from which they can obtain a high margin by offering in-store freshness. Obviously, this trend further erodes Landberg's position.

4. *The likelihood of a competitive crunch*, as today's healthy retailing margins, over three times higher in the UK than for most European equivalents, attract new entrants and as labor shortages and costs depress retailer profits. The multiples have put immense pressure on Landberg's margins by playing it off against Heller; with both companies stuck with production overcapacity and the need to produce as close to capacity as possible, the multiples have been able to minimize price increases over the past five years.

Each of these factors pushes the leaders among the multiples in the same direction: toward closing as many open loops in the supply chain as they can and significantly shortening planning, ordering, and delivery cycles. Landberg needs to find a way to benefit, not lose, from this trend.

Landberg's customers in tier 1 provide 60–70 percent of its revenues. Relationships are more antagonistic than cooperative in several instances, with one Landberg manager commenting that "Heller Select isn't the enemy; ABC [a major supermarket multiple] is." Imagine calling one of your main customers the enemy! The leading retailers dominate the dairy industry, with electronic data interchange being just the most recent addition to their armory for dealing with their dairy suppliers. They are able to put pressure on Landberg along the entire business chain, from pricing to packaging to distribution, which has created the sense of conflict revealed in the comment concerning Landberg's "enemy."

Interviews with Landberg managers quickly highlighted the following problems in dealing with the top tier:

> "Customer expectations are far higher in terms of price, concessions, and service than is justified by the realities of the industry and our own organizational abilities. It's over the top."

> "We have no leverage anywhere."

> "EDI should bring us relief. Instead, it's a hassle and provides absolutely no benefits to us. The customer takes them all."

> "Retailer power is massive."

> "XYZ initially wanted *us* to pay *them* half a million pounds for our so-called 'savings' on EDI!"

We heard many examples of how tier 1 customers use this massive power to squeeze Landberg and play the major and independent suppliers off against each other. The overall picture is one of relationships that are at best edgy and are often overtly antagonistic.

Several managers and outside experts provided a different slant on these relationships, suggesting that they can be made much more cooperative by removing many of the sources of error and lack of communication along the supply chain. Comments included the following:

> "The customer has better information than we do. We don't know how we are doing until they tell us."

> "The customer tells us about our errors. We don't know what we made, what they ordered, and what we delivered."

> "All ABC wants is that we deliver what they ask for to the right place at the right time and to the right quality. They don't trust us to do that."

> "We can't track the product through the process, so we can't trap problems. The customer finds them first."

> "If you have a problem and tell XYZ *before* they find out, they are very reasonable and understanding."

> "Customers *must* have accurate information on products and status."

> "Shared forecasts [between customer and supplier] is the name of the game."

> "The replenisher's biggest beef is always the unreliable service of the suppliers." (The supermarket replenisher is responsible for handling day-to-day store inventory levels; the multiple's buyer handles long-term contracts and pricing.)

"My [a replenishment manager for a multiple] biggest problem is talking to the suppliers as I reduce my own lead time."

"HIJ wants to work with its suppliers, give them forecasts, and streamline everyone's operations. They are very tough negotiators, but they are skilled and anxious to help their suppliers use their skills. Procter and Gamble started out seeing EDI as a dangerous area. Now, their attitude with suppliers is, 'Hey, how quickly can we get moving together?' "

"Joint optimization is the basic logic of EDI. It should not be seen as zero-sum—that the retailer wins and the supplier loses or the other way round."

"The store never talks to the dairy."

In no way can information technology or anything else suddenly turn Landberg's relationship with a difficult customer into constant sweetness and light; the nature of the business, history, the buyers' job pressures, and the like all get in the way of that. However, the study team saw many ways in which improved information and communication can at least remove many of the current sources of tension and complaints. At some stage in this highly commoditized market, in which there are few sources of differentiation in products and prices, service must become the differentiator. Landberg today lacks direct operational communications with its tier 1 customers. Mobile computing, fax, direct hotline phone links from the customer, and perhaps electronic mail offer opportunities here, many of them at low cost and with relative simplicity of operations.

The most obvious and (for Landberg Dairies) far-reaching trend among tier 1 customers is toward central distribution. The issue with respect to this trend is not if it will occur but how fast. The economics of distribution across industries in the UK and Europe favors a hub-and-spoke strategy: moving goods into central hubs and then moving them out along the spokes. Technology, especially electronic data interchange, will provide the just-in-time information base for just-in-time management. The current uses of EDI are embryonic; evidence from many other industries strongly indicates that as the systems become fully embedded in operations, using the information they provide will become a priority.

From the viewpoint of suppliers and customers, central distribution has many pros and cons, but overall it is an inevitable competitive necessity in responding to increasingly shortened ordering and selling cycles, to pressures to reduce inventories, to consumer demands for increasing variety, to rationalization of manufacturing, and to labor and administrative costs.

None of the managers and experts interviewed doubted that central distribution will become the norm for tier 1 retailers. The consequences for Landberg are immense, with the exploding volumes and costs of trunking, which have risen from 5 percent of production to 45 percent in four years, creating major cost strains and inefficiencies. Consider the following comments:

"Central distribution offers the ideal. . . . Happy Shopper now does its own breakdowns. . . . "

"Fillers already has 40 specialized dairy goods lines on central distribution."

"By 1999, everything will be via central distribution."

"It's clearly going to come, though they [multiples] may find it harder to make it work than they assume."

"It's 12–18 months away. They all want it."

"With central distribution, you don't need me [a dairy sales manager]."

"It gives the chains control over the stores. They can then pick on small independent suppliers. It renders obsolete our own strength to service every store in the country."

"The independents will push it ahead. They can pick off the plums and offer the multiples central distribution."

Central distribution, increased demands for product variety, and rationalization of plant all drive the trend toward more trunking. Trunking costs are growing rapidly; today there is no way to optimize trunking across the system of dairies because it is basically handled bottom-up, dairy by dairy, and as needed.

The study team saw trunking as Landberg Dairies' potential Achilles' heel. With central distribution an inevitability, the team considered it vital that all the dairies be linked electronically to a central ordering and scheduling point, so that decisions on production scheduling and outbound logistics can be made from the viewpoint of the system, not of the individual dairies. This requirement can be met only through a comprehensive telecommunications capability that provides a new degree of reach: from the dairies to the head office and from customers to the head office.

Tier 2: The Smaller Stores

The second tier of Landberg's customers provides high unit prices, with little need to offer discounts. These customers are small stores, some of which buy less than the equivalent of $50 of goods a day. There is more loyalty, albeit a loyalty often based on the truck driver's salesmanship. The truck drivers, who are paid an hourly wage, collect cash from many of the smallest customers, handle returns, and in many instances advise them on what to stock.

Service and administrative costs (delivery and administration) for tier 2 customers are prohibitively high. Over 1,000 of Landberg Dairies's staff of 10,500 are in finance and administration, mostly in the dairies. Data entry costs an estimated £1–2 million a year. It takes two weeks to invoice a new customer. Tickets (records of purchases) are up from around 90,000 a week four years ago to 112,000 now, with 25,000–30,000 weekly invoices sent out that take around 24 hours of batch processing to produce.

Delivery costs are similarly high, with all the attendant issues of labor management and the risks of "fiddles" and of drivers having to handle up to £20,000 in cash a week. (Fiddles are frauds and/or manipulations of the system to earn a little money on the side; one industry expert described the truck drivers' operations as "legalized banditry.") Sales are in single units, although the demands of efficient logistics are making it increasingly clearer that they should be by the crate.

It is not an option for Landberg to reduce its efforts to attract, retain, and serve the second tier because of the nature of the upstream production process, with its constant need to sell to capacity. At any given time Landberg has a fixed capacity of output, of which 70 percent goes to tier 1 customers. If it reduced production to 70 percent of capacity, the fixed costs would reduce operating profits even further. Landberg must sell its production "downstream," even if only at break-even prices.

The ideal, but seemingly impractical, option for serving tier 2 customers is to transform the role of the truck drivers from employees who deliver goods to independent agents who handle their own small business. This transformation would displace many aspects of administration, accounting, logistics, returns, invoicing, and so on. We raised this possibility in several of our interviews; the general consensus of managers is that it is not practical, even if it were desirable. Some of their comments included the following:

"It's inevitable that we will have to get rid of small calls."

"Franchising is looked at each time a dairy closes but business drifts away."

"We do use merchandisers. It's a cushy job for them. I don't think we would know how to set up and run a real franchising operation."

To the research team, franchising made immense sense, with telecommunications providing a major new opportunity to run the business better by linking field staff to the head office. Mobile computing plus mobile telecommunications (which is a growing and basic tool in many package delivery businesses, most obviously Federal Express and UPS) offers many simple and practical ways of making the strategy of using an independent agent logistically and administratively practical. Mobile computing is the thin edge of a large wedge for change, for it makes it practical to provide the truck drivers with all the tools they need to handle billing, with real-time directions for deliveries and the ability to call in orders as received. The time window for Landberg's operations is around six hours, so the earlier dairies can get orders in, the more time they have to schedule production runs, set up bottling and label assembly lines, and schedule next-day deliveries.

The milk market in the UK over the past decade has been fairly predictable in terms of consumer tastes and trends. The UK market is conservative, with no signs of the fads that drive just about every U.S. market. That said, the study team's analyses of the retailing and dairy industries in the two markets suggest that this may well change.

Landberg can afford to favor scale over flexibility only if it can be sure that major shifts from today's patterns in demand will not occur. The study team's view was that volatility in consumer habits is likely to occur and that the marketplace of 1999 will be highly varied and fragmented. These conditions favor the well-positioned independent at the expense of Landberg.

These assumptions reinforce the importance to Landberg of (1) scheduling and optimizing its entire dairy system through electronic links to a central coordination point that was part of the recommended strategy for IT, and (2) positioning itself to handle a two-tier market. Many of the factors just listed are most likely to affect tier 2, the smaller delivery to a store, whether a dairy, a new specialty shop, or a gas station forecourt.

Most of the Landberg managers interviewed expected diversity to be the norm:

> "We will see diversification of products, so there will be diversification of dairies, with own label, branded, and mixed, and that means further dairy rationalization."

> "ABC has 360 dairy outlets with 360 production managers and 360 quality problems. If it's hard for us, it's harder for them. To make dairy outlets work, the smart retailers will simplify operations and drastically limit the range to a basic line, with no service counter. They will buy in dried and frozen products; this is only just starting. For us, it means more production complexity and logistical variations."

> "The UK market does not move that quickly but still is becoming more and more diverse."

> "Specialty goods and added-value milk are up since 1985 from 14 percent to 22 percent of what is otherwise a very stable and slow-changing market. Skimmed milk is now 10 percent."

> "We are losing market share in specialty goods, which is where the margins are— and where consumers' tastes are."

> "The Molait brand [Landberg's largest selling product] is just a cultural legacy. Our brands are overpriced and more and more costly to protect."

> "It's the independents who can exploit changes in demand patterns, not us."

Predicting the Future Business Environment

Although no one can reliably forecast the business environment of 1995, let alone 1999, the study team predicted several inevitabilities, most obviously:

1. A growing divergence between tier 1 multiples and tier 2 individual stores and smaller chains

2. In tier 1, central distribution, increasing reliance on electronic data interchange, barcoding, and scanning

The team also predicted many likelihoods, less than inevitable but more than merely possible:

1. Diversity and variety everywhere, with new outlets, specialty stores, in-store dairies using a range of new materials and technologies, and shifting consumer purchasing patterns

2. A labor crunch that will dominate the strategies of more and more firms in both basics of operations and the trade-off between technology and traditional ways of working

3. Shortened cycles in every area of marketing, planning, scheduling, and distribution

4.   A margin crunch for every player in the industry, with the retailers squeezed by labor costs, the consequences of overexpansion, capital shortages, and new competition

The marketplace of the rest of the 1990s will surely be even harsher from Landberg's viewpoint than today's marketplace. Revenues will be difficult to find and profits will come under constant strain.

| | | | | | |

Landberg's Cost and Profit Structures

The view of every manager interviewed was that Landberg Dairies must find ways of improving its cost structures. Consider the following comments:

"We must be the most efficient producer. We can't afford to lose 2 percent market share and so have to buy it back again."

"We are at a crossroads. We must examine every cost. Trunking is becoming the significant one, but we have so many areas where costs are either not known or not acceptable."

"The independents' overhead costs structures are 2–7 percent lower than ours."

"We are trying hard to come to grips with cost and profitability planning."

"Costs have to be taken out to make money."

Production and Distribution

Orders take too long to arrive at Landberg Dairies; even EDI orders from a multiple can take six hours to reach the dairy. Trunking has a 40-hour lead time. Last-minute adjustments are commonplace:

"Until you get the final 24-hour figures, you have to work on last week's actuals plus any seasonal adjustments."

"If we could bring the time back from 6 P.M. to 1 P.M., we could certainly save money."

Trunking is a growing cost and problem, and there is no coordinated information resource that managers can tap into on-line, as needed:

"If we make a national promotional launch from 21 sites, we have to phone the dairies to find out what they are making tomorrow."

"The receiving dairy doesn't know what's coming in on the next truck. . . . Some trucks go back empty."

Variety and flexibility in sales and distribution heavily penalize production efficiency, especially in packaging:

"There's high variance everywhere—mix, skills, disruptions. There is nothing standard anywhere, and that makes it impossible to keep on top of. There is waste and errors."

"Selling in physical blocks—pallets, layers or vehicle loads—instead of individual items is a solution whose time needs to come as soon as possible."

There are far too many adjustments and mistakes in production and distribution:

"We do not know what we make."

"The salesman goes out not knowing what he's got."

"Unaccounted losses in manufacturing, despatch, on road amounts to perhaps 2–3 percent of sales. . . . We don't know our real losses."

"We do not know what we make. . . . There's an inventory muddle. For intercompany goods in and out, there's no information, only documents."

"We don't know what 'returns' really mean."

"There's a 10–20 percent despatch adjustment at most sites."

More consistent, up-to-date, and easily accessible information will not in itself transform this situation, but without it there can never be a solution. The long-term answer is to move information with the goods themselves through the entire process of production, delivery, and accounting. Barcoding and scanning make this fully practical, but the initial costs are very high, and such systems rely on a literate, skilled, stable, and loyal employee base.

Management Information Needs and Resources

A constant theme that runs through the comments of Landberg's managers is their lack of information to run the business:

"We don't have the information for central coordination."

"The receiving dairy doesn't know what's coming in on the next truck."

"The salesman goes out not knowing what he's got."

"One hundred four people sitting at HQ don't have information. Decisions are made on-site."

"We lack useful information for taking action throughout the chain."

"At any point in time, we don't know which or how many people we have actually at work."

These "we don't know" problems can be fixed through a *suitable* IT strategy, which requires a complete shift from the "applications" approach that has historically dominated Landberg Dairies' use of computers to a data "repository" and "network" approach.

In the applications approach, data is specific to individual systems and is maintained in files designed to facilitate efficient transaction processing; reports are derived from these systems. Reports almost invariably lead to new forms—the complaint "more bloody forms!" runs through the interviews in discussion of special studies, information systems, planning, and reporting.

The concept of a repository is that data should be managed like a library: It should be created and updated in one place, eliminating duplications. Application programs—for sales, accounting, and payroll, for instance—access data through database management software that contains indexes to the items in the repository.

IT in Landberg
Dairies

Landberg Dairies' information technology investments have been focused mainly on accounting applications. Its managers have mixed views on the current value and future importance of IT:

> "We have spent huge amounts of money on new sales and product reporting systems. I don't believe we know whether they are useful or useless."

> "I don't even know what information's available, let alone useful."

> "What's the point of spending money on technology when you still have to copy figures off reports? We spend money on systems that generate mountains of paper. It does not appear to me to provide any value, and we cannot afford to waste money."

> "We are 25 years behind where we ought to be with IT."

> "[Company X] that I came from was ahead ten years ago of where Landberg is now."

Landberg's existing IT systems are inadequate to meet the business trends and competitive needs identified in the preceding sections of this chapter. That is not a criticism of Landberg's IT group; the systems are very typical of those built in the late 1970s in most large firms. Time and technology have moved on, with a rapid shift from automating clerical procedures through computing to streamlining business processes through telecommunications, data repositories and access tools, and management alerting systems.

Landberg has not invested heavily in IT. It lacks development staff, and there seem to be strong cultural barriers that are not unusual in any company. Relevant comments include the following:

> "Some dairy general managers hid the machines—they were embarrassed."

> "They see information systems as taking away ways of hiding causes."

> "The dairies see information as 'head office.'"

The IBM System/36 computers that are the base of the dairies' computing capabilities are overloaded and operate mainly in a batch mode. Landberg has no core on-line processing and information capability. None of its systems are cross-linked directly to others to share data resources. At one dairy, three people work close to full-time transferring data from personal computers to the S/36. Systems and information are frequently duplicated, with a mainframe version at the head office and an S/36 version in the dairy; the personal computer and S/36 systems are not synchronized and mutually up to date; and there is too much rekeying and transcribing of data.

| | | | | | |

Recommendations for Landberg's IT Strategy

The business priorities that must drive Landberg's telecommunications and computing strategy cover a wide range of areas and issues but point to the same needs in the technology platform. The recommendations are summarized next, with a review of (1) why this is a business opportunity for Landberg, (2) the technical priorities it implies, and (3) its implications for the firm's telecommunications platform.

Recommendation 1: Position for Central Distribution

Why This Is a Business Opportunity The trend is already strongly apparent, and Landberg's management and industry experts agree that it is inevitable. Trunking, which is a correlate of central distribution and rationalization of production facilities, is an Achilles' heel for Landberg Dairies today, and the costs, coordination, complexity, and lead time central distribution imply will increase it.

The Technical Priorities It is vital to link the dairies and the head office to a "location-independent" central point—it can be virtually anywhere—for receipt of orders electronically via any suitable and convenient medium. This may be simple fax, phone, electronic data interchange, in-truck mobile communications, personal computers, or S/36 computers. The aim is to get orders in early and directly. When customers make an 800 number call, they neither know nor care where the physical center is; thus, Landberg's center could be in London or in Wales.

This central location will handle production scheduling and trunking across the entire dairy system instead of site by site with limited central coordination. Landberg needs to be managed logistically as a single unit, which will require new analytic models and planning systems. The telecommunications network is the key element both in making it as easy to send an order to the center as to send it to a dairy (but much earlier and faster) and in getting schedules out to the dairies and responding to late changes.

The technology needed to make this electronic management network practical is proven and widely available. It is expensive for Landberg to build and manage its own network capacity, and it makes no sense to do so. A variety of outside providers with the skills and facilities to do this are available by **outsourcing**—by using an outside company to handle functions that were originally handled in-house. In this instance, outsourcing may involve systems development, network operations, and even running the entire IT function. The basis for choosing an outsourcing firm must be its skills in integrating telecommunications, data management, and application development.

Implications for the Telecommunications Platform The telecommunications network, the infrastructure for the use of computers and information management, provides the reach of the platform. Database management systems at the center create and manage the repository of shared data that is the core for coordinating operations. The network provides inputs to scheduling and trunking models, which will take time and effort to develop, but even simple models and planning systems that look at trunking needs from the viewpoint of the Landberg Dairies national system (instead of from the perspective of each dairy) will bring potentially runaway costs under greater scrutiny and control. (Work was already underway in several studies of trunking options.) Dairy S/36s and personal computers will link to these central repositories and planning systems, instead of being self-contained, stand-alone facilities.

These components of the platform—telecommunications links, the data repository and database management systems, and workstations—are cited again and again in the recommendations for the IT strategy. They create a shared information and communications base, as well as tools to access shared data, that are independent of specific transaction processing, reporting, accounting, and information systems. The telecommunications links are from the dairies to the head office and from customers using EDI; later, they are likely to include links from on-truck computing and perhaps from customer fax machines or voice response phone systems.

Recommendation 2: Position for a Two-tier Customer Base

Why This Is a Business Opportunity Demands for service, the nature of logistics, and customer relations are all diverging rapidly, with the top tier of multiples substantially raising the level of service they insist on and with the costs of distribution and administration becoming excessive for many tier 2 customers. It is difficult to see how Landberg Dairies can afford to manage these two tiers through the 1990s in the same way as today. The tiers are so different in cost dynamics and service requirements that they need distinctively different strategies that allow a flexible response; new options for delivery, management of milk crates, and pricing by crate; and possibly major changes in both the role of the truck drivers and their formal relationship with Landberg. The tiers involve very different information needs for operations, management control, and cost accounting and accountability.

The Technical Priorities The main technology priority is to create the base for these flexible options. To serve tier 1 customers both to their satisfaction and to Landberg Dairies's reasonable profit, the telecommunications infrastructure needs to provide both direct electronic links to key customers for EDI and problem alerting and status information. It must also dramatically improve the quality and ease of access to customer-relevant information.

The long-term ideal is the strategy that has made Federal Express so successful: Information moves with the goods. Federal Express promises to be able to tell customers exactly where their package is within half an hour of the query by using scanning, barcoding of documents and packages, mobile computing with mobile communications links to databases and processing systems, and careful attention to helping the customer communicate with the supplier. Federal Express even gives a free personal computer, costing around $6,000, to customers who send over 25 packages a day via Fedex.

The Federal Express strategy cost over a billion dollars to implement and operate, and it is obviously not possible for Landberg to fully implement it. But the principles apply for Landberg, and parts of it can be implemented in low-cost, high-payoff phases. A major technology imperative in serving tier 1 customers is to make positive use of their moves toward EDI. With a suitable telecommunications infrastructure, the technical issues of EDI are simple to implement, and the information and service by-products are very valuable.

For service to tier 2 customers, the technology imperative is to use IT to help cut costs. The simplest costs to cut are in administration. Cutting distribution costs is a far more complex problem that depends on factors that IT cannot directly influence, especially "franchising" the truck drivers.

Implications for the Telecommunications Platform The concept of a technology platform emphasizes flexibility. The aim is to make sure that standard personal computers, the S/36s, selected customer locations and EDI systems, mobile computers, fax machines, and portable computers can be linked to the firm's communications resource simply and easily and as needed.

"As needed" does not mean that they *must* be so linked, but that they *can* be, on a phased basis or to meet new business opportunities and necessities. For instance, the mobile computers under consideration today are stand-alone devices with no communications capability. There is a strong likelihood that at some future time some of these will be directly linked to Landberg and even to customers. Such linkage can be achieved through the mobile data communications technology that is emerging rapidly as one of the next large waves of IT innovation.

Recommendation 3: Get Administration and IT Costs Out of the Dairies

Why This Is a Business Opportunity This recommendation is linked to the preceding one, and is especially relevant to tier 2 customers. Administration and accounting inevitably depend on IT in any large firm. Thus the nature of the IT base, its costs and quality of operations, and the number of staff it displaces significantly affect the firm's overall costs of doing business. Landberg can get much better value out of IT than it is getting now and should be able to get a better deal in terms of price, quality, service, and applications development.

With margins and revenues under continuing strain and the costs of serving small customers very high in relation to the revenue they provide, this is an obvious target of priority. Landberg Dairies must find any and every way of cutting unnecessary costs without creating organizational strains and expensive capital investment.

The Technical Priorities Let the dairies concentrate on what they are good at—making dairy products; IT is not among the skills of the dairy managers. The current proliferation of independent IT systems and personal computers with different information bases, the need for rekeying and transcribing, and the many multiple accounting systems and highly inefficient computer operations are far more expensive than they need to be.

Approximately 1,000 employees are tied up in administrative and IT-related tasks:

300 operate dumb terminals

100 handle ledgers

100 deal with weekly management acting and costing

150 handle clerical tasks

100 handle secretarial and security functions

100 deal with payroll

150 (estimated) are scattered around the larger sites

The cost of this staff is around £10 million a year. The study team could not yet estimate by how much staff costs can be reduced, but other companies' experience suggests that at least 20 percent can be cut, with additional improvements in quality possible through streamlining and central consolidation of IT and IT-based administration.

The key to accomplishing this is to create a central IT service, run by an outside firm, with telecommunications links to the dairies. This service is built around a data repository that uses one of the powerful and proven new generation of relational database management systems. Data is captured whenever possible at point of event; EDI data goes directly into the center, as do orders. Personal computers access data through the network, and new application systems extract and update data in the repository.

The central administrative group would work with interactive systems that update the data repository directly, instead of key-punching information that is processed through batch systems overnight. Many transactions will come in to the operators by telephone, and more and more will be immediately completed on-line, with few (if any) additional steps, paperwork, and people involved.

Direct data entry simplifies business processes by removing "handoffs." Processing systems can be designed to check for errors and consistency through preformatted screens. The center and its staff can be located anywhere—including Ireland, where more and more U.S. firms are electronically sending "paperwork" to be processed by relatively low-cost, high-skilled labor. (The Irish government provides many tax incentives for this practice.) Landberg can locate its center on the basis of labor costs and quality, tax incentives, and grants.

Over time, documents would be eliminated through CRT data entry, electronic data interchange, and scanners. This cannot be done in a single step, obviously, nor can the data repository be built easily and quickly. Some existing systems on the dairies' S/36s and personal computers and on the head office computer will need to be kept in place for some time. The sooner the S/36s are migrated to a single data center, the faster the current expensive and cumbersome dairy IT operations and the staff they tie up can be displaced.

Developing the data repository is as much an organizational and political task as it is a technical task. It is fully feasible to create the repository, providing there is strong management drive from the top and provided Landberg outsources development to an absolutely top-rate IT service provider. Once the core business systems, including ordering, budgeting, and accounting, are organized as part of the data repository, key ratios can be kept on-line—fully up to date literally to the second—and simply accessed from personal computers in the head office and the dairies. It will be practical to allow selected customers to tap into subsets of the data repository, such as delivery status. Although many network components may need to be upgraded or replaced, this can be handled in phases, and the existing systems can still coexist with new ones developed within the technical architecture for the platform.

Implications for the Telecommunications Platform The combination of the telecommunications network and data repository entirely repositions IT in Landberg. It moves Landberg firmly into the mainstream of management and customer service computing, rather than remaining stuck in clerical and administrative computing. It provides the dairies with decentralized access to IT services and information, and it gives the head office coordinated information access.

The software needed to manage the data repository is complex, requiring specialized technical expertise, particularly in ensuring that it runs efficiently. Landberg does not have these skills in-house; they will need to be provided by an outside company.

Recommendation 4: Schedule Dairy Production and Trunking Across the Entire System

Why This Is a Business Opportunity This recommendation follows from just about every finding of the business review; it is the only way to deal with both the immediate problems and costs of trunking and the future problems that are likely when market forces lead to a combination of reduced scale and increased demands for flexibility.

Today, Landberg's production and distribution are highly decentralized, reflecting the historical need to locate the dairy as close to its customers as possible. Orders came into a single dairy, to be scheduled by that dairy and delivered from it; specialized dairy goods, which were a small fraction of total volume, could be trunked on an as-needed basis. That mode of operations is passing fast. Indeed, it is already obsolete, as is evident from the high rate of increase in trunking and in its approaching the volume of nontrunked goods. At the same time, the old narrow time windows for production remain as tight as before, but key customers are demanding just-in-time service and are shortening their replenishment cycles. Landberg is pushing an old production and delivery system to its limits. The business opportunity enabled by telecommunications is to optimize the entire system rather than scheduling its individual components and using trunking to cover the gaps.

Landberg must optimize its entire system, not just run the parts. If orders can come into a central point at several times a minute, analytic models and skilled people can carry out the balancing act between production and trunking. Today, for instance, dairies A, B, C, and D make trunking decisions largely independently of each other, with 48-hour advance notice needed. A limited experiment in coordinating trunking centrally relies on a few individuals' rules of thumb; the sheer size and complexity of the Landberg Dairies system makes this at best a short-term approach.

Optimizing production and distribution rests both on having an up-to-date database and on getting orders into the system fast and early; EDI is part of this. The dairies do not feel comfortable with the very limited EDI in use today. Rather than push them in an area in which they lack skills and in some cases interest, there are many advantages (in expertise and clarity of coordination) in having EDI managed and used from the center.

The Technical Priorities A telecommunications network now makes getting orders in faster and earlier a practical opportunity; without the network, it is a fantasy.

Having an up-to-date, complete data repository on today's orders, trunking costs, and production capacities also provides the base to apply scheduling and trunking models that optimize the entire system. There are many practical problems in developing such computer models. Landberg's current model, for instance, cannot handle a context as volatile as the one it faces every day. The mathematical combinations, predictive rules, and necessary data are too many and too complex to fit existing methods. No current off-the-shelf software package meets Landberg's needs.

That said, there are almost certainly simple ways of improving the existing scheduling and trunking process through central coordination. Just having an overall picture of the system helps. Making sure that orders are matched to individual dairies' capacity and that trunking needs are looked at across the national system is likely to improve efficiency and costs.

Landberg's distinctive strength is also potentially its distinctive weakness: It can serve every store in the country. That becomes prohibitively expensive if more and more of its production capacity is far away from those stores and more and more of the value-added goods it produces are made in just one or a few of its dairies. If Landberg has the equivalent of the airlines' reservation and scheduling systems (in a far simpler environment than the airlines face), it will be able to serve a national market while exploiting local production and distribution facilities.

Today, the independents enjoy the best of all options. They can choose high value-added products, exploit the tier 1 customers' moves to central distribution, and offer heavy discounts. Their business strategies pick off plums and look for niche markets. Landberg is a full-service, full-geography provider; its production strengths rest on lack of variety—if the market were only for standard whole milk at a premium price, it would lead the industry.

Landberg's telecommunications platform will be a critical element in positioning it to maintain its full-service, full-geography capabilities and at the same time be able to meet the demands of a marketplace in which production and distribution will no longer involve single locations. There is a distinct possibility that many of the independents will come under pressure in the 1990s from giving too many discounts. The recommended platform would enable Landberg

Dairies to acquire independents and add them directly into its highly flexible electronic scheduling and distribution system. An independent in this context is just another source of trunking.

Implications for the Telecommunications Platform Telecommunications is the central element in getting information into the center and back out to the dairies reliably, quickly, and continuously. Neither the network alone nor data is enough. A new generation of scheduling models will be needed, together with accurate cost data, particularly for trunking. Several such systems, now in use in the airlines and in the freight industry, are built on powerful new personal computers that have both the number-crunching power of a mainframe and far more flexibility of use.

Recommendation 5: Meet the Individual Information Needs of Managers

Why This Is a Business Opportunity Landberg's senior managers are information-poor in a business that has increasingly heavy demands for just-in-time information. A common theme in the interviews was, "We don't know"— costs, causes of errors, production, work force, and so on. Managers cannot simplify or streamline parts of the business without precise, detailed, and accurate data in the form they need. The information they receive is mostly formatted in fixed ways and at fixed levels of detail. There are some tools for generating special reports, but they are not tools managers can use directly, and they take time to produce results. In general there are too many forms, too much paper, and too many variations in data definitions and accounting systems for managers' information needs to be met.

Lack of relevant, timely, and flexible information pushes Landberg's management into a reactive mode that frustrates managers and blocks their ability to address many evident problems. Managers are also frustrated by the extent to which lack of data makes it difficult to pinpoint responsibilities and creates too many opportunities for excuses.

The Technical Priorities Landberg needs to move as fast and as comprehensively as possible into a technical environment in which information is organized, stored, and accessed as a shared corporate resource. Ideally, all operational and financial data should be on-line and cross-linked through relational database management software. Transaction processing systems then feed data into the data repository, and managers use personal computers to access information in a variety of ways and forms, including the following:

1. *Key figures* held on-line: These are preformatted graphs and numbers that individual managers want to look at first thing in the morning. They are not elaborate reports, but they answer the question "If I as a manager could see only one screen of data to reassure me that the business is under control or to immediately alert me to a problem, what would that screen contain?" Landberg's head of operations loved this idea. He pointed to a three-inch stack of reports on his desk and commented that "this only tells me what happened last week or last month. I *must* know about last night.

Give me three figures and I can be on top of everything." (The figures are customers whose deliveries were late or incomplete, production break-downs, and scheduled output versus actual output.)

2. *Flexible inquiry tools*: These tools allow managers or their staff to zero in on data at any level of detail, moving, for example, from a single screen that shows a graph of this week's returns, to a screen with the returns for individual dairies, and then to returns by specific customers in a given dairy yesterday. This is practical because one of the main principles of relational database management systems is to maintain data in molecular form—at the very lowest level of detail. Software used in a reporting system can always aggregate data on, say, individual daily deliveries, into a weekly total. If, however, the software stores only the total, then obviously it cannot immediately disaggregate the data back to the individual deliveries, and the molecular information is lost.

3. *Cross-referenced data*: Relational database management essentially means cross-referenced data. Instead of cost data being organized by customer profitability *or* product line *or* dairy—and in separate (and often inconsistent) files—cost data at the molecular level of detail can be cross-referenced as needed. Simple (from the user's perspective) commands generate powerful software routines that work out the cross linkages. Most of the information problems Landberg's managers reported reflect a lack of ability to cross-reference data as they want and when they want and, more specifically, to relate costs to activities along the entire supply chain.

4. *Management alerting systems*: These systems extract information from the repository of on-line operations data. The more timely that data is, the earlier the systems can be alerted to trends, exceptions, and problems. Once order data comes in electronically, managers can get a snapshot of the entire system at any point in time. If goods can be barcoded and tracked, the level of responsiveness that Federal Express provides becomes a practical option in giving customers status information or in alerting managers to a delivery delay or despatch adjustment. In this context EDI becomes a supplier of data to Landberg rather than a logistical nuisance.

Implications for the Telecommunications Platform The principle of capturing data at point of event, moving it via telecommunications into the repository, and accessing it as needed underlies the uses of IT by the firms that are recognized as receiving a sustained advantage from it. The technical limitations of the 1970s and early 1980s prevented progress except at enormous cost and risk. Computer systems were almost entirely limited to overnight batch processing because there was no source of widely available and reasonably inexpensive telecommunications services. Information was organized inflexibly because database management software was slow, expensive, and inefficient. Reports were paper-based and derived from accounting files because personal computers with the software and telecommunications to access, massage, and display information were not available.

All these tools are widely available and proven now. There will be many practical difficulties in defining and building a data repository, but a platform based on a central data repository, a network with reach to all dairies and some customers, and personal computer-based access tools is fully practical and provides the blueprint for providing the information management needs to manage.

Recommendation 6: Develop a "No-Surprise" Service Capability

Why This Is a Business Opportunity Landberg's tier 1 customers are highly demanding and in the driver's seat. They have plenty of reasons to be dissatisfied with the level of service they receive, which is marked by errors, conflict, and unreliability. Some of them are developing along the supply chain far better information capabilities and logistical systems than Landberg has. They will all increase their use of EDI and scanners over the next few years, and they will also become more, not less, demanding in dealing with their suppliers. The business priority for Landberg Dairies in this context is to find ways of improving service without increasing costs and of shifting the relationship with selected suppliers to a mutually more satisfactory basis.

In many commodity markets, electronic service has become the key competitive differentiator across the entire industry. In each instance, the innovation is at first an opportunity for a few firms, which may or may not make money or establish a competitive edge. Within a five- to seven-year period, the innovation redefines the base level of service for the entire industry and becomes a competitive necessity for all major players. Customers drop suppliers that fall below this base level and shift more and more of their business to the suppliers that provide service.

This is really what electronic data interchange is about. It will surely set the base level of service for the dairy industry well before 1995. It will put pressure along the entire supply chain for quality, responsiveness, and information in service. it will make poor service even more expensive than it is today, in terms of the cost of finding and repairing errors and the consequences of mistakes and delays.

The Technical Priorities Historically, IT has been used mainly to automate a firm's internal processes; now it is increasingly being used to link a company's processes to its customers'. The technology priority for Landberg is to move in that direction as cost-efficiently as it can. Communications and on-line alerting information are the key enablers of this move. Most of the complaints about service by key customers relate to surprises. They understand that there will be problems in any business with very short time windows, but they want to be informed about problems early, and they want to be able to deal with suppliers simply and conveniently. Unfortunately, they want this without having to pay for it, knowing that they control the suppliers, not the other way around.

Implications for the Telecommunications Platform The concept of an IT platform is that it is a shared resource and hence a reusable one. The network Landberg Dairies will build to link its own operations and to handle customer

EDI can also be used to send information to customers. The data repository it uses to meet its own management needs can provide information to help meet customers' management needs as well.

Simple examples of providing "no surprise" service include a direct fax link from the national accounts department to customers or replenishers and standard electronic mail messages to customers' personal computers or back along their EDI communication links. Messages might include status alerts, consolidated invoices, or confirmation of special orders. The U.S. dairy industry is already extending EDI to provide head office-to-head office invoicing.

Responding to the business imperative of no-surprise service is mainly an issue of routing and reusing data. The recommended technical architecture for the platform extends Landberg Dairies' IT base out to its customer base.

The Recommended Telecommunications Platform

Only after all six recommendations are presented and justified as business priorities are the technical priorities identified. If all the technical priorities were different, there might be no need for an integrated platform. Taken together, however, the priorities point to common needs for reach and range, particularly in linking the bakeries to the head office and the head office to tier 1 customers and in creating a shared data repository for scheduling, management alerting, and customer service. The business recommendations themselves in effect define the technical recommendations for the platform, which is described below.

Reach

The proposed platform has the following levels of reach:

1. *From the dairies and the head office to a central electronic service point,* which may be in the head office but is location-independent (there is no need to link the dairies; they will communicate with each other via the center)

2. *To key customers* from the electronic center

3. *To trucks*

4. *To Landberg's own suppliers* as they move toward EDI, as they will and must

The technical architecture for the platform will allow Landberg to implement and extend both reach and range over time. Initially, for instance, there may be neither any need to increase reach to trucks nor any supplier EDI capability. The option will be there, though, at no added cost, because the technical standards for the platform will be chosen to ensure this. The telecommunications standard will include X.25, which is the established base for interorganizational communications through public data networks provided by British Telecom and Mercury, and value-added networks offered by GEIS (General Electric Information Services), Infonet, IBM's Information Network (IN), and others. The architecture will also need to accommodate the main EDI protocols for inter-firm electronic transactions, including the UK Tradacom standard. It will also use the X.400 standard that is becoming the accepted base for electronic mas-saging, electronic mail, fax, telex, and many simple forms of EDI.

Range

Range is more complex than reach, both to define in the platform and to imple-ment. Today, Landberg cannot share information across applications; all appli-cations are independent data files, with a few limited inquiry tools that can access each single file but cannot select and combine information directly across them. Achieving this means committing to a core database management system, of which there are several practical options, such as Oracle and IBM's DB2. Oracle is popular in companies that use a wide range of different vendors' equipment and have no single central repository of data, but databases are spread across many locations. The team's strong recommendation was DB2, which has become the workhorse for companies that are building the type of shared corporate resource that is the core of the proposed IT strategy. XDB supplements the mainframe-based DB2 with a powerful personal computer capability.

Ad hoc access to information that complements the regularly produced management alerting systems will be handled through the SQL standard, which has become the accepted, though slightly cumbersome, language for creating reports from relational databases. Many powerful tools are available for data manipulation and display that can easily meet senior managers' needs and ac-commodate their personal modes of work and their preference for how infor-mation is presented. The critical issue in choosing such tools is that they fit into the overall platform—that they are part of an integrated capability. Selecting them on the basis of their specific features is a common mistake that leads to additional incompatibility among systems.

Responsiveness

Landberg operates 24 hours a day, except Sunday morning. Its time windows are very narrow; if the proposed communications systems were down between 6 P.M. and 10 P.M., the company would face a complete disaster. Many of the bakery managers may argue that the current system works well enough, so why put the company at risk by relying on a telecommunications network? Land-berg's existing IT base is weak and the firm lacks expertise, especially in telecommunications.

These concerns mean that Landberg's platform must ensure guaranteed perfect service, the highest level of responsiveness. The best way for Landberg to provide this is to use an outside service, such as a value-added network and/or PDN. The providers of these services have the expertise, capacity, and expe-rience in network management and disaster recovery to resolve the concerns.

Figure 13-2 summarizes the business functionality of the telecommunica-tions platform and highlights the business priorities that it makes practical and to which it most directly relates. Those priorities are disabled if either reach or range is reduced. For example, if Landberg were to continue with its current

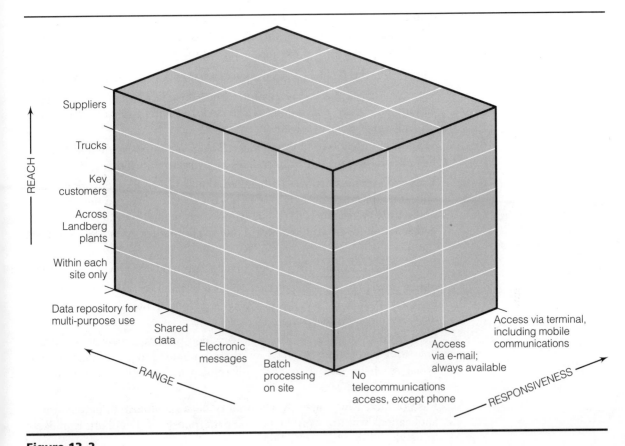

Figure 13–2

The Recommended Telecommunications Platform for Landberg

approach to IT and did not add telecommunications links through X.25, the option of providing no-surprise service to key tier 1 customers essentially disappears. If Landberg stays with an application approach to information and reporting systems development, instead of moving to a data repository strategy, the aim of comprehensive management alerting systems is disabled.

Network Design Variables

Although the research study did not include detailed network design, it was the basis for the RFP (request for proposal) sent to vendors to bid on designing and implementing both the platform and the key applications. Landberg's top management team accepted the recommendations in their entirety and authorized immediate action on them. The following list contains the authors' assessment of the relative priorities and trade-offs among the variables:

▸ *Capability*: the technical applications and services the facility provides.

Variety of applications: The platform must be able to accommodate applications that share a common data repository. Thus the core DBMS is a critical choice and its selection a very high priority.

Degree of integration of the services: This is not a priority in and of itself, but integration of the data resources is a very high priority.

Range of volumes that can be handled: the traffic on the network includes almost no on-line transactions between 6 A.M. and 6 P.M. and then a continuous stream of bursty traffic, mainly messages to and from tier 1 customers and scheduling and trunking instructions. There must be no delays at peak times.

▸ *Flexibility*: how easily the platform can accommodate growth and change.

Speed of change in adding capacity: This is not a priority because there will be plenty of lead time in dealing with the closing of dairies, the addition of locations, and so on.

Ease of increasing reach: This is a very high priority, especially in adding links to tier 1 customers. However, there is no need to provide for international reach because Landberg will remain a domestic firm due to the nature of its industry.

Ease of increasing range: A low priority because almost all potential new applications will draw on the central data repository. Landberg is unlikely to need multimedia applications.

Ease of increasing responsiveness: The required responsiveness is already at the highest extreme; Landberg cannot consider any option that does not ensure it.

▸ *Certification*: how reliable the platform is and what guarantees of service it provides.

Operating reliability: Ultra-high priority.

Ease of access: Not a priority, except for tier 1 customers. Access by tier 2 customers can be by fax and phone into the location-independent order and scheduling center.

Security: A relatively low priority, though Landberg is anxious that competitors not learn about either its response time or its price terms for tier 1 customers. (The industry is cutthroat in respect to the latter, and firms often try to hire key national account managers to ascertain the company's prices so that they can be undercut.)

▸ *Cost*: fixed and variable development and operating costs.

Fixed costs of equipment and software: It is an absolutely dominating priority for Landberg not to need to divert very scarce capital to IT investments. Even a $5 million expenditure is unacceptable to Landberg and to its parent company, Waterfield.

Development costs: Keeping these low is a high priority. IT has not to date provided any apparent payoff.

Operations costs: Cutting the current costs of operating the dairy and head office S/36 systems is a high priority.

Maintenance, support, and education: It is a relatively high priority to keep these costs low. The new systems must be easy to understand and use and must not involve adding specialized staff.

Economic Benefits to Landberg

The specific benefits to Landberg Dairies from implementing the proposed strategy need much more study before they can be reliably quantified. The following analysis summarizes the sources of the benefits without attempting to quantify them:

▶   *"Hard" cost savings*
    Reduction in administrative, accounting, and finance staff in head office and dairies
    Reduction in IT staff in the dairies, especially for key punching, forms, copying, storing and locating documents, correcting errors, and rekeying data into the computer
    Dramatically improved trunking costs
    Reduction in the costs of errors and repairs

▶   *Organizational benefits and "soft" savings*
    Much improved ability of senior managers to track problems, respond quickly, and get the information they need when they need it
    Tighter head office control over operations, with more accountability and fewer excuses
    Additional value from on-truck computing and EDI
    More service options and improvements in quality of service
    Better value for money from existing IT expenditures and wider choice of services
    The opportunity to streamline the head office as the network and data resources create new management tools and reduce delays, administrivia, and so on
    The managing director and his top management team will have far greater control of operations decisions
    A solid base for managing distribution and trunking that can only be provided through IT

| | | | | | | |

The Outlook for Landberg

Landberg was in the process of implementing the recommendations when this book was published. The first step was the key one: acknowledging that Landberg must outsource the network because of capital costs, lack of in-house skills, and the need for a level of responsiveness Landberg cannot afford to provide through a private network. The RFP specified the core standards for the platform, including X.25, X.400, and SQL-compliance for DBMS. Landberg's future remains uncertain, but telecommunications has for the first time opened up entirely new avenues of direction. Indeed, it is no exaggeration to say that telecommunications has created the practical possibility of freeing the company from many constraints and burdens that threaten its survival: cost, service, management, and organization.

Summary

Landberg Dairies faces an increasingly harsh environment. Its market is not growing, and the major multiples—large chains of supermarkets—are squeezing it in terms of prices and are demanding higher and higher levels of service. Smaller "independents" offer a wide range of specialty goods. Landberg, which operates 21 dairies, cannot adapt its production schedules and facilities to produce these goods in every location; about half its production must be shipped between dairies at growing expense; this process is called trunking.

Landberg makes limited use of information technology, with each dairy running an independent mid-sized computer system and using personal computers. The head office has its own capability, and there is no telecommunications network linking it to the dairies. Landberg is experimenting with hand-held computers in its delivery trucks. It also makes use of electronic data interchange, but only because of pressure from the multiples. Several earlier information technology plans have not provided payoffs, and senior managers are very skeptical of its value.

The chapter reports the recommendations of a research study carried out to apply two of the core frameworks of this book (the telecommunications services platform map and quality profit engineer-

ing) to a company whose management has not been receptive to the use of telecommunications. The study used interviews and secondary sources of information to identify a range of business opportunities, most of which related to running the business better. The main recommendations were (1) position Landberg for central distribution of its goods by the multiples; (2) position for a two-tier customer base, with the multiples being handled differently than the smaller second-level tier of customers; (3) get administration and information technology costs out of the dairies, using a location-independent center and telecommunications links from the dairies to the head office; (4) schedule production and trunking across the entire system, instead of on a dairy-by-dairy basis; (5) meet the individual information needs of managers; and (6) develop a "no-surprise" capability for tier 1 customers.

Each of these business opportunities has similar technical priorities, most particularly reach from the head office to the dairies and to tier 1 customers and range to provide needed management alerting systems and to greatly improve customer service. The telecommunications platform offers Landberg a new source of service, coordination, and improvement in its cost and profits structures.

Review Questions

1. Briefly describe the business conditions and structure of the dairy products industry in England after World War II, and discuss Landberg's position in the industry.

2. Briefly assess Landberg's business situation at the time of the study. Focus on its profitability, cost structure, prospects for future growth, and management team.

3. Describe Landberg's use of information technology at the beginning of the study, and discuss the extent to which it was adequate to support the dairy's business needs.

4. Briefly describe the multistep process for carrying out the research study on information technology for Landberg Dairies, and explain why the study does not focus explicitly on the technology and its capabilities.

5. Describe the "tone" of the reports summarizing the interviews, and explain why this approach was important to the future success of the project.

6. Briefly describe Landberg's situation and competitive positioning with respect to the following:
   a. Dairy product retailers
   b. Flexibility in production capability
   c. Operating margins
   d. Small and large competitors

7. Should Landberg use telecommunications and information technology to try to expand its market and market share? Explain.

8. Seventy percent of Landberg Dairies's revenue comes from six customers, and yet 70 percent of its delivery calls are for very small orders in which the cost of the milk crates and transportation are greater than the cost of the product sold. What does this suggest about the focus of a strategy for the use of information technology to enhance business performance?

9. Outline the basic telecommunications strategy for improving relations with the large, tier 1 customers.

10. Discuss the implications for Landberg Dairies of the trend toward central distribution for the multiples.

11. Briefly describe the problems Landberg encounters in serving the smaller, tier 2 customers, and explain why the tier 2 customer base must be maintained for Landberg to be successful.

12. Briefly outline the major trends in consumer requirements and behavior with respect to the purchase of dairy products in the UK, and state the implications of these trends for Landberg, assuming they continue to operate as they have done in the past.

13. Describe the role of EDI in the UK dairy industry, and explain its current and projected future impact on Landberg Dairies in particular.

14. Describe Landberg Dairies's cost problems, and identify the principal cause of these problems.

15. Briefly describe some generic telecommunications and information systems strategies that would help solve Landberg's cost problems.

16. Explain the difference between the "application" approach and the "data repository and network" approach to an information technology strategy that would help Landberg solve its problem of insufficient information to run the business effectively.

17. Describe Landberg's current information technology system, and evaluate the system's adequacy in meeting the needs of the company.

18. Identify and briefly describe the components of the proposed telecommunications platform for Landberg Dairies. Indicate how this platform both helps Landberg adapt to central distribution of its products and improves the efficiency of operations.

19. Describe the differences in information needs for operations, management control, and cost accounting for tier 1 and tier 2 customer segments that result from differences in cost dynamics and service requirements for the two groups of customers.

20. Explain the meaning and importance of the following statement in the context of satisfying

Landberg Dairies's information requirements: The concept of a technology platform emphasizes flexibility.

21. Describe the proposed approach for reducing the costs of administration and related information technology costs for Landberg Dairies.

22. Explain why it is important for Landberg to schedule dairy production and trunking across the entire system, instead of continuing to follow the current approach of decentralized production and distribution.

23. Explain how EDI can be used by Landberg as a source of data for its management alerting systems, and how this can translate into improved service to the customer without increasing cost.

24. Explain how the concept of an information technology platform (that is, a shared, reusable resource) can help meet the business imperative of "no-surprise" service for Landberg Dairies.

Assignments and Questions for Discussion

1. Analyze the competitive forces that affect Landberg's relationship with its large customer base (the multiples). Include the following in your analysis: (a) the relative power of supplier vs. buyer and its implications, (b) trends in buyer operations and purchasing practices that affect the relationship, and (c) entry conditions into the market and the threat of substitute products and services.

2. Develop a plan for franchising the distribution and delivery of Landberg Dairy products to small, tier 2 customers. Build into your plan mechanisms to ensure quality customer service and adequate responsiveness, and explain how the plan uses information technology and telecommunications to achieve the franchising goals.

3. Federal Express has pioneered the use of information technology to ensure that information moves with the goods through a "hub and spoke" centralized distribution mechanism in the overnight package delivery business. Using these same principles, develop a telecommunications and information systems strategy appropriate for Landberg Dairies (taking into account that Landberg does not have the financial resources or organizational sophistication of Federal Express).

4. Lack of relevant, timely, and flexible information pushes Landberg's management into a reactive mode that frustrates them and blocks their ability to address many of their problems. Develop a proposed set of generic types of information that would enable Landberg to improve its problem-solving ability and its management decisions. Justify your choice of the types of information included.

5. Describe the degree of business functionality needed by Landberg for each of the three dimensions of the telecommunications platform (range, reach, and responsiveness) based on the generic types of information identified in question 4. Indicate differences in the business functionality requirements of the platform for the near term and for the longer term.

Four

Managing the Telecommunications Resource

14

The Telecommunications Industry

Objectives

- Learn about the components of the telecommunications industry and the main competitive forces driving it
- Understand the main emerging trends in the industry
- Learn the implications of these trends for telecommunications choices

Chapter Overview

| | | | | | |

Chapter Overview

The telecommunications industry is as about as stable as the former USSR. Every month we hear major new announcements from companies concerning new products and promises of new products. The aim of this chapter is to describe the main forces driving telecommunications competition and innovation and to consider their implications for the telecommunications designer's options and opportunities. The chapter classifies the industry under five main headings:

▶   U.S. local services, by far the largest part of the market and until recently a competitive oligopoly

▶   The long-distance carriers

▶   Differentiated services that either provide some added value to transmission or are confined to wireless transmission only

▶   International services

▶   Nontransmission facilities, including switches, LANs, and telecommunications software

In each of these areas, there is both rapid and often volatile change in the technology and its providers. Choice of vendors and relationships with them are thus an important consideration in design, implementation, and operations. Understanding the industry is as important as understanding the technology; although technical innovations pace the development of telecommunications, identifying supplier trends and capabilities is crucial in exploiting those innovations.

| | | | | | |

Telecommunications: The New Equivalent of the Automotive Industry

The sheer size of the telecommunications industry now makes it the new equivalent of the automotive industry. It is highly capital-intensive, is often the major employer in a country, and contains the largest trade union.

Before January 28, 1982, there was no real telecommunications industry, but rather an almost entirely regulated worldwide monopoly. There were pockets of regulated competition; the United States had (and still has) over 1,400 phone companies, most of which were covered by state regulation, and a few companies were licensed to provide international telex and special-purpose data communications services. There were a few, limited competitors to AT&T's services and equipment. Ten years later, competition is everywhere, and in the United States, the last bastions of monopoly are under attack. The Regional Bell

Operating Companies ("Baby Bells") have had almost total control over the key part of the telecommunications system—the wall plug for the phone. No matter how a long-distance call travels, it ends up there, and the RBOC gets a fee from the long-distance carriers that amounts to about half of the cost of the call. The situation is now changing, and within a very few years that same phone call may end up in a cable television company's equivalent of the RBOC's central office and delivered via fiber optics, or perhaps the local loop may be bypassed by other carriers using a shared cellular resource. At the same time, a video may come to you down your phone line, not through a cable company. In the emerging competitive climate, cable TV companies are competing in the phone market and phone companies are competing in pay-per-view. The shift in competition and the breaking down of boundaries between local and long distance, and among the cable, phone, and cellular industries, is paralleled by shifts in technology.

The United States was the first country to allow full competition in telecommunications services. The announcement of the divestiture of AT&T occurred in January 1982 and took effect in 1984. Great Britain followed in 1986 with "liberalization" (limited competition), and the old quasi-government agency run by the UK Post Office became a private company. Between 1984 and 1990, there were few dramatic innovations in the deployment of technology beyond LANs. Fiber optics and digital switches were installed by the long-distance carriers for major routes, leading to continued but incremental reductions in costs for both consumers and large businesses; small and medium-sized businesses did not receive much benefit because they lacked economies of scale. One of the main arguments PTTs in such countries as France have made in preserving their monopoly is that the fragmentation of telecommunications in the United States has resulted in the lack of a comprehensive national infrastructure that offers small and large users alike the benefits of advanced technology. This argument was one of the main forces driving ISDN.

Competition and technical and product innovation in the LAN market were far more dramatic than in the long-distance market, but until the late 1980s the LAN market was an entirely separate area from wide area communications dominated by computer vendors, proprietary systems, and PC-to-Lan connectivity. The technology was immature, and standards were lacking. LANs were also almost entirely outside the corporate telecommunications unit's area of responsibility and authority in the organization; they were handled either by business departments as part of their development of a PC-based capability or according to information systems functions. In large organizations telecommunications was an overhead function; voice communications constituted the main cost element and management concern. The major industry issue telecommunications managers focused on was the choice of long-distance provider, made mainly on the basis of price and quality. As the quality of the three main companies converged, price increasingly dominated the decision. For the network designer, the crucial choice was between public and private networks, with price again the main criterion and the volume of traffic determining the trade-off.

This first decade of the "new" telecommunications industry is now over, and change is everywhere. In 1992, major events, new alliances and international consortia, and technical breakthroughs occurred monthly. AT&T stirred up the immature but fast-growing cellular communications market by acquiring Mc-Caw in 1993, the leading company; this move was quickly countered when MCI promoted a coalition of providers to create a national wireless communications infrastructure. In 1993, RBOCs restructured themselves as legal entities to position for the new era of competition. Ameritech offered to give up its regulated monopoly and allow free competition, provided it was also allowed to compete in entertainment and information services. New names joined the familiar long-distance providers: LDSS and WilTel created new niches in the 800-number market and in data communications. British Telecoms applied for a license to become a U.S. long-distance carrier, and Sprint applied for a license to operate in Great Britain.

New high-bandwidth technologies leapfrogged ISDN and stimulated new industry competition in 1991 and 1992. ISDN, defined in the late 1970s, was only just becoming available in the early 1990s, with nationwide tests of interoperability held in late 1992. By then, WilTel had become the first firm to offer frame relay, a huge leap ahead of ISDN. Motorola had announced its Iridium project, which will launch into a low orbit (413 nautical miles above the earth) 77 minisatellites that can be reached by low-power signals from cellular phones. Iridium will create a worldwide mobile phone system, if it can attract enough investors.

In 1993 two related and pivotal decisions made by the FCC opened up an intense new competition: the "Fresh Look" and 800-number portability. The Fresh Look specified a 90-day period in early 1993 during which companies that had signed long-term contracts with AT&T were permitted to reconsider and accept counteroffers from other carriers. These contracts were written under the rules of **Tariff 12**, which provided major discounts that competitors were unable to match, with the condition that the customer had to give AT&T all of its long-distance voice and data traffic. Tariff 12 users, among the very largest of U.S. companies, have massive telecommunications budgets, so that the Fresh Look offered AT&T's competitors a huge new opportunity.

Transmission speeds have increased dramatically in the 1990s. In the second half of the 1980s, ultrafast LANs had to link to slow WANs, with T1 the practical limit of transmission links for most firms and 64 kbps the standard channel transmission rate. SONET is already operational, providing speeds that would have seemed fantasies in the 1980s. There are now over five million miles of fiber optics in the United States; their capacity has barely been used to date. SMDS is providing the RBOCs with the first type of service in which they are the technology leader over the long-distance carriers, with MANs (metropolitan area networks) bridging the gap between LANs and WANs and opening up a flood of new applications, especially in health care. LANs are being brought into the RBOC, VAN, and long-distance providers' sphere of operations, as they all offer LAN-to-MAN, LAN-public WAN, and even global LAN connectivity.

ATM-based transmission promises to provide LAN-rate speeds over public and private WANs.

The combination of ferocious competition among suppliers and a wealth of new technology offers the client new applications and the designer new options. It also creates new challenges for designers and implementers, especially because there is often a two- to four-year gap between the preliminary announcements of new products and services and their becoming fully available and reliable. Moving immediately to new products and services might mean betting on the wrong technology, but waiting until matters became clearer might mean a loss of both competitive opportunity and the ability to cut costs.

As the technology dramatically shifted in the early 1990s, regulation, which had been a rather closed issue since divestiture, also became highly volatile. States such as Illinois and New York began to allow competition with the RBOCs, whose reputation for service and innovation has been largely poor. Judge Harold Greene, who oversaw the antitrust case that led to divestiture and who maintained oversight through the 1980s, allowed the RBOCs at last to enter the potentially profitable equipment and information services market. Most important for industry competition, cable television companies were allowed to offer phone services, putting RBOCs into direct competition with cable companies that have fiber-optic capacity to spare but are also highly undercapitalized.

To complete the stirring up of the fairly stable long-distance industry, in 1992 the FCC ordered full portability of 800 numbers. The implications of this are enormous, because a firm with a distinctive number (like British Airways's 1-800-AIRWAYS or Pet's Sakes's 1-800-FOUR PAW) was very unlikely to shift carriers and risk losing recognition, ease of memorization, and its payoff from advertising. AT&T had 90 percent of the 800 number market. Given that about half of AT&T's long-distance traffic during peak business hours is 800 number calls, which have high margins for the company, the new competition strikes at its profitability, not just its volume. Some 80 million 800-number calls occur each day in the United States; all of them became up for grabs.

U.S. firms have moved abroad, with Sprint applying for a license to compete against British Telecoms in the UK. BT had already become a powerful presence in the United States, through both its acquisition of Tymnet, a leading value-added network, and its alliance with IBM to create Syncordia, which offers to manage large firms' entire network.

In mid-1993, British Telecom announced that it planned to acquire 25% of MCI. If approved by regulators, BT will be positioned in the United States, the world's largest and most competitive telecommunications market, and will pose a strong threat to AT&T. In addition, the BT-MCI alliance will become a dominant force in the world's busiest areas of telecommunications traffic flows: New York City and London. MCI will get $4 billion of new capital that it will certainly use to enter emerging markets, such as multimedia, cable television, or cellular communications.

Twenty foreign telecommunications service providers have offices in New York City. Cable and Wireless, a United Kingdom firm, has sought a license to provide a U.S.–UK–Hong Kong end-to-end service that would attack such giants as AT&T and BT on the world's busiest international route, New York to London. GTE led a U.S. consortium that owns 40 percent of the newly privatized Venezuelan CANTV. AT&T has signed major contracts with PTTs in such countries as China. Virtually every major player in every country has international plans, joint ventures, and subsidiaries. The large PTTs and private companies are looking to find markets, and the smaller companies and/or those in smaller or less developed countries are seeking partners with capital and technology.

Time to market was very slow before the new age of cutthroat competition. It took AT&T 15 years to launch its Princess telephone receiver to offer an alternative to the standard black handset; now, a $1 billion project to create a new switch takes about 18 months. Installation of equipment and services, which used to take about six weeks, now often occurs in under a day.

The degree of dependence on telecommunications across business and society was made very clear in a series of AT&T network crashes that badly damaged the company's reputation for reliability. Now, AT&T guarantees that 800 service will be back up inside an hour and for some customers will refund all their total monthly bill for every *hour* the network is out of service.

Our discussion of the telecommunications industry only hints at what is happening, suggesting in the process the following issues that cause many problems for designers and implementers:

▸   Telecommunications is a young industry with continuing innovation, uncertainty, political shifts, risks, and changes in competition. Many more shakeouts will come, and the pace of change is unpredictable. The telecommunications designer is put in a very uneasy situation, for the rapid rate of change suggests closely following the wave of change, but uncertainties often equally strongly suggest a tendency toward conservatism.

▸   The old, well-established standards-setting process is in danger of breaking down, for the pace of change moves faster than the standards-setting committees can keep up. The more volatile the technology and the faster the growth rate of the market, the fewer the standards. This is hardly surprising; technology and competition do not wait for standards but either assume or ignore them.

▸   The economics of telecommunications are being transformed by competition and technology, opening up many opportunities for skilled managers to cut costs (without cutting quality of service), to add new services, and to achieve ever higher levels of reliability.

▸   It is becoming increasingly difficult to keep up to date on key technical and competitive trends. The 1990s have already seen many totally new terms, products, standards, and companies, and there is every reason to expect more of them in the coming years.

▸   It is also more difficult to sort out fact from hype and hope. As in any technical field that changes quickly, there are many bandwagons and many excessive claims, and anyone can make predictions. If you skim any of the leading telecommunications trade magazines, you will find wildly varying views of such important developments as cell relay, SNMP, OSI, MANs, and ISDN, ranging from euphoric approval to dogmatic dismissal.

It is vital in this context for designers and implementers to understand the forces that drive the industry. The details will continue to change, making it even more important to be able to sort through hype and focus on key trends. In addition, designers and implementers must be able to explain to clients why their own views of the industry lead them to make recommendations about technologies or about vendors and providers. For such a large industry, business executives (and people in general) are surprisingly unfamiliar with telecommunications. Name recognition is low for even such major players as MCI, whose 1990 revenues were larger than those of the entire industry in 1978. Few people realize, for instance, that the revenues of the RBOCs are three times those of AT&T, that close to half of a long-distance provider's revenue from a call goes to an RBOC for local access charges, or that on average 75 cents of every dollar you pay when you make an international phone call goes to the PTT in the country you are phoning.

The rest of this chapter summarizes key trends in the telecommunications industry. It spends little time looking back at the evolution of the industry and at divestiture because the new dynamics of the industry no longer rest on the past. The key issues once were if and how AT&T's competitors could build an effective competitive counterpart to AT&T, but MCI now has close to 20 percent of the long-distance market, well above the 12 percent estimated to be the minimum for long-term survival. The development of the local telephone market was greatly slowed by lack of competition for the RBOCs; if the 1980s were the era of the long-distance wars, the 1990s will be that of the challenge to the Baby Bells and the Bells' aggressive expansion.

The technology base of 1984, when divestiture took effect, no longer drives the industry, for then the following key components of the new industry either did not exist or were in early development or testing: fiber-based LANs, broadband transmission, bridges, smart hubs, image processing, fiber networks, electronic data interchange, VSAT, and (nonmilitary) cellular data radio.

There are many ways to categorize the elements of the telecommunications industry. In this chapter, we use the following categories:

1. *Local, national,* and *international core services*: Basic transmission for public and private networks plus the broad range of services that most businesses routinely need.

2. *Differentiated services*: These services either provide specialized services (such as VSAT networks), use specialized technology (especially wireless), or are formally regarded as providing value-added network capabilities.

3. *Nontransmission facilities*: Switching equipment, LAN technology components, and software. This category is the most recent and volatile in terms of innovation, market entry (and exit), and standards.

Together, these categories add up to a worldwide industry with over $700 billion of revenues a year ($450 billion for services and $250 billion for equipment). The total is increased by at least $100 billion when telecommunications-related computer software is included. The rate of overall growth of the industry has been between 15 and 20 percent a year; growth is greater outside the United States than inside. The U.S. market has been marked by ferocious price competition, so that, even though the long-distance phone companies increased their volumes by about 12 percent in 1992, the revenue growth was just over 5 percent.

Unless otherwise stated, the figures given for volumes, revenues, and prices are for 1992, the most recent data available at press time.

| | | | | | | |

The U.S. Telecommunications Service Industry: Core Local Services

Historically, the long-distance market has been viewed as the core of the telecommunications industry, with AT&T at the center. From a business perspective, however, it makes more sense to focus on the local loop, which is by far the largest part of the domestic market, with revenues of over $100 billion. The Regional Bell Operating Companies, which have over 98 percent of this total, face growing competition from teleports, cable television firms, and companies that use private local fiber-optic facilities to connect their customers to large transmission hubs, most usually ones with large satellite earth stations. The RBOCs are already in place; they have access to capital, although their eroding growth and high cash payouts of annual dividends make them less attractive to investors than they were five years ago.

The Regional Bell Operating Companies

The RBOCs consist of Ameritech, Bell Atlantic, BellSouth, NYNEX, Pacific Telesis, Southwestern Bell, and US West. After divestiture in 1984, they owned a cash cow because the long-distance carriers had to connect to the **local exchange** at each end of the connection—and had to pay to do so; in 1986 over 60 cents of every dollar of long-distance carriers' revenue was paid to RBOCs. This has dropped to 45 percent, but it still adds up to around $30 billion.

One of the main reasons for divestiture was that the old AT&T system handled both long-distance and local phone calls, with heavy cross-subsidization of local services. Business and residential long distance was paying about $10 billion more than it should. MCI, the force behind the attacks on AT&T, had exploited this by offering long-distance service at a discount of over 20 percent. AT&T blocked MCI from local access, which led to MCI having a negative net worth of $600 million. When federal intervention forced the old Bell system to allow local access to competitors, the question became "At what price?" The creation of the RBOCs resolved the situation by clearly separating local access and long-distance elements. A new subscriber access charge was created; ini-

Figure 14–1

Sales and Profits for Corporate Operations for Three Long-distance Carriers and the Regional Bell Operating Companies During 1991–92

	Sales (billions)	Profits (billions)	Sales growth (%)	Profit growth (%)
Long-distance carriers				
AT&T	$64.9	$1.4	2.9	500.0
MCI	10.6	0.6	11.3	10.5
Sprint	10.0	0.4	5.1	16.2
RBOCs				
Ameritech	11.1	1.3	3.1	24.0
Bell Atlantic	12.6	1.4	3.0	3.8
BellSouth	15.2	1.7	5.2	10.1
NYNEX	13.2	1.3	−0.6	118.0
Pacific Telesis	9.9	1.1	0.4	12.5
Southwestern Bell	10.0	1.3	7.3	12.6
US West	10.3	1.2	−2.8	113.0

tially it was a dollar a month, but it rose to $3.50 by the end of the 1980s. Because local services had been subsidized prior to divestiture, the RBOCs' case for rate increases was largely supported by the state regulators who supervise the services that produce 75 percent of the RBOCs' revenues. Within a few years, however, regulators began to restrict increases in rates. In 1987 the RBOCs had to return $550 million they had overcharged their subscribers.

The Economics of Local Telecommunications Services

Even when rate increases become difficult to justify, the RBOCs continue to flourish. They faced no competition and their subscriber base is huge, as is shown by their 1992 revenues and profits as compared to those of the long-distance carriers (see Figure 14-1). Only in the first quarter of 1992 did MCI overtake the smallest RBOC in revenues, and that figure includes the half of its official revenues that it had to pass on to RBOCs.

The long-distance carriers' profit margins are below 4 percent (AT&T's is 7.3 percent for all its operations, with manufacturing providing the addition). Among the RBOCs, only NYNEX had a 1991 margin below 10.6 percent, and that largely reflected a long and damaging employee strike during 1989–90. MCI's and Sprint's sales in 1990 were larger than the entire long-distance industry's revenues at the time of divestiture; that is an indication of the competition and technical innovation that divestiture fueled. The revenues and profits of the four leading non-RBOC alternative providers of local services—Centel, Rochester Telephone, Cincinnati Bell, and Southern New England Telephone—were close to $6.5 billion, with a margin of 7.6 percent. Each of these providers competes against the RBOCs on price and service.

These figures show the relative profitability of local versus long-distance services. The evidence suggests that lack of competition has meant lack of innovation and service. The real cost of a long-distance phone call has dropped by over 40 percent since divestiture; the cost of a local business call has dropped by just 10 percent after adjusting for inflation.

Figure 14–2	Total sales	Total profits	Sales per employee (rank)	Profits per employee (rank)
Rankings of Three Long-distance Providers and RBOCs According to Sales and Profits, 1992				
AT&T	1	1	$206,000 (3)	$12,100 (9)
BellSouth	2	2	$157,000 (9)	$17,100 (6)
NYNEX	3	5	$159,000 (7)	$15,800 (8)
Bell Atlantic	4	3	$170,000 (4)	$18,600 (3)
Ameritech	5	4	$154,000 (10)	$18,400 (4)
MCI	6	9	$358,000 (1)	$21,000 (2)
US West	7	7	$159,000 (7)	$16,100 (7)
Southwestern Bell	8	6	$166,000 (5)	$21,600 (1)
Pacific Telesis	9	8	$161,000 (6)	$18,400 (4)
Sprint	10	10	$215,000 (2)	$9,900 (10)

The RBOCs' wealth has subsidized fairly notorious inefficiencies and over-staffing. Figure 14-2 ranks the RBOCs and the three largest long-distance providers by revenues per employee and profits per employee. Bell South, for instance, is second in sales and profits but ninth and sixth in sales per employee and profits per employee, respectively.

In 1991, MCI's sales were smaller than those of any of the RBOCs; it was the only firm to increase its sales in 1992 by more than 10 percent. NYNEX and US West lost sales but still increased profits by over 100 percent. Without the protection of regulation and the guaranteed revenues from local access charges paid by long-distance firms, the RBOCs look weak and unproductive.

Competition in the local loop makes the RBOCs' low levels of productivity untenable. The RBOCs are scrambling to innovate. Their main problem has been that their core market revenues are growing at just 3–5 percent a year. Volume growth is 7 percent; advances in technology and new competition are eroding margins. (By contrast, the data communications market is growing by 15–30 percent.) Because the RBOCs need new revenue sources, they have lobbied aggressively to be allowed to enter all telecommunications and information services markets. In October 1991 they were granted permission to do so, although the legislative issues will take years to resolve. Newspaper publishers and television programmers strongly oppose any moves by the RBOCs into their territory until there is equal competition in the local telecommunications market.

Although bypass technologies and smaller competitive carriers have cut the RBOCs' advantage, by 1992 competitors had gained only 1 percent of the $5 billion business for local access. A bill introduced into Congress prohibits the Bells from providing information content over their networks until 50 percent of their customers are able to access an alternative supplier and until 10 percent actually do so. It may be five to ten years before the competitive picture is clear and the new rules redefined. In the meantime, many of the RBOCs have made moves and joint ventures in electronic directories and on-line yellow pages; this is their most obvious market as information providers and one that they have eyed ever since divestiture. BellSouth has entirely restructured the company by splitting it into two parts; one will continue to handle the old core business while the other is positioned to make acquisitions and alliances and issue stock to compete in cellular and entertainment markets. Bell Atlantic has even sued the federal government, arguing that because the Supreme Court agreed that video is speech, blocking the RBOCs from entering the TV and video programming market amounts to a violation of its constitutional rights to freedom of speech.

The RBOCs have applied their cash base and expertise to international joint ventures. For example, BellSouth owns a European paging company; US West is implementing Hungary's first cellular network and owns part of UK Cable Corporation, the largest cable firm in the English-speaking world. BellSouth and Cable and Wireless (UK) spent over $3 billion to set up a telecommunications system to compete with the PTT in Australia when that country moved to **privatization** and competition. Until Judge Greene limited the amount of their revenues that could come from outside their regulated services to 10 percent, many of the Bells also diversified, largely ineffectually. For instance, Bell Atlantic bought a chain of computer stores in 1985 and sold it off at a loss in 1988.

The technical infrastructures of the RBOCs have until recently failed to keep pace with the innovations among the long-distance and value-added network providers. There are an estimated 122 million access lines in the United States; 85 million of these are lines into residences. Bellcore (the research and development arm of AT&T that also provides services for the RBOCs and helps coordinate the linkages across the national phone system) predicts that around one-fifth of these will be converted to ISDN service by 2001. In the local public networks that are the RBOCs' base, 50 percent of lines use digital switching; the United States lags behind France, Great Britain, and Scandinavia here. Almost all the existing access lines are copper cable; ISDN will increase the carrying capability of these by two and a half times. Copper to the home is already in place. There has been little incentive for the RBOCs to replace this with fiber, and before the early 1990s they did not reinvest their profits in upgrading their networks. It is estimated that it will cost around $120 billion for them to bring fiber into the 95 million households that have phones.

The Bells have begun to focus on ISDN, on the often very profitable and fast-growing cellular business, and on metropolitan networks. The old profitability is eroding, though. There may be little incentive for consumers to pay the extra for ISDN services and equipment. The preliminary tariffs filed for ISDN in 1992 were about twice the price for equivalent analog services. ISDN offers literally hundreds of new features; the demand for them is entirely unclear. In addition, the RBOCs' early ISDN operations were marked by expensive and highly visible network failures, largely created by the complexity of the new software needed to implement SS7.[1] The costs of connecting subscribers often exceeds the revenues acquired. New York Telephone claims that it costs $21 to

---

[1] Signalling System 7 is at the heart of ISDN. An SS7 error that brought down Bell Atlantic's network for nine hours in 1991 was caused by a single incorrect line of code that was the equivalent of one mistyped word among five copies of Herman Melville's *Moby Dick*. There were six such SS7 problems in 1991, each of which cut off service to millions of homes.

install one basic service, for which the charge for installation is just $7. The estimated average cost for a new subscriber connection went up from $35 to $43 between 1983 and 1989. (A caveat is needed here: The calculation of costs and returns are extremely complex in the telecommunications services industry, with elaborate rate formulae, allocation of shared costs, jargon, and decisions about what should be excluded from the RBOCs' cost base.)

In the business services market, the RBOCs face powerful competition from the value-added network providers and from companies' own uses of LANs and private WANs. As a result RBOCs have become more aggressive in marketing and technical innovation. They are reducing the time required to install new circuits for business customers, cutting weeks down to days and in some cases hours; Pacific Telesis has cut the time for "provisioning"—scheduling and installing—from 30 to seven days for most services. They are also rapidly installing fiber-based MANs in large cities, the technical base for which is switched multimegabit data service (SMDS). The developments in the RBOCs' strategies involve standards that are either largely missing, still emerging, or highly varied in the details of implementation.

The sluggish response that marked the RBOCs in the late 1980s created the same sort of aggressive competition in the local market as did AT&T's complacency in the 1970s. The revenues of one local competitor, Telport Communication Group Inc. of New York, are only around $60 million, but it is attracting customers for its services and is being helped by state legislation that now requires New York Telephone to interconnect Telport's customers and Telport's fiber local network. In 1986, a group of financial service organizations were told by NYNEX that the waiting time for a new digital circuit would be nine months; Telport completed the installation in weeks. New York Telephone, the local NYNEX unit, soon cut its installation time to a month and its costs for this circuit from $900 a month to $500. Telport's local network links locations in Long Island, New York City, and Newark, New Jersey, and also provides long-distance links to seven major cities, including Los Angeles, Boston, Dallas, and Chicago. Given that 21 million phone calls a day originate from lower Manhattan, the potential market for Telport is huge.

Only Nebraska had fully deregulated local telecommunications by mid-1993, but Illinois and New York had already begun to break up the RBOC monopoly. Illinois created a "free trade zone" and, like New York, requires the RBOC to provide access to rivals. Metropolitan Fiber Systems (MFS) has exploited this by installing a fiber-optic network in 12 major cities. The long-distance **interexchange carriers** typically locate international gateways in major cities, such as Dallas, New York, Chicago, and San Francisco. Now, they can reach these gateway cities, where MFS operates, seamlessly and much more cheaply.

The RBOCs as a whole have had a bad press and seem to deserve it. (The RBOCs' main defenses are that they have been required to provide universal service, that their costs of installing phones are far higher than what they can charge, and that they have been locked out of information markets that would have allowed them to reduce costs by increasing their revenue base.) A nationwide 1991 study of 200 large businesses reported that 45 percent of firms would switch local phone companies, given the chance. The RBOCs are seen by many technical specialists and industry observers as very poor in marketing, uneven in service, and high in price; many are skeptical concerning their ISDN plans and actions. But that was yesterday; tomorrow looks very different.

The Long-distance Carriers

In contrast to the RBOCs, the competitors in the long-distance market have made innovation their norm. The industry consists of three traditional competitors—AT&T (70 percent overall market share), MCI (17 percent), and Sprint (10 percent)—several emerging upstarts, about 20 alternative common carriers (ACCs) and alternative operator services (AOS), 300 resellers, and several emerging new providers. The ACCs have revenues of around $23 billion (including local services), with GTE the leader. The main measure of growth in the long-distance market is the number of call minutes rather than dollars, for prices drop rapidly and continuously. In 1984, the total traffic was 38 billion call minutes; in 1990, it was 90 billion.

It may seem eccentric to talk about "traditional" competitors in an industry that began just over a decade ago, but for the first years after divestiture only three companies really mattered. AT&T began with a market share of 85 percent, which has dropped overall as the market becomes much more differentiated. AT&T has well over 90 percent of the private network business but only 60 percent of the residential market. MCI has eroded AT&T's position through aggressive marketing of its Friends and Family program, probably the single most successful service innovation since divestiture. Sprint has around 10 percent of the business and consumer markets and has had to struggle to maintain solvency, even though it has been a frequent innovator in technology and services. Its May 1992 proposal to merge with Centel in a stock swap of close to $3 billion would create the third largest long-distance firm, a local phone company with six million customer lines, and a cellular company with a potential customer base of more than 20 million subscribers.

The traditional Big Three now face aggressive and innovative competition. In March 1991, WilTel announced its frame relay service. (It moved into voice services in 1992.) WilTel, a subsidiary of an oil company, built its 11,000-mile fiber network through the imaginative use of old oil pipelines. To move fiber, it used "pigs," devices that push oil along pipes. The company has added switched data services and voice services to its offerings. WilTel may or may not flourish, but the fact that an oil pipeline firm was the technical innovator in frame relay shows that the traditional industry is changing fast. In 1992, the first full year in which frame-relay services were available, WilTel's market share was 38 percent, with AT&T in second place with 21 percent.

Figure 14–3

AT&T's Costs and Revenues during Three Years under Option 58 (figures in millions of dollars)

	1991	1992	1993	Total
Revenue	4.14	4.14	4.14	12.42
Costs	4.25	3.88	3.88	12.01
Profit/Loss	–.11	.26	.26	.41

In late 1992, a new competitor emerged. When LDDS Communications Inc. acquired Advanced Telecommunications Corporation, it created a company with revenues of close to $800 million. Both firms focus on business customers who use daytime phone service. (Residential customers typically use less expensive night and weekend rates.) LDSS thus buys capacity in bulk from the Big Three long-distance providers and resells it. It is exploiting its geographic coverage of the southern United States to attack the rapidly growing U.S.-to-Latin America and U.S.-to-Caribbean markets, which have both grown by over 50 percent in ten years.

Long-distance voice and data services are a buyer's market. Each of the major players is aggressively wooing new customers, with price a key issue in the consumer telephone sector, small and medium businesses, and large public and private sector corporations. The industry has moved through four main phases of competition. The first was price-dominated and immediately followed divestiture. In the second phase (1986–88) network quality dominated. The third phase was largely product-based and lasted through the early 1990s. Each of these three phases addressed both business and consumer markets. The new, fourth phase is marked by integrated service for integrated technologies, with a focus on small, medium, and large businesses rather than on consumers.

The Business Market for Long-distance Providers

The long-distance carriers compete for voice and data traffic from large customers, with a growing focus on getting small businesses' voice and dial-up traffic. AT&T's stronghold has been large companies, for which it has over 90 percent of the market for private networks. It has used its size to offer low prices if companies commit to using AT&T for all their major communications. The average cost per minute of long-distance service in early 1993 was 17 cents (versus 0.5–1.5 cents for local calls and 30–35 cents for cellular calls); AT&T's volume discounts can cut this to 6–8 cents.

Tariff 12 offers 3–5-year deals that can undercut MCI and Sprint by 30 percent and lock them out of the customer account. (AT&T is the only interexchange carrier required to file tariffs, because of its dominant size and market share.) Tariff 12 is a point of contention, as are many multimillion or even multibillion dollar telecommunications deals. Lawyers thrive in the litigious world of the AT&T–MCI battle. AT&T asked the FCC to provide an escape clause for Option 58 of Tariff 12 in 1992. This offers attractive opportunities for resellers who sign up for Option 58, which includes 800 numbers and other switched services. Option 58 was not profitable for AT&T in its first year. The minimum charges and costs for a three-year deal are presented in Figure 14-3.

The FCC has issued rules that allow all Tariff 12 customers to discontinue service without any penalties in 1993, when 800 numbers became fully portable. This creates a scramble as AT&T tries to hold its business and others (especially MCI) try to capture the business of the 100 largest users of telecommunications. In the first week (starting May 1, 1993) that firms were allowed to move their 800 numbers away from AT&T, MCI gained 2 percent market share; that percentage is not trivial, for it amounts to $160 million of immediate new revenue.

AT&T faces Option 58 pressures, too. In 1992, eight resellers were using Option 58, and many others wanted to purchase the service so that they would then be able to resell 800 services. A firm like British Airways could keep its 1-800-AIRWAYS distinctiveness, but could drop Tariff 12 and buy on favorable terms from a specialist reseller, who makes small margins by focused pricing and limited extra services. AT&T asked the FCC to allow it to charge a $1 million installation fee.

In the first years after deregulation, AT&T was at a price disadvantage but had a major quality advantage over MCI, the number two in the industry. MCI's network was highly unreliable, and many businesses were reluctant to trade price for quality. Sprint moved rapidly to make digital communications a differentiator against MCI; its television commercials were built around the ability to hear a pin drop over the phone. MCI responded by investing heavily in its own network. In the third quarter of 1990, it wrote off $550 million of analog equipment, and spent $1.1 billion on capital investment in 1991. Sprint spent over $3.5 billion between 1989 and 1991. AT&T wrote off over $1 billion in 1990 and had its only after tax loss in 103 years. The drive for quality and capacity pushed the leaders to write off obsolete but undepreciated plants and to invest in the state of the art.[2]

As a result, there is very little price difference outside the Tariff 12 options across the three providers, and very little quality difference as well. The FCC no longer monitors the quality of interstate communications. The Big Three rank well on price and quality, with the customer generally being able to drive a hard bargain. Pricing increasingly centers on such issues as whether the provider charges in ten-second or six-second increments. In many cases quality comes down to subsecond call connect time, with AT&T advertising its ability to provide two-second access and 5.1-second call setup time. For JC Penney, shaving half a second matters; when it was able to get the time between a caller dialing and a customer service agent answering down below ten seconds, it reduced the number of impatient customers who hang up to under 4 percent.

The pace of technical change and fast response to that change makes it difficult for the leaders to differentiate their products. They all aggressively target emerging niches, most recently 800 and 900 numbers, virtual private

---

[2] AT&T spent $3 billion in 1992 to upgrade its network. Since divestiture, it has invested over $20 billion. Its main priority has been to add fiber; it now has over 30,000 miles of fiber-optic link on its 63,000-mile domestic network.

networks, and fast packet-switching services. An example is Sprint's "Live Broadcast 900," a 900-number service with fees of $1,500 for sign-up, $100 a week per line, $900 a month for billing, and 35 cents a call minute. The purchaser marks up the call minute charge.

What is the "product" here? In more and more instances in both the business and consumer market, it is a billing system. The same transmission capabilities can be sold in many forms. Hence, Tariff 12 is a product. The intelligence in the switches in a carriers' network become crucial here. Virtual private networks—which offer the economics and control of private networks plus the guarantees of capacity and accessibility on demand of public ones—have become a new battleground, with AT&T's advantages of features, MCI's quality of billing and management-control information, and Sprint's lower prices providing a wide range of choices for the large user.

Serving Consumers of Long-distance Services

In the consumer market, the three major carriers have all built the basics of their strategy around **equal access**—the set of rules, still in force, that requires that all customers choose AT&T, MCI, or Sprint as their long-distance carrier, on an equal-opportunity basis; AT&T was not allowed to automatically retain its existing base.[3] Equal access was the key practical outcome of the divestiture in terms of competition. Equal access also means that phone bills have become very complex because there is now a part showing the long-distance carrier's charges and another showing the local carrier's charges.

The main problems faced by long-distance carriers are price erosion and low margins. Competition has cut the cost of a call minute, the basic unit of business, to as low as four cents. The average revenue per minute before divestiture was $0.40; by 1992, it was $0.15. There are special deals everywhere, including AT&T's Reach Out America and MCI's Friends and Family. Sprint led the introduction of personal 800 numbers, a facility available before divestiture only to companies with very large volumes of calls and, later (through MCI) to small businesses. Calling cards, issued by both long-distance and local carriers, have become a standard product. More and more of them can now be used abroad. By 1992, MCI's card could be used in over 70 countries. Many of the calling cards offer discounts and even frequent flier mileage with particular airlines.

One of the most radical moves in the consumer market was AT&T's entry into credit cards, which immediately led Citibank to move its business to MCI. AT&T is now one of the three largest card providers in the United States. Even though this is a sideline for AT&T's core business, it has provided the carrier with a large database of consumer information that should be of great value in targeted marketing.

The tight margins in the voice market have severely damaged the prospects of the other companies providing long-distance services. These companies,

---

[3] Customers who did not specify a provider were assigned on the basis of a formula to one of the Big Three. There are still a few areas of the nation where equal access has not yet been implemented.

which hold about 5 percent of the market, operate mainly by reselling capacity provided by the Big Three to hotels, hospitals, colleges, and the like. Some have a specialized role in rural markets. MCI acquired Telecom°USA in 1990, a move that cemented its position as a strong number two to AT&T. Alternative operator services (AOS) provide switchboard services for hotels, schools, and other institutions, with whom they share revenues. The AOS have a poor reputation for price-gouging; some of them have very high markups for calls. AT&T's advertising, which shows an outraged customer either complaining about getting a huge bill from some operator service he or she has never heard of or slamming down the phone on an operator, is targeted at the AOS, though the implication is that AT&T's two main rivals also provide poor operator service.

The Next Ten Years in the Long-distance Market

Competition is obviously here to stay and is likely to increase. Basic transmission services are now a commodity, and the major players will find it difficult to gain a sustained advantage. In the business market, smaller-sized firms are a target of opportunity, mainly by offering attractive packages that "bundle" many services. The consumer market will be driven by price and packaging. AT&T's 1986 Reach Out America promotion added one billion call minutes; MCI's Friends and Family program added ten million customers and around 4 percent market share to MCI's 16 percent of the residential market.

The large customer business market is now driven by four main factors; (1) reliable and timely information on network traffic and performance, (2) knowledgeable and responsive sales and service staff, (3) fast and responsive support, and (4) quality and level of detail of billing systems.

Differentiated Services

Value-added Networks

Value-added networks were originally a regulatory distinction. Licenses to operate networks that sold services were granted only to firms that did not offer basic services that competed with the PTT, and the service had to add value to raw transmission. Electronic mail, electronic data interchange, and funds transfers are examples of value-added services. Today, VANs generally are designed to meet an industry's needs to share transmission, particularly for intercompany transactions. VANs are thus a core element in international electronic data interchange and in customer-supplier links in the manufacturing, automotive, and retailing industries, especially in Asia and Europe, where geography and a more cooperative business climate has led to VANs being adopted at a faster rate and earlier than in the United States.

Most VANs are standard X.25 networks that provide an efficient and easily accessed service. Generally, prices are higher than for public data networks, but the network is tailored to the type of message traffic VAN users typically send and receive. More importantly, VANs provide far more security than PDNs and increasingly aim to differentiate their offerings from the main alternative option of private networks through network management, reliability, flexibility, and (of course) cost. They compete against public data networks largely on the basis of the enhanced services they offer, which has led more and more VAN providers to shift from simple X.25 networks. Frame relay is the natural direction of innovation for X.25-based VANs, but one barrier is the poor quality of circuits in many countries. VAN providers must lease lines from PTTs and may not be able to get lines that are sufficiently error-free.

The VAN market is attracting many players. Sears operates an SNA VAN. BT's Global Network Service, which combines access to its VAN services and management of the customer's own network, grew from an initial 473 customers to 1,200 customers in its first 18 months of operations. Much of the new action is international for the obvious reason that many companies lack the size and resources to justify private global networks but want more than country-by-country access to PDNs. The growth of VANs in the Pacific Rim since 1980 has been very rapid, reflecting the fact that within a few years trade between Asia and the United States will be greater than that between Europe and the United States.

The market shares for X.25-based international VANs for U.S. carriers are: Sprint, 36 percent; BT North America, 35 percent; Infonet, 11 percent; GEIS (GE Information Services), 11 percent; and Compuserve, 7 percent. BT estimates worldwide X.25 public and private service revenues as follows: public, $567 million in 1991 and $615 million in 1992; and private, $718 million in 1991 and $777 million in 1992. BT expects the figures to be reversed in 1996, with just $604 million for private and $849 million for public X.25 services.

Information
Providers

Related to (and sometimes including) VANs are networks that provide information services. Electronic delivery of information has been a target for many providers in many countries. The results have been varied, ranging from the huge successes of Reuters in providing the foreign exchange rate information that got it occupancy in every trader's and corporate treasurer's office, to the many flops in U.S. efforts to reach the consumer market through videotext—the established (though obsolescent) term for information delivered via a terminal or PC through simple menus.

In the United States, the casualty rate in electronic information services has been high. The successes have mainly been providers of financial information, most obviously Reuters. No firm has yet created a large consumer market; Prodigy is the latest of many videotext services trying to do so. The main success has been the France Telecom's Minitel, but this increasingly looks like a special case. It succeeded because terminals were provided free, Minitel provided directory assistance, and the project was highly subsidized. The average use of a Minitel terminal is only two minutes a week, and the service lost $800 million in 1991. That said, it stimulated the creation of many business and consumer services offered via Minitel by a wide range of providers.

The United States is very different. Unlike in France, where the PTT had complete monopoly over all telecommunications, there are thousands of information providers and many networks to access them. There has as yet been no takeoff application that has led the owners of the millions of PCs in use today to videotext. On-line consumer services earned only $300 million in 1990; by contrast, home alarm monitoring was a $2.5 billion business, and business use of financial information earned over $3 billion.

Part of the problem providers have faced has been telecommunications itself. Dial-up access to, say, Prodigy, to check news, send and receive electronic mail, or initiate shopping orders ties up the family phone line and is very slow. The videotext displays are slow and poorly formatted in comparison to the quality of multimedia displays on PCs; in addition, most home PCs lack memory, color, and high-resolution screens.

All this may be changed when fiber-optic cable brings unlimited multimedia transmission into the home. Such service will not occur, except in a few cities, within the next five years.

Satellite Services in the United States

Most aspects of satellite transmission are international and operate under international conventions (see the section "The International Services Industry" later in this chapter). The fastest emerging area of domestic satellite competition is VSAT, the very small aperture terminals that are increasingly the base for retailers to send and receive information from all their locations. The main advantages of VSAT are ability to broadcast data from the head office, not having to provide expensive cabling to small stores, speed of installation and operation, and high reliability. The leader in VSAT is Hughes, with 50 percent of the market. (GTE has 30 percent and AT&T's Tridom service 7 percent.) There were an estimated 190,000 VSATs in use in 1992; retailing accounted for 35 percent and the automotive and financial industries for 17 percent each.

VSAT is growing rapidly, but the U.S. satellite services market has been overtaken by the many alternative high-capacity transmission options. Video-conferencing generally relies on satellites for the main reason that it is an occasional and ad hoc service. Users typically go to a conference room that has an uplink to and a downlink from a satellite. Time can be rented on a transponder by the hour.

Wireless Services

The most volatile area in telecommunications in terms of technology, regulation, and competition is wireless communications. Originally confined to radio broadcasting, this now includes cellular phones, infrared wireless LANs, PCNs (personal communication networks), and other uses of the high end of the electromagnetic spectrum. In the main area of growth—cellular communications—Europe and Canada lead the United States. In the UK, Cellnet and Vodafone services have over a million subscribers, with revenues growing at 25–30 percent a year; sales are almost $1.5 billion. The total U.S. market is currently about that size as well. In 1991, the leading three companies—McCaw Cellular, Metro Mobile, and Vanguard Cellular—had total revenues of $1.5 billion but losses of over $600 million. McCaw is betting on growth and is adding capacity as fast as it can. Its subscribers to its cellular services grew 42 percent to 1.6 million; that is still a tiny proportion of the U.S. population. Cellular has largely been a local service; McCaw aims to build a national infrastructure. The infusion of capital from AT&T in 1992 will strengthen its solidity. It is expected to break even in 1994 and be very profitable thereafter.

So, too, does Motorola, which already offers radio services and is the industry leader in portable phones. Its Iridium project, mentioned earlier, aims at a global personal communications network. Its ARDIS data communications service is well established, and it offers a mobile electronic mail capability called EMBARC. Motorola is also involved in a joint venture with IBM and Apple, just one of many speculative ventures underway to exploit the as yet underdeveloped wireless market. Chapter 19 summarizes the potential high-growth opportunities, but it is too early to discern the shape of the wireless industry. It is at roughly the same stage as the PC market was in 1982, with few established players or clear direction. The FCC is encouraging innovation by opening up access to the wireless frequency spectrum.

Cellular PC Technology	Mobile cellular phones have been commonplace for years. The race is now on to apply wireless technology to personal computers. Only an estimated 2,500 out of the 2.7 million laptop and notebook computers sold in the United States in 1991 used wireless communications. Based on the rate of growth of the cellular telephone market in the 1980s, wireless technology should increase to around 180,000 out of 4.5 million at the end of 1993, a rapid rate of increase but still a small percentage of the total sold.

The opportunities are immense. One is the liberation of the PC from phone lines and from hotels, where the phone wire cannot be unplugged from the handset and connected to a laptop; some veteran users of laptops have left dismantled wall plugs and a disarray of cables behind them at checkout times. With cellular, you can connect your PC to a network from your car, the beach, or any hotel. Cellular has an advantage over the other two main wireless technologies—infrared and radio frequency (RF)—in that cellular networks cover almost the entire United States. Infrared is best suited to LAN communications, and RF networks are not yet fully deployed or standardized.

There are many barriers to **cellular PC** connections. The technology is still largely analog, making it difficult to ensure reliable and noise-free circuits. Cellular is designed for voice communications, in which pauses and interruptions are not critical; such interruptions often garble data or cause a modem to disconnect. And the cost is high. In 1992, sending a megabyte of data over a cellular link cost $600 and a page of fax cost $12, well over ten times the average price for terrestrial links. Cellular phones do not have a dial tone, so that special interface units must be added to the laptop machine—and to the user's briefcase. Setup times for cellular calls can be up to a minute.

Progress is rapid, however. Manufacturers of PCs are offering PCs with built-in cellular modems, and prices are dropping fast. The 9.6 bp modem of 1991 had a list price of $1,245, whereas in 1992 it listed for about $600; the modem of 1995 should cost $200. The cost of sending 1 megabyte of data is expected to drop from $600 to $25 by the year 2000 and that for a page of fax is expected to fall from $12 to 50 cents.

| | | | | | |

The International Services Industry

The world's largest international carriers are, in descending order of size, AT&T (five billion minutes of international traffic a year), DBP Telekom (Germany), Telecom UK, France Telecom, Telecom Canada, the Swiss PTT, ASST/Italcable (Italy), the Dutch PTT, the Belgian PTT, and Japan's KDD. Reliable figures in international traffic often take years to compile, but these relative rankings seem likely to hold through the mid-1990s.

The U.S. telecommunications industry is open and intensely competitive. That is not true in most countries in the world. Until the late 1980s, telecommunications was generally viewed as either a "natural monopoly" or a social good and was managed as a quasi-government agency called a PTT. The German, French, and Belgian PTTs have even sued the Commission of the European Community in opposition of its requirements that telecommunications becomes competitive as part of the 1992 Open Market initiative.

Through the 1980s, the Deutsche Bundespost kept private networks illegal. This was to the huge advantage of the United Kingdom, where telecommunications was liberalized in 1986. London processes 40 percent of the world's daily foreign exchange, its stock market is bigger in volume than all the other exchanges of Europe, it is a hub for international telecommunications, and it handles close to 40 percent of Europe's international EDI traffic. Germany was a nonplayer in all these areas simply because its firms could not access telecommunications services that were manageable, cost-effective, and reliable. The Bundespost moved to privatization and limited competition in the early 1990s, as have most PTTs worldwide.

There have been two general patterns of liberalization. The first is for the PTT to retain basic services and license competition in other services. Great Britain set up a duopoly; British Telecoms was privatized, and a new company, Mercury, was licensed. In 1992, the UK Office of Telecommunications (OFTEL), which oversees telecommunications policy, allowed application for licenses from any source; Sprint quickly applied. The British model ensures some competition with the new privatized PTT but maintains order and preserves the national public network. Australia and South Africa have followed the duopoly model. In Japan, NTT and KDD, two noncompeting organizations, were privatized, with three new long-distance competitors permitted for the domestic NTT and one for the international KDD. Prices plummeted, achieving the goal of better service and lower costs while preserving the national infrastructure.

The other pattern of liberalization is straightforward privatization, with monopoly being maintained and prices set to ensure innovation. France Telecom and the Belgian RTT fought hard against privatization but accepted its inevitability, largely because PTTs now need to seek international business, which is impractical for a government agency. France Telecom, one of the most innovative telecommunications providers in the world, has one of the highest percentages of digital switches in the world. Its public packet-switched X.25 public data network, Transpac, led the world in volumes of traffic for a PDN. In each of these areas, France was ahead of the United States. The logic of France Telecom's strategy has been to be aggressive in technical innovation as a defense against pressures to privatize or allow competition. Ironically, while arguing against competition it has been a fierce and effective competitor in international markets.

In general, countries that have historically been trading nations have moved quickly to privatize telecommunications, for they recognized that far from being a "natural monopoly," telecommunications is a core national resource vital to businesses' ability to compete internationally. The Netherlands and Sweden were thus early participants in privatization. Hong Kong similarly took a lead; all local calls there are free, a stimulus to fast and easy communication that is part of its frenetic business culture.

The monopolistic PTTs once faced no domestic price competition. Now, the large differences in prices across countries have led many large firms to either seek out or avoid them in their choices of location of factories, offices, data centers, and communication hubs. This explains why London outperforms Frankfurt in electronic trading and why it has remained one of the three main financial centers of the world (along with Tokyo and New York). Singapore has the lowest communication costs in Asia and aims at exploiting this in competition for trading services against Hong Kong, Taiwan, and other Asian cities. One large Swiss firm moved its main data center from Basel to the UK because the PTTs costs were so dramatically out of line; a leased line from New York to Basel was one-tenth the cost of one from Basel to New York, for instance.

Alliances among international firms are commonplace. The main problem for large international companies is the lack of "one-stop shopping." They want to have one point of contact for ordering and billing, instead of having to deal with, say, the Swiss PTT for a link from Basel to Paris and with France Telecom for the very same link back from Paris to Basel. VAN providers provide some help here, and many of the major PTTs, in alliance with each other and a U.S. carrier, offer such services. However, restrictive practices are still a strong legacy from the old monopolies.

International Satellite Services

For many years satellites were the only source of high-speed international transmission. In 1964, INTELSAT was set up, largely through U.S. government initiatives, to improve global communications. It is owned by its member countries, which now number over 120, under an international treaty and mostly serves PTTs and organizations that resell services to carriers and users. Fiber optics has greatly reduced the importance of INTELSAT since 1988, when the first international cable was installed. By the middle of 1991, the bandwidth leased on fiber had overtaken that on satellites. In the mid-1980s, over 60 percent of AT&T's international calls were via satellite; today, it is closer to 40 percent. INTELSAT now gets about 60 percent of its revenues of about $600 million from switched international services and just 10 percent from private lines; 15 percent comes from television services. INTELSAT remains the key, however, to reaching the many countries that do not have fiber links, most obviously developing nations.

Competition from fiber has caused INTELSAT to cut prices in return for long-term commitments. Customers can get a dedicated voice-grade circuit for $27 a month, if they contract to buy in bulk for 15 years. Of course, technology has also cut prices. A leased circuit that cost $22,800 when INTELSAT began business now costs just $80. Competition is also coming from new players in satellite transmission. INTELSAT has agreed to allow any private satellite system with less than 30 36-MHz transponders to compete, provided it offers less than 100 international switched circuits. An Argentinian entrepreneur has been fighting (and largely losing) in the world courts to allow his tiny PANAMSAT to compete in the closed international market.

The operations of other providers of international satellite communications are more special-purpose than those of COMSAT, a private company licensed by the U.S. government. For instance, INMARSAT handles marine communication needs, ranging from the obvious navigational needs of ships to the less obvious transaction needs. Between the time an oil tanker leaves, say, Abu Dhabi and arrives in Rotterdam, its contents may have changed ownership 20 times and its route may have been altered even more frequently to minimize time and cost and avoid bad weather. ARINC and SITA similarly provide satellite communications to airplanes; until recently, these services were analog and limited in capacity. A pilot of a jumbo jet knew the plane's position only within 100 square miles when it was halfway across the Atlantic; now it can be pinpointed to within a few square feet, and data communications augments voice to and from the cockpit. Federal Express now begins the process of clearing Brussels customs when it is several hours away from Europe; when it lands, customs agents can list the few packages they wish to inspect.

The Growth Opportunity of International Telecommunications

Telecommunications, transportation, and education are the three major infrastructures of a modern nation that are crucial to economic and social growth. International telecommunications is a guaranteed growth industry, given that (1) each 1 percent increase in investment in a nation's telecommunications system creates $3 of gross domestic product, (2) many advanced countries have telecommunications infrastructures and regulations that badly need updating, and (3) rates of penetration of basic services are still very low in most of the world.

The United States is by far the world's largest and most mature telecommunications market. The average number of phone calls per person is three times that of Germany, the United Kingdom, France, and Japan. One-sixth of the world's population lives in countries with less than four phones per 100 people; Canada has 80 per 100 people. The cost of raising Eastern Europe to 50 per 100 people is estimated to be $116 billion.

Every country, whether developed, semideveloped, or developing, recognizes the importance of telecommunications to its future. Every major provider equally recognizes the importance of an international presence. Whereas in the United States the new agenda is likely to be dominated by competition, joint ventures and alliances among cable TV companies, RBOCs, long-distance providers, cellular phone providers, and VANs, the international market will remain focused on the core of telecommunications policy issues: the role of the PTT, privatization and liberalization, basic phone and data services, and national infrastructures. It may be 20 to 40 years before we see the world wired (and/or wireless), but it is coming, and every year will be a growth year for international communications.

| | | | | | |

Nontransmission Facilities

The market for telecommunications equipment and software is very much like that for computers: a wide range of options, competitors, products, and features exists, as does constant innovation. This makes it far simpler to explain than the services industry; regulation is minor, and history is largely irrelevant. There are few barriers to entry and exit. In any list of small growth companies, at least 10 percent will be in the telecommunications equipment market.

Large-scale switching equipment is the only exception to this picture, a direct result of the nature of the international services market. Each PTT's operations depend on the quality of the switches it uses to handle call setup, routing, billing, security, error-checking, and many other functions. They must be fast and reliable. The more intelligence that can be built into them through software, the more features and services the PTT can offer. It costs over a billion dollars to develop a new generation of switch; to recover that investment, the manufacturer must capture around 15 percent of the world market.

As a result, only about six manufacturers will survive; small players cannot afford to get into the competitive game, and weak ones will get out. More importantly, the major PTTs' decisions on which switch providers to use determine the fate of the competitors. For that reason, there is often a strong nationalistic element in their choice.

Apart from large-scale intelligent switches, the part of the telecommunications market that does not provide transmission services outside the LAN is very open. It has three main components: switches, including PBX (private branch exchange); LAN equipment; and software.

Switches

Because the equipment market is open, it is heavily international, apart from the PTT nationalist bias. A major problem for developing countries is that they must pay hard currency for switching equipment. That underlies many countries' decisions to deal with U.S. long-distance carriers and RBOCs; it also underlies the U.S. deficit in imports and exports of equipment. Before divestiture, AT&T's switches were made by AT&T's Western Electric, and the United States ran a surplus of as much as $6 billion a year; by 1989 this had become a deficit of $2.5 billion. In 1988, U.S. sales of switches were $4.3 billion, with imports amounting to $461 million, around 20 percent; the sales of telephones were $1.2 billion in 1982 and $1.6 billion in 1988. However, in an area of U.S. strength—mobile phones—imports were just $278 million out of total sales of $2 billion.

In the U.S. market for switches and PBX, AT&T's market share has changed substantially. In 1970, 1980, and 1989, its share of PBX was 80 percent, 46 percent, and 26 percent, respectively. Northern Telecom's market share was 13 percent in 1976, 8 percent in 1980, and 23 percent in 1989. For digital switches, AT&T's market share grew from 1 percent in 1982 to 46 percent in 1988; Northern Telecom's share declined from 66 to 32; and GTE's share grew from 3 to 10 percent.

Overall, prices in the telecommunications equipment market are dropping by 7–10 percent a year. The obvious reason for this is that the costs of computer hardware are dropping by an even faster rate, as much as 30 percent. Telecommunications equipment is increasingly based on microprocessor technology. A router, bridge, concentrator, or small switch is really a computer that processes a bit stream instead of transactions. The hardware provides much of the needed intelligence, although software may be required to provide additional intelligence.

In the days of analog technology, electromechanical devices determined equipment design and performance. Digital communications exploits computer technology, which is fundamentally digital, which is why prices will continue to drop and performance improve across the nontransmission markets.

The LAN-related Equipment Market

The switch market is mature and WAN-based. The marketplace for LANs and for the interconnection of LAN to LAN and LAN to WAN have been the main areas of innovation and growth in the 1990s. The competition, like competition in the PC market, is fierce, with continued price-cutting and improved performance. The main battlegrounds have been routers, bridges, and smart hubs. Progress is hardware-driven, so that prices have dropped at the same pace as for computer chips. The market is unstable in terms of products, standards, technology, and prices.

In May 1992 *Network World* conducted a reader survey of products and services that identified the leaders in various categories. The results are summarized below, together with comments on the nature of the industry competitors:

1. *LAN adapters* for Ethernet, token ring, and FDDI: Except for IBM (token ring) and Digital Equipment (Ethernet and token ring), every one of the top 20 firms listed is a relatively small company, and almost all have been founded less than ten years ago. The Ethernet standard was created by Xerox, which is no longer a player in this market.

2. *Bridges:* The situation is the same as for LAN adapters; IBM is the only established firm in a market dominated by new names.

3. *Routers:* Every company here is a fairly recent startup company, and several of them have either become favorites of stock market tipsters or—when they fail to grow at 50 percent or more a year—stocks they have

dumped. This is an area that has moved from premium innovation to commodity very quickly. Cisco and Proteon have set the pace, very much in the same way that Digital Equipment, Data General, and Prime did with minicomputers in the 1970s. It is worth noting that each of the latter were fast-growing innovators for a decade, but only DEC is still a leader, and it is facing many of the same problems as IBM faces. The field of information technology moves so fast that today's small, high-growth firm may be tomorrow's giant—or tomorrow's loser.

4.   *Smart hubs:* After routers, they have become the fastest growing and most technically innovative area of the LAN interconnect market. SynOptics Communications was the pacesetter; its growth was over 80 percent a year from its inception. AT&T has a presence through its acquisition of NCR.

5.   *Diagnostic and LAN management tools:* The players here include established computer vendors like Hewlett-Packard; established LAN software and hardware vendors, most obviously Novell; and many newcomers.

6.   *LAN servers and superservers:* The established computer companies are very active here, for the obvious reason that servers and superservers are PCs and minicomputers optimized for networked rather than stand-alone use. Compaq, IBM, Sun, and many other manufacturers of PCs and workstations naturally target the high end of their products to the server market.

7.   *Wireless LANs:* The technology here is very new, so that the firms competing are almost all new entrants—with the exception of Motorola, one of the U.S. manufacturing firms that takes no back seat in quality or innovation to any Japanese company.

The Telecommunications Software Market

Many types of software are relevant to telecommunications but are not limited to it. For instance, today's computer operating systems must include many communications facilities. Software packages for PCs must have versions applicable to LANs to have any chance of succeeding in the business market; database management systems must similarly accommodate distributed databases.

The three types of software that are specific to telecommunications (that is, there is no stand-alone version) are electronic mail, network management software, and groupware. E-mail and groupware have no dominant vendor, although Lotus Notes has established a strong position. Network management software is emerging as one of the most important competitive fields. In LANs, Novell has built a commanding lead through its Netware product. IBM, AT&T, and Digital Equipment are fighting to establish their own systems as the base for end-to-end network management of computers, applications, and communications.

A new term in the telecommunications industry is **middleware**—software that resides on a device such as a router and enables interoperability and provides network management and other services; in contrast to it are the operating systems that reside on client and server workstations and/or mainframes. The term is not well defined, and its very existence signals the growing role of software in what has historically been a market driven by hardware and equipment.

The overall network software market is in flux. Every major player in the information technology industry must provide a telecommunications capability if it is to remain a major player. That is why Lotus and Microsoft, whose main base is PC software, are providers of groupware. IBM, DEC, and HP, whose core markets are in computer hardware and software, are active in every area of telecommunications for the same reason; so, too, are the leading providers of database management systems, image processing, and just about every other business-centered area.

Summary

The new telecommunications industry has five main components: (1) U.S. core local services, by far the largest single market and one dominated by the regulated Regional Bell Operating Companies; (2) long-distance carriers, the most competitive of all the markets; (3) differentiated services, most obviously value-added networks; (4) international services; and (5) nontransmission facilities, including switches, LANs, and telecommunications software.

Before the divestiture of AT&T in 1982, there was no real telecommunications competition. This deregulation of the long-distance market created a new environment that has cut prices and stimulated innovation in technology, products, and services. Deregulation is currently the main trend in other markets, replacing monopoly control by PTTs abroad and by RBOCs in the United States.

The local communications market has been the least innovative, largely because of control by RBOCs, whose flow of profits has been constant, encouraging overstaffing and lack of innovation. Now, locals face competition from many sources, including cable television firms, and in turn they are being permitted to enter new markets. Their core market—local phone services—is not growing fast, and they see entertainment, cellular phones, publishing, and information services as major targets of opportunity.

Competition in the long-distance market has cut the real cost of a call by around 40 percent in under ten years. AT&T's dominant position has been challenged, first by MCI and Sprint and now by several new niche players. The carriers have invested billions of dollars in new facilities and have innovated aggressively in the 1990s following a period of relatively little technical innovation in the industry. Today, the carriers are positioning to offer virtual private networks, SDMS, frame relay, and nationwide cellular services.

Value-added network providers offer differentiated services that are mainly targeted to specific industries and types of user. Many are configured for EDI traffic or international communications and use X.25 as the dominant protocol. Information providers offer a wide range of business and consumer services that as yet have not created a large base of customers—with the exception of financial information, such as Reuters' foreign exchange rate quotations. Satellite services in the United States have grown primarily with respect to VSAT, but satellites play a key role in many international aspects of business and broadcasting. The international marketplace has been dominated by monopoly PTTs, but liberalization, privatization, and alliances are growing rapidly.

Nontransmission facilities, especially in LANs, have been an area of intense innovation and competition. Many of the fastest growing small and medium-sized firms are new entrants, offering routers and bridges, network operating systems, and software for telecommunications.

Review Questions

1. Briefly explain the notion that before January 28, 1982, there was no real telecommunications industry.

2. Consider the following statement: Competition and deregulation in the U.S. telecommunications market have resulted in the lack of a comprehensive national infrastructure that offers the benefits of advanced technology to both small and large users. Do you agree or disagree? Support your answer.

3. In the latter half of the 1980s, corporate telecommunications development was limited because ultrafast LANs were linked to slow WANs. Describe recent developments that are alleviating this problem.

4. Explain why "800 portability" is such an important issue. Discuss it from the standpoint of the end-users and the carriers supplying the service.

5. Explain why telecommunications designers are put in a very uneasy position with respect to rapid changes occurring in the industry, and why it is critically important to understand the key forces that drive the industry.

6. The profit margins for the major long-distance carriers are significantly lower than for the RBOCs. Discuss the reasons for this, indicating whether this trend is likely to continue or not, and explain your reasoning.

7. Identify the principal markets and services that the RBOCs are targeting for future growth, and briefly explain why these markets/services make sense from a strategic and competitive positioning perspective.

8. Describe and assess the competitive threat to the RBOCs posed by alternative local (bypass) carriers such as Telport Communications Group (in New York) and Metropolitan Fiber Systems (which operates networks in 12 major cities).

9. Who is "WilTel" (what services/products does it provide); how did it get into the long-distance market; and what competitive threat does it pose to the "Big Three" long-distance carriers?

10. Describe the key elements of price and quality of service competition in the highly competitive market of providing long-distance services to businesses.

11. Explain how the billing system is the product in long-distance services.

12. Describe how the major long-distance providers are attempting to combat low and continually eroding profit margins in long-distance services provided to the individual consumer market.

13. How do VANs differentiate themselves from both public data networks and the alternative of private corporate networks?

14. Explain why subscriber information services (such as Prodigy, Dow-Jones Retrieval Service, and so on) have largely failed to catch on and take off in the United States, compared with France's Minitel service.

15. Explain the difference between the "duopoly" model and full privatization for liberalizing international telecommunications in countries that have historically relied on a government or quasi-government monopoly to provide equipment and services. Discuss the strengths and weaknesses of the two approaches.

16. Why have countries that historically were trading nations moved quickly to privatize telecommunications?

17. What single factor is likely to have the biggest impact on reducing the prices of international telecommunications services offered by non-U.S. PTTs and carriers and aligning them more closely with prices for outbound calls from the United States to other countries? Explain.

18. Explain why the market for central office switching equipment exhibits a strong nationalistic bias and is not nearly as open and competitive as the markets for other types of telecommunications equipment.

19. Describe the market for LANs and LAN interconnection equipment in terms of the size and stability of markets, the extent of competition, principal products, and the role of standards.

20. Identify the three types of software that are specific to telecommunications (that is, that have no stand-alone version), and describe the general market conditions and trends for each.

Assignments and Questions for Discussion

1. Describe the differences between the evolution of data communications and voice communications in U.S. companies, and discuss the implications of this evolution for achieving a completely integrated information technology function within a company.

2. Put yourself in the position of a federal judge who is preparing to rule on requests to permit the Regional Bell Operating Companies to offer long-distance services and cable television services, and to manufacture and sell telecommunications equipment. Discuss the arguments for and against granting this request, and indicate the conditions under which you would permit other entities to offer local telephone service.

3. Devise a strategy for the RBOCs to compete effectively in a largely deregulated telecommunica-tions equipment and service market (long distance and local service). Be sure to include in your plan a realistic approach to dealing with a bloated bureau-cracy and a corporate culture that is largely inexpe-rienced with a globally competitive business environment.

4. The global satellite consortium INTELSAT was created to provide affordable global connectivity to all countries of the world. The underlying premise of the organization is that INTELSAT's competi-tors are harmful to global welfare because needless duplication of facilities (satellites) uses scarce fre-quency spectrum and orbital slots, and such com-petitors undermine the inherent economies of scale in satellite transmission by siphoning off the high-volume, high-margin traffic and leaving the higher-cost, thin-route traffic for INTELSAT. Eval-uate and assess these arguments for prohibiting en-try of alternative satellite systems.

5. Design a profitable, on-line information service that will attract a significant number of U.S. con-sumers. Include in your design a description of the target market for the service, its key features, and how it will be accessed and priced.

Minicase 14-1

The Gang of Five

100 mbps to the Desktop

The Gang of Five is a group of vendors who worked together to produce a proposal in 1990 to help jump-start FDDI (fiber distributed data interface). Their goal was to accelerate the use of FDDI on the shielded twisted-pair (STP) cabling that has been the base for Ethernet and token-ring LANs. They published a *Green Book*, which they made available to any vendor without charge, to speed up the standards-setting process by helping vendors move quickly to implement FDDI in real products. In 1990, FDDI on fiber cost up to eight times as much as Ethernet on cable. The Gang of Five saw that if FDDI could be run over cable, installation costs would be greatly reduced because twisted pair is much cheaper. In addition, there was a huge base of IBM STP cabling systems; users could upgrade to FDDI running at up to 100 mbps.

The ANSI X3T9.5 committee, which oversees FDDI standards-setting, established a working group in late 1990, after the publication of the *Green Book*. The group had the cumbersome title of the Twisted Pair Physical Layer Medium Dependent Group. Its goal was to draft a proposal for FDDI running over twisted pair wiring systems, primarily unshielded ones. Crescendo Communications, a vendor that developed a working unshielded twisted pair (UTP) technique, collaborated with the working group to set up the UTP Development Forum (UDF) in June 1991. The members of the forum, which included AT&T,

Hewlett-Packard, British Telecom, and providers of LAN equipment, used laboratory tests to prove that Class 5 (data grade) unshielded twisted pair could indeed run FDDI along cables up to 100 meters in length (and up to 75 meters with Class 4 cables). The UDF technology would also run on shielded twisted pair.

Thus there were two totally incompatible and competing proposals for the same standard. IBM had joined the *Green Book* supporters and was marketing a well-received product. The UDF proposal had the advantage that it could run over a higher percentage of twisted pair installations, whether shielded or unshielded. Meanwhile, Cabletron and National Semiconductor had developed their own proposals for FDDI on unshielded twisted pair.

The Gang of Five proposed that there be two separate standards, one for STP (their own) and one for UTP. They argued that this would both speed up adoption of FDDI as a standard and rapidly expand the market. UDF and the Cabletron and National Semiconductor consortium argued for a single standard, so that users would not have to pick among incompatible products that were all the same "standard." In October 1991, the ANSI working group voted to reject the Gang of Five's STP proposal. Only after mid-1993 can the differences between the two UTP proposals be resolved. Products based on the Gang of Five's *Green Book* are already on the market. Leading firms like Digital Equipment and SynOptics offer concentrators, and hundreds of sites are using them.

Of course, none of the UTP versus STP issues matter if firms replace twisted pair cable with fiber. The main reason for the Gang of Five's initiative was to address the needs of companies that had installed IBM token-ring cabling.

Questions for Discussion

1. Why was the "Gang of Five" vendor consortium formed, and what is contained in their jointly developed *Green Book*?

2. Identify the other major industry players in the FDDI standards struggle, and describe their respective positions with respect to FDDI cabling.

3. Do you agree with the decision by the ANSI standards body to reject the Gang of Five's proposal? Explain your reasoning and discuss the implications of the decision for FDDI.

Minicase 14-2

Hong Kong and Singapore

Competition in Southeast Asia

The economic growth of Southeast Asia in recent years has been one of the marvels of the modern age. The gross domestic products of countries such as Thailand and Indonesia have increased at a rate of 7–10 percent a year for almost a decade. Malaysia is now the world's largest exporter of computer chips, and Singapore is the world's largest port in terms of cargo volumes; Hong Kong is no longer just a trading city but the key hub for Guandphong province in South China, an area half the size of California with a population of 50 million. In both 1991 and 1992, Guandphong's economic growth was over 20 percent.

In this context, telecommunications carriers are moving very aggressively to improve their networks and services. In particular, Singapore and Hong Kong are extending their rivalry from shipping and banking to telecommunications. Both want to be the center for trading in the region and seek to attract multinationals to build hubs in their city. Conservative estimates are that their telecommunications spending will grow at 20 percent a year between 1993 and 2000.

All the region's PTTs are modernizing quickly. It is estimated that over 90 percent of all lines in Singapore, Hong Kong, and Thailand will be digital by 1995, up from 48 percent, 60 percent, and 75 percent, respectively, in 1990. Both Singapore and Hong Kong dropped the price of leased lines by an average of 23 percent between 1988 and 1991.

Their 1991 average availability for leased international lines exceeds that of France Telecom, one of the most modern telecommunications providers in the world, at over 99.94 percent uptime.

Fears of China's 1997 takeover of Hong Kong have led many companies to move to Singapore. For example, SITA (Societe de Telecommunications Aeronatiques), the Paris-based organization that runs a global network for international airlines, set up its third telecommunications center in Singapore. Reuters and Telerate, the two leading financial electronic information providers, moved from Hong Kong to Singapore.

Other firms are playing the two trading city-states off against each other. GEIS (General Electric Information Services) has kept its main regional hub in Hong Kong but has increased the size and scope of its Singapore unit. According to its country manager, "More and more operational headquarters of companies are moving from Hong Kong to Singapore. Once the decision-making people such as the chairman and officers are here, the communications hub usually is not far behind." The Hong Kong and Shanghai Bank, a major force in both Asia and Europe, keeps careful score of Hong Kong Telecom's and Singapore Telecom's performance, service, availability, and prices.

Both PTTs are slow in delivering leased lines as compared to U.S. carriers, but they are up to par with Japan and well ahead of Europe. Hong Kong is more competitive on tariffs and more responsive to customers, partly because making deals is part of the culture. Singapore Telecom is more paternalistic. One customer commented that "what they offer is technically very good, but they offer what they want to offer." As with Hong Kong's culture of deal making, this fits the stereotype of Singapore as a highly disciplined environment in which it is illegal to chew gum in public and police may arrest any male with long hair and have him shorn. Although Hong Kong will sign quality-of-service agreements with key customers, Singapore Telecom ignores requests to do so. It is, however, investing over $250 million between 1992 and 1995 in fiber optics under the Pacific Ocean to link it to other major world centers. It is carefully privatizing while retaining a monopoly over all service except value-added networks and terminal equipment. It provides virtual private networks and ISDN service.

Hong Kong and Singapore match European levels of quality, cost, and service. A problem for firms operating in the region is the dreadful quality of telecommunications elsewhere. Whereas Hong Kong has 57 telephones per 100 population, Malaysia has 9, Thailand 2, the Philippines 1, and Indonesia less than 1. Thus although circuits for international gateways are widely available in all countries in the region, it is a major problem to build domestic networks. The waiting list for phone lines in Thailand numbers 1.2 million customers. The World Bank, Asian Development Bank, and other agencies are pouring funds into telecommunications development. VSAT technology is as important here as fiber optics because of the massive distances and scattered populations of such countries as Indonesia. It would be far too expensive to provide fiber links across these distances; VSAT is the simplest and fastest way to upgrade telecommunications and to reach remote locations.

The problem here is lack of satellite transponder capacity. It will be 1994 before there is any reduction in the estimated shortfall of 60 to 70 transponders needed to meet today's demand. There are only two regional satellite services: Asiasat, operating out of Hong Kong, and Palapa, owned by the Indonesian government. About 25 percent of Asiasat's capacity is used for VSAT. There are a number of VSAT networks and network providers, most of them operating networks with 150 to 250 VSAT receivers. Rates are attractive. Globe-Mackay, an aggressive carrier in the Philippines, offers a 9.6-kbps data link plus 16-kbps voice link for under $2,000 a month; an analog terrestrial leased line from Hong Kong or Singapore Telecom costs close to twice that amount.

Questions for Discussion

1. Describe the implications for telecommunications of both China's 1997 takeover of Hong Kong and companies' strategy of playing the two Asian trading cities off against each other.

2. How do Singapore and Hong Kong compare with the United States, Europe, and Japan both in terms of the availability of telecommunications equipment and services and with respect to quality and price? Do you think some or all of these differences are attributable to the monopoly provision by the PTT of telecommunications in Hong Kong and Singapore? Explain.

3. Describe the role played by VSAT satellite systems in domestic networks of countries in Southeast Asia, and explain why these systems are being deployed instead of higher-capacity, state-of-the-art fiber-optic cables.

Minicase 14-3

Cisco

Keeping Product Leadership

Cisco Systems Inc. is one of the most successful companies in the telecommunications equipment industry. Its market share for routers is about 40 percent, and it is consistently the technological innovator in internetworking devices. Its revenue and profit growth since its founding less than ten years ago has been at least 40 percent a year. An example of its innovativeness is its Communications Server product, announced in May 1992, which combines four functions that once needed separate equipment. The 500-CS combines a 38.4-kbps asynchronous router, a terminal server, a protocol translator, and a remote communications server. Cisco was the first company to combine terminal server hardware with routing software. Because the cost of the 500-CS is about the same as for a conventional server, it cuts users' costs substantially. More importantly, it simplifies network design and management.

In late 1991 the CEO of Cisco, John Morgridge, was interviewed by *Network World*, a weekly telecommunications magazine. The following summary of the questions put to him and his answers provide a practical perspective on progress in internetworking from the viewpoint of a company whose entire business depends on it.

Question 1: It is often said that internetworks are growing to the point of chaos. What does Cisco plan to do to address network management?

Answer: Our strategy is to build a basic set of tools for the equipment we know and understand and then work with other vendors to provide interfaces to their products. The management area of LANs is still in an experimental phase. I don't see any quick or early breakthroughs because it requires a lot of software—and a fair amount of hardware. . . . Certainly, this is a horizon we're further away from than we'd like to be. It's unclear how that will evolve. . . . The industry today doesn't have adequate management tools that can give the user comfort in terms of being in control.

Question 2: A key challenge today for vendors of routers is IBM's Systems Network Architecture. Who's going to gain control of the single internetwork backbone?

Answer: SNA is not as well understood in Silicon Valley as in IBM's own labs. People thought there would be a single magic solution from some vendor. There won't be. We are tackling the problem in phases.

Question 3: Analysts claim that service and support is weak in vendors like Cisco that grew up serving academic and technical customers. Commercial customers may prefer to rely on products from firms like IBM or DEC because of the bigger resources they have for support.

Answer: That's a fair comment. Internetworks are multidimensional. We have to work in a dozen different environments. In addition, corporate networks are becoming very diversified. It's one thing to support a ten-router network and another to have 450 routers in a global network. Our industry lacks trained people. We will have to invest heavily here.

The IBMs and DECs may not have any real edge here, though. It's not enough to know SNA and the hardware side of this business is a very small problem. It's software and interoperability that are the key issues. That demands in-depth knowledge.

Question 4: Will routing and hub technologies merge?

Answer: Absolutely. We will play a leading role here because we are primarily a software company and the hub vendors tend to be hardware firms.

Questions for Discussion

1. Use information provided in this minicase to give a plausible explanation for Cisco's tremendous success in the industry, and explain how it is able to compete successfully against established rivals that have huge resources at their disposal.

2. Some users claim that it is not enough to have innovative products. They argue that small companies do not have the resources necessary to adequately support their products in the field. Evaluate this statement and include Cisco's position on the issue.

3. Briefly describe the key issues and problems associated with internetwork management, and evaluate Cisco's approach to dealing with these issues.

Minicase 14-4

Washington International Telport

A New Communications "Retailer"

Washington International Telport (WIT) is the telecommunications equivalent of a seaport; it is a gateway for voice, data, and video traffic into and out of the mid-Atlantic region. Most other telports rent space and facilities to users, but WIT owns each of the 23 satellite earth stations on its eight-acre site. The stations, which range from two meters to 18.3 meters in size, access all domestic U.S. satellites and most INTELSAT ones. WIT has made joint agreements with foreign providers such as Germany's to route traffic from their country through WIT into and out of the United States, linking to WilTel's fiber-optic frame-relay network and operating a 2,000-mile microwave network.

Washington, D.C., is the news capital of the world. Until the mid-1980s, the high cost of studios in and transmission from there meant that local stations across the United States had to rely on feeds from the three major networks, ABC, CBS, and NBC. Stations were unable to customize their broadcasting to local needs, personalities, and issues. European broadcasting was largely a quasi-government national monopoly that provided limited coverage of Washington.

When COMSAT, the U.S. satellite monopoly, was deregulated in 1984, foreign broadcasters took advantage of direct access to INTELSAT; previously they were required to use COMSAT as a gateway. Competition and improved technology cut transmission rates. WIT was an outgrowth of

the new environment. In 1987 it introduced its Capitol Coverage service, which offered computer-controlled switched-video transmission over fiber optics from multiple downtown locations, including the State Department and Capitol Hill. Small broadcasters now had access to top-rate facilities; their reporters in effect had their own studio. WIT's facilities could be booked by the hour.

In late 1987, WIT added a new service that provided 24-hour two-way TV transmission between Washington and London. The D.C.-to-UK link used C-band to C-band connection. European digital Ku-band traffic was linked to C-band domestic services. Central and South America, France, Italy, Germany, and Poland were added to the countries that could directly hub into WIT.

The C-band to C-band and Ku-band to C-band links were needed to deal with the fact that although many TV stations have Ku-band capability, by far the majority of radio and television broadcasting operates in the C-band. C-band dishes are large because the signal is more difficult to pick up than that of the higher-frequency, more focused Ku-band. The dishes are almost impossible to locate in crowded urban areas (no one wants to have a giant dish on the top of his or her building or apartment complex, and zoning laws generally forbid it), so broadcasters either had to go to rural and suburban areas or use a telport. WIT's 23 dishes offered both C-band capacity, for which the de-

547

mand has grown rapidly, and Ku-band capacity, which offers more comprehensive and higher-quality services.

By the end of the 1980s, government agencies, trade associations—a mainstay of the D.C. economy— and corporations in the D.C. area were rapidly increasing their use of business TV and videoconferencing. Few of these users needed or could afford their own private facilities. WIT offered top-of-the-line services on an as-needed basis. A trade association might want to hold a two-hour broadcast training program every three months, for example, linking to its regional members across the country.

WIT has continued to add services that can be bought on this retail basis. It added voice/data transmission when it built a 13-meter earth station in 1987. In 1988 it added an 18-meter one (56 feet in diameter) that provides a dedicated circuit via INTELSAT to WIT's Telemetry, Tracking, and Command Center in Italy. It added a United States-to-Switzerland 800 service for both voice and fax.

WIT is entrepreneurial; it markets widely and moves quickly to add new services as regulation and technology open up new opportunities. Telports are neither a public, a private, nor a value-added network. They are a type of transmission broker; they manage the building of facilities, contracting with INTELSAT and others and providing guarantees of service availability and quality. Like good retailers, they are opportunistic in the best sense of the term. They are demand-driven, whereas the PTTs are largely supply-driven. PTTs are in the utility business, whereas WIT draws on utilities to provide a buy-as-you-need service to a growing range of customers. Given the rapidly growing demand for occasional use of international telecommunications services—whether for television, videoconferencing, 800 number services, or training programs—Washington International Telport may become as important a communications gateway as Washington National Airport and Washington Dulles Airport are for the physical movement of senators and journalists.

Questions for Discussion

1. What market niche does WIT serve that is not adequately covered by either the established satellite companies, such as Comsat/INTELSAT, or the traditional local or long distance/international service providers?

2. The WIT has been likened to a good retailer that buys from telecommunications utilities (e.g., PTTs) to serve the changing needs of end-user customers. Discuss the meaning of this statement and state the characteristics of a good retailer that are exhibited by WIT.

3. How much of WIT's success is due to its strategic location in the Washington, D.C., metropolitan area? Would you project another success if WIT were to build a similar telport in another city, either in the U.S. or abroad? Explain and justify your answer.

15

Managing Network Design

Objectives

- ▶ Understand the role of the designer in network design
- ▶ Demonstrate the trade-offs among the four principal design variables
- ▶ Identify the major policy issues influencing design choice
- ▶ Demonstrate the applicability of total quality management methods in network design

Chapter Overview

| | | | | | | |

Chapter Overview

This chapter examines the role of the designer in network design. Complete network design encompasses everything from analyzing the trade-offs among key network variables, such as **capability, flexibility, quality of service,** and cost, to determining the specific size and type of circuits and equipment that constitute the network system.

The chapter begins with a discussion of the designer's role in this process. This is followed by a description of the four principal network design variables and a discussion of the trade-offs among them. We then turn to an investigation of the major policy choices affecting design decisions, including such factors as the company's tolerance for risk, planning horizon, vendor preferences, telecommunications architecture, and so on. Network design is also bounded by several factors that constrain choices, including the availability of technology, the firm's organizational capability, regulation, and industry practices. The chapter concludes with a discussion of how to manage the network design task, including an introduction to the concept of total quality management and its applicability to network design implementation.

The focus of the chapter is on managing design, not on design itself. The technical details of design can be extraordinarily complex. Before they reach a decision, companies frequently seek outside expert opinion, spend substantial time meeting with key vendors, and carry out detailed technical and economic analyses. The designer is often the coordinator of these activities and may not have the in-depth technical knowledge of other experts. But experts' knowledge is only of real value if the designer provides the management framework. The telecommunications business opportunity checklist facilitates the client's business-centered input to the design dialogue, and the telecommunications platform map translates these into criteria for design. Managing network design involves ensuring that these are then reviewed in terms of the following four key network design variables:

- ▸ The capability of the network: What business functions does it deliver?
- ▸ Its flexibility: How easy will it be to add and extend business functions?
- ▸ The quality of service: How reliable is it?
- ▸ Cost: What's the price tag, and is it worth it?

| | | | | | | |

The Designer's Role

A complete network design process includes (1) a high-level analysis of client preferences and tradeoffs with respect to important design features that frame the capabilities and limitations of the network, and (2) the detailed translation of these strategic requirements into specific circuit sizes, equipment, software, and standards needed to actually deliver the chosen capabilities. The designer's role is to focus on both the determination of client preferences with respect to strategic variables that determine the network's ability to meet the needs of the business and the trade-offs among these variables. The outcome of this analysis is input to the implementer, whose role is to translate these choices into actual network specifications. The implementer undertakes detailed traffic analysis studies and applies mathematical models to determine the proper circuit capacities and the optimal location of switches, multiplexers, and other equipment. There will be discussions—and probably arguments—about which vendors to use, standards to adopt, and products to test. All of this requires that the designer constantly ask, "How do these relate to the business opportunities and the platform requirements?"

In many organizations, no one is performing the designer's role; consequently, the entire design process is left to the implementer, with very little client input regarding high-level trade-offs among key network parameters that affect business results. The implementer generally has just two guidelines: Meet a specific need, such as connectivity among locations, devices, or applications, and minimize cost. As a result, many networks are ineffective in supporting the achievement of key business goals because these goals were never stated in terms of telecommunications business opportunities and business criteria for the telecommunications business platform. This is how the "network from hell," described in Chapter 11, grows from case-by-case design decisions to multitechnology chaos.

To create an effective telecommunications network, the design process must be clearly linked to the overall business planning process, starting from the identification of telecommunications opportunities. In Chapter 13, which described the Landberg Dairies study, this was a process of funnelling; the study team began with a very broad business scan that looked at Landberg's industry, customers, and company. The funnel was very wide; the team poured in findings from interviews, outside experts, the business and telecommunications trade press, customers, and many other sources. The team then narrowed down its options to the priority business opportunities and in turn to the business criteria for the telecommunications platform. At that point, the management of network design can begin.

Key Design Variables

Capability

Capability refers to the range and volume of services that can be handled by a network. The range of services can include voice, image, and video applications, as well as every type of computer related traffic, including electronic mail, customer transactions, access to databases, word processing, videotext, electronic data interchange, voice-mail, funds transfer, facsimile, and so on.

The required range of services must be based on application opportunities and requirements that support the strategic business goals of the organization. If the designer's role is not performed properly, bottom-up design based merely on technical opportunities and availability of a given technology can result. Of course, the designer does need to look ahead and ask which new technology vehicles might open up new application opportunities. The business vision is the reality test for answering that question.

Technology solutions that ignore the wider business vision typically occur as a result of technology-generated problems. For example, in 1992 one of the world's largest international economic development agencies initiated a new budgeting process for IT. Each department was required to submit an IT plan as part of its business plan. Department after department indicated that LANs were a priority. The total cost would be over $13 million. In-depth review of the proposals revealed no apparent reason for the unexpected, dramatic, and expensive demand. The agency already had a WAN, an organizationwide e-mail system, and backbone campus LANs in its main offices. Why the need for so many departmental LANs?

The explanation was simple: problems with printers. The agency had encouraged the use of PCs by all its professionals. The wide variety of their work had led to a wide variety of choices; there were Macs, Suns, Grid portables, IBM clones, DOS machines, Windows machines, and even machines that used NEC's Japanese operating system. These machines could not interconnect, so that output from one could be made input to another only through that wonderfully user-friendly medium called paper. That meant using printers, but the agency had not provided funds for a printer on every desk to go with a PC on every desk (or in every briefcase). The demand for LANs was driven almost entirely by a need for shared printers.

An outside consultant who was asked to help develop the architecture for the new LAN complex calculated that every desktop machine user could be provided with a high-quality laser printer for a total cost of under $1 million. The issue was not network design; the capability needed was simply stand-alone printing. Behind the problem, though, was the question of whether the agency should now focus on integration of information and outputs from multi-technology desktop machines and begin to design a platform with reach, range, and responsiveness across its departments. That is a very different design issue from simply installing a LAN to share printers.

Capability has many facets. As in the previous example, it mainly relates to realizing a given opportunity. Here, integration is not the main concern. Integration of separate capabilities must be viewed in a broader context. In the development agency, the need to share printers narrowed the focus to LANs. Addressing the needed reach and range for the agency to do its work effectively would have shifted the issue from printers to service integration.

This is not to say that it is a foregone conclusion that all network services will be integrated. Some services may require separate terminals and use different transmission facilities and be entirely independent of each other. For example, a word-processing software package may also send and receive electronic mail, but it does not need to process transactions. A PC may retrieve inventory data, place orders, and check credit information, but it will not process digital voice messages.

The concept of integration in telecommunications often means different things to clients and implementers. From the perspective of the senior manager as client, the issue is not integrated technology, but integration of customer service. What is the reach of the network service? What range of services can the customer get access to from a single workstation? Can separate transactions be cross-related and handled together?

From the implementer's perspective, the concept of integration means shared network facilities. For example, voice/data integration means that the same transmission links and switches handle both telephone and computer data traffic. The implementer deals with the realities of the network—essentially, do the bits arrive accurately, reliably, securely, and cheaply? Given that for an international electronic funds transfer those bits may have traveled across ten network nodes, been converted into ten different protocols, been packetized and depacketized ten different times, moved from LAN to private Wan to PDN to WAN to LAN, and transmitted through fiber, microwave, and satellite links, it is a miracle that networks achieve 99 percent reliability. The miracle maker here is the implementer.

Telecommunications clients need to know nothing about the adventurous life of a bit. Their interest is in the services that can be delivered quickly and efficiently to the workstation and combined together to create a new business opportunity. There may or may not be sound technical reasons for a firm to move ahead with the network equipment integration necessary to provide these services over the network's lines and switches, which are independent of the business reasons for integrating the services.

It is generally far cheaper and easier to design a network that provides specific services than one that plans ahead for full integration. It may also be that customers do not need integrated services delivered to the same workstation now, but what about in the future? The implementer's main responsibility is to ensure that a stated technical need can be met. The designer's responsibility is to ensure that today's and tomorrow's needs do not conflict. Fundamentally, the proper degree of integration in network design is a business question that should be driven by customer requirements, not technical convenience.

It is also important to recognize the key role played by the designer in ensuring that the client's need for integrated services drives integration of network transmission and switching facilities, and that these needs become part of the overall network architecture to accommodate growth and future changes in requirements. The client is in the position of the typical user of electricity. How many of us who use a VCR "understand" electricity? In what way would explaining what happens behind the wall in any way empower us here? The electrical grid is an integrated delivery capability very analogous to a telecommunications services platform. If it used different voltages in different states, the impacts of systems disintegration would be very apparent to users—your VCR would blow up as one state's 976 volts surged through a machine designed to operate at only 102 volts.

Integration is never apparent. Indeed, users see it only when it is missing. Much of the multitechnology chaos in the "network from hell" came from the fact that business clients never realized that integration mattered or even existed. It is the designer's responsibility to be thinking ahead about it and looking at a given need in this wider context. That is why the telecommunications platform map can be so useful in building the client/implementer dialog. It is in practice a blueprint for integration expressed in business terms.

Another aspect of network capability is the overall capacity or the volume of traffic it can accommodate. Telecommunications capacity is expensive, and its level for the network is determined by the equipment component with the smallest capacity (that is, the weakest link in the chain determines the strength of the overall chain). For example, an airline reservation system may have ample transmission link capacity to handle the required volume of transactions, but if the switches are unable to accommodate enough terminals at the same time, this is the limiting element in network capacity.

Forecasting capacity needs is very difficult. Often, telecommunications network capacity (supply) creates its own demand. Companies typically underestimate the rate of growth in traffic volume, and their short planning horizons and emphasis on controlling communications costs often put them in a reactive position. Unfortunately, there is no simple answer to the capacity planning problem under uncertain demand conditions and rapidly changing technology, cost of facilities, and services. Adopting an architecture with substantial capacity flexibility is certainly one approach, but that often involves paying a premium for short-term availability. And the amount of capacity built into a network must be evaluated relative to the other design variables (and their respective costs) in order to arrive at sensible trade-offs with respect to client preferences and needs.

Flexibility

Flexibility refers to the ease and speed with which changes can be made to any part of the telecommunications network platform and to the range of changes that can be made without having to replace, redevelop, or throw out existing network components. One way to build flexibility into a network architecture to permit the addition or deletion of small capacity increments is by relying on public network services or virtual private network arrangements with carriers. Public network service offerings have built-in flexibility because they consist of dial-up connections established as needed and have the ability to change the capacity of the service through a simple order change with the carrier that provides the service.

For example, if a bank subscribes to MCI's fractional T1 (1.544-mbps) service offering, it is charged for the fraction of the T1 (measured in 64-kbps increments) it actually uses. The bank can start out by activating only 64 kbps of the total capacity available for transferring customer account information among two or three of its smaller branches. However, as headquarters comes on line with its new mainframe computer and all customer account data is centralized on this machine, the bank merely contacts its MCI sales representative and orders the activation of additional capacity in increments of 64 kbps up to the full T1 speed.

Flexibility in the form of rapid network modifications to accommodate additional locations, types of services, and service integration is determined by the network architecture that has been adopted. In general, the most flexible architectures are the ones with the most overhead and fixed cost. Telecommunications managers will often pay a premium for a major vendor's components because the underlying architecture ensures flexibility to accommodate future system growth and changes.

Quality of Service

The third key design variable is quality of service, which is determined by how well the network design meets customer requirements. Such requirements should dictate the level of capability and flexibility designed into the network. In addition, they should also determine the level of reliability, responsiveness, and accessibility to the network, which are the principal aspects of quality of service. The responsiveness dimension of the telecommunications services platform map helps clarify the business importance of reliability. If the platform requires guaranteed on-demand service, reliability is a requirement, not an option to be traded off against cost. Similarly, if "anyone/anywhere" reach is the business imperative, accessibility is a central priority in design.

Reliability is a measure of error-free continuity of service. How often will the customer get a busy signal (especially at peak hours), or find the network "down," or encounter noise on the lines that creates costly errors, delays, and disruptions? Reliability is essential when the business is on-line; it is also expensive to ensure.

Network customers may also have preferences with respect to the distribution of network interruptions and downtime. For instance, the same average downtime can result from one or two long outages at off-peak periods or from a larger number of very short outages during peak use periods. The type of work being done and the individuals involved will determine which of these two downtime patterns is more onerous.

Responsiveness addresses how quickly the network customer's message is processed and an answer provided. A given average response time can also be achieved in different ways, with different implications for customers. For example, a response time of three seconds or less for 95 percent of the time can include some periods of very long response times in the other 5 percent. Some customer groups may prefer a higher average response time with less variability, such as six seconds or less for 99 percent of the time.

Furthermore, it is not always reasonable to assume that a faster response time is important. One major bank spent millions of dollars to give international customers a five-second response time for electronic funds transfers and later learned that customers were perfectly happy with 20 minutes. The simple (and yet often violated) rule of thumb is that the response time of a network should match the speed requirement for real-time information. If there is a real business advantage connected with a faster response time (and its benefits outweigh the costs of providing it), then it is justified; otherwise, it is a waste of expensive and scarce network resources.

The third major component of quality of service is accessibility, which refers to how simple and convenient the procedures are for customers to get access to the network and its services. Do they need special equipment or software? Can they use the facility only during given times and from specific locations? Must they use a complex set of procedures?

Many corporate employees use neither their company's new telephone system with callback, auto dial, and other features nor the special long-distance service that cuts phone bills by 20 percent because they have to dial too many digits first. Many network services that appear on the surface to be attractive are inconvenient and cumbersome to use. Employees are unlikely to choose to complicate their lives merely to save the company some money.

Many aspects of accessibility relate to the design of the terminals, the procedures for logging on to the system, and the quality of the network facilities, which generally fall under the responsibility of the telecommunications unit. Other aspects depend on the design of the system software and on human engineering, which are generally handled by the information systems department. These two units must cooperate in the design and delivery of services if true accessibility is to be achieved.

Electronic delivery of services relies on accessibility, convenience, ease of use, and freedom from worry. Computers and telecommunications networks are still difficult to use. Gaining access to a network can appear to be an intelligence test or a form of Russian roulette. The telecommunications implementer does not usually have the time, interest, or sufficient understanding of the customer's requirements to ensure that adequate accessibility is provided; consequently, this function must be performed by the designer. It is important to keep in mind at all times that the customer should never be distracted by the network, and that if the service is difficult to use, it is unlikely to be used. That is why the old cliché about most people being unable to program their VCRs is a fact, not just a joke. In general, the desktop operating system determines ease of access, but the network reach determines ease of accessibility.

Finally, the fourth aspect of quality of service is network security. Secure telecommunications requires attention to the whole chain of activity involved in electronic delivery. The signal may be encrypted, which means that it is sent in scrambled form that makes it look like garbage if it is intercepted in route.

Companies and customers have, on the whole, been remarkably casual about security. Simple passwords, encryption, and cards with magnetic stripes cannot provide adequate security. Visa found that about 30 percent of its customers forget their password. There is clearly a trade-off between convenience and security.

How much security should be built into network design? The answer to this question depends on the answers to another series of questions: When is security a differentiator, or even a product itself? How much are we at risk from errors, omissions, and fraud? Clearly, the implementer can install and maintain a security system and its procedures, but it is up to the designer both to evaluate the full range of security requirements based on current and future client needs and to integrate security into network design. Ensuring good security is very expensive. Today's telecommunications network services are not at all secure.

Cost

The fourth key category of network design variable is cost. Even though it would be ideal to have a network with infinite capability, total flexibility, and perfect quality of service, that is no more practical than guaranteeing all U.S. residents access to the best health care facilities, whenever they need them and with every service matching the very best provider's offerings. It is not that the designer—or doctor— does not want to ensure such service, but that there is always a cost, and money spent on A must come from not spending it on B. In making the trade-offs between, say, flexibility and cost or reliability and cost, it is thus crucial to define cost elements as carefully as defining capability and quality of service.

There are three main categories of telecommunications costs:

▶ Variable: These costs increase directly with volumes. For example, your phone bill increases the more calls you make.

▶ Step-shift: These costs are variable but only up to a limit of capacity. For example, a switch can handle a given number of ports, with transmission over, say, an X.25 PDN being a variable cost, but at a given number of users and mix of traffic it must be upgraded; this is a step-shift in cost, in that the variable cost of the PDN remains the same but additional investment is needed.

▶ Fixed: The cost is the same, regardless of volume of traffic or degree of use. A LAN is largely a fixed cost, as is a leased line or file server. The cost is the same when it is not used as when it is used at full capacity.

At the network design stage, the costs have not yet been fully defined; that comes once a complete design has been selected and a detailed, formal business proposal has been created. The aim in design is to highlight trade-offs and assess when a given level of capability, flexibility, and quality of service affect (and are affected by) choice of costs and choice of a technology, which is a variable cost. A few rules of thumb here are:

▶ The narrower the range of capabilities, the simpler it will be to select the most cost-efficient network design. By contrast, the harder it will generally be to adapt that design to add new capabilities. Choose the design that most simply, efficiently, and cheaply provides the needed capabilities, but first make sure that those needed capabilities are not likely to change within the next few years.

▶ The more predictable the volumes of traffic, the more likely it is that the designer can select a low-cost, high-efficiency design. Wide variations in traffic at different periods easily lead to slow response time, congestion, and even network crashes. For this reason, the designer needs to look at potential factors that may increase volumes and not just at estimated average volumes. It is vital to analyze a wide range of scenarios of traffic growth.

▶   Unless quality of service is defined as a priority *for business reasons*, cost is likely to become the main factor the client considers. Make sure that the trade-off between quality and cost is explicitly addressed by the client.

Trade-offs Among the Design Variables

More capability, flexibility, and quality of service in the design of a given network is clearly better, but it generally requires more resources to achieve (that is, it costs more). Therefore, the essence of the designer's role in network design is to manage the trade-offs among the design variables such that the resulting network provides needed business support at an acceptable cost. This is, in turn, determined both by client trade-off preferences among the design variables and by the client's view on the business value of the network.

For instance, a designer can choose to invest in a network capability that provides innovative services but may be less reliable for a number of reasons: Overload at peak times leads to poor response times, or the technology is new and complex and hence subject to errors and failures. The trade-off between reliability and performance—or between reliability and cost or between reliability and speed of response—requires a clear business model. It is impossible to answer the question "How reliable should our network service be?" in a vacuum. The appropriate questions must be: "Given a set of application opportunities, policies, and external constraints, what is the relative emphasis we should place on reliability? How much should we pay to improve it by a given amount?"

The literature on network design emphasizes methods for network optimization. Generally, this means meeting given requirements (locations, number of workstations, volume of messages, and so on) at lowest cost. However, minimizing cost is not always the key objective, and using telecommunications to support business strategy may involve multiple, often conflicting objectives. In some instances the goals are to maximize service, control costs, guarantee ease of expansion, ensure complete security, and so on.

Thus there is no "optimal" network design as defined by purely technical considerations. The trade-offs among aspects of capability, flexibility, and quality of service can be complex, which is why many companies have as many as 50 different network facilities, each dedicated to a particular application and all of them largely incompatible. When the customer workstation becomes the delivery point for a range of services—and when more and more mail, meetings, reports, and conversations depend on digital communications—then high capability, flexibility, and quality of service become crucial. But what premium is the firm willing to pay? What must we have in these areas? What can we ease back on or postpone? What premiums or savings in investment and operating costs are involved? These questions obviously require business directives. They illustrate why telecommunications planning must be done from the top down; they are impossible to answer if the business criteria are not already well defined.

The cost range of comparable networks in comparable firms can vary considerably. Some of the differences relate to management skills and economies of scale resulting from using fixed-cost leased facilities instead of incremental volume-sensitive ones, whereas other differences are due to technical and operational efficiency. Many of the cost differences, however, relate to design trade-offs. For instance, if a firm wants to be able to interconnect to customer networks and installs protocol converters that can convert messages from the format used in IBM's System Network Architecture to that used by public "packet switched" data networks, approximately 20 percent more message overhead is required in the form of additional transmission bits. This can significantly affect response times, capacity needs, and the like. Is it worth it? It certainly is for any company whose application opportunities largely depend on being able to connect to the processing base of customers or suppliers.

The dialogue between designer and client (and that between designer and implementer) should be about trade-offs. The designer needs to know what the firm wants to optimize and cannot afford to get locked into a technical, operations-centered mindset and lose sight of the customer-driven design considerations. Finally, to be effective, the network design dialogue must take place within the context of a broader policy framework that provides guidance on a number of important issues that influence design choices. This is the subject of the next section.

| | | | | | |

Policy Issues Influencing Design Choices

Policy is an explicit set of corporate mandates and directives. It defines boundaries of planning and design, and it clarifies responsibilities and authority. It thus sets the criteria for the architecture and establishes the role of the designer. Policy is partly a set of choices—"This is the way we will do things around here"—and partly self-imposed constraints—"We will consider only those options that fall within these limits." In some areas no policy choice is needed, but it should be an explicit decision, not something that happens by default. As the client, designer, and implementer move through the design dialogue, they will often have a list of existing policies that constrain their options, but they also need to build their own list of policy issues that need review.

The elevation of a business issue—say, security— from feature to policy will surely change many aspects of an expected network design. The most likely option—encryption devices in network nodes—will be expensive, whereas most of the network design process aims at reducing cost. Policy decisions remove options to make trade-offs among network design variables in the interest of the wider enterprise priorities.

The following list identifies the major policy agenda that all firms must address if they are to effectively link network design to the support of a business strategy;

- ▸ The degree (and pace) of change acceptable to management
- ▸ Choice of vendors
- ▸ Platform architecture and the range of services accommodated
- ▸ The planning horizon, including funding and cost recovery
- ▸ The level of risk considered acceptable

Other items can be added to the list. In fact, one question for senior management to ask its designer is simply "Is this list of policy items complete, or are there other items we should add to the agenda because they require a directive from the top?" Let's consider each of these policy issues in turn.

Degree of Change

Many firms overlook the need to link the pace of technical change with that of organizational change. How fast must a firm move for competitive reasons? How fast can it move without disrupting the existing organization and operations? What degree of change will it accept in terms of business activities, culture, and shifts in jobs and skills? What resources must it commit to smoothing the process of change?

In financial services, for instance, the business and supporting technical strategy has often run way ahead of needed changes in selling skills and in the sales force's attitudes toward making those changes. Citibank's and Merrill Lynch's aggressive account officers blocked, rather than supported, those firms' moves into the electronic marketplace. There can be a conflict between the need to accelerate change to meet competitive pressures and demands to slow change down to meet cultural constraints. Which factors should set the pace?

The policy on the scale and rate of change that management wants must be complemented by a decision to provide resources to make the policy effective. Increasingly, this means a commitment to education, and education requires a set of strategic goals that include changing attitudes toward technology and its uses within the organization, building the skills needed to participate in the process of change, providing a forum for people to express and resolve their concerns, sharing information, and ensuring that there is a shared business vision that staff understand in terms of what it means for their jobs.

Education to support a change in policy must begin early and be pervasive. In fact, there are those who believe that the only true sustainable competitive advantage for an organization is the ability to learn and adapt to change more quickly than others. This gives rise to the concept of the "learning organization," which is committed to the application of organizational change principles and information technology to help the company adopt a systematic, organizationwide learning process.

Choice of Vendors   The idea of choice of vendors as a strategic policy issue may seem strange at first. Shouldn't the company decide what it wants, create the specifications, and then choose the supplier based on the supplier's proposed price, service, delivery, and product design features? This approach works for purchasing individual pieces of equipment when the technical specifications ensure compatibility among equipment supplied by different vendors. However, in designing telecommunications networks, the adoption of an overall vendor policy is very important because of a lack of widely accepted standards to ensure compatibility. Furthermore, vendors become partners in the network design process by offering technical solution options and a migration path to accommodate future needs.

One option is to adopt a single primary vendor for both telecommunications and computing. Other vendors' products may be used only if they are fully compatible with the primary supplier's standards. In practice, this means using IBM or the merged AT&T/NCR combination. Historically, IBM was the only real choice for a single vendor to supply telecommunications and computing systems for a company, but many companies that adopted the IBM architecture were not completely satisfied with the telecommunications network side of the architecture. In its early form, IBM's SNA (discussed in Chapter 7) was rigid, cumbersome, and tied to a mainframe computer at the center of a star configuration. Recently, SNA has been made more flexible with peer-to-peer distributed processing capability, and more open systems connectivity is available. IBM purchased a controlling interest in Rolm to be able to integrate voice communications with its SNA computer networking capability; unfortunately, the marriage never worked effectively, and IBM sold its interest in Rolm.

Clearly, AT&T has been the dominant vendor for voice communications, but it has not been able to supply a company's complete computing needs. Its merger with NCR signals a desire to become a primary vendor for both telecommunications and computing, but a complete architecture similar to SNA has yet to be developed.

The principal argument for a single supplier is the long-term strategic importance of integrating all aspects of the information technologies; the arguments against it are that no one firm, including IBM, combines leadership and the best products in all areas, and that there is some risk in being tied to a single supplier. Furthermore, as we move closer to worldwide standards and compatibility among different vendors' equipment, the case for a single vendor policy is less compelling.

Another approach is to have a primary vendor for telecommunications and another primary vendor for computing. This used to mean AT&T plus IBM. AT&T, like IBM, has some competitive advantages: staying power and standards. AT&T has the financial clout to fund the huge investments in product development that will be essential for the rest of this century and beyond, and it owns many of the key telecommunications transmission and switching standards. Deregulation and the breakup of the Bell System has altered this entirely; now the decision involves the choice of a long-distance carrier for telecommunications, and in some cases a choice between the local RBOC or a local bypass carrier for local service. MCI, Sprint, and newer companies like WilTel have invested heavily in long-distance transmission equipment and subscribe to open standards. Although AT&T's market share is still the largest by far, it is shrinking, and there is now a viable choice among several primary telecommunications vendors.

Interestingly, although the dominant group of long-distance carriers is capable of supplying all wide area communications needs and can interface with any computing vendor's equipment, only AT&T provides a LAN system, the fastest growing segment of the data communications market. AT&T's Starlan product is not a market leader, however, and therefore it is difficult as a practical matter to adopt a single primary vendor for all telecommunications requirements.

Because no one company has solved all the problems of meeting a large company's telecommunications requirements (let alone those involved in integrating communications and computing), there are strong arguments against using any single primary vendor. However, defining the criteria for using multiple vendors is much more difficult than sticking with a few established leading suppliers. There is the obvious problem of incompatibility among the various vendors. Despite claims of adequate protocol conversion, IBM-compatibility, full integration, and adherence to standards, often the pieces still do not fit together.

Any multivendor plan must be built around standards of compatibility. The policy may require using only vendors that can provide full compatibility with IBM's SNA or OSI standards. Digital Equipment Corp., Tandem, Amdahl, Northern Telecom, AT&T/NCR, and Wang are just a few of the communications and computer suppliers that have products that meet this criterion. In fact, it is possible for a firm to adopt IBM's architectures without having to use any of its products, by combining (for instance) Amdahl mainframes, Tandem switches, Compaq personal computers, and MCI's virtual private network.

In addition to technical merit and price, two other key criteria for selecting vendors are increasing in importance: service and support. Telecommunications products become part of a utility service; when there are problems, no matter how minor, we want repairs and a return of service immediately. As more telecommunications products become commodities, support and service are the competitive differentiators, not price.

Rapidly changing technology and the changing support needs of a corporation require a much closer working relationship between supplier and customer to ensure long-term success in meeting business goals. It no longer is a question of simply buying products off the shelf; it is a matter of building mutual understanding and joint development efforts. In fact, many companies adopting a total quality management philosophy are eliminating large numbers of suppliers selected on short-term criteria (dominated by price) and are entering into long-term relationships with only a few vendors to work as partners in designing and implementing networks. The criteria for selecting long-term vendor partners should include staying power, a strong balance sheet, funds allocated to research and development, proven technology, trust, and compatible organizational cultures.

Unfortunately, general policy prescriptions do not always apply to many advanced products in specialized areas supplied by firms that are newer and smaller than the established giants. Vendor policy must be shaped to cover these situations as well.

The absence of an overall vendor policy—laissez faire—is a surprisingly common practice, especially when local divisions have substantial autonomy over purchase decisions. Quite often, a corporate telecommunications group defines equipment standards, but these decisions are not backed up by any degree of authority. In some instances the central recommendations conflict with local needs, the scale of operations, existing facilities, or (in the international arena) with regulation and government telecommunications monopolies.

Once the content of the policy is established, the central issue is who decides. If authority is ambiguous, the result is often political conflict between a "bureaucratic" central telecommunications staff unit and the field staff. Responsibility without authority has too often been the burden of the beleaguered telecommunications manager. The traffic on the corporate network is increasingly decentralized in terms of decision making, development, and operations; it should not be recentralized under a monolithic communications agency. Nor, though, can telecommunications be left to laissez faire planning with no vendor policy.

The main issue for the policy agenda is to ensure consistency. If the policy establishes strong and tight criteria (such as reliance on a primary vendor), there must be a central coordinator who oversees conformance with the policy and resolves the many exceptions that will inevitably arise. If, on the other hand, the policy is a loose one, there is less need for ensuring conformance with policy.

Telecommunications Platform Architecture

Another major policy issue is the choice of a telecommunications architecture for the platform. Conceptually, it would be sensible to decide first on the architecture and then on the vendor policy. If there were a well-established set of universally adopted standards, that would be practical because the choice of equipment and services would be relatively vendor-independent. Managers could shop around for the products that offered the best price or performance.

At present, specific architectures are determined more by vendors than by general standards. Even though the Open System Interconnection (OSI) and ISDN standards have been adopted by many vendors, there are differences in the details of implementation of these standards, which means that equipment made by different vendors may not work together even if technically they conform to OSI or ISDN standards. Also, ISDN is being passed over by many users in favor of waiting for the higher capacities available through the Broadband ISDN (B-ISDN) standard still under development.

Standards are emerging, and major vendors are conforming to them, which will eventually provide the integration of the many islands of information technology that now exist. That defines the goal; the starting point, however, generally means multiple facilities and protocols. This leads to the following main architectural policy issues:

> ▸ Do we commit to a single overall architecture?

> ▸ If so, what are the timetable and major phases for integrating the existing incompatible components now in use?

> ▸ Do we need different architectures for particular applications, countries, or business units? Why exactly?

> ▸ What range of services should be anticipated in the architecture? ("Anticipated" is used instead of "included" because the architecture is a blueprint for developing an increasingly integrated resource over changes in technology, locations, users, and applications.)

The business vision and application opportunities should determine the policy for the architecture and the services it should be able to accommodate, not the reverse. The overall issue is what level of integration is the firm's goal.

A financial services organization, such as Citibank or American Express, must push for more rather than less integration because its business vision rests on delivering a growing range of services through the same electronic delivery base to each workstation. Many manufacturing firms do not need this. They may, for example, want to operate special high-speed networks for computer-aided design, videoconferencing, electronic mail, or dealer transactions and inquiries. Each of these is very different in terms of length of messages, requirements for speed of transmission and response time, and trade-offs between cost and performance.

Business strategy guides the definition of the architecture. To effectively blend business strategy with the telecommunications architecture, the following questions must be answered:

> ▸ Are the plans for integration or for maintaining separate facilities explicit and consistent with our business vision and principal application opportunities?

> ▸ Is there a target architecture as the end point of our investments for the next five to ten years? What are the business and technical assumptions underlying our choice?

> ▸ Does senior management need to give more direction, clarify business issues, or make policy decisions so that a statement of the principles on which the architecture will be based can be developed?

These questions will be answered in the natural course of business activity if there is an ongoing dialogue between client and designer.

The Planning Horizon

The planning horizon for telecommunications is a key policy issue. Most senior executives want in a network innovation that leads to growth and competitive advantage, but also a rapid cost recovery of the investment. Unfortunately, these rarely go hand-in-hand. There is generally a three- to five-year planning horizon for intermediate range innovation in telecommunications, and the architecture must be designed to allow evolution over a five- to ten-year horizon.

However, numerous business and technical uncertainties generally limit specific technical plans to three years or less.

Investment payback and tangible bottom-line results are demanded over an even shorter period. This mismatch between the planning horizon needed for innovation and that required for cost recovery and benefit payoff simply will not work. Building or substantially expanding the communications infrastructure as part of an aggressive business strategy requires spending money now to get benefits later. The budgeting process for telecommunications in most firms means that the planning horizon is too short to make innovation possible.

Ford Motor Company's World-wide Engineering Release System network is a rare exception. It involved an investment commitment of over $70 million, with a possible financial break-even point occurring in 15 years. This represents a long-term commitment to the corporate vision of building a "world car" and the resolve to construct a worldwide network to support the vision without the expectation of a short-term financial payoff.

A one-year planning horizon is too short to do more than make tactical investments in telecommunications. For instance, many LAN projects, upgrades of PCs, or leasing of additional T1 capacity have short lead times. Designing and implementing a comprehensive platform or developing major software applications takes two to five years. Unless a firm's planning horizon focuses on the three- to five-year horizon, it is extremely unlikely that it will fund key platform infrastructures.

Level of Risk

The last main area of policy in which management must make sure it has an explicit position is the question of how much risk to accept. Several different facets of risk need to be addressed:

▸   *Technological risk*: Do our application opportunities, vendor policy, criteria for architecture, and planning horizon commit us to serious risk that the technology will not work as planned, will be delayed, or will involve unpredictable and substantial extra costs? Are the potential gains from using cutting-edge technology worth the associated risk, or are we better off with proven technology and more modest expected returns from its application? The policy should involve an explicit risk/reward trade-off, and it is senior management's call. The role of the designer is to make sure senior management has the proper information on risk and reward to make the call.

▸   *Financial risk*: Can we limit the cost involved, even if only approximately—that is, not less than $X million and not more than $Y million over Z years? A true ceiling on costs avoids unpleasant surprises when actual costs deviate significantly from a single estimate that was used to justify the network during the budget review process.

▸   *Organizational risk*: Can we handle the organizational and cultural changes with our existing management and technical team? How much change will be required in the way we work on a day-to-day basis to take full advantage of the network? What level of training and support and how much time are needed to undertake the required organizational changes? Again, senior management must often accept a trade-off between the degree and speed of technological change and the extent of accompanying organizational change needed to fully benefit from the power of the network.

▸   *Business risk*: How confident are we that customers will respond positively to our application objectives? How long will it be before we have a clear idea of the profitability of the innovation? In telecommunications, as with most things in life, high return and high risk generally go together. Innovation involves unavoidable gambles, and the edge generally comes from doing it sooner than someone else. Senior management must set policy regarding the degree of business risk it is willing to accept in its quest to gain a competitive advantage through information technology.

The intersection of these various policy choices—subject to limitations imposed by regulation, available technology, and accepted industry practices—defines the working domain of the network designer. Key policy choices provide guidance in the trade-offs that must be made in choosing and implementing telecommunications technology to meet the company's goals and objectives in both the short and longer term. Without clear-cut policy guidance in these areas, there is a reasonably good chance that the network design will fall short of expected effectiveness, or cost more than management is willing to spend, or generate unacceptable levels of organizational change. The client and the designer must be on the same wavelength, and the client must set policy to guide the design effort.

Improved Network Design Through Total Quality Management

Total Quality Management (TQM) is increasingly being recognized as a thought revolution in management whose goal is complete satisfaction of the customer. This is attained through continuous improvement of all processes in the organization based on decisions arrived at analytically. It also involves viewing the organization as a true system with many interdependencies among its various departments and units.

TQM was developed during the 1950s primarily by two Americans, Dr. W. Edwards Deming and Dr. Joseph Juran. However, Americans were not interested in this new management thinking approach during the post–World War II boom period. Deming and Juran were invited to Japan in the early 1950s to help Japanese industries recover from the devastation of the war and to overcome their reputation for making poor-quality products. Today, the Japanese have a worldwide reputation for very high-quality products, and their total quality management methods are being adopted by companies in other countries.

Today, many U.S. companies are looking to TQM as a quick fix or as a prescription for healing a sick "bottom line." The reality is that the successful implementation of TQM generally requires a radically different approach to managing the organization that involves a cultural transformation and a new set of values. Most U.S. companies are focused on a "quarter-to-quarter" earnings mentality, and Wall Street is the primary customer. There is no loyalty to employees, who are there merely to serve the interests of top management and the shareholders.

In contrast, a true TQM company (as exemplified by some Japanese companies and a few exceptional companies in other parts of the world) has as its primary goal the satisfaction of the customer (and, as a close second, job growth, continuity, and fulfillment). If these two goals are sought correctly, profits will follow as a by-product, which naturally leads to a longer-term view of the organization and to such values as respect for employees, which drives fear out of the workplace and stimulates continuous improvement.

A true systems view of an organization leads to a focus on the processes that generate the final results, rather than on the results themselves. From a systems perspective, improvement of the processes will inevitably lead to better results because this directs attention to the causal factors responsible for generating the results and takes into account the many interdependencies that inextricably link the various parts of the organization. In contrast, a focus directly on results often leads to weaker overall performance because individual units of the organization will work to optimize their own performance in isolation, which can lead to suboptimization of the whole by ignoring the interdependencies across units.

Although there are differences in the various TQM systems and in the methods proposed by various consultants and gurus, there are some key elements of commonality. Almost all TQM approaches include the following elements to a greater or lesser degree:

▸ A holistic, systems view of the organization

▸ A continuous process-improvement focus that is based on a recognition that variation in processes and outcomes is pervasive, and that statistical thinking and methods are required to separate systematic variation from random variation

▸ Management by fact, rather than by opinion and intuitive feel

▸ An obsession with customer satisfaction

▸ A true respect for people and a strong commitment to invest in employee development and fulfillment through training and supportive management practices

With this background, we turn to the real issue of what TQM can contribute to network design. Its primary contribution is the concept of building quality of network performance and the capability to build customers' requirements into the design itself, rather than ensuring quality through extensive inspection and alteration after the network is constructed and operating. Evidence gathered from companies practicing TQM has shown that it is much less costly to eliminate failures and defects by altering the design of a production or service process than it is to correct problems after they are created. For example, IBM conducted a study of the costs of correcting defects in its computer systems after the product and its manufacturing process is designed, as compared with eliminating the defects through better design of the product and production process itself. The results showed that it was 45 percent more costly to correct a defect after production but before shipping a product to customers, and 85 percent more costly to correct the problem after delivery to customers, as compared with improving the process upstream in the design phase.

The conclusion is simply that overall cost will be minimized and quality of service enhanced if customer requirements for network features, performance, and reliability are incorporated into the design itself, rather than trying to manage and control these factors after the network is designed. TQM uses a structured method to transform customer requirements into design features and engineering characteristics; it is called quality function deployment (QFD).

QFD is accomplished through a mapping process that systematically links customers' requirements directly to product/service features and engineering characteristics using a matrix system that tracks interrelationships and trade-offs. This system, nicknamed the "House of Quality," is shown in Figure 15-1.

The left side of the matrix contains customer requirements for network performance obtained through surveys, focus groups, and observation of people actually using network services. The QFD matrix in Figure 15-1 includes customers' requirements related to only one aspect of the network—availability—to keep the illustration simple. Notice that the customers' requirements are stated in terms that are meaningful to the network user, but they do not contain any technical design information to operationalize these requirements.

On the other hand, the columns of the House of Quality matrix contain specific service and engineering characteristics that operationalize the user requirements. For example, if users require a three-second response time to access the network, this is achieved by specifying appropriate technical parameters for the network: the extent of circuit utilization, the degree of buffer utilization, the lengths of queues, and data-processing cycle time. These are shown in the matrix as positively related to user response time.

The symbols in the matrix cells indicate the strength and direction (positive or negative) of the relationships between customer requirement variables (the rows) and engineering/service characteristics (the columns). Note that, although the customer requirements are stated as specific target levels (for instance, 3-second average response time), the symbols in the matrix cells refer to relationships with the underlying variable itself (in this case, average response time).

The roof of the House of Quality is used to designate the interrelationships and trade-offs among the various engineering or service characteristics shown in the columns of the matrix. In Figure 15-1, circuit utilization is positively related to buffer utilization and queue length. Thus an increase in circuit utilization resulting from heavy traffic on the network will result in increased buffer use and longer queues for a given data processing cycle time. Processing time is

Figure 15–1

The "House of Quality" matrix

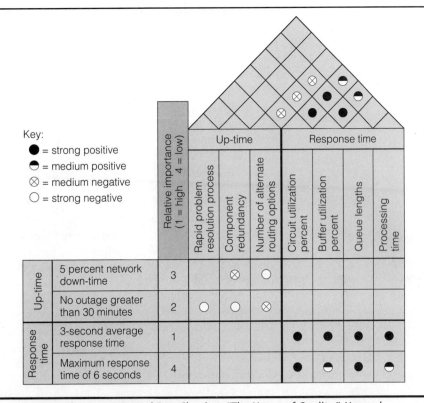

Source: John R. Hauser and Don Clausing, "The House of Quality," *Harvard Business Review*, May–June 1988.

also positively related to buffer utilization and queue length. Consequently, reduction of data-processing cycle time will also tend to decrease queue lengths and buffer utilization because the data is processed and distributed in less time.

Using the QFD process in network design ensures that the complex interrelationships among the full range of customer requirements and engineering/service characteristics embodied in the four primary design variables (capability, flexibility, service quality, and pricing/cost recovery) are considered and that appropriate trade-offs are made. This approach also permits key policy variables, such as technical risk, overall architecture choices, and preferred vendors, to be explicitly included in the analysis of engineering alternatives (by identifying which engineering features are consistent with which policies) and allows us to see where some of these policy choices may conflict with the ability to satisfy customer requirements at minimum cost.

The alternative to a comprehensive QFD type approach to network design is a much narrower focus on optimizing the network to meet a narrowly defined performance criterion, such as the projected level of busy hour traffic with a specified average response time. This approach generally does not consider the full range of customer requirements, particularly the ones that are more difficult to quantify (such as policy issues), and it does not examine the full range of engineering and service feature alternatives and the trade-offs among them. Network design software programs are available to select a least-cost configuration, but the explicit trade-offs among the design variables are masked by the program, and much of the richness of information visually available through a QFD matrix is lost to the designer.

QFD imposes a structured discipline on the network design process, and it forces the designer to consider all relevant trade-offs and the full range of customer requirements. This approach does necessitate a greater up-front investment in time and effort, but if done properly it will reduce the overall cost of network operations because it eliminates alterations and redesign of the network after it is placed in operation.

Summary

Network design involves many complex processes. This chapter has addressed the process of managing design and ensuring that the detailed technical simulations, sizing, and traffic analyses are carried out in a context of business needs and priorities. According to the network design variables framework, which defines four categories of variables: (1) capability—the types of service and the volumes of traffic the network provides; (2) flexibility—ease of change and extension of services and locations; (3) quality of service—availability, reliability, response time, and security; and (4) cost.

Trade-offs between, say, flexibility and cost or between security and capability must be made in the context of the business criteria developed through the earlier steps in the telecommunications decision sequence, most particularly the telecommunications services platform map, which identifies the degree of importance of platform features. In addition, policy decisions constrain design choices. An obvious example is that if security is required as a policy, there is no longer any option for the designer to trade it off against, say, cost.

Policy issues that must be explicitly addressed include (1) the degree and pace of technical and organizational change, (2) choice of vendors, (3) platform architecture, (4) planning horizon, and (5) level of risk.

The purpose of any network is to provide service; quality of service must thus drive design. The total quality management approach to network design focuses on (1) a holistic and systemic view of the organization for which and in which the network is to provide quality service, (2) continuous improvement in every area of process and outputs, (3) management by fact and metrics, (4) an obsession with customer satisfaction, and (5) commitment to and respect for people.

TQM in network design can minimize cost and maximize service through such structured methods as quality function deployment, which systematically links customer requirements directly to product/service features and engineering requirements using a matrix of interrelationships and trade-offs.

Review Questions

1. Consider the following statement: The designer is responsible for the following in the network design process: (a) a high-level analysis of client preferences that determines the overall capabilities and limitations of the network, and (b) the detailed translation of these strategic requirements into specific circuit sizes and equipment. Do you agree or disagree? Explain.

2. What is necessary for the design process to create an effective telecommunications network?

3. Briefly describe what is meant by "capability" in a network design context.

4. What is wrong with a bottom-up design of the network based on technical opportunities and the availability of a given technology?

5. Describe the different perspectives of client, designer, and implementer on the concept of integration in telecommunications.

6. What is meant by the following statement: "Telecommunications network capacity supply creates its own demand"?

7. What does "flexibility" mean in a telecommunications network context, and how is it incorporated into the network through the use of public service offerings from carriers?

8. List and briefly describe the principal quality of service factors that are relevant to network design.

9. Explain the following statement: Two different networks with the same average outage or downtime do not necessarily have the same reliability.

10. Consider the following statement: A faster network response time is always better. Do you agree or disagree? Explain.

11. Describe a situation in which encryption alone would provide essentially no security.

12. Evaluate the following statement: The fully allocated cost of network services should be charged directly to users.

13. What do you need to know in order to answer the question "How reliable should our network service be?"

14. Isn't the essence of network design simply meeting given requirements at lowest cost? Explain your answer.

15. The cost of comparable networks in comparable firms can differ by a large amount. What are some of the factors that explain these cost differences?

16. Why is corporate "policy" an important factor in network design? Explain its role.

17. Consider the following statement: Limitations on the pace of change in information technology essentially determine the rate at which the organization can change. Do you agree or disagree? Explain.

18. Discuss the following statement: A company should decide what it wants, create the specifications, and then choose the vendor based on the proposed price, service, delivery, and product design features.

19. Briefly discuss the strengths and weaknesses of choosing a single primary vendor for both telecommunications and computing.

20. Discuss some ways to build a close working relationship between supplier and customer to ensure long-term success in meeting business goals.

21. Does the following statement reflect reality? "Specific telecommunications architectures *should* be determined by general standards and not by what particular vendors have to offer."

22. Make a list of the key architectural policy issues.

23. Explain the mismatch between the planning horizon needed for innovation and that required for cost recovery and benefit payoff.

24. Consider the following statement: If the level of technological risk is known and is within the limits set by corporate policy, then there is no further need to consider the risk factor in designing the network. Do you agree or disagree? Explain.

25. For most companies, the primary goal is to maximize profits or shareholder wealth, and the focus of management is on results. How does this differ from a company that adopts a total quality management philosophy?

26. List the key elements that are common to the various schools and approaches to TQM.

27. Sketch a simple 2×2 House of Quality matrix and explain the use of each element of the house: the rows, the columns, and the roof.

Assignments and Questions for Discussion

1. A large machine-tool manufacturer has just hired you as director of telecommunications for the company. In the initial review of the company's communication facilities, you discover that the company relies exclusively on dial-up service from the local RBOC and a chosen long-distance carrier. The company clearly needs a network design. Describe your role in the network design process, and outline the process steps in developing an overall network design for the company.

2. The range of costs for outwardly similar networks in comparable firms can vary by a large amount. Some of the differences relate to management skills and possibly economies of scale (substituting fixed-cost facilities for volume-sensitive ones) and to technical and operational efficiency. However, much of the cost difference is due to design trade-off decisions. Explain and discuss.

3. Company X has ignored many of the changes taking place in the information technology environment and holds steadfastly to its teleprocessing network consisting of dumb terminals linked through cluster controllers and connected to IBM mainframe computers with low-capacity dial-up and leased lines. Describe a set of corporate policy directives consistent with this situation.

4. The director of telecommunications for a large, multinational company said that "the best way to ensure a quality network design is to simply build the network, and then deal with problems and changes to the network as the need arises. There is entirely too much emphasis on trying to figure everything out ahead of time; just build it and work things out as you go along." Discuss the weaknesses of this approach, and briefly describe a way to avoid the problems that this approach creates.

Minicase 15-1

Bank of the West

Downsizing to LANs

Banks have relied mainly on mainframe computers in developing electronic services such as automated teller machines, electronic funds transfers, and foreign exchange trading. The reasons for this are the high volume of transactions involved, the need to centralize customer data and use powerful database management software, and issues of telecommunications management, security, and guaranteed availability. Only the most powerful mainframe computers could combine all these functions efficiently. Personal computers and LANs are increasingly used to distribute functions and front-end processing, such as error checking and message formatting. Many new applications—for budgeting, electronic mail, word processing, financial analysis, and management reporting—are built on LANs. However, the mainframe remains the core of most banks' information systems and telecommunications.

Bank of the West, headquartered in California, expanded rapidly in the late 1980s and early 1990s, acquiring another bank. It controlled costs tightly and kept its information systems staff lean. Expansion increased the need for new systems, however, and its mainframe computer was increasingly overloaded in handling the additional processing this created. Bank of the West decided to use LANs to remove reliance on the mainframe. It already had several LANs in place for departmental computing and communications. The acquired bank also had some in use, but these LANs were incompatible with the others.

The most immediate need was in the controller's department, which heavily used the mainframe system for budgeting, accounting, regulatory reporting, and management reporting. This use requires the department to integrate data from the mainframe into reports that go to a variety of people inside and outside the bank. The process was time-consuming, clumsy, and expensive. Staff downloaded data onto their stand-alone PCs, collected other data from many sources, and used PCs to organize the information and produce reports. A number of people might manipulate the data before it was finalized in a report; errors were frequent. Floppy disks were passed around, as were hard-copy reports. The same numbers were entered and reentered in different systems; information used in the central mainframe systems then had to be "massaged" into the format these required. It could take up to two days to schedule mainframe processing plus another day to get the outputs.

This is hardly a satisfactory way to manage in the 1990s. Bank of the West brought in a systems integrator, PC Edge, to create a LAN-based solution for an initial 12 users. The service would build on PC software already in use, including Lotus 1-2-3. The first problem PC Edge encountered was cabling. Bank of the West's wiring conduits were filled to capacity, and new cables could not be added. PC Edge recommended replacing the existing ARCNET cables with twisted-pair Ethernet products to ensure adequate capacity. The bank decided to standardize on 10Base-T, the standard that is an extension of and is compatible with Ethernet. The electrical contractor installed this in a few days; adding new PC hardware and software took around the same time. The PCs selected were 386-based, operating at 25 megahertz, with 1 mb of memory. The file server had 600 mb of disk storage. This adds up to a fast and powerful capability at low cost.

It is also a simple system. Several of the workstations have an IRMA board that lets the PC emulate an IBM 3270 terminal so that they can communicate directly with the mainframe, but the controller department's data now resides on the file server. Lotus 1-2-3 is the standard software for budgeting, analysis, and reporting. The LAN users also handle investment tracking, asset and liability management, and treasury operations. Branches submit their budget data on floppy disks; this data is stored on the file server. Only when all inputting and relevant analysis, verification, and approval are complete is the data uploaded to the mainframe general ledger system for regular budget variance reporting.

Turnaround time for developing and finalizing the budget has been halved, and the resulting figures are far more accurate and consistent. The strain on the mainframe has been greatly reduced; Bank of the West has not had to add capacity. It has outsourced many mainframe applications, such as payroll. It continues to add LANs across the bank, with 10Base-T its standard. Many software packages are now available for banking applications at low cost. The LAN file server ensures that these share a common database and can also share such devices as high-speed color laser printers. In 1993, Bank of the West was to create an internetwork for the departmental LANs. There is no LAN-to-LAN link at present.

Questions for Discussion

1. Describe the key issues and problems that cause the Bank of the West to bring in a systems integrator to redesign its system.

2. Evaluate the LAN-based solution developed for the bank in terms of the four design variables for networks (capability, flexibility, quality of service, and cost/price), compared with the mainframe system currently in place.

3. Indicate which of the five major policy issues (degree of change, choice of vendor, telecommunications architecture, planning horizon, risk)—if any—had a significant impact on the new network design, and discuss the role played by the relevant policy factors on the new design.

Minicase 15-2

WHDH-TV

Getting the News Out

WHDH-TV, CBS's Boston affiliate, is probably the most technologically advanced television company in the world, which helps explain why it has moved from a poor third in its market in the 1970s to a strong contender for first place. In the spring of 1992, it introduced PCs linked by a LAN into its newsroom. It took the 50 reporters and news staff just a week to master the technology, with dramatic effects. According to the producer of the 5 o'clock news, the installation of the LAN was "like I suddenly got my arm out of a cast and sling. Before we had the PCs and the LANs, it was a real headache to change the format of a story during a live broadcast. . . . [Now] we can make all changes on-line, electronically—we get exact timing and fewer mistakes." An anchorman says, "I love access to the wire service. I can sit down without having to walk anyway or interrupt what I'm doing. . . . It amazes me how many stories I can crank out with my computer and tape deck side by side."

Tape libraries are accessible on-line from the LAN; footage in libraries can be viewed instantly. Cameras are controlled by the network in real-time, too. All editing is done on-line and stored immediately—and hence is immediately accessible by another PC. The producer of a newscast can even read a story on-line as the reporter is writing it while also monitoring wire services and police, fire, and weather-emergency scanners.

The network evolved in two phases. The first was the installation of nine file servers connected by thick-wire Ethernet cable via Ethernet concentrators. The network operating systems were Novell's NetWare 286 and NetWare 386. These networks linked newsroom staff and around 150 employees in the sales, accounting, promotion, and research departments.

These LANs were upgraded and consolidated into three LANs using file servers built on Intel 486-based hardware. The LANs are linked by the NetWare 386 network operating system. The new system adds a far higher level of security; the mixed 286 and 386 environment made it impossible to encrypt passwords on the 286 personal computers. Security is a priority for WHDH: According to the station's network administrator, "We view our networks the same way ordinary people view electricity, the telephone, or their cars. That is, they're just there, you don't even think about them until something goes wrong." Hackers and computer viruses are a major concern for WHDH-TV, given its visibility and responsibility as a news provider up to five times a day in real-time.

The choice of concentrator equipment was made largely because of its ease of use and flexibility. The Ethernet Series 4000 concentrators are rack-mountable; thus, in the event of disaster, a reporter or janitor can be "walked through" making an adapter swap over the phone. The Series 4000 is also very fast, with a throughput of up to 500 mbps via multiple Ethernet segments. Each concentrator has five such segments. This speed is needed to control cameras electronically (electronic robots move them during a broadcast) and to load video-cassettes from the table library.

The value of the LANs in improving news operations is very clear. In addition, there have been many economic benefits for the station. The accounting department has been halved in size while adding accounting functions for WHDH's radio stations. The LANs have contributed to this improvement in productivity by reducing paper and administration. New England has been in a very severe recession, perhaps even a depression, and the broadcasting industry is in a tight cost squeeze, with margins plunging. One manager commented that the station cannot afford to lose even a single commercial.

The telecommunications team is already planning the next generation of technology for the station. It aims at implementing the "open-slot" concept. Here, reporters edit their own videotape at a PC (this is practical now on both Macintosh and multimedia-based PCs, including MS.DOS, Windows, and OS/2-based systems). The station is experimenting with software that lets news staffers automatically insert special instructions for individual news stories directly onto the TelePrompTer. "Our dream in a few years is that a reporter will be able to walk into a newsroom with a story on his notebook computer and attack it to the network via a wireless LAN. So there's plenty of stuff on the drawing board to keep us excited."

The network administrator adds that "we run the business on our network—the entire business. Ed [his boss] and I used to joke that we would stage a 'network appreciation day' for ourselves by unplugging the file servers and going home sick just to underscore everything that we and the networks do around here—because the people just have no concept of the scope of what we do. But we can't afford to do that because if we shut off the networks, the station would be off the air. Our LANs are up and running 24 hours a day, seven days a week."

Questions for Discussion

1. Briefly describe the LAN-based network used by the WHDH-TV news department, and identify the key benefits of the system.

2. Describe the importance of security to WHDH-TV, and evaluate the security aspects of the quality of service features incorporated into the LAN-based network.

3. Describe the next generation of technology planned for the station, and explain how the "open slot" concept will change both the way a video news story is developed and the role of support personnel in the process.

Minicase 15-3

The National Retail Federation's 1992 Convention

Future Technology for Stores

When new information technology products are first announced and demonstrated, it typically takes around five years before they are widely implemented across an industry—and, of course, some of the most promising ones turn out to be duds, because of design flaws, overpricing, poor support, or customers' lack of interest.

Industry trade shows are generally a showcase for new products and new technology. A 1992 trade show such as The National Retail Foundation's (NRF) convention gives a good sense of the standard technology we will see in use in shops in the mid- to late-1990s. Not all the products shown will be successes, and given the pace of the technological innovation, we can expect each new convention to include new breakthroughs. That said, here is where the mainstream trends in retailing are moving.

Many of the products shown at the NRF convention use wireless technology. Several companies showed new radio frequency devices that offer stores more options in choosing mobile equipment that can communicate within the store, replacing hard-wired terminals and eliminating the cost of cabling. Portable point-of-sale (POS) registers allow salesclerks to process a sale anywhere in the store, including making credit card authorization and printing out a receipt. One company, Symbol Technologies, offers a two-pound scanner RF terminal that can update inventory information and send data via the firm's Spectrum One RF network. It has published a 30-page booklet entitled *The Wireless Store, the Strategic Environment for Retailing's Future.*

Telxon showed a POS terminal with touchscreen menu display that can accept a signature electronically for credit sales. It has an add-on scanner and can communicate over narrow-band and spreadspectrum networks. Many vendors offered portable scanners, including ones that can be positioned in many ways, can even be activated by movement, and do not require being held and precisely aimed.

Several vendors demonstrated interactive kiosks. Kmart tested one marketed by IBM; it wanted to see if the kiosks improve customer service and help boost sales. Lechmere showed one it is using at its Express Self-Service Centers: Customers order and pay for goods at the kiosk and get them at a pick-up desk. Montgomery Ward has tested credit kiosks, where shoppers can get credit information and make payments.

Siemens Nixdorf and IBM were demonstrating the retailers' kiosks. IBM has invested heavily in making kiosks easy to use and is the provider of the Info/California kiosks described in Minicase 4.2. It sees as its strength in what it expects will be a massive long-term market its research on human factors and ergonomics in multimedia applications.

Siemens Nixdorf is betting on its high-quality engineering and reliability.

Other innovations included POS systems that can communicate in an open environment. HP announced that its UNIX POS devices can now support IBM's token-ring standard. Security Tag Systems showed anti-shoplifting tags that break apart and spread two dyes if forcibly removed. 3M has shrunk its security tag to .6 by .9 inches, the smallest on the market. Both types of tag can be printed in a store and scanned.

A major announcement at the NRF convention was SpecNet, a service sponsored by NRF for specialty stores. The idea had been created in 1990 by a small group of specialty stores that face large competitors that have the economies of scale to justify on-line systems within their stores and large-scale private networks linking them to its head office and supply centers. NRF agreed to sponsor the development of a consortium that would have the same buying power as the large retailers, provided that all users of SpecNet also became members of NRF.

SpecNet began with shared voice communications because this presents no technical challenge and is a simple matter of negotiating voice discounts. The plan announced at the 1992 convention was an agreement with MCI that provides up to 30 percent discounts for a range of services, including 800 numbers, virtual private networks, and private networks. The initial 36 specialty store chains include Barnes and Noble, The Gap, Stride Rite, Crate & Barrel, and Waldenbooks.

The next step, which is expected to take up to five years, is to extend the simple voice service to a shared network for data communication, including POS, credit authorization, and video. The technology for this may be VSAT. Some NRF members feel that 1997 is too long to wait; one is heading a separate consortium, formed in 1987, called CARDS (Cooperative Access Retail Data System). CARDS planned to negotiate an agreement with a VSAT network provider, either Hughes or AT&T

Tridom, by the end of 1992. A CARDS spokesman said that "if and when SpecNet rolls out a data network, we may merge ours into theirs. But then again, we may not. There may be differences between the groups." CARDS is concerned about NRF's control of SpecNet decisions, including choices of consultants and requiring fees and membership in NRF. The spokesman also pointed to the hurdles SpecNet faces in handling the administration of multiple independent companies.

Muzak announced at the NRF 1992 convention its ZTV Retail Channel, a combination of music videos, fashion, sports, cartoons, and "lifestyle shorts." It will provide five broadcast programs, one for each of five separate demographic groups.

So the store of 1997 may well be one of wireless POS registers, self-service kiosks, MTV overhead, a far smaller number of staff, and first-rate data communications.

Questions for Discussion

1. According to the National Retail Federation's convention, one of the major trends in retailing is expected to be the "wireless store." To what extent and in what manner does the "wireless store" concept affect each of the four key network design variables (capability, flexibility, quality of service, and cost/price) and the trade-offs among them?

2. Discuss the impact of self-service kiosks on the process of retail sales and customer service. Specifically, indicate how the steps in the sales and customer service processes, including the role of personnel and merchandise display, are likely to change as a result of the use of kiosks.

3. Briefly explain the SpecNet concept, and discuss the major hurdles to successful implementation within the context of the major policy issues influencing network design: (a) the acceptable degree of change; (b) choice of vendor; (c) telecommunications architecture; (d) the planning horizon; and (e) acceptable risk.

Minicase 15-4

Southwestern Bell

Designing a 125-Mile Fiber-optic Single-building Network

When Southwestern Bell Corporation (SBC) moved into its new headquarters in 1985, it took the opportunity to establish IT standards from scratch and to develop a 125-mile fiber network in a single building so that the wiring will not need to be replaced or moved as new technology and applications emerge. It set up a steering committee of managers with IT expertise to establish priorities and recommend standards. For instance, the committee decided that DOS- and IBM-compatible machines should be the workstation base, IBM PC LAN the networking software, and Ethernet the underlying protocol. Ungermann-Bass was chosen to provide the LAN backbone because of its proven ability to support several alternative network architectures and for its support of IBM 3270 terminals in a PC network.

Each work area in the One Bell Center headquarters building in St. Louis was connected by a four-fiber cable to one of two wiring closets located at a corner of each floor. There could be as many as 21 four-fiber cables attached to a closet. A fiber-optic modem translates fiber-optic light signals into Ethernet electrical voltages. The modem is connected to multiplexers that can each handle eight PCs.

SBC's mainframe computers use IBM SNA and 3270 terminals. Ungermann-Bass provides the complex hardware and software to connect PCs on the LAN to the hosts, using 3270 emulation where needed. Files are routinely uploaded and downloaded between the PC and the mainframe. ISDN gateways connect to the telephone company's 550-node Ethernet LAN and into its wide area packet-switched network. A permanent network monitor handles trouble-shooting, diagnosis, and security. There are 320 network stations on the LAN. Only a small fraction of the available bandwidth is used, around 10 percent, but that is expected to increase rapidly as image processing, videoconferencing, and other high-bandwidth applications are added. Network support, hardware maintenance, equipment installation, and user education are provided by outside contractors; SBC has only two specialists on its staff to support the LAN users. The DOS software and IBM LAN are robust, and few problems have been reported.

SBC regards four factors as having made the installation and operation of the LAN a success:

1. A reliable, flexible wiring medium

2. Access by users to trained outside specialists, by phone or by pager

3. The SBC steering committee, which encourages and coordinates connection of compatible systems to the network

4. "Undoubtedly the most important"—the wide variety of available application software packages that take advantage of the net-

worked PCs, including standard DOS packages, print servers, and graphics transfers.

The network will evolve to the client/server model. SBC has implemented OS/2 on PCs and is migrating to OSI and TCP/IP. A client/server network will greatly reduce disk input/output across the network and will make it practical to access data on distributed relational databases within and outside the LAN, opening up many new sources of on-line information for customer service, operations, and internal reporting. SBC has discouraged multivendor hardware on the network, imposing tight systems standards in order to keep costs down.

Questions for Discussion

1. Explain the role of choosing standards in the design of SBC's single-building network.

2. Two of the four factors that SBC regards as having made the installation and operation of the LAN a success do not relate directly to the technology (hardware, software, protocols, and so on) itself. Explain both the role of these two factors in the network design process and how they relate to one or more of the four network design variables (capability, flexibility, quality of service, cost/price).

3. Assess the positive and negative aspects of SBC's decision to outsource network support, hardware maintenance, equipment installation, and user education.

16

Total Quality Operations

Objectives

- Know the scope of telecommunications operations responsibilities
- Understand the designer's role in managing network operations
- Identify the key quality-of-service issues for network operations
- Show how total quality management (TQM) methods are used to manage network operations

Chapter Overview

| | | | | | | | ## Chapter Overview

This chapter begins with a brief description of the scope of telecommunications operations responsibilities and continues with a discussion of the role of the designer in managing network operations. Most treatments of network operations focus on the role of the implementer, not the designer, and little emphasis is placed on linking customers' requirements directly to management of network operations. Business has entered an era of unprecedented change, marked by global competition, that has significantly raised the standard of performance and quality of service necessary to remain competitive. These new standards of quality and performance have been extended to all parts of the company, including telecommunications. Thus increasing emphasis is being placed on managing network operations, and the standards for acceptable performance have increased by an order of magnitude.

Following this introduction, we turn to a discussion of the key quality-of-service elements that are the focus of network operations: network availability, reliability, security, and accessibility. **Availability** consists of two components: the amount of time the network is operational and response time. **Reliability** is a measure of both the durability of the network and its components and freedom from transmission errors. **Security** includes protection of information sent over the network and assurance that only authorized individuals have access. Finally, accessibility refers to ease of access to the network over a range of conditions and locations.

The last part of the chapter describes how a total quality management system can be used to manage these quality-of-service factors effectively. First, approaches to discerning customers' requirements are described; then we introduce a method for linking customer requirements to key telecommunications operational accountabilities. This topic is followed by discussions of the role of teams in managing operations and the development of a daily management system that drives continuous improvement of the aspects of network operations that are most important to customers. Finally, the chapter concludes with a discussion of the importance of building long-term relationships with vendors and of comparing key operational processes with other organizations.

| | | | | | | | ## The Scope of Telecommunications Operations Management

A large firm's telecommunications operations are extraordinarily complex to manage, roughly equivalent to managing one's own electric utility and telephone company. The scope of network operations management encompasses a wide array of activities that are necessary to avoid business failure resulting from failure or inadequate performance of the network. Network operations management activities can be classified into four main areas:

▶ *Telecommunications Operations and Control Centers (TOCC)*: Every network facility needs a control and command center that oversees the operation of the entire system (often including multiple, nonoverlapping networks). This is the key decision-making unit at peak traffic times.

▶ *Software and Equipment Monitoring and Testing*: Real-time operations make it essential to be able to monitor what is happening anywhere across the entire system for the purpose of diagnosing faults and troubleshooting problems. These faults and trouble spots can be related to either hardware or software or a combination of both.

▶ *Backup*: If any major part of the system fails, there must be a provision for backup (redundancy and alternate routing) and restarting.

▶ *Installation and Field Support*: Network systems must grow and change to meet changing user requirements; this requires installation of new equipment, new software releases, reconfiguration of network elements, and so on. Some of this function can be handled through the TOCC, whereas the remainder requires installation in the field. When users experience problems or when a piece of equipment in a remote location fails, field support personnel must respond quickly.

Well-managed network operations are critical to business success in today's hypercompetitive, global business environment; a clear telecommunications policy and a well-designed network architecture become irrelevant if network operations fail. One of the key decisions concerning the appropriate scope of a firm's network operations is the extent to which these activities are managed internally (versus outsourcing them or relying on public network facilities). The following points should be considered in this decision:

▶ If you can't manage it effectively, don't build it.

▶ If you build it, allocate sufficient resources to manage it well.

▶ Provide 24-hour guaranteed service and immediate field assistance.

There are advantages to relying on a public network, or to outsourcing this activity altogether and having someone else manage the company's network operations. A number of firms are choosing the outsourcing route, usually because of cost savings and the argument that "our business is not telecommunications."

However, there are corresponding disadvantages to outsourcing and relying on a public network; concern about security and loss of control are two of the most important considerations. With a public network, a company cannot be sure that traffic from other users will not degrade its own service, and the company must still acquire and operate a lot of equipment on its own premises. Also, managing security and controlling access is much more difficult when using a public network.

With outsourcing, concerns include the facts that the vendor managing network operations may not be as committed as the company to high levels of service quality, and that the vendor is not as likely to strategically innovate and strive to meet changing customer requirements. Furthermore, the economic disadvantage of outsourcing depends on the vendor's ability to realize substantial economies of scale and/or to more efficiently manage network operations than a firm can do on its own. The outsourcing vendor must recover its cost and earn a profit, whereas a firm's telecommunications department is not required to earn a profit on its internal operations. The counterargument is that the vendor or systems integrator can focus effort and attention on network management and build specialized expertise.

There is a growing body of literature on reasons for outsourcing, reasons for not outsourcing, and on successes and failures. There is no clear consensus on principles for going outside versus keeping functions in-house. That said, case studies have revealed one key management issue that may seem surprising: Even though outsourcing requires fewer in-house staff, such staff must be better than those needed for in-house operations alone. This surprises many business executives who hope that outsourcing will remove any need for complex planning and decision making; they hope to dump the problems on someone else.

In practice, a small group of internal planners must focus on priorities, policies, relationships with the outside firm, contracting, and quality assurance. The issues related to a firm's business directives—the focus of the telecommunications decision sequence—cannot be outsourced. The most effective outsourcing thus aims at focusing the internal group's skills, attention, and time on business goals while ensuring that the outside unit focuses its skills, attention, and time on providing first-rate operations for its client.

The role of the designer in telecommunications operations is to develop and implement a coherent network operations strategy that focuses on managing quality-of-service issues. Few firms have such a coherent strategy; instead they generally rely on vendor support and ad hoc servicing and repairs. This is unacceptable in a multivendor, multitechnology environment in which the firm's entire cash flow is on-line.

The industry is moving rapidly toward automated network management, with products such as IBM's Netview and AT&T's Unified Network Management Architecture (UNMA) systems as key elements of the strategy. However, these and similar products are not comprehensive enough to permit complete automation of operations. Moreover, these systems do not address the issue of monitoring changing customer requirements and incorporating them into network performance parameters. There are several emerging network management standards that, though not sufficiently comprehensive to ensure end-to-end, LAN-WAN-LAN capabilities, offer the same as TCP/IP; they are available, proven, practical, and do the job adequately, if not always elegantly or efficiently. These standards include SNMP (Simple Network Management Protocol) and CMIP (Common Management Information Protocol). Ironically, the more open a protocol or standard is, the less likely it is to address network management because open systems enthusiasts focus on ease of access, not on management and control. Neither the original OSI reference model nor TCP/IP considered network management. Most LAN network operating systems are primitive in network management, however sophisticated they may be in either ease of use or GUIs.

Network management is still an art form. However, given its growing importance to the success of an on-line business, it must move quickly from art to science, and another role of the designer is to facilitate that transformation. Toward that end, the next section discusses the key quality-of-service elements that are the focus of network operations management.

| | | | | | | |

Quality-of-service Factors

The network design process, described in Chapter 15, incorporates into the network such key features as overall throughput capacity and the range of services provided. These features are based on the three dimensions of the telecommunications platform: reach, range, and responsiveness. The network is designed such that the quality-of-service levels required by customers can be achieved with proper management of network operations. In turn, proper network operations management consists of both identifying the key causal determinants of quality of service and establishing a process for ensuring that these factors are properly managed and controlled.

Availability

There are two different aspects of network availability: network uptime and response time. *Uptime,* the portion of total time that the network is up and operating, can be more precisely defined as follows:

$$\text{Uptime} = \frac{\text{MTBF}}{\text{MTBF} + \text{MTTR}},$$

where MTBF equals *mean time between failures* and MTTR equals *mean time to repair.* This formula makes it clear that operational availability of a network is a function of both the average time between network failures and the average time it takes to repair the problem and restore service when it does fail.

One of the major reasons for a firm to outsource network operations is to increase MTBF, a measure of efficiency, and reduce MTTR, a measure of expertise. The MTBF of the network is determined by the underlying reliability of its component parts and the level of preventative maintenance that is performed. The MTTR is determined by the level of component redundancy built into the network design, the extent of alternate routing capability used to bypass failed segments of the network, and the inherent speed and efficiency of a department's problem-resolution process. The dramatic improvement in the reliability of hardware and in diagnostic tools has greatly increased MTBF across network equipment. However, the complexity of software and the variety of network services often increases MTTR also, especially for unproven tools.

Careful analysis of data on network outages and the time needed to restore service can help pinpoint where management should focus its attention for maintaining and improving network availability. (The process Texas Instruments used, described in Chapter 9, is a model for collecting and using such data.) The data often indicate whether the principal effort should be directed toward either working with suppliers to obtain more reliable parts, striving to improve the maintenance or trouble-resolution process, or building greater component redundancy and alternate routing capability into the network.

The trouble-resolution process must address two major problem categories: routine problems and disasters. *Routine problems* consist of a malfunction or failure of a network or individual component that occurs during the regular course of business activity, and the trouble is generally restricted to only a few locations or pieces of equipment. Management must design and implement efficient processes for quickly detecting and correcting such routine problems, including processes for identifying the cause of the problem and for quickly restoring service through either alternate routing or substitution of a spare or duplicate piece of equipment for the failed or malfunctioning unit. Once such processes are in place, continual efforts should be made to improve them to achieve a resulting decrease in the MTTR. Many companies provide diagnostic hardware and software that can quickly detect such routine problems as a failed node or a lost line.

Disasters are failures that are generally widespread, often with multiple points of failure, and require emergency measures to manage and restore service. Examples of telecommunications disasters include Hurricane Andrew, which completely destroyed private and public networks in a large part of South Florida; a fire that destroyed an entire telephone central office switching center, completely shutting down all telephone service in metropolitan Chicago; and a software problem that caused AT&T's eastern-regional network to crash completely. It is the magnitude of service disruption and the number of points of failure that distinguish disasters from routine problems. With a routine problem, traffic can be rerouted, and most of the network continues to operate; in a disaster the entire system collapses. Separate disaster-recovery plans must be developed and reviewed periodically. Because of the widespread nature of disaster damage, recovery operations must generally rely on mobile and wireless facilities. Wireless LANs, personal communications networks, cellular telephones/data terminals, wireless PBXs, and portable satellite dishes should be considered part of the disaster recovery equipment arsenal, and their use should be carefully planned, coordinated, and practiced.

The second aspect of network availability, response time, is a measure of how quickly a user's query, message, or computation request is processed and a response or acknowledgment is received. Network response time can significantly affect worker productivity in large organizations that rely on the network for on-line business operations; the difference between, say, one second and three seconds can drastically affect entire activities (entering orders, for example). Also, rapid network turnaround time in booking reservations, placing purchase orders, or querying a customer account database is essential for the provision of world-class customer service.

Network response time is determined principally by traffic levels relative to design capacity, and by the processing time of intelligent devices on the network. Recall that the throughput capacity of the network is established in the design process, and when operational utilization approaches the design capacity on any segment of the network, traffic congestion will increase the response time. Processing time for intelligent devices on the network is determined partly by the operating cycle time of the device itself and partly by the number and magnitude of requests for processing at any given time. Slower cycle times and a larger number and magnitude of processing requests will slow network response time. The network design parameters (for overall capacity and other features) establish an average response time, but careful management of the network is needed to consistently attain the average and to minimize variation around it. Traffic can grow or change in mix without warning. The designer cannot anticipate all of this; the network operations function must monitor traffic and provide alerts to management.

Reliability

A reliable network is one in which errors are absent and failures are a rare occurrence. Error-free operation is attained by designing transmission links with low noise levels and by using protocols with effective error detection and correction schemes. The trade-offs in attaining networks with low error rates are usually higher costs (low-noise lines cost more) and sometimes longer response times. Response times can increase if extensive error checking and correction are performed at each node in the network, as in the case of the X.25 packet-switching protocol. Newer high-speed protocols such as frame relay rely exclusively on end-to-end error detection and correction because it is assumed that the underlying transmission lines are reliable and have relatively low noise levels.

Network failure is the telecommunications manager's nightmare, and there is increasingly less tolerance for failures as more of a business moves on-line. Failure rates are determined in large part by the underlying durability of the individual components that make up the network. The MTBF of the least reliable component determines the average reliability of the entire network. In turn, the MTBF of individual components is a function of both the quality control process used in manufacturing the equipment and the limits of the technology itself. The reliability of individual components can be improved by working with vendors that supply parts that improve the quality-control process, and possibly to support R & D efforts to push back the technological barriers to higher reliability.

Another factor that affects the overall reliability of the network is following specified maintenance practices. Some network equipment requires regular maintenance in order to meet specifications for reliability. Neglecting to perform these procedures as specified will result in a lower MTBF for the component.

One other factor plays a significant role in overall network reliability: following proper procedures for network access and operation. Network failures can result if users fail to follow specified procedures for gaining access to the network and instead attach pieces of equipment that are incompatible with the network protocol system or move workstations to new locations without modifying network address codes. Trying to access a network with an incompatible terminal device can have the effect of radio jamming signals; it can interrupt traffic flows and create chaos on the electronic highway.

Security

One design function activity is to develop a network security strategy and an implementation plan based on its importance to the organization and trade-offs between cost and security. The role of the designer in ensuring effective daily security is to oversee the development of a standardized process to ensure that the designed levels of security are consistently realized. This process must include clearly defined policies and responsibilities for various aspects of security, as well as a control/feedback mechanism to determine how well design standards are being met.

The essence of telecommunications security operations is managing and controlling access to equipment and facilities and to the network and its information databases. The crux of the security problem is providing simple and inexpensive access on a wide-reach basis (ideally from anywhere) while protecting the physical facilities from harm and sensitive information from unauthorized users.

Security of physical facilities is a lock and key issue and should be managed as such. Network access security is a more complex problem and requires answering the following questions: How do we know who is trying to gain direct access to the network from a known terminal device? How do we limit this person's activity to the functions he or she is authorized to perform and to the information she or he is permitted to access? How do we prevent indirect access by an unknown terminal device (for example, wiretapping, pirating satellite signals, and so on)?

Log-on procedures that require the use of personal identification numbers (PINs) and passwords help ensure that the person requesting access is, in fact, the person authorized to gain access. Unfortunately, PINs and passwords can be stolen, often quite easily, because the vast majority of users adopt simple, easy-to-remember codes, such as names of family members, phone numbers, birthdays, and so on. Access to dial-up lines is controlled by special terminal identification protocols that are made part of the "handshake" connection procedure, or by a callback system whereby the network computer verifies the terminal requesting access by calling it back at a predetermined number. This is clearly not appropriate for portable terminals that may be calling from locations that cannot be predetermined.

It is relatively easy to limit a user's access to particular network functions and information databases through software-defined gatekeeper arrangements. Network devices such as PBXs, LAN servers, and front-end processors all possess this programmable capability. However, a much more difficult problem is to ensure that once access to the function or information source is obtained, it is used in the intended manner.

For instance, John Doe may be authorized to access a dial-up voice line for business-related long-distance telephone calls. The PIN or password ensures that the user requesting access is John Doe, or at least someone who has his PIN or password, but it does not prevent John Doe from making a social call to a friend in a distant city. A detailed analysis of call logs can help uncover acute abuses of the social-calling restriction, but this is costly for a large organization. Many companies report theft of employee phone passwords; a major area of contention between companies and long-distance providers is who pays for this.

Encryption of network information streams is generally used to prevent un-authorized access by an unknown terminal. If the information bit stream is coded effectively, wiretapping or the use of an unauthorized satellite dish or radio receiver to gain access will be thwarted. However, any encryption scheme can be broken, given enough computing power and enough time. Therefore, the choice of an encryption method is essentially the selection of a price level (time and money) that the unauthorized user must pay to obtain access. If the price is set sufficiently high, unauthorized access will be prevented in practice.

In the end, the security dimension of managing network quality of service comes down to cost. It costs more for tighter security, both in terms of money and greater difficulty of access. The use of PINs, passwords, and user restrictions makes access more complex and time-consuming, and costly modifications to network software are needed to limit access, verify PINs and passwords, and encrypt data streams. Typically, network security is not sufficient to meet the business needs in most organizations today. The design is inadequate, and security operations are not managed efficiently. A high level of security is needed to protect the core business activities that are increasingly on-line, and senior management must recognize that this carries a price that must be paid.

Accessibility

Network design should incorporate the key aspects of accessibility appropriate to a firm's business need: simple and convenient access procedures, flexibility in the times and locations of permissible network access, and no need for special equipment to access the network. Accessibility is often a key determinant of whether or not information technology is used effectively within an organization, and it is seldom given attention worthy of its importance. Stories are legion about expensive networks and information systems that do not gain widespread use and acceptance within an organization. Often, a primary cause is that access procedures are complex and time-consuming and that there are severe restrictions on time, location, and the type of equipment needed to gain access. For example, the simple telephone access system that requires dialing 1 plus area code plus local number is often replaced by a more "efficient" and low-cost access system that involves learning or looking up many codes and entering as many as 30 digits. People avoid using it, and the efficiency and cost savings are lost.

The need in daily operations is to establish and manage an effective process for delivering the specified level of accessibility in a dynamic environment. This process involves continually reviewing actual network access procedures and restrictions placed on users to verify that design targets are attained (or at the very least not thwarted) by procedures. It also requires a means for managing changes in network access stemming from the addition or deletion of people, workstation terminals, and new locations for work groups, and from new information flows resulting from external forces and the reengineering of business processes to make more effective use of information technology.

As previously indicated, the designer's role in managing the operational aspects of quality of service is the development, implementation, and control of processes that consistently deliver appropriate performance levels for these factors. In turn, the appropriate performance level targets themselves are established in the network design process through an assessment of users' requirements. To this point we have discussed what needs to be done, but not how to do it. The remainder of the chapter is devoted to a discussion of how total quality management methods can be used by a designer to effectively manage these operational aspects of quality of service.

| | | | | | |

Ascertaining Customer Requirements

Total Quality Management of Network Operations

Total quality management of network operations begins with discerning customers' requirements for each of the quality-of-service elements. Recall from Chapter 15 that TQM uses a systematic and scientific approach to management and problem solving. It is management by fact, rather than by opinion. Therefore, ascertaining customers' requirements for quality of network service requires an objective method for fact gathering and analysis.

The following list (Deming, 1986, pp. 156–66) contains some key questions for management regarding customer requirements and their satisfaction. Note that the questions seek more than customer needs; they also seek a method and a *factual* justification for obtaining results and drawing conclusions.

Key Questions for Management Concerning Customer Satisfaction

1. Who are your customers, and how do you know?
2. Can you demonstrate how you are satisfying your customers?
3. How do you get feedback from your customers?
4. How are you doing based on that feedback?
5. What are your customers' priorities, and how do you know?
6. How do you help customers meet their priorities?
7. How do you prioritize work to improve customer satisfaction?

The first step in determining customer requirements is establishing the identity of your customers. There are both internal and external customers. *External customers* purchase a company's products and services and have a major impact on company decisions. For example, a regulatory body such as the state public utility commission is a customer of the electric power companies operating in the state. *Internal customers* are the individuals and groups within the organization that receive and use the output of your work. Dr. Kaoru Ishikawa, a respected quality expert, coined the expression, "The next process is your customer." This captures both the notion of the business organization as a large system with a number of interdependent subsystems and the concept that the downstream department in the production chain is the internal customer of the adjacent upstream department that supplies them with goods or services. Sales is a customer of accounting, which is a customer of MIS, for example.

This notion of internal customers is often difficult for organizations to embrace. Frequently, the next department in the production chain is viewed as the enemy and the primary cause of the company's ills. Therefore, the notion of spending the workday finding ways to please and satisfy the members of that department requires a complete transformation in attitude and work relationships. Imagine the impact on productivity and efficiency if everyone in the organization started treating the next process as the customer. It would eliminate the wasted energy and inefficiency resulting from noncooperation, bickering, and turf battles that, if not checked, can lead to organizational gridlock.

Federal Express has adopted a formal system for implementing the concept of "the next process is your customer." The system consists of a series of meetings and exchanges between two departments that have a customer/supplier relationship in the production or service chain to arrive at a written service contract that states specifically what Department A will do for Department B by when, and what measures are to be used to determine whether A is meeting its commitments to B. The written contract becomes both a set of specific goals and targets for Department A and the operational definitions of what Department B can and should expect.

The following list shows the set of customers for the corporate telecommunications department at Florida Power & Light Corporation (FPL).

Florida Power & Light's Telecommunications Customers

- General office telephone systems users
- ACD, PBX, and key system sites
- Regional data centers
- Data network users in Trouble Offices
- Sites with data network outages greater than eight hours
- All voice network users
- All data terminal users

There are two important points to glean from this list: First, most of the customer segments are internal to the corporation, and this is likely to be the case for most corporate telecommunications operations. Second, some of the customer groups are overlapping. For instance, "Sites with data network outages more than eight hours" is a subset of "All data terminal users." Overlapping customer segments are appropriate when there are particular problems or needs to be addressed for particular subgroups of customers.

Once the key telecommunications customer segments and groups have been identified, the next step is to assess their needs and priorities. Several methods have proven useful for identifying customer requirements. The first and most popular is the use of survey questionnaires. Surveys are straightforward and permit data to be gathered from large customer populations; however, most individuals are surveyed to death, and their tendency is either not to respond at all or to fill out forms quickly without concern for accuracy or completeness. Low response rates and low accuracy rates call into question the validity of many surveys. Therefore, surveys should be used sparingly, and preferably in conjunction with other methods that can corroborate the survey results.

Interviewing focus groups is another method of assessing customer needs and priorities. A focus group consists of a representative sample of the target customer group. The group is assembled and asked a series of questions designed to elicit preferences and priorities with respect to the given product or service. Considerable skill is needed to conduct a successful focus group session, but the output is generally very useful. The informal nature of the session allows probing of specific points and issues that are not possible with a formal survey. Ironically, one of the weaknesses of focus groups is their lack of formality and structure; it is often difficult to generate a sufficient amount of unbiased, objective data upon which to base sound decisions from these small group discussions.

A newer method for discerning customer needs, called systematic observation, is being used extensively by Japanese TQM companies to complement information obtained from surveys and focus groups. These companies realized that customers often are not conscious of what they want, particularly if the product or service does not exist. For example, one of the Japanese camera companies, using traditional surveys and focus groups, attempted to find ways to improve its cameras to increase customer satisfaction. Surprisingly, the results indicated that customers think the cameras are great and that there is no need to improve them. Nevertheless, the camera company probed further by using trained teams of observers to watch customers taking pictures and to gather data at photo development laboratories. After careful analysis of large quantities of processed film, the teams found that a high proportion of the first few frames on each roll were overexposed. Further observation and analysis revealed that the problem was caused when users opened the back of the camera after loading a new roll of film to determine whether the film leader had attached properly to the winder stem. The results of this systematic observation study led to the development of the automatic load and wind feature that is found on most cameras sold today.

Progressive companies are learning to apply the methods of cultural anthropology and other techniques for scientific observation and measurement of be-

Figure 16–1		Ranking*
Elements Important to FPL's Customer Segments	Sales and Service Quality	
	Accurate Answers/Timely Actions	7.2
	Accurate Bills	6.8
	Considerate Customer Service	5.4
	Energy Management Assistance	2.1
	Continuity of Service	8.3
	Understandable Rates/Bills	3.9
	Delivery	
	Character of Service	4.2
	Capacity	5.1
	Safety	
	Public Safety	7.9
	Employee Safety	6.5
	Cost	
	Price	7.5
	Rate Options	3.5
	Financial Integrity	6.7
	Corporate Responsibility	
	Prevent Pollution/Protect Public Health	4.2
	Protect Property and Equipment	3.3
	Concern for Community	2.1
	Visual Appeal	1.1
	Protect Natural Environment	4.3
	Reporting & Filing Requirements	5.4

*On a scale of 1 to 10; 1 = least important, 10 = most important

Source: J. M. Cummins and E. C. Stonebraker, "Total Quality Management of Telecommunications," *Business Communications Review,* December 1989, Vol 89, No. 12, pp. 36–41.

havior in an attempt to elicit a true and complete "voice of the customer." Many telecommunications units would benefit by observing how their customers use PCs, for example, for it would reveal opportunities to improve ease of access or uncover other problems such as user difficulties in configuring modems.

Finally, another way to ensure that the voice of the customer is incorporated into key decisions and activities in a telecommunications department is to include customers on teams within the department. Invite customers to team meetings to discuss operational problems or to brainstorm on how service can be improved. Customers will provide useful input, and the meeting will demonstrate that the department cares about their needs and is interested in serving them—the first, and most important, step in building a long-term relationship.

The Florida Power & Light Corp. has used all of these methods to identify and analyze their customers' requirements. The results of this extensive analysis are shown in Figure 16-1, which is a detailed listing of elements important to FPL's various customer segments and a ranking of their relative importance. FPL's customers range from residential and commercial users of electricity to the Nuclear Regulatory Commission and the Florida State Public Utility Commission.

Linking Operations Processes to Customer Requirements

Once a company's customers have been clearly identified and their overall needs determined and prioritized, the next step is to link these requirements directly to key operations processes within the company to ensure that resources are focused on things that are important to customers. This may seem like obvious common sense, but surprisingly few companies follow this practice; and as a result, many departments within the company carry on activities that bear little relationship to what is important to customers.

Each department must first identify its major accountabilities to customers and management and the key processes used to carry out its mission. Although every telecommunications department will undoubtedly have a superficial understanding of its major areas of responsibility, such knowledge is generally not sufficient to determine the extent of customer satisfaction and whether or not the designed level of service quality is being achieved on a consistent basis. The following list, adapted from Deming (1986, pp. 156–66), contains important questions for management regarding telecommunications operations. The effort required to answer these questions should provide management with a deeper understanding of the department's critical activities.

Telecommunications Operations Questions for Management

1.   What are the major operations accountabilities of the department, and how do you know?

2.   What are the major work processes for the department, and how do you know?

3.   What is the process used to decide operational priorities?

4.   Can you demonstrate your process for managing daily work?

5.   How do you build long-term relationships with vendors and customers?

A useful method for linking overall customer requirements to the major accountabilities of the telecommunications department is to create a correlation matrix that lists the quality elements important to customers down the left-hand side; the columns of the matrix represent the key departmental accountabilities (see Figure 16-2).

The relative rankings of customer needs, shown in the second column of the matrix (which is taken from Figure 16-1), represent a priority ordering of what is important to customers. The telecommunications department at FPL then did a careful assessment of the relationship between each of the customer quality elements and the four major telecommunications responsibilities: (1) designing and maintaining data communications networks, (2) designing and maintaining fiber-optic networks, (3) intelligent tandem network design and maintenance, and (4) voice communication system design. Each combination of a customer quality element and one of the telecommunications responsibilities is assigned a number from 1 to 3, with "1" indicating a weak relationship and "3" a strong relationship. The number shown in each cell of the matrix is the product of the relative ranking of customer quality elements (column 2) and the

Quality Element	Weight	Rank	Telecom Responsibilities				Total Quality Element Weight	Overall Quality Element Rank
			Design and Maintain Datacom Networks	Design and Maintain Fiber-Optic Networks	Intelligent Tandem Net Design and Maintenance	Voice Comm. System Design		
Sales and Service Quality								
Accurate answers/Timely actions	7.2	4	S 21.6	S 21.6	W 7.2	W 7.2	57.6	1
Accurate bills	6.8	5	W 6.8		W 6.8	W 6.8	20.4	5
Considerate customer service	5.4	7						
Energy management assistance	2.1	16			W 2.1		2.1	7
Continuity of service	8.3	1	S 24.9	M 16.6	W 8.3		49.8	2
Understandable rates/ bills	3.9	13						
Delivery								
Character of service	4.2	11						
Capacity	5.1	9						
Safety								
Public safety	7.9	2	M 15.8		W 7.9	W 7.9	31.6	3
Employee safety	6.5	7				W 6.5	6.5	6
Cost								
Price	7.5	3		S 22.5	W 7.5		30.0	4
Rate options	3.6	14						
Financial integrity	6.7	6						
Corporate Responsibility								
Prevent pollution/ protect public health	4.2	11						
Protect property and equipment	3.3	15						
Concern for community	2.1	16						
Visual appeal	1.1	18						
Protect natural environment	4.3	10						
Reporting and filing requirements	5.4	7						
Total department/activity weight			69.1	60.7	39.8	28.4		
Overall department activity rank			1	2	3	4		

Figure 16-2

How Telecom Responsibilities Relate to Quality Elements at FPL

Note: The values are obtained by multiplying the quality element weights by a strength of relationship factor (Strong = 3, Moderate = 2, Weak = 1, none = 0). The values are summed vertically to obtain a weighted ranking for each telecom department responsibility and summed horizontally to obtain a weighted ranking for each quality element. Value numbers in the chart have been altered to protect the confidentiality of FPL's data.

Source: J. M. Cummins and E. C. Stonebraker, "Total Quality Management of Telecommunications," *Business Communications Review,* December 1989, Vol. 89, No. 12, pp. 36–41.

strength-of-relationship rating (1, 2, or 3) provided by the telecommunications department.

The sum shown at the bottom of each column is a measure of the relative overall importance of each of the telecommunications accountabilities in meeting overall FPL customer requirements. The sum shown at the end of each row represents the relative ranking of the various customer quality elements in terms of how closely they are linked to the company's telecommunications functions. This matrix is a useful guide for setting priorities and focusing limited departmental resources on things that directly relate to customer satisfaction. It also helps identify long-standing activities that have virtually no effect on meeting the needs of customers. These activities are prime targets for reduction of effort or outright elimination, thereby freeing up time and resources to allocate to other activities that do have a major impact on customer satisfaction.

Imagine the power of this simple exercise conducted in every department, from accounting to building maintenance. Many well-meaning, but irrelevant (to customers) activities would be eliminated, and the organization's resources would be focused directly on those activities that have the largest impact in satisfying the needs and desires of its customers. We can think of no better recipe for success.

Teams and Continuous Process Improvement

One of the underlying principles of total quality management—that teamwork is superior to a collection of individuals undertaking the same task—is based on the concept of synergy and complementary skills among team members. However, effective teams are made, not born, for team members need to learn and practice team skills in addition to developing their individual professional capabilities (see Scholtes, 1988).

In a TQM environment, there are departmental teams, cross-functional teams, and generally an overall lead team or quality council. Departmental teams, made up of members from a single department or organizational unit, are usually the easiest to initiate because the members have the same reporting channel and typically work together on a daily basis. Cross-functional teams include individuals from different departments or units within the organization. They are generally more difficult to create because the members report through different channels and have different allegiances, and they do not work together or cooperate on a regular basis. However, most of the operational bottlenecks and quality problems in an organization stem from inefficient and ineffective cross-functional work processes. The handoff and exchange of products, services, and paper between different departments or divisions are frequently poorly executed. It is common to hear exchanges such as, "Planning is supposed to handle that; it is not our responsibility in network operations." Personnel and planning usually respond that "It is not our job; network operations handles that."

A lead team or quality council, made up of senior executives, guides the efforts of individual teams and sets priorities for the work undertaken. Unfortunately, true teamwork is rare in most organizations, largely because a culture of individualism is prevalent, and formal schooling (except for athletics) and the work environment discourage teamwork and reward individual accomplishment. Today's modern enterprises competing in global markets are quickly recognizing this deficiency, and many are moving quickly to correct this problem. These organizations realize that it is no longer a choice, but a competitive necessity.

In addition to basic teamwork skills, TQM organizations also provide their teams with a standard problem-solving process. At Florida Power & Light, this process is called the **Quality Improvement (QI) Story** and consists of the following seven steps:

1. Statement of the reason for improvement
2. Assessment of the current situation
3. Analysis of root causes
4. Development and testing of solution countermeasures
5. Analysis of countermeasure test results
6. Standardization of the countermeasures
7. Statement of future plans for further improvement

Adopting a formal problem-solving process (such as the QI Story) for the entire organization has several important benefits. If each member of the team is trained in the standard problem-solving method, the team can begin immediately to address the problem at hand, rather than devoting considerable time and resources to agreeing on a method for attacking the problem and then making sure each member understands the approach and his or her role in it. If an individual is transferred to another part of the organization or is asked to serve on a cross-functional team with individuals from another division, everybody is trained to use the same approach to address the problem, and no time is wasted in getting started. Finally, a complete QI Story can be represented in the form of a storyboard layout, with each step shown graphically. This helps managers who have not been working on a particular problem to quickly grasp the issues and the solution approach that the team has adopted. At FPL, QI Story storyboards are often posted on bulletin boards where the teams are working.

The QI Story method is an operationalized version of one of TQM's basic concepts—the **Plan, Do, Check, Act (PDCA) cycle**, which is the approach used to implement continual improvement through managing by fact. The four-step cycle was created by Walter Shewhart at Bell Laboratories and was popularized in Japan and the United States by W. Edwards Deming. It is now often referred to as the Deming wheel or cycle, and in more recent treatments of the cycle Deming has substituted the word "study" for "check" as the third part of the cycle, but the meaning is essentially the same (Deming, 1986, p. 88). The essence of the method is to *plan* a change that will lead to an improvement; *do* the proposed change on a test or trial basis; *check* the results of the test to determine whether the improvement was realized; and *act* on those results to refine the proposed change or standardize the improvement. The PDCA cycle is really a shorthand version of the scientific method, and it is embodied in the

QI Story: The first three steps are the *plan* part of the cycle; development and trial testing of countermeasures is the *do* part; analyzing the results of the countermeasure test is the *check* part; and standardizing the countermeasures into daily work processes is the *act* part of the cycle. The statement of future plans for improvement starts the cycle again, which is necessary for continual improvement to take place.

Daily Management System for Operations

In a TQM system, teams are formed to solve problems that arise on an ad hoc basis in the workplace, but there is also a need to manage and continually improve the normal, day-to-day functions and activities of the telecommunications department. This requires a **daily management system**, which should include the following tasks:

▸ *Identify and document key processes* associated with the major accountabilities of the telecommunications department.

▸ *Devise statistical measures* to determine whether each process is consistent and in control.

▸ If the process is inconsistent or unstable, *determine the causes and make corrections to bring the process under control.*

▸ *Assess whether the output of the process meets customer requirements and expectations.*

▸ If not, *apply quality improvement tools and methods to improve the process.*

▸ *Continue to monitor the process* and apply the PDCA cycle to realize continual improvement.

For each of the major accountabilities of the telecommunications department, several key processes generally constitute the daily work related to those accountabilities. For example, one of the four major accountabilities of the telecommunications department at Florida Power & Light is designing and maintaining corporate data networks. FPL has identified five important processes that constitute the daily work related to designing and maintaining corporate data networks: (1) data network engineering; (2) data network trouble resolution; (3) the repair cycle of spare data communications equipment; (4) data network moves, adds, and changes; and (5) LAN trouble resolution.

After identifying the key work processes of the department, the next step is to document them by developing a detailed flow chart describing each significant step in the process and identifying the entity responsible for that step. Most companies that have undertaken this flow charting exercise have come to a remarkable conclusion: Management and the workers don't understand or agree on how work is performed on a daily basis. In many instances, employees in the same department that have been working together on the same job for years do not agree on the steps in the flow chart that depict their daily work activity. Therefore, one of the initial outcomes of a flow charting exercise is reaching agreement among the work team as to exactly how work is performed.

In addition, the flow charting exercise will often identify an obvious inefficiency or useless activity in the process that can be changed or eliminated to immediately improve the process.

For example, the telecommunications department of a large corporation discovered through the flow charting process that their procedure for handling the failure of a telephone set was to dispatch a repair person to the trouble location, remove the faulty telephone set, and proceed to repair the instrument on-line while the trouble location was out of service. This led to an average outage time from failed telephone sets of 4.5 hours. The flow chart the telecommunications department created drew attention to this absurd situation, which led to an obvious revision of the process: Immediately replace the faulty set with a spare to restore service, and then repair the faulty set off-line. This simple change in procedure reduced the average outage time for failed telephone sets from 4.5 hours to ten minutes. The point of this example is that much of what we do on a daily basis goes unnoticed, is generally determined by how things have been done in the past, and often bears scant relationship to things that are important to the people that buy the company's products and services.

Figure 16-3 contains a simplified flow chart of the data communications network trouble resolution process at FPL. The beginning and end of the process are represented by circles. Rectangles are used to represent activities or handoffs between departments or entities; diamonds are decision points that lead to a branching of the work flow, and arrows connecting the symbols represent the flow of activities. Listed across the top of the chart are all of the entities (both within and outside the company) that have a significant involvement in the process. The symbols placed under each heading (viewing them as columns) represent work performed by that entity, and movement down the flow chart from top to bottom generally coincides with the passage of time.

A flow chart can include only the very major activities involved, or it can be quite detailed. Generally, detailed flow charts are divided into segments, and subprocesses or activities are detailed on separate connecting charts.

In many instances, a cursory examination of the flow chart will reveal obvious opportunities for process improvement; needless redundancy and duplication of activity are immediately apparent. Also, a large number of arrows connecting activity flows between various departments or entities may be an indication of wasted activity in the form of unnecessary cross-functional handoffs.

After flow-charting key work processes, the next phase in the development of a daily work system is to establish for each process indicators and targets that are linked to customer requirements. The flow chart of the process is a good place to look for opportunities to make measurements to serve as indicators of how well the process is doing in satisfying customers. A good indicator is observable, measurable, and closely related to what is important to customers. For example, the duration of a network outage is an appropriate indicator for the network's trouble resolution process; it is observable, measurable, and closely related to network availability, which is important to users (customers) of the network.

Process flow chart

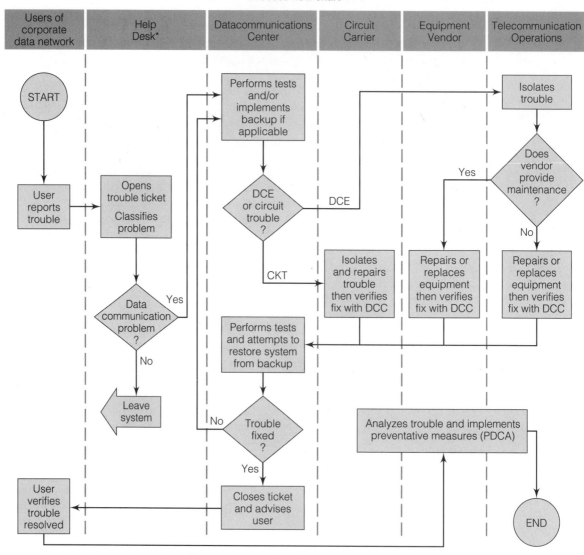

*After hours, Help Desk performs limited Datacommunications Center functions

Figure 16–3

**Data
Communications
Trouble Resolution
Process**

Source: J. M. Cummins and E. C. Stonebraker, "Total Quality Management of Telecommunications," *Business Communications Review,* December 1989, Vol. 89, No. 12, pp. 36–41.

Next, measurement data on these indicators are collected and analyzed to determine whether the process is stable and in control. The concept of **process stability and control** is based on a statistical theory for quality control developed by Dr. Walter A. Shewhart of Bell Telephone Laboratories in the latter half of the 1920s. Shewhart noted that almost every observed phenomenon exhibits variation about a target or ideal value, and that this variation has two components: a steady, random element that is attributable to chance and unassignable causes, and an intermittent element attributable to identifiable causes. He concluded that identifiable causes can be diagnosed and removed without making basic changes to the process. However, the random variation component can be altered only by making basic changes to the process itself. Under Shewhart's theory, a process is stable and in control if data generated from that process exhibit a random pattern of variation that is within plus or minus three standard deviations of the mean of the data. Any data points lying beyond three standard deviations above or below the mean or exhibiting a non-random pattern are attributable to identifiable and assignable causes (Juran and Gryna, 1988). Figure 16-4 shows examples.

The importance of this theory is that the approach taken to improving quality is different depending on whether the underlying process is stable and in control or not. If the process is not in control, the approach is to focus on determining the root cause(s) of those observations lying outside the three standard deviation control limits or exhibiting a non-random pattern and to take corrective action without altering the process itself. If the process is in control, improvement can be achieved only by changing the process itself.

Once corrective action has been taken on the special causes of variation and the process is stable and in control, improvement targets for the indicators are established. Then teams collect and analyze data on the process and recommend changes in the process that are designed to achieve the target objectives for the indicators. The PDCA cycle, together with the application of various data-analysis and problem-solving tools, is used to develop and test modifications to the process (see Ishikawa, 1982; Ozeki and Asaka, 1988). The new, improved process is then standardized and implemented as the current best-practices-method only after the evidence from the check phase of the PDCA cycle demonstrates that the modifications do, in fact, yield improvement in the designated indicators. The new process is then monitored periodically to ensure that it continues to be stable and in control and that it is performing at an acceptable level in terms of meeting customer requirements.

This is a radically different approach to managing operations on a daily basis than the all too familiar "Don't mess with it unless it's broken or somebody complains; then, we'll fix it when we get a chance." The TQM approach explicitly recognizes that the organization is a huge system with many interlinked subsystems, and that the goal is to optimize the overall system in the delivery of superior performance and customer service. Optimization is achievable only if the interdependencies among the various subsystems are recognized and incorporated into daily operations management, and if changes to the various subsystem processes are based on sound analysis of facts. Otherwise, chaos reigns and

Figure 16–4

Examples of Shewhart's Quality Control Theory

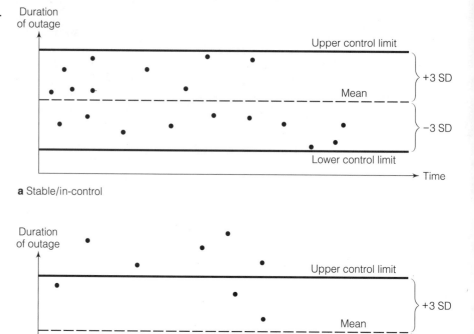

a Stable/in-control

b Unstable/not in-control

Chart **a** shows a stable process, where all instances of network outages exhibit a duration that is within ± three standard deviations of the mean. Chart **b** shows an unstable process: a number of data points fall outside the three standard-deviation control limits.

General framework from J. M. Juran and Frank M. Gryna, eds. *Juran's Quality Control Handbook,* 4th ed., New York: McGraw-Hill 1988, pp. 24.2–24.3.

acts of local suboptimization dot the corporate landscape. Theory is very clear that a collection of local optima does not generally lead to an overall system optimum; yet, most executives stubbornly persist in managing their companies this way, ignoring crucial interdependencies among departments and functions.

One of the crucial interdependencies among subsystems that is seldom given the serious attention it deserves is the link between the telecommunications department and key vendors supplying equipment and services. To optimize the overall delivery of high-quality telecommunications services to users, there must be a close working relationship with vendor organizations. Instead of the strained courtship that exists today, with the user playing one vendor off against another for short-term financial gain, it makes much more sense to select the smallest practical number of vendors on the basis of criteria important for long-term success, and then build a lasting partnership. Include vendor personnel in key planning and decision meetings, help them to improve service delivery processes, invite them to training sessions, and so on.

Florida Power & Light developed a formal vendor certification program that includes various levels of achievement: quality vendor, certified vendor, and FPL excellent vendor. FPL staff work with vendor organizations, both informally and through formal training sessions, to help them implement their own quality management systems and better understand and meet FPL requirements. FPL has recognized that quality is limited by the weakest link in the value chain, and that continuing to achieve customer satisfaction is critically dependent on a working partnership with vendor organizations.

Process and Performance Comparisons

Most telecommunications departments assess their operational performance against some form of internal standard, if at all. Seldom does Company X's telecommunications department either compare its performance and management of operations with Company Y's telecom department or seek out the worldwide "best-in-class" telecom department for a comparison. TQM is changing this view because it strongly encourages "best practices" comparisons, or **benchmarking**—essentially a systematic observation and comparison of how somebody else does something that you also do. Benchmarking activities can lead to improvements in process methods, and it can also provide measures of achievable target levels for key quality indicators. It is always easier to strive to attain something that you know is achievable because someone else has demonstrated that it can, in fact, be done.

The growing interest in benchmarking among telecommunications departments has led to the creation of comprehensive databases containing network cost and performance information for a wide range of companies in many different industries. Also, the Malcolm Baldrige National Quality Award in the United States requires that all applicants undertake benchmarking comparisons of their key activities (see Figure 16-5). Many companies have also found that some of the best benchmarking information is found outside of telecommunications departments. The idea is to find the best example of the process you are trying to improve, regardless of the type of business or operation in which it is found. For instance, the "best-in-class" process for customer billing may be found in a retail catalogue business, such as L.L. Bean or Lands' End, rather than in a telecommunications department or company.

Finally, a word of caution about benchmark comparisons of telecommunications network operations is in order. Any comparison should be between similar networks or network components. This sounds simple, but it is actually quite difficult in practice to find two similar networks in terms of overall reach (number of nodes and geographic distribution), capability (transmission capacity and range of services provided), and quality of service (availability, reliability, level of security, and so on). Consequently, it is very easy to draw erroneous conclusions based on a comparison of dissimilar networks.

Figure 16–5	Areas to Address
Malcolm Baldrige Quality Award Benchmarking Criteria, 1993	Describe the company's processes, current sources and scope, and uses of competitive comparisons and benchmarking information and data to support improvement of quality and overall company operational performance.

a. how the company uses competitive comparisons and benchmarking information and data to help drive improvement of quality and company operational performance. Describe: (1) how needs are determined; and (2) criteria for seeking appropriate comparison and benchmarking information—from within and outside the company's industry.

b. brief summary of current scope, sources and principal uses of each type of competitive and benchmark information and data. Include: (1) customer-related; (2) product and service quality; (3) internal operations and performance, including business processes, support services, and employee-related; and (4) supplier performance.

c. how competitive and benchmarking information and data are used to improve understanding of processes, to encourage breakthrough approaches, and to set "stretch" objectives.

d. how the company evaluates and improves its overall processes for selecting and using competitive comparisons and benchmarking information and data to improve planning and company operations.

Notes:

(1) Benchmarking information and data refer to processes and results that represent superior performance and set a "stretch" standard for comparison.

(2) Sources of competitive and benchmarking information are of several types, and could include: (1) information obtained directly from other organizations through sharing; (2) information obtained from open literature; (3) testing and evaluation by the company itself; and (4) testing and evaluation by independent organizations.

Source: Malcolm Baldrige National Quality Award, 1993 Award Criteria, Gaithersburg, MD: National Institute of Standards, 1992.

Suppose that Company A is comparing network performance and cost with Company B's network, which has the same number of nodes and a similar geographic coverage (average distance between nodes). The results show that for the dimensions of network performance that were analyzed, performance was essentially the same, but Company B's cost is significantly lower. Without additional information, Company A may draw the erroneous conclusion that its network operations are less efficiently managed than Company B's, and that severe cost-cutting measures are required. However, further investigation would reveal that the reason for the cost differential is that Company B does not provide the same level of component redundancy and alternate routing capability as Company A. Although this generates the higher cost level, it also yields a higher level of network availability; consequently, severe cost cutting by Company A is not justified by the comparison. It is clearly very important to compare similar networks, or at least to understand their differences, in a benchmarking exercise to avoid drawing the wrong conclusions from the data.

The TQM methods described here represent an integrated, coherent approach to managing network operations. They are applicable to all of the operations functions (TOCC, monitoring and testing, backup, and installation/field support) and can serve to coordinate and link them together. In addition, these methods ensure that the efforts of a telecommunications department are focused on things that are important to customers. For in the end, the acid test of any commercial organization is its ability to obtain and keep customers, which means continually satisfying them at a price they are willing to pay. Finally, in the globally competitive environment of the 1990s, both internal and external customers are much more demanding, and it is no longer sufficient for telecommunications operations to keep the network operating reasonably well most of the time. In surviving companies, telecommunications operations will provide extraordinary service levels that are essential for company success. Ad hoc management methods are not capable of delivering such a high level of performance on a consistent basis.

Network Management Tools

Of course, managing a complex, geographically dispersed, "mission-critical" network requires tools, including automated network management systems that combine hardware and software that monitor circuits and terminals, provide diagnostic messages, and generate alerts and statistics. The systems also maintain data on network configuration, which may change daily as new nodes and terminals are added and switches, servers, and lines are upgraded. The total quality metrics previously described are the obvious base for deciding which statistics to collect and review.

The main trend in network management tools is toward comprehensive problem detection and resolution in a multitechnology environment. The systems increasingly adopt a GUI interface so that the network operations staff can quickly make sense of the flood of data that comes into the network control center. Previously, device-specific proprietary systems generated simple alarm messages about the status and failures of that device; the information could not be consolidated or coordinated. Standards were almost entirely missing; the original OSI reference model omitted network management entirely.

Today, the OSI-based Simple Network Management Protocol (SNMP) is widely adopted as the base for what are termed *agents* to send data in a common format. The Common Management Information Protocol (CMIP) is a more recent OSI development. IBM's NetView and SystemsView provide the base for managing SNA networks. The leaders in providing tools for complex LAN-based platforms are Hewlett Packard, with its OpenView, and Sun, whose SunNet manager is widely used in UNIX environments. Many vendor groups are cooperating to define a comprehensive Integrated Network Management Platform (INMP), including the Open Systems Foundation (OSF); this is a priority for both users and vendors because "open" is meaningless if the open elements of the network cannot be managed together. The OMNIPoint set of standards has the support of OSF, the Corporation for Open Systems, and several other leading consortia. It builds on CMIP and SNMP and defines an object-oriented "request broker" architecture.

Managing proprietary networks has always been a challenge. Managing complex networks in an open systems environment is a very recent development, and there are few proven tools. Although it is the implementer who has the most responsibility for network management tools, the designer must ensure that the need is addressed in the design process; it is part of the quality-of-service category of network design variables. Quality of service is what network operations is about. Of course, the client should not know or care about network management; all clients want is quality.

Summary

A large firm's telecommunications platform is extraordinarily complex to manage. There are four main areas of focus: (1) the telecommunications operations and control center (TOCC), (2) monitoring and testing, (3) backup and disaster recovery, and (4) installation and field support.

The designer's role in network operations is to ensure during the design stage that the strategy focuses on quality of service. The main criteria for assessing and monitoring quality are defined by the levels of reach, range, and responsiveness that capture the client's statement of business priorities. Availability, reliability, security, and accessibility are the main general measures of quality; these network design variables must be translated into network operations capabilities.

The TQM approach to network design also applies to network operations. It requires a focus on the customer's priorities and on measures of quality. The requirements of internal and external customers must be explicitly identified and addressed and performance must be monitored in relation to these requirements.

Total quality operations needs a team approach that aims at continual improvement. Florida Power & Light adopted a seven-step process called Quality Improvement Story in problem solving and in handling the inevitable failures and breakdowns in a complex telecommunications network. The process requires a daily management system for documenting, monitoring, and resolving problems. The goal is process stability and control, using Deming's Plan, Do, Check, and Act (PDCA) cycle.

There are many network management tools to assist in effective network management, which remains an art form. The Simple Network Management Protocol (SNMP) and Common Management Information Protocol (CMIP) are widely adopted. However, integrated network management tools emerge only slowly.

Review Questions

1. List the four main areas of network operations activities.

2. Explain how an outsourcing firm is able to profitably manage another company's network for a price that is lower than the company's cost of managing it themselves.

3. Briefly explain the role of the designer in network operations.

4. Explain the concept of network availability.

5. Define network uptime, and indicate how its two components can be managed to increase the level of network uptime.

6. Briefly explain the key causal determinants of network response time.

7. How is network reliability measured, and what are its key causal factors?

8. What is the role of the designer in security operations, and how does this role differ from the designer's responsibility in the security aspects of network design?

9. What is the essence of telecommunications security?

10. Indicate what is wrong with the following statement: "We have all the security we need; you need a personal identification number (PIN) or password to get access to our network."

11. Explain the following statement: "The choice of an encryption method is essentially the selection of a price the unauthorized user must pay to get access."

12. What is the role of the designer in the operational aspects of obtaining access to the network?

13. Give an example from your personal experience of a decision not to use an available network service because of difficulty in obtaining access.

14. What is the starting point for implementing total quality management of network operations?

15. Give a brief explanation of the idea that "the next process is your customer."

16. In identifying an organization's customers, is it ever appropriate to have overlapping customer groups or segments (in which some of the same customers are included in two or more different groups)? Explain.

17. Briefly describe how customer requirements are ascertained using the method of systematic observation, and explain why this technique is complementary to more traditional ways of determining customers' needs and wants.

18. Consider the following statement: It is dangerous to include customers on teams that are designing, developing, and improving products and services because they may provide competitors with sensitive information that would put the company at a disadvantage. Do you agree or disagree? Explain.

19. How are the requirements of a company's customers linked to the major accountabilities and responsibilities of its telecommunications department?

20. Name three different types of teams found in a company using total quality management methods.

21. What are the advantages of having a formal problem-solving process that is used throughout a company?

22. Briefly explain the Plan, Do, Check, Act cycle and its role in a total quality management system.

23. List the key elements or steps in a "daily management" system designed to manage operational activities.

24. What is meant when a process is said to be "stable and in control"?

25. What is benchmarking, and why is it important for effective management of telecommunications operations?

Assignments and Questions for Discussion

1. A large manufacturer of consumer products has determined that information systems and telecommunications networks are a crucial factor for the company's success in the 1990s. However, the company is concerned with its ability to manage this resource effectively. Consequently, you have been retained by this company to evaluate the pros and cons of outsourcing telecommunications planning and operations functions. Because you do not have access to specific information about the company, focus your analysis on key factors that should influence the choices for the company.

2. Create a correlation matrix diagram, similar to the one in Figure 16-2, with the rows consisting of the four quality-of-service elements (availability, reliability, security, and accessibility) and the columns consisting of the four principal functional activities of the network operations department (TOCC, monitoring and testing software/equipment, backup, and installation and field support). Then, fill in the cells of the matrix by determining whether the relationship is strong = 3, moderate = 2, or weak = 1, assuming that the relative weights for the quality-of-service elements are availability = 5, reliability = 2, security = 6, and accessibility = 3.

3. Using the flow chart describing the data communications trouble resolution process in Figure 16-3, determine the number of times that cross-department or cross-functional handoffs or activities occur. Then, redesign the process (redraw the flow chart), eliminating as many of the cross-functional exchanges as possible while still maintaining an effective and efficient trouble-resolution process.

4. The steps in a TQM daily management system (with the possible exception of assessing the statis-tical stability of the process) seem like common sense that should be applied regardless of whether or not the company has implemented TQM. Yet, most telecommunications departments do not have a structured daily management system. Explain why, and identify what is needed to ensure that such a system is implemented.

5. Discuss the following statement and its implications for the long-term success of a corporation: Building relationships with vendors and suppliers should consist of the following: (a) managing a request-for-proposal process, (b) ensuring that vendors compete against each other to yield the lowest price to the buying company, and (c) negotiating a contract that contains strict performance standards with little flexibility and room to maneuver.

Minicase 16-1

American Airlines

The Great Fare Wars of 1992—Down Goes the Network

American Airlines's Sabre reservation system, one of the most famous IT systems in the world, is the core of marketing, customer service, operations, and profit management for the airline that is widely recognized as the industry leader. It is worth well over $1.5 billion—Delta offered to purchase half of it for $650 million in 1989. Some 23,000 travel agencies use Sabre, which typically handles 2,000 transactions a second. The data and network control center is an underground fortress, with six physical layers of security, including sets of reinforced steel doors. Sabre has earned more profits for American than the airline has made from flight operations over the past ten years.

Sabre also crashes. In mid-1992, American took the industry by surprise in introducing a simplified set of air fares, aimed mainly at helping stimulate demand during a recession that has seen the industry lose billions of dollars. When American cut some fares drastically, competitors responded quickly with fare cuts and discount deals, such as taking a companion with you for free. American escalated the war in turn, and prices dropped by the day; phone calls also increased by the day.

As a result, Sabre was completely overloaded, as were travel agents' switchboards. Some travel agents had their offices down for up to two-and-a-half hours, during which they could not make any bookings. Crashes of half an hour became routine. AT&T reported its heaviest day of calls ever; histor-ically, Mother's Day set the record. In many regions, AT&T blocked calls from going through to avoid its own network crashing. American Airlines took Sabre down in some areas of the country for the same reason. Sabre processed 3 million new bookings in just a few weeks, with transaction rates increasing from 2,000 to 3,000 a second at one point.

For travel agents, the situation was a disaster. Many of the calls were from passengers wanting to cancel tickets they had ordered at the old prices and to rebook them at the new price, which could be as much as 50 percent lower. The travel agents lost the commission on the original booking and had to go through additional work to earn half of what they had previously earned. One CEO of a major agency with 300 Sabre locations complained that his people were sitting around unable to do any business and then had to be paid time-and-a-half for the overtime required to catch up. Another agency had to redo hundreds of tickets, but because Sabre does not have the ability to print invoices for discount ticket credits, the agency had to do double the work and lose half the commission as well.

Ironically, in the very same week that Sabre and its competitors' reservation systems overloaded, American announced both a new information management system that will make travel agents' work much easier and an upgrade of its network that should increase transmission rates by 300 percent. This is a shift from Sabre's proprietary, mainframe-centered Agency Data System (ADS), which has been in use for over 15 years and is an obsolete technical architecture that has fallen behind its leading rival in computerized reservation systems, Covia (United Airlines and European partners). The new system, TravelBase, is a much-needed move to a client/server architecture and open systems. It is based on UNIX servers with client PCs running OS/2 on Novell NetWare LANs. The graphical interface is IBM's Presentation Manager; the database interface is SQL. New applications include many back-office sales entry systems, accounting, and customized report generation. There are additional features for security, electronic mail, and network administration. The architecture enables users to add a range of off-the-shelf software and desktop devices instead of having to adopt proprietary Sabre terminals. The system was being piloted in late 1992, with orders being taken in November for 1993 delivery. The new service will cost 10–15 percent more than ADS. Users can upgrade directly to TravelBase; American will continue to provide ADS indefinitely. The new telecommunications network, Sabrenet, migrates transmission from shared leased lines to a private packet-switched capability.

Major travel agents welcome the announcement of the new service, for they are still very angry over the costs and disruptions of the great fare wars. Other airlines' CRS crashed, too, but as the largest single system, Sabre's crashes had greater impact. Covia, the number two in the industry, has already implemented a client/server architecture and consequently has had far fewer problems.

Questions for Discussion

1. Describe the causes of the Sabre reservation system crashes that occurred in mid-1992, and relate these causes to one or more of the quality-of-service elements associated with telecommunications operations.

2. Explain how the move by American to implement a new information management system based on a client/server architecture and open systems will reduce the likelihood of similar network crashes in the future.

3. Evaluate American's new client/server information management system with respect to the four quality-of-service elements of telecommunications operations, and indicate how the new system compares with the original Sabre system using these evaluation criteria.

Minicase 16-2

Motorola

Making Internetworking Secure

Motorola's worldwide network is huge. It includes over 150 T1 lines plus some T3s, three satellites and many earth stations, 500 routers, FDDI LANs, and over 30,000 workstations. Communications systems are built to make access easy and to facilitate adding new users and uses; almost by definition, then, it is difficult to ensure security. Client/ server technology compounds the problem; when you use your ATM card, the system checks your password before allowing you to access the bank's processing system, but client/server technology moves between systems and databases, where it may be essential to reauthenticate the user at various stages. Each server may need to check user authorization. Individual software packages do not address these needs; they assume that security has been handled before they are accessed.

Laptop computers can access the Motorola network from just about anywhere. Dial-in access over public phone lines poses extra security challenges. The software available to Motorola for authentication is limited; it prevents hackers from tapping into the system and can ensure that anyone trying to penetrate the layers of software and communications services is quickly detected. The basic problem Motorola addresses is that interoperability and ease of access and use—the business goal for the internet—directly conflict with security and control. On the one hand, the company is trying to extend who can access it; on the other hand it is trying to limit access.

The sheer scale of Motorola's network makes both interoperability and security difficult to achieve. In general, off-the-shelf software packages work well only up to a certain limit. For example, with 50 or even 500 workstations, it is fairly easy to exchange files while maintaining compatibility and security, but there are over 50,000 electronic mail users on the network. According to Motorola's director of enterprise integration, control is relatively easy in a mainframe-centered system just by adding more computing horsepower, but this does not work for a networked system; there is no single point of control.

Motorola has a director of systems security who works closely with the director of enterprise integration. Their mutual aim is to extend the reach of the firm's network. They feel that its traffic is only half what it could be; "What we are trying to do is make it so attractive that people will not want to be left out."

Questions for Discussion

1. Identify and briefly discuss the major problems faced by Motorola in managing its network.

2. Discuss the difficulties associated with achieving both high levels of interoperability and adequate security and control in a network based on the client/server architecture, and contrast this with a mainframe-centered environment.

3. Discuss both the importance of accessibility in attaining greater use of the Motorola network and the trade-off between accessibility and other quality-of-service elements associated with network operations.

Minicase 16-3

European Public Data Networks

Service Quality

Public data networks (PDNs) play a far more prominent role in large firms' international use of telecommunications than in their use in the United States, mainly because PTTs have until recently remained monopolies. In many countries, there was no choice except the PTT's PDN. The German Bundespost placed extremely tight restrictions on use of private networks through the 1980s. France Telecom even made it illegal to use packet switching over a private network, retaining this technology for its own Transpac network, one of the biggest and most efficient networks ever built.

Service quality on PDNs is regularly surveyed by the European Association of Information Services and the European On-line User Group. Their 1991 report shows very wide variations in reliability. In Northern Europe, between 9 percent and 12 percent of transmissions fail due to problems in PDN facilities. For Southern Europe, which includes Spain, Italy, and Greece, the figure is 25–30 percent. Spain and Italy are spending billions of dollars to upgrade their notoriously inefficient infrastructures, so the figure will improve significantly year by year. When the survey was first carried out in 1986, the overall failure rate was 31 percent; in 1991 it was 20 percent. (Some of these failures reflect user error.) The survey estimates that 14 percent of 1991 failures were attributable to the PDN. The most common problem is busy signals at local PDN nodes, which is an almost cer-

tain indication that traffic levels are exceeding the PDN provider's forecasts. Congestion, equipment failures, and line noise are other typical causes.

Austria had the lowest failure rate, 3.2 percent, but is a very small provider. Of the giants, the United Kingdom has a surprisingly high figure, 11 percent; Germany's rate is 8.6 percent and France's is 16 percent. Spain offers the second least reliable and one of the most expensive services in Europe, with a failure rate of 29 percent.

In almost every country, transmissions routed through dedicated local lines into the PDN node had a far lower failure rate than those sent via dial-up lines: 7 percent versus 17 percent. Only a quarter of all calls use dedicated local access lines. The survey commented that the standard justification made by PTTs for restricting private networks is that they are the "champion of the little guy." This is the argument that telecommunications is a natural monopoly and that the PTT provides a social service; France Telecom has argued this viewpoint strongly. The survey concludes that "it still pays the user to be a bigger guy and to invest in better communications facilities."

AT&T reported about the same time that it was having major problems in many countries in provisioning international toll-free 800 numbers. In one unnamed country, it takes a day, whereas it may take as much as 60 days in others. The average across the 60 countries in which AT&T provides international 800 numbers is ten days. The main problem is insufficient capacity.

The fiercely competitive U.S. market and massive technical innovation means that firms can expect more than a 99 percent level of reliability. Kmart reported just half an hour of outage in its VSAT network for all of 1991.

Questions for Discussion

1. Explain why European public data networks exhibit lower reliability and availability than their U.S. counterparts. Also, explain the wide variability in PDN failure rates among the countries in Europe.

2. The standard justification made by European PTTs for restricting the use of private networks is that the PTTs support and protect the small user of telecommunications. Explain the reasoning behind this concept, and assess whether or not data presented in this minicase supports the claim made by the PTTs.

3. What do you think would happen to PDNs in Europe if the PTTs lifted restrictions on the availability and use of private networks?

Minicase 16-4

The Industrial Bank of Japan International

Linking the Back Office to the Dealing Room

The Industrial Bank of Japan International (IBJI) is a subsidiary of the giant Industrial Bank of Japan. IBJI operates across the globe, including London, the financial center of Europe and handler of 40 percent of the world's daily foreign exchange trading.

IBJI is the Industrial Bank of Japan's international investment banking arm. The parent bank's assets are over $200 billion. IBJI grew very rapidly in London in the late 1980s, taking advantage of the deregulation of securities and trading markets. Its staff grew from 130 to 180 in two years; it increased profits by around 20 percent a year and moved from number 20 to number 12 in volumes of Eurosecurity revenues. It expanded its activities in many areas, including U.S. treasury bonds, Japanese government bonds, futures, equities, and warrants, in a highly competitive market.

One of its key initiatives was to install a dealing room in London in 1985. The system was initially designed around workstations that used "digitizer tablets" instead of the standard keyboard. These tablets eliminate the need to type; the dealer simply presses a finger on the square on the tablet that indicates the relevant bond, rate, currency, and so on. The tablet then inputs the data to the computer system. Dealers can access price and interest rate information and can review current and historical data on bonds, contracts, and positions; they can also connect to a LAN to run analytic software

packages and to a central minicomputer that contains a database on market information.

The initial system has been fairly continuously upgraded, mainly to ensure rapid response and on-line updating of information. The dealing market is extremely time-sensitive, with prices changing almost by the second in many instances. The first system, built in 1985, lacked real-time data input; handwritten tickets recording trades were keyed in as quickly as possible, with frequent bottlenecks at the end of a busy afternoon. A core element of the system was the use of a fault-tolerant computer. Fault tolerant means that if there is any computer "crash," the system does not lose any transactions or data that are being processed at that time. This is very different from a typical PC, for instance, for if it loses power or has a hardware failure, all the items in memory are lost.

The central database used a complex relational software system. A relational database management system cross-references information, so that a dealer can make such requests as "Give me all the bonds that were traded by Customers A and B in June that have dropped in price by more than 10 percent, and that were handled by Dealer C."

There is substantial processing overhead as the software interprets the query and accesses the relevant records. Response times were far too slow to be acceptable to dealers seeking instant information in a time-critical market; they were often as high as 20 seconds. The software was replaced with a less powerful but faster system, and the digitized tablet has also been replaced. As dealers got used to the system, they were perfectly comfortable keying in their own data; most trades required only a few items of input. The digitizers helped dealers at the start, though, so they were of value in reducing fears, embarrassment, and input errors.

Volumes of transactions expanded by 50 percent between 1985 and 1988. The number of PCs linked to the dealer system local area network grew to 85, with another 40 in use elsewhere in IBJ. (This is close to one PC per employee.) There are now three fault-tolerant minicomputers; one of these is used for software development. The original LAN could not handle the growth in traffic and has been upgraded. The main new needs were for network management software and for a network operating system that provides improved file server performance. (File servers are computers that store databases and software needed by PC users linked by a LAN; instead of each PC having to store a copy, they share the file server. Because this may mean many users are trying to access the same database just as it is being updated, file server technology is complex in terms of hardware, software, and telecommunications demands.)

Many stand-alone systems in IBJI are not part of the dealing system; these include customized software for investment analysis, fixed asset accounting, and a multicompany, multicurrency general ledger. The in-house team of programmers has developed special dealer spreadsheets that run on the central minicomputers. These are large and complicated in structure and cannot be handled easily on a standard PC; they access large data files stored on the minicomputer.

The IBJI systems are complex because the business is complex and covers so many countries, currencies, and types of instrument. In addition, the bank was among the first in London to combine the "front office" dealing system with the "back office" settlement system. Most companies developed an on-line trading system but handled settlements through overnight processing. In 1989, one of the five top banks in London discovered that an error

in paperwork input to the settlement system had overstated its profits on foreign exchange by $21 million. The error was not discovered until after it had published its earnings results.

IBJI's system combined the front and back offices; the dealing system links directly to the settlement system and updates the accounting information in real-time. The director of IBJI in charge of developing the system sees this as a direct source of competitive advantage. IBJI had to develop its own general ledger system for this; there were no available accounting software packages for the fault-tolerant minicomputer "because the machines are pitched primarily at mission-critical environments, where most people don't need general ledgers." The many available standard accounting packages did not handle accounting transactions in real-time. Building the system was expensive and had many pitfalls, but "we got the systems we wanted. When you buy a package, you never get what you want, just a set of compromises."

The bank's rate of growth through 1990 required them to expand their offices and prepare to move to a new location in the early 1990s. The telecommunications facilities needed expansion, too. The existing base was the LAN that supported the dealer systems and the minicomputer, with many additional terminals that accessed such services as Reuters; the dealers' desks were crowded with devices. IBJI planned to link the LAN to its Tokyo offices at some future time. It also saw a potential need for national links to other offices, perhaps including customers.

Questions for Discussion

1. Today, the situation described in this minicase looks like a fairly standard use of software and LANs, but in 1988, IBJI was a leader in the London market, which includes every major international bank worldwide. Most other banks had an on-line telecommunications-dependent front office trading system and an off-line telecommunications-independent settlement system. What competitive advantages do you see in having both systems on-line and linked directly?

2. Why did so many banks rely on batch processing? Why do many still rely on it?

3. What potential competitive advantages do you see in IBJI extending its network to link to Tokyo?

4. Fault-tolerant machines are widely used in retailing point-of-sale systems, banking ATM systems, and foreign exchange trading systems, as in IBJI. Why? What difference (if any) would it make if IBJI did not use them, in terms of business operations, telecommunications, administration, and software?

5. Map the reach, range, and responsiveness of the platform described in this minicase.

17

Managing Costs

Objectives

- ▸ Understand the main components of telecommunications budgets
- ▸ Learn about methods for ensuring both high-quality service and cost effectiveness
- ▸ Review six areas of opportunity for managing costs

Chapter Overview

| | | | | | |

Chapter Overview

As telecommunications capital expenditures and operating expenses become an ever larger component of most businesses' core operations, managers are naturally concerned with its costs. This chapter addresses how to manage those costs by first reviewing the main components of the cost base and then discussing six key ways to ensure that needed services are provided at the best cost: (1) choosing standards for the telecommunications platform that reduce the costs of change and of incompatibilities, (2) choosing public, virtual private, or private networks that identify where a variable cost (public network) or large fixed-cost base may be most effective, (3) multisourcing: making intelligent choices between in-house operations and outsourcing, joint ventures, or use of shared facilities, (4) monitoring the timing of investments in new technologies to avoid moving too early on unproven or high risk innovations without waiting too long to exploit the advantages of new developments, (5) choosing vendors with whom you can create a real and mutual partnership and dialogue, and (6) exploiting the many opportunities opened up by distributed computing and location of facilities.

| | | | | | |

Telecommunications Budgets

The amount of money organizations spend on telecommunications depends on a wide variety of factors, including the industry, size of the organization, business priorities, and the number and nature of locations. The following figures, taken from a late 1991 survey carried out by *Network World* concerning budgets for 1992, provide a rough guide to the nature of telecommunications expenditures. Although the figures are somewhat distorted by widespread cutbacks created by recession, the allocation of money to specific areas, such as LANs and WANs, has been relatively stable since 1990.

- Operating budgets are split 50–50 between voice and data, whereas in 1981 voice expenditures typically amounted to 85 percent of the budget.

- In capital expenditures, LANs dominate; 45 percent goes to LAN and LAN interconnection equipment (22 and 23 percent each). Twenty percent is for voice-related equipment (for example, PBX, voice processing equipment), 17 percent for WAN transmission equipment (for example, T1 multiplexers), and 15 percent to network management equipment.

- The average network operating budget, which includes salaries and long-distance access charges, is $6.6 million and grew 8 percent even in the recession when most firms' revenues were flat and profits were falling. This figure shows the growing role of telecommunications as a core business resource.

▸ The average capital budget was $4 million for 1992, an increase of 16 percent. Telecommunications continues to be one of (and often the) fastest growing components of business investment.

▸ Twenty-four percent of operating budgets goes to salaries. This figure has been constant from 1990 to 1992. Line charges amount to 17 percent; in 1990, they were 26 percent and in 1989 were 37 percent. Just over half of this is for long distance. Companies are exploiting both the highly competitive nature of the long-distance carrier market and the wealth of options and are waiting to see real competition in the local market.

There are three main areas of telecommunications cost: capital investment, mainly for equipment; carrier charges (payments to transmission providers for public and private network services); and other operating costs. Carrier charges make up the largest single component of budgets, greater even than salaries; telecommunications managers view this component as the area of largest potential savings. In mid-1992, large companies were paying about 17 cents per minute for long-distance calls; the largest virtual private network users pay as little as 5 cents. In turn, the carriers pay 3–4 cents per minute at the terminating end of a call to the local exchange carrier.

In the equipment market, prices have dropped substantially, just as they have for computers. Most telecommunications equipment is special-purpose computer hardware; because the price and performance of microelectronic chips improve at an average rate of 25–30 percent a year, it is certain that prices of routers, hubs, PBX, and other equipment will continue to fall.

The Service–Cost Double Bind

A 1992 survey of 150 U.S. network managers asked them to identify the most crucial issues they face. The issue most frequently cited was compatibility and interoperability—the new key to service—and the second was cost. Issues 3–10 were lack of network management tools, managing staff, product reliability and performance, network availability and keeping systems running, standards, keeping up with technology, and training end-users. Five out of the ten relate to ensuring service, but cost was among the three most critical issues for almost every respondent.

At the heart of the telecommunications professional's job is balancing service demands with costs. The demand side requires adding flexibility, capacity, and guaranteed access, making network management one of the most important aspects of telecommunications operations, but equally important in most firms is cost management. Telecommunications designers, implementers, and managers of operations face constant pressure from clients to cut costs; this is so partly because telecommunications is a growing and highly visible expense—for a small firm it is often one of its largest costs outside salaries and rent—and partly because the rapid pace of development in the technology and the resulting aggressive vendor strategies raise expectations of continued cuts in unit cost. A third reason is that unless telecommunications is justified in terms of the business logic and business-focused dialogue represented by the telecommunications/business decision sequence, many clients see it as a utility, like heat, light, and power. Thus all they are interested in is cutting its costs; it is a business necessity but only as a portion of overhead.

Clients also want service, though. Telecommunications is fundamentally about service; it is intended to provide convenience; access to people, information, and transactions; and timeliness. Thus it is demand driven. Its users can unwittingly drive up costs and degrade performance through sudden increases in the volumes of messages and transactions. This happened in mid-1992, when American Airlines first introduced its new four-tier fare structure and then responded to competitors' price cuts by halving prices.

Such major network crises are rare, of course, but there are many smaller crises that result in the typical business network being out of service several hours a month, at an estimated $5,000 to $50,000 per hour of lost business. Whenever a public network fails, the effects are nationwide and can effectively bring public and private activities to a halt. As described in Chapter 16, organizations that invest in backup, redundancy, automated network management, and disaster recovery capabilities stay in business when the others are down, but such features cost money to provide.

Providing both best service and acceptable (or even lowest) cost is not necessarily a conflict, but it is a difficult challenge. The network design variable checklist can help anticipate trade-offs early in the business decision sequence. The more "mission-critical" the services on the network, the more businesses recognize that they must pay for flexibility and quality. Unless this is done, the likely priority in design is to provide a given level of capability for a given cost, relegating subsequent needs for flexibility and exposure of inadequate quality of operations to a matter of crossing your fingers and hoping.

Even when flexibility and quality are given added weight in design trade-offs, cost is still crucial. Why pay more when a combination of management decisions can improve costs by as much as 30 percent? Costs can be minimized in the following ways:

1. *Choose standards for the telecommunications platform*. These choices can reduce the later costs of change and can reduce the likelihood of systems disintegration and multivendor, multitechnology chaos.

2. *Choose among public, virtual private, and private networks*. Public networks are largely a variable cost—the more you use, the more you pay—whereas private networks are a fixed cost. Virtual private networks offer the advantages of extra bandwidth on demand and thus variable costs and guaranteed capacity. In general, the choice of private versus public network depends on volumes; the economics of private networking favor the large user.

3.   *Make choices for multisourcing*, ranging from total outsourcing of development and operations; to selective outsourcing of subcomponents; to joint ventures, use of value-added networks, and other forms of sharing; to in-house development and operations. Telecommunications "strategy" increasingly comes down to choices of sourcing.

4.   *Select key technologies.* Steering between moving too quickly on unproven technology and too slowly exploiting the many cost advantages of new developments makes keeping up to date a necessity (and a difficulty) for designers and implementers. Such emerging developments as SMDS, ATM, and frame relay offer many potential cost savings—and headaches; the savings are lost when the costs of implementation turn out to be far higher than anticipated.

5.   *Choose vendors and management of vendor relationships.* This applies to negotiations about prices and contracts and to support and service. The leading vendors like to talk about being "partners" with their customers; customers talk about wanting suppliers to understand their business and to be responsive. In many ways, achieving this parallels the dialogue *Networks in Action* presents, with the roles of designer and (often) implementer being shared by staff from the client organization and the vendor. Given the complexity of today's high-performance networks and the many long-distance, international, and local services and equipment providers they involve, vendor relationships can be a major element in cost management.

6.   *Exploit distributed computing options and location of facilities.* The entire basis of modern information technology is to balance low-cost, intelligent devices, most obviously PCs, with central high-capacity ones, and to balance low-cost LANs with high-performance shared wide-area resources. In general, the economics of a network are improved the more processing is distributed and the more communications are kept local. However, there are many areas of operation in which central coordination is more effective and shared large-scale resources is less costly overall. Consolidation is part of distributed systems, not contrary to it. In particular, the economics of location for international data centers and communication hubs frequently favor recentralization.

| | | | | | |

Instituting Standards: Avoiding Multitechnology Chaos

The concept of a telecommunications platform—instead of a case-by-case choice of technology to meet individual application and business unit needs—is a key to both service and cost. Unfortunately, telecommunications managers and their staff often inherit a telecommunications function rather than starting from scratch. Thus they inherit incompatibility, much of which comes from three sources:

1. Explosive growth of departmental LANs when there were few standards and many technical innovations. Although bridges and routers are easing the difficulties of interconnecting LANs and connecting LANs to WANs, the general situation in large organizations is one of widespread incompatibilities and high costs of administration and support.

2. The development of transaction processing systems in which decisions about telecommunications were made largely on the basis of specific computer vendors and operating systems, and in which proprietary rather than open standards have been the norm.

3. A situation in which technology has outpaced the slow standards-setting process, resulting in many differences in the ways product standards are implemented.

Firms must move toward a platform architecture that ensures compatibility and and interoperability, for business reasons as well as technical ones. In the end it is cheaper to have an overall platform blueprint than not. Companies are spending far more on fees to systems integrators and on equipment to link dissimilar systems than it would have cost to plan the platform in the first place.

Careful choice of standards is a major way to avoid future costs, even if it does not solve the problem of today's multitechnology chaos. Because choosing standards obviously puts a premium on open standards, "open" must be carefully defined. A true open standard (from the perspective of managing telecommunications costs) meets four criteria:

1. It is fully-defined, with no ambiguities or gaps, and is implemented in proven products.

2. It is stable and the interfaces are published, so that equipment and software vendors are not "aiming at a moving target."

3. No single provider or regulator can control the standard, and there is a range of providers from which to choose.

4. It does not require jettisoning existing investments and is consistent with other mainstream trends in standards, so that there should be no irresolvable problems in future interoperability.

Defining Versus Implementing a Standard

Many of the disappointments with the progress of OSI products result from incomplete definitions and gaps in network management; as a result, more and more network designers adopted the proven TCP/IP, which was incompatible with OSI, and SNMP (Simple Network Management Protocol) became a de facto standard. Defining a standard is no guarantee that it will be implemented or that systems using it will be able to interoperate; this was a problem with early implementations of frame relay. *Communications Week* reported in January 1992 that "many vendors have improperly implemented not only the frame relay protocol, but even long-established communications interfaces" that were uncovered in laboratory tests sponsored by the Frame Relay Forum, a consortium of 15 organizations that provide test beds; field trials are still to come. This is why vendor groups' tests of compliance and trials of interoperability can take several years and why they are so necessary.

An example of how tiny differences in implementation can create problems and costs is T1. On a standard T1 link from the United States to Mexico, sending a "1" as the first bit indicates "available to receive"; on a standard T1 link between the United Kingdom and France, it means "busy (that is, "unavailable to receive").

Throughout the 1980s there was a strong tendency among many academics, consultants, and vendors to talk as if a standard was self-implementing, and neglecting the practical issues that must be addressed once it is defined. They ignored lead times. In practice, it takes from eight to 15 years for a standard to be defined and fully implemented. As a result, many effectively "open" standards emerge from outside the committee-based standards-setting process. A proprietary standard may thus turn out to be an open one, just as a standard that is open in definition may be effectively proprietary in implementation (such as was the early situation in frame relay). The ubiquitous MS.DOS became a de facto standard in this way; it was highly proprietary and would not have fared so well had IBM picked one of its two strong competitors for its first PC. Users adopted it; software developers made it the base for their offerings, which further increased adoption; hardware vendors used it to create IBM "clones." This loosened IBM's hold on the market. DOS drove the PC market and made it as fully open as any area of business in the world.

Understanding the differences between definition and implementation helps explain why many very skilled designers believe that some proprietary standards will still play a major role even as the mainstream pushes toward open systems; increasingly, too, providers add OSI products and features, or equipment manufacturers add protocol converters and other means of ensuring interoperability. These designers are looking for the option that is both the most cost-effective and moves in a *practical* direction toward interoperability. Implementers who adopt the same perspective look for products that are the most reliable and are "vendor-neutral."

The literature on open systems throughout the 1980s largely played down or ignored the practical economics of standards. Apart from the issues already raised, they implied that once we have open standards, problems disappear. A survey of 170 users and 59 vendors at a 1991 conference reports a very different view. Respondents were asked to rank "What we want to happen by 1997" and "What we think will happen" on a scale of 1 to 5, with 1 being most desirable (for question 1) or most probable (for question 2):

	1. What we want to happen		2. What we think will happen	
	Users	Vendors	Users	Vendors
Open systems prevail	1	2	4	3
Network-based computing takes hold	2	1	5	5
Technology advances drive market	3	3	3	4
Users stuck with incompatible systems	4	4	1	1
Users locked into proprietary systems	5	5	2	2

What users want is the opposite of what they expect—and of what most vendors' ads and many academic articles promise.

Proven open standards clearly help network designers, implementers, and operators cut costs. Once a standard meets the four criteria previously listed, the market guarantees that prices will fall. Consider the following examples:

► MS.DOS- and MS.Windows-based PCs, for which prices per unit of processing power have dropped by at least 25 percent per year for almost a decade, and software quality and prices have similarly improved, though not at such a sustained rate.

► The growth of the intelligent hub and multiprotocol router markets. This has been practical only because the protocols they support are established and stable, the interface specifications published, and the market open to any firm, old or new. In 1990, both markets more than doubled, to a total of $640 million. By 1992, growth had dropped but was still over 50 percent.

► Ethernet 10BaseT smart hubs. The price per port for these dropped from $233 in 1989 to $54 in 1993. The development of 10BaseT was faster and was less flawed and fragmented in terms of problems of definition or implementation than were most standards developments, because it built on the well-established Ethernet standard. By contrast, FDDI products were rushed to market even before the full definition of the standard was complete.

► Routers. Prices at the low end of the market, which typically support only the most widely established standards, have fallen rapidly and continuously. Prices are below $7,000. However, prices rose in 1991 and 1992 for top-of-the-line routers, which provide more features and are nonstandard.

Standards also greatly reduce training and support costs because they reduce the range of knowledge needed, streamline procedures, and make problem diagnosis easier. Much of the cost of support for PCs and LANs comes from the frequently out of control software and word processing and spreadsheet packages, the plethora of operating systems, cables, and boards, and the like. Given that most telecommunications managers are under continual pressure to cut staff levels and costs, narrowing down the range of standards to be supported can offer large savings.

Obviously, choice of standards affects and is affected by many issues. From the perspective of managing costs, standards are an issue of products and implementation, not of concepts and definition.

Public Versus Private
Networks

Choosing between private and public networks was relatively easy in the 1980s; if an organization had high capacity needs, "big was beautiful" and private networks were the main option, except in countries in which they were not permitted or where the PTT kept costs high to discourage their use. The choice is much more complex now. Three new forces have emerged: (1) the new option of virtual private networks, (2) high-bandwidth services provided over the public network, and (3) the growth of intercompany electronic transactions that span networks.

Private networks are point-to-point; the user leases circuits such as T1 or T3. The Gartner Group, a firm that specializes in forecasts and analyses of the information technology industry, predicts that from 1991 to 1995 the price of private voice-grade lines will drop 8–10 percent a year and that high bandwidth prices will fall 30 percent a year. Fiber optics is the reason; it increases a carrier's transmission capacity at a rate far faster than either its own operating costs or demand for services (though investment costs are heavy). Telecommunications is marked by immense economies of scale, and large users have exploited this. Fiber is central to carriers' abilities to offer such services as SMDS, ATM, and frame relay over the public network as an alternative to private lines. For this reason, most carriers are implementing SONET as fast as they can. Although their users may not need 2.4 gbps, carriers make their money by squeezing out every possible percentage increase in traffic from the bandwidth on a given link.

Until recently, the public network mainly offered convenience and ease of access. For low-volume users, it was the only practical option. Now, switched 56-kbps services are widely available from all major carriers at around 40 percent of the cost of T1 lines. Switched public network services—such as SMDS (Switched Megabit Data Services), which is a primary target for the RBOCs—are closing the price gap between private and public network services to under 10 percent. Frame relay and cell relay allow a carrier to provide bandwidth on demand.

Virtual private networks have the flexible economics of public networks and the management control—or controllability—of private networks. For firms that spend over $25,000 a month on long-distance service, they are an obvious opportunity. The key to operating them is software in the switches. Thus each provider offers very different features in terms of pricing, network management, access and egress facilities, applications that can be supported, and billing. In general, they are like buying a car: The standard model comes cheap but the cost of the features adds up quickly. The distinction between virtual private and public networks is diminishing; SMDS and frame relay, for instance, are offered as both of these by different carriers. The main distinction now is in service level agreements. Virtual private networks generally involve a contract for a minimum given level of usage, with discounted prices for additional volumes plus guarantees of availability. Public networks may have an installation and monthly service charge, but the pricing is on a pay-as-you-use basis.

The growing extension of networks from mainly linking internal locations to linking suppliers and customers means that the straightforward private network is no longer the single base for many large organizations, which generally need to connect their own network to the public network and/or a value-added network.

No one option stands out among public, virtual private, and public networks. The economics are highly situational and are a key element in managing the 17 percent of the typical operating budget that goes to line charges. Generally, the two key issues in choosing among them and getting the best value from the many vendors are building an effective relationship between vendor and customer and having a clear picture of the trade-offs among network design variables. Without these issues, the only real issue is cost, not service, support, reliability, flexibility, or being able to adopt new technology. The less that telecommunications is a considered business resource and the more that it is considered operational overhead, the more cost dominates over all other factors. The more it is a business resource, the more the attention must be paid to noncost issues.

Relationships with Vendors	Vendors increasingly talk about wanting to be "partners" with their customers. A cynic might argue that this partnership adds up to "You write the check, partner, and I'll cash it," but the best companies agree that the type of dialogue Texas Instruments built with its main vendors (see Chapter 9) can benefit them in reducing their own costs and can benefit the vendor through long-term customer commitments. Such a partnership requires that both parties provide the other with information about their business plans and service developments. In addition, it demands that the vendor provide a high level of technical support and fast, responsive service. Given that more and more of the challenges of telecommunications relate to support, support alone can be one of the major ways to manage costs effectively. A top-rate vendor that emphasizes quality of service saves itself and its clients time, trouble, and money.
Multisourcing	The term *outsourcing* has become a common word in the vocabulary of both information systems and telecommunications personnel. Outsourcing is the contracting of activities to an outside firm instead of conducting them in-house. There are three main justifications for outsourcing: (1) The cost is lower, (2) the operation does not require that knowledge and expertise be built and kept in-house, and (3) management time and effort is reduced. Outsourcing is controversial, with many debates—about whether it risks losing control over a strategic resource; whether contracts can be written that accommodate all the many inevitable contingencies of volumes, services, upgrades, and the like that business changes will create over the many years that outsourcing contracts often

cover; and whether the cost really is lower and the expertise as great as the firm's in-house operations.

In practice, the debate is irrelevant. All large (and many small) firms use a combination of sources, with joint ventures, use of shared resources and value-added networks, and cooperative alliances offering opportunities between the extremes of everything in-house and everything outsourced. Many companies are multisourcing selectively—outsourcing, say, network operations but keeping network management oversight in-house. Merrill Lynch did this; it saw no competitive, economic, or organizational advantage to developing skills in network operations, which is the expertise and specialty of the long-distance carriers, so it outsourced operations to MCI. However, it viewed network management as a key issue and insisted on maintaining control of it. The logic was to outsource overhead and keep strategic functions in-house.

Most analyses of outsourcing conclude that the key element is contracting. The contract must address all the uncertainties of future change that may alter the basic assumptions, commitments, and measures of performance on which the original agreement was based. It must specify in great detail what those measures will be and how to deal with support and service issues. Outsourcing may look simple, but is not at all so. It requires careful planning and a continuing relationship and dialogue with the outsourcing firm. Then it can both save money and enable the in-house telecommunications unit to focus its efforts and skills on leveraging the business without having to worry about managing operations.

Choosing Technologies

In mid-1993 the designers and implementers of a large firm's network could choose among the following technology options for a WAN: private fiber, VSAT, SMDS at first at the local and then the national level, ISDN, long-distance switched 56-kbps services, T1, T3, virtual private networks, frame relay, cell relay, SONET, and X.25 value-added networks. For a LAN, options included FDDI fiber, token ring, wireless LANs, 10BaseT, smart hubs, bridges, and routers—the list may not be endless, but it is very long.

All these options involve trade-offs in network design variables, mainly between capability and cost, flexibility and cost, and quality and cost. Cost is the anchor measure; if it is all that a client's management is interested in, then there is little to discuss. Otherwise, the issue is how to provide a given level of service, flexibility, and quality at the best price. A useful rule of thumb is that the best managed telecommunications units are 30 percent more cost-efficient than the average. Timing the choice of adopting an emerging technology is crucial here.

Generally, a new technology favors particular types of traffic and levels of volume. For instance, T1 private networks provide the best cost-per-unit for businesses whose traffic flows are fairly predictable and do not vary much within a given period. The user can fill up the T1 link on an even flow. Virtual private networks or SMDS are far more cost effective for large-volume users with less predictable traffic peaks and troughs. Frame relay suits those that mix short, bursty message traffic with longer, megabit and gigabit data such as engineering designs and databases. There is no tidy rule about what is the "best" technology; the key is that the best for a given type of traffic and type of variation in traffic is very much less costly than for those of another mix.

Thus we return to the basics of the telecommunications decision sequence: It is impossible for a designer and an implementer to select a technology without knowing both the type of business traffic and the relative priorities for the network design variables.

Choosing Vendors   Standards mean little without products, and products mean little except in relation to the quality of the vendor in terms of price, commodities, or service, support, and ability to work with the client. Just ten years ago a PC was a premium "high-tech" product; now it is something to order over the phone. LANs are highly commoditized, as are T1 circuits, but just a decade ago looked like a fantasy. End-to-end network management is not commoditized; nor is vendor support for a multitechnology, multivendor, multinational network.

Commodity products require little dialogue between supplier and customer; premium products demand it. With commodities, it may not matter much if the supplier goes out of business or does not spend enough money on research and development to keep up with its competitors. For plug-in-and-go components, it is irrelevant whether or not the vendor understands the customer's business. When the network is crucial to the business, however, and responsive support and service are essential, that understanding may be vital.

In the ferociously competitive long-distance marketplace, it is common for large customers to play competitors off against each other, looking for lower and lower prices. They can afford to do so. After all, transmission is now just a commodity, and no carrier can hope to establish a technical advantage it can hold for even two years. Fiber optics is creating massive overcapacity; the customer is in the driver's seat.

Interestingly, firms are frequently spending several million dollars more than the difference between the bid prices of two suppliers in a consulting firm's study of strategic options for telecommunications. There is no contradiction here. In telecommunications, the added value is increasingly business expertise, not product features. In general, the telecommunications industry's sales and marketing staff are far more knowledgeable about products than about the use of products. Customers want more understanding of business and more of the skills of business consultants than people who can just spout the details of bandwidth, enhancements, and special features. In many instances, though, they are not willing to pay for these skills, putting the sales reps in a difficult situation; margins are low for commodity products and services, yet they are expected to combine the lowest price with the best service, support, and expertise.

Obviously, every customer is looking to get the best price, but just as in trading off cost and flexibility or capability, there is a similar trade-off between price and value-added expertise. In general, the less important expertise, support, and service are, the more this area is a candidate for outsourcing, and the less critical is the choice of vendor versus the choice of product.

Exploiting Distributed Systems and Location-independence

The last major area of opportunity for managing costs is the most technical: exploiting location-independence. Before the era of high-speed, cost-effective communications, there were few ways to configure a network or to choose the location of data centers. Now the options are almost infinite. Instead of having to decide between centralized mainframes with high costs of accessing them over slow and expensive lines or over departmental minicomputers with limited interconnection (the main choice just ten years ago), companies can use a combination of LANs, which by themselves reduce communications costs; client/server computing, which reduces the costs of accessing remote central machines; and consolidation of data centers to place them in the location that offers the best combination of labor costs, quality, and telecommunications costs. For example, one large European pharmaceutical firm headquartered in Switzerland relocated its main data center to the west of England, where it also consolidated its other ten European data centers. Swiss telecommunications costs were very high and service and support very poor. The company exploited Great Britain's low international costs and was also able to recruit and retain a small group of top-level personnel in such areas as network design and operations, network management software, and operating systems. Its smaller data centers had been unable to attract such people; they could not offer salary, scale of operations, or opportunity for career growth and promotion.

By contrast, many firms are aiming to move as much traffic off the central mainframe and WAN as they can; they rely on client/server computing built around LANs. Routers replace many functions of switches and backbone WANs. The difference between the move to consolidation via the WAN and distribution via LANS is primarily the firm's economics. In the international field, where UK prices for leased circuits are often half to a third of those of Spain's Telefonica and Italy's Italtel, consolidation offers many cost opportunities (and also service and quality; it is an ironic outcome of PTT monopolies that the most expensive providers often have the worst quality). By contrast, if it were located in the United States, the European pharmaceutical firm might well have decided to distribute its operations instead of consolidating them.

Fitting the Pieces Together

The price trends of most aspects of telecommunications are fairly predictable; the only exception is international costs, which depend heavily on PTTs, regulation, and international agreements. U.S. businesses and consumers are penalized by over $10 billion a year by distortions in the formulae by which the provider in the country from which a phone call is made pays the international equivalent of an access charge to the receiving country's provider. On average, U.S. carriers thus pay out 74 cents for each long-distance dollar of their revenue for international calls. In the case of a call from the United States to Brazil, the figure is 99 cents. Many lesser developed countries regard *incoming* international phone calls as their main source of revenue and, not surprisingly, oppose changes or private network facilities that bypass the public network. There is little that an individual company can do to affect this situation or manage the relevant costs better. The U.S. and UK governments have taken the lead to end the distortions, but it may take years.

In most other cases, price per unit for services and equipment can be fairly well extrapolated from the trends over the past five years. Depending on federal and state legislation and regulation, local transmission costs should begin to fall closer to those of long-distance costs than has been the case since divestiture; the more competition the RBOCs face, the more they and the new competitors will use price as leverage. Equipment costs should drop by at least 5 percent, and up to 30 percent a year. Fiber optics guarantee that unit costs of transmission will not increase and ensure that there will be no shortage of transmission capacity.

However, even though the costs of the components of a firm's network can be predicted, the costs of the network cannot easily be. The reasons for this are mainly that business change leads to changes in the type and volume of traffic, and that technical change may do the same. For example, today's networks carry mainly voice, messages, and transactions; by 1995 they will be flooded with image and multimedia. One large insurance company calculates that the industrywide shift to managing electronic documents through image processing will increase the number of bits transferred over its WAN by at least 19,000 percent by 1995. Electronic data interchange and exchange of digital designs between manufacturer's and customers' engineers similarly exponentially increase the bits moving through a high-bandwidth network.

Managing the flow of bits is the challenge for implementers and operations staff; anticipating the flow of costs is a job for clients and designers. New business initiatives must be reviewed in terms of what they may mean for traffic mix, new bottlenecks such as lack of fast switching or slowed response times, and added support and service costs. Some of these may change the basic design principles and priorities and require major new investments in the network. For instance, the insurance firm facing a 19,000 percent increase in traffic over the WAN is actively looking at making as much of its image-processing operations LAN-based as it can; this may not only change the design of the network but the design and location of work as well.

Historically, managing the costs of telecommunications has been viewed in terms of cost containment achieved while maintaining required levels of service. At the extreme of the early-1980s internal telephone utility, in which the "manager" was little more than a supervisor, his job (there were few women in the field) was to cut costs while ensuring the phones worked. He made the case for new investment on the basis of cost savings. That mentality still lingers in some companies, but in more and more firms the new priority is service and business/technology integration at the best cost. Cost remains very important in most firms, but it is now part of an often complex trade-off among many aspects of current service, provision for future service, and guarantee of business quality through network quality. The service-cost double bind is at the core of telecommunications planning, and the trade-offs rest on dialogue.

Summary

Telecommunications budgets are split 50–50 between voice and data communications, with data by far the faster-growing expenditure. LANs dominate WANs as a percent of capital investment costs. Salaries amount to about one-fourth of the operating cost base.

The three main elements of telecommunications costs are capital investment, mainly for equipment; carrier charges, the largest single component of the budget; and operating expenses. Telecommunications managers rank compatibility and interoperability as the most critical issue they must address and cost as the second most important. Cost is one of the top three issues for almost every manager surveyed.

The challenge for the telecommunications function is to balance service and cost. There are six main ways to achieve this: (1) choosing standards; (2) choosing among public, virtual private, and private networks; (3) multisourcing, including options for outsourcing; (4) selecting key technologies; (5) choosing vendors and management of vendor relationships; and (6) exploiting distributed computing and location of facilities.

Review Questions

1. Evaluate the following statement: Typical telecommunications operating budgets for 1992 were dominated by expenses for voice communications, and almost half the budget was allocated to WAN equipment. Indicate whether or not telecommunications budget allocations have been stable over the past few years.

2. In the opinion of network managers themselves, what are the two most critical issues they face today?

3. Why are telecommunications managers faced with the difficult bind of having to provide high levels of service at reduced cost? Doesn't senior management understand there is a trade-off between these two elements? Explain and discuss.

4. Identify the means at management's disposal to significantly reduce the costs associated with delivering a high level of network service quality and flexibility.

5. Explain the cost trade-offs among public, virtual private, and private networks for both small and large user organizations.

6. Explain how the choice of vendors and the management of vendor relationships can be a major factor in cost management.

7. Explain the role of distributed computing architectures and the significance of location of facilities in cost control and management.

8. Telecommunications managers often inherit multiple, incompatible networks developed to solve a particular problem, rather than an overall platform architecture. Identify the primary sources of incompatibility among networks.

9. What must be included in the definition of a true open standard from the perspective of a manager responsible for telecommunications costs?

10. Explain and discuss the meaning of the following statement: Defining a standard is no guarantee that systems built to that standard will be able to interoperate.

11. What is the impact of well-defined, implemented standards on training and support costs (short-term and long-term)?

12. Explain how the new high-bandwidth public network services and virtual private network capabilities are changing the trade-offs between public and private networks.

13. Consider the following statement: Telecommunications should either be totally outsourced or totally managed in-house—there should be no in-between. Do you agree or disagree? Explain your reasoning, and if you disagree, indicate the appropriate criteria to use in deciding which telecommunications activities to outsource and which to continue to manage internally.

14. Certain types of technology are optimal (that is, the most efficient for the least cost) for particular types and volumes of traffic on a network.

Therefore, given a variety and mix of types and volumes of traffic, how does one choose among the variety of technologies available for both local and wide area networks?

15. Under what conditions does it make sense to pay a price premium for a telecommunications product or service in order to obtain vendor expertise and support?

16. In Europe, it often makes sense for a company to consolidate its data center operations in a single location, but if the user were operating in the United States, the best choice is more likely to be distributed databases and computing architectures, rather than consolidation. Explain and discuss why these two different approaches can both be correct.

17. Explain and discuss the meaning of the following statement: The costs of the components of a firm's network can be predicted fairly easily, but the costs of the network cannot.

18. Which role in the dialogue among client, designer, and implementer should focus on the impact of new business initiatives on traffic mix, on new network bottlenecks, and on added service and support costs? Explain.

Assignments and Questions for Discussion

1. You have been hired as a telecommunications consultant by a Fortune 500 company to prepare a brief report that evaluates the cost impact (short-term and long-term) of having a unified telecommunications platform architecture, compared with maintaining and managing a portfolio of individual networks that are largely incompatible and cannot share information directly. Prepare such a brief report.

2. Discuss the current cost implications (short-term and long-term) for an end-user that adopts the OSI open system standard, instead of the TCP/IP closed system standard, for choosing internetworking equipment.

3. Historically, managing the costs of telecommunications has been viewed in terms of containing costs while maintaining required levels of service. The case for new investment was made on the basis of cost savings. Discuss the changing role of positioning and managing telecommunications costs in a progressive company of the 1990s.

4. Review the data on the relative allocation of money in telecommunications budgets in 1992 among the following categories: (a) overall operating budget expenditures; (b) proportion of operating budget split between voice and data traffic; (c) overall capital budget expenditures; (d) proportion of capital budget split among LANs, WANs, and voice-related equipment; and (e) proportion of operating budget allocated to salaries. Using this information together with trends in corporate management and the use of technology, project the relative allocation of money in telecommunications budgets among the five categories for the year 1997, and explain your reasoning.

Minicase 17-1

Laura Ashley

Outsourcing Distribution

Laura Ashley lost $16 million in 1991, with a 10 percent decrease in sales to $453 million. This represented a significant *improvement* over the previous year. This fashion retailer had been very successful with its distinctive fabrics in the 1980s but failed to keep up with changing tastes. It has had to make many changes. Some 3,000 jobs were cut at a cost of $8.5 million, and the U.S. headquarters relocated, costing close to $2 million.

Laura Ashley faces many problems in basic operations and logistics; only 65 percent of merchandise arrives on time. An executive described its delivery systems as "complex, costly, and inefficient." It operates stores in 28 countries and has suppliers in 40. It needs to change its product mix, add new stores in North America (where it is reasonably profitable), and improve its profit margins instead of discounting in an effort to increase volumes. Laura Ashley has neither the experience, skills, or money to try to match the electronic logistical capabilities of a Wal-Mart, a Toys "R" Us, or a Circuit City, so it has looked elsewhere for help.

In the spring of 1992, the firm signed a ten-year contract with Federal Express amounting to around $300 million. Fedex will manage Laura Ashley's entire inventory and shipping through its Business Logistics Service. This major new thrust by Fedex had sales of over $400 million in its first full year of operation. Business Logistics manages its customers' entire supply chain, including their warehousing, building on Federal Express's almost legendary global electronic tracking systems, experience in time-based logistics, and unmatched customer service.

Laura Ashley will save 10–12 percent of its total distribution costs. Because it gets access to Federal Express's global systems, it will save at least $5 million on investment in basic improvements to its existing software systems. It should also get substantial reductions in inventory and hence improvements in working capital. It will be able to expand its important global mail-order business.

Laura Ashley has made a similar deal for international freight and supplier logistics with the San Francisco-based Fritz Companies. Fritz will assume responsibility for coordinating worldwide shipments of raw materials and manufactured goods; it will handle freight consolidation, vendor notification, purchase-order tracking, freight forwarding, and customs brokerage. As a Laura Ashley executive commented in early 1992, "we wanted to strip out as much as we could from our lead times and to do that you need visibility in the flow of goods and information around the world. We retain the responsibility for communicating with our suppliers and buyers, but this approach has relieved us from having to communicate with shipping agents in overseas offices."

Fritz's worldwide communications system provides the visibility. Its director of integrated logistics systems argues that the old "we move it, you catch it" style of forwarding is obsolete: "Anytime you interrupt the process, there's a delay that creates uncertainty, as well as inventory."

Laura Ashley's CEO is Jim Maxmin, who turned around THEi (see Minicase 1.1). Maxmin's program for Laura Ashley is called "Simplify, Focus and Act." The outsourcing of end-to-end logistics to Federal Express and Fritz will "enable us to focus our own resources on activity where we can add to the core business."

Questions for Discussion

1. Describe Laura Ashley's problems in basic operations and logistics that, in part, are responsible for reduced earnings.

2. Describe what Laura Ashley gives up, if anything, by outsourcing its inventory and shipping operations to Federal Express (in the United States) and outsourcing freight and supplier logistics with Fritz Companies for its international shipments. Is the reduced cost and improved efficiency from outsourcing worth the trade-off?

3. Explain how Federal Express's Business Logistics Service is able to maintain or increase service quality while reducing cost.

Minicase 17-2

Missouri Division of Data Processing and Telecommunications

Saving Money by Never Buying a Switch

In early 1989, the state of Missouri commissioned a study on ways to reduce telecommunications costs over a five-year period while upgrading service and performance by beginning the move to voice/data integration. Today, Missouri is a highly cost-efficient user of telecommunications and an innovator in many areas of telecommunications application. The secret of the state's Division of Data Processing and Telecommunications's (DDPT) strategy is not owning switching equipment.

It got rid of its old and inefficient analog PBX that provided Centrex service to 13,000 lines and contracted with United Telephone (UT) to install and run a new Northern Telecom DMS 100 digital central office switch. That saved $600,000 a year. By consolidating the state network with other customers of United Telephone (the local carrier), the state spread UT's overhead and thus got a lower price from it; previously, UT had charged a special tariff for the private state network run through the PBX.

DDPT requested competitive bids to upgrade its WATS and long-distance lines to tariffed data circuits that would piggyback over voice on a T1 backbone. Major needs were to link this network to the state university system's MORENET and to networks used by the Department of Corrections and by Department of Mental Health. Southwest-

ern Bell and United Telephone made a cooperative proposal and won the bid. The network was completed in August 1991, with ten of the state's 23 sites linked to it. The first phase focused on voice and the second on data. Southwestern Bell set a tariff that accommodated the rapidly growing number of LANs that needed to access the state's remote mainframe computers.

Originally, the LANs were IBM token-ring networks transmitting at 4 mbps and accessing the mainframes over 9.6-kbps circuits. When the LANs were upgraded to 16 mbps, these circuits were far too slow. Southwestern Bell upgraded transmission across the WAN to 56 kbps. The carrier manages and owns the switches and has added many valuable features for quick line reconfiguration, bandwidth on demand, and automatic tracking of calling records of state clients. The state saves over half a million dollars a year through the new network; much of this saving comes from reduction in staff needs and the ability to add services as needed, when needed. DDPT has added video to the network, made recommendations on sites, and requested competitive bids for equipment. It was able to add the needed bandwidth just by ordering additional T1 services. In 1992 it began to develop another five-year plan. According to the assistant director of DDPT, "the good news is that we don't own any equipment so that we are free to make changes as we want. We can move in any direction."

Questions for Discussion

1. Explain how not owning a switch saved the state of Missouri money on telecommunications and avoided the service-cost double bind.

2. Explain how the Missouri Division of Data Processing and Telecommunications actually integrated its voice and data requirements, and how this led to reduced cost for the overall network.

3. In addition to the benefit of reduced cost, which of the main network design variables is significantly enhanced by the new system adopted by the Missouri DDPT? Explain why.

Minicase 17-3

Electronic Mail

Counting the Hidden Costs

The hidden costs of PC network electronic mail are analyzed in detail in a 1992 report, published by Ferris Networks, entitled *Integration of PC Network E-mail: Planning, Product Evaluation, Implementation*. It shows very clearly how in a networked environment the direct software and hardware costs of the PC are but a small fraction of the whole. Ignoring the cost of the PC itself (because it is already in place and therefore a sunk cost) and ignoring the LAN cost for the same reason, adding a mailbox costs $350 to $450 a year.

The workstation software costs around $25 per PC (the Ferris report calculates costs for a 1,000-PC network electronic mail service at a single site); extra communications software brings this up to $60–100. Message servers must be added to the LAN; generally, each server can handle about 100 mailboxes, which can be standard low-cost PCs with a network adapter card and cable. That adds another $20 per mailbox. In addition, message transmission PCs must be added to the message servers; these need a modem and phone wiring in order to access outside users over a public electronic mail service, such as MCI Mail. That costs $25 per mailbox. Adding fax servers, a public mail gateway, and a gateway to the corporate WAN brings the total installation cost up to $174 per mailbox, a large jump from the $25 for the basic software.

The major costs, however, are the annual oper-

ating expenditures, which amount to twice the installation cost—well over $300 per year. Local technical support requires one full-time equivalent personnel (FTE) for every 40 PCs in a typical simple LAN environment. About 10 percent of this network support staff's time goes to supporting electronic mail; that means one FTE per 400 mailboxes. Central technical support is needed to ensure LAN-WAN integration, maintenance, upgrading, synchronization and propagation of organization wide electronic mail directories, and other functions that can be complex in a distributed environment and immensely complex in a client/server one. Together, technical support adds an annual cost of $225 per mailbox.

Communications to remote locations via X.25, SNA links, fax, or dial-up adds $70 per mailbox per year, and product maintenance brings the annual operating cost up to $313 a year. Amortizing the installation costs over three years results in an annual cost per mailbox of $386. Technical support is almost 60 percent of this total, and product costs, including maintenance, make up just 25 percent.

The purchase and operating cost of a PC LAN varies widely, but Ferris estimates that an annual expenditure of $5,000 per workstation is common. The fraction of the cost to be allocated to electronic mail for purposes of planning should be around 10 percent, if that is the fraction of network staff's time spent supporting e-mail. Add the resulting $500 to the $386, and $25 of e-mail software has generated close to $900 a year of expense. It may well be worth it, but the business case is very different depending on how many of the hidden costs are overlooked.

Questions for Discussion

1. Classify the distribution of the full costs of e-mail operation among the following telecommunications budget categories: (a) the portion of the operating budget that does not include salaries and benefits of employees; (b) the employee salary and benefit portion of the operating budget; (c) the portion of the capital budget allocated to LANs and LAN interconnection; (d) the portion of the capital budget allocated to WAN equipment; and (e) the portion of the capital budget allocated to message-processing equipment.

2. Describe several methods for extending a company's e-mail system to remote sites outside the company, and assess the cost impact (both operating and capital costs) of such an extension.

3. Propose a method for identifying and measuring the benefits of an organization's e-mail system that can be compared against the full cost of such a system, to determine whether e-mail is truly worth the cost.

Minicase 17-4

Massachusetts Department of Public Utilities

Setting ISDN Rates

In February 1991, New England Telephone (NET) filed its first proposal for its ISDN tariff. In 1992, the Massachusetts Department of Public Utilities (DPU) launched a comprehensive investigation of the proposal that showed the practical and often very different problems regulators must address and the many legal conflicts, uncertainties, and calculations rate-setting involves. The basic concern of the DPU was whether ISDN is likely to become a mainstream service to businesses and consumers or a specialized niche service. Its social responsibility is to encourage the diffusion of telecommunications services to the benefit of the community. If those services are underpriced, there is no incentive for carriers to market them aggressively; if they are overpriced, customers will not buy them.

NET's initial proposal was to offer ISDN to customers served by 18 of its central offices by the end of 1993, and to extend it to 22 other COs by the end of 1994 (NET has just under 300 central offices). There would be a monthly charge of $5 for the digital subscriber line, plus $8 to $75 a month for basic services. Costs per unit of ISDN voice would be set at the same rate as for existing voice services; data rates would similarly be the equivalent of switched 56-kbps and packet-switched tariffs.

The DPU may approve a tariff as filed, in which case the provider may bring the product to the market, or it may suspend it for up to six months of

investigation. "Intervenors" can then join the investigation. In the case of the NET ISDN proposal, intervenors included many competitors of NET, who objected to the local exchange carrier's potential exploitation of its regulated monopoly position. MCI and Sprint (long-distance carriers), Prodigy (on-line information services), and the New England Cable Television Association were key intervenors. Prodigy argued that ISDN could allow it and other information providers to greatly expand and accelerate their services, especially by making multimedia services practical and cost-efficient; a major limitation of Prodigy is its slow speeds of transmission and unattractive presentation format.

NET estimated that only 2–3 percent of its customers would adopt ISDN by the year 2000. Its proposal set its prices to recover investment at that level of usage. It also argued that ISDN is price inelastic and that cutting prices would not significantly increase demand. Prodigy countered strongly that NET's rates were artificially high and that it should set its prices at its marginal cost, not its average cost at the assumed number of subscribers. If Prodigy is correct, NET's demand forecasts are far too conservative; if it is wrong, NET could lose substantial amounts of money. MCI and Sprint took the same position as Prodigy.

The crux of the argument was not so much price as whether or not ISDN is a basic service that fundamentally affects how people communicate.

NET's Proposed Rates

	NET Proposal	NET marginal cost	DPU approval	New York rate	Maine/ Vermont rate
Digital subscriber line	$5.00	$7.40	$8.40	$10.00	$5.00
Circuit-switched data	22.00	2.50	5.00	2.00	22.00
Low-speed packet data	8.00	3.30	6.60	2.00	8.00

NET argued that ISDN is not a basic service. Other commentators strongly argued that this becomes a self-confirming prediction. Just as the mass profusion of PCs created a market for new software products, mass diffusion of ISDN as a basic service would generate spin-offs; they pointed to the explosive growth of portable fax machines that was stimulated by the combination of low-cost hardware and deep price cuts in long-distance phone rates. New customers create new uses and new users; if libraries or doctors, for instance, subscribe to ISDN, so too may patients, schools, and even households.

No one can yet reliably predict whether ISDN will flourish. It is late, being built on obsolescent or even obsolete standards, and has not yet generated much excitement in the marketplace. On the other hand, every commentator accepts that at some stage universal digital services to businesses and homes will be the norm and that this may generate services and demand for services that do not exist but that can be brought into existence only if the ISDN platform is in place.

The DPU accepted that the forecasted demand for ISDN was a critical element in setting rates and that there was no reasonable and reliable forecast. It tried to calculate a rate that should help the market grow and enable NET to recover its own costs "within an acceptable time period." It used the ISDN rates already established in New York and Maine/Vermont to help guide it. The following table shows examples of NET's proposed rates, its marginal cost, the other states' tariffs, and the DPU's recommendation:

The differences are substantial. For example, New York's tariff is $2.00 for circuit-switched data per tariff unit, whereas NET proposed $22.00, the same rate that applies in Maine and Vermont. NET's marginal cost for this is just $2.50—its average cost would be high if demand meets its low estimate. The DPU adjusted NET's figures to provide volume discounts and to keep them closer to its marginal costs. In a few instances, it set the tariff below the marginal cost. For example, NET proposed that the first minute of a basic data call be 10 cents in the western part of Massachusetts; its marginal cost is 6 cents. The DPU approved a rate of 2.6 cents.

Questions for Discussion

1. Explain why the prices for ISDN service approved by the Massachusetts Department of Public Utilities depend on whether ISDN is likely to become a mainstream service to businesses and consumers or a specialized niche service. Should ISDN prices depend on how the service is used? Explain your reasoning.

2. Explain why you agree or disagree with the following statement: Tariff rates for telecommunications services should never be greater than the marginal cost of providing the service. (Note: "Marginal cost" means the additional or incremental cost incurred in supplying one additional unit of service.)

3. Explain why the Massachusetts Department of Public Utilities is involved in determining the price of ISDN services to the public, and describe what is likely to happen if the department decided to let the free market determine prices for ISDN service.

Five

The Telecommunications Career Agenda

18

Telecommunications Management and the Organization

Objectives

▶ Learn how telecommunications has evolved within organizations and what that evolution means for career opportunities

▶ Understand the main principles for organizing telecommunications and the main roles involved

Chapter Overview

| | | | | | | |

Chapter Overview

Historically, telecommunications and information systems developed along separate technical, managerial, and organizational paths. Now, they are more interdependent. This chapter reviews the evolution of the telecommunications function in large organizations and identifies the principles for establishing a unit that provides effective service and makes telecommunications a business resource.

This evolution has occurred in three main stages. First, in the telecommunications operations era, it was little more than an overhead function. Second, the internal utility stage, analogous to a heat, power, and light utility for which cost control and reliability were the main priorities. Third, the stage wherein telecommunications is an integrated business resource, the era that is the topic of this book. Each of these eras requires different skills, and each builds on the others.

This chapter reviews those skills and management priorities the telecommunications unit needs in moving to the era of the integrated business resource. It needs to balance central coordination of the platform with decentralized business unit autonomy in using the platform and ensure that charges to users of the telecommunications resource do not impose unreasonable costs, stifle innovation, or create bureaucracy.

| | | | | | | |

The Evolution of Computers and Telecommunications in Organizations

Telecommunications has moved through three fairly distinct eras in large organizations (Keen, 1986):

1. *The operations era*: This was the period of "hands-on" supervision of a telephone and telex unit. Planning was a matter of ordering from the telephone company; operations was a matter of ensuring nothing went wrong, and, when it did, phoning the phone company. Telecommunications was little more than an overhead function. In the United States this era lasted until divestiture, but it lasted far longer in countries where the PTT maintained total control and the technology remained analog telephony. Note that most of the procedures for managing telecommunications as a business resource were relatively new in many organizations and that managers now in their forties and fifties learned their skills in the operations era.

2. *The internal utility era*: In the early 1980s, even before divestiture, many organizations recognized that telecommunications was becoming a growing cost and that there were now two very different types of telecommuni-

cations activity: analog voice, which remained primarily an issue of operations, efficiency, and cost control, and limited data communications, which were an extension of computer operations and other telecommunications-dependent applications. Planning, cost allocations, and service became more of a priority than before and required a more coordinated approach. In the operations era, a firm might have literally dozens of voice communication operations that needed to be rationalized scattered across locations and business units; the reaction was largely to build a centralized unit that too easily became a bureaucracy. Business management's priority was to get telecommunications under control, especially when divestiture transformed the previously tidy world of AT&T as the only vendor to a world with multiple vendors, new types of negotiation, growing demand for data communications, and far more responsibility in the hands of the telecommunications manager.

3.   *The integrated business resource era*: The transition to the third era depended far more on business management's enlightenment about the new role of telecommunications as a key part of the core fabric of the business, rather than as an expensive subset of administration and operations. In many instances the shift was led by the head of information systems, who first built relationships with senior business executives that put "IS and competitive advantage" higher on their agenda and then either formally or informally added data communications to the IS function. In some companies, this left the traditional voice telecommunications unit as an increasingly separate and subordinate unit, with IS more and more in charge of key network decisions. The increasing recognition in IS of the importance of both LANs and integration of computing and communications accelerated this process.

Today, a number of organizations are in the third era, in which telecommunications is in itself either a strong unit or plays an important role in a strong information services unit; a far larger number consider telecommunications more than a utility but less than a true business function; and some remain stuck in the old organizational models and modes of management. It is difficult to estimate the relative distribution among the Fortune 1,000 firms and similar-sized public-sector organizations, but the weight of evidence from cases, surveys, and research studies suggests that it fits the standard bell-shaped curve, as shown in Figure 18-1.

From the perspective of helping you build a career in telecommunications, the focus of *Networks in Action* is to provide you with the knowledge and techniques to help shift the curve to the right in an Era 2 organization and to be a real contributor in an Era 3 one.

Managing firms in each of the eras requires different skills and roles and different managerial priorities. Each builds on the other. For instance, solid basic operations skills are still essential, even when the main focus of the telecommunications organization is on building the business/technology dialogue;

Figure 18–1

The Distribution of Today's Large Organizations Among the Three Eras of Telecommunications

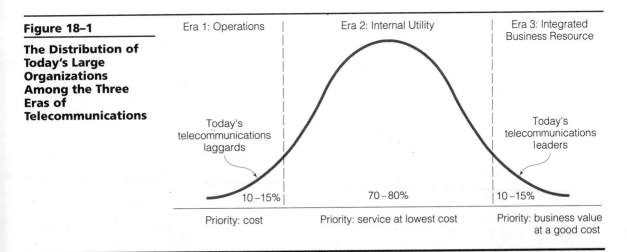

Era 1: Operations | Era 2: Internal Utility | Era 3: Integrated Business Resource

Today's telecommunications laggards

Today's telecommunications leaders

10–15% | 70–80% | 10–15%

Priority: cost | Priority: service at lowest cost | Priority: business value at a good cost

when the network is down, the company's business strategy may suddenly become irrelevant. Similarly, the administrative skills of the internal utility provide the base of efficiency needed to ensure that the telecommunications platform is well managed both fiscally and relative to the business service.

However, being in one era should not block movement to the next. In many companies it has been difficult or even impossible to move on from the internal utility stage to the integrated business resource stage. In almost every instance, the main causes were lack of business/technology dialogue, either because the telecommunications managers' mind-sets were too focused on technical efficiency and operations (which prevented looking for business opportunities), or because the business managers lacked interest in actively viewing telecommunications in business terms (which also prevented seeing business opportunities). Most organizations in 1993 are a mix of the internal utility and the business resource modes, and telecommunications managers are often most frustrated by the lack of dialogue.

The Evolution of Information Services and Telecommunications

Until the late 1980s, telecommunications and computers evolved along separate technical, professional, and organizational paths. Now, all three of these paths and both the fields have come together, to the extent that it is increasingly difficult to discuss them separately. In addition, the barriers of culture, language, and interests that divide telecommunications and computer groups on one side and business units on the other have been breached, if not removed entirely. In 1970, technical specialists and businesspeople were almost invariably physically and psychologically isolated; telecommunications was part of operational overhead, responsible for telephones and telex, and the data processing (DP) department built transaction processing systems and ran large computer centers.

Both the IS and telecommunications fields have made major shifts in roles and knowledge. IS faced challenges much earlier and thus in many ways is better positioned to enter the business/technology dialogue. In 1980, business units were beginning to create their own experience base in end-user computing and office technology; there was no similar movement toward end-user telecommunications, which was impossible in the era of slow analog voice, regulation, and high costs.

The new computer hardware and software vendors generally bypassed the central DP fortress, creating a new internal market and set of skills in, for instance, word processing and financial modeling. When PCs began to appear on the office landscape in the late 1970s, many DP units tried to either block or control them. By the mid-1980s, the better ones had recognized that PCs were not little computers for amateurs, but a key new element in business. IS professionals began to focus on service and support, to link to business units through management education programs, and to develop a wide range of business-oriented and organizational skills. By the second half of the 1980s, conferences in the information services field looked like strategic planning meetings. The programs focused on such topics as information technology and competitive advantage, the new role of the chief information officer, and building management awareness. At times it seemed almost fashionable for IS managers to boast about their lack of technical knowledge in their zeal to be business-centered.

As a result, the IS field is now far more sensitive to business needs than before. This sensitivity did not develop in telecommunications. At the start of the 1980s, as the IS field opened up to business inputs and responsibilities, only a handful of telecommunications professionals were thinking in the same terms. Upon the divestiture of AT&T, telecommunications managers faced a variety of new problems, none of which had to do with understanding the business. In the period between the announcement of divestiture in 1982 and its implementation in 1984, managers faced many uncertainties, especially about how tariffs would be set. They knew that parts of their communications costs would increase and parts would decrease, but they did not know which or by how much. They also knew that they would have to deal with a new AT&T plus a new local RBOC, and that they would have the opportunity to use one of the new competitors, Sprint and MCI. "New" here meant no previous experience and guidelines.

The result of this situation is that in general telecommunications managers and specialists took a very narrow approach to their work, almost as if they had decided to concentrate on today's problems and shrug off the broader long-term issues. To be fair, their internal focus had led many business managers to be concerned only with cutting telecommunications costs (or complaining about them when they could not be cut). Most telecommunications professionals had little if any business background, for the old world of telecommunications had been highly hands-on and operational.

As the 1980s progressed, telecommunications increasingly came out of the room in the basement where phone lines came into the building. In rough order of sequence, wide area voice and data communications became a core part of business services; LANs grew explosively to connect PCs initially and then as a powerful technology in their own right; telemarketing, 800 and 900 numbers, and enhanced voice services such as ANI became key elements in customer service; and, most recently, electronic data interchange, electronic payments systems, and customer-supplier links have moved to being the norm, not the exception, in almost every industry.

All these trends created a climate for a new style of telecommunications operation that was less focused on operations and efficiency than in the previous eras. The new style needed some degree of corporate coordination and was closely interlinked with issues of corporate infrastructure. As a result, telecommunications is one of the only business functions in which the trend has been toward centralization rather than decentralization. (The need for integrated network management is the main driver here. Note that the more effective the platform in terms of integration and degree of reach, range, and responsiveness, the easier it is to combine centralization of platform planning and network management with decentralization of business planning and network use.) The growing importance of computers, databases and information systems has led business units to play a stronger role in decisions and to have much more control over development and operations. Distributed systems and the astonishing, continued improvement in the cost, power, and quality of PCs facilitated decentralization. As a result, more and more aspects of central IS have been moved out to business units. Indeed, in many firms, corporate IS plays a minor advisory role and mainly develops and operates transaction systems.

Telecommunications has tended to move in the other direction. Every large firm needs a coordinated backbone network strategy because key wide area facilities cut cross locations, functions, and even companies. A private network, electronic data interchange capability, a funds transfer network, and the like cannot be built up locally and piecemeal without generating redundancy, fragmentation, duplicate facilities, incompatibilities, and extra costs. That occurred frequently in the 1980s, when telecommunications was just a utility. Typically, manufacturing firms had as many as 40 separate and incompatible networks.

Just about every development in WAN technology and costs moves priority toward rationalizing such a multitechnology muddle. Hence, there is generally a strong central telecommunications group with responsibility for planning and running "the" network. The group's primary focuses are on vendor selection and relationships (with contract and price negotiation a constant activity), on network management and operations, and on meeting service levels for the business units. In other instances, this unit is in effect a self-contained business that interacts closely with both IS and business groups. It is more service-centered than cost-centered, though it usually sets the prices for its internal users at an attractive rate, typically around 80 percent of what an outside firm would charge. It aims at exploiting developments in technology and new vendor offerings, advancing end-to-end network management—an emerging new art form in the field—and increasing quality of service while also reducing costs.

This unit may be organized in any one of three ways: as a separate company or autonomous profit center, as part of IS, or as a service unit. The first alternative is often used by very large firms. Sears, for example, runs telecommunications through a company it set up; Westinghouse runs it as a profit center. In each instance the logic is that companywide telecommunications is a major operation in its own right that needs a skilled manager with the autonomy to select vendors, design and run the network, develop new types of service, and set prices instead of allocating costs on a formula basis. As a profit center, it has its own market test; if it cannot provide a premium price and level of service and support, business units can go outside the firm and telecommunications becomes a candidate for outsourcing. Prices are generally compared to those of leading outside suppliers.

More typically, telecommunications has been moved from being a low-level internal utility function to being part of the information services group. The reasons for this are obvious; it is increasingly difficult to handle data communications, database management, and information systems as separate activities. The old organizational split was between information systems and telecommunications, each of which had both a development and an operations arm; now the split is more typically between development and operations, each of which has an IS and a telecommunications arm. The skills needed for development are more business-centered, and those for operations are more technology-centered. The development side typically addresses the client-designer part of the business-telecommunications dialogue and the operations side the designer-implementer part.

These observations apply mainly to WANs and corporate facilities. LANs, which are very different in terms of organizational responsibility, are more an IS issue than a traditional telecommunications issue. Their growth has been fueled by a combination of distributed computing, business unit autonomy and decentralized decision making, PCs, and the thousandfold improvements in bandwidth per dollar of investment. Many business units make their own LAN decisions and support them themselves through a divisional IS unit. This extreme of decentralization views the LAN as an extension of the personal computer; it sees no need for either corporate oversight or any form of corporate standards.

More and more firms find this too extreme and are trying to get a better balance between coordination and local autonomy. This obviously creates for a central unit that oversees LANs the same role as those created by the trends in wide area networking, with the exception that LANs generally fall under IS. An additional difference is that management issues of vendor negotiations, operations, and pricing or cost allocation that are crucial for corporate telecommunications are much less so for LANs. The major issue for LANs, by contrast, is technical support. User departments that install a "simple" LAN soon add new applications; software, servers; links across LANs, including routers and network operating systems; new types of peripheral such as image servers or high-speed color printers; and so on. This becomes more complex than a typical data processing department of the 1970s. At the very least, the business unit needs its own LAN administrator; more generally, it needs a help desk and a hotline to IS.

It is not at all unusual for a large organization to have a corporate telecommunications group that is concerned with LAN-WAN integration and that thus seeks to impose corporate standards, plus corporate and divisional IS units with varying views on standards, plus business departments with little if any interest in corporate needs for standards. The rate of growth in LAN installations has been so fast and the technology so volatile that both management issues and technical and political issues have generally not been resolved. In a 1989 survey by *Datamation*, 56 percent of respondents reported that their company uses LANs, up from 33 percent a year earlier; 64 percent of the companies had a plan for interconnecting LANs to corporate computer systems, but only 22 percent had any stated formal policy or strategy for LAN use. By 1991, a frequent topic in the IS trade press was the growing problem of LAN chaos and the impossibility of establishing coherence across the organization; this was before routers, bridges, and smart hubs began to simplify the situation for IS managers. By 1992, a plangent theme in the LAN trade press was that the good old days were over, that IS was intruding, and that LAN administrators were becoming bureaucrats who impose complex procedures for user directories and security. The reason is that the days of amateur LAN management are over. LANs have become complex, and LAN-WAN-LAN integration is a new business priority that makes some degree of central coordination essential.

It also makes some degree of conflict inevitable. Business units can often entirely ignore the central unit in areas relating to PCs and LANs. According to one LAN consultant in Hughes Aircraft's corporate communication and data processing group, "from the corporate standpoint, I see a major need for implementing control and management processes, but we have people who literally are going down to [a computer store] and buying equipment. We are the overseers, and they don't have to listen to us if they don't want to."

Hughes has over 30,000 PCs in use, with 7,500 of them connected to a backbone Ethernet complex of LANs. Hughes management gives its seven business groups autonomy; the role of corporate IS/TC is thus mainly advisory. By contrast, the corporate group in any firm has very strong operational control over WANs.

The tension between central coordination (especially to ensure interoperability and ease of support) and local business options is a complex one requiring good will, collaboration, and incentives. The situation directly parallels the related issue of PCs, which began as amateur activities outside IS and led to proliferation, incompatibility, and problems of support. Little by little, completely local options generally gave way to local autonomy within a small set of standards, committees to coordinate and review key decisions, and business units accepting central IS oversight so long as IS in return provided first-rate support and service.

There is no "typical" telecommunications organization with respect to choices of degrees of business decentralization, IS-telecommunications fusion or separation, levels of outsourcing, and strategies for WANs and LANs. The organizational form shown in Figure 18-2 is becoming fairly common, however.

Regardless of the specific organizational form, the two dominant principles for telecommunications and information services in general are to make service and support to users the key principle, and to ensure that there is a strong business reason for any central control, and that it is backed by mechanisms for

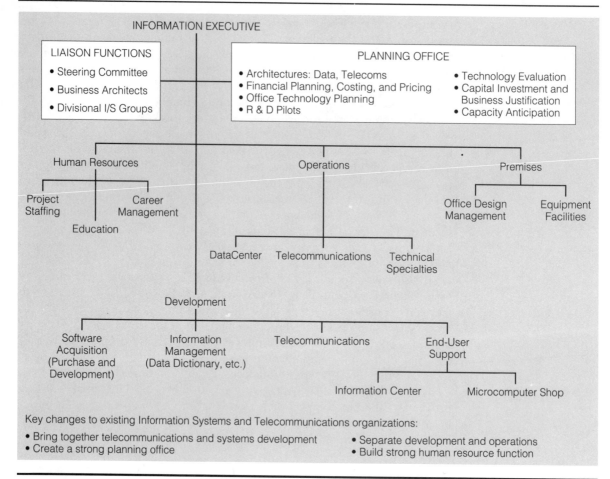

Figure 18–2

**Organizational Chart
for Typical Tele-
communications
Department**

coordination. These two issues can be in conflict, for coordinated operation of the corporate network can easily become unresponsive to local needs. Local service can get in the way of economies of scale, can lead to reinventing the wheel, and can result in too many vendors, equipment, software, and services.

*Pricing and
Allocations*

A very difficult management task is handling recovery of costs for shared resources, most obviously the corporate network. There are three options:

1. *Charge nothing*; absorb costs as part of corporate overhead, rather like the company parking lot.

2.   *Allocate all costs* to users via a formula that aims at fully recovering them every year; the formula may be based on a fixed budget or on usage.

3.   *Use a pricing mechanism* that is not directly cost-based but is more like a regulated power utility's rate.

The first two options have been the main ones used by companies to handle telecommunications as an internal utility, with cost allocations the most established—and the most unsatisfactory—option. Any pricing scheme must answer the following five questions:

1.   Does it fully recover costs, with no operating deficit or egregious price gouging?

2.   Does it minimize the cost of costing—administration, overhead in network operating systems to capture, process, and store information; billing; and so on?

3.   Does it encourage users to experiment and innovate?

4.   Does it motivate users to act in the corporation's best interests?

5.   Does it help flatten peaks and troughs in demand and maximize use of capacity?

Charging Nothing   The logic for having no charges for use of a telecommunications facility are that this strategy minimizes the cost of costing, stimulates innovation, and emphasizes service over bureaucracy. That approach is reasonable if (and only if) the cost of the telecommunications resource is relatively small. There is, for instance, little reason to charge people for using a simple departmental LAN; the equipment has already been paid for out of the capital budget. The operating costs are very small, and to handle billing would cost more money than is being recovered. By contrast, however, more and more aspects of telecommunications are variable costs and substantial ones. An analogy here is the electrical utility: If electricity were provided free, why would anyone bother to turn lights out? Giving a valuable resource away free means no cost discipline.

However, charging nothing encourages innovation and experimentation. Supporting a LAN may be expensive, but if users must pay a fee to get information or help, they are unlikely to do so, even when they are in difficulty. If the operating cost for accessing data or sending messages is high, users may be unwilling to produce prototypes and pilots and may overlook high-payoff business opportunities. If the initial volumes of users of a system are low, the average cost will be high; to allocate this cost is to discourage additional use, which keeps the cost high. An electronic mail service, for instance, might cost $50,000 a month to provide for 1,000 users, but with 10,000 users the cost only goes up to $65,000. The full cost recovery approach requires a charge of $25 a month with 1,000 users, but only $6.50 for 10,000 users. For a department of 50 people, the difference is significant: $15,000 a year versus $3,900. Its proponents can reasonably argue that the system's value is very limited with just 1,000 users and that it is in the firm's interests to stimulate demand. They may then just as reasonably add that at $6.50 a month, it's not even worth sending out the bill, which costs at least that to process. Hence, the recommendation is to absorb the cost centrally and give the electronic mail service away for free.

The equally valid counterargument is that if it is free, it is not valued and may be misused, accidentally or deliberately. The cost may be "only" $65,000 a month, but add up just five of these giveaways and you have spent almost $4 million a year. Proponents of the view that there must be some charge to ensure rational decision making about usage and responsible behavior could point to the international bank whose usage of e-mail in its New York head office went from 400 users to 4,000 in four months. This looks like a success; the e-mail system has increased communication and hence productivity. In fact, however, there was a curious pattern to the new messages: They were almost all sent to Manila and contained such requests as "I'd like to put $100 on the Giants versus the 49ers. The *New York Times* says the line is SF +4." Of course, this is a football bet, not a funds transfer, and because it is illegal to bet on games in New York, a friendly colleague had arranged to place bets in Manila. The staff were not being irresponsible; if a resource is free, people will use it as often and however they want.

Allocating All Costs   The e-mail advocate could counter this counter-argument, though. In another division of the very same bank the head of the business unit sent a memo to his staff, with the instruction that they should always try to communicate internationally by mail. If they need faster turn-around, they should use telex or fax; if the matter was urgent, they should use the international direct-dialing telephone system. Only if there were no other alternative should they use the bank's private network.

That network was operating at 15 percent of capacity. Allocating its costs over the small volume of traffic meant a very high unit charge—exactly the case if the e-mail system had just 200 users to share the $50,000 monthly cost instead of 1,000. The full recovery charge is now $250 each, even though the next 800 users have a marginal cost of zero dollars. The main problem with full cost recovery is that the early users must often bear a heavy burden. This discourages investment in infrastructures, such as a private network, a new switching facility, customer databases, and so on.

The main reasons for full cost recovery schemes are the historical legacy of data processing. Most DP costs were expensed, including software development. Full cost recovery helped keep costs from getting out of control. Debits equaled credits. It also provided a rational base for making sure that heavy users of a system paid their fair share. It is obviously important to find a formula that is equitable, but there must be a formula. Financial control staff who have seen how easily IS costs get out of control and development projects overrun their budgets are not enthusiastic about letting IS off the hook. They want to ensure that users pay for what they use and that IS makes neither a profit nor a loss. By focusing on cost recovery, the firm has a reliable metric for evaluating performance. If users find the allocated costs too high and outside providers can offer a better deal, it may make sense to outsource the activity.

Setting Prices, Not Costs Proponents of a price-based rather than a cost-based approach accept the argument that there should be no free giveaways but make the following arguments about full cost recovery:

1. It is expensive. One company's telecommunications manager estimates that 10 percent of his staff are devoted full-time to internal billing; another calculates that it costs 18 cents in software, disk storage, and administration to allocate 16 cents of WAN usage, so that he must charge 34 cents. He reports that the firm's corporate accountants are unmoved. Let debits equal credits.

2. It discourages investment in infrastructures, because the initial cost is spread over a small user base.

3. It generates behavior that is not in the corporate interest.

4. It does not provide users with incentives to use the resource in a way that helps them get the best value and meet corporate needs.

The last point is the key one. AT&T provides many incentives to phone at off-peak hours; this gives people choices. They can phone during business hours or wait until after 7 P.M. and save money. AT&T can set the discount to a level that reduces peak traffic; this cuts AT&T's costs and the customer's as well. In the electronic mail example of 1,000 users costing $50,000 and 10,000 costing just $65,000, an outside vendor would set prices to encourage growth, taking a loss initially but quickly building volume and hence reducing average cost.

Imagine going into McDonalds and ordering a hamburger and then being asked to pay $23.50. You would consider that crazy. Would you accept that price as reasonable if the owner explained that he or she had just had the restrooms renovated and that you as a user had to bear your share of the cost allocation? Of course not. The problem with rigid allocations that recover costs on an annual basis is that no matter how clever the formula, allocations too often generate something unfair and silly. Proponents of unit value pricing argue that the internal telecommunications and computing utilities must operate as real external utilities; they must quote a fair price. Thus if the corporate telecommunications unit cannot offer prices that at least match those of the best outside providers, business units must be free to shop around. And if economies of scale, security, or the need to ensure volumes lead senior management to insist that business units use the corporate facility, and if the telecommunications unit cannot match outside prices, then telecommunications should probably be outsourced.

The price must be predictable; a given unit of service should always cost the same. A Big Mac should not cost $1.23 on Monday and $2.69 on Friday. Cost-based allocation schemes often lead to a higher charge for bad service than for good service. For example, a convenient and frequently used charge formula for using a public or private network involves connect time, a cost per minute for the time your PC is connected to the service you want to access. On Sunday morning at 3 A.M. you are able to make transactions over a public data network in, say, 45 seconds at a connect time cost of $1.20 ($1.60 a minute); on Friday at 3 P.M., when the network is overloaded, response time degrades from three seconds to three minutes, so that the transaction now takes 40 minutes. The cost for appallingly slow service is now $64.

A business that charges in this way is likely to lose customers quickly. The advocates of cost-based allocations would argue that the corporate telecommunications unit is not a business because it has a captive market. If it is allowed to set its own prices, it could charge whatever it wanted because business units are not allowed to go outside. The advocates of price-based charges would respond that all that is needed here is to ensure that the telecommunications unit must offer a good price as well as recover its costs in the longer term, just as the owner of the McDonalds restaurant who renovated the restrooms must and will. They would treat the telecommunications unit as a business within a business. Both the extremes of full cost recovery and free services consider telecommunications as an overhead function; it must be run like a business.

Businesses have customers, and customers have choices and can go elsewhere. They also expect value. Wherever a product they buy is provided by a monopoly, as in the case of public utilities, regulatory agencies try to ensure that rates are fair to provider and customer alike. When a monopoly becomes bureaucratic, arrogant, lazy, and fat, pressures for deregulation and competition increase. (This is illustrated by the RBOC's complacency in their highly profitable local marketplace and by once-complacent AT&T's increasingly aggressive business strategies in its new competitive environment.)

Just about every aspect of telecommunications organization and management discussed in this chapter—corporate coordination of wide area telecommunications, support of LAN users, creating mechanisms for coordination without intruding on business unit autonomy, and charge procedures—pushes in the same direction. The telecommunications organization lives, dies, or gets outsourced by its ability to provide service. If it offers first-rate operations and support, business units may be willing to accept corporate standards; if it offers a price that matches those of outside vendors, business units may invite it to contribute to their planning meetings.

*Development,
Service, and
Operations*

There are many specific telecommunications jobs and job titles; Figure 18-3 lists examples. They vary in balance among the technical, business, and organizational skills needed.

Regardless of their specific responsibilities and skills, the jobs fall into the three main categories that comprise the telecommunications organization: (1) planning and development, (2) service and support, (3) operations. In Era 1, most staff were in operations. The internal utility added far more mechanisms and people for planning and development. The distinguishing feature of telecommunications as part of the integrated business resource, whether it is a

Figure 18–3

Telecommunications Job Titles

Planning and Development
Director, telecommunications planning
Manager, network planning
Data network design technician
Voice network design specialist
Business applications development specialist

Service and Support
Director, network services and support
Data communications service manager
Help desk technician
LAN service manager
Office automation applications specialist

Operations
Director of network operations
Network security manager
Data network operations manager
LAN manager
Voice systems technician
Voice network operations technician

separate function, profit center, or component of IS, is its focus on adding service and providing new types of support to its clients.

Planning and development cover topics ranging from identifying opportunities, to planning and implementing the telecommunications services platform, to designing specific applications. In addition, a major part of the work here relates to selecting vendors for key services and building relationships with them that lead to a mutually beneficial partnership. Operations involves a wide range of functions, most of which are highly specialized and technical, such as network operating systems, network management, and expertise in particular types of switching equipment or software. Billing is an additional core function.

Service and support roles are mainly those of internal consultancy, including help desks, to which PC users can call to get answers to problems, and experts in such areas as installation of new software. Using a PC in a networked environment is not a simple activity, and there are many types of problems, particularly problems of incompatibility. Help desks use telecommunications to maximize the use of limited telecommunications and IS staff. Some firms and many vendors of software and hardware use voice mail menus to route calls to the relevant expert, who accepts calls as they come in. The importance of LANs and PCs to more and more aspects of a business unit's activities will generally lead it to create its own support and service group, which may be just a few people. The LAN administrator, for instance, handles issues of adding new users, trouble-shooting, and handling queries on a daily basis.

The emerging principles for ensuring first-rate support include:

▶   Make sure the support system is absolutely simple and reliable from the client's perspective. This means providing a single point to contact for assistance, automating tracking and monitoring of problem resolution, encouraging standardization to simplify diagnosis, and building in-depth expertise. Users want a quick response, not excuses, explanations, and delays.

▸   Get away from extremes of either decentralization or centralization. The complete decentralization that some companies adopted (or just fell into) has given us LAN chaos, systems disintegration, and fragmentation of data. The extreme of centralized corporate data processing and centralized telecommunications operations gave us bureaucracy and very frustrated users. The overall pattern today is to coordinate the IT infrastructure from the center, to decentralize as much application development and local PC and LAN operations as practical, and to provide nodes of training, service, and support.

▸   Take an entrepreneurial role to encourage innovation and keep clients well informed. The old utilities of IS and telecommunications were passive and reactive. At the extreme, the main goal of the telecommunications unit was to control costs while meeting a minimum service level. There was little reason to identify new technology options or business opportunities, or to work closely with business units. Many IS groups use management education to build awareness, share ideas, communicate what IS is doing and can offer, and position people for new developments.

▸   Greatly enhance network management capabilities. The increasing dependence by more and more parts of the business on a firm's networks has raised network management from a fairly simple process of monitoring equipment and transmission and reacting to problems to a major new discipline in itself. The goal is automated end-to-end management using advanced software and hardware tools to anticipate and diagnose problems, to ensure maximum security and reliability, and to place as much intelligence as far out as possible in the network to reduce reliance on central systems. The new approaches to network management focus on "self-healing" networks that can recover automatically if a link is down.

In summary, the effective telecommunications organization is strong in all areas of activity. Planning and development establish business relevance and build the business-technology dialogue. Service and support make the dialogue work at the level of daily interactions and ensure that telecommunications is not just a utility. Operations makes sure the utility part of daily activities is reliable and efficient. To have only two out of the three in place is not enough.

Summary

Telecommunications has moved through three distinct eras of management: (1) the operations era, in which the main issues were hands-on supervision of a telephone and telex unit in a regulated environment; (2) the internal utility era, in which cost control and efficiency of voice communication was the priority and data communications was mainly a peripheral issue handled by the information systems unit; and (3) the era of telecommunications as an integrated business resource. Each of the first two eras required important skills, mainly in the areas of operations.

Until the mid- to late 1980s, telecommunications and information systems were separate functions, skills, and professional knowledge bases; now they are increasingly interdependent. That separation contributed to the multitechnology muddle and the mass of incompatible systems that companies are now struggling to rationalize, coordinate, and integrate. Effective information-services units are now bringing together their telecommunications and computing groups, are focusing on service, and are moving away from traditional charge schemes for allocating costs.

There are three main approaches to handling prices and allocations of the information services' unit's costs: (1) charge nothing, treating telecommunications as part of corporate overhead; (2) allocate all costs; and (3) use a pricing mechanism that creates a quasi-profit center. The first two methods create many problems and do not address the five most important issues that ensure that users, the company, and the telecommunications function provide best value at least cost: (1) full recovery of costs with no operating deficit or price gouging; (2) minimization of the cost of costing—administration, billing, and network overhead; (3) encouragement of user innovation and experimentation; (4) motivation of users to act in the corporation's best interest; and (5) reduction of peaks and troughs in demand.

Service and support have become for telecommunications units a higher priority that adds to and reinforces its role in development and operations.

Review Questions

1. Briefly describe the three eras of the telecommunications function in large organizations.

2. Describe the skills and roles of the telecommunications executive that are most appropriate for each era.

3. Explain why it has been so difficult for companies to move from the internal utility stage to that of the integrated business resource.

4. Explain why IS department personnel are much more likely to be sensitive to the business needs of an organization than are their telecommunications counterparts.

5. Why is telecommunications one of the only business functions that has moved toward centralization, rather than decentralization? Explain and discuss.

6. Describe the primary focus of a strong, central telecommunications group with responsibility for planning and running a company's network.

7. Explain why companies have evolved from an organizational split between IS and telecommunications, each with a development and operations branch, to a split between development and operations, each with an IS and telecommunications arm.

8. Are LANs treated differently than WANs in terms of organizational responsibility in most organizations? Explain.

9. Explain the meaning and the rationale behind the following statement: The good old days are over; IS is intruding on LAN turf, and LAN administrators are becoming bureaucrats who impose complex procedures for user directories, security, and so on.

10. Is there a conflict between ensuring adequate service and support to end-users and maintaining central control and coordination? Explain.

11. What are the criteria that should be met by any acceptable scheme for recovering the costs of a shared telecommunications network?

12. Under what conditions does it make sense to have no charge for the cost of telecommunications services and to provide network usage essentially free of charge?

13. Under what conditions would it make sense to recommend using a charge mechanism that allocates all costs of telecommunications services to the users?

14. Describe a set of conditions under which it is appropriate to have a charge for telecommunications services that is less than the fully allocated cost of providing the service.

15. Cost-based allocation charge schemes often lead to a higher charge for bad service than for good service. For example, this phenomenon occurs when the charge is based on network connect time. Explain why this happens and discuss its implications.

16. Classify the following list of telecommunications activities under one of the three major roles of the telecommunications organization (planning and development, service and support, and operations): (a) selecting and building relationships with vendors of key services, (b) personal computer help desk, (c) network management, and (d) adding new users to an office LAN.

Assignments and Questions for Discussion

1. Telecommunications departments and the professionals working in them are often thought of as the poor stepchildren of information systems and data processing. Identify the root causes of this situation and discuss them in the context of the parallel evolution of telecommunications and IS in the business organization.

2. Discuss the strengths and weaknesses of organizing the telecommunications function in each of three alternative ways: (a) as a separate company or autonomous profit center, (b) as part of the IS function/department, and (c) as an independent service unit. Also, indicate the conditions under which each might be appropriate.

3. Discuss the issue of choice between central coordination and control and local business unit autonomy with respect to telecommunications activities. Indicate whether or not the appropriate choice differs depending on the type of network being considered (that is, LAN or WAN).

4. The CEO of the Merit Widget Co. strongly believes that all business units should pull their own weight and meet the overall corporate target profit rate of 15 percent after taxes. You have been hired by the director of IS to prepare for the CEO a consulting report that will convince him that operating telecommunications as a profit center may be detrimental to the long-term success of the company. Prepare such a report.

5. Discuss the changing importance of the three main roles in the telecommunications organization as they have evolved over time, and develop a forecast of the key roles in the effective telecommunications organization of 1998.

Minicase 18-1

Westinghouse

Merging the Voice and Data Telecommunications Cultures

David Edison, the executive vice president of Westinghouse Communications Systems, comments on his reorganization of voice and data communications into a single organization: "One is surprised at how hard one has to work to merge the two cultures successfully. We thought they would combine more thoroughly than they did."

Westinghouse Communications, set up initially to run the company's worldwide internal telecommunications worldwide, has grown to provide commercial services to many other firms. It controls more than 300 T1-level (or above) high-capacity circuits. It has seven switching centers in the United States and another in Europe. About half its traffic on the WAN that absorbs a third of its operating expenses comes from outside Westinghouse. The WAN is a volume-driven business, so that extra traffic cuts unit cost; the per-unit cost for data communications is half that of the 1982 peak.

The merging of the voice and data communications groups freed up an estimated 25 percent of staff resources. It was, however, very difficult. Edison comments that the voice culture is very cost-driven and the data culture service-driven: "Voice people believe that it's always possible to work around a fault or break in a circuit. The data world is very service-driven. . . . They are extremely quality-conscious because they have to be error-free. A minor problem on a voice network is easily handled, while a minor problem on a data network may in fact cause total disruption."

Edison and his managers had to build a new team culture, getting the two groups to understand how similar and linked their two previously separate worlds now are. They also had to create a customer-focused culture that could compete effectively in an open market. "Sensitivity to customers doesn't come easily to people who are very much dedicated to resolving problems. They see themselves as problem resolvers, not as being customer communicators."

Edison also found that people tended to be very territorial. When he merged the voice and data communications control centers, some staff lost positions of power. He decided to "skillfully promote the idea of freeing up resources for new challenges." The control room merger allowed one of his staff to create the new position of engineering manager. There was no overt resistance to change but plenty of passive resistance, filibustering, and pet views about how things should be handled.

Edison's summary of the change is very positive, but he advises firms to "take your time with the human side of it. It's well worth your time to get round pegs into round holes."

Questions for Discussion

1. Describe the differences between the voice and data communications cultures at Westinghouse, and explain the underlying causes for these differences.

2. How did Westinghouse overcome these differences and build a new team culture during the merger of the two groups?

3. Given the difficulty involved in combining two different cultures, would it have made sense to continue to operate telecommunications at Westinghouse with separate groups for voice and data? Discuss the strengths and weaknesses of this approach.

Minicase 18-2

Business Communications Review

Survey of Telecommunications Outsourcing

Business Communications Review is a monthly telecommunications publication that covers industry and technology in depth and with insight; it is one of the most useful periodicals for keeping up to date. In May 1992, *BCR* published the results of a survey of its subscribers about outsourcing, a subject that "everybody is talking about . . . [but about which] nobody seems very sure what is actually occurring." Close to 75 percent of 211 respondents were in organizations with budgets under $5 million. They span a very wide range of industries, and over half are managers or directors of telecommunications. Their answers to *BCR's* questions give a good idea of just what is happening with outsourcing.

Seventy percent of the respondents' organizations outsource some aspect of telecommunications. This figure is a bit misleading. Hardware maintenance is the only aspect in which half the respondents outsource (voice 53 percent and data 49 percent). Second in frequency is network management software development (15 percent). In order of frequency, the rest of the list is:

1. Total voice facilities management and end-user data communications software development: 10–15 percent.

2. Voice network design, total data facilities management, network management help desk and troubleshooting, LAN inter-

networking and data network design, data communications data center operations, and LAN internetworking total facilities management: 5–10 percent.

Relatively few firms report a "boomerang" effect, in which they regret outsourcing and bring it back in-house. The average is 10 percent, although in health care it is 22 percent. The main factors in support of outsourcing are cost savings, especially in personnel; quality of service; and lack of in-house skills. The main arguments against it are its "mission-critical" nature, lack of proven cost savings, and lack of confidence in the supplier. Talent and track record are the key factors in choosing an outsourcer, with price third in importance. Only one reported factor (mission-critical nature of the function) is listed by even 30 percent of the respondents; however, 52 percent picked quality of the contract as crucial to success in outsourcing, a clear statement of responsibilities, definition of performance criteria, and the like.

Over half the firms using outsourcing said it did not free up their staff to work on new projects, suggesting that overload is a major stimulus for it. All in all, the survey adds up to a "definite maybe." Firms do not have clear criteria for outsourcing, are wary of suppliers' claims, and are unsure of the benefits—but they actively and continually evaluate the option of outsourcing.

Questions for Discussion

1. According to the *BCR* survey, what aspects of a firm's telecommunication function are most likely to be outsourced, and why?

2. What are the main reasons given in the survey for firms *not* outsourcing some or all aspects of the telecommunications function?

3. What criteria are used by firms to evaluate outsourcing contractors, and what factor is considered most critical to having a successful outsourcing experience?

Minicase 18-3

Sears

A View from the Chief Networking Officer

This minicase summarizes a 1991 *Chief Information Officer* interview with John Morrison, a vice president in Sears Technology Services, which is responsible for providing voice communications to Sears and its subsidiaries, including Allstate Insurance, Dean Witter Reynolds, and Coldwell Banker. Sears has faced many problems in its core retailing business, losing industry leadership to Wal-Mart. Its main strength is its Discover credit card, which is one of the top three in the industry. In late 1992 it decided to sell off most of its nonretailing assets.

Morrison is widely respected as a manager and as a telecommunications professional. In the interview, he reviews a major organizational, technical, and economic shift he made in deciding to dismantle Sears's private network and move to a public switched network. The interview provides useful insight into the thinking of a large firm's chief network officer.

Question 1 Why did you make the switch?
Answer When we first formed Sears Communications/Sears Technology, our long-distance costs were around 31 cents a minute. Large firms like us could use private networks to provide a much lower cost. When we first centralized corporate telecommunications in 1986, we found that our 800 number services were costing us $30 million a year. We hadn't yet added the Discover card or our centralized telecatalog operation. We realized our private network did not have the capacity to handle the growing 800 traffic, so we would

have to use public services anyway. We decided to leave data on the private network since we have a humongous network we can't easily move to a public network.

We went to AT&T, Sprint, and MCI and said, "We would like to be able to pick up the phone, dial 9-1-area code and phone number. You take that call and complete it however you want, but we want that call to go through [and fast]. We would also like customers to dial 1-800 and the catalog number and be answered very quickly at the point they're trying to call." We didn't care how they handled those calls. We felt, though, that they would gain some economies of scale by adding all our traffic and by our not caring how they routed the calls. They ought to be able to offer us an attractive rate, and we made it very clear that we would choose the lowest bidder.

That was Sprint. This was the first time AT&T ever lost a customer as large as Sears. They didn't believe we really would move our long-distance inbound and outbound traffic to Sprint. We did, but we still work closely with AT&T and do not give 100 percent of our traffic to Sprint. We have what we call strategic locations, including the Sears Tower, where we have some AT&T facilities for backup and contingencies.

Question 2 Was your staff skeptical about the change?
Answer At the start, there were just two people who thought this was a good idea—my boss and myself. We almost had a rebellion when I announced we were moving to Sprint. At that time,

Sprint had an appalling reputation for its chaotic billing [this forced its CEO to resign, in fact]. We knew that, but we weren't buying the billing. We were buying a network. The network was stable; the billing was lousy. It took a lot of time to convince my people I knew what I was doing.

I showed them the economics and told them none of them would lose their job. Many of them had come up through the ranks because of their technical knowledge of private networks; we would use their expertise in other ways.

Then I got lucky—I always call it foresight but it really was luck. The corporation announced a major reorganization in the merchandise company. Just about 100 percent of the staff moved to new jobs. Many of the locations moved, too. If we had had a private network, I would have had to double my staff just to react to putting in and taking out circuits, and installing PBXs. Instead, my people could concentrate on working with the business to understand its needs and how they wanted to communicate. All of a sudden, my staff became very supportive.

Question 3 How did you reassign your private network specialists?

Answer I have some technical people who are as good as the best in AT&T, Sprint, and MCI. These folks work closely with Sprint and AT&T on the intelligent network designs that I think will be the base for public as well as large customers' networks in the future.

We made a strong effort to keep our best technical people partly because our contract with Sprint will run out at some time. We told the carriers we still have the expertise to build and run our private network. If costs get out of line, we will do so.

No one resigned.

Question 4 How do you think telecommunications will change Sears's competitiveness?

Answer By exploiting intelligence in the network. To do that, we need to have in-depth understanding of what Sprint is planning and doing in this regard. For instance, if you phone 1-800-366-3000, our catalog number, the call will go through to an ACD [automatic call distributor] in one of ten locations. The ACDs are managed from Sears Tower in Chicago. There are people there who can monitor calls on a dynamic basis and reroute them, if there's a snowstorm in Louisville, for ex-

ample; the call will be sent to San Antonio or Provo without the customer knowing the difference.

That's today, but if you look at Sprint's intelligent network plans and Sears's database, we should be able to develop a capability neither company has by itself. Calls can be routed automatically to cut any "hold" time by looking at where they originate and where can they be answered in the shortest time. We have to do some software development to mesh our needs with Sprint's capabilities and plans.

We have a phone number today, 1-800-DIS-COVER, which we don't advertise. We published the number to do a test and found that people called it for around 20 reasons. Some wanted account information, others wanted to talk to Greenwood Trust, a small bank that is really part of Discover. Most callers reached an office in Columbus, Ohio, and had to be told to call another number. We ought to be able to transfer the call to the person who can handle the customer's request, using the intelligence in the network. We can't do that cost-efficiently today.

We believe we can get a competitive edge here because some companies we compete with haven't the foggiest idea of what the intelligent network means for them.

Question 5 Has telecommunications increased Sears's productivity?

Answer Absolutely. The catalog operation is 10 ACDs, but you can now run the human resources as if it is one. We have calculated that this gives us a 5–8 percent improvement in productivity. Given our payroll, that adds up to a lot.

Question 6 How many of your 120 staff have a formal voice telecommunications background?

Answer Probably half.

Question 7 You came from data processing. How did you pick up voice communications expertise?

Answer By accident. I was assigned to find someone with a technical background to put in a telecommunications and telephone system in Sears Tower. Every time I came up with a candidate, either that person didn't want the job or the VP didn't want the person. My boss called me in one day and asked what was wrong with the job. I told him it was a great opportunity and anybody would be crazy to turn it down. He said, "Great,

I'm glad you feel that way because you've got the job."

I educated myself by going to the vendors. I went to some Bell Labs classes, and AT&T was very good. I told them I knew nothing about telephone systems. Even today, I get much of my education from the vendors, including IBM.

Question 8   How do you set priorities for your staff?

Answer   I don't think I do that as such. I'm trying to instill in my staff the idea that we must become a service-driven organization. I sit down with the senior managers of our business units and they set the priorities for me. That translates into priorities for my people.

However, you have to be careful when you are trying to be a user-driven organization in the technology field, because many users are uninformed and narrow in their view of technology. We have meetings every six months with Sears's chairman and the top managers of the business groups to keep them informed of where the technology is moving. We want to feed into their business thinking so that they can exploit the technology for a competitive advantage by being ahead of other companies' thinking.

Question 9   How did you mesh voice and data skills in your organization?

Answer   We assumed in 1983 that the integration of voice and data was just around the corner. We thought we needed to set up an integrated organization for integrated technology. Progress was much slower than we expected. We found we needed different skills for voice, particularly in commodity areas where getting the best technology was not an issue; we needed the best price. We split our organization back along the lines of voice and data in 1988. Now, though, voice functions are becoming more software-driven. Parts of my voice organization are being coupled with the data folks for applications that are voice-data driven.

Question 10   How has telecommunications changed in Sears over the past few years?

Answer   It used to be an overhead function. The Discover card changed that. Sears could not have launched it without telecommunications. We had many separate networks at the time and had to pull a network together that would provide credit card transactions within our company and link to other businesses. In addition, we had to take a broader business perspective on telecommunications operations. In 1990 we spent $614 million on telecommunications, so that we have to operate as a business resource. When the catalog organization wanted to get a closer focus on its costs, it came to us. In 1986, our voice costs were 30 cents a minute; now this is down to single digits. We are part of the business, not overhead.

Question 11   How did you adapt your organization to meet the business responsibility?

Answer   We appointed a Board of Directors composed of senior officers in the business groups. We present our plans to them. We developed a detailed strategic plan that we presented to Sears's chairman. He decided that the business groups should submit their annual five-year plan for our review, so we could assess if their plans were feasible and make sure we adjusted our plans to ensure the technology was in place. The chairman of our company got us off the ground.

Question 12   Does this support from the chairman make it easier or more difficult to get cooperation from the business units?

Answer   There's always a political element in implementing our programs. We are always resisted at various levels, no matter how much top management support we have. We need to keep our activities in front of the chairman and his staff. That's one reason why we carry out telecommunications audits that are professional, candid, and rigorous. We can show where we've saved the company millions of dollars. That gets us listened to.

The telecommunications bottom line is that you have to do things that are truly, truly productive. Building the best network, having the brightest people, or creating the best organization doesn't do anything if nobody knows it or sees the payoff. In that case, telecommunications is just an operating expense.

Question 13   There seem to be two very different cultures in our industry: data processing and telecommunications. How do you merge the two?

Answer   It's a very, very difficult task. I spent some time in DP and found that people there are driven by the technology, not the business case or user needs. They also believe they understand the technology and the need better than the people who use the technology. I try to extend our communication with and understanding of users by bringing people from the business who have no knowledge of either DP or telecommunications into our organization at fairly high levels, so that we can teach them things they need to know in the telecommunications world. They can bring in the user's perspective even though they don't know the technology and can't explain it to their peers.

Questions for Discussion

1. The Sears telecommunications staff was initially skeptical of the switch from a private to a public voice network with Sprint as the primary vendor. Explain why, and indicate what occurred that convinced the staff that it really was a good move.

2. At Sears, telecommunications changed from an overhead function to a competitive necessity with the launching of the Discover charge card. Explain how the telecommunications organization adapted to meet this new business responsibility.

3. Explain how senior management of the Sears information systems group both handles resistance to telecommunications initiatives from the business units of the company and deals with the difficulties of merging the two separate cultures of data processing and telecommunications.

Minicase 18-4

Forrester Research

A View of the Network Manager of the 1990s

Janet Hyland, director of network strategy research for Forrester Research, Inc. in Cambridge, Massachusetts, presented her views of the new skills and organization needed to be an effective telecommunications manager in the 1990s. She reported that the major challenges identified by telecommunications executives in 50 Fortune 500 companies listed organizational changes as the primary need, including centralized LAN management and improved coordination between information services and end-users. Choosing and implementing standards was the second main factor and smart hubs the third.

Hyland views telecommunications as an integral part of IS and LAN internetworking as the foundation network, which will coexist with SNA networks but demand an entirely new style of thinking. She argued that network managers must understand this shift to internets or "they will be replaced by someone who does. . . . Managers cannot take what they know about 3270 multidrop lines and apply it to the internet. If they do, they'll be woefully constrained in this new environment and wind up with a network that's too hierarchical."

She believes that information systems organizations lack these skills. A 1991 survey of network executives in the largest 1,000 firms in the United States concluded that there is a dearth of skilled LAN and LAN internet managers. Forty percent of the companies had installed routers, but of these

only half were installed by IS, which mainly still operates "traditional" networks. End-users are playing a strong role in decisions about internets, bringing in outside specialists and systems integrators.

"Managers must get involved and take control of these LAN internets that have grown in a haphazard, anarchic fashion. It's too important to an organization's business objectives to be done otherwise." The network needs a central architecture and management oversight to increase IS visibility and contribution to coordination, to obtain volume purchases and economies of scale, and to trim redundant internet links.

She argues strongly for IS drafting of formal plans for a corporate network architecture: "Many network offerings cannot be delivered unless they are coordinated by IS." The network manager must build coalitions with end-users and must build internal support, mainly by providing a first-rate help desk and central technical support capability. Planning horizons must be shortened because the internet market is dominated by change, multiple vendors, and dramatic technical innovation that often moves faster than standards. Users must make pragmatic decisions within a maximum two-year planning horizon.

She believes the transition will be very difficult, with much controversy and many "uncomfortable decisions."

Questions for Discussion

1. What are the major challenges faced by tele-communications managers of the 1990s?

2. Describe the new skills needed by telecommunications professionals to be effective in meeting the challenges identified in question 1.

3. Outline the organization of a telecommunications department that is designed to meet the challenging requirements imposed by companies shifting from a terminal-mainframe architecture to a client/server, internetworking system.

19

Telecommunications and Your Career

Objectives

- ▸ Understand the different types of telecommunications career paths
- ▸ Learn about likely trends in technology and its uses that provide new career opportunities in telecommunications
- ▸ Review the telecommunications priorities of leading companies as a guide to potential career opportunities

Chapter Overview

| | | | | | |

Chapter Overview

The main aim of this book has been to help you prepare for your career in a world where telecommunications will permeate every sector, private and public, and strongly influence organizations' efficiency, effectiveness, service, and innovation. It reviews the four career quadrants discussed in Chapter 1, each of which require a mix of (1) technical experience and currency of knowledge and (2) business and organizational skills. The Business Services quadrant requires very strong business and organizational skills, with only minimal knowledge of telecommunications. At the other extreme, Technical Services demands top-rate technical skills. Between these two are the hybrid roles of Business Support, and Development Support. Business Support, the new mainstream for information services and telecommunications, primarily requires business skills but adds the requirement for sound technical skills to translate business insight into technical planning and implementation. Development Support focuses on technical issues but adds the understanding of business needed to create the type of dialogue that is the main theme of this book. This chapter illustrates each of these roles.

This chapter—and thus the book—concludes with an assessment of the emerging technological developments that are most likely to have a dramatic impact on business and organization in the coming years and relates these developments to career opportunities for readers who are comfortable with both telecommunications and business.

| | | | | | |

The Four Career Quadrants

In the first era of telecommunications, just about every job involved hands-on operations, with the main skill being knowledge of the relevant technology. Today, telecommunications spans business, organizational, and technical skills. Consider a hypothetical bank that is developing a new service for merchants that offers electronic payment services at point of sale, including direct debit for purchases made by either credit card or ATM card, electronic cash management for each store, and netting and zero balance accounts that ensure that the retailer gets its funds as fast as possible. Figure 19-1 depicts the service.

This is an ambitious project that many banks are considering, but none of them has put all the pieces together. It is a major opportunity for this hypothetical bank, which we will call First Transamerican (FT). A task force has analyzed the business opportunity, designed the platform, and made the business case, and top management has approved the project. FT wants to move fast, so it sets up a new unit called FirstTrans FastTrade, or FT^2, which will be part of the Corporate Services division.

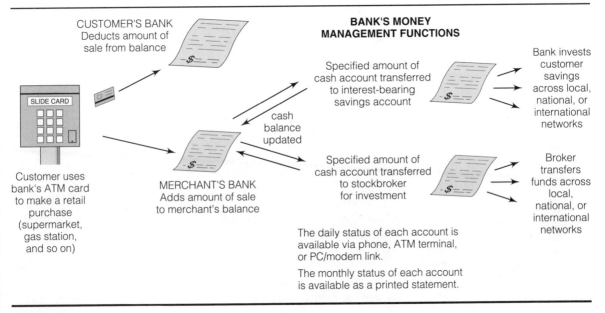

Figure 19–1

Electronic Point-of-Sale Payment and Cash Management

The newly appointed head of FT², Joan Makkia, has identified four key jobs to be filled:

1. Marketing director: This person handles all aspects of positioning FT² in the marketplace, developing relationships with key potential customers, and arranging promotion and advertising, pricing, and contracts.

2. Head of systems development: He or she will work with IS to develop new transaction processing systems and databases, as well as links to existing processing systems and data resources, such as those for credit-card authorization and funds transfers.

3. Manager of network operations: FT² will use FT's existing data communications and data centers. In addition, however, FT will set up a new switching center and virtual private network that will connect FT² customers to the bank's other networks, transaction systems, and databases.

4. Customer service director: This individual will oversee all aspects of education, support, and services for FT² customers.

Each of these jobs involves a different mix of business and organizational skills on one hand and technical skills on the other. For instance, the marketing director must have broad experience in retailing to be credible with customers but does not need in-depth knowledge of technology. Makkia's "only worry here is that I may have to pick between someone who knows retailing and merchandising backwards and can charm even the toughest customer but is a Luddite, and someone else who is not quite as sharp on the business side but at least has a sense of what the technical people have to deal with. I used to be in corporate banking at FT, and I saw far too many instances of credit people who now had to sell electronic banking products but didn't have a clue what they were. They could talk traditional banking but had nothing to say about new banking. My instinct is that while I want a first-rate business track record for the marketing director, I won't take anyone who I feel can't work well with the IS and telecoms folk."

Makkia's views on what she needs from the manager of network operations are the opposite: "I want someone with the business imagination of a clam—but the very best in technical operations I can find. If World War III breaks out, I want that person worrying about what that means for the 10 P.M. shift. I talked with Federal Express's top network staff, and they told me they have a 99.975 percent reliability record on transactions, but that's not good enough because it means 150,000 errors a year. The network team at the Royal Bank of Canada told me they are at 99.95 percent uptime on the network and apologized—they promise to do better next month. FT^2 pushes the telecommunications state of the art. We've met with several vendors who tell us that the requirements for fast switching and security are a challenge for even their newest products. When a store's crowded the week before Christmas, a tenth of a second makes a difference in transaction processing time."

Makkia then summarized what she needs: "I want a team of top technical experts and top business experts. In the middle, I want people who combine the two. The head of development is a technician first but must have solid business knowledge. This isn't just carving computer program code. It needs lots of discussion with marketing and sales about how to make sure the computer and communications systems fit with how our customers run their businesses. The customer service director is a businessperson first but had better know the technical issues well. I expect we'll find that more and more of our resources for customer service will go into training in the use of the systems."

Figure 19-2 places these key roles into four career categories (Keen, 1991). It shows two dimensions of career trajectory: business/organizational and technical. Historically, these have been at cross purposes to each other: People moved up a progressive career ladder but rarely crossed over from business to technology or vice versa. They might make changes along the way, such as leaving a large firm to join a startup firm, or becoming a consultant, but business and technology were very separate worlds in terms of recruiting, qualifications, training, and skills, and little communication occurred between them. The technical career was firmly unidirectional, with information systems and telecommunications clearly requiring separate skills. The main career path in telecommunications was through operations, first to supervisor and then to manager of the phone utility; this was a very limited area of opportunity and growth. In the information systems field, the standard career path was either project-based—programmer to analyst to project leader to head of systems development to IS manager—or focused on technical specialization—technical

Figure 19–2

**Telecommunications
Job Categories**

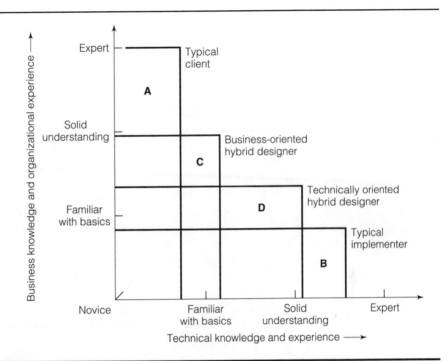

specialist to senior specialist to consultant to senior consultant to manager of other specialists.

This tidy world has been turned upside down to an extent that there really are few career "paths" or career "ladders" now. The very terms imply an orderly progression with fairly predictable job titles, skill needs, and promotion opportunities. In the field of telecommunications and information systems, there is a host of new technical specialties and jobs that need a mix of business and technical expertise. These have no ladders or paths because they never existed before; they have at best a general trajectory, a direction for personal growth and opportunity in which innovators in effect create the new jobs and promotions. For business careers, it is increasingly difficult to find any area of an organization that is not deeply affected by information technology and that thus does not urgently need a new mix of staff, again with some level of technical as well as business expertise.

A brief discussion of the four career quadrants in Figure 19-2 follows:

1. *Business services:* The key need here is for business skills, combining functional and/or company knowledge and experience with organizational skills, especially in communication. Technical knowledge need be only minimal.

    Historically, almost every job outside IS and telecommunications fell into this category. A marketing manager in a bank, a financial analyst, or a procurement supervisor needed to know nothing about IT to perform well and to move up the job ladder. It is still a matter of some pride for business executives to boast about being computer-illiterate. Even in the mid-1980s, a student could graduate from a top university with an MBA without a single required course on information systems; there were no required telecommunications courses in any major business school's curriculum.

There are still a few areas of business in which some degree of understanding of telecommunications and computers is irrelevant, but they are difficult to find. Business services is no longer the mainstream of careers and jobs.

2. *Business support:* This is one of the emerging hybrid roles that will increasingly become the mainstream for management, staff, and professional positions in more and more organizations. It requires strong business and organizational skills, plus an adequate understanding of information technology, especially telecommunications, because of its central role in customer service, coordination, and logistics. A business support role in banking might be a product manager who combines knowledge of customers and the market with understanding of electronic service delivery and use of customer databases for targeted marketing. In telecommunications, it might be a specialist in applying groupware to teams and project activities.

3. *Development support* reflects a major and often difficult shift in information systems from its mainstream in applications development, with programming the core skill and with little need for business knowledge and organizational skills beyond those relevant to designing and implementing a specific system. In the 1970s and most of the 1980s, information systems departments were largely staffed with people whose lack of business interest and insight were fairly notorious. Telecommunications staff were even less interested and even more isolated in most instances.

Now, although the core of both information systems and telecommunications remains (and will always remain) technical in expertise and emphasis, more and more roles require solid business understanding, and consulting and communications skills. This applies to most support roles, to business analysis, and to jobs involving direct contact with clients to help build the business/telecommunications dialogue.

4. *Technical services:* Telecommunications needs top-rate technical specialists, as does information systems. In addition, it needs an entirely new generation of specialists who are capable of functioning in a climate of increasing interdependence among telecommunications, data management, and computers. Network management, LAN management, switched data services, frame relay, image processing, distributed relation databases, multimedia, SNMP, CMIP, and voice recognition are just a few of the fields in which there are no specialists with 15 years of experience simply because they did not exist at that time. In several instances, there are none with even three years of experience.

An effective IS organization must of necessity include people from all four quadrants. Consider Joan Makkia's needs to staff FT^2:

▶   Marketing director: Business and organizational skills dominate, with only limited technical understanding needed. In terms of Figure 19-2, business/organizational skills needed exceed technical skills needed. (The technical skill needs are modest.) The role here falls into the category of business services.

▶   Customer service director: Here the priority is business and technical skills, but the technical skills need to be far sharper. This job falls into the category of business support.

▶   Head of systems development must have first-rate technical skills and experience, including experience in managing large projects. He or she will be supervising project managers in many areas of information systems and telecommunications development and must earn their respect and make sound technical judgments. FT^2 needs a top technical expert in charge; however, that person must also have a shrewd understanding of business issues, must be able to work closely with marketing and customer service, must understand the retailers' needs and mode of operation, and must know how to explain technical issues in business terms. Using the categories in Figure 19-2, the job falls into the development support quadrant.

▶   Manager of network operations is in the opposite quadrant from the marketing director, which is in business services. Technical services requires very strong technical skills, and Makkia will choose a superb technical "nerd" who lacks business sense and interest over an average technician with business savvy.

The FT^2 example focuses on fairly senior-level jobs. Many of the most difficult jobs to staff are at more junior levels, at which there are few experienced people and those who are technically knowledgeable and up-to-date tend to be young and lack "hands-on" business experience. (And those with such experience usually are neither up-to-date nor recently qualified.) A June 1992 survey of salary guides by the Robert Hall organization, a large and well-respected recruiting firm, demonstrated the shortage of and demand for LAN/WAN skills. Salaries in 1992 were almost 7 percent higher than those in 1991, whereas for all other areas of telecommunications and information systems, the 1990s recession, downsizing, and outsourcing held growth in earnings to 1–4 percent. Comments on the findings of the survey include the following:

▶   "There are more openings than there are qualified personnel; therefore, companies are willing to pay a premium. . . . Once you have 30 to 40 users onto a LAN, you need someone to administer it."

▶   "The future of corporate computing is the network platform. . . . There are the design skills necessary to set up LANs, such as the wiring and internetworking, bridges and routers, and backbones. That part alone is very complex. Then you have all the issues involved in managing and maintaining the LAN once it is up and running."

▶ "A LAN administrator needs to be able to communicate effectively with both technical and nontechnical users. In addition, knowledge of standard off-the-shelf office automation packages is very useful."

▶ "Because this kind of skill is so recent, it's difficult to find candidates who have formal training or education. . . . Much more important is actual hands-on experience."

Commentators agree that as people move higher up in the information services organization—which increasingly integrates telecommunications and computing staff instead of keeping them in separate departments—hands-on experience becomes less critical, although sound knowledge of ever-changing technical basics remains vital, right up to the position of chief information officer. The chief networking officer must be sharp technologically because of the astonishing pace of technical change. In addition, "moving to a networked environment is one of the biggest changes in the history of the computer industry. . . . This is really an entire science that requires a special set of skills." Technical skills are crucial here. However, "a lot of people have basic networking skills, but we've found that our clients have trouble getting people who can see the big picture and help with long-term planning." Those are key skills for the client-designer dialogue.

It is easy to identify many growth areas for telecommunications careers in all four quadrants of the career map, with the possible exception of business services, in which, by definition, technical knowledge and experience are irrelevant. In business support we need a massive infusion of business-focused people who are familiar with point of sale (retailing, banking), image processing (insurance, health care), electronic data interchange (just about every public and private sector organization, large or small), and payments systems (again, every organization), to list just a few obvious examples. A particular shortage for business services are people who understand international telecommunications, especially regulatory issues.

In development support, it is increasingly difficult for the average technician to progress in his or her career without adding business knowledge and being able to talk in business terms with clients. Only first-rate specialists can move ahead in technical services; their roles are more circumscribed, and if they do not keep up-to-date, their knowledge has an increasingly short half-life. Equally, as the Robert Hall survey shows, technical change plus business change opens up continued opportunities for those who learn and keep learning.

Three themes run through just about all articles on IT employment needs and trends, although in many ways they are contradictory. One theme is the vital importance of business understanding in both telecommunications and information systems professionals. The second theme is at the opposite extreme of the career map: the vital importance of creating a new cadre of technical professionals for the era of client/server and object-oriented methods. The third theme is that it is in telecommunications that firms are most lacking in key skills.

This point appears in just about every survey of the IS field and of job opportunities in it.

It is difficult to identify any industry in which in ten years telecommunications will not be a critical aspect of business innovation and operations. Today, more and more core aspects of most industries' business activities depend on people with skills in the hybrid career trajectories.

"Wham" Technologies: Trends and Possibilities

Forecasting the impact of new technological developments in telecommunications is always difficult and often foolhardy. Enthusiasts frequently overestimate the pace of change in the adoption of technology because they assume it will match the pace of the technical change. Notorious instances are the heralding of the paperless office and the cashless society in the 1970s; neither has yet happened.

At the other extreme are the skeptics who just as often pooh-pooh new technology and overlook how quickly and completely telecommunications can change the basics of business and organization. The chairman of a major insurance firm told one of the authors in 1988 that telecommunications would never be important in financial services, thus ignoring one of the new drivers of the industry—which made USAA a competitor his firm cannot match.

He and the predictors of the paperless office were equally mistaken: Successful business managers and successful technical professionals both face the challenge of how to identify which aspects of technology to ignore and which to focus on.

This final section presents a checklist that might be termed "watch this space." It is not a prediction but a short review of what we, the authors, feel are likely to be the "wham!" technologies of the next five to ten years. By "wham" we mean that once these take off, they will have a massive impact on business and/or society, in the same way that ATMs had in the 1970s, fax machines in the 1980s, and mobile communications has started to have in the 1990s.

Telecommunications is a business resource, as much as money or people are. At the same time, it is a technical resource that changes at astonishing rates. There can be no question that the technology of 2004 will be dramatically different from that of 1994, in ways that we cannot predict. The telecommunications field is full of expert predictions, many of which turn out to be wildly wrong. Sometimes they err in their optimism; the progress of ISDN and OSI, for example, has badly lagged behind expectations. In 1990, articles in the trade press spoke of OSI as the only choice for telecommunications and berated any manager who was not on the bandwagon. One of these articles claimed that by the end of 1991, all seven layers of OSI would be fully implemented "right down to the local area network protocol stack."

That is downright silly but not atypical. By 1993, more and more commentators were writing about the disappointing progress of OSI. Again and again, the hype of the 1980s has become the "whatever-happened-to?" of the 1990s. It takes a combination of people with real business savvy to spot the technical innovations that offer real opportunity and people with sound technical judgment to sort out which innovations are most likely to turn into practical products. Without such acute judgment, there will always be wild overestimates and forecasts.

In trying to make sense of technical and industry developments there is also much excessively conservative prediction. Very few industry experts anticipated that ISDN—designed in the 1970s with the expectation that 64 kps was a practical upper limit on basic public network transmission services—would be at the *trailing* edge of 1990s applications. No major industry commentator at the time of its definition rapped ISDN by stating that T3 speeds of 45 mbps and LAN transmission at 100 mbps (FDDI) and at 2.4 gbps (SONET) would be operational at the very time ISDN was being tested in the United States.

It is extraordinarily difficult to keep up with, let alone anticipate, progress in telecommunications in terms of either the industry or the technology. There are, however, the following obvious trends that directly affect telecommunications career opportunities for the mid- to late-1990s:

▶   *Multimedia networking*—"the singing PC," video, speech, image, CD-ROM, digital photography on disk instead of film, and even live television at the desk. The names of some new PC products introduced in 1992 indicate the pace and direction of multimedia: the Sound Blaster from Creative Labs, Multimedia Extensions for Microsoft Windows, PC-SoundPort, and Music Notation. However, according to a 1992 article in *Business Communications Review,* "the state of multimedia today is woefully incomplete; almost no systems integration is going on. . . . From a networking point of view, most of these systems are in the dark ages."

▶   *Voice recognition and voice processing*: New generations of chips, particularly digital signal processors, have the capacity to digitize voice, and new types of software are able to interpret its meaning, not just its sound. Such systems are largely experimental but are gradually moving into commercial use. Voice and spoken language are the basic media of human life; screen menus, typing, and graphical user interfaces are a limited substitute for them. It may be a long time—even centuries—before you can phone your bank and make a request in words that can be immediately and accurately interpreted and processed, but by the mid-1990s advanced voice technology will be as hot an issue as advanced data communications.

▶   *Massive bandwidth internets*: This is a given. The public networks of today are built on transmission and switching systems that are exponentially slower than the best available technology. These networks are being replaced and/or upgraded, but largely incrementally, because the carriers need to preserve their existing investments. Well before the end of the 20th century, however, today's capabilities will be widely deployed, and tomorrow's will make these look slow.

Fiber optics guarantee that. It is likely that massive bandwidth networks that span many organizations and countries will initially be used for specialized applications. Health care and research are two obvious examples; they have an immediate need for transmission of high-resolution images, for radiography in health care, and for scientific data in research.

▶   *Optical switching*: Telecommunications advances that remove a bottleneck of speed, compatibility, or cost always reveal the next bottleneck. For decades the bottleneck was transmission speed; electronic switches were faster than the bit stream they processed. A first-generation PBX was basically a minicomputer that processed bits instead of, say, payments and orders. The incoming message at, say, 1,200 bits per second could be handled with hardware computing cycles to spare; those cycles could be used for protocol conversion, accounting, or error checking.

When transmission speeds are measured in millions or billions of bits per second, software and hardware in switches become the new bottleneck. These switches are electronic, with time units measured in microseconds for operations. Computer performance is measured in instructions per second, with 300 a current (but temporary) upper limit on commercially available mainframes. A link operating at 100 mbps (FDDI) sends 100 bits in a microsecond; SONET at 2.4 gigabits per second sends 2,400. A simple 1,000-bit e-mail message thus travels in one one-hundred-thousandths of a second through an FDDI link (100 mbps) but may take up to a tenth of a second to move through the switch. This is why the end-to-end "latency" or additional flow time over and above transmission on an X.25 network adds up to around 100 milliseconds. If the switch is the bottleneck, improving transmission speed does not increase message speed.

The low cost of today's electronic switches, including devices such as routers and smart hubs, has made possible many of today's uses of telecommunications in large organizations, but fiber optics push the limits of their capabilities. T1 multiplexers often have problems in handling synchronization and message framing. SONET compounds the problems. Optical switching is as yet an idea, not a reality.

▶   *Wireless communications*: This field, rapidly emerging as one of the most promising areas of telecommunications, includes cellular phones, which will move from analog to digital within a very few years, new services in the parts of the electromagnetic spectrum opened up by the FCC, new international services created through negotiations of relevant agencies and committees, a new generation of satellites, international use of VSAT, personal communications networks, direct broadcast satellite, infrared LANs, and many other innovations. It is difficult at present to predict the exact pace and direction of industry developments and regulatory issues.

There are many reasons to adopt wireless transmission, most obviously their ability to bypass local exchange carriers and to remove the need for expensive cabling, and the mobility they allow. The next generation of transmission systems may involve a competition between or combination of fiber optics (which offer unbeatable speeds and cost per bit) and wireless communications.

► *Fiber into the home*: At some point before the end of the 20th century, at least a few selected municipalities will have fiber optics and thus virtually unlimited bandwidth into the home. This will provide a test bed for creating new consumer services. So far, electronic consumer information services and videotext has largely been a bust, with many millions of dollars poured into "great ideas" that customers ignore. It is clear that video entertainment is the driver for consumer markets, and with fiber new forms of view-when-you-want entertainment will open up. Whether or not these support and encourage electronic shopping, educational services, financial and travel services, and the like is unclear. IBM's acquisition of a minority share of Time Warner in 1992 is a signal of new likely combinations and ambitions.

► *Total image and document management systems*: Image processing is already an area of explosive growth, but firms have only just begun to exploit its potential. The combinations of high-bandwidth networks, optical scanners and storage, data compression, and multimedia technology will accelerate what is happening now. The 1980s was the era of the PC; the 1990s looks like the era of image.

There are many other areas of likely innovation, and many that we do not yet know about. For telecommunications as a career opportunity, the main issues concern dialogue. Technology in itself does not generate innovation; it is the dialogue between business and technology, between end and means, that produces progress. However creative the results of that dialogue, making new technology work and integrating it with existing technology is always a challenge, and the implementer is a key contributor to the most business-centered dialogue.

| | | | | | | ## The Priorities of Some Leading Companies

Although companies are looking ahead at ways to exploit new technologies, they must deal with today's issues. At any one time a large organization's information technology base will span at least 20 years of technical progress. It will have "legacy" systems that were based on 1970s computer hardware and telecommunications, with 3270 terminals, X.25 as a major standard, and analog phone systems. Its 1980s legacy will include PCs, stand-alone LANs, private T1 links, and database management software; its 1990s base will have added routers, high-speed LANs, client/server applications, image processing, and many other innovations. The knowledge and skills of the past remain important, even as the pace of the technology changes; most of the challenges of integration reflect the need to mesh the old with the new.

The priorities of some companies, identified by *Information Week* as leaders in the use of telecommunications, provide a reliable picture of the current practical state-of-the-art, and thus of those topics and themes presented in *Networks in Action* that will be of most relevance to you in your own choice of career trajectory. The companies are not all leaders in their business, and one topic *Information Week* does not address is how well the organizations have meshed their business thinking with their technical planning through dialogue. You can make your own judgments.

As you read about these companies, think of what you have learned about telecommunications, and about those aspects of the many topics we have covered that most interest you. If it is the technical and analytic side, you may want to consider a career as an implementer, in which case the most relevant features of these companies' plans and experience for you may be how they deal with challenges of integration and network management. If you are most interested in the link between the business and the technology, your career direction may be as a designer or as a consultant who either brings technology strategy into business planning or business strategy into technology planning. If the technology does not grab you and you now breathe a sigh of relief, knowing that there will be no more acronyms and standards to learn about, you may well become a client, not a designer, of telecommunications. If that is the case, ask yourself the following question: Can anyone be an effective business client in the coming decades without some insight into telecommunications?

Norwest Mortgage Inc.

Norwest is the third largest issuer of home mortgage loans in the United States, with 440 branch offices. Its main priority for information technology is to remove its IBM mainframe computers by the end of 1994 and replace them with LANs. The main reason for this move is the huge increase in volumes of transactions as homeowners refinance their loans because of the plummeting interest rates that marked the early 1990s. Currently, the branches use IBM 3270 dumb terminals in an SNA architecture, with some UNIX workstations. In head office and operations centers, large LANs handle distributed applications.

Norwest's plan is to install LANs as fast as possible. By early 1993 it had built a 1,100-user token-ring LAN complex. To do this, it relied on wireless communications because cable and equipment takes too long to install when the staff needs immediate and continuous access to information. Norwest thus installs an NCR WaveLAN card in a PC that acts as a bridge to the wired LAN. Once cable has been installed, the WaveLAN card is removed, and the PC is connected directly to the LAN. The wireless capability has saved several years in making the massive move from mainframes and dumb terminals to a completely distributed environment and is yet one more indicator of how important wireless communications are likely to be in the coming decade.

Rhone-Poulenc

Rhone-Poulenc is a chemical and pharmaceutical firm with operations in 140 countries. It has five separate business sectors and ranks fifth in the world among suppliers of chemical and pharmaceutical products. Not surprisingly, its computing and communications systems evolved on a decentralized basis, and its new priority is to create an integrated global platform.

Rhone-Poulenc's strategy for doing this has been to steer a careful path between too much and too little standardization. The variety of equipment and applications is very wide, with IBM and DEC dominating on the computing side and PCs using Novell's NetWare dominating on the LAN. An X.25 international data network connects countries. There are also seven national networks shared across the five business sectors that provide electronic mail, data and file transfer, and many shared applications, especially financial ones. The X.25 international data network extends these applications to other countries.

This relatively simple strategy provides a high degree of flexibility and autonomy for countries and lines of business, as well as coordination among them. The X.25 network is managed from Paris; the national networks are in the United States, UK, France, Japan, Germany, and Brazil. Some 19,000 terminals can now connect to the platform. The major difficulties reported by the central group that runs the X.25 global network are the lack of network management tools and the difficulties in handling large-scale data exchange in a heterogeneous environment. According to one manager, "Due to nonstandardization, you have to spend too much time on IT issues instead of user issues."

Rhone-Poulenc's main business priority that drives its entire networking strategy is research and development. To optimize its R&D, it must be able to share information and tools across its entire geography because "it's not practical to put everyone in one room."

Kennedy Space Center

Kennedy Space Center has three to five satellite launches in the planning phase at any one time. It must assemble experiments, check out equipment, put experiments into racks and racks into modules, and coordinate thousands of steps. The size and complexity of projects has increased greatly over the past few years; so, too, have the communications requirements. In 1989 the Payload Operations Network (PON) had only 100 users; now there are 3,000 users spread across 27 facilities. There are 20 different operating systems running on 33 different types of computer, ranging from PC and Unix workstations right up to supercomputers. There are dozens of different telecommunications architectures and standards.

Planners admit that they have constantly underestimated the rate of growth of the network. The 3,000 users are twice the number forecasted in 1989 and are expected to increase to 4,500 by 1994. The Center has been able to handle the growth through three means:

▶ The use of TCP/IP as the platform protocol that links LANs and workstations regardless of their protocols and operating systems. A mainframe thus appears to users to be just another server.

▶ The use of shared server-based software to cut costs and create a level of standardization that does not intrude on user autonomy.

▶ The use of fiber. The Center has thousands of miles of fiber underground. It has in place sufficient bandwidth—and can add more as needed—to make network efficiency far less of a constraint on its design choices. The use of TCP/IP as the mechanism that creates a platform out of a morass of systems is highly inefficient but very effective.

Carnegie-Mellon University

Carnegie-Mellon University has been a leader in the use of telecommunications in education and research. Its current activities continue to keep it a technological pace-setter, and not just among universities. It is installing a 2.5-gbps ATM campus backbone network that will link over 4,000 terminals; it will connect close to 200 Ethernet LANs with over 4,000 workstations and Macintosh and DOS/Windows PCs. The main impetus for the move is to prepare for the expected flood of multimedia applications and to ease the burden of maintenance. Real-time video for teaching is a main target of opportunity.

In 1992 Carnegie-Melon introduced one of the nation's first distributed electronic libraries. Students and faculty can access page images from books on high-resolution workstation displays. The images are stored on distributed servers and can be accessed from UNIX-based workstations or via Macintosh, DOS, and Windows machines. By distributing the libraries, the university should be able to handle virtually unlimited growth, and departments can publish their own papers and/or add specialized journals without going through the library system.

Pizza Hut

Pizza Hut has solved one of the longest-standing telecommunications challenges: creating efficient and effective large networks of Macintosh machines. It has linked 30 AppleTalk networks that could link to each other only via IBM mainframes, Unix, and DOS machines. It is rare to find an AppleTalk network with even 50 users; Pizza Hut now connects over 600. It uses Ethernet LANs and is now able to combine the wonderful ease of use of the Mac with powerful networking capabilities, including image processing, which it mainly uses for EDI and management of invoices. This has improved customer service and generated large cost savings. The Mac can now use the same communications gateways as any other workstation or PC and can even access mainframe databases.

McDonnell Douglas

McDonnell Douglas has signed a ten-year, $3 billion outsourcing contract with IBM's Integrated Systems Solutions Corporation to move a huge mass of machines off the premises. Space is at a premium for the firm; there was not enough room for it to install fiber-optic cable to link its head office's 11,000-node Ethernet LAN to its eight assembly buildings. The solution has been to install a wireless LAN instead, which operates at 5.7 mbps using spread spectrum transmission. The link carries Unix-based graphics among 11 Digital Equipment machines and Sun Microsystems Unix workstations. With the outsourcing, McDonnell Douglas's computing and communications platform is effectively invisible.

Lawrence Livermore Labs

Lawrence Livermore Laboratories operates over 8,000 Apple Macintoshes, 5,000 PCs, and 2,000 Unix workstations. Over 350 AppleTalk LANs connect workstations to Cray supercomputers. Apart from AppleTalk, the lab's networks support TCP/IP and DECnet. The LANs are Ethernet 10 mbps, running over fiber, with extra room left along the cabling channels to upgrade to higher-speed networks.

The labs need both speed and protocol-independence on the LANs. Scientists routinely transfer large volumes of databases and images across four research networks. The next step for Lawrence Livermore is a shift to asynchronous transfer mode over fiber, which will provide 1-gbps speeds across the LANs and to the desktop. This will be a joint development project with AT&T.

Microsoft

Microsoft's network connects 12,000 employees via 1,000 servers and 40,000 nodes. Not surprisingly, the platform is built entirely on Microsoft's own architecture: its Windows and NT operating systems, LAN Manager, and SQL Server software. All employees can communicate to anyone in the company over e-mail. Microsoft is upgrading the network and installing a 100-mbps backbone that will run over untwisted shielded pair cable, not fiber optics, and will improve performance by a factor of between two and five.

The State of Texas

The state of Texas has over 200 agencies and departments—and multivendor, multitechnology chaos as a result of the uncoordinated explosion in acquisition of PCs in the 1980s. The state spends $1.4 billion a year on information technology. Its new priority is to create a platform based on open systems, with demonstrations and testing in 1993; migration will begin in 1994, and full implementation will take ten years or more. The key standards are TCP/IP, OSI, X/Open, and Unix.

Texas is also designing a new integrated platform based on TCP/IP and OSI that will allow a wide range of agencies to share access to large databases. Judges, state executives, environmental specialists, and welfare workers, among many others, will thus be able for the first time to use their PCs to get information that has been unavailable to them because of the many incompatibilities of equipment, software, and communications.

Northrop

Like other aerospace and manufacturing firms previously described (see, for example, Minicase 2-4 and 5-2) Northrop, the developer of the Stealth bomber, needs massive bandwidth to transmit and receive large databases and high-resolution images of design documents. In 1988 it was one of the first firms to install fiber-optic backbone networks, using the then-unproven FDDI draft standard. In order to maximize reliability and uptime, it installed two routers per LAN and doubled the redundancy on the backbones. One backbone links workstations in its main plant with fiber to the desktop; the other links 30 Ethernet and token-ring LANs at its other plant to a complex of mainframe computers three miles away.

This is an expensive strategy, but one that is fully justified by the cost incurred when the network is down. The redundant links and routers mean that downtime is measured in minutes per year, not hours per month.

Caterpillar

Caterpillar, the world's leading maker of earth-moving equipment, needs to improve the information flow to and from its 75 dealers that sell and repair the machinery. Such data and transactions include equipment history, parts availability, product information, invoices, and warranty claims, all of which until recently have been handled on large mainframes, some of which handle software applications that are over 20 years old and require the use of dumb terminals.

Caterpillar is well into its program of a phased distribution of its main business systems. It is using mid-range IBM AS/400 machines as servers that are connected to IBM PS/2 PCs. Dealer LANs, mostly token ring, link to the mainframes from the servers via a private WAN. This evolutionary strategy buffers Caterpillar and its dealers both from degraded performance created by the growth in volumes that dumb terminals cannot efficiently handle and from having to undertake a massive conversion process in one giant step. By maintaining a clean IBM-based architecture, Caterpillar is able to improve the performance of the existing mainframe applications and gradually rebuild the old systems in a distributed AS/400 and PC environment. Over time, it will front-end more and more functions and positions for client/server computing.

There are many lessons you can draw from this collection of examples. One is that you already know all of this. In many ways, these companies merely summarize what you have learned from this book, and by now you should be very comfortable in understanding the technology and the uses it illustrates. This is not a trivial point. For a variety of reasons, relatively few people in the field of information technology (and even fewer in business) know what you now know.

Summary

Historically, business and telecommunications were very separate career, knowledge, and skills areas. Now, more and more aspects of business require some degree of understanding of telecommunications, and vice versa. These hybrid roles are two of the four career quadrants that establish your career trajectory: (1) business services, in which only a minimal understanding of telecommunications is needed; (2) business support, in which the key skills are business-centered, but a relatively strong technical understanding is essential; (3) development support, in which technical skills are key but business ones are also essential; and (4) technical services, which requires top-rate technical skills and minimal business ones. Each of these career quadrants is needed to create an effective business resource in an increasingly telecommunications-dependent world.

Your business and/or telecommunications career will be affected by future changes in technology that will be at least as fast-paced and dramatic in the next ten years as in the previous ten. The most likely areas of developments are multimedia networking, voice recognition and voice processing, massive-bandwidth internets, optical switching, and wireless communications.

Review Questions

1. In the First Transamerican Bank example, describe the differences in the skill requirements for the head of systems development and the manager of network operations. How likely is the same person to serve in both positions in his or her career?

2. Describe the traditional career path in telecommunications and information systems.

3. Identify and describe the two main forces that have shattered the traditional career paths in telecommunications and information systems.

4. Describe the requirements of the new business support role in organizations. Specifically indicate the type of telecommunications and information systems skills required.

5. How has the set of skills required for a successful development support professional changed in recent years?

6. Describe the skills and qualifications necessary to be a LAN administrator in an environment in which interconnected LANs and client/server computing represent the primary information systems architecture for the organization.

7. Moving to a networked environment is one of the biggest changes in the history of the computer industry. Describe the qualifications needed to become the chief information officer in a large company committed to a networked environment.

8. Describe the particular skill requirements for a telecommunications professional responsible for managing multimedia networking for a telecommunications department in a Fortune 500 manufacturing company whose operations are limited to the United States.

9. Describe the particular skill requirements for a telecommunications professional responsible for managing advanced voice recognition and voice-processing activities for an airline reservation system.

10. Describe the particular skill requirements for a telecommunications professional responsible for managing wireless communications for a multinational retail products distributor with worldwide operations.

Assignments and Questions for Discussion

1. Develop a plan for hiring, training, and supporting technical services specialists for a telecommunications/information systems department. Assume that new technologies will continue their rapid proliferation and that well-trained technical specialists will be valuable assets to the company. Be specific about the types of skills required and how the company will ensure an adequate supply to meet its needs.

2. Assess your own current skill set and place yourself in one of the four career quadrants shown on the map in Figure 19-2. Compare your skill set with the skill set and quadrant of the map you would like to be in to meet your career goals. Outline the skills/knowledge base that you need to reach your goal, and indicate how you propose to achieve it.

3. Identify which of the technology trends discussed in the chapter will require a radically different managerial skill/knowledge base than is available today, and describe the new skills/knowledge that will be required to manage these technologies effectively.

Minicase 19-1

Korn/Ferry International

Staffing Global Networks

Korn/Ferry International is a recruitment firm with seven regional offices in Asia and its Asian headquarters in Tokyo. John Harlow, a director of the company, spoke in late 1991 to *Network World* about the skills needed to manage global telecommunications networks. The article includes comments from other recruiters about skill needs and availability of skilled personnel in Europe and North America.

All the people interviewed agreed that there are "extraordinary challenges" for managers trying to staff global nets, starting with finding candidates with international experience and those willing to relocate. Currently, the lack of people with up-to-date technical knowledge and the ability to help firms apply that knowledge is a problem, but Harlow and others see this as temporary. The explosive worldwide growth in the use of telecommunications is fueling an increase in the pool of skilled technical personnel. As international protocols and standards stabilize and more countries adopt advanced technology, there will be less variation in skill needs. Vendors will be able to provide more technical support, and outsourcing firms will offer alternatives to building skills in-house.

The recruiters also expect more movement across companies and countries, so that technical practices will become more homogenous. This is already the case in the information systems field, in which a COBOL programmer or UNIX expert can move from, say, France to Canada to Singapore with little need to pick up new technical knowledge.

Harlow identifies as the key skills for getting a job in Asia "soft business skills, such as cross-cultural and language abilities . . . the ability to function in unfamiliar cultures, to be sensitive to business environments . . . and to cope with sometimes autocratic and unstable governments." He is emphatic that these skills take precedence over technical ones. If candidates have the soft business skills, "they can quickly be brought up to speed on the technology by comparison."

Fundamental engineering skills are strong in most Asian countries, according to Harlow, but each country has its own distinctive capabilities. The Chinese are "usually well-skilled no matter where they live." Australians who are frustrated by the country's notoriously lazy, unionized, and politicized business environment are increasingly open to relocation in Asia. More and more Japanese telecommunications professionals in the 40–45 age range are willing to leave and work for a foreign firm; this was unheard of 20 years ago. Many of them have been trained outside Japan, getting a masters or doctorate from a top U.S. technical university. They are ready for and qualified for a senior management position but are blocked by the seniority system.

Global network nodes are heavily concentrated

in Japan, but staff are often pulled in from across the region. English is a key language, even in Japan, both because of the presence of many U.S.-based firms and the fact that a large proportion of telecommunications equipment was purchased from English-speaking nations. (The UK's Cable and Wireless is a major presence in the region, for instance, exploiting its historical role as provider of telephone equipment in the old British Commonwealth plus its strong presence in Hong Kong.) Taiwan has many skilled telecommunications and electronic engineers, but little English is spoken there; that is a serious disadvantage for them in finding jobs.

In Singapore, education and language skills are high. The government has made technical education a major priority, establishing its influential National Computer Board. It is also very aggressive in using IT. Its deployment of electronic data interchange through its TradeNet was a major factor in moving it from the tenth busiest to the busiest port in the world. The nation-state's plans to make Singapore the trading hub of the region, including electronic trading. As a result, however, Singapore is short of qualified staff; its own success has moved it from a low-cost producer to a high-cost one, and skilled telecommunications managers are snapped up quickly.

About 20 percent of the telecommunications managers Korn/Ferry recruits to Asia are American or European expatriates. Typically, their key job is to transfer understanding of new technologies to local executives and to help establish effective coordination and communication with headquarters and branches. It is often difficult for Harlow to find people willing to relocate, and he feels that the demand for expatriates will drop, mainly because Japan is rapidly catching up in technical training and skills. He believes that by 1994 to 1996, Westerners will not have any edge in knowledge of products or technology. By 1999, he sees the rest of Asia catching up.

In Europe, there is a far higher demand for than supply of both telecommunications specialists with strong technical knowledge and telecommunications professionals with proven management skills. The shortage is such that recruiters must either interview candidates across Europe or look to the United States. The massive investment Germany is making in rebuilding the telecommunications infrastructure of eastern Germany has created a very large shortage of engineers. Although the former East Germany has many technical staff who can be relatively quickly updated about the technology (because they have sound engineering skills and training in fundamentals), they lack experience in applying advanced technology and thus are not well-positioned for senior management jobs. Network management, for instance, is absent from their background, even though it is central to any telecommunications manager's responsibilities.

The United Kingdom provides the best combination of education, technical skills, and management experience, partly because the liberalization of telecommunications in the early 1980s put Great Britain well ahead of the rest of Europe in managing in a competitive environment. Great Britain has 40 percent of volumes or installations in most key European markets, including LANs (which were impeded by regulation in most countries), electronic data interchange, and private networks. The strong position of London as one of the world's three electronic financial centers, along with New York and Tokyo, has added to the combination of strong technical and managerial skills. London trades 40 percent of the world's foreign exchange, and its stock market trading volumes exceed the rest of Europe's combined.

France is the most difficult country in which to find skilled telecommunications managers, mainly because France Telecom held out for so long to maintain its monopoly; this stranglehold impeded the emergence of telecommunications management, because there was nothing to manage—France Telecom handled it all.

PA Consulting, Europe's top professional service agency, often looks to the United States for candidates. It finds many U.S. telecommunications professionals and managers "overexperienced" for Europe and asking for far too high a salary. PA often hires them as consultants on a one- or two-year contract. It feels that as Europe increasingly uses the same telecommunications standards as the United States, Americans often have all the technical skills needed; up to a few years ago, the telecommunications environment—in terms of regulation, equipment vendors, costs, and standards—was so different that it could take years to become adequately knowledgeable. The only additional knowledge that is now essential is understanding the PTT, which in Germany still owns all network lines and controls all satellite connections. Until 1992, the Swiss PTT determined what switches would be installed for its customers.

One major factor leading U.S. telecommunications professionals to move to Europe is the downsizing of large U.S. companies, fueled by the recession and, to a lesser extent, by the flurry of mergers and acquisitions that led to consolidation of telecommunications operations and to outsourcing, and thus to elimination of many functions. As a result, although Europe is short of skilled staff, the United States has a glut. The North American Free Trade Agreement and the Canadian government's decision to deregulate telecommunications should create a new job market that keeps more of them on this continent instead of looking to move to Europe.

The *Network World* article identifies the following priorities for companies whose business is becoming more and more tightly coupled across nations and increasingly dependent on telecommunications:

1. Managers and technicians who combine international experience with an understanding of international and regional protocols and standards.

2. Senior executives who not only have exposure to the broad range of international technologies that exist now, but who can look ahead five to ten years and forecast and plan accordingly. This is "the most pressing demand."

3. Managers who can link the multiplicity of technologies and platforms; the need here is for people who can take a broader view of the technologies of tomorrow.

Today, the recruiters must constantly choose between candidates who are strong in either technical or management, but not in both. Business skills are increasingly favored by U.S. companies: "The right person who is well informed will become the better manager than a technician who lacks an understanding of the big picture."

Questions for Discussion

1. Identify the key skills recruiters are looking for in telecommunications professionals hired to manage global networks, and discuss the reasons why the current shortage of available talent is likely to be reduced or eliminated.

2. Compare and contrast the demand and supply of well-trained telecommunications professionals in Europe and the Far East, and indicate any likely changes in this situation in the near future.

3. Describe the impact of the downsizing of U.S. corporations on the skill level of telecommunications organizations worldwide, and indicate which countries or regions of the world are likely to be most affected.

Minicase 19-2

Kash n' Karry

Jobs of the Future in the Food Store of the Future

Kash n' Karry, a 112-food store chain, is in the middle of a five-year program that is as radical in terms of business, organization, and technology as any business anywhere. It is moving to implement distributed object computing and to entirely replace its hardware, software, data, and network resources. The risk is huge, but the business environment demands radical moves. Margins are very low in the industry—1–2 percent of sales before tax. "Spoilage," which is mainly shoplifting and employee theft, is typically 2–4 percent of sales. New and powerful competition is everywhere, including mass merchandisers, price clubs, and specialty retailers who are beginning to sell groceries. Wal-Mart, the destroyer of small town merchants, department stores, and just about every competitor it has met to date, is targeting groceries as a major opportunity. You can already buy Sam's Cookies and Sam's Cola (named for Wal-Mart's founder, Sam Walton).

Kash n' Karry's goal is to implement by 1996 a business capability built on information technology that will make it nimbler than any competitor, so that it can launch new lines of business and add promotions in such departments as floral shops, pharmacies, and bakeries without being held up by the slowness of systems development or the cost and time involved in making systems changes. Kash n' Karry's head of MIS defined the rationale as follows: "We need systems that can respond immediately to different marketing decisions. Our business is moving so fast that traditional systems development cycles are completely inadequate."

The key to Kash n' Karry's plan is object-oriented thinking, which is rapidly emerging as what many experts see as the long-awaited solution to the software productivity and quality crunch. It adopts not so much a new style of programming—although object-oriented programming (OOPS) is a specialized field in its own right, with debates about whether the C++ language is better than Smalltalk—as a new way of thinking about design. Kash n' Karry sees its business as a set of "views" of the company that can be pieced together in real-time by the interaction of objects exchanging messages across the network. These objects form a library—often called a repository—that represents the fundamental components of the firm's operations. Anything can be an object: a price list, scanner, manager, supplier, register, or stock item.

The goal is to make systems development and systems use so simple that anyone can access the objects and combine or update them: "Ultimately, we will have one large network in which stores and offices become transparent and data resides somewhere on the network and is available to anyone who needs it in a format they can readily use."

Creating this "transparency" pushes the state of the practical art. Few companies have implemented distributed object systems. Few vendors offer proven expertise and proven tools, though many of them have ongoing development projects—and plenty of promises to go with them.

Kash n' Karry's information services culture lacks not just expertise but also willingness and/or ability to adapt. Most of Kash n' Karry's software is very old and mainframe-based. It replaced 70 percent of its information systems staff in the first year of the project, for they had trouble unlearning old skills and approaches to systems development and did not want to move away from the mainframe methods.

The technology base for the new strategy is UNIX processors linked by a fiber-optic network, plus a development language based on C++ as the object creation tool, an object database, an object request facility, and a specially designed user interface. The current telecommunications network is a 10-mbps Ethernet backbone, which will be replaced with a 100-mbps FDDI dual-ring network that will support six UNIX-based object repository servers and 15 UNIX object servers. The object servers will support 250 X-terminals, powerful UNIX-based workstations linked by Ethernet subnets. The object servers are Sun Microsystems SPARC 2 stations; they support object request software that processes user queries or business events and sends messages to an object manager.

The object manager is a software layer on a relational database stored on a Sun UNIX processor. It locates objects in the relational database and ties them back to the object requester so that the user can interact directly with them. It also stores objects created by business events or transactions and sends messages across the network when objects in its own domain need the services of objects on other servers.

Questions for Discussion

1. Describe how Kash n' Karry's plan to adopt object-oriented thinking will improve their competitive position relative to mass merchandisers, price clubs, specialty retailers, and Wal-Mart.

2. Identify and discuss the reasons why 70 percent of Kash n' Karry's IS group was replaced when the distributed object-oriented system was adopted. Could this have been avoided? Explain.

3. Identify and describe the skills/knowledge requirements needed to manage the telecommunications part of the distributed object environment that is planned for Kash n' Karry, and recommend how these skills could be obtained by the company.

Minicase 19-3

Rohm and Haas

Building Skills for the New Era of Client/Server Computing

Chic Sailes is the manager of personal computing services at Rohm and Haas, the $3 billion company that makes Plexiglass and other chemicals and related products. He describes his job as a "delicate balancing act." Rohm and Haas has over 1,000 PC users in its head office, roughly half of them with Apple Macintosh and half with MS.DOS and Windows. The growing availability of packaged software has made support for development far easier, but not the shifting of applications from mainframe computers to PCs as part of the company's adoption of a client/server architecture. "Moving to this new architecture will be a culture shock to both the technical staff and users. When you start changing a way of doing things that has been around for years, you can't just plug it in. You must demonstrate its worth and value to the organization."

The six-person help desk in Sailes's unit receives over 4,000 phone calls a month. To complement this overloaded service, Rohm and Haas buys products in bulk from a supplier that provides technical support and draws on its own Corporate Computing group for evaluating new products, including "connectivity products." The client/server architecture involves a joint project between his group and the mainframe development staff, a first for them both. Previously, Sailes's team was an autonomous service group with no systems development responsibility; it also did not have to deal with

anything other than local area networking in the area of telecommunications.

Sailes, who is in his 50s, joined Rohm and Haas as an accountant. He was introduced to PCs as a user. He moved to his new job in 1989 in order to challenge himself professionally. "The technology is changing so rapidly that it provides you with a tremendous opportunity to really make a difference in your organization." He feels that his background is a major asset in his work: "I bring a slightly different perspective to the table because I've been in the business community."

He reads everything he can get his hands on to keep up-to-date with technology and products. At the same time, he sees it as essential that he meet regularly with individuals and departments in his user community to make sure he knows about their needs and concerns. "I'd rather be more involved with applying the technology to the business and handling analyses of how a business process can be improved through technology than in studying the nuts and bolts of the computer."

Sailes is representative of a new type of professional in the information services field; he did not come up through either telecommunications or information systems, and he lacks much of the technical background needed for meshing standard end-user computing on PCs into a new company-wide client/server platform. He has a level of knowledge that few people in either telecommunications or IS can easily build: business experience, hands-on experience as a user, and knowledge of a wide range of PC software, much of which has never existed in either of those communities; examples are desktop publishing, multimedia, and software for graphics presentations.

Client/server computing is a growing force in the information technology field. Many companies report difficulties in finding good people, in retraining technical staff, and even in deciding exactly what skills and jobs are required to plan and implement it. Sailes represents part of the skill base.

Questions for Discussion

1. Chic Sailes, manager of personal computing services at Rohm and Haas, describes his job as "a delicate balancing act." Explain what he means.

2. Describe the job skills needed to successfully implement the new client/server architecture at Rohm and Haas through a project team consisting of professionals from personal computing services and the mainframe development staff.

3. Explain how Chic Sailes's unusual interests and career track have served him well in the current information technology environment at Rohm and Haas. Would you recommend a similar path for others? Explain.

Minicase 19-4

Baxter Healthcare Corporation

Moving from Network Efficiency to Business Effectiveness

Baxter Healthcare Corporation is a leader in its industry—some might say a killer. It makes and distributes medical supplies worldwide and has used its telecommunications and information systems as a major force in attracting business, cutting costs, and streamlining operations. Baxter built on the pioneering innovations of American Hospital Supply, which it acquired in the mid-1980s, in using telecommunications to capture over half the hospital market. Well over 75 percent of medical product distributors disappeared when electronic customer-supplier links became the base for industry logistics. Baxter has 60,000 employees in 40 business units, operates in 33 countries, and has revenues of over $8 billion a year coming from 120,000 products.

In 1985 Baxter consolidated its voice and data operations into a single organization that has worldwide responsibility for telecommunications. It sets standards for services and equipment and negotiates major contracts with such key suppliers as AT&T, MCI, Northern Telecom, and ROLM. It tracks regulation, evaluates technology, and manages the corporate backbone network. It is mainly domestic and relies on T1 and fractional T1 links. It administers a virtual private voice network linking 240 U.S. sites. The telecommunications organization is part of corporate information services. Data communications in Baxter is growing at 20–30 percent a year while voice communications is static. There are limited opportunities to integrate the two because 85 percent of voice traffic is outside the company; adding voice to the data communications backbone would thus address only 15 percent of volumes.

In 1990 the telecommunications organization was restructured, and several functions were moved from the corporate group to business units. Maintenance activities and switchboard operations were outsourced. The changes reflected success; since 1985 the priority had been to reduce telecommunications costs and increase efficiency. Overall expenditures were cut by $26 million in five years, despite the 20–30 percent growth in data traffic. Baxter's corporate telecommunications unit replaced and optimized hardware, especially PBX and switches, and it negotiated new contracts with suppliers, including a 1990 deal with AT&T for a Tariff 12 contract and a volume purchase agreement with MCI. Together, these two deals saved $50 million in three years.

In mid-1991 the telecommunications organization was realigned. The previous emphasis had been on technical, economic, and operational efficiency; the telecommunications unit aimed at building a first-rate utility. It worked with business units on a project-by-project basis. Staff were selected for assignment because of their technical expertise in either voice or data communications. There was little continuity of relationship; the telecommunications staff were technical advisers to business units and were not business-focused.

Now, Baxter's corporate telecommunications group assigns people to business units, not to projects. They work with the units on an ongoing basis to learn their communications needs; when they need additional or specialized technical expertise, they bring in the relevant individuals. They aim at being proactive rather than operating in the old mode of reacting to stated project needs. According to the director of corporate telecommunications, "the job of telecommunications personnel assigned to business units is first to understand their business needs. Then, they focus on new voice and data technologies and present them to the units as ways of more effectively meeting those needs. Networks today are providing many new functions and features which business units can use to improve operations such as customer service or order entry. ISDN, for example, can be used with database look-up to provide better customer service."

Baxter is making strong moves to develop a network strategy based on business needs in the international marketplace. It is also exploiting the new generation of LAN technology to integrate the customer service and order entry systems that are backbone-based with office productivity tools such as electronic mail and spreadsheets. This will require more organizational shifts and new skills.

Questions for Discussion

1. In 1985 Baxter consolidated its voice and data operations into a single organization, but the two networks were kept separate and were not really integrated. Explain.

2. Describe the key skills required to operate in an environment that emphasizes technical, economic, and operational efficiency and that produced an overall reduction in telecommunications expenditures by $26 million in five years by replacing and optimizing PBX and switching hardware and by obtaining more favorable new contracts with service suppliers.

3. Explain why Baxter now assigns telecommunications professionals directly to business units, and describe the differences in skills needed to operate in this new mode, compared with the efficiency mode outlined in question 2.

Product Summary Appendix

This appendix is intended mainly for browsing. It describes approximately 100 telecommunications products listed under 19 headings. They are a representative sample of the range of products on the market in 1993. The purposes of this Telecommunications Product Summary are to help you get a sense of the available technology and to make the terms of telecommunications more concrete. Whereas you can get an understanding of both the functions of a router or multiplexer (from the text of this book) and their uses (from the many minicases), the differences in features among competing products, typical prices, and standards and protocols supported can be found in this appendix.

The Telecommunications Product Summary is very selective. There were an estimated 800 vendors and 5,400 products on the market in early 1993, up from 600 and 3,000 the year before. The summary lists the main features of each product, many of which refer to standards and terms that are not only outside the range of this book but outside the range of knowledge of most telecommunications professionals. Many of these are special-purpose or proprietary; they are all part of the very complex technical environment with which implementers must deal.

Browse the Product Summary. You will see common patterns, particularly the frequency of references to particular standards. You will get a picture of prices, equipment, and technology. The summary focuses on LAN and LAN-to-WAN products, mainly because these are the products with which implementers are most concerned.

The products are listed under the following headings:

1. Adapters
2. Bridges
3. Bridge/Routers (Brouters)
4. Cabling
5. Communications Servers
6. E-mail Gateways
7. File Servers
8. Gateways
9. Hubs
10. Modems
11. Multiplexers
12. Network Management Tools
13. Network Operating Systems
14. Network Printers
15. Repeaters
16. Routers
17. Storage Devices
18. WAN Access
19. Wireless Products

Adapters

3Com Corp. EtherLink II
An eight-bit adapter for connecting PCs as workstations to networks. Supports ISA/EISA bus types (IBM PC, XT, and PS/2 models 25, 30, 35, 40, and compatibles). It is a half-size board with thick (AUI) and thin (BNC) coax connections. $250.

Accton Technology Corp. EtherCoax-HP/EN 1620
A 16-bit coaxial Ethernet adapter for PC, XT, AT, and compatible computers. Provides a 64-kb memory buffer with local memory management. BNC and AUI connectors are provided. $250.

CNET Technology Inc. CN200E/2
An Ethernet Card built for IBM PS/2 Micro Channel and compatible machines. IEEE 802.3-compliant and uses carrier sense multiple access/collision detection (CSMA/CD) packet passing with a 10-mbps transfer rate. $250.

Gateway Communications Inc. G/Ethernet MC
A 16-bit Ethernet adapter for use in IBM Micro Channel computers as either a file server or a workstation. Features a wide range of operating system software

drivers. Operates at 10 mbps over standard coaxial, RG-58, and fiber-optic cable. $450.

Codenoll Technology Corp.   CodeNet 9440
Singlemode FDDI Single Attached Station Adapter
A Category One single-mode SAS 100-mbps ANSI FDDI adapter. Connects to a single-mode glass fiber port on a FDDI concentrator at distances of up to 12 miles. The card includes the SMF-MIC (Media Interface Connector) connector, diagnostic indicators, software support for leading network operating systems, and FDDI SMT (Station Management). $9,000.

Accton Technology Corp.   RingPair 4/16/TR1607
A high-performance adapter for 4-mbps to 16-mbps token-ring networks. Compatible with the AT, AT 386/486, and compatible computers. Supports both Type One and Type Three cabling. $800.

Network Generation   Pocket LAN Adapter
Laptop, notebook to LAN, for IBM PC, XT AT, and compatibles. With 8 kb of memory, for thin and thick Ethernet and 10Base-T; data transfer rate of 10 mbps. $3,000.

Plusnet Corp.   FDDI/MCA 16/32
A 16-bit or 32-bit, 100-mbps NIC (network interface controller). 8 mbps throughput for Micro Channel bus computers. Fiber-optic, Coax, STP, and unshielded twisted-pair (UTP) cabling supported. Connectors are ST, BNC, DB9, or RJ45. Includes software drivers for NetWare 2.x, 3.x, and Open Data-link Interface (ODI) drivers. $1,500.

Bridges

V.C. Advanced Relay Communications Inc.
XBridge
X.25 bridge for LANs that use Novell's NetWare network operating system. Package supports fractional T1 speeds. Connects up to 11 remote LANs to a WAN. LANs connect point-to-point, or if connected through PDNs, point-to-multipoint. 256 virtual circuits (VCs) are supported. Physical connection through modem or CSU/DCU. $2,500.

Madge Networks Inc.   Smart 16/4 MC Bridge/58-02
Bridge for Micro Channel buses; connects two token-ring networks and utilizes source routing technology to forward data between the rings. Package consists of two Smart 16/4 Bridgenodes and the Smart source routing bridge software, which includes Madge's Fastmac Mi-

crocode. As well as providing the bridging function, these products enable users to monitor hardware and software errors, manage the network configuration, and establish links to network management utilities. Both products fully support IBM bridge management features. $4,000.

Plexcom Inc.   PLEXNETLocal Token-Ring Bridge
Links two independent 802.5 compatible networks together. The 8039 supports source routing protocol, and optional firmware is available to support SRT (Source Routing Transparent) protocol. 4-mbps to 16-mbps operation is strap-selectable, and networks do not need to be operating at the same speed. The 8039 includes a Simple Network Management Protocol (SNMP) agent and also supports NetView (IBM LAN Manager). Additional features include load balancing and Spanning Tree Algorithm. The 8039 occupies two slots in a Plexnet Chassis. $4,500.

RAD Data Communications Inc.   MBE–Ethernet Bridge/MBE-1, MBE-8
Extends a central Ethernet network to a remote cluster of up to eight workstations. The remote connection is accomplished using a serial communication link that operates at data rates up to 256 kbps over private lines or public networks, including Integrated Services Digital Network (ISDN). Operation is in pairs; the MBE connected to the main LAN filters and forwards to the remote unit only those packets whose destination address belongs to the remote site. The unit connected to the remote LAN is self-learning; it filters and transmits to the main LAN packets whose destination address is not in the remote site. $800–$1,700, depending on interface.

Applitek Corp.   LANcity Ethernet Bridge/N110/E
A high-speed Metropolitan Area Networking (MAN) bridge available at 10 mbps that connects standard Ethernet subnets over a single standard 6 MHz broadband channel up to 35 miles away. LANcity Bridge is a protocol transparent remote bridge with no limitation on any protocol that connects Ethernet-to-Ethernet via long-distance MAN. $14,000.

Bridge/Routers (Brouters)

Magnalink Communications Corp.   ROUTE-3000
Frame Relay Bridge/Route-3000 FR
Provides LAN-to-LAN connectivity over frame-relay networks. Examines 14,800 packets per second and forwards over 7,000 packets per second, with a sustained

throughput of 2.048 mbps. Redundant link with auto-restoral is standard in all models. Compression model uses proprietary technology to increase real throughput without the need to upgrade transmission media. Addresses of up to 8,192 local and remote devices are learned for packet routing, and only packets destined for the remote LAN are passed in order to minimize LAN-to-LAN traffic. Traffic customization package provides flexible filtering of unwanted traffic and prioritizing of time-sensitive data. Comprehensive diagnostics, including noninvasive loopbacks, threshold alarms, and performance statistics, can be activated and monitored either through a user terminal, a LAN station, or MUX-View-Telco Systems's network management systems. Privacy and integrity of transmitted data is guaranteed through optional DES (Data Encryption Standard) encryption. All models have RS-449, V.35, RS-422, EIA-530, RS232 interfaces. $7,000.

Microcom Inc. Microcom Bridge/Router/MBR/6000
Connects up to three LANs when configured as a local bridge/router. When configured as a remote bridge/router, an MBR/6000 simultaneously links a local network (Ethernet or token ring) with up to four remote networks. Each MBR/6000 supports all leased line services, including analog (9.6–19.2 kbps), 56–64 kbps, Fractional T1 (64 kbps to 51.536 mbps), T1 (1.544 mbps), and E-1/Conference Europeene des Administration des Postes et Telecommunications (CEPT) (2.048 mbps). $4,000–$8,000.

Teleglobe Communications Inc. ConnectLAN/CL 5200
Remote high performance bridge/router connects geographically dispersed LANs over wide area network services at rates from 9.6 kbps through T1/E1. The CL 5200 connects to a 4–16 mbps token-ring and/or Ethernet LAN and supports up to two WAN serial ports. WAN protocol options include HDLC for point-to-point connections, plus X.25 and variable bandwidth frame relay for packet network connections. Where required, the CL 5200 will perform the conversion between token-ring and Ethernet LANs. The CL 5200 supports multiprotocol routing (TCP/IP, DECnet, XNS, IPX, and AppleTalk) and bridging of nonroutable or selected traffic simultaneously. Ethernet bridging complies with the IEEE Spanning Tree standards and Token-Ring to the Source Routing Bridging standard. The CL 5200 can be supported by Teleglobe's Con-

nectView 5800 Network Management System and is SNMP-compatible. $7,500–$9,500.

Cabling

MUX LAB SWITCHEX Premises Wiring Management System
A PC-controlled electronic patching system that manages the physical connections of a network. Comprises a hardware switching module that is protocol-transparent up to 16 mbps and uses 50-pin telco or RJ-45 connector for the input/output pairs. Can switch up to 48 host ports or up to 192 office outlets. Also comprises a connections software and user-defined database that keeps track of the connections and allows moves, adds, and changes to be made to a network. $2,000–$6,000 depending on configuration.

Plexcom Inc. Plexstar/8040/8043
Provides 16 RJ-11 ports, including two RJ-11 ports to cascade modules. The 8043 module provides 16 ports on a 50-pin telco. The 8040-8043 can support up to 16 physical segments, each up to 5,000 feet of cable (80,000 feet per PlexStar 8040). Linked 8040 modules can be used to create campus and building networks. The PlexStar 8040-8043 can use the existing telephone cabling or dedicated wiring in a building, connecting into the building telephone wire in the wiring closets. $1,300.

Unicom Electric Inc. Unicom-Dyna A400 HUN
An active hub designed to convert IBM 3X and AS/400 twinaxial "daisy chain" wiring to a more practical "star-wired" configuration via unshielded twisted pair (UTP) cabling. Has both twisted pair and twinax host hookups for total flexibility. Design permits each device to be connected up to 4,000 feet from the host to workstations using UTP. A single hub supports up to seven terminals, features both terminal and host-activity monitoring. Has corresponding yellow/red/green LEDs to indicate host port/power/device port status. Bidirectional repeating capability ensures accurate regeneration of the device signal. Proprietary digital fiber circuits on all device connections prevent repeating line noise, and a polarity switch enables different polarity UTP input from twinax baluns. $1,000.

Bytex Corp. Ring Out Tester
A troubleshooting tool for 4-mbps and 16-mbps token-ring networks. Transmits, receives, and decodes IEEE 802.5 token-ring signals to detect and isolate high-frequency cable faults. Supports both unshielded and

shielded twisted pair and offers built-in DB-9 and RJ-45 connectors. $1,500.

Fotec Inc.   Fiber Optic Cable Testing Software
Accurate characterization of fiber-optic cables requires determining loss and bandwidth for future high-speed networks. This program supports fiber-optic installation and testing by calculating variations in cable loss and in source wavelength and by calculating multimode fiber-optic cable bandwidth with LED sources. Runs on an IBM PC or IBM-compatible using MS.DOS. $7,500.

Communications Servers

American Data Corp.   NetBlazer/NST
A full line of integrated communications servers. Connects to Ethernet or token-ring networks and provides IP routing over leased or dial-up lines. Additionally, provides terminal service for local terminal users or dial-in users. Gives network hosts and workstations access to a common pool of modems for modem sharing. As a dial-up router, allows an organization to expand its network over the public telephone system to any location in the world. This flexibility saves leased line charges by automatically creating networks on demand over dial-up connections. $2,400.

Digital Equipment Corp.   DEC Commserver 100
Allows Digital Equipment systems to connect to financial market data feeds such as Telerate International Quotations (TIQ), Standard and Poor's Tickers III and IV, and Reuter's Integrated Data Network (IDN). Can also interface to SWIFT and CHIPS and market broadcasts from SIAC, NYSE, and NASD. Military and government users can use the DEC Commserver for communications connections to DDN/ADCCP NRM and ADCCP ABM. $7,300.

Network Products Corp.   ACS2/SA-Asynchronous Communications Server/ACS2
A high-speed network gateway providing asynchronous connections for network users. Can be used to share modems, communicate with a remote computer, dial out to professional services, and provide connectivity for remote dial-in applications. When used with Network Products Communication Controller (NPCC), offers four ports of high-speed access supporting transmission rates of up to 38.4 kbps. Four NPCCs can be installed in a single dedicated workstation providing 16 ports in a single gateway. $1,300.

Shiva Corp.   LanRover/L
A dial-in server that provides PowerBook and Macintosh users with fast, easy, dial-in access to all network services, and shared dial-out to electronic forums and bulletin boards. With LanRover/L, users can connect to an AppleTalk network and access file servers, Electronic mail, and network applications. $200.

E-mail Gateways

Computer Mail Services Inc.   S-Bridge for Higgins
An electronic mail gateway between Enable Software Higgins and Simple Mail Transfer Protocol (SMTP). Connection with UNIX is possible through SMTP. Requires a dedicated PC with 640-kb RAM. The Higgins Exchange PC requires 640-kb RAM. Price for one Higgins domain, with upgrades available for additional domains and remote Higgins users. $2,500.

Computer Mail Services Inc.   V-Bridge for MHS
An electronic mail gateway between a Banyan VINES mail service and Novell NetWare MHS or other Message Handling System (MHS) based software and LANs. ASCII/Binary attachments, forwarding, answering, and return-receipts are among the features supported. Runs as a "server based service" on a VINES server and is fully integrated with Banyan StreetTalk. $800.

File Servers

Intergraph Corp.   InterServe 6400
A deskside RISC-based server designed to provide larger workgroups with database, file, compute, and plot serving in distributed and client/server environments. A 36 mips CPU and a separate IO subsystem for increased network performance. Base configuration includes 16 mb of memory, a 426-mb internal disk drive, three RS-232 ports, a 15-pin AUI Ethernet interface, a parallel port, and 3.5-inch floppy disk drive. Is configurable with up to 256 mb of memory, and up to 5 gb of disk. Additional cabinet space allows the addition of optional devices such as add-on RS-232 boards, an SNA gateway, and a IEEE-488 interface. $23,000 and up.

Gateways

3Com Corp.   CS/1-SNA
Provides connectivity between Ethernet-attached de-

vices and SNA front-end processors. Ethernet users gain access to IBM data centers. Emulates 3278 and 3287 IBM host sessions. Ethernet-attached PCs running either TCP/IP or Xerox Networking System (XNS) can be used to access IBM hosts with the CS/1-SNA Gateway. $9,500.

Digital Communications Associates IRMALAN/EP
SNA gateway that provides IBM host connectivity via 802.2, SDLC, or DFT; allows simultaneous IBM 3278/79 terminal emulation or IBM 3287 printer emulation for any Macintosh or PC on the same gateway. Supports SDLC link speeds up to 72 kbps, five sessions per workstation, and 128 LUs per gateway. DOS client terminal emulation software included. IRMALAN Windows client and IRMALAN Macintosh client are separate add-on software available on a gateway license. $1,500–$7,500.

Interlink Computer Services Inc. PCaccess
Gives IBM PC/XT/AT or compatible computers multivendor interoperability via TCP/IP. Also, provides FTP file transfer rates of up to 60 kbps and Digital Equipment VT220 and IBM 3270 terminal emulation. Also supports internet standard LPR and LPD utilities for printing across the network. PC access can be loaded co-resident with NetWare or LAN Manager and supports token-ring, Ethernet, and SL/IP connections. $350.

Cleo Communication HS/3770LINix
Enables UNIX systems to emulate IBM 3776 and 3777 RJE (remote job entry) stations for remote batch transfer of files and data in SNA networks. Communicates with IBM host Job Entry Subsystems, such as MVS/JES2, MVS/JES3, and VSE/POWER. Users can submit jobs to the SNA host for processing and receive output generated by the host. They can also issue commands to the host Job Entry Subsystem and receive output from these commands via the host console. A single installation can be configured to provide multiple RJE stations. For each station, users get up to five simultaneous host sessions, plus support for up to nine printer and nine punch output devices. $1,000–$2,000.

Computone Corp. LYNXWARE/Netcom II X.25 Gateway
A full-function X.25 communications gateway that allows UNIX users to transfer data across packet-switched data networks, X.25 dedicated links, or X.25 dial-up (X.32) links. Includes an X.25 "C" language ap-

plication program interface that can be used to develop and test Open Systems Interconnection protocols and applications; 16 to 112 virtual circuits. $1,140–$2,450.

Microdyne Corp. EXOS X.25 Extended Adapter for PC
Communications card that features a Motorola 68008 microprocessor, 1 megabyte of memory, and a RS-232C (V.24) port. Combined with either EXOS Link/X.25 or EXOS X.25 WAN Gateway software, allows users to gain access to either private or public X.25 packet-switching data networks. $1,800.

Andrew Corp. NetLynx/5400e
Designed to allow up to seven DECnet Local Area Transport (LAT) users access to IBM System/3X or AS/400 sessions. Users appear as locally attached twinax devices. Features include simple installation; access to AS/400 PC support; and IBM display station and printer emulation. Supports over 80 ASCII terminals, printers, and PC terminal emulators; flexible node addresses are assigned on screen; and has user-defined security passwords, multiple session capability, and auto-dialing. $6,000.

Hubs

Plushnet Corp. FFDI/IMR 4 or 8
An intelligent, active, four- or eight-port hub used in a star topology for connecting workstations utilizing the PlusNet FFDI 100-mbps ISA, EISA, or Micro Channel network adapters. No limit to the number of IMRs that can be interconnected within a network. Maximum of 65,534 workstations. Fiber-optic, coax, shielded twisted pair (STP), and unshielded twisted-pair (UTP) cabling supported. Connectors are ST, BNC, DB9, or RJ-45. Maximum distances between repeaters or workstations are 100 meters using fiber; 100 meters using coax or STP; 30 meters using UTP. $2,000–$2,500.

3Com Corp. LinkBuilder 10BT and LinkBuilder 10BTi
Low-cost wiring hubs that provide chassis flexibility at fixed-port prices. Customers can add 10Base-T nodes, Simple Network Management Protocol (SNMP) management, and BNC or fiber backbone connections and can form one logical 12–48-port repeater. One LinkBuilder 10BTi hub provides SNMP management for up to three LinkBuilder 10BT hubs. $1,000–$2,000.

Penril Datacomm Networks Series 2000
Modular, internetworking hub that supports mul-

tiprotocol routing and concurrent bridging. The four-slot chassis supports WAN, Ethernet LAN, and 10Base-T device interfaces. Can be configured as a multiport Ethernet router/bridge, a 96-port 10Base-T routing concentrator, a WAN router/remote bridge, or any hybrid combination of these. Routing and bridging functions are selectable on a per-port basis. All traffic between interface modules is routed or bridged at wirespeed over the high-performance internal backplane. Maximum performance speeds of 30 kbps are supported. Filtering capabilities are extensive, supporting network security and enhanced throughput configurations. Routing functions support open shortest path first (OSPF) and EGP to enhance interoperability. Protocols routed includes: TCP/IP, DECnet, IPX, and Xerox Networking System (XNS). WAN protocols supported are PPP, Frame Relay, and X.25. Multiple WAN interfaces are supported: T1/E1, RS-232, V.11, V.24, and V.35. Comprehensive Simple Network Management Protocol (SNMP) management is also available on a PC-based, menu-driven system. $10,000–$32,000.

Synoptics Communications Inc.   LattisNet Model 3000 Premises Concentrator/Model 3000-01
An intelligent hub that supports up to 132 ports of Ethernet, 144 ports of token ring, and 42 ports of FDDI in a physical star topology using unshielded twisted-pair, shielded twisted-pair, coaxial, or fiber-optic cabling. Hubs can be connected to create very large networks. Modular local and remote bridges and routers are also supported, as are SynOptics's network management system. $3,000.

IBM Corp.   IBM 8230 Model 002 Base Unit
Intelligent controlled access unit. Self-healing and managed by IBM's Lan Manager NOS. Supports 4–16 mbps on shielded and unshielded twisted-pair media. Provides connection for up to 80 connections on a token ring by allowing up to four 20-lobe access modules to be attached. $3,500.

Optical Data Systems Inc.   294-RC
Module supports out-of-band token-ring network management. By placing a single 294-RCB in a chassis, this manager can control up to 32 ports. Communication is via RJ-45 or fiber-optic connectors on each module. One RJ-45 or fiber connector is used as an input, with the other provided as output to additional cascaded ODS modules because distances between chassis are often in excess of the 200 feet customary for RS232 and RS422. RS422 allows connections up to 2,000 feet between the ODS 294-RCBs. $3,000.

Modems

Hayes Microcomputer Products Inc.   ULTR 96-08-00863
A V.32/9,600-bps stand-alone modem. Dial-up or leased line operation and throughput to 38,400 bps with V.42bis data compression and V.42 error control. Also provides X.32 (Dial X.25 PAD) and MNP 2-5, as well as automatic feature negotiation. $1,000.

Intel Corp., PC Enhancement Div.   SatisFAXtion modem/200/model 200 fax modem
A group 3-compatible internal faxmodem card combining 9,600-bps fax transmission with a 2,400-bps V.42bis modem. Available as a two-thirds length bus board. Model 200 includes DCA's CROSSTALK Communicator for modem connections and software for faxing from DOS and Windows applications. $350.

Plexcom Inc.   PLEXLINK Fiber-Optic High-Speed Modem
A full-duplex, fiber-optic modem designed to be compatible with BELL 306/CITT V.35 and RS-422 interfaces. Has both BELL 306/V.35 and RS-422 interface circuitry built in. Supports all common types of clock configurations. Has front panel status indicators and is compatible with the Plexview Network Management System. Plexlink 8306 available with either ST or SMA style optical connectors. Additionally, has an optical budget of 16 dB (decibels) with 62.5/125 fiber. The Plexlink is also modular and can reside in a number of different chassis. $1,200.

Shiva Corp.   NetModem V2400
A communications device designed to meet two-LAN communication needs. Allows remote Macintosh users to dial in and have access to network services and lets Macintosh users dial out to remote information services or hosts. Consists of a Hayes-compatible modem with V.22 and V.22bis support and a built-in LocalTalk interface. $500.

U.S. Robotics Inc.   Courier HST
Modem with ASL 16.8-kbps, 14.4-kbps, 12-kbps, 9.6-kbps, 7.2-kbps, 4.8-kbps, 2.4-kbps, 1.2-kbps, and 300 bps speeds, with CCITT V.42/MNP 2-4 error control and V.42bis/MNP 5 data compression for error-free throughput up to 57.6 kbps. Available as internal, external, or rackmount. $4,000.

Multiplexers

ADC Kentrox   DataSmart T1 and 45 SMDSU
Provides subscriber access to switched multimegabit data service (SMDS) at T1 and T3 rates. SMDS offers connectionless, packet-switched service to cost-effective transport data, such as LAN-to-LAN communication, over a wide area network. This intelligent SMDS connects to SMDS-compatible router or other data terminal equipment to the SMDS network through a T1 line. SMDSU is designed to conform with Bellcore SMDS specifications and related standards. SMDSU T1 and T3 are members of the DataSMART family of Intelligent DSU/CSUs (IDSUs) products. $6,450–$14,950.

Ascend Communications Inc.   BRI-3, BRI-8, T/RI, T/P2
Multiband bandwidth-on-demand controller family of products uses Ascend's inverse multiplexing technology to provide bandwidth on demand for applications with high throughput requirements such as video-conferencing, imaging, and LAN bridging and routing. Provides ISDN basic rate, ISDN primary rate, and T1 access to worldwide switched digital services, including AT&T Accunet Switched Digital Services (56–64–384–1536 kbps), MCI VPDS 56–64 kbps, Sprint VPN 56 kbps, local exchange carrier 46–64-kbps services, and PTT services in the European Community and the Pacific Rim. Multiband creates point-to-point virtual connections at data speeds for 56 kbps to 3 mbps in 56-kbps or 64-kbps increments. $5,000–$11,000.

Ascom Timeplex Inc.   TIMEPAC X.25 Packet Switching Family
Family of X.25 nodal processor packet switches and packet assembler/disassemblers (PAD) includes the NP1000, NP500, and NP100E and is managed by the TIME/VIEW 2500 Network Management System. The NPs provide advanced networking capabilities, such as packet-by-packet routing for optimum response time with maximum resilience and bandwidth utilization, security features, automatic routing table generation, and a frame-relay option. $5,000–$30,000.

Digital Access Corp.   Fracdial inverse multiplexor
Installs at the customer's premises and provides dial-up, high-bandwidth (to 1.544 mbps) data and video transmission on any existing digital telephone network, public or private, including voice networks. Bandwidth is user-selectable. System looks and functions like a dial-up modem. In use for video, saves 70 percent or more in costs over dedicated, broadband facilities and provides the flexibility to connect to corporate partners, vendors, and customers. Is compatible with all video codec makes and supports RS-366 dialing from the codec. Dialing from the Fracdial front panel is executed through menu-controlled, stored directory. For WAN connections, interfaces to many LAN bridges and provides dial-up bandwidth for large file or batched file traffic that must be routed to different locations on demand and that requires high bandwidth. For data rates to full T1: $11,500; for data rates to half T1: $6,995.

Multi-Tech Systems Inc.   MultiX.25 PAD
Packet assembler/disassembler (PAD) for accessing X.25 public data networks (PDNs) and private packet networks (PPNs). Converts native mode (non-X.25) protocol to CCITT X.25 standard packets, then transmits the packets to the host terminal or PC. Has eight asynchronous channels, one X.25 trunk port, and a command port. The command port connects the PAD to any async terminal for use in configuring the PAD operating parameters. $1,700.

Newbridge Networks Inc.   3604 Frame Relay PAD
Full-function intelligent packet assembler/disassembler supporting standardized frame relay. It supports up to two network ports capable of speeds up to 64 kbps (V.35, V.24, V.11). Up to eight user ports can support frame relay, X.25, X.32, any bit-oriented sync protocol, and async protocols up to 19.2 kbps. Full SNMP management is available through the object-driven 4602 Network Manager. Price starts at $6,500.

Telenetics Corp.   Frame Relay Multiplexor and Node Switch II/FRM II
Series of frame relay internetwork nodal processors (INPs) provide a flexible and cost-effective means to use public frame-relay services with local switching and concentration capabilities, or for provision of private or hybrid public/private frame-relay networking. Provide bandwidth on demand for inter- and intrasite connections and up to 1,024 frame-relay virtual circuits per node. Support Ethernet LANs, SNA, X.25, transparent, and asynchronous protocols, and provide frame-relay interfaces to router, bridges, and PADs with frame-relay capabilities. Equipment allows network managers to provide seamless, mixed-vendor and mixed-technology interworking on consolidated frame-relay networks with very high utilization. These networks provide very high throughput, low delay, reduced number of network nodes, and reduced number of physical access lines or

data rates needed to transport all the firm's data and voice traffic. $25,000.

Dowty Communications Inc.   Subrate Data Multiplexor/DCP9072
A five-channel subrate data multiplexor (SDM). Standard features include a built-in customer service unit data service unit (CSU/DSU), console and front panel control, built-in diagnostics, and up to five sync inputs. Input speeds are to 384 kbps. Fully compliant with the SDMS services offered by AT&T and other carriers. Benefits include end-to-end diagnostics, network management, and, where applicable, single-ended multiplexing and individual channel switching. $1,800.

Netrix Corp.   ISS Integrated Switching System
A wide area networking, X.25 packet-switching, and time division multiplexor. Netrix is the first company to offer products that simultaneously support these three technologies. Provides the ability to consolidate traffic as diverse as X.25, Systems SNA, DECnet, LAN-to-LAN traffic, video, and voice into a single network. $15,000–$40,000.

Optical Data Systems   ODS 300/800 Series
Full line of fiber-optic multiplexors, including internal, external, asynchronous, synchronous, LED, and laser. Speeds to 2.048 mbps. Interfaces include RS-232, CCITT, RS-422, RS-449, V.21, V.35, T1, and others. $400–$5,000.

Dowty Communications Inc.   Frame Relay Access Device (FRAD)/FPX2195
Provides direct access to frame-relay networks for users of equipment supplied by IBM, Bull, and other leading manufacturers. Can support 80 remote synchronous terminals and printers. Consists of eight user ports that can operate at speeds up to 19.2 kbps. Provides dual links that can be used for communications with both frame-relay and X.25 networks for load balancing or redundancy. Will automatically reroute calls to the surviving link in the event of a link failure, without interrupting calls. Other features include printer sharing, sync/async operation, and user-controlled switching. $6,000.

Network Management Tools

Dolphin Networks   LAN Command Professional/2.1
Network diagnostic system that monitors and troubleshoots IBM, Microsoft, 3Com, Banyan, and NetBIOS networks in real-time. Proactively analyzes every node across all known bridges, routers, and gateways without additional software or hardware. From a single PC, polls every station and gathers hundreds of statistics to diagnose common problems that occur in a LAN/WAN environment. Using pulldown menus and extensive on-line help, it provides expert-driven network management system that ensures support of Arcnet, Ethernet, and token-ring topologies, auto-populating databases, extensive diagnostics of both file-servers and clients, English-text alarms, English-text reports and suggestions, custom alarm forwarding, benchmarking tools, remote PC control, and custom report generation. $1,000–$8,000.

Intel PC Enhancement Division   Intel NetSight Sentry Monitor
Hardware/software monitor for Ethernet and token-ring networks. Lets the network administrator supervise network traffic, evaluate station activity, and detect network faults. It consists of software and an intelligent PC add-in board, which installs in any IBM-compatible PC or AT that accepts a full-length expansion card. $1,995 for Ethernet, and $2,995 for token-ring version. Also available: NetSight Professional Enhancement, a software upgrade that adds a full-featured multi-protocol analyzer to the NetSight Sentry monitor. Price is $6,495 for the Ethernet version and $6,495 for the token-ring version.

Integraph Corp.   Network Management System/NMS
A standards-based network platform for managing communications equipment, lines, and hosts on large, heterogenous networks. Includes a discovery facility that automatically identifies all TCP/IP devices residing on the network. Performance management facilities track transmitted packets, loss, and load at individual LAN components. Customizable alarm feature supports automated performance management. Fault management facilities define, set, and monitor alarm conditions, network outages, and packet loss. Configuration management facilities allow devices to be reconfigured and objects to be enabled or disabled. An interface to Informix databases enables users to create queries, generate reports, and graph trends or usage. Runs on an Intergraph workstation or server under the X Windows System with the Open Software Foundation (OSF)/Motif graphical user interface. $15,000.

Legent Corp.   LANSpy
Reports token-ring LAN performance, configuration, and error statistics to a central site IBM or compatible

computer. Provides end-to-end problem determination and capacity planning capabilities, real-time performance and utilization statistics, SNA response time statistics, and central site command and display facilities. LANSpy data is forwarded to NetSpy, LEGENT's enterprise network performance manager, providing an integrated, centralized view of WAN and LAN performance. Runs as a background program under DOS, monitoring all LAN stations, including those using DOS, OS/2, and UNIX. No dedicated PC is required. Network operating systems supported include Novell NetWare, IBM LANServer, and Microsoft LAN Manager. $20,000–$100,000.

Novell Inc.   NetWare Communication Services Manager 1.0
Windows-based application that provides network management capabilities for configuring, monitoring, and maintaining NetWare Communication Services, including NetWare for SAA 1.2 and NetWare Asynchronous Communication Services (NACS) 3.0 anywhere on a NetWare Network. Helps minimize service downtime through real-time fault management, and provides performance management through the collection of statistical information. Can access and manipulate audit trail logs of communication services usage, and an event monitor can be configured to receive real-time notification of potential security violations. Based on the NetWare Management System core technology, allowing it to operate concurrently with other Novell network management tools. $3,500.

Synetrix Inc.   Net-Toolbox
Features tools for Novell NetWare management; allows quick access to information to simplify network tasks and optimize efficiency in LAN operations; includes functions to track user log-in and log-out information, establish interactive communication between two workstations, or display connection usage data; and offers a total of eight utilities for tracking, communications, and control. Operating system and software requirements: Novell NetWare 2.15 and above, DOS 3.1 and above. Hardware requirements: IBM PC XT, AT, PS/2, or compatible; minimum 512 kb of available RAM; CGA monitor or above is recommended; compatible with ARCNET, Ethernet, token ring, and 10Base-T. $100.

Network Operating Systems

Banyan Systems Inc.   Banyan VINES
Fully distributed PC network operating system based on UNIX and other industry standards. Enables an organization to interconnect an unlimited number and variety of PCs, minicomputers, mainframes, LANs, WANs, application programs (DOS, Macintosh, OS/2, and Windows), and other computer resources into one easy-to-use, easy-to-manage global network. Banyan's Directory, Management, Security, Messaging, Time and Communications network services interoperate across a distributed environment to create a single logical system. The VINES system product family features five fully integrated product offerings: VINES 5, VINES 10, VINES 20, VINES Unlimited, and VINES SMP. Each includes Banyan's full array of network services. $600–$6,000.

Novell Inc.   NetWare 3.11
A 32-bit, multitasking network operating system that integrates diverse computing resources into a single, enterprisewide system. Allows DOS, Windows, UNIX, Macintosh, and OS/2 workstations to connect to the same server and share a consistent set of network resources. Supports separate products that provide AFP, AppleTalk, Network File System (NFS), Open Systems Interconnection (OSI), File Transfer, Access, and Management (FTAM), communication, database, and network management services. Security features let users control who logs into the network, who has access to files and directories, how much disk space a user can use, and so on. Built-in reliability features include read-after-write verification, Hot Fix, disk monitoring, disk duplexing, resource tracking, the Transaction Tracking System, and uninterruptible power supply (UPS) monitoring. NetWare Loadable Modules (NLMs), software modules that link dynamically to the operating system, enable server-based applications to be added to the server while it is running. NLMs are provided by Novell and third parties. NetWare 3.11 is available in various configurations up to 250 users. NetWare 3.11 requires a dedicated server with a minimum of 4 mb of RAM. Five-user: $1,095; 10-user: $2,495; 20-user: $3,495; 50-user: $4,995; 100-user: $6,995; 250-user: $12,495.

Novell Inc.   NetWare Lite 1.0
DOS-based peer-to-peer network operating system. Supports up to 25 PCs and features easy installation, file and printer sharing, CD-ROM support, read-ahead disk caching, single log-in to network, location-independent network resources, security system with passwords and access control, audit and error logs, NetBIOS support, auto-reconnection between work-

station and server PCs, message transmission, network management utilities, and coexistence with server-based NetWare operating systems. Requires 28 kb RAM on workstation and 56 kb on a server. No dedicated server is necessary. Runs on most network boards that have a Novell IPX driver. $99 per node.

Quantum Software Systems Ltd.   QNX/Version 2.15 and 4.0
POSIX-compliant, real-time, multiuser, multitasking, message-passing, networked OS for Intel-based microprocessors. Consists of a 7-kb microkernel and a team of cooperating processes. Environment is configurable—can support up to 32 or more serial devices, hundreds of concurrent processes, and as many networked PCs as the network platform can handle. Provides responsive context switch speeds (17 used per context switch on a 33 MHz 80486) and offers true distributed processing, allowing all programs, data, and resources to be shared transparently across the network. Provides transparent fault-tolerance as well as load balancing, and handles multiple networks (for example, Arcnet, Ethernet, and so on) through individual drivers. $525 and up.

Network Printers

QMW Inc.   QMS-PS2000
20 pages per minute, RISC-based network laser printer. Optional direct Ethernet (NetWare, TCP/IP, or DECnet) or token-ring (NetWare) interfaces plus standard parallel, serial, LocalTalk. Built-in printer-server, simultaneous I/O; emulation switching; resident spooling. Resident PostScript, HP PCL, HP-GL. 11″ × 17″ duplexing, collation, 1,500 sheets in/out. $16,000.

Talaris Systems Inc.   TALARIS 5093 PRINTSTATION/T5093
50 page-per-minute network ion deposition printer; simultaneously accepts multiple protocols on Ethernet (TCP/IP, DECnet, Ethertalk). Multiplexing I/O (serial, parallel, optional Ethernet) for multihost connections, 500–2,000-sheet total capacity, including legal, 45 fonts, 5.5 mb RAM, 300 dpi (dots per inch), 250,000 page/month duty cycle. Includes HP LaserJet, PostScript optional. $32,000.

Hewlett-Packard Co.   HP JetDirect Cards
Network peripheral interface allowing an HP printer to be installed as a node anywhere on the network. Instead of putting printers near the file server or print server, they can be positioned where the users are. Because the printer connects directly to the network, the

parallel port is bypassed, speeding up printing and overall network performance. HP Jet Direct cards available for the HP DesignJet plotters. Functional for every type of PC-LAN as well as for systems running SCO UNIX, HP-UX, and Sun OS. $500.

Nu Data Inc.   FASTPORT Ethernet/TCP/IP PRINTER SerOver
Unix printing made easy and fast. Provides direct access to two printers via direct network spooling. LAN connection via AUI or BNC. Features a fast 16-bit microprocessor; supports both printers simultaneously; built-in support for PostScript. $900.

Repeaters

Farallon Computing Inc.   PhoneNET Repeater/ PN200
Hardware that regenerates and reclocks LocalTalk and FlashTalk signals, allowing LANs to extend longer distances. Also links a variety of network topologies, including backbones, daisy chains, and stars, as well as StarController hubs and networks in different buildings. For very long networks, users can link PhoneNet Repeaters together with over a mile of ordinary telephone cabling between each repeater. Repeater includes built-in PhoneNet Connectors to attach one or two wire pairs directly to the repeater. $500.

MUX Lab   Etherease 16 Multi-port Repeater
Powered wiring hub for use in larger Ethernet IEEE 802.3 10Base-T networks. Operates on networks at 10 mbps over unshielded twisted-pair cable by creating a star rather than a bus topology. Supports 16 RJ-45 UTP ports and one BNC and AUI for 10Base-2 (thinnet) or10Base-5 (thicknet) connections. Features automatic preamble generation, jabber function, external collision detection on all ports, and link and partition status LEDs for each UTP port. $990.

Routers

Neon Software Inc.   RouterCheck/Version 1.1
AppleTalk router monitoring and management package for the Macintosh. Tracks and monitors AppleTalk routers from a Macintosh located anywhere in the internet. Query and report features build lists of router type, name, location, and network configuration. Simple Network Management Protocol (SNMP) support provides detailed information for a broad range of AppleTalk routers. Analysis capability checks the in-

ternet for misconfigured routers and recommends corrective actions. Real-time monitoring features watch
the network for changes in performance and network
configuration, and for statistical information such as
traffic and error levels. Runs on all types of AppleTalk
networks including LocalTalk, Ethernet, and token
ring. $900.

Cisco Systems   AGS+ Router
A nine-slot, rack mountable chassis assembly with nine
system bus connections. Includes a 68030-based pro-
cessor card. Delivers an aggregate forwarding rate of
70,000 packets per second. Protocol support includes
TCP/IP, SDLC Transport, Novell IPX, AppleTalk, Ban-
yan VINES, DECnet (Phase IV and V), 3Com 3+/3+
Open, ISO CLNS (OSI), Xerox XNS, Ungermann-Bass
Net/One, and Apollo Domain. WAN support includes
HDLC, HDH, PPP, X.25, DDN X.25, frame relay,
SMDS, IP head compression, and priority output
queuing. Optional bridging support includes concurrent transparent bridging, IEEE 802.1d Spanning
Tree, DEC Spanning Tree, concurrent transit bridging,
and source route bridging. AGS+ can internetwork one
FDDI interface with as many as 22 Ethernet networks.
$12,300.

Storage Devices

**ADIC/Novell Co-Developed Disk Subsystems/
Various models**
Certified by Novell and designed specifically for
NetWare applications. Capacities range from single
disk 335-mb units to dual-disk 2-gb units with performance ranging from 15.5 ms to 14.0 ms average seek
time. Dual-disk units (DP) feature dual power supplies
to allow disk duplexing in system fault tolerant configurations. Has external SCSI address switches and
external termination. ADIC BusMaster DCBs are recommended for top performance. $4,500–$12,500.

Hammerman Associates Inc.   Data-Safe RAID
Subsystem
Redundant Array of Inexpensive Drives (RAID) subsystem provides speed and security for large Novell and
UNIX networks. Consists of five SCSI-2 drives connected to a specialized controller. Data is written, block
by block, across the drives, rather than to any drive.
Parity information is also spread across the drives. If
one drive should fail, the remaining drives will support
the network. When the failed drive is replaced, the sys-

tem will restore the information that was previously on
the failed drive. The controller is based on the Intel
RISC chip for fast operation. Available in floor-
mounted and rack-mounted enclosures. Parallel, redundant power supplies. User-selectable alarms and
dial-in diagnostics are provided. Can be incorporated in
the same case as a file server. 1.2-gb to 8-gb models
available. $20,000–$50,000.

Micropolis Corp.   RAIDION/680, 1340, 2060,
2680, 3500
A modular disk array subsystem that operates under
NetWare 3.11. Can be configured as a 680-mb "mirror"
(two-drive) array and can be expanded by adding disk
modules—up to 28 drives with a capacity of 47 gb. Hot
spare and hot swap drive options available in selected
configurations. Novell-certified and includes NetWare-
ready SCSI disk drives and device drivers. $5,000–
$49,000.

Pacific Micro Date Inc.   MAST VI Disk Array
Supports industry standard, nonproprietary SCSI technology with RAID-5 fault tolerance. "Hot Fix" remov-
ability supported by six removable half-height drive
shuttles and one internal power supply, monitoring
LEDs, shock mounts, and circuit board connection to
SCSI bus. Operates under Novell NetWare 3.11 on ISA
or EISA bus platforms and available with or without
drives. List price without drives: $7,000.

WAN Access

Fastcomm Communications Corp.   FastComm/
Time Machine
Eliminates network bottlenecks that are the cause of
poor internetwork performance. With compression ratios of from 2:4 to 6:1, lets network users access data
across a WAN 500 percent faster. Fully compatible with
all major bridges and routers. Plugs easily into all
existing CSU/DSU equipment. Built-in automatic fall-
back and fault tolerance protects the network, minimizing downtime while maximizing throughput. $12,495
for fractional and $14,495 for full T1.

Gandalf Systems Corp.   Infotron 2000 Series
T1/E-1 networking multiplexor that also provides
frame-relay concentration for up to 60 remote Infotron
2120 multiplexors. Provides full WAN networking capabilities, including switching and network management. Prices from $8,000.

Wireless Products

Cardinal Technologies Inc.   WAVEcomm Wireless
Network
Peripheral sharing, e-mail, and file-transfer device for
use with PCs. The size of an external modem; utilizes
spread-spectrum radio technology. Sharing capabilities
provided via the coded radio signals that transmit
through walls, floors, and other traditional office ob-
structions. A single WAVEcomm can communicate up
to 10,000 square feet and connect with other systems
in nearby buildings. Up to 100 users can exchange files.
$400.

Motorola Inc.   Altair Vista Point Wireless LAN
Link
A wireless LAN link providing connectivity to both
interbuilding and intrabuilding networks. Connects
Ethernet networks ranging up to 500 feet apart, both
between and within buildings, and operates with a
3.3-mbps full-duplex maximum throughput. $11,500.

Telesystems SLW Inc.   ARLAN 620 Wireless
Ethernet Bridge
Enables high-speed data connections between separate
Ethernet cable LANs up to 6 miles apart on a line-of-
sight radio path. A protocol-independent MAC level
Ethernet bridge, which automatically forwards packets
intended for the remote LAN, it automatically monitors all local Ethernet traffic and can filter up to 1,000
node addresses. $5,000.

Glossary

Adapter Small boxes or boards that incorporate microprocessors in the design; typically used to connect a PC to a specific type of local area network.

Advanced peer-to-peer networking (APPN) An additon to IBM's Systems Network Architecture (SNA), APPN allows nodes to establish and manage sessions on a peer-to-peer basis, without the need for a designated host to provide centralized support and control.

Alternative operator services Operator-assisted telephone services provided by entities other than AT&T, MCI, Sprint, and the Regional Bell Operating Companies. They generally provide special-service, low cost transmission, niching their products; some, for example, offer hotels or schools bulk discounts for the right to handle calls by guests and students.

American National Standards Institute (ANSI) A U.S. federation of standards-making and standards-using organizations created for the purpose of establishing voluntary standards.

Analog transmission A transmission method that creates a replication of the sound or image to be sent (i.e., a speaker's voice—literally an analog) by varying the electrical signal to correspond with the variation in the sound wave or image.

Application program interface (API) Software that manage the linkages between the user and the server in a client/server computing environment. It passes information between two software elements that is necessary for them to work together.

Architecture The technical strategy of the organization. It is the design blueprint for evolving the telecommunications network platform over time, technology, uses, volumes, and geographic locations.

ASCII code (American Standard Code for Information Interchange) One of the oldest standard coding schemes for digital information. The letter "c" is represented in ASCII as "0010011," for instance.

Asynchronous transfer mode (ATM) A fast packet-switching protocol based on the cell relay concept of fixed size cells sent at speeds up to 150 Mbps. ATM is the basis for the broadband ISDN architecture.

Asynchronous transmission A communications mode that involves sending one character at a time, with start and stop bits included to signal the receiving terminal when a character begins and ends. Error control is limited to a single parity bit check (odd or even) accompanying each character.

Automatic number identification (ANI) A controversial feature that is part of the evolution of ISDN. It displays the caller's phone number on the receiver's handset. Some commentators see this as a threat to privacy and a discouragement to users of emergency phone lines, such as the Samaritans or phone numbers for runaway children to call.

Availability There are two different aspects of network availability—network uptime and response time. Uptime is the portion of total time that the network is up and operating. This can be more precisely defined as: Uptime = MTBF/(MTBF + MTTR), where MTBF = Mean Time Between Failures and MTTR = Mean Time To Repair. This formula states that operational availability of the network is a function of both the average time between failures of the network and the average time it takes to repair the problem and restore the network service when it does fail.

Backbone network The central wide area network that links a firm's distributed operations and which may link to outside WANs (suppliers, banks, public networks, and so on). More recently, the term has been applied to the central LAN to which other LANs interconnect. Think of the backbone as the spinal column of the communications resource, with many joints and nerves connecting to it.

Backward compatible Works with existing and earlier versions of software or hardware, even though the newest version may not be able to exploit all the features of the earlier version.

Bandwidth The basic measure of the information carrying capacity of a transmission link. Bandwidth is literally the width of the frequency band, or the range of usable frequencies. It is computed by taking the difference between the highest and lowest frequencies.

Bar code An electronic tag consisting of a group of parallel, vertical black lines. The width of a black line on the tag indicates a number.

Baseband LAN A LAN consisting of computer workstations connected directly to a shared transmission medium, such as a twisted wire pair, coaxial cable, or fiber optic cable, configured in either a star, bus, or ring topology. LAN transmission is in digital form, and the entire capacity of the medium is allocated to each signal on the network. This means that there is potential for congestion and interference among separate transmission signals vying for the same shared medium capacity.

Benchmarking A systematic observation and comparison of a similar activity or process performed by someone else, generally outside the organization. Benchmarking "best practices" examples of a process or activity can lead to improvements in process methods, and it can also provide measures of achievable target levels for key quality indicators.

Bit The fundamental unit of coding for computers, where an "on" signal indicates a 1 and "off" means 0.

Bit error rate (BER) The basic measure of quality of digital transmission measured by the frequency of errors in the bit stream. A BER of 1 in 10^{-6} means an average of one error for every million bits (10^6) transmitted.

Bit mapping A method for mapping a digital bit stream representing images directly to specific locations called pixels (picture elements) on the video display screen. This process, without the intermediate step of translation to characters or symbols, results in a higher resolution image, but also requires more memory in the terminal to attain this result.

Bridge A switching device used for interconnecting local area and regional subnetworks that use identical protocol structures.

Broadband High speed, high capacity transmission systems (literally broad bandwidth) used to send large volumes of traffic over long distances. It generally refers to capacity ranges from T1 (1.544 mbps) to T3 (45 mbps).

Broadband Integrated Services Digital Network (BISDN) A public switched telecommunications platform that will support transmission rates greater than primary access ISDN. The plans include multigigabit (billions of bits) transmission rates capable of worldwide retrieval and sharing of information from multiple sources in multimedia format, including on-demand, real-time distribution of educational and entertainment programming to the home or office.

Broadband LAN A LAN that uses analog transmission techniques. The total capacity of the cable medium forming the network is subdivided into separate circuits or channels through multiplexing. Separate subnetworks can be created on the individual multiplexed transmission channels, and all forms of information (voice, data, and video) can be accommodated on a broadband LAN.

Broadcast radio An omnidirectional, over-the-air transmission technique that uses a lower frequency band than microwave transmissions. Unlike microwave systems, broadcast radio methods do not require a dish-shaped antenna, or line-of-sight alignment. The frequency band from 30 MHz to 1 Ghz covers the FM, VHF, and UHF broadcast bands. The transmission signals in this range of frequencies are not sensitive to rain, but they are subject to interference from other signals and from reflection off buildings and other physical objects.

Bus topology A bus is a high speed link directly connecting nodes or terminal devices. In a bus network, the terminal devices or nodes are attached directly to the bus medium (e.g., coaxial or other type of cable), and all terminals share the same transmission link.

Business process reengineering The redesign of business functions and activities to improve efficiency and quality of products and service. The base of reengineering is to take a fresh look at work processes and remove as many documents, people, administrative procedures, and delays as possible. The redesigned processes usually rely heavily upon telecommunications and information systems to achieve the productivity and efficiency gains.

Business television The use of one-way video broadcast capability to inform company employees about corporate policy, new products, and other relevant events and activities.

Byte A group of seven bits plus a parity bit used to represent a single character, such as "1," "q," "Q," or "@." It is the basic unit of the digital language of computers.

C band The frequency band used by a class of communications satellites known as C band satellites. A portion of the electromagnetic spectrum in the 4 and 6 gigahertz band has been allocated for use by these satellites.

Capability The range of telecommunications services and the volumes of each service that can be handled by the network. The range of services can include voice, image, and video applications, as well as every type of computer related traffic, including electronic mail, customer transactions, access to databases, word processing, videotext, electronic data interchange, voice-mail, funds transfer, facsimile, and so on.

Carrier Sense Multiple Access with Collision Detection (CSMA/CD) A contention protocol system used in the Ethernet LAN specification to manage access to the shared medium. A contention system relies on each terminal device on the network listening to the network to determine if the bandwidth is available or occupied with a transmission. The protocol also includes a collision detection and retransmit feature.

Cell relay A fast packet-switching transmission system that uses a fixed-length frame that encapsulates an existing LAN packet, without altering its format, for transmission across the backbone network. Cell relay systems operate at speeds ranging from 1.5–30 mbps. This speed range is suitable for public fast packet service offerings that multiplex together packets from a number of different user sessions.

Cellular PCs A wireless transmit and receive capability incorporated into personal computers based upon digital cellular radio technology. Cellular radio efficiently uses the limited radio frequency spectrum assigned to mobile radio by spatially dividing a metropolitan area into small geographic areas, called cells. As users move from cell to cell, they are reassigned different channels from within the allocated group.

Channel A one-way communications path. It can refer to the entire bandwidth (capacity) of the transmission link, or a specified part of it, such as the bandwidth needed to carry a single voice or data signal. A high-speed transmission link may be split up into many slower-speed channels.

Channel service unit/data service unit (CSU/DSU) An interface device that synchronizes a terminal's digital signal with the requirements of the transmission link facilities of the network. It performs synchronization and clocking functions, digital signal regeneration and reshaping, line conditioning, and some testing activities.

Circuit switching A switching method where the information signal is routed through the various nodes and switches along a temporary fixed path connecting the point of origin to the destination point. This path is dedicated for the duration of the call, session, or message. Circuit switching is contrasted with packet switching.

Client/server computing A dynamic distributed processing system that establishes a variable and changing work flow between a client application that needs data or software and a server that is a source for some or all of the needed data and software. Client/server is replacing cooperative processing as the general term for the new generation of distributed systems. They are essentially the same in their aims and principles.

Closed architecture An architecture that is not compatible with other architectures. The interfaces to other architectures are either not permitted or not practical.

Cluster controller A device that manages the communications operations (including congestion and competition for the transmission link) for a group of terminals linked to the network through a single interface. In a stand-alone configuration, each terminal must contain its own control logic and interface to the network. In contrast, in the cluster configuration, the cluster controller includes the control logic and provides the single interface to the network for all of the connected terminals.

Coaxial cable A cable constructed with two conductors—one an outer cylinder of conducting material acting as a shield, and the other an inner conducting wire separated from the outer conductor by insulation. This type of construction permits a wider range of transmission frequencies, and hence more bandwidth, which means more information carrying capacity.

Communications mode The method used by terminals to coordinate and synchronize the data stream sent across the network. A terminal may operate in asynchronous or synchronous mode, depending upon the design of the network.

Conformance testing The process by which laboratories, PTTs, and other agencies certify that a product meets a standard. In the United States, the National Institute of Standards and Technology handles conformance testing for Federal Government procurement. The Corporation for Open Standards (COS) provides seals of approval for Open Systems Interconnection (OSI) standards-compliance.

Connection establishment A technique for making a connection between two terminals on the network that desire to communicate and releasing this connection upon termination of the session.

Connection-oriented service A service that designates a specific path through the network for the duration of the communication session, and all data follows the same path through the network for that session. The public telephone network operates as a connection-oriented system.

Connectionless service A communication session between two terminals that does not establish a specific path through the network to carry all of the session traffic. Individual packets of information constituting a message or conversation will likely follow different paths through the network and be re-assembled at the final destination. Although a connectionless service protocol does not establish a specific path through the network, it includes logic in each switching node in the network that interprets addresses and routes information to adjacent nodes closer to the ultimate destination such that traffic congestion in the network is minimized.

Consultative Committee on International Telegraphy and Telephony (CCITT) An international standards body, with over 150 countries represented. It is one of four units of the International Telecommunications Union, which was founded in 1865 under a United Nations treaty.

Control Point (CP) A software entity in IBM's Systems Network Architecture (SNA) that directs activation and deactivation of links in the system and provides network management and control services.

Cooperative processing A processing method that extends the client/server concept to complex transaction processing systems that utilize software distributed among different hardware units that are cross-linked and automatically work together across the network.

Cost and hop A method for choosing a path through the network that has the lowest cost and traverses the fewest number of nodes—hops.

Customer Premise Equipment (CPE) An industry term for the switches and other equipment that act as the interface between the customer's location and the network.

Cyclic Redundancy Checking (CRC) An error detection method that uses a variety of often complex mathematical formulae that typically add the remainder from a division operation to the transmitted message as a check sequence that is transmitted with a data packet. The receiving terminal does an identical calculation and compares the result with that found in the packet received. If the results match, no error has occurred during transmission. CRC techniques ensure only about three bits per hundred million are received incorrectly.

Daily Management System A quality management method to manage and continuously improve the normal, day-to-day functions and activities of a business unit. It involves identifying and documenting key processes of the business unit, systematically working to improve the performance of those processes.

Data Circuit Terminating Equipment (DCE) A device whose function is to move and manipulate transmission signals to and from a terminal. It provides all the functions required to establish, maintain, and terminate a connection between the data termination equipment and the transmission line.

Data compression A technique for reducing the number of digital bits needed to transmit an image across a network. It essentially screens the image, and instead of sending "red dot, shade 109, red dot shade 109, red dot shade 109," it transmits "next 1200 dots all red shade 109." This can reduce the number of bits to be transmitted for video and image by a factor of up to 200.

Data rate The speed at which a bit stream travels or the speed of operation of a terminal device. The data rate of a terminal ranges from approximately 75 bits per second for a teleprinter to 64,000 bits per second for an intelligent terminal.

Data Terminal Equipment (DTE) A digital terminal that is the source and/or destination for data streams and provides some control functions for the movement of data in and out of the terminal.

Datagram routing method A packet-switching routing method whereby the first switching node determines a route for each packet, based upon current network congestion status. This means that packets representing various parts of the same message or block of data may travel on different paths through the network and arrive at different times and out of sequence. Upon arrival, they are reassembled in correct order.

DBMS (data base management system) A set of computer programs used to define, process, and administer a set of data and applications associated with that set of data.

Digital coding The use of patterns of bits ("0" and "1" combinations) to represent information.

Digital transmission Combinations of bits sent as pulses at high speed through a transmission medium.

Direct broadcast satellite A satellite system used to distribute television signals, which are received directly from the satellite using a small earth station antenna on the roof of a home or apartment building.

Distant vector algorithm A routing algorithm that chooses paths through the network on the basis of the lowest cost path that traverses the fewest number of nodes in the network ("cost and hop").

Distributed processing Sharing (distributing) the workload in computing and communication between the local PC and remote mainframe or network server computer. Some computing and communication tasks are performed by the local PC, and others are performed by the mainframe or server computer.

Divestiture The breakup of AT&T, which was the outcome of an antitrust suit. It created competition in long distance and the splitting up of AT&T's local phone services into seven monopoly regional phone companies; it took effect in 1984.

EDIFACT The international standard for electronic data interchange and electronic trade based upon the ANSI X12 standard for the United States.

Electromagnetic spectrum The invisible radio and infrared rays and visible light whose frequencies range from low to ultra-high. It extends from the sound we can hear (low frequency) through to AM radio, then FM, then UHF television, radar, microwave, infrared, and finally visible light.

Electromechanical devices Machines or devices whose moving parts are powered by electricity. This makes them slower than microelectronic-based ones, which have no moving parts.

Electronic cash management Software systems and telecommunications links that process corporate banking transactions electronically. These services include funds transfers, letters of credit and foreign exchange. The bank provides software and network access. The workstation will generally be a standard personal computer that exploits available software to add reporting features, spreadsheet analysis, and access to corporate information.

Electronic data interchange (EDI) The transmission of electronic documents in a form that allows another firm's computer to convert the sender's message format for, say, a purchase order, into its own formats.

Electronic mail (E-mail) A widely used type of messaging, where the message is transmitted to a computer that stores it, and then forwards it on when the intended recipient next links to the electronic mail service.

Emulation A software and/or hardware technique that converts the protocols used by a given terminal or workstation device into another set of protocols used by another device (computer, terminal, workstation, etc.) connected through the network.

Encapsulate The act of enclosing a packet or frame of data and control information formatted for a given protocol completely inside a packet or frame formatted for another protocol. This is a less efficient alternative to a complete protocol conversion of the original packet or frame.

Equal access The set of rules developed as an outcome of the AT&T divestiture decision, still in force, that required all customers to choose AT&T, MCI, or US Sprint as their long distance carrier, on an equal opportunity basis; AT&T was not allowed to automatically retain its existing base. Equal access meant that phone bills became very complex, since there was now a part showing the long distance carrier's charges and another showing the local carrier's charges.

FDDI (fiber distributed data interface) A standard for transmission of data over a fiber optic cable medium. The standard specifies a 100 mbps transmission rate for a token ring local area network configuration.

Fiber optic A transmission medium that sends light signals along glass cables thinner than a human hair via laser beams that pulse signals at sub-sub-subsecond speeds. This permits communications transmission rates and the corresponding ability to move large volumes of information cheaply and rapidly that ten years ago were unimaginable.

File server A computer on a local area network whose central function is to provide shared data resources to the LAN user PCs or workstations.

Flexibility The ease and speed with which changes can be made to any part of the telecommunications network platform and what range of changes can be made without having to replace, redevelop, or throw out existing network components.

Forms mode An operating mode for a video display terminal that shows a blank form covering the entire screen. The user manipulates a pointer on the screen to a position on the form where information is to be entered. The user then types text or data into the appropriate blank space on the form.

Fractional T1 service A service offered by local exchange and interexchange carriers that provides for the leasing of any fraction of a full T1 data rate transmission capacity, down to a minimum of a single 56 kbps channel.

Frame A grouping of bits that constitutes a discrete unit for transmission and is part of a message or transaction. A message may be composed of multiple frames, each of which includes actual message data, and additional information for the destination address, control, and error detection/correction.

Frame relay A fast packet-switching transmission system that encapsulates LAN packets into special frame relay packets of variable length and transmits them across the backbone network at speeds ranging from 56 kbps to 1.5 mbps.

Frequency division multiplexing (FDM) A multiplexing technique that divides the bandwidth of the transmission medium into partitions or slots, each consisting of sufficient bandwidth to carry the required information and have a little left over between each slot to provide a buffer against interference. The separate frequency band slots assigned to each voice conversation or data stream are called channels.

Front end processor (FEP) A stand-alone mini or microcomputer (or an internal board in a multipurpose computer) that handles the routine communications tasks for a host computer. It performs a terminal control function by acting as a communications network interface for a host mainframe or minicomputer and as a controller for remote terminals accessing the host system. These tasks include data formatting, character or message assembly or disassembly, code conversion, message switching, polling of remote terminals, error checking, protocol support and conversion, automatic answering and outward calling, as well as compiling network operating statistics for management.

Full duplex Signals carried both ways simultaneously along the same transmission link. A full duplex transmission uses two parallel links: one for transmissions from A to B, and the other for transmissions from B to A.

Gateway A switching device that connects different types of local area and regional subnetworks with incompatible protocol structures, such as one based on a proprietary architecture and one conforming to the OSI model.

Generic services The functions that a lower protocol layer provides to its adjacent higher layer in a modular, layered protocol system.

Geosynchronous Stationary and fixed with respect to a point on the earth. Communications satellites are placed in orbit 22,300 miles above the earth. At this distance, they maintain a fixed position relative to the earth as it rotates.

Graphical user interface (GUI) Displays users' options on the screen in the form of a menu and pictures (icons) that show what the menu choice represents, such as a document, a printer, and so on. Instead of having to type in—and remember—commands, the user moves a cursor around the screen via a pointing device (a mouse), and then selects an icon.

Graphics mode An operating mode for a video display terminal designed to display high resolution graphics images (engineering drawings, pictures, and so on). Graphics mode terminals use a method for direct mapping of digital bits to specific locations on the screen, rather than the technique of first translating digital bit streams into characters (text and numbers) and then moving these characters one by one to specific positions on the video screen.

Group A basic unit of public phone system capacity, made up of twelve 4 Khz voice channels multiplexed together; five groups (of twelve channels each) multiplexed together constitute a supergroup (60 voice channels); and a master group consists of ten supergroups multiplexed together (600 voice channels). Different combinations of groups, super groups, and master groups may all be placed on the same high capacity medium, such as a fiber optic cable, for long distance transmission to another region of the country or the world.

Groupware A catchall term for software designed to help business teams work together across locations. It coordinates schedules, messages and workflows among group members and usually includes modules for group analysis and decision making, as well as group document preparation.

Guard band A range of unused frequencies between channels that is needed to prevent interference between them.

Guided media Media that operates with an enclosed path, such as a wire or cable, and with the transmitter and receiver terminals connected directly to the medium. Wireless communications are unguided. Both guided and unguided media can carry all forms of information—voice, video, data, image.

H.261 An international standard for videoconferencing systems. It defines the protocol for coding and decoding a high speed digital video signal, such that systems developed by different vendors with their own proprietary protocols can communicate with each other.

Half-duplex A two-way alternate method of transmission where the signal can travel only one way at a time, with the sending and receiving terminal sharing the same transmission channel.

HDTV (high definition television) A television transmission method that provides a high resolution picture with over a thousand lines on the screen (versus today's 525 lines).

Hertz The number of complete cycles per second of the electromagnetic wave carrying information over the air or through a cable-like medium (one hertz equals one cycle per second). It is a measure of the transmission signal's frequency; the higher the frequency (number of hertz), the more information that can be encoded onto it. In the higher frequency bands, a prefix is used to indicate order of magnitude of the frequencies: 1 kilohertz is a thousand hertz; 1 megahertz is a million hertz; 1gigahertz is a billion hertz; 1 terahertz is a trillion hertz.

Hierarchical configuration (network) A configuration of nodes and switches in a network designed for efficient routing and distribution of the traffic. The hierarchy is based upon geographical proximity to the final destination location. Local traffic is routed through local central office switches, and long distance and international traffic will be transferred from local switching centers to a series of regional switching nodes that constitute a path to the final destination.

High-level Data Link Control (HDLC) A commonly used data link layer protocol system. Its principal function is to provide flow, error, and other transmission link control functions. The protocol operates by first formatting the data received from the user into a frame. A typical HDLC information frame will include: synchronization information, general control information (indicating the type of frame and its function), flow control codes, user data, and error detection information.

Hot swappable The ability to swap equipment modules in and out without disrupting network operations. If the equipment is not hot swappable, the change will require stopping transmission.

Image processing The scanning of documents and storing them in electronic form for transmission, access, and processing by computers.

Information technology (IT) A term that covers computers, data base management, and modern telecommunications. The entire field of IT is built on an astonishingly simple base: the ability to represent any form of information as a combination of representations of two conditions—the presence or absence of an electrical signal.

Integration The complete compatibility and working together of the elements of an information technology resource: hardware, software, communications, and data. Voice/data integration means that voice and data signals are combined and transmitted simultaneously over a single transmission link.

Intelligence The ability of a terminal or other device to process or alter the data that it is sending over the network or receiving from the network.

Intelligent workstations A general description of a personal computer with built-in telecommunications capability that manages many aspects of a transaction or computation through its own software, instead of it all being handled by a computer at the other end of the telecommunications link.

Interexchange carriers The name given to the long distance service carriers, such as AT&T, MCI, and U.S. Sprint following the divesture of AT&T. The term refers to those entities that carry traffic between the local exchanges or switching centers.

Interface The point of interconnection between two devices, such as a printer and a personal computer, or a telephone receiver and phone line.

International Standards Organization (ISO) An organization established to promote the development of standards, to facilitate the international exchange of goods and services, and to develop international cooperation in science and technology.

Internet The union of interconnected separate networks, where each retains its own identity.

Interoperability The ability of different systems to work together reasonably efficiently.

ISDN (Integrated Services Digital Network) A blueprint for the public networks of the mid-1990s that combines voice, data, text, and image on the same digital transmission line into the home or office. The basic ISDN service consists of two 64 kilobit-per-second transmission channels and a separate signalling channel used to set up and terminate the call. A higher capacity channel of 1.544 megabits per second (2.048 megabits per second in Europe) is also available.

Ku band The frequency band used by a class of digital communications satellites known as K band or Ku band satellites. A portion of the electromagnetic spectrum in the 14 and 11 gigahertz band has been allocated for use by these satellites. The higher frequencies used by the Ku band satellites means that the wavelength of the signal is shorter and therefore smaller earth station antennae can receive it; however, it also means that rain drops can distort the signal.

Link Access Protocol, Balanced (LAP-B) A data link layer protocol system used to initialize and terminate the connection, to acknowledge receipt of frames, to indicate errors, and so on. It is used in the X.25 packet switching protocol standard.

Link state algorithm A network routing algorithm that selects paths through the network on the basis of many factors, including reliability of a link, delay, and

maximum data rate, rather than just cost and the number of network nodes traversed.

Local area network A network that links together personal computers, printers, and other devices within the same building or local area building complex.

Local exchange The local telephone company switching center that interconnects the end-user subscriber line from home or office to other local switching centers and to the long distance carriers. Following the AT&T divestiture, the Bell Operating Companies that provided the local service were called the local exchange carriers.

Local loop A twisted wire pair that connects a residence or office building to the telephone company's local central switching office. A large number of these wire pairs are bound together in thick cable sheaths; individual pairs are consolidated from adjacent office buildings and residence locations to efficiently link them to the central office.

Logical Link Control (LLC) A data link layer protocol system used to send and acknowledge receipt of frames, to indicate errors and so on in some local area network configurations.

Logical Unit (LU) A software representation of end users (either terminals or application programs) in IBM's Systems Network Architecture (SNA). Every end user is represented to the network by a logical unit that enables it to communicate with other end users. There can be more than one logical unit at a network node, because application programs are defined as end users.

Lotus Notes A PC software package that organizes and coordinates the messages, schedules, and specified workflows of individuals and groups. It is becoming a quasi-standard for groupwork software.

Low earth orbit satellite (LEOS) Satellites that are placed in orbit 100–200 miles above the earth. In contrast to fixed satellites that are stationary in orbit 22,500 miles above the earth over a single location, LEOS are continually moving across the earth's surface. Consequently, a whole series of them are needed to provide continuous coverage of a given geographical area. Low earth orbit satellites are cheaper to launch, smaller, and easier to operate than the high orbit geosynchronous ones that are the core of international satellite services. Motorola has designed an ambitious service called Iridium that, as of the end of 1992, had not attracted enough investors for its very practical $3.4 billion project to go ahead.

Medium access control (MAC) A protocol that appends a terminal station address to the frame received from the logical link control sublayer protocol. It then manages access to the network to avoid congestion and interference among transmissions vying for the shared medium.

Mesh network (topology) A configuration of the network nodes such that there is a direct connection between every pair of nodes in the network.

Metropolitan area network (MAN) A collection of LANs linked together in a corporate or university campus area (sometimes called a campus area network), or a wideband, high capacity network extending throughout a metropolitan area.

Microwave system A directional radio broadcast transmission method operating in the 2–40 Ghz super high frequency band. This band requires a line-of-sight relationship between transmitting and receiving antennae, and the typical microwave antenna is a parabolic dish attached to a tall tower within sight of the next tower relay station. It is similar to a wireless, high-tech telephone pole system.

Middleware Software that resides on a device such as a router and that performs functions of interoperability, network management, and other services. It is contrasted with the operating systems that reside on client and server workstations and/or mainframes. The term is not well-defined, and its very existence signals the growing role of software in what has historically been a market driven by hardware and equipment.

Modem A device that converts digital to analog signals and analog to digital signals. The modem detects a binary 0 or 1 signal sent from the terminal and modulates a self-generated carrier wave such that it represents the pattern of 1s and 0s present in the original signal. The analog carrier wave is transmitted through the network and received by a companion modem at the other end of the link. The destination modem demodulates the carrier and retrieves the binary 0s and 1s; it then forwards them to the destination terminal in the form of a serial bit stream.

Monomode fiber Optical fiber cable that has a core which is so small that the light has only one possible transmission path, and thus there is no divergence along its path and consequently no slowing down.

Multitasking The ability of an operating system to run many software programs simultaneously.

Multidrop line A single transmission circuit to which several terminal devices are directly attached. A familiar example of a multidrop line system is the almost extinct "party line" telephone.

Multimode fiber Optical fiber cable that has a wide core, which makes it cheaper to produce than monomode fiber. Light rays propagate slightly, reflecting off the boundary between the core and the cladding of the fiber strands.

Multiplexing A technique used to combine smaller, lower capacity data or voice information streams into one large stream to increase capacity utilization and take full advantage of efficient high capacity transmission links (similar to an airline feeding passengers from small commuter planes to central hubs where they are combined to fill a fuel-efficient jumbo jet for long distance transport). Time division and frequency division multiplexing are the most prevalent forms of the technique in use today.

Network A set of devices or entities that share a directory and can thus directly access each other. The directory is directly analogous to the standard telephone directory, except that it provides an address for each component of the network. If the device is not included in the directory, it is not part of the network.

Network addressable unit (NAU) A software component of a network node in IBM's Systems Network Architecture (SNA) that supports communication between end-users (terminals or application programs residing on computers), and between end-users and other entities that provide network services (control and management activities). Each NAU has an address on the network, and is classified as either a physical unit, a logical unit, or a control point.

Network interface unit (NIU) A microprocessor device that performs the communications functions necessary to link a PC workstation to a local area network: it accepts and buffers data from the terminal (PC); addresses and transmits packets; scans the network for packets with a matching address and reads them into the terminal; and transmits data from the terminal to the network at the proper data rate of the LAN.

Network operating systems (NOS) Software that manages the basic communications and processing functions of a local area network. The NOS controls the operations of the network, coordinating the communication between the devices on it. It also determines which devices can communicate via the network and how.

Network server A micro- or minicomputer that acts as the central routing and control node for a local area network. The network server is the central storage repository for files, application software programs that are used by other devices on the network, and electronic mail and messaging services. It also controls access to the various files, manages file transfers, and ensures data integrity.

Node A point to which a group of devices and transmission lines connect; it will usually contain a switch or series of switches. Hundreds or even hundreds of thousands of devices may at any point in time be sending messages across the network via its nodes.

Noise A distortion that makes a telecommunications signal either fuzzy or inaccurate. Noise is encountered in everyday life, on telephone lines, and in television reception. Transmission errors are routinely caused by noise.

Object-oriented thinking The concept of thinking in terms of "objects," where an object can be defined as a data base, a piece of music, a transaction processing routine, or a picture, and so on. Objects can operate on or be accessed by other objects. Each object is entirely self-contained; it includes knowledge about the characteristics it inherits from the broader class of objects to which it belongs. For instance, the object "car" inherits features of the class "vehicle." Objects are instances of classes. Each object packages data and related procedures together in a self-contained module, which can be combined with other objects to form more complex entities, without having to recreate the basic objects again. Object-oriented thinking has led to "object-oriented programming systems" (OOPS), which generally use one of the two most common object-oriented program development languages—Smalltalk and C++.

On-line Data needed to answer a question or carry out a transaction can be immediately accessed by a computer, instead of having to incur a time delay to locate and make the data available, as with a filing cabinet or book.

Open system A vendor-independent specification or standard that ensures compatibility among different manufacturers' equipment and software.

Open Systems Interconnection (OSI) model A standards-based architecture "reference" model intended to guide the development of products, not to define a product in itself. The purpose of the model is to establish complete and consistent communication rules that permit the exchange of information between dissimilar systems. The OSI model is known as a layered architecture, or modular approach to communications standards. Each of its seven layers represents a necessary element in the electronic communication process.

Operating system Complex software that manages all aspects of a computer's applications. These include basic utilities such as file management; running software such as spreadsheets, word processing, and transactions; input–output of data; error-handling; and many other functions. The operating system determines the capabilities of a computer, especially the variety of software it can run.

Optical character reader (OCR) A photoelectric cell that senses light and dark and converts patterns of light to a character code such as ASCII. A barcode reader is a similar device found in many retail outlets. It is generally connected to a point-of-sale terminal, such as an electronic cash register, and is used to scan a small strip attached to the outside of an item that contains a series of parallel lines that represent the Universal Product Code scheme. This is a 10-digit code that identifies a particular manufacturer and its specific products.

Optical fiber cable Thin glass fiber strands, which look like one-pound test monofilament fishing line, surrounded by cladding. Cladding is a different type of glass that is reflective, and thus directs stray light rays back to the core fiber strand. The cladding is usually covered by an outer protective shield of plastic. The total diameter of the fiber strand is less than that of a human hair. Individual fibers are bundled together in groups to form a cable configuration. A light source (light-emitting diode or laser) is the carrier that transmits information along the glass fibers. Because of their very large information carrying capacity, optical fiber cables are now used extensively for long distance transmission links to carry voice and data in the public telephone network.

Optimization Theoretically optimization means maximizing the performance of the network through mathematical models that analyze the flow of traffic under various conditions. In practice, today's networks are so complex that modeling techniques have to make simplifying assumptions, which make it close to impossible to reliably predict exact performance.

Outsourcing The use of an outside company under a contractual arrangement to perform functions that were previously handled inside the company; it may include systems development, network operations, and even running the entire information technology function.

Packet A small, structured block of digitally coded information containing a message or other data, along with addressing and error-detection information.

Packet assembler/deassembler (PAD) A device that frames data and destination address information into blocks called packets. These packets are sent through the network from node to node; the destination address of the packet is read as it arrives at each node, and the packet is then forwarded on a path through the network toward its final destination. Upon arrival at a PAD at the distant end, the destination address and error-detection bits are stripped from the packet, and the data is passed in serial form to the host computer or other destination terminal.

Packet switching The well established alternative to the circuit-switching method of transmission, where information to be sent is divided up into small blocks called packets. Each packet is sent separately and may travel over various routes before being reassembled in correct order at the final destination.

Page mode An operating mode for a video display terminal that shows an entire page of text or data on the screen at one time, and the full page is stored in the terminal prior to transmission through the network.

Parallel transmission A transmission method that sends all bits in a given grouping (a byte) simultaneously, each over its own separate wire. The receive end, computer or printer, processes the byte intact. A typical byte sent using parallel transmission will consist of 8, 16, or 32 bits. The advantage of parallel transmission is speed: 8, 16, or 32 bits are sent in the same time it would take to transmit 1 bit using a serial transmission method. The disadvantages associated with parallel transmission methods are limited distance capability and difficulty in control and synchronization of the transmission. Most parallel transmission links are limited to 15 feet or less.

Parity checking A technique used to detect errors in transmission. The parity bit is set to 1 if the number of 1s in the transmission is odd, and to 0 if the number of 1s is even. The receiving device checks the parity bit; if there has been an error in transmission, the value of the parity bit will not correctly match the pattern of the information bits.

PBX (private branch exchange) Essentially a scaled-down version of a telephone company central office switch. It is designed primarily to switch voice traffic, but it is also capable of switching moderate-to-slow–speed data traffic. (PABX is the term for a private automatic branch exchange).

Peer-to-peer Refers to direct communication between similar devices across a network: For example, two personal computers exchanging information directly across the network, without the need for an intervening host computer to manage the communications session between the two PCs.

Personal Communications Systems (PCS) A wireless mobile system with portable terminals capable of providing the user with all the services he or she gets from owning a phone at home, from using cellular mobile phones, and from having a laptop computer with a modem. The user communicates to anywhere, from anywhere, and is accessed through a single number, regardless of location. Generally, the PCS networks provide wireless connectivity for their mobile terminals within a very narrow geographic area, such as a single building, but these wireless cells can be interconnected to wide area networks via regional cellular systems land-lines, or satellite.

Physical unit (PU) A software module used to manage the hardware resources in IBM's Systems Network Architecture (SNA) at a given node and to allocate their use to particular communications tasks.

Plan, Do, Check, Act (PDCA) A quality management method used to implement continuous improvement through managing by fact. It is really a shorthand version of the scientific method: **Plan** a change that will lead to an improvement; **Do** the proposed change on a test or trial basis; **Check** the results of the test to determine if the improvement was realized; and **Act** on those results to refine the proposed change or standardize the improvement.

Platform A particular combination of hardware technology, software, and protocol rules capable of delivering a range of information services to a wide variety of user groups. In contrast, a nonplatform system generally has a limited information service capability available to a single user group, and multiple systems or networks are needed to provide the full range of information services to multiple user goups that is available through a given platform. A telecommunications platform refers only to the networking services component of information technology, and an information technology or technical platform encompasses a broader range of components and information services.

Polling In a multidrop line configuration, the host computer polls each terminal in turn to determine if it has something to transmit. This process avoids contention for the single line if more than one terminal tries to transmit data at the same time.

Portability The ability to run a given software package in different hardware environments.

Primary Access ISDN An ISDN service offering that consists of 23 64-kbps bearer channels and 1 16-kbps delta channel, or 23B+D. The 23 bearer channels have a transmission capacity almost equivalent to a T1 circuit (24 64-kbps channels). The service is also called primary rate access.

Private network A network where the transmission facility is leased by a firm for its own dedicated and exclusive use.

Privatization The act of converting government-owned and -operated enterprises into private or quasi-private companies. The deregulation movement that began in the United States has spread to many other companies in the form of privatizing the government-owned and -operated telecommunications entities (PTTs).

Process stability and control A statistical concept of stability that exists when the pattern of variation of a process about its mean or target value is attributable to random chance. This occurs when all process data points observed are within a prespecified upper and lower control limit (usually three standard deviations from the mean).

Proprietary systems Systems that are vendor-dependent, and require vendor-specific products to operate.

Protocol The strictly defined and precisely coordinated set of rules and conventions by which two machines talk to each other and exchange information. This includes procedures for establishing a telecommunications link, interpreting how the data is represented and coded, how it is transmitted, and how errors are to be handled. Transmission methods use a variety of ways of coding both messages and the instructions and formats needed for the sending and receiving devices to interconnect and process information, all of which require specific protocols.

Protocol converter A piece of equipment that enables data to flow from one type of network to another or from one piece of equipment to another in a manner that is transparent to the user. Protocol converters modify the bit stream passing through them to ensure that it conforms to the rules and procedures of the new network or piece of equipment that it is entering. They check an incoming message to identify its protocol and reformat it.

PTT (Poste Telegraphique et Telephonique) The government-controlled agency that operates a nation's telecommunications monopoly. It generally includes the postal service and telecommunications equipment and services.

Public network A network available to a wide range of subscribers, on a pay-as-you-go basis; the most obvious and widely used example is the public phone system.

Pulse code modulation (PCM) A modulation method that is the core of digital transmission. The signal is transmitted by sampling its amplitude x times a second and transmitting it as a bit, instead of in continuous analog form. Each bit sent can be distinguished from another. This means that the signal need be only accurate enough to tell 0 from 1.

Quality Improvement Story A seven-step problem-solving process adopted by the Florida Power & Light Co. as part of its Total Quality Management system. The QI Story method can be represented in a visual storyboard format, which facilitates a quick understanding of the problem and its solution.

Quality of service A measure of how well the network design meets customer requirements. The principal quality of service factors are the level of reliability, responsiveness, and accessibility to the network.

Quality profit engineering A framework for thinking creatively and practically about how to use telecommunications and information systems to improve company profits by focusing on four key elements: (1) delivering higher quality products and services at the same or lower costs; (2) delivering higher value information just-in-time at the point of decision; (3) enhancing revenues without commensurate increases in cost; (4) reducing cost through favorably altering the underlying cost structure.

Quick Response systems (QR) A general term for uses of computers and telecommunications to streamline the entire chain of business transactions from point-of-sale to retailer to manufacturer to shipper to warehouse to store.

Range The property of telecommunications networks that represents the scope of information directly and automatically shared across business functions and processes; the extent of cross-linking information over different vendor systems, different technologies, and different applications. Low range is cross-linking standard messages; high range is cross-linking through cooperative processing.

Reach The property of telecommunications networks that represents the extent of interconnection among people, locations, and organizations. Low reach is interconnection of a few people within a single location; high reach is interconnection to anyone, anywhere.

Real-time Insignificant time delay between initiating an action at one point on the network and attaining the result of the action at another point on the network.

Redundancy The provision of duplicate or alternate equipment and/or transmission links that are not needed for operation but that can be immediately activated if there is any problem with the equipment or transmission links that are operational.

Reliability A measure of continuous, error-free operation of the network. Perfect reliability over a given time period is no failures and no errors.

Remote program call (RPC) Software that manages the linkages between the user and the server in a client/server computing environment. It initiates the activation of one software program from an instruction issued by another software program residing in another location on the network.

Repeaters Devices placed at intervals along a transmission link that check, regenerate, and amplify the signal, which can fade and become distorted as it travels along the transmission medium.

Response rate (time) A measure of how quickly input and output is sent and received over the network.

Responsiveness The property of telecommunications networks that represents the level of service (in other words, speed, reliability, security) provided on a consistent basis; the quality and reliability of network service. Low responsiveness is non-immediate service with some interruptions and failures; high responsiveness is on-demand service that is devoid of errors and failures.

Responsiveness bottleneck The slowest device or link in the network that reduces the throughput of other elements or parts of the network.

Ring topology A configuration of the network nodes such that all nodes are connected to a common circular or oval loop medium (cable or wire).

RJ-11 jack connector A standard for connecting a phone jack to the wall outlet. The wall outlet is in turn connected to a copper wire pair that runs to the telephone company's local central office, where it feeds into the telephone company's network. The standard specifies the configuration and number of wires in the connecter and the functions of each wire.

Router A switching device that connects different types of local area and regional subnetworks that use different protocols, but that belong to the class of open networks whose architectures are compatible with the OSI model structure. As the technology for these devices evolves, hybrids have been developed that combine multiple internetworking functions into a single device. For example, a "brouter" is a device that combines the features of both a bridge and a router.

RS-232 A standard for connecting a computer to an external device, such as a printer. RS-232 defines the electrical features and timing of signals for a 25-pin interface cable.

Satellite system A microwave relay system in the sky. It is essentially a repeater or amplifier that receives a signal, from a ground (earth) station, amplifies the signal, and retransmits it to one or more receiving earth stations on the ground. Most communications satellites are located in geosynchronous orbit over the equator (they are fixed in one place) 22,300 miles above the earth.

Scroll mode An operating mode for a video display terminal that shows text one line at a time on the screen (similar to a hardcopy printer). When a new line of text is displayed, the previous line scrolls off the screen and disappears.

Security Protecting the network's physical facilities from harm and sensitive information from unauthorized users, while providing simple and inexpensive access on a wide-reach basis.

Serial transmission A transmission method that sends one bit at a time in a serial sequence. Its advantages are simplicity, ease of control, synchronization, and its reliability over long-distance transmission links.

Shielded twisted pair A twisted pair that has a layer of protective insulation, which reduces attenuation and interference, making it possible to transmit high data rates over short distances—typically up to a maximum of 100 meters.

Signal compression A method used to reduce the amount of information that is actually sent across the network, such that less transmission capacity is required to send a given voice conversation or block of data (similar to dehydrating bulky materials by removing the water content, transporting it in compressed powder form, and then adding water and reconstituting it at the destination).

Slave/host An arrangement where dumb terminals are connected directly to a mainframe computer in a configuration that characterizes a centralized computing environment. Under this arrangement, one terminal can communicate with another terminal only indirectly through the host mainframe computer, and the communication session is controlled by the host computer.

Smart hub A single device that replaces a multitude of separate hardware components connecting the elements in a local area network and interconnecting separate local area networks. It helps simplify cabling and the interconnection of incompatible systems and provides network management diagnostics necessary to efficiently operate a multitechnology telecommunications environment.

SNA See System Network Architecture.

Software defined network (SDN) A wide area virtual private network service provided by long distance carriers. Customer organizations lease a certain transmission path capability between designated locations, which is made available on demand by the carrier. However, specific transmission and switching facilities are not dedicated to the customer full time; instead, the specific transmission/switching resources required for any communications session are allocated from a pool of available capacity on the network at the time service is needed, and these facilities are released upon termination of the session.

SONET (Synchronous Optical Network) The standard for a fiber optic-based public communication network. It provides speeds of up to 2.4 gigabits per second and is the basis for the next generation of advanced networks.

SQL (Structured Query Language) An English-like grammar and syntax for a user to access information from a computer database without having to know how the information is organized or where it is stored.

Standard A specification that describes the physical configuration, performance requirements, and interface parameters for a piece of equipment, a network, or a software program such that it will work on and with any other system following the standard.

Star configuration (topology) A configuration of the network nodes such that all nodes connect directly to a central hub, and all communications between the various nodes are routed through this hub, making it easy to coordinate and manage the network.

Statistical (asynchronous) time division multiplexing A form of multiplexing that eliminates predetermined time slots assigned to particular devices or information sources. Inputs from source terminals are buffered in a queue until the multiplexer device scans the input buffers and assigns data to available time slots, filling the frame (no empty time slots are permitted). Then the multiplexed signal is sent through the network.

Subnetworks Individual local area and regional networks linked together to form an internet.

Switched Multimegabit Data Service (SMDS) A cell relay, fast packet-switching transmission system service provided by carriers that permits an organization to provide high speed interconnection among a number of LANs by simply providing a local interface connection from a LAN bridge or router to a carrier's local point of presence.

Switching devices Typically microprocessors or specialized computers located at nodes in the network that direct information streams over alternative paths through the network toward their final destination point. They have evolved through a primary emphasis on supporting voice communications requirements; however, modern digital switches are capable of handling both voice and data traffic, and some specialized switching devices are designed exclusively for data.

Synchronous time division multiplexing A multiplexing method used to consolidate and combine multiple digital information streams when the data rate (bps) of the transmission medium is greater than or equal to the sum of the data rates of the individual digital streams being multiplexed. The multiplexer device combines the separate bit streams by interleaving bits from each of the separate source signals into pre-arranged time slots.

Synchronous transmission A communications mode that involves transmitting multiple characters, along with control and error detection information, in frames or blocks that are synchronized in time, so that transmitter and receiver are working together and know precisely when a frame of information begins and ends.

System integrators Companies that provide integrated information systems solutions to business needs, ensuring that multi-vendor software, hardware, and telecommunications equipment work together. Leading systems integrators include Andersen Consulting, EDS, and computer vendors such as IBM and DEC.

System Network Architecture (SNA) A proprietary architecture designed to help customers integrate new IBM product offerings into existing IBM system networks. SNA was originally a closed architecture but has recently incorporated support for open systems standards, such as the X.25 packet switching interface standard TCP/IP and OSI.

Systems programmers Highly specialized technical experts who generally focus their computer programming skills on a single operating system, or at most on the operating systems of a single vendor.

T1 link A digital transmission link service offering from AT&T and other carriers with a transmission capacity of 1.544 megabits per second. A fractional T1 link service is also available, which permits users to activate and pay for fractions of the total 1.544 mbps capacity in increments of 56 kilobits-per-second channels.

T3 link A digital transmission link service offering from AT&T and other carriers with a transmission capacity of 45 megabits per second.

Tandem-switched networks A network of transmission links and switches directly linked together in a non-hierarchical fashion. Most private line networks are configured in this manner.

Tariff 12 The price structure and terms and conditions for AT&T's long distance, volume discount service. A tariff is the description of terms and conditions (including price) of a service offered to the public by a regulated common carrier. AT&T is the only interexchange (long distance) carrier required to file tariffs, because of its dominant size and market share.

TCP/IP (Transmission Control Protocol/Internet Protocol) A fairly simple, low overhead transport and internetwork protocol designed to interconnect a wide variety of networks and the computer equipment connected to those networks. TCP/IP is the dominant standard for UNIX-based environments.

Telecommunications architecture The technical strategy of the organization. It is the design blueprint for evolving the telecommunications network platform over time, technology, uses, volumes, and geographic locations.

Teleprinter terminal A traditional hard copy printer device located in a place remote from the computer generating the printed output. The basic unit contains the printer mechanism and a set of electronic controls, including the communications interface. It may also include a keyboard for an operator to enter information to be sent to the distant computer.

10BaseT A high capacity, copper wire-based transmission cabling system that is used as a LAN transmission medium; it is based on the Ethernet standard.

Terminal An input/output device that acts as the interface between a user and the telecommunications network resource. A "dumb" terminal is one that has no processing intelligence of its own and is limited to simple input and output (display) functions.

Time division multiplexing (TDM) A multiplexing method that divides a circuit's capacity into time slots. Each time slot is used to carry a signal from a different terminal, which can be voice, data, image, and so on.

Time-sharing The processing of multiple transactions from multiple terminals, using a powerful host computer. The computer is fast enough to simultaneously accommodate multiple transactions and to buffer and sequentially process any that cannot be handled on a simultaneous basis. Airline reservations and automated teller machines at banks are the classic examples.

Topology The physical structure of the transmission links; the manner and shape of their arrangement in the network.

Total Quality Management (TQM) An approach to management that views the corporation, its customers, and its suppliers as a true system whose goal is complete customer satisfaction. The goal is achieved through optimizing the overall system by the continuous improvement of all processes in the organization based upon decisions arrived at through fact and analysis.

Transmission The process of sending information from a point of origin to a reception point. It can also refer to a set of characters, messages, and other data.

Transmission link The medium or channel path that carries the message from sender to receiver. More formally, a transmission link is defined as the physical path between terminal devices that carries the coded electronic signal from origin to destination. The transmission link is also often referred to as a circuit (two-way path) or channel (one-way path); it consists of the physical medium through which the signal travels and includes any intermediate hub points, or nodes, that act as a switch or relay points to direct the signal on its way to the final destination point.

Transmission medium Any substance through which a propagated electronic signal travels from origin to destination. A transmission link can use any of a wide range of media available to support electronic transmission. Within this range of available media, each has its own characteristics that make some better suited for certain types of transmission applications than others.

Twisted wire pair A twisted wire pair consists of two insulated copper wires twisted into a spiral pattern. The twisting is used to prevent the wires acting as antennae and inadvertently picking up electromagnetic waves in the air. It is the most prevalent guided medium in use today and accounts for a very high percentage of the outside wiring associated with the world's public telephone networks. Shielded twisted-pair wire has extra insulation in addition to the twisting of the wire to prevent interference.

Unguided media Nature's natural carriers of electromagnetic waves, including air, water, and the vacuum of space. The transmission is generally a multidirectional broadcast signal that uses antennae to send and receive the information.

Universal Product Code (UPC) A printed bar code, stamped or taped onto retail goods and magazines, that contains the vendor's identification number and the product number based upon a widely accepted standard coding scheme.

UNIX operating system A computer operating system with an open design compatible with multiple central processor designs, which has evolved independent of both Apple and IBM, with Sun Microsystems and Apollo leading the way. UNIX is also a powerful operating system capable of running many software programs simultaneously and accommodating multiple users at the same time.

Unshielded twisted pair A twisted pair that has no insulation layer surrounding the copper wires. It is the cheapest guided medium in use, and the one with the lowest data rate. It is also the medium most widely used for standard telephone lines.

Very Small Aperture Terminals (VSAT) Satellite–earth station terminals whose antennae are less than 6 feet in diameter. VSAT networks consist of hundreds or thousands of these terminals connected through the satellite to a central hub station on the ground, which controls the traffic distribution for the network.

Videoconferencing Meetings or sessions involving people in different locations linked through two-way video and audio. Until recently, special rooms were equipped with video and audio technology to support the conferencing. Currently, desktop workstations can be equipped to support video conferencing in a local and wide-area network environment.

Video display terminal (VDT) A terminal with a video screen that shows characters and graphics. Most personal computers and remote non-intelligent data entry terminals connected to mainframe computers are examples.

Virtual circuit routing method A routing method that establishes a single path through the network connecting the communicating terminals at the beginning of a session based upon current congestion conditions. All packets will travel the same path for the duration of the session. Packets arrive in the order in which they are sent, because they stay in sequence following the same path through the network.

Virtual path A designated path or route through the network along which information will flow between two terminals. The path is not dedicated exclusively to these two terminals and may be shared by other communication sessions connecting other terminals.

Virtual private networks (VPN) Networks that provide specific transmission link and switching capability on demand to the subscribing customer provided from a dedicated pool of facilities shared by a group of VPN and public network customers. They do not include a specific set of equipment and facilities dedicated full time to a single customer (as with true private networks).

Virtual terminal service A software package that is an abstract representation of a real terminal, including all of its basic functions and characteristics such as display mode, character code, data rate, and so on. The virtual terminal service interface is designed to solve the problem created when there are many different types and makes of terminals remotely accessing multiple host computers made by different vendors.

Voice mail The digital equivalent of telephone answering machines and very much a voice version of electronic mail. Because people are very used to both phones and answering machines, many find voice mail far more "natural" than e-mail. Messages are stored and forwarded, just like e-mail. Advanced systems include features for notification of receipt, redirection, message sending priority, time stamps, and interworking of different vendors' voice-mail systems.

Voice messaging A system for recording audio messages that can be retrieved on demand by the recipient. In a sophisticated system, the caller may select from a menu of options by pressing, say, a 1 or 8 in response to recorded voice instructions; a computer then handles the instructions and response to the caller's selection.

VSAT network (Very Small Aperture Terminal) VSATs are the tiny satellite earth station dishes that brought news from Saudi Arabia to CNN in real-time in the Gulf War. A VSAT network consists of large numbers of these terminals linked through the satellite to a central control hub on the ground.

Wide area network A network that covers a broad geographic area.

Wide Area Telecommunications Service (WATS) A bulk-rate, switched, semiprivate network service, priced at a discount versus regular long distance service. It is provided by the long distance carriers primarily as a voice service, and there are two versions—inbound and outbound service. This is the widely used 1-800 number service.

X12 An American National Standards Institute (ANSI) standard for electronic data interchange (EDI) that describes how firms can send documents to each other in a common format, without having to adopt one or the other party's specific format.

X.21 An international standard that specifies the physical connection of terminals to the X.25 network. It is equivalent to the RS-232-C, 25-pin–connector standard in the United States.

X.25 A worldwide standard protocol for packet-switched data networks, most public data networks, most value-added networks, and almost all international networks. The standard actually specifies the interface between the terminal equipment devices and the network.

X.400 The international standard for electronic messages, including e-mail, fax, and telex. It links different electronic mail services with proprietary protocols.

X.500 The emerging international standard for managing directories of addresses across networks.

References and Bibliography

The authors of *Networks in Action* built a library of well over 4,000 articles as part of writing this book. These were scanned and stored on optical disk. Most date from 1991 to late 1993 and comprise material from over 120 trade press publications and academic journals, including North American, Asian, European, and African sources. Rather than provide an exhaustive bibliography, we thought it would be more useful to offer selected references whose titles give a sense of the major issues, attitudes, concerns, and conventional wisdom of the telecommunications field as of mid-1993. In addition, where a chapter refers to an article or book, it is listed in the bibliography, as are the sources for each chapter's minicases.

The following books are the original sources for the core frameworks of *Networks In Action*:

- ▶ Keen, Peter G. W., *Competing In Time: Using Telecommunications For Competitive Advantage, Second Edition* Cambridge, MA: Ballinger, 1988. This is the source for the Telecommunications Business Opportunity checklist (Core framework 1, Chapter 10), Network Design Variables (Core framework 3, Chapter 14), and the Career Quadrants framework (Chapter 19).

- ▶ Keen, Peter G. W., *Shaping The Future: Business Design Through Information Technology*, Cambridge, MA: Harvard Business School Press, 1991. This is the source for the Telecommunications Services Platform (Core framework 2, Chapter 11) and Quality Profit Engineering (Core framework 4, Chapter 12).

In each case the frameworks have been updated and extended for *Networks in Action*, which therefore presents the authors' newest thinking and experience.

Sources For Minicases

1.1 THEi O'Leary, Meghan, "Rethinking The Organization." *CIO*, Jan., 1990, p. 44.

1.2 Shearson, Lehman Dyson, Esther, "Shearson's Image Of Groupware." *Datamation*, Apr. 1, 1990, p. 61; McMillan, Tom, "Shearson Lehman Brothers: Savvy Investment Strategy." *Integrated Image*, June 1992, p. 6.

1.3 The State Of California O'Leary, Meghan, "Home Sweet Office." *CIO*, July 1991, p. 30; Fritz, Jeffrey N., "Telecommuting Using ISDN and Remote LAN Applications." *Telecommunications*, July 1991, p. 55.

1.4 Telestroika Gilhooly, Denis, "Hungary Embraces Telestroika." *Communications Week International*, Nov. 10, 1991, p. 1; Horvath, Paul, "Telecommunications In Hungary." *Transnational Data And Communications Report*, Jan./Feb. 1991, p. 13; Shetty, Vineeta, "Hungary's 'Goulash' Monopoly." *Communications International*, Nov. 1992, p. 22; Shetty, Vineeta, "Priming The Primary Networks." *Communications International*, Feb. 1993, p. 37.

2.1 Terra International Stratigos, James, "VSATs: Far-Out Communications For Remote Sites." *Telecommunications*, Sept. 1990, p. 23.

2.2 Dun And Bradstreet "D&B's Big Play In Client/Server." *Datamation*, Dec. 15, 1992, p. 28.

2.3 Saving New York $14 Million Curran, Lawrence, "Systems Integration: Defining A Big Business Opportunity." *Electronics*, Apr. 1, 1990, p. 85.

2.4 The View From Volkswagen Tate, Paul, "Users Burn While EC Fiddles." *InformationWeek*, Aug. 9, 1993, p. 1; "Telecommunications Survey: Coping With The Globe Girdlers." *Economist*, Oct. 5, 1991, p. 25.

3.1 Conrail Part A Ginsberg, Stanley, "Putting Information Technologies On Track At Conrail." *IN Magazine* Winter 1988, p. 2.

3.2 Conrail Part B "Planes, Trains, Systems." *InformationWeek*, Oct. 14, 1991, p. 18; Bowman, Robert J., "Keeping Track Of Tracking." *World Trade*, June 1992, p. 31; Pope, Gregory T., "The Iron Horse Enters The Space Age." *High Technology Business*, Apr. 1989, p. 18.

3.3 Rosenbluth Travel Clemons, E. K., and M. Row, "Information Technology At Rosenbluth Travel: Competitive Advantage In A Rapidly Growing Service Company." *Journal Of Management Information Systems*, Vol. 8, Fall 1991, p. 53; Miller, David B., E. K. Clemons, and M. C. Row, "Information Technology and The Global Virtual Corporation." In Bradley et al., 1993, p. 283; Pastore, Richard, "The Mad Doctor Of Corporate Travel." *CIO*, May 1, 1992, p. 44; Rosenbluth, H., and D. Mcferrin, *The Customer Comes Second*, Philadelphia: Morrow Press, 1992.

3.4 Whirlpool Corporation Juneau, Lucie, "The Education Of A Line Manager." *CIO*, June 1, 1992, p. 38; Rothfeder, Jeffrey, "Crossed Wires." *Corporate Computing*, undated, p. 159.

4.1 The Internal Revenue Service Cooke, Catherine J., "Taxation By Automation." *Bank Technology News*, Nov./Dec. 1992, p. 40; Messmer, Ellen, "Voice Response Tax Filing Could Offer IRS A Big Return." *Network World*, Apr. 19, 1991, p. 32; Venkatraman, N., and A. Kambil, "The Check's Not In The Mail: Strategies For Electronic Integration In Tax Return Filing." *Sloan Management Review*, Winter 1991, p. 11.

4.2 Info/California Gould, Russell S., "Info/California Clears Path Through State Bureaucracy." *Contra Costa Times*, Nov. 19, 1991, p. 1; *Sacramento Bee* editorial "Government Service By Fingertip." Dec. 6, 1991; Moody, Ken, and David Catzel, "Providing Public Information Quicker And Better." *California County*, July/Aug. 1991.

4.3 Health Link Johnson, Maryfran, "Service Delivers Painless Insurance Claims." *Computerworld*, July 15, 1991, p. 1.

4.4 The National Cargo Bureau *Computer Buying World*, April 1992.

5.1 European Payments System Services Crockett, Barton, "Credit Firm To Build Pan-European Net." *Network World*, Sept. 2, 1991, p. 19.

5.2 Rockwell International Gareiss, Robin, "Rockwell Pulls Plug On Frame-Relay." *Communications Week*, June 22, 1992, p. 1; Sutter, James, "Getting Behind The Technology: The Rockwell Story." *Journal Of Business Strategy*, July/Aug. 1993, P. 44.

5.3 Analog Devices Crockett, Barton, "Net Lets Firm Unify Sales Office Abroad." *Network World*, June 7, 1991, p. 1.

5.4 Reliance National company sources

6.1 University Of Miami Wallace, Bob, "Miami School Plans Ahead With SONET." *Network World*, June 15, 1992, p. 31.

6.2 Chrysler Messmer, Ellen, "Chrysler Extols Benefits Of Move To Radio-Based Nets." *Network World*, June 16, 1991, p. 22.

6.3 Adaptive Corporation "Adaptive Introduces ATM Switch."

Broadband Networking News, July 15, 1992, p. 2; "New Adaptive STM/S Cost-Effectively Extends Broadband Networking To Smaller Sites." *Edge,* Feb. 8, 1993, p. 1; "New Adaptive ATM Switch Solves LAN Problems And Enables New Applications." *Edge Work-Group Computing Report,* July 13, 1992.

6.4 Covia Schultz, Beth, "Covia Test-Flies Hybrid Frame-Relay Network." *Communications Week,* Sept. 1991, p. 33.

7.1 Dana Corporation McWilliams, Brian, "The Electronic Connection." *Enterprise,* Oct. 1992, p. 31.

7.2 Department Of Commerce Cavedo, Robert F., "The Federal Government Gets A Handle On Gossip" *Networking Management,* Nov. 1990, p. 91; Fisher, Sharon, "Group Puts Final Touches On Government Standard." *Communications Week,* June 22, 1992, p. 36; Molloy Maureen, "Department Of Commerce Issues EDI Mandate." *Network World,* May 13, 1991, p. 27.

7.3 IATA (International Air Transport Association) Feldman, Joan M., "Going Global, Air Transport World." Apr. 1991, p. 67; Horwitt, Elisabeth, "Air Transport Industry Flying The OSI Banner." *Computerworld* May 25, 1992, p. 14.

7.4 Compression Labs and PictureTel Compression Labs, Inc., "Compressed Digital Video And The Video Communications Revolution." Apr. 1992; Schwarz, Jeffrey, "Video Interoperability Tests." *Video Forum,* Fall 1991, p. 3; Crockett, Barton, "Net Lets Firm Unify Sales Office Abroad." *Network World,* June 7, 1991, p. 1; Wexler, Joanie M., "PictureTel Makes Videoconferencing Cheaper." *Computerworld,* May 17, 1993, p. 58.

8.1 Redstone Arsenal Jackson, Kelly, "Gulf War Net Still Dug In." *Computerworld,* June 8, 1992, p. 32.

8.2 Health One Alter, Alan E., "Health Care's Band Leader.", *CIO,* May 15, 1992, p. 48.

8.3 The Travelers Richter, M. J., "Infrared LANs Shine Brightly At Travelers." *Computerworld,* Aug. 19, 1991, p. 1.

8.4 Byer California "Setting The Trend With A Frame Relay VAN." *Networking Management,* Jan. 1992, p. 64.

9.1 Texas Instruments Dryden, Patrick, "TI Tries To Manage Its Past and Future." *LAN Times,* June 18, 1993, p. 25; Eckerson, Wayne, "TI Sharpens Its EDI Edge." *Network World,* Nov. 25, 1991, p. 37; Eckerson, Wayne, "TI's Global Net Is Instrument Of Success." *Network World,* Nov. 25, 1991, p. 35; Magnet, Myron, "Who's Winning The Information Revolution?" *Fortune,* Nov. 30, 1992, p. 110; Pantage, Angeline, "TI's Global Window." *Datamation,* Sept. 1, 1989, p. 49; Thyfault, Mary E., and Bob Violino, "And Then There Was One." *InformationWeek,* July 29, 1991, p. 12; Vacca, John, "Users Easily Cost-Justify EDI; Indirect Savings Are The Big Bonus." *Communications Week,* Sept. 18, 1991.

9.2 The Royal Bank of Canada "Internetwork Provides Royal Savings." *Token Ring Internetworking,* Apr. 1992, p. 24; "Banking On Technology." *Canadian Business,* Mar. 1991, p. 79; Cukier Wendy, "Timely Strategies." *Canadian Business,* Mar. 1991, p. 75.

9.3 British Airways North America Stanislevic, Howard, "Diverse Access To The Public Network For High-Speed Data Circuits." *Telecommunications,* Sept. 4, 1990, p. 43.

9.4 Digital Equipment Brown, Peter E., "Business Productivity Through Networking." *Inter Net,* Apr. 1992, p. 8; Dubinskas, Frank A., "Managing Complexity In The RA90 Project." *Digital Equipment Product Marketing Strategic Programs,* FY90–1, Sept. 1989; Lorsch, Jay, "The Elements Of Engineering." *Enterprise,* Summer 1990, p. 35; Schweizer, Susanna, "Increasing Profitability Through Distributed Networking." Digital Equipment Corporation, 1993.

10.1 U.S Defense Department and CIA Carpenter, Betsy, and Tim Zimmerman, "Spying On The Earth." *U. S. News & World Report,* June 29, 1992, p. 66.

10.2 Fidelity Investments Smith, Geoffrey, "Fidelity Jumps Feet First Into The Fray." *Business Week,* May 25, 1992, p. 104.

10.3 JC Penney "Penney Brings Far East Suppliers Closer." *Chain Store Executive Age,* Oct. 1989, p. 71; Meeks, Fleming, "Live From Dallas." *Forbes,* Dec. 26, 1988, p. 112.

10.4 Mexico Moffet, Matt, "Rewiring A Nation: Telefonos De México Makes A Promising Start on a Daunting Task." *Wall Street Journal*

Europe, Feb. 25, 1992, p. 1; Schwartz, Jeffrey, "Mexican Firm To Construct Hubless VSAT Net." *Communications Week,* Mar. 16, 1992, p. 19.

11.1 Sea-Land Foster, Ed, "Desert Storm Pushes Sea-Land's Enterprise Into Action." *Infoworld,* Sept. 9, 1991, p. S66; Ruffin, William, "Sea-Land, Leaner, Faster and Wired For Service. *IN Magazine,* Spring 1988, p. 12.

11.2 Towers Perrin Rittle, Mary, "Masters Of Time." *ROLM Customer,* Number 4, 1991, p. 23.

11.3 Maryland State Government Messmer, Ellen, "Feds, States Advance With Benefits Distribution Nets." *Network World,* Mar. 30, 1992, p. 1; Thyfault, Mary E., "Providing Benefits Through Electronic Funds Transfer. " *InformationWeek,* June 10, 1991, p. 15.

11.4 Progressive Insurance King, Julie, "Re-engineering Puts Progressive On The Spot." *Computerworld,* July 15, 1991, p. 58.

12.1 American Standard Strassman, Paul A., et al. "Case Studies: American Standard, Inc." *Measuring Business Value Of Information Technologies,* Washington DC: ICIT Press, 1988, p. 144.

12.2 United Parcel Service Booker, Ellis, "UPS To Deploy Wireless Network." *Computerworld,* May 18, 1992, p. 6; Crockett, Barton, "UPS Telecom Enters International Resale Market." *Network World,* Dec. 30, 1991, p. 19; Eckerson, Wayne, "Hand-Held Computers Will Help UPS Track Packages." *Network World,* May 6, 1991, p. 15; Flint, Perry, " Maybe They Should Call It United Profit Service." *Air Transport World,* Feb., 1990, p. 91; Kindel, Sharen, "When Elephants Dance." *Financial World,* June 9, 1992, p. 76; Schultz, Beth, "UPS Improves Info Delivery With New Backbone." *Communications Week,* Feb. 24, 1992, p. 19.

12.3 Nordstrom Corporate Computing, "Nordstrom E-Mail At Your Service." Case Study, reprinted from *Corporate Computing,* Oct. 1992; Littman, Jonathan, "At High-Fashion Nordstrom, Pricey EDI Is Out, Well-Worn Technology Is In. It Sells." *Corporate Computing,* July 1992, p. 155.

12.4 City Of Hope Cummings, Joanne, "Hospital Puts Hope In FDDI-based Net." *Network World,* Mar. 30, 1992, p. 13; Appleby, Chuck, "The Network Cure-All." *InformationWeek,* Mar. 23, 1992, p. 53.

13 Landberg Dairies International Center for Information Technologies, St. John, U. S. Virgin Islands, unpublished research study, 1991.

14.1 The Gang Of Five "Kudos To X3T9.5 Committee." *Communications Week* Editorial, June 29, 1992, p. 46; Eugster, Ernest, "Users Would Be Better Off Waiting For ANSI Standard." *Network World,* July 20, 1992, p. 42; MacAskill, Skip, "IBM Dealt Blow By ANSI's FDDI-Over-Copper Decision." *Network World,* June 29, 1992, p. 1; Schneier Bruce, "CDDI Breathes Life Into FDDI Standard." *Network World,* Sept. 10, 1992, p. 35; Semilof, Margie, "100-Mbps Copper Choices Down To Two." *Communications Week,* June 22, 1992, p. 12; Wexler, Joanie M., "Copper FDDI Rivals Ally Against ATM Threat." *Computerworld,* June 29, 1992.

14.2 Hong Kong And Singapore King, J., and B. Konsynski, "Hong Kong Tradelink: News From The Second City." Harvard Business School Case 1-191-026, 1990; King, J., and B. Konsynski, "Singapore TradeNet: A Tale Of One City." Harvard Business School Case 1-191-009, 1990.

14.3 Cisco "Cisco Exec Discusses Internet Strategies." *Network World,* Dec. 2, 1991, p. 21; Mulqueen, John T., "Cisco's Story: Like Nothing Else." *Communications Week,* June 8, 1992, p. 48.

14.4 Washington International Telport Nicholson, Paul J., "Telports Come Of Age." *Telecommunications,* July 1989, p. 60.

15.1 Bank Of The West Adicoff, Sam, "Bank Of The West Downsizes To LANs." *LAN Times,* Mar. 23, 1992, p. 29.

15.2 WHDH-TV Didio, Laura, "WHDH-TV News Adopts Netware For Leading Edge." *LAN Times,* Jan. 12, 1992, p. 12; Semilof, Margie, "LAN Puts Boston TV Station On The Air." *Communications Week,* Aug. 3, 1992, p. 13.

15.3 The National Retail Federation's 1992 Convention "NRF Highlights Future Tech." *Discount Store News,* Feb. 3, 1992; "Retail Distribution and Logistics: The Impact Of Information

Technologies." *Chain Store Age Executive,* May 1993, Section 2; Tahminicioglu, Eva, "Convention Focus: Enhancing Existing Technology." *Women's Wear Daily,* Jan. 10, l991, p. 9.

15.4 Southwestern Bell Carter, John and Dave Stein, "A Fiber-Optic LAN At One Bell Center." *Telecommunications,* May 1989, p. 57.

16.1 American Airlines Schultz, Beth, "Airfare War Strains Data, Voice Nets." *Communications Week,* June 8, 1992, p. 1.

16.2 Motorola Rinaldi, Damian, "Motorola's Massive Internet Poses New Type Of Security Challenge." *Software Magazine,* Client/Server Computing Special Edition, May 1992, p. 32; Weber, James A., "Motorola's Rapidly Developing Peer-To-Peer Network Helps Achieve A Worldwide 'Wall-Less Workplace'." *Networking Management,* July 1992, p. 14.

16.3 European Public Data Networks Schenker, Jennifer L., "European Data Links Are Still A Struggle." *Communications Week,* Apr. 19, 1993, p. 51.

16.4 The Industrial Bank Of Japan McLening, Maggie, "Japanese Trader Makes A Second City Splash." *Banking Technology,* Mar. 1988, p. 20.

17.1 Laura Ashley Caldwell, Bruce, "Halfway House—The Warehouse Is Dead; Long Live The Automated Distribution Center." *InformationWeek,* Sept. 16, 1991, p. 80; McPartin, John P., "Laura Ashley's Fashionable Move." *InformationWeek,* Mar. 23, 1992, p. 12; Thomhill, John, "Future Based On A Maxmin Maxim." *Financial Times,* Aug. 22/23, 1992, p. 21.

17.2 Missouri Division Of Data Processing And Telecommunications Greenstein, Irwin, "For Missouri, Not Owning Switching Gear Means Savings and Flexibility." *Networking Management,* Jan., 1992, p. 18; Compression Labs Inc., "Videoconferencing Helps University With The Business Of Education." *Teleview,* 4th Quarter 1991.

17.3 Electronic Mail Ferris, David, "Count Hidden Costs Of E-Mail." *LAN Times,* Feb. 10, 1992, p. 69.

17.4 Massachusetts Department Of Public Utilities Baldwin, Susan, "ISDN Rate-Setting In Massachusetts." *Business Communications Review,* June 1992, p. 47.

18.1 Westinghouse Kronstadt, Paul, "Matching A Pair." *CIO,* May 25, 1990, p. 62.

18.2 *Business Communications Review* Survey Thobe, Deborah, "Who's Minding The Shop? BCR's Survey On Outsourcing." *Business Communications Review,* May 1992, p. 22.

18.3 Sears Dickey, John R., and K. Koriath, "Why Sears Towers In Telecommunications: Interview With John Morrison." *Chief Information Officer Journal,* Spring 1991, p. 5.

18.4 Forrester Research Molloy, Maureen, "Landscape Changing For Net Execs Of '90s." *Network World,* Dec. 16, 1991, p. 17.

19.1 Korn/Ferry International "Changing Net Landscape Impacts Global Hub Selection." *Network World,* Mar. 9, 1992, p. 31; Knauth, Kristin, "Staffing Your Global Network." *Network World,* Sept. 23, 1991, p. 67.

19.2 Kash N' Karry Eckerson, Wayne, "Intrepid User Braves Risks Of Distributed Object World." *Network World,* June 8, 1992, p. 1; Johnson, Maryfran, "Kash N' Karry Shops In New Technology Aisles." *Computerworld,* Mar. 1, 1993, p. 1.

19.3 Rohm and Haas Laplante, Alice, "Moving To A Client/Server Architecture Is A Major Task." *Infoworld,* June 8, 1992, p. 361.

19.4 Baxter Healthcare Corporation Girishankar, Saroja, "Baxter To Use SNA Server." *Communications Week,* Aug. 23, 1993, p. 10; Short, James E., and N. Venkatraman, "Beyond Business Process Redesign: Redefining Baxter's Business Network." *Sloan Management Review,* Fall 1992, p. 15; Weber, James, "Focus Switches From Network Efficiency To Business Effectiveness." *Network Management,* May 1991, p. 26.

General References

Abelson, Alan, "Why The Baby Bells Own The Future." *Barron's,* Aug. 10, 1992, p. 16

"Accounting For Technology Costs." *Enterprise,* Special Report, Winter 1990, p. 9

Adams, D. A., M. C. Lacity, and J. R. Mullins, "Telecommunications Research In Information Systems: An Investigation Of The Literature." *Database,* Summer 1991, p. 35

Adams Eric J., "Sizing Up EDI." *World Trade,* June 1992, p. 102

Alter, Alan A., "EDI Release 2.0." *CIO,* June 15, 1993, p. 73

Andrews, Edmund L., "A Baby Bell Primed For The Big Fight." *New York Times,* Feb. 21, 1993, p. 3–1

Audin, Gary, "How Bridges and Routers Choke Today's WANs." *Business Communications Review,* Nov. 1992, p. 37

"Banks Reach Middle Market With Expanded EDI Offerings." *Bank Technology News,* June 1992, p. 1

Baum, David, "Middleware: Unearthing A Software Treasure Trove." *Infoworld,* Nov. 30, 1992, p. 46

Berg, J. L., *An Analysis Of The Information Technology Standardization Process,* Amsterdam: North Holland, 1990

Blenheim Online, *The Networked Economy 1991,* Conference Proceedings, Mar. 1991

Bochenski Barbara, "Database Mix Poses A Sharing Challenge." *Software Magazine,* Mar. 1992, p. 63

Bolles, Gary A., "Enterprise-To-Enterprise Computing." *Network Management,* May 1993, p. 88

Bolles, Gary A., "Gearing Up For EDI: A Primer On Electronic Data Interchange." *Network Computing,* Sept. 1991, p. 88

Bonatti, M., F. Casali, and G. Popple (Eds), *Integrated Broadband Communications,* Amsterdam: North Holland, 1991

Booker, Ellis, "Motorola E-Mail Service Bows." *Computerworld,* July 27, 1992, p. 16

Bovoni, Stephen J., "Joint Ventures: Networking's Final Frontier." *Beyond Computing,* Aug./Sept., 1992, p. 28

Bowman, Robert J., "Keeping Track Of Tracking." *World Trade,* June 1992, p. 30

Bradley, Stephen P., J. A. Hausman and R. L. Nolan, *Globalization, Technology and Competition: The Fusion of Computers and Telecommunications In The 1990s,* Cambridge, Mass: Harvard Business School Press, 1993

Braue, Joseph, "The Future Of Enterprise Networking." *Data Communications,* Sept. l992, p. 22

Briere, Daniel, "Public Or Private?" *Network World,* July 20, 1992, p. 55

Briere, Daniel, "The Secret To Success With Virtual Nets." *Network World,* Mar. 23, 1992, p. 1

Brown, Bob, "A&P Setting Up VSAT Net To Register Service Gains." *Network World,* June 29, 1992, p. 17

Brown, Bob, "A Crop Of Feature-Rich X.400 Products Emerging." *Network World,* Mar. 23, 1992, p. 1

Brown, Ronald O., "What You Need To Know To Plan For Disaster." *Networking Management,* Apr. 1993, p. 25

Business Communications Review, *A Guide To Frame Relay,* Oct. 1991

Caldwell, Bruce, "CRSs Still Reeling." *InformationWeek,* July 13, 1992, p. 14

Caldwell, Bruce, "Security Slack At Exchanges." *InformationWeek,* Sept. 2, 1991, p. 14

Caldwell, Bruce, and D. Bartholomew, "Banc One's Competitive Edge." *InformationWeek* Sept. 9, 1991, p. 42

Callan, James E., "Long-Term Vision For Corporate Network Management." *Journal of Networking Management,* Summer 1990, p. 6

"The Changing Role Of The Mainframe." *I/S Analyzer,* Jan. 1993, p. 1

Chismar, William G., and L. Chidambaram, "Telecommunications and The Structuring of U. S. Multinational Corporations." *International Information Systems,* Oct. 1992, p. 38

"Chrysler Steers Net Management." *InformationWeek,* Sept. 2, 1991

Cikoski, Thomas R., and J. S. Whitehill, "Integrated Network Management Systems: Understanding The Basics." *Telecommunications,* June 1993, p. 41

Conning, Sue, "Image-Enabled Software Stores History." *Systems 3X/400,* Mar. 1993, p. 40

Coy, Peter, "The Baby Bells Learn A Nasty New Word: Competition." *Business Week,* Mar. 25, 1991, p. 96

Coy, Peter, "The New Realism In Office Systems: Computers Can't Take The Place Of Good Management—But They Can Help." *Business Week,* June 15, 1992, p. 128

Coy, Peter, "Your New Computer: The Telephone." *Business Week,* June 3, 1991, p. 126

Crandall, Robert W., *After The Breakup: U. S. Telecommunications In A More Competitive Era,* Washington, D. C.: The Brookings Institute, 1991

Crandall, Robert W., and K. Flam, Eds., *Changing The Rules: Technological Change, International Competition, And Regulations In Telecommunications,* Washington D. C.: The Brookings Institute, 1989

Crawford, Philip, "Minitel Spread Its Wings." *Credit Card Management,* June 1992, p. 64

Crockett, Barton, "Gillette Hands International X.25 Traffic To BT Unit." *Network World,* Feb. 17, 1992, p. 23

Crockett, Barton, "PDN Service Quality Mixed Across European Countries." *Network World,* Sept. 9, 1991, p. 34

Cummins, J. M. and E. C. Stonebraker, "Total Quality Management of Telecommunications." *Business Communications Review,* December, 1989. p. 36–41.

Cummings, Joanne, "Videoconferencing Gives Recruiter Edge." *Network World,* Mar. 1, 1993, p. 23

Cummings, Joanne, "Another Bright Idea." *Network World,* November 23, 1992, p. 31

Cummings Joanne, "Interactive TV Gains Support In Industry." *Network World,* Aug. 17, 1992, p. 29

Dagres, Todd, "Frame Relay's Day Will Dawn." *Business Communications Review,* Apr. 1992, p. 28

Davenport, Thomas H., *Process Innovation: Reengineering Work Through Information Technology,* Cambridge, MA: Harvard Business School Press, 1993

Davidson Peter, "Multimedia Finally Appears Within Networks' Reach." *Network World,* Apr. 6, 1992, p. 39

Davis, Dwight B., "Hard Demand For Soft Skills." *Datamation,* Jan. 15, 1993, p. 28

Davis, Dwight B., "Marrying Wireless Communications To Mobile Computing." *Electronic Business,* May 1993, p. 83

Deering, Ann M., "The Fate Of Your Network Is In Your Hands—Not The Telcos'." *Chief Information Officer Journal,* Winter 1992, p. 51

Deloitte and Touche, *Annual Survey Of North American Telecommunications Issues,* 1993

Deming, W. Edward, *Out Of The Crisis,* Boston, Mass: MIT Press, 1986

Dern, Daniel, "Internet System Experiencing Meteoric Growth." *Infoworld,* Sept. 21, 1992, p. 56

Desjardins Richard, "OSI Really Is A Good Idea." *Communications Week,* June 29, 1992, p. 47

Desmond, John, "Building Around Unknowns Of Client/Server." *Software Magazine,* Mar. 1992, p. 68

Dickey, John R., "Reliability Is No Accident." *Chief Information Officer Journal,* Winter 1991, p. 44

Dixon, Hugo, "U. S. Telecom Authorities Target Phone Cartel." *Financial Times,* July 16, 1991, p. 1

Dizard, Wilson P., *The Coming Information Age,* Second Edition. New York and London: Longman, 1984

Doll, Dixon R., "The Spirit Of Networking: Past, Present, and Future." *Data Communications,* Sept. 1992, p. 25

Donlan, Thomas G., *How America Can Win The Technology Race,* Homewood, Ill: Business One Irwin, 1991

Dos Santos, B. L., K. Peffers, and D. C. Mauer, "The Impact Of Information Technology Investment Announcements On The Market Value Of The Firm." *Information Systems Research,* Vol. 4, 1993

"Don't Write Off X.400." *Communications Week,* Editorial, June 8, 1992, p. 48

Duffy Caroline A., "14.4 Kbps Modems Set Comm Pace." *PC Week,* July 13, 1992, p. 83

Duffy, Tim, "T-l Keeps Shaw's On Top Of Grocery Game." *Network World,* Aug. 3, 1992, p. 9

"Dumb Terminals Remain The Preferred Choice Of Many." *Digital On The Desktop,* 1992, p. 30

Dutta, Amitava, "Telecommunications Infrastructure In Developing Nations: Rural Coverage." *International Information Systems,* No. 3, July 1992, p. 31

Earl, M. J., and D. J. Skryme, "Hybrid Managers—What Do We Know About Them?." *Journal Of Information Systems,* No. 2, 1992, p. 169

Economic Development Board, State of Washington, "Advanced Telecommunications For Economic Development in Washington State." Working paper, 1988

Ethem, Mel, "Migrating From Ethernet To FDDI: A Case Study." *Business Communications Review,* Mar. 1992, p. 38

Feder, Barnaby J., "Frito-Lay's Speedy Data Network." *New York Times,* Nov. 8, 1990, p. Dl

Ferris, David, "Work Flow Applications Simplify Office Processes." *ABA Banking Journal,* Feb. 1, 1992, p. 79

Fisher, Sharon, "Users Pay High Price For LAN Management." *Communications Week,* Jan. 4, 1993, p. 32

Fitzgerald, Jerry, *Business Data Communications*, Third Edition, New York: Wiley, 1990

Fitzgerald, Jerry, *Business Data Communications: Basic Concepts, Security, and Design, Fourth Edition.* New York: Wiley, 1993

Fitzgerald, Michael, "Users Agree: Support Takes Center Stage." *Computerworld,* July 20, 1992, p. 87

Flanagan, P., "Intelligent Hubs: A Marketplace Survey." *Telecommunications,* Oct. 1993, p. 33

Flanagan, P., "VSAT: A Market and Technology Overview." *Telecommunications,* Mar. 1993, p. 19

Flanagan, Patrick, "Multiprotocol Routers: An Overview." *Telecommunications,* Apr. 1993, p. 19

Foley John, "EDI Innovations Coming To Users." *Communications Week,* Sept. 16, 1991, p. I

France Telecom, *ISDN: A User's Guide.* Booklet, 1992

Freece Alan L., "Majority's Needs Met By VSAT." *Communications Week,* June 22, 1992, p. 41

Freedman, David, "Retailing In Real-Time." *CIO,* Aug. 1992, p. 34

Freedman, David, "A Virtual Company." *CIO,* Sept. 1, 1991, p. 42

Fulk, Janet, and Charles Steinfield, *Organizations and Communications Technology,* Newbury Park, Calif: Sage, 1990

Galbraith, Michael, "Japan Opens Up." *Communications International,* Nov. 1992, p. 14

Gantz, John, "Outsourcing: Threat Or Salvation?" *Networking Management,* Oct. 1990, p. 22

Gareiss, Robin, "Chip Company Tests AT&T Frame-Relay." *Communications Week,* June 29, 1992, p. 1

Gareiss, Robin, "X.25's Popularity Remains High." *Communications Week,* Aug. 3, 1992, p. 21

Gerwig, Kate, "Management, Recovery Become Software Problems For Carriers." *Communications Week,* Oct. 19, 1992, p. 30

Gilder, George, "Cable's Secret Weapon." *Forbes,* Apr. 13, 1992, p. 80

Gilder, George, "The New Rules Of Wireless." *Forbes,* Mar. 29, 1993, p. 96

Glass, Brett, "The Well-Connected Notebook." *Infoworld,* Nov. 16, 1992, p. S88

Glass, Brett, "High-Speed Modems." *PC World,* June 1992, p. 220

Gleckman, Howard, "The Technology Payoff." *Business Week,* June 14, 1993, p. 57

Goff, Leslie, "Getting A Line On Telemarketing." *CIO,* Jan. 1991, p. 61

Goldberg Aaron, "Why Mainframes Aren't Extinct." *PC Week,* July 18, 1993, p. 224

Gore, Al, "Infrastructure For The Global Village." *Scientific American,* Sept. 1991, p. 150

Greenstein, Irwin, "Users Talk About Implementing OSI." *Networking Management,* Oct. 1992, p. 46

Gregg, Lynne, "Wireless LANs: The Next Frontier In Network Computing." *Infoworld,* June 22, 1992, p. 95

Gross, Neil, "It's 'The Elephant Against The Rats' For Japan's Phone Business." *Business Week*, Nov. 9, 1992, p. 112

Hammer, Michael and J. Champy, *Reengineering The Corporation*, New York: Harper Business, 1993

Hausman, Jerry A., "The Bell Operating Companies Venture Abroad While British Telecom and Others Come To The United States. " In Bradley et al. 1993, p. 313.

Head, Beverley, "Banking From The Outback." *Banking Technology*, May 1991, p. 27

Hector, Gary, *Breaking The Bank: The Decline Of Bank Of America*, Boston: Little Brown, 1988

Higgins, Steve, "Groupware: Getting A Grip On Work-Group Computing." *PC Week* Special Report, Oct. 26, 1992

Himwich, H.A., "Cooperative Network Management: SNA and SNMP." *Business Communications Review*, Oct. 1992, p. 51

Hoffman, Thomas, "Amex Seeks Wireless Trades." *Computerworld*, May 17, 1993, p. 6

Hoffman, Thomas, "Imaging Cures Hospital's Paper Woes." *Computerworld*, June 29, 1992, p. 74

Horwitt, Elisabeth, "Organization, Not Tools, Key To LAN Management." *Computerworld*, Sept. 7, 1992, p. 8

Horwitt, Elisabeth, "Canadian Bank Finds Needed Link." Computerworld, July 20, 1992, p. 32

Horwitt, Elisabeth, "Nestle To Blend E-Mail Systems Into Corporate Net." *Computerworld*, May 18, 1992, p. 51

Huber, Peter, "Telephony Unbottled." *Forbes*, Jan. 18, 1993, p. 94

Iacobuzio, Theodore, "Network Fretwork." *Bank Systems And Technology*, May, 1992, p. 4

IBM, *Systems Network Architecture: Concepts and Products* fifth ed. IBM Corp., Mar. 1991

Institute For Information Studies, *A National Information Network: Changing Our Lives In The 21st Century*, Aspen Institute, 1992

"Interop Showcase Award Winners." *LAN Times*, Nov. 23, 1992, p. 19

Ishikawa, K., *Guide To Total Quality Control*, White Plains, NY: Asian Productivity Organization, 1982

Janson, Jennifer L., "Technologies, Budgets Keep IS Managers On Their Toes." *PC Week*, Feb. 10, 1992, p. S/19

Jazwinski, Andrew, "Applying Simulation To Network Planning." *Network World*, May 17, 1993, p. 33

Johnson, Jim, "A Survival Guide For Administrators." *Software Magazine*, Dec. 1992, p. 54

Johnson, Johna T., "Rebuilding The World's Public Networks." *Data Communications*, Dec. 1992, p. 60

Johnson, Johna T., "ISDN Goes Nationwide, But Will Users Want It?" *Data Communications*, Nov. 1992, p. 93

Johnson, Johna T., "SMDS: Out Of The Lab and Onto The Network." *Data Communications*, Oct. 1992, p. 71

Jordan, Michael, 1991, personal communication

Juneau, Lucie, "Smart Hubs: Head Of The Class." *Datamation*, Jan. 1, 1993, p. 86

Juran, J. M. *Juran on Leadership for Quality: An Executive Handbook* New York: Free Press, 1989

Juran, J. M., and F. M. Gryna, eds., *Juran's Quality Control Handbook*, Fourth Edition New York: McGraw-Hill, 1988, p. 24.2–24.3

Keen, Peter G. W., "Business/Technology Fusion: Making The Management Difference." *IBM Systems Journal*, Apr. 1993, p. 17

Keen, Peter G. W., "Telecommunications and Organizational Choice." In Fulk and Steinfield, p. 295

Keen, Peter G. W., "Wa$te Not, Want Not." *Computerworld*, Feb. 25, 1991, p. 77

Keen, Peter G. W., and Ellen M. Knapp, "Business Process Investment: Getting The Right Process Right." Coopers and Lybrand and International Center for Information Technologies, monograph, 1993

Keller, John J., "AT&T Sues MCI, Saying Competitor Violated Patents." *Wall Street Journal*, Jan. 12, 1993, p. B12

Kiely, Thomas, "Looking For Networking's Mr. Big." *CIO*, Feb. 1992, p. 28

Killette, Kathleen, "Ford Adds Dealers To Custom Net" *Communications Week*, June 17, 1991, p. 13

King, William R., and V. Sethi, "An Analysis Of International Information Regimes." *International Information Systems*, No. 1, Jan. 1992, p. 1

Kirkpatrick, David, "Could AT&T Rule The World?" *Fortune*, May 17, 1993, p. 54

Kronstadt, Paul, "Ship 54—Where Are You?" *CIO*, May 1990, p. 30

Korzeniowski, Paul, "RPCs Help User Put Mainframe Apps On Unix LANs." *Communications Week*, Sept. 14, 1992, p. 1

Korzeniowski Paul and Paul Strauss, "Wireless WANs Have A Ways To Go." *Datamation*, May 15, 1992, p. 52

Kraemer, Joseph S., "Local Competition: All Against All." *Business Communications Review*, Mar. 1993, p. 35

Kupfer, Andrew, "The Race To Rewire America." *Fortune*, Apr. 19, 1993, p. 42

Kushnick, B., "The State Of The Public Network: Why We Need Another Divestiture." *Telecommunications Management*, Jan. 1993, p. 17

Lacity, Mary C., "The Information Systems Outsourcing Bandwagon: Look Before You Leap." University of Missouri, Aug. 31, 1992

LAN Times, *Buyers Directory*, July 24, 1992

Laplante Alice, "Federal Express Gives Clients On-Line Access To Tracking System." *Infoworld*, Nov. 16, 1992, p. 108

Laplante, Alice, "LAN/WAN Skills Sought After In IS Shops." *Infoworld*, June 8, 1992, p. 558

Laplante, Alice, "Taking A Second Look At The Concept Of Outsourcing." *Infoworld*, May 13, 1991, p. 58

Lindstrom, Annie, "McCaw To Deliver Cellular Data Service. " *Communications Week*, May 3, 1993, p. 6

Loh, Lawrence, and Venkatraman N., "Diffusion Of Information Technology Outsourcing: Influence Sources and The Kodak Effect." *Information Systems Research*, Vol. 3 No. 4, Dec. 1992

Lombardo Nicholas, "Routers May Test Frame Relay Users' Patience." *Network World*, May 25, 1992, p. 35

Longsworth, Elizabeth, and J. Montgomery, "Network Puzzle." *Corporate Computing*, May 1993, p. 99

Lucas, Henry C. Jr., and R. A. Schwartz eds. *The Challenge Of Information Technology For The Securities Markets*, Homewood, Ill: Dow Jones Irwin, 1989

Lynch, Karen, "Canada Deregulates Long Distance Market." *Communications Week*, June 29, 1992, p. 33

Macaskill, Skip, "California Network Takes Aim At Deadbeat Parents." *Network World*, July 6, 1991, p. 6

Macaskill, Skip, "Hub Vendors Mixed On Use Of Wireless." *Network World*, June 8, 1992, p. 21

Macaskill, Skip, "Cost Issues Still Loom As Economy Brightens." *Network World*, May 18, 1992, p. 47

Malamud, Carl, "TCP/IP—A Dependable Networking Infrastructure." *Networking Computing*, Apr. 1991, p. 83

Malone, T. W., J. Yates, and R. I. Benjamin, "Electronic Markets and Electronic Hierarchies." *Communications Of The ACM*, Vol. 30, 1987

Malone, Thomas, and J. Rockart, "Computers, Networks and The Corporation." *Scientific American*, Sept. 1991, p. 128

Manheim, Marvin L., P. G. W. Keen, and J. Elam, "Strategic Assessment: The Use Of Telecommunications As An Element Of Competitive Strategy By The City Of Amsterdam." working paper, Cambridge, MA: Systematics, Inc., May 1988

Mason, Richard O., Joyce Elam, and Daniel Edwards, "Creating An Urban Advantage: How Cities Compete Through Information Technology." working paper, International Center For Information Technologies, 1988

McClusker, Tom, "Networks That Manage Themselves." *Datamation*, June 1, 1992, p. 53

McDonald, Joe, and S. Sherizen, "Is Your LAN Data Secure?" *Software Magazine*, Nov. 1992, p. 66

McGraw-Hill, Data Communications, 1993 Product Selection Guide Issue, *Data Communications*, Special Report, Oct. 15, 1992

MCI Communications, "Frame Relay Technical Description," MCI Data Services, Technical Paper, 1992

McQuillan, John, "Why ATM?" *Business Communications Review*, Supplement, Feb. 1993

McWilliams, Brian, "The Electronic Connection." *Enterprise*, Oct. 1992, p. 31

"Measuring Up: Users Rate The World's Carriers." *Communications Week International*, Oct. 5, 1992, p. 1

Meier, Edwin, E., "The Decline and Fall Of The Novell IPX Protocol." *Communications Week*, Nov. 9, 1992, p. 29

Meier, Edwin E., and Betsy Yocom, "Is Fiber Cabling The Future?" *Communications Week*, Aug. 17, 1992, p. 35

Meijer, Anton, ed., *Systems Network Architecture: A Tutorial*. Pitman, 1987.

Messmer, Ellen, "The Daring Few Lead Mobile Data Charge." *Network World*, Apr. 5, 1993, p. 25

Metcalfe, Bob, "Economists and The Productivity Paradox." *Infoworld*, Nov. 16, 1992, p. 72

Miles, R. E., and C. C. Snow, "Network Organizations." *California Management Review*, Spring 1986, p. 62

Mische, Michael, "EDI In The EC: Easier Said Than Done." *Journal of European Business*, Nov./Dec. 1992, p. 19

Mohen, Joe, "OSI Interoperability: Separating Fact From Fiction." *Data Communications*, Jan. 21, 1992, p. 4L

Montgomery, Leland, "Paperless Claims: A Miracle Cure." *Finance Weekly*, June 9, 1992, p. 32

Moss, Mitchell L., "Telecommunications, World Cities And Trade." *Urban Studies*, Vol. 24, 1987, p. 534

Mott, Stephen C., P. G. W. Keen, and J. D. Sulser, "Electronic Information Services: Looking Ahead By Looking Back." ICIT Briefing Paper, 1988

Mulqueen, John, "Imaging Lights Up The Network." *Communications Week*, May 11, 1992, p. 37

Multitech Systems Booklet, "Local Area Networking: Opening Gateways In The LAN Maze." 1993

Musich, Paula, "Novell Upgrades Netware 3270 Emulator For DOS Clients." *PC Week*, Dec. 7, 1992, p. 51

Nelson, Fritz, and T. Hinders, "Making X.400 Work." *Network Computing*, Jan. 1993, p. 118

Network World Staff, "IEEE To Weigh Fast Ethernet Alternative." *Network World*, Nov. 9, 1992, p. 1

Newman, Mark, "Europe's Calling—But At A Price." *EuroBusiness*, July/Aug. 1993, p. 70

"New Momentum For Electronic Patient Records." *New York Times*, May 2, 1993, p. 8f

Nicholson, Paul J., "Teleports Come Of Age." *Telecommunications*, July 1989, p. 60

Nolle, Tom, "Groupware: The Next Generation." *Business Communications Review*, Aug. 1993, p. 54

Nolle, Tom, "X.25's Underrated Staying Power." *Business Communications Review*, Aug. 1992, p. 19

Norvell, Bruce, "The Forgotten Fourth In Long Distance." *New York Times*, Dec. 27, 1992, p. 6

Novell Research, *Power and Grounding For Distributed Computing*, Provo, Utah: Novell, Inc., 1991

Oberst, Gerald E. Jr., "Liberalising The Satellite Market." *Networking Management Europe*

O'Brien, Timothy, "Oracle Glues Desktops To Net Services." *Network World*, Nov. 9, 1992, p. 1

O'Leary, Meghan, "Next-Generation Retailing." *CIO*, Mar. 1991, p. 28

O'Reilly, Brian, "Novell Faces The Battle Of Its Life." *Fortune*, Aug. 9, 1993, p. 81

Ozeki, K., and T. Asaka, *Handbook Of Quality Tools*, Cambridge, MA: Productivity Press, 1988

Pastore, Richard, "Networking On The Fast Track." *CIO*, Mar. 1992, p. 20

Perry, Linda, "Handle With CareNet." *InformationWeek*, Apr. 6, 1992, p. 12

Pierce, John R., and M. Noll, *Signals: The Science Of Telecommunications*. New York: Scientific American Library, 1990

Plesums, C.A., and R. W. Bartels, "Large-Scale Image Systems: USAA Case Study." *IBM Systems Journal*, No. 3, 1990, p. 343

Powell, Dave, "The Hidden Benefits Of Data Compression. "*Networking Management*, Oct. 1989, p. 46

Preston, Robert, "Taiwan Revamping Telecoms—Plan Calls For Eased Rules, New Networks." *Communications Week International*, Nov. 9, 1992, p. 34

Quint, Michael, "4 Banks Plan Automated Teller Setup." *New York Times*, July 23, 1992, p. D5

Radding, Alan, "Superservers To The Rescue." *Infoworld*, Nov. 9, 1992, p. 73

Rajendran, Joseph, "India's Public Telephone Authority Setting Up Offices In Singapore." *Singapore Press Holdings*, Mar. 31, 1992, p. 3

Rash, Brian, "Preparing For The Worst." *Communications Week*, Nov. 2, 1992, p. 45

Rash, Wayne, "The Good and Bad Sides Of Vendor Alliances." *Communications Week*, June 29, 1992, p. 45

Reinhold, Bob and Harris Allen, "Imaging Makes Its Mark On Nets." *Network World*, June 8, 1992, p. 43

Richter M. J., "Toppling The Local Monopoly." *Communications Week*, June 15, 1992, p. 55

Roach, S. S., "America's Technology Dilemma A Profile Of The Information Economy." *Special Economic Study*, New York: Morgan Stanley & Co., 1987

Roesler Paula, "Building For The Future: A New Jersey Initiative." *NRF Show Daily*, July 17, 1992, p. 11

Rosaire, Claire, "FDDI: The Promised LAN." *Computerworld*, Aug. 10, 1992, p. 42

Rosenberg, Robert, and Steve Valiant, "Electronic Data Interchange—A Global Perspective." *Telecommunications*, Oct. 1992, p. 50

Roussel, Anne-Marie, and Peter Heywood, "Pacific Carriers Get A Head Start." *Data Communications*, Dec. 1992, p. 95

Roussel, Anne-Marie, "Southeast Asia Telecom." *Data Communications*, May 1992, p. 92

Rowe, Stanford H. *Business Telecommunications*. New York: Macmillan, 1990.

Runge, David A., *Winning With Telecommunications: An Approach For Corporate Strategists*, Washington, D.C.: ICIT Press, 1988

"A Rush To Modernity For East German Phones." *New York Times*, Mar. 11, 1992, p. D7

Rutz, Dave E., "Texas Commerce Bancshares: Image Of A Winner." *Document Management and Windows Imaging*, Mar./Apr. 1993, p. 37

"Sailor's Distress Call Unheeded For Weeks." *The San Juan Star*, Nov. 17, 1992, p. 32

Salamone, Sal, "Sizing Up The Most Critical Issues." *Network World*, May 18, 1992, p. 63

Salamone, Salvatore, "Standard For Fiber Ethernet Advances." *Network World*, Aug. 19, 1991, p. 19

Salz-Trautman, Peggy, "DBP Telekom: Dancing As Fast As It Can." *Communications International*, Oct. 1992, p. 26

Santiago, Paul, "The International Telecommunications Union (ITU)." *Business Puerto Rico*, Feb. 1990, p. 62

Santosus, Megan, "Merced County's Magic Act." *CIO*, Jan. 1993, p. 104

Santosus, Megan, "The Taxpayer Is Always Right." *CIO*, Aug. 1993, p. 84

Santosus, Megan, "Where Are They Now?" *CIO*, Apr. 1, 1993, p. 28

Schatz Willie, "The Race To Wire The Home Will Yield A $100 Billion Market." *Electronic Business*, June 6, 1993, p. 21

Scholtes, Peter R., *The Team Handbook*, Madison, WI: Joiner Associates Inc., 1988

Schriftgeisser and Levy, "SMDS vs. Frame Relay: An Either/Or Decision?" *Business Communications Review*, September 1991.

"Shoot-Out Served Up Many Bests, But No Winner." *PC Week* editorial, Nov. 12, 1990, p. 21

Shuilleabhain, Aine, "VANs Flood In To Asia-Pacific Market." *Communications International*, May 1993, p. 53

Simpson, David, "Uniting DECnet and SNA Networks: Goal Is Clear, Approaches Are Many." *Digital News and Review,* Oct. 26, 1992, p. 53

Singleton, Loy A., *Telecommunications In The Information Age,* Cambridge, MA: Ballinger, 1986

Slutsker, Gary, "The Tortoise And The Hare: How AT&T Is Succeeding Where IBM Faltered." *Forbes,* Feb. 1, 1993, p. 66

Smalley, Eric, "Time To Join The FTAM Team?" *Datamation,* Sept. 1, 1992, p. 53

Smith, Laura Brennan, "Mobile Computing." *PCWeekly* Supplement, Apr. 19, 1993

Smith, Gail, "Planning For Migration To ATM." *Business Communications Review,* May 1993, p. 53

"SNMPv2: The New Direction In Network Management." *Network Computing,* July 1993, p. 140

Sproull, Lee, and Sara Kiesler, *Connections: New Ways Of Working In The Networked Organization,* Cambridge, MA: MIT Press, 1991

Stallings, William, "SNMP-Based Network Management: Where Is It Headed?" *Telecommunications,* June 1993, p. 57

Stallings, William, *Business Data Communications,* New York: Macmillan, 1990

Stallings, William, *Data and Computer Communications,* Third Edition, New York: Macmillan, 1990.

Stamper, David A., *Business Data Communications.* Redwood City, CA: Benjamin/Cummings, 1991

Stevenson, J. G., "Management Of Multivendor Networks." *IBM Systems Journal,* No. 2, 1992, p. 189

Strassman, P.A., et al., *Measuring Business Value of Information Technologies,* Washington, D. C.: ICIT Press, 1988

Strehlo Christine, "Navy Automates On-Ship Documentation." *Infoworld,* Aug. 17, 1992, p. S72

Sweeney Terry, "Two Bells Frustrate National ISDN Effort." *Communications Week,* Nov. 23, 1992, p. 1

Szabat, M. M., and G. E. Meyer, "IBM Network Management Strategy." *IBM Systems Journal,* No. 2, 1992, p. 154

Taber, Mark, "Call Growing For Open Systems Management." *Software Magazine,* Mar. 1992, p. 51

Taff, Anita, "Carriers Plot Strategies At Dawn Of War Over 800 Users." *Network World,* Nov. 9, 1992, p. 1

Taff, Anita, "Users, Carriers Fret About 900 Portability." *Network World,* Aug. 3, 1992, p. 19

Taff, Anita, "U.S. Lead In Network Technology Slipping." *Network World,* Oct. 28, 1991, p. 13

Tanenbaum, Andrew S., *Computer Networks,* Second Edition, Englewood Cliffs, NJ: Prentice-Hall 1989.

Terplan, Kornel, "Should You Outsource Network Management?" *Chief Information Officer Journal,* Winter 1991, p. 23

Thiel, Thomas J., "Floating Lighter, The Navy's Paperless Ship Project." *Document Management,* May/June 1992, p. 12

Tolly, Kevin, "Testing The New SNA." *Data Communications,* May 21, 1992, p. 58

Torgersen, Don A., "Diginet Diversifies Its Fibernet With Microwave." *Telecommunications,* Oct. 1989, p. 51

Tucker, G., "Global ISDN Analysis from Physical Layer To Network Layer." *Telecommunications,* Aug. 1991, p. 43

"Tough Budgets For Tough Times." *Network World,* Sept. 23, 1991, p. 47

USAA, "Fact Sheet: Image Systems At USAA." USAA Building, San Antonio, Texas, May 1992

"USAA: Insuring Progress." *InformationWeek,* May 25, 1992, p. 4a

"Usable Flying Objects (Bouncing Radio Waves Off Meteors)." *The Economist,* Oct. 21, 1989, p. 96

Vanirk, Doug, "IBM, Sears Combine In Giant Network Venture." *Computerworld,* Aug. 21, 1992, p. 60

Varhol, Peter D., "Keeping Data On Tap." *Corporate Computing,* Mar. 1993, p. 47

"WAN Connections: The New WAN." *Communications Week* Supplement, 1993

Zachary, G. Pascal, "IBM-Apple Operating System, Taligent, Is Ahead Of Schedule." *Wall Street Journal,* Jan. 12, 1993, p. B12

Zadra, Randy, "The Telecommunications Payoff In Latin America." *Telecommunications,* July 1993, p. 33

Zaheer, Akbar and N. Venkatraman, "Determinants Of Electronic Integration In The Insurance Industry: An Empirical Test." *Center For Information Systems Research,* Sloan School Of Management, MIT, working paper, Dec. 1992

Zboray, "How Fast Packet May Affect X.25 Data Networks." *Business Communications Review,* November, 1990

Index

Numbers set in *italic* indicate an illustration or table on that page.